2nd International Conference on Fluid Mechanics and Industrial Applications (FMIA 2018)

AA006864

Journal of Physics: Conference Series
Volume 1064

Guilin, China
12 – 14 July 2018

ISBN: 978-1-5108-6866-3
ISSN: 1742-6588

Printed from e-media with permission by:

Curran Associates, Inc.
57 Morehouse Lane
Red Hook, NY 12571

Some format issues inherent in the e-media version may also appear in this print version.

Copyright© (2018) by the Institute of Physics
All rights reserved. The material featured in this book is subject to
IOP copyright protection, unless otherwise indicated.

Printed by Curran Associates, Inc. (2018)

For permission requests, please contact the Institute of Physics
at the address below.

Institute of Physics
Dirac House, Temple Back
Bristol BS1 6BE UK

Phone: 44 1 17 929 7481
Fax: 44 1 17 920 0979

techtracking@iop.org

Additional copies of this publication are available from:

Curran Associates, Inc.
57 Morehouse Lane
Red Hook, NY 12571 USA
Phone: 845-758-0400
Fax: 845-758-2633
Email: curran@proceedings.com
Web: www.proceedings.com

Table of contents

Volume 1064

2018 the 2nd International Conference on Fluid Mechanics and Industrial Applications

12–14 July 2018, Guilin, China

Accepted papers received: 19 July 2018
Published online: 10 August 2018

Preface

Papers

Thermodynamics and Heat Transfer

Combustion and Emissions

Modeling and Simulation

Fluid Properties, Performance and Applications

Fluid Theory and Related Algorithms

IOP Conf. Series: Journal of Physics: Conf. Series **1064** (2018) 011001 doi:10.1088/1742-6596/1064/1/011001

Preface

2018 the 2nd International Conference on Fluid Mechanics and Industrial Applications (FMIA 2018: http://www.icfmia.org) includes original and peer-reviewed research papers and held on July 12-14, 2018 in Guilin, China. This conference was sponsored by Asian Union of Information Technology, and was attended by academicians, researchers, entrepreneurs, government agencies and policy-makers, postgraduates.

FMIA 2018 is the annual international conference devoted to the discussion of recent developments and applications of parallel computing in the field of fluid mechanics, CFD, fluid power, fluid flow and related disciplines. The field of fluid mechanics is vast and covers numerous and diverse applications. The conference covered a wide range of topics, including basic formulations and their computer modelling as well as the relationship between experimental and analytical results. The meeting provided a forum for discussing new work on fluid mechanics and in particular for promoting the interchange of novel ideas and the presentations of the latest developments in this field. This conference will also provide an ideal environment to develop new collaborations and meet experts on the fundamentals, applications, and products of the mentioned fields.

FMIA 2017 received 386 manuscripts, and 77 submissions have been accepted by our reviewers. By submitting a paper to FMIA 2018, the authors agreed to the review process and understood that papers would undergo a peer-review process. Manuscripts were reviewed by appropriately qualified experts in the field selected by the conference committee, who took detailed comments and-if the submission was accepted-the authors would submit a revised version that took into account this feedback. All papers were reviewed using a double-blind review process: authors declared their names and affiliations in the manuscript for the reviewers to see, but reviewers did not know each other's identities, nor did the authors receive information about who had reviewed their manuscript. The committees of FMIA 2018 invested great efforts in reviewing the papers submitted to the conference and organizing the sessions to enable the participants to gain maximum benefit.

With our warmest regards,

Jun Xiao

Wuhan, China

Content from this work may be used under the terms of the Creative Commons Attribution 3.0 licence. Any further distribution of this work must maintain attribution to the author(s) and the title of the work, journal citation and DOI.
Published under licence by IOP Publishing Ltd

IOP Conf. Series: Journal of Physics: Conf. Series **1064** (2018) 011002 doi:10.1088/1742-6596/1064/1/011002

Committees

Chairmen

Prof. O. Somei, Hokkaido University, Japan
Prof. J.J. Xu, Northeast Petroleum University, China
Dr. J. Xiao, Wuhan University of Technology, China

Members

Prof. P.L. Liu, National Chung-Hsing University, Taiwan
Prof. J.P. Wang, Beijing Institute of Technology, China
Prof. S.D. Tan, Beijing Institute of Technology, China
Prof. J.J. Xu, Northeast Petroleum University, China
Prof. F. Liu, Huazhong University of Science and Technology, China
Prof. C.J. Jia, Huazhong University of Science and Technology, China
Prof. F. Yang, Huazhong University of Science and Technology, China
Prof. X. Long, Zhejiang University, China
Prof. C.H. Li, Zhejiang University, China
Prof. L.L. Deng, Zhejiang University, China
Prof. Y.D. Shi, Zhejiang University, China
Dr. X.L. Li, Peking University, China
Prof. H. Chen, Shanghai University of Engineering Science, China
Associate Prof. S. Zhu, Wuhan University, China
Prof. G.S. Liu, Hunan University, China
Prof. P. Wang, Guangxi college of Education, China
Dr. Yuan Xi, Panjin Industrial Technology Institute of DUT, China
Dr. H.Y. Su, Central South University, China
Dr. T.Q. Liu, Central South University, China
Dr. J.P. Fang, Chongqing University, China
Dr. G.P. Pan, Chongqing University, China
Dr. X.B. Wang, Chongqing University, China
Mr. W. Jiang, Xin Wen Sheng Conference Co., Ltd, China
Mr. C.P. Wang, Xin Wen Sheng Conference Co., Ltd, China
Mr. G.F. Tan, Xin Wen Sheng Conference Co., Ltd, China
Ms. Z.L. Liu, Xin Wen Sheng Conference Co., Ltd, China

Content from this work may be used under the terms of the Creative Commons Attribution 3.0 licence. Any further distribution of this work must maintain attribution to the author(s) and the title of the work, journal citation and DOI.
Published under licence by IOP Publishing Ltd

IOP Conf. Series: Journal of Physics: Conf. Series **1064** (2018) 011003 doi:10.1088/1742-6596/1064/1/011003

Peer review statement

All papers published in this volume of *Journal of Physics: Conference Series* have been peer reviewed through processes administered by the proceedings Editors. Reviews were conducted by expert referees to the professional and scientific standards expected of a proceedings journal published by IOP Publishing.

Content from this work may be used under the terms of the Creative Commons Attribution 3.0 licence. Any further distribution of this work must maintain attribution to the author(s) and the title of the work, journal citation and DOI.
Published under licence by IOP Publishing Ltd

Exergy analysis of water cooling cycle system for scramjet engine

Bangyun Wang[1], Yu Pan[2,*] and Ning Wang[2]

[1]School of National University of Defense Technology, Changsha, China.
[2]School of National University of Defense Technology, Changsha, China.

*Corresponding author e-mail: nudt_wangbangyun@163.com

Abstract. Thermal protection technology is one of the key technologies for scramjet engines. The traditional regenerative cooling technology using hydrocarbon fuel as coolant has some limitations, such as insufficient cooling capacity and easy pyrolysis coking. This paper presents a water cooling cycle system for scramjet engine based on Rankine cycle, and introduces the system composition and working process. On the basis of the thermodynamic analysis, we have made the exergy analysis of each process and obtained the exergy loss and exergy efficiency. And the influence of the system operating parameters on the exergy loss and exergy efficiency has been analysed.

1. Introduction

Studies have shown that when the sramjet engine is flying at high Mach number, the total temperature of the combustion chamber can reach more than 3000K [1], which far exceeds the temperature resistance of the material of combustion chamber wall and will cause the engine deforming or even burning. Therefore, settling the thermal-production question of combustion chamber is a significant technology of the study and development of scramjet.

At present, there are mainly two methods used to realize thermal protection of scramjet engine: passive thermal protection and active thermal protection [2, 3]. The onefold passive thermal protection scheme is very demanding on the capability of materials [4]. It is not suitable for prolonged work or repeated use in the working conditions of the scramjet combustion chamber. Regenerative cooling technology applies the fuel as a coolant, and it has gained researcher's prominence. The United States X-51A machine performed an regeneration cooling experiment using J-P7 hydrocarbon fuel as a coolant, and GDE-2 scramjet engine achieve combustion time 300s under the Mach 5 flight conditions [5].

Moreover, the flowing characteristics of kerosene in the supercritical state are complicated, and hydrocarbon fuel easily cracks and forms coking at a higher temperature, resulting in local heat-transfer deteriorating or even clogging the pipeline [6], which limits the maximum Mach number of the aircraft. Gascion et al. [7] studied the characteristics of pyrolysis and coking of hydrocarbon fuels under supercritical conditions and analyzed the effects of coking on the cooling effect and pressure loss.

However, with the limit of the thermal properties of fuel, there are lots of problems in the utilizations of regenerative cooling [8]. Under the basis of the thermal balance of regenerative cooling, researchers consider adopting heat collection processing or distribution processing. This thermal management system can indirectly improve the fuel heat sink and effectively improve the thermal protection system performance. Bowen et al [9, 10] designed a scramjet engine secondary cooling cycle system, using

Content from this work may be used under the terms of the Creative Commons Attribution 3.0 licence. Any further distribution of this work must maintain attribution to the author(s) and the title of the work, journal citation and DOI.
Published under licence by IOP Publishing Ltd

high-temperature fuel to drive turbine work after decalescence through the engine wall, and the cooling fuel can cool the wall again. Qin et al. [11] analyzed the performance of RCC and introduced some thermodynamic parameters that characterize its performance. The results show that this open recirculation performance is superior to the traditional regenerative cooling technology. Wu Xianyu and Chen Xuefu [12, 13] proposed an expansion cycle scheme of ethylene-assisted start-up to transfer waste heat to other parts which requires energy.

Based on the theory of Rankine cycle, this paper constructs a water cooling cycle for scramjet engine. This system utilizes the heat flow of engine wall and converts it into electrical energy and preheats kerosene to achieve efficient energy utilization. This paper will also analyze the exergy of the cooling circulatory system, obtain the exergy loss and efficiency of each process, and study the effects of different operating parameters.

2. Working principle and thermodynamic analysis of water cooling cycle system

The cooling cycle system consists of four main parts: the engine cooling passage, the turbo expander, the condenser, and the water pump, which is shown in Figure1. The specific working process is as follows. First, the circulating water enters the cooling channel and absorbs the heat flow of the engine wall. After absorbing heat, it vaporizes into the superheated steam. Then, the superheated steam pushes the blade through the expander to do work, and the rotating shaft of the blade is connected to generate electricity. The temperature and pressure of the steam will reduce. After that, the steam enters the condenser and exchanges heat with the cold kerosene. The water vapor condenses into liquid water. The preheated kerosene is injected into the combustion chamber. Finally, the circulating water flows into the engine cooling channel after being pressurized by the water pump to finish the next path.When receiving the paper, we assume that the corresponding authors grant us the copyright to use the paper for the book or journal in question. Should authors use tables or figures from other Publications, they must ask the corresponding publishers to grant them the right to publish this material in their paper.

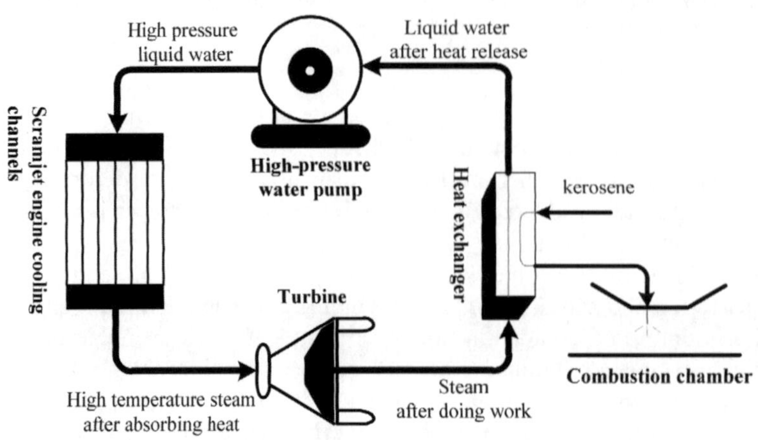

Figure 1. Concept scheme of cooling cycle

During the entire cycle, the circulating water absorbs huge heat from the wall of the engine. Through thermal cycling, heat is transferred from the high-temperature heat source to the low-temperature cold source, and the potential work between the heat source and the cold source is utilized. The turbine expander does work to drive the generator to generate electricity and realize the conversion of thermal energy into electrical energy. The generator can generate electrical power for use in water pumps and other electronic devices. The entire cooling cycle not only achieves cooling of the high-temperature wall, but also preheats kerosene fuel, and realizes waste heat power generation. The research value and significance are enormous.

IOP Conf. Series: Journal of Physics: Conf. Series **1064** (2018) 012001 doi:10.1088/1742-6596/1064/1/012001

The water cooling cycle system is designed based on the Rankine cycle. Figure 2 shows the operation of the cooling cycle system. It is divided into four-step thermodynamic processes: 1-2 is the isobaric endothermic process of water in the cooling channels; 2-3 is the adiabatic expansion process of superheated steam; 3-4 is the isobaric exothermic process of steam; 4-1 is the adiabatic compression process of water. Figure 3 is the T-s diagram of the cycle.

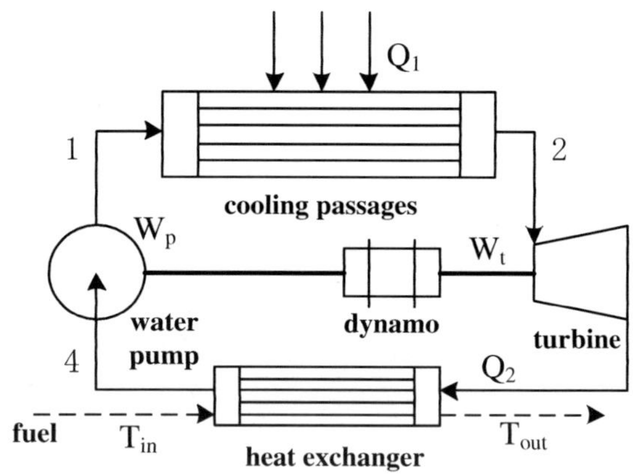

Figure 2. Operation scheme of the water cooling cycle

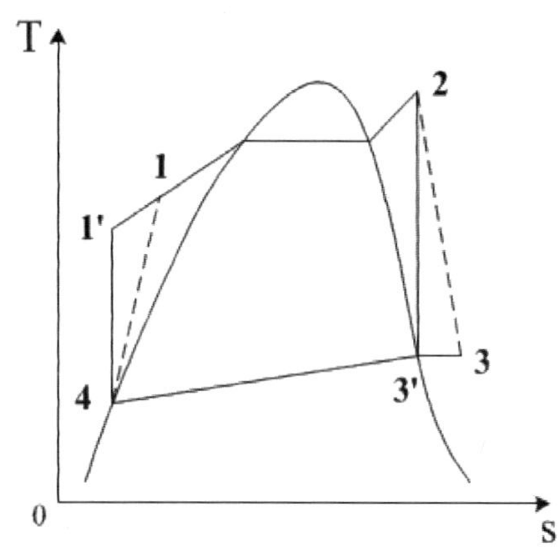

Figure 3. T-s diagram of the cycle

2.1. Endothermic process in cooling channels
The heat absorbed by the circulating water in the cooling channel is

$$Q_1 = \dot{m}_w \cdot (h_2 - h_1) \tag{1}$$

In the formula, Q1 is the heat to be absorbed of the engine; mw is the flow of the circulating water; h1 is the enthalpy of water at the inlet of the cooling channels; h2 is the enthalpy of the steam at the outlet of the cooling channels.

2.2. Expansion in the turbine
In the process of expansion, the reduced energy of the steam is Q_2

$$Q_2 = \dot{m}_w \cdot (h_2 - h_3) \tag{2}$$

Then the power generated by the turbine expansion is calculated as the formula,

$$W_t = \eta_t \cdot Q_2 \tag{3}$$

in the formula, h_3 is the enthalpy of the steam at the outlet of the turboexpander.
The expansion ratio of the circulatory system is defined as follows,

$$\pi = p_2/p_3 \tag{4}$$

2.3. Heat transfer process in the condenser
When the fuel coolant flow rate is$\dot{m}_f$, the fuel coolant absorbs heat in the cryogenic heat exchanger can be expressed as the formula,

$$Q_f = \dot{m}_f \cdot C_{pf} \cdot (T_{out} - T_{in}) \tag{5}$$

During the cooling system, the heat absorbed by the fuel coolant isQ_3,

$$Q_3 = \frac{Q_f}{\eta_e} \tag{6}$$

And it can be expressed as

$$Q_3 = \dot{m}_w \cdot (h_3 - h_4) \tag{7}$$

In the formula, η_e is the heat exchange efficiency of the low-temperature heat exchanger.

2.4. Compression process in the pump
In the compression process, the energy of cycling water increases Q_4

$$Q_4 = \dot{m}_w \cdot (h_1 - h_4) \tag{8}$$

The energy consumed in the compression process is calculated as follows,

$$W_p = \frac{Q_4}{\eta_p} \tag{9}$$

According to the Rankine cycle, the gas expansion causes the turbine to rotate to do work, and at the same time, the turbine feeds the water pump (compressor) through the coaxial rotation. The compression cycle fluid consumes a part of the work and the residual work outputs, and the useful power output of the cooling cycle system is calculated as

$$W_{sys} = W_t - W_p = \dot{m}_w \cdot [(h_2 - h_3) - (h_1 - h_4)] \tag{10}$$

Therefore, the energy conversion efficiency of the entire cooling cycle system is as follows,

IOP Conf. Series: Journal of Physics: Conf. Series **1064** (2018) 012001 doi:10.1088/1742-6596/1064/1/012001

$$\eta_{sys} = \frac{W_{sys}}{Q_1} = \frac{(h_2 - h_3) - (h_1 - h_4)}{(h_2 - h_1)} \quad (11)$$

3. Exergy analysis of water cooling cycle system

Due to the limitations of the scramjet engine's own environment and conditions, the quality of thermal energy in the process of energy transfer and conversion of the cooling cycle system is continuously reduced. In order to improve the output power and conversion efficiency of the system scheme, the exergy transfer model and the exergy loss model of the cooling circulatory system were established from the perspectives of heat transfer and thermodynamics, and the change of exergy of the different positions or components of the heat transfer and conversion process of the system were analyzed.

For a steady-state equilibrium thermodynamic system, ignoring the effects of kinetic energy and potential energy, the mass conservation and exergy balance equations can be expressed in each thermal process as

$$\sum m_{in} = \sum m_{out} \quad (12)$$

$$Q + W = \sum m_{out}\, h_{out} - \sum m_{in}\, h_{in} \quad (13)$$

$$E_{x,Q} + W = \sum E_{out} - \sum E_{in} + I \quad (14)$$

The “in” and “out” indicate the import and export of the selected thermal system. Q is the heat absorbed by the thermal system from the external environment. W is the work done by the external environment to the system. m is the mass flow rate of water. E is the exergy value, equal to me, where $e = h - h_0 - T_0(s - s_0)$. In the formula, h_0, s_0 are the specific enthalpy and specific entropy of the circulating water at ambient temperature. T_0 Is the ambient temperature. I Is exergy loss. $E_{x,Q}$ is the heat enthalpy, which is input by a heat source.

The following describes the exergy analysis method for the four processes of the cooling cycle, and gives the calculation formulas for the exergy loss and exergy efficiency.

3.1. Heat transfer process in cooling channels

1-2 thermal process is the absorption of heat from the engine wall by the working fluid, which ideally is an isobaric heat absorption process. The amount of exergy input into the system is the amount of heat exergy input to the wall of the engineE_H, and the exergy change E_{1w} of the circulatory water is the difference amount of exergy between point 2 and point 1. The exergy balance equation during heat transfer can be expressed as

$$E_H - E_{1w} - I_{eva} = 0 \quad (15)$$

The exergy efficiency $\eta_{e,eva}$ is as follows,

$$\eta_{e,eva} = \frac{E_{1w}}{E_H} \quad (16)$$

3.2. Expansion process in the turboexpander

2-3 is the expansion work process of the expander. The amount of exergy E_{2w} input into the system is the exergy change of the circulating water. The exergy balance equation for the process is as follows,

$$E_{2w} - W_t - I_t = 0 \tag{17}$$

The exergy efficiency $\eta_{e,t}$ can be calculated as,

$$\eta_{e,t} = \frac{W_t}{E_{2w}} \tag{18}$$

3.3. Heat exchange process in the condenser

3-4 is the isobaric heat release process. In the heat process, the input amount of exergy equals to the heat exergy E_{3w} which is transfered into the condenser carried by the circulating water. The exergy change of the cold kerosene is E_L. The exergy balance equation for the process is as follows,

$$E_{3w} - E_L - I_{con} = 0 \tag{19}$$

The exergy efficiency $\eta_{e,con}$ of this process can be calculated as,

$$\eta_{e,con} = \frac{E_L}{E_{3w}} \tag{20}$$

3.4. Compression process in the pump

The pump pressurizes the working medium to increase the pressure and consumes energy. The heat balance equation and the exergy balance equation for the 4-1 thermal process are as follows,

$$W_p = m_w(h_1 - h_4) \tag{21}$$

$$E_{4w} + I_p - W_p = 0 \tag{22}$$

The exergy efficiency $\eta_{e,p}$ is calculated as,

$$\eta_{e,p} = \frac{E_{4w}}{W_p} \tag{23}$$

3.5. Total exergy loss and exergy efficiency of the cooling cycle system

The total exergy loss of the cooling cycle is the sum of loss in each process. The total exergy loss of the cooling cycle is as follows,

$$I_{sys} = I_{eva} + I_t + I_{con} + I_p \tag{24}$$

The total effective energy E_{sys} of the cooling cycle system is as follows,

$$E_{sys} = E_H + W_p - I_{sys} \tag{25}$$

The total exergy efficiency $\eta_{e,sys}$ of the cooling cycle is as follows,

$$\eta_{e,sys} = \frac{E_{sys}}{E_H + W_p} = 1 - \frac{I_{sys}}{E_H + W_p} \tag{26}$$

In the following, the calculation of the system is based on the actual conditions of a known scramjet engine combustion chamber. When the scramjet engine is working, the wall temperature can reach 2000K, which will seriously damage the engine. After regeneration cooling, the wall temperature can be reduced to 1000K. The heat to be absorbed in the combustion chamber is$Q_1 = 609.7MW$. The temperature and pressure of the cooling water at the inlet of the cooling channels are$T_1 = 25°\text{C}, p_1 = 6MPa$, and the temperature and pressure of the water vapor at the outlet of the cooling channels are$T_2 = 700°\text{C}, p_2 = 6MPa$. The value of water in a certain state is calculated using IAPWS-IF97 [14] software, and we can get that $h_1 = 110.38kJ/kg, h_2 = 3894.47kJ/kg, s_2 = s_{3s} = 7.425kJ/kg/K$. So the cooling water flow is as followed,

$$\dot{m}_w = Q_1 \div (h_2 - h_1) = 0.161\text{kg/s}$$

We suppose that the adiabatic efficiency is$\eta_t = 0.85$, the expansion ratio is $\pi = 3$, and the specific heat ratio is $\text{r} = 1.3$, then the pressure of the water vapor at the outlet of the expander is $p_3 = 2MPa$. According to the formulas which are as followed,

$$h_{3s} = \text{REFPROP}(p_3, s_{3s}) \tag{27}$$

$$h_3 = h_2 - (h_2 - h_{3s})\eta_t \tag{28}$$

$$T_3 = \text{REFPROP}(p_3, h_3) \tag{29}$$

We can get the temperature of steam at the outlet of the turbine is$T_3 = 790.18K$. The output power of the turbo expander is (ignoring the adiabatic efficiency of the expander)

$$W_t \approx Q_2 = \dot{m}_w(h_2 - h_3) = 62.38\text{kW}$$

Similarly, the pressure of water at the inlet of the pump is$p_4 = 2MPa$, we set $T_4 = 297.15K$, then the power consumption of the pump is as followed,

$$W_p = \dot{m}_w(h_1 - h_4) = 1.29\text{kW}$$

Because of$W_p < W_t$, the mechanical work generated by the turbine can meet the power consumption of the feed pump.

In the condenser, the value of heat release of water vapor Q_3 can be calculated as,

$$Q_3 = \dot{m}_w(h_3 - h_4) = 548.61\text{kW}$$

Considering the heat transfer efficiency is $\eta_{\text{con}} = 0.8$, the average constant-pressure specific heat of kerosene is $\text{C}_{pf} = 2.6\text{kJ/kg/K}$[15], the initial temperature of cold kerosene is $\text{T}_{\text{in}} = 15°\text{C}$, and the vaporization temperature of kerosene is $\text{T}_{\text{out}} = 450°\text{C}$, so the flow rate $\dot{m}_f$ of kerosene is as followed,

$$\dot{m}_f = \frac{Q_3\eta_{\text{con}}}{\text{C}_{pf}(\text{T}_{\text{out}} - \text{T}_{\text{in}})} = 0.388\text{kg/s}$$

Under the given initial conditions, with the designed cooling cycle, the matching of energy and power can be achieved. The heat absorbed from the wall of the combustion chamber can preheat the kerosene to vaporization temperature, and the mechanical work generated by the turbine can satisfy the work of the water supply pump. The whole cycle conditions are shown in Table 1.

After knowing the process parameters of the circulatory system, the physical properties of the working fluid at each state point were found and calculated by REFPROP software. The results are shown in Table 2.

Table 1. Parameters of the entire cycle

T_1[K]	p_1[MPa]	T_2[K]	p_2[MPa]	T_3[K]	p_3[MPa]
298.15	6	973.15	6	790.18	2
T_{in}[K]	T_{out}[K]	W_t[kW]	W_p[kW]	$\dot{m}_w$[kg/s]	$\dot{m}_f$[kg/s]
288.15	723.15	62.38	1.29	0.161	0.388

Table 2. Thermodynamic parameters at each state point of circulatory system

h_1[kJ/kg]	h_2[kJ/kg]	h_3[kJ/kg]	h_4[kJ/kg]	$\bar{T}_H[K]$	$Q_3[kW]$
110.38	3894.47	3507.13	102.39	1442.7	548.61
s_1[kJ/kg/K]	s_2[kJ/kg/K]	s_3[kJ/kg/K]	s_4[kJ/kg/K]	$\bar{T}_L[K]$	$T_0[K]$
0.366	7.425	7.513	0.353	472.76	298.15

Based on the formulas for the exergy loss and exergy efficiency of each process, the exergy loss and efficiency of the components of the cooling cycle are calculated, as shown in Table 3. During the exergy transfer process, the exergy efficiency of each component was different. The exergy efficiency in the cooling channel, turbo expander, condenser, and water pump was 55.94%, 93.64%, 99.02%, and 51.64% respectively. It can be seen that the exergy efficiency of the high temperature heat transfer process and the pump compression process is relatively low.

Table 3. Exergy loss and exergy efficiency of each process

Process	Exergy loss [kW]	Exergy efficiency [%]	The proportion of process exergy loss to the total exergy loss [%]
Heat transfer process in cooling channels	213.12	55.94	96.88
Expansion work process in the turboexpander	4.24	93.64	1.93
Heat exchange process in the condenser	2.01	99.02	0.91
Compression process in the pump	0.622	51.64	0.28
The total cycle system	219.98	53.35	

4. Effect of circulatory system operating parameters on exergy loss and exergy efficiency

4.1. The influence of the maximum temperature of the working fluid during the cycle

Under the condition of satisfying the heat change requirements, the maximum temperature of the working fluid during the cycle is changed from 933.15K to 973.15K. The changes of exergy loss in the different thermal processes are shown in Figure 4. As the maximum temperature increases, the exergy loss in the cooling channel, the turbine expansion process, and the water pump compression process all decrease. However, the exergy loss in the condenser increases because the temperature of the fluid

IOP Conf. Series: Journal of Physics: Conf. Series **1064** (2018) 012001 doi:10.1088/1742-6596/1064/1/012001

entering the cryogenic heat exchanger increases as the maximum temperature increases, which causes the increase of the heat exchange temperature difference and the exergy loss.

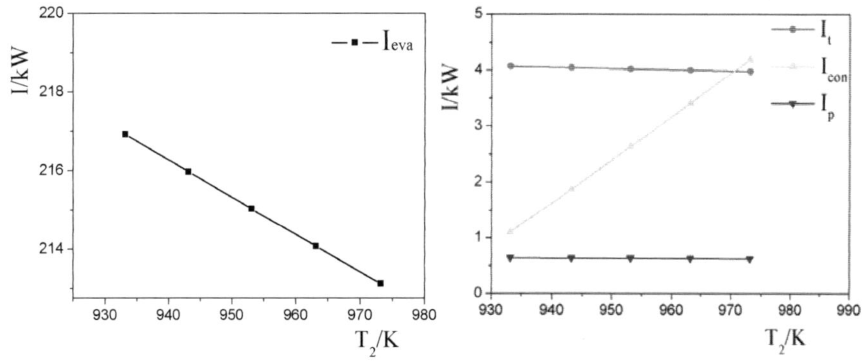

Figure 4. Variation of exergy loss of different thermal processes under different maximum temperature during the cycle

The exergy efficiency of each thermal process is shown in Figure 5. The change of exergy efficiency of the turbo expander and the pump is very little and remains basically constant with the increase of the maximum operating temperature. The exergy efficiency in the heat transfer process of the cooling channels increases with the increase of the maximum operating temperature. However, the exergy efficiency of the condenser decreases as the maximum operating temperature increases.

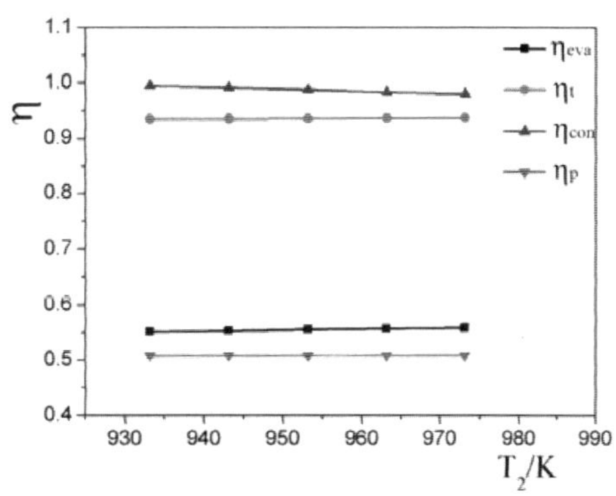

Figure 5. Variation of exergy efficiency of different thermal processes under different maximum temperature during the cycle

4.2. Effect of expansion ratio

The different expansion ratios of the cooling circulatory system change the parameters of each state point, and then affect the exergy changes of each thermal process. Under the conditions of satisfying the heat exchange requirements, assuming that the expansion ratio of the cycle varies between 1.2 and 3.2, and the change of exergy loss in different process is shown in Figure 6. With the increase of the expansion ratio, the exergy loss in the heat exchange process of the cooling channels is almost unchanged, the exergy loss of the turbo expander is slightly increased, and the exergy loss in the condenser and the water pump is reduced. At the same time, the exergy loss in the condenser is reduced to a large extent, and the exergy loss in the pump is reduced to a small extent. In general, as the expansion ratio increases, the system exergy loss decreases. The exergy efficiency of each thermal process changes with the expansion ratio as shown in Figure 7. As the expansion ratio increases, the exergy efficiency changes in each process are as follows: the heat transfer process in the cooling channel is almost unchanged, and the expansion in the expander decreases slightly. There is a significant increase in both the condenser and the pump.

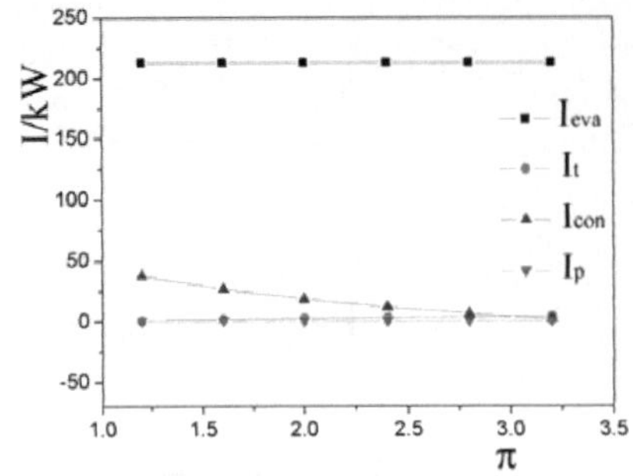

Figure 6. Variation of exergy loss of different thermal processes under different expansion radio during the cycle

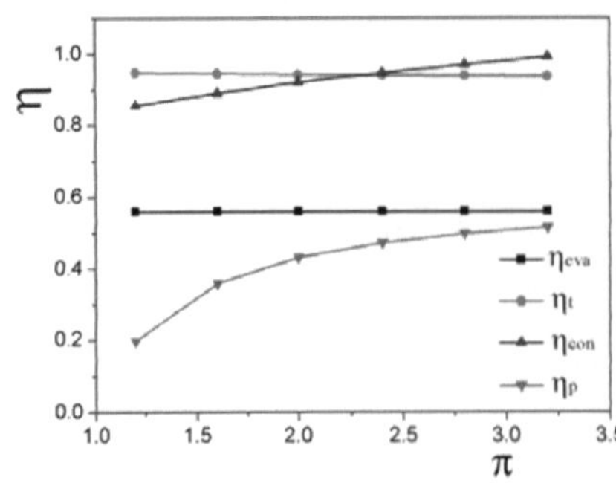

Figure 7. Variation of exergy efficiency of different thermal processes under different expansion radio during the cycle

4.3. The effect of the flow rate of cold kerosene

Cold kerosene absorbs the heat of water vapor in the condenser. The change of its flow rate affects the temperature of the cold source, and then affects the exergy change of the process. Under the condition of meeting the requirements of heat exchange, the flow rate of kerosene was changed from 0.5kg/s to 1.0kg/s. The calculations of the changes of exergy losse and exergy efficiency of the processes were shown in Figure 8. With the increase of the flow rate of kerosene, the exergy loss in the condenser increases significantly and the exergy efficiency decreases significantly, and other processes do not change significantly.

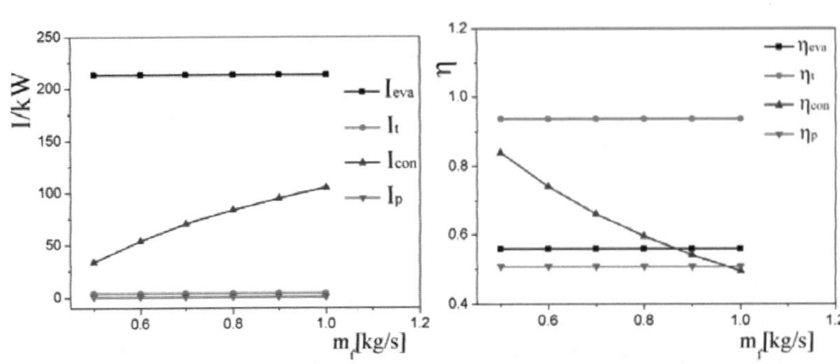

Figure 8. Variation of exergy loss and exergy efficiency of different thermal processes under different flow rate of cold kerosene during the cycle

IOP Conf. Series: Journal of Physics: Conf. Series **1064** (2018) 012001 doi:10.1088/1742-6596/1064/1/012001

4.4. Influence of ambient temperature

Ambient temperature is the reference point temperature in the exergy analysis process, and its change affects the calculation of the input exergy and the changes of exergy of the system, and then affects the exergy loss and exergy efficiency of each process. Under the condition of satisfying the heat exchange requirements, the ambient temperature is changed from 398.15K to 348.15K. The calculation of the changes of exergy loss and the exergy efficiency are shown in Figure 9 and Figure 10, respectively. With the increase of the ambient temperature, the exergy loss of all processes increase, and the exergy efficiency of all processes decreases.

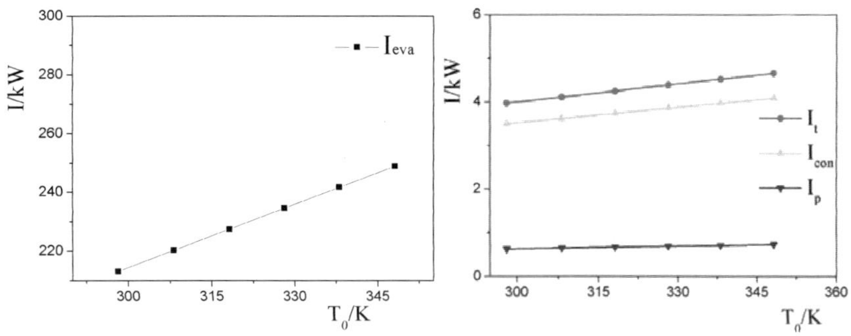

Figure 9. Variation of exergy loss of different thermal processes under different ambient temperature during the cycle

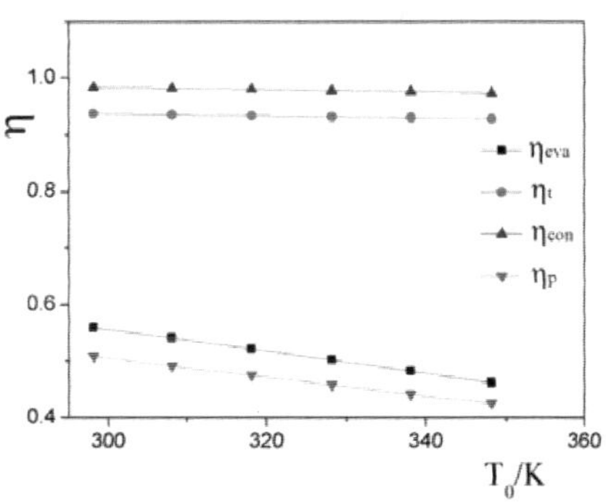

Figure 10. Variation of exergy efficiency of different thermal processes under different ambient temperature during the cycle

5. Conclusion

This paper describes the composition and working principle of the water cooling cycle system of the scramjet engine. Based on the thermal analysis of the system, the exergy in various processes of the cooling cycle system are analyzed, and the exergy loss and exergy efficiency of each process are obtained. Furthermore, the influence of the cyclic operation parameters on the exergy loss and exergy efficiency of each process was analyzed, and the following conclusions were obtained.

- ➢ The biggest exergy loss is at the endothermic process in the cooling channels, which is caused by the huge heat exchange temperature difference;
- ➢ With the increase of the maximum temperature of the working fluid in the cycle, the exergy loss of the endothermic process in the cooling passages, the turboexpansion process, and the compression process is reduced, and the exergy efficiency is increased; the exergy loss in the condenser is increased, and the exergy efficiency is reduced.
- ➢ With the increase of the cyclic expansion ratio, the exergy loss of the turboexpansion process increases and the exergy efficiency decreases; the exergy loss of heat transfer process in the condenser

and the compression process in the pump decreases, and the exergy efficiency increases; The exergy loss and exergy efficiency in the cooling channels is basically unchanged;

➢ With the increase of the flow rate of cold kerosene, only exergy loss of the heat exchange process in the condenser increases, and the exergy efficiency decreases; the other processes all remain basically unchanged;

➢ With the increase of the ambient temperature, the exergy losse of the four process all increases, and the exergy efficiency all decreases.

Under the guidance of the above conclusions, appropriate expansion ratio, lower ambient temperature, adjustment of the cooling channel configuration, etc., can be used to reduce the system damage and improve the cycle thermal efficiency.

References

[1] Kelly, H. N., and Blosser, M. L., Active Cooling from the Sixties to NASP, NASA TM 109079, July 1994.
[2] S Stoleriu, E Volceanov, A Volceanov. Ceramic composite based on silicon carbide and vitreous silicon dioxide obtained under unconventional sintering treatment[J]. Romanian journal of materials, 2013, 43(2):174-184.
[3] Stephen J Scotti, Carl J Marthn. Active cooling design for scramjet engines using optimization methods. AIAA 1988.
[4] Zhi-Yong L I, Li-Han L I, Zheng R H, et al. Effects of delamination of needled preform on thermal insulation of C/C-SiC chamber[J]. Journal of Propulsion Technology, 2011, 32(5):698-705.
[5] Hank J, Murphy J, Mutzman R. The X-51A Scramjet Engine Flight Demonstration Program[C]// Aiaa International Space Planes and Hypersonic Systems and Technologies Conference. 2008.
[6] JIN Xuan, SHEN Chibing, WU Xianyu, et al. Progress of regenerative cooling technology for scramjet[J]. Journal of rocket propulsion, 2016, 42(5):66-73. (in Chinese).
[7] Nicolas Gascoin, Philippe Gillard, et al. Characterisation of coking activity during supercritical hydrocarbon pyrolysis[J]. Fuel Processing Technology, 2008, 89: 1416-1428.
[8] Yu Gang, Fan Xuejun. Supersonic combustion and hypersonic propulsion[J]. Advances in Mechanics, 2013, 43(5):449-471. (in Chinese).
[9] Qin J, Bao W, Zhang S, et al. Comparison During a Scramjet Regenerative Cooling and Recooling Cycle[J]. Journal of Thermophysics & Heat Transfer, 2015, 26(4):612-618.
[10] Bao W, Li X, Qin J, et al. Efficient utilization of heat sink of hydrocarbon fuel for regeneratively cooled scramjet[J]. Applied Thermal Engineering, 2012, 33(1):208-218.
[11] Qin J, Bao W, Zhou W, et al. Performance cycle analysis of an open cooling cycle for a scramjet[J]. Proceedings of the Institution of Mechanical Engineers Part G Journal of Aerospace Engineering, 2009, 223(6):599-607.
[12] Wu X, Yang J, Zhang H, et al. System design and analysis of hydrocarbon scramjet with regeneration cooling and expansion cycle[J]. Journal of thermal science, 2015, 24(4):350-355.
[13] Chen Xuefu. Investigation on Design and Performance of Scramjet Feed System Scheme[D]. National University of Defense Technology, 2013. (in Chinese).
[14] Wang Pei-hong, Jia Jun-ying, Cheng Mao-hua. General Calculating Models of Water and Steam Properties (IAPWS-IF97), Power Engineering, 21 (6), 2001, pp. 1564-1567. (in Chinese).
[15] Deng H W, Zhu K, Xu G Q, et al. Isobaric Specific Heat Capacity Measurement for Kerosene RP-3 in the Near-Critical and Supercritical Regions, Journal of Chemical & Engineering Data, 57 (2), 2011, pp. 263–268.

Critical relative humidity of evaporative cooling in CCPP based on thermodynamic and economic analyses

Hao Lyu[1], Jianhong Chen[1, *] and Huanyan Huang[2]

[1]College of Energy Engineering, Zhejiang University, Hangzhou, China.
[2]China Energy Engineering Group Zhejiang Electric Power Design Institute CO., LTD. Hangzhou, China.

*Corresponding author e-mail: 21627035@zju.edu.cn

Abstract. A calculation and analysis to determine the feasibility of inlet air evaporative cooling systems for combined cycle power plant (CCPP) is carried out. According to the psychrometric chart, the effects of ambient temperature and relative humidity (RH) on the temperature drop of the inlet air across an evaporative cooling system are analyzed. The power differences between CCPP with and without an inlet air evaporative cooling system in different environmental conditions are investigated. Based on the thermodynamic and economic analyses, a new concept, the critical RH, is proposed to evaluate the feasibility and economics of CCPP with inlet air evaporative cooling system considering different ambient conditions. If ambient RH is lower than the critical value, an inlet air evaporative cooling system is suitable and economical for regions considered. Sensitivity analysis of the factors affecting the critical RH is also carried out. The critical RH varies with climatic condition, electricity price, fuel price and expected payback period (PBT).

1. Introduction

Gas turbine and CCPP are widely applied in power generation with the advantages of high power-to-weight ratio, short duration of construction, quick-start, strong peak load regulation performance, low greenhouse gas emissions and so on. Despite these advantages, a certain disadvantage of gas turbine is that the power output and efficiency both decrease as ambient temperature increases [1-4]. According to statistics, as the compressor inlet air temperature rises 1 °C, the power output of gas turbine declines about 0.5-0.9% [5, 6]. So the power output of gas turbine and CCPP falls considerably in summer. However, the electricity power demand on these days rises dramatically because of the operation of air conditioners and cooling systems. Strictly speaking, the performance profile of gas turbine and the electric power demand profile go in opposite directions.

Performance losses of gas turbine and CCPP because of atmospheric conditions can be prevented by cooling inlet air before entering the compressor. Since this reduces the power consumption of compressor and increases the air density. Various inlet air cooling technologies are used to enhance the peaking capacity of gas turbines and CCPP, such as evaporative cooling, compression chiller, absorption chiller and thermal energy storage. The power output and efficiency of gas turbine and CCPP with and without an inlet air cooling system are analyzed [7-9]. Mohapatra [10] compared compression cooling with evaporative cooling integrated to a gas turbine power plant. Khaliq and

Content from this work may be used under the terms of the Creative Commons Attribution 3.0 licence. Any further distribution of this work must maintain attribution to the author(s) and the title of the work, journal citation and DOI.
Published under licence by IOP Publishing Ltd

Choudhary [11] provided an exergy analysis of gas turbine with absorption chiller. The analysis indicated that the exergy losses in different components of gas turbine varied with the compressor inlet air temperature and ambient RH.

Compared with other inlet air cooling technology, evaporative cooling is the most economical system in hot and dry areas because of the ease of installation and its low initial investment cost [12, 13]. De Lucia et al. [14] proposed that evaporative cooling could improve the power output by 2-4% according to atmospheric condition. Chaker et al. [15] concluded the potential cooling hours of direct evaporative cooling in 122 different locations.

Most of the previous studies concerned about the effects of evaporative cooling system on performance of gas turbine and CCPP. However, in which environmental conditions the evaporative cooling system are suitable and economical haven't been carried out. Therefore, a new concept, critical RH, is proposed to evaluate the feasibility and economics of evaporative cooling system for CCPP according to real atmospheric conditions. For a given period, the economic benefit of the evaporative cooling system may be exactly equal to total investment cost in some condition. The RH of this condition is the critical value. If ambient RH is lower than critical value, the evaporative cooling system for CCPP is suitable and economical for regions considered. Otherwise, the evaporative cooling system is not economical or inlet air needs to be dehumidified before entering the evaporative cooling system. In addition, the sensitivity of critical RH for ambient temperature, electricity price and fuel price is carried out. With the concept and criteria proposed in our study, gas turbine operators can easily utilize it to determine the feasibility of the evaporative cooling system for their own gas turbine or CCPP.

2. Modelling and analysis

2.1. Description of the system investigated

Figure 1 shows the schematic diagram of CCPP with evaporative cooling system. Ambient air is cooled in the evaporative cooling system because the heat is absorbed by the latent heat of vaporization involved in phase changes. Then the cooled air is sent to compressor for compression. The CCPP is composed of a GE PG9351 (FA) gas turbine, a triple pressure reheat heat recover steam generator (HRSG) and HP, IP and LP steam turbine. The main parameters and performance indexes of this CCPP without inlet air cooling at ISO ambient base load condition are shown in Table 1.

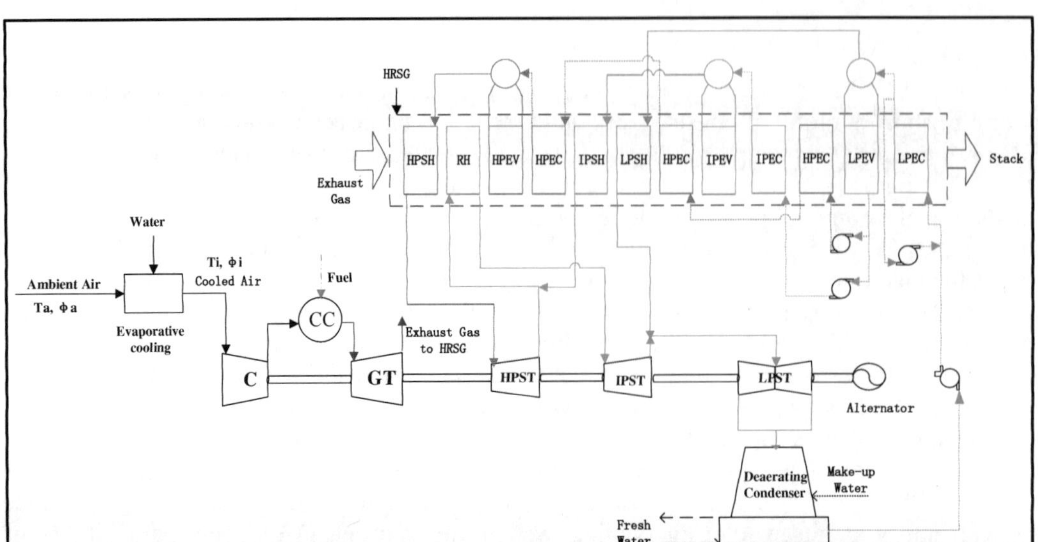

Figure 1. Schematic diagram of CCPP with evaporative cooling system

IOP Conf. Series: Journal of Physics: Conf. Series **1064** (2018) 012002 doi:10.1088/1742-6596/1064/1/012002

Table 1. Parameters of CCPP (ISO ambient base load)

Parameter (unit)	Value	Parameter (unit)	Value
Ambient air temperature(°C)	15	Natural gas input (MW)	686.53
Ambient RH (%)	60	GT power output (MW)	251.43
Ambient air pressure (kPa,a)	101.35	ST power output (MW)	143.12
Natural gas flow (kg/s)	14.10	Gross power output (MW)	394.55
Air flow (kg/s)	639.81	Gross thermal efficiency (%)	57.48
Air/fuel mass ratio	45.38	Gross Heat rate (kJ/kWh)	6262.80
Natural gas-LHV (MJ/kg)	48.69		

2.2. Thermodynamic analysis

2.2.1. The evaporative cooling system. Figure 2 illustrates the inlet air cooling process on the psychrometric chart. The cooling process follows a constant enthalpy line. In the cooling process, demineralized water is injected into the evaporative cooling system to cool inlet air from ambient condition (Ta, φa) to a lower temperature. Part of the injected water evaporates, and then it cools the inlet air due to the absorption of latent heat. In ideal condition, the compressor inlet air temperature (Ti,c), can reach wet bulb temperature as RH rises up to 100% (point 's'). However, the outlet temperature is higher than wet bulb temperature in reality. The bold line (a-i) and (a-s) represent real cooling process and ideal theoretical cooling process, respectively.

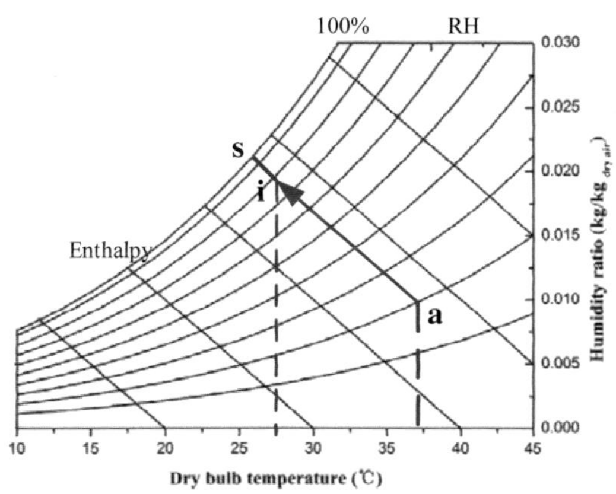

Figure 2. The inlet air cooling process on the psychrometric diagram

The temperature drop (ΔT), which is the temperature difference between the inlet and outlet of the evaporative cooling system and is given by

$$\Delta T = T_{a,dry} - \mathrm{T}_{i,dry} \quad (1)$$

The ambient air consists of dry air and water vapor. The humidity ratio (d) is identified as the mass of water vapor contained in per unit dry air and can be obtained from

$$\mathrm{d} = 0.622 \frac{p_v}{p_a - p_v} \quad (2)$$

$$p_v = \varphi p_s \quad (3)$$

IOP Conf. Series: Journal of Physics: Conf. Series **1064** (2018) 012002 doi:10.1088/1742-6596/1064/1/012002

Where p_a is the atmospheric pressure, p_v is the partial pressure of water vapor, p_s is the saturation pressure of water vapor, φ is the RH of air.

2.2.2. CCPP performance. The performance parameters, including power output and efficiency of gas turbine, steam turbine and CCPP are calculated with some governing equations given as follows:

The compressor work is given by

$$W_c = \dot{m}_{a,i} \cdot c_{p,c,i} \cdot \left(T_{e,c} - T_{i,c}\right) \quad (4)$$

Where $\dot{m}_{a,i}$ is the mass flow rate of inlet air, cp,c,i is the specific heat at constant pressure of air across compressor, Te,c and Ti,c are the air temperatures of the exit and inlet of compressor, respectively.

The power output and efficiency of gas turbine are given by

$$W_{gt} = \dot{m}_{g,i}\left(h_{g,i} - h_{g,e}\right) - W_c \quad (5)$$

$$\eta_{gt} = \frac{W_{gt}}{\dot{m}_f \cdot LHV} \quad (6)$$

Where $\dot{m}_{g,i}$ is the amount of burnt gas entering to the turbine, $\dot{m}_f$ is the amount of natural gas, LHV is the lower heating value of fuel, hg,i and hg,e are the enthalpies of burnt gas entering and exiting to the turbine, respectively.

As shown in Figure 1, steam turbine is composed of HP, IP and LP stages. The steam from HRSG is firstly brought to HP steam turbine for expansion to do work and generate electricity. Then the steam returns back to the HRSG for reheating, and goes to IP steam turbine. Finally, the steam from LPSH and IP steam turbine is brought to LP steam turbine for further expansion. The power output and efficiency of steam turbine are expressed as

$$W_{st} = \sum^{stages} \dot{m}_{s,i}\left(h_{s,i} - h_{s,e}\right) \quad (7)$$

$$\eta_{st} = \frac{W_{st}}{Q_0} \quad (8)$$

Where Q0 is the heat energy entering into HRSG, $\dot{m}_{s,i}$ is the amount of steam entering to each stage of the steam turbine, hs,i and hs,e are the enthalpies of steam entering and exiting to each stage of the steam turbine. Therefore, the power output and efficiency of CCPP are given by

$$W_{plant} = W_{gt} + W_{st} \quad (9)$$

$$\eta_{plant} = \frac{W_{plant} \cdot \eta_{alt}}{\dot{m}_f \cdot LHV} \quad (10)$$

Where η_{alt} is the efficiency of alternator.

2.3. Economic analysis

2.3.1. Electricity production enhancement. The evaporative cooling system causes an increase of the power output of CCPP and, consequently, electricity generation is improved. With the installation of the evaporative cooling system, the power output of CCPP will change, including: 1) the power output of gas turbine increase since compressor inlet air temperature decrease (ΔWgt); 2) the power output of gas turbine and steam turbine increase since air density rises as Ti,c declines (ΔWd); 3) the inlet air pressure drop rises as a result of the installation of the inlet air evaporative cooling system, which leads to a reduction of the power output of CCPP (ΔWp); 4) the energy consumption of auxiliary equipment of the evaporative cooling system increase (Wa). Therefore, the increase of the net power output of CCPP is:

$$\Delta W = \Delta W_{gt} + \Delta W_{\mathrm{d}} - \Delta W_{p} - W_{\mathrm{a}} \tag{11}$$

The additional income of CCPP with the evaporative cooling system is

$$\mathrm{I}_{add} = \Delta \mathrm{W} \bullet \eta_{alt} \bullet p_{e} \tag{12}$$

Where p_e is the price of electricity.

2.3.2. Extra fuel consumption. With the application of the evaporative cooling system, the gross power output increases and, as a result, the fuel consumption increases, so fuel consumption increases. The extra fuel consumptions of CCPP with the inlet air evaporative cooling system (ΔF, kg/s) are given by

$$\Delta F = (W_e \bullet \beta_e - W_0 \bullet \beta_0) / (3.6 \bullet LHV) \tag{13}$$

Where β is gross heat rate (kJ/kWh); the subscript e and 0 are the values of CCPP with and without the inlet air evaporative cooling system, respectively.

2.3.3. Initial investment cost. The initial investment cost of evaporative cooling system mainly depends on the maximum capacity of cooling load, and is given by

$$I = k_i \bullet Q_{CL,\max} \tag{14}$$

Where ki is the initial investment cost for per unit cooling load, $Q_{CL,max}$ is the maximum capacity of cooling load.

2.3.4. Critical RH. The annual net profit by installing the evaporative cooling system to CCPP consists of the following three parts: the additional income from electricity production enhancement, the increase of fuel consumption cost and annual operation and maintenance fees of evaporative cooling system.

$$S = \left(\Delta W \bullet \eta_{alt} \bullet p_e - \Delta F \bullet p_f\right) \bullet H - M \tag{15}$$

Where p_e is the price of electricity, pf is the price of natural gas, H is operating hours, M is annual maintenance and operation fees of the evaporative cooling system. So the payback period (PBT) is

$$PBT = \frac{I}{S} \tag{16}$$

To gain a profit (S>0), the additional income should be greater than the sum of fuel consumption cost and operation and maintenance cost.

The critical RH is relevant to local environmental conditions, design parameters of CCPP, fuel price, electricity price and expected PBT. The calculation flow of the critical RH is shown in figure 3.

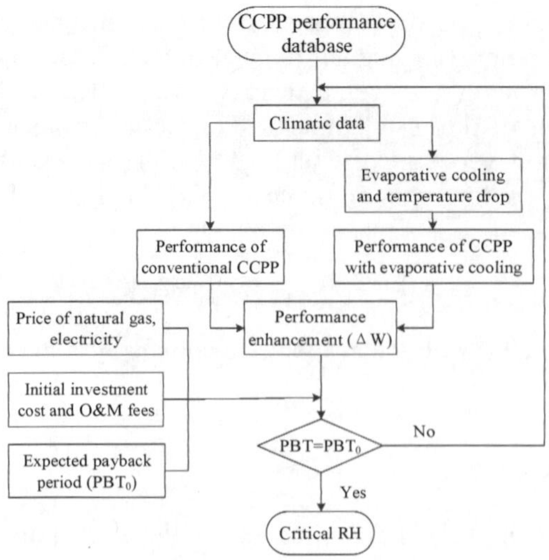

Figure 3. Calculation flow chart of the critical RH

3. Results and discussion

In this study, the temperature drop of inlet air in different environmental conditions is obtained. The power outputs of CCPP with and without the evaporative cooling system are also compared. The critical RH of the evaporative cooling system for CCPP is investigated as well.

3.1. The temperature drop with the evaporative cooling system

Figure 4 illustrates the effect of environmental conditions on temperature drop (ΔT) of air at the inlet and outlet of the evaporative cooling system. It can be observed that ΔT rises continuously as ambient temperature rises. The higher of ambient temperature and the lower of RH, the difference between dry bulb temperature and wet bulb temperature is larger, which means the greater ΔT can be achieved by evaporation cooling.

For a given ambient RH, ΔT increases as ambient temperature rises. However, the increasing rate of ΔT goes down as ambient RH rises. ΔT even changes slightly as ambient RH reaches a turning point. For example, when RH is 60%, ΔT increases slowly as ambient temperature rises. It is obviously that at high RH situation, the effectiveness of the evaporative cooling system is weak and ΔT is not as large as that in arid conditions. This phenomenon indicates that the evaporative cooling system is not suitable in humid regions. For example, when ambient temperature is 35 °C and RH is 60%, ΔT is merely 5.95 °C, however, as the RH decreases to 10%, ΔT dramatically goes up to 18.58 °C, which is double larger than the former.

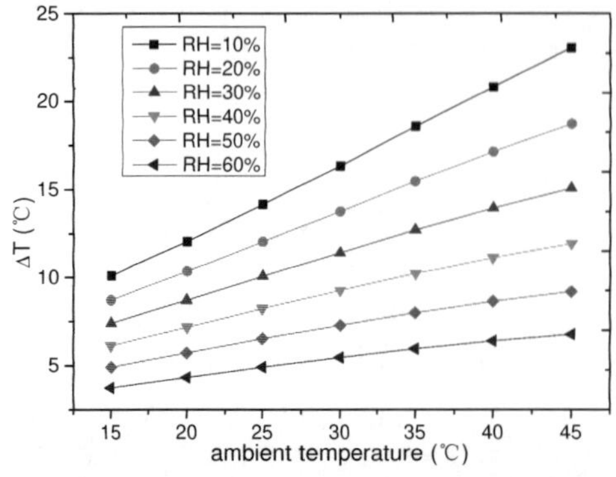

Figure 4. Temperature drop (ΔT) at various ambient conditions

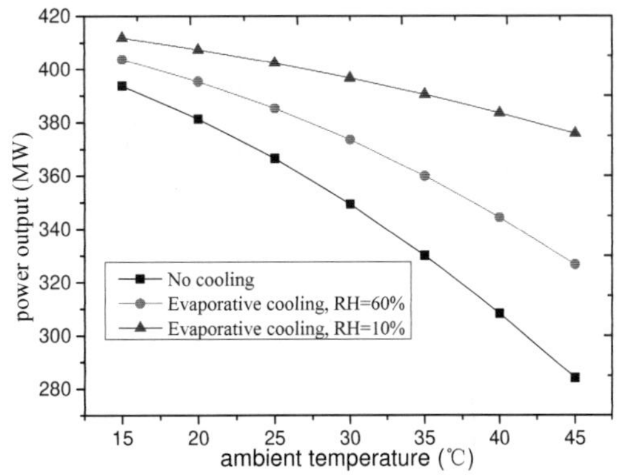

Figure 5. Power outputs of CCPP with and without the evaporative cooling system

3.2. Comparison of power outputs of CCPP with and without the evaporative cooling system

The power outputs of CCPP with and without the evaporative cooling system are compared as shown in figure 5. It is observed that power outputs of CCPP with and without the evaporative cooling system both decrease almost linearly as ambient temperature increases. However, the power output of CCPP without the evaporative cooling system decreases more quickly than that of CCPP with the evaporative cooling system. In the state of no evaporative cooling, when ambient RH is 60%, as ambient temperature increases from 25 °C to 30 °C, the power output decreases by 4.89%. For CCPP with the evaporative cooling system, the decrease is only 3.16%.

The results also show that the effectiveness of the evaporative cooling system for CCPP significantly depends on RH. For CCPP with the evaporative cooling system, the power output decreases more slowly in lower ambient RH. For example, as ambient temperature increases from 25 °C to 30 °C, the power output decreases 3.16% at 60% RH and 1.42% at 10% RH respectively.

3.3. Effect of ambient conditions on power output of CCPP with the evaporative cooling system

Figure 6 shows the variation of power output of CCPP with the evaporative cooling system in different ambient conditions. Although the temperature drop increases as ambient temperature rises, the power output of CCPP with the evaporative cooling system reduces. This is because for a higher ambient temperature, Ti,c is still high, though ambient air is cooled by the evaporative cooling system. As Ti,c rises, the mass flow of inlet air reduces and power consumption of compressor increases correspondingly. The increase in power consumption of compressor because of higher Ti,c leads to a reduction in CCPP power output. For a given ambient RH of 20%, as ambient temperature increases from 25 °C to 35 °C, the power output of CCPP with the evaporative cooling system reduces 15.61MW.

It is also observed that as ambient RH is at a low level, the power output of CCPP with the evaporative cooling system decreases slowly, however, as the ambient RH rises, the CCPP power output decreases more quickly. The outlet RH of the evaporative cooling system is assumed to be 95%, a lower ambient RH means more water evaporates and absorbs more latent heat. So the effect of the evaporative cooling system is improved and, consequently, temperature drop rises and Ti,c falls. At ambient RH of 80% and ambient temperature of 35 °C, the temperature drop is 2.36°C, while it goes up to 15.47°C at an ambient RH of 20%. This indicates less power is consumed by compressor, which results in the enhancement of CCPP power output.

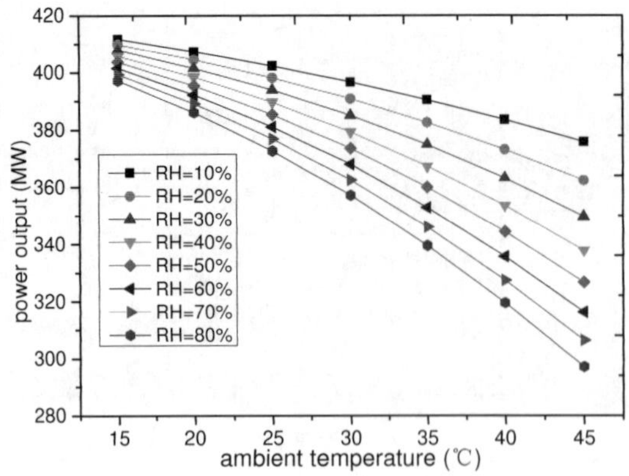

Figure 6. Effect of ambient conditions on CCPP power output

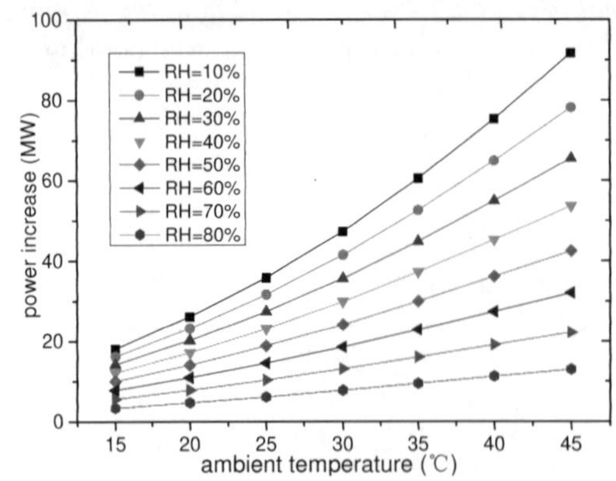

Figure 7. Effect of ambient temperature and 3.4. Effect of ambient conditions on CCPP power increase with the evaporative cooling system

Figure 7 shows the power increase (ΔW) of CCPP after the application of the evaporative cooling system in different ambient conditions. The result illustrates the advantage of CCPP with the evaporative cooling system over that without inlet air cooling. For a given RH, as ambient temperature rises, the additional power generation due to the installation of the evaporative cooling system increases. At a higher ambient temperature and lower RH, the difference between wet bulb temperature and dry bulb temperature is higher, so a greater temperature drop can be achieved and as a result, less power will be consumed by compressor. Therefore, the difference between power output of CCPP with and without evaporative cooling system is greater, namely power output increase is greater. When RH is very low, ΔW increase sharply as ambient temperature rises. While RH rises near saturation, the increase tendency becomes slow. When ambient temperature is 35 °C, the power increase is 9.45MW at ambient RH of 80%, while it rises dramatically to 52.56MW at ambient RH of 20%.

3.4. The critical RH varies with ambient temperature, electric price and PBT

The critical RH is a value depending on economic analysis of PBT to determine the feasibility and economics of the evaporative cooling system for CCPP according to different ambient conditions. When the ambient RH is lower than the critical value, the actual PBT of CCPP unit with inlet air evaporative cooling system installation is shorter than the expected PBT. On the contrary, when the ambient RH is higher than the critical value, the actual PBT is longer than expected PBT. The critical ambient RH can be used to evaluate the feasibility of implementation of the evaporative cooling system for CCPP.

The variation of critical RH with different annual average temperature and PBT is shown in figure 8. The annual average temperature is the average temperature of 5 months (May-September) in a year. It is obviously that the shorter of PBT, the critical RH is lower for a given annual average temperature. This is because in a lower RH, the difference between the wet bulb and dry bulb temperatures is greater, which means a larger temperature drop is possible. The larger temperature drop leads to more power increase and electricity generated. As the additional generated electricity increase, the net income for the year rises and the PBT becomes shorter. It is also observed that as the payback become longer, the critical RH for different annual temperature gradually tends to be close. This is because as the RH increases to a high level, the temperature drop decreases and tends to the same.

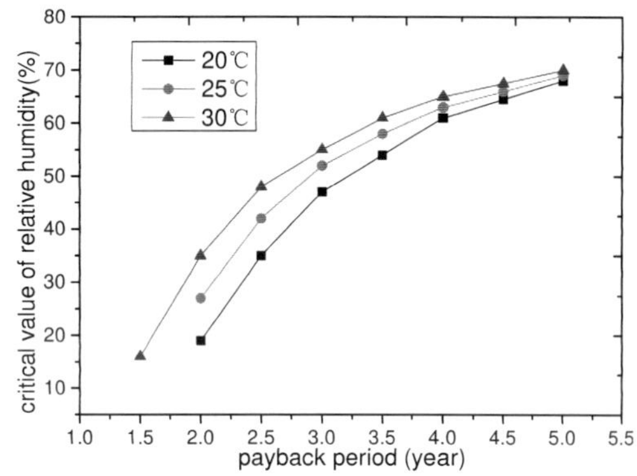

Figure 8. The critical values of RH vary with annual average temperature and PBT

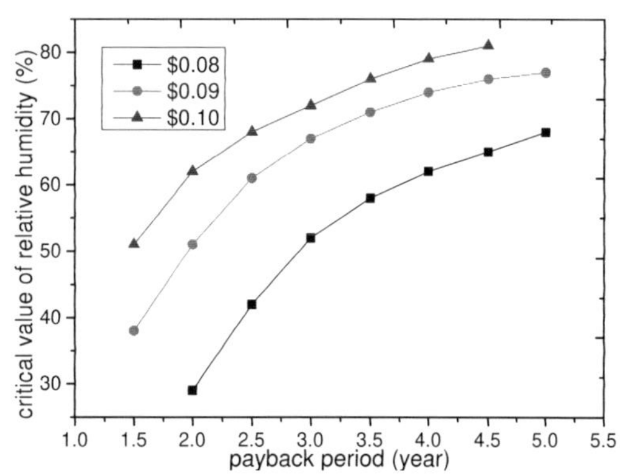

Figure.9 The critical values of RH vary with electricity price and PBT

The sensitivity of critical RH varies with electricity price and PBT is shown in figure 9. It illustrates that the shorter of PBT, the critical RH is lower for a given electricity price. The lower of RH, the temperature drop and power output increase is greater as explained earlier. For the same temperature drop and additional generated electricity, the higher of electricity price, the more of net income is gained. Therefore, as the electricity price increases, the critical RH for the same PBT is higher. For example, for a given PBT of 2.5 years, the critical RH is 42% for electricity price of 0.08 \$/kWh and 61% for electricity price of 0.09 \$/kWh, respectively.

4. Conclusions

Based on the thermodynamic and economic analyses and calculation of critical RH, the following conclusions have been drawn:

(1) The temperature drop (ΔT) increases as ambient RH declines. At ambient temperature of 35 °C and RH of 60%, ΔT is merely 5.95 °C, while ΔT is 18.58 °C at 10% RH.

(2) As ambient temperature rises, the power outputs of CCPP with and without evaporative cooling system both decrease almost linearly. In the state of no cooling, when ambient RH is 60%, as ambient temperature increases from 25 °C to 30 °C, the power output decreases by 4.89%. For CCPP with the evaporative cooling system, the decrease is only 3.16%.

(3) The application of the evaporative cooling system improves the CCPP power output. The power increase is greater at higher ambient temperature and lower ambient RH.

(4) The calculation and analysis of critical RH in different ambient conditions are carried out. It is observed that the shorter of expected PBT, the critical RH is lower for a given annual average temperature.

(5) The sensitivity of critical RH varies with electricity price and expected PBT is analyzed. For a given expected PBT of 2.5 years, the critical RH is 42% for electricity price of 0.08 $/kWh and 61% for electricity price of 0.09 $/kWh, respectively.

References

[1] De Sa A, Al Zubaidy S. Gas turbine performance at varying ambient temperature, Applied Thermal Engineering, 2011, 31(14): 2735-2739.

[2] Arrieta F R P, Lora E E S. Influence of ambient temperature on combined-cycle power-plant performance, Applied Energy, 2005, 80(3): 261-272.

[3] Mohapatra A K. Thermodynamic assessment of impact of inlet air cooling techniques on gas turbine and combined cycle performance, Energy, 2014, 68: 191-203.

[4] MA Ehyaei, M Tahani, P Ahmadi, M Esfandiari. Optimization of fog inlet air cooling system for combined cycle power plants using genetic algorithm. Applied Thermal Engineering, 2015, 76, 449-61.

[5] Cortes C R, Willems D E. Gas turbine inlet air cooling techniques: an overview of current technologies, POWER-GEN, Las Vegas, Nevada, USA, 2003.

[6] Mahto D, Pal S. Thermodynamics and thermo-economic analysis of simple combined cycle with inlet fogging, Applied Thermal Engineering, 2013, 51(1): 413-424.

[7] Mohanty B, Paloso G. Enhancing gas turbine performance by intake air cooling using an absorption chiller, Heat Recovery Systems and CHP, 1995, 15(1): 41-50.

[8] Hariri R, Aghanajafi C. The influence of recovering wasted energy and air coolers on gas turbine cycle performance, Energy Engineering, 2009, 106(1): 24-39.

[9] Boonnasa S, Namprakai P, Muangnapoh T. Performance improvement of the combined cycle power plant by intake air cooling using an absorption chiller, Energy, 2006, 31(12):2036-2046.

[10] Mohapatra A K. Comparative analysis of inlet air cooling techniques integrated to cooled gas turbine plant, Journal of the Energy Institute, 2014.

[11] Khaliq A, Choudhary K. Exergy analysis of the regenerative gas turbine cycle using absorption inlet cooling and evaporative aftercooling, Journal of the Energy Institute, 2009, 82 (3):159-167.

[12] Gareta R, Romeo L M, Gil A. Methodology for the economic evaluation of gas turbine air cooling systems in combined cycle applications, Energy, 2004, 29(11): 1805-1818.

[13] Erdem H H. Thermodynamic and economic assessments of gas turbine inlet air-cooling by evaporative technique, International Journal of Exergy, 2009, 6(5): 605-619.

[14] De Lucia M, Carnevale E, Falchetti M, et al. Performance improvements of a natural gas injection station using gas turbine inlet air cooling, ASME 1997 International Gas Turbine and Aeroengine Congress and Exhibition. American Society of Mechanical Engineers, 1997: V002T07A001-V002T07A001.

[15] Chaker M, Meher-Homji C B, Mee T, et al. Inlet fogging of gas turbine engines detailed climatic analysis of gas turbine evaporation cooling potential in the USA, Journal of engineering for gas turbines and power, 2003, 125(1): 300-309.

Flow field simulation of the spiral-wound heat exchanger

Guanghui Chen, Jie Bao, Shu Guo, Jianlong Li*

College of Chemical Engineering, Qingdao University of Science and Technology, Qingdao, China.

*Corresponding author e-mail:ljlong321@163.com

Abstract. The flow conditions and heat transfer processes in the spiral-wound heat exchanger were studied. The flow characteristics at different sections were also analysed. Numerical calculation results demonstrate that the velocity gradient of the fluid on the shell-side of the heat exchanger increases with the inlet flow velocity increasing, and the increased velocity gradient enhances the turbulence of the shell side fluid.

1. Introduction

As a new type of highly efficient and compact heat exchanger, the spiral-wound heat exchanger fits the development trend of heat exchange equipment well. The spiral-wound heat exchanger also has many advantages over other heat exchangers. For instance the tubes twisted on the center cylinder, and the parts of the exchanger such as spacer and pipe clamp affect the flow state of the shell-side fluid. In the meantime, due to the special spiral winding structure, the fluid in the tube forms a strong secondary flow, which greatly improves the efficiency of the heat transfer. In order to investigate the heat transfer coefficient on the shell-side of the spiral-wound heat exchanger, Sam pour [1] established an experimental model of the coiled tube heat exchanger, and calculate the heat transfer coefficient on the shell side of the heat exchanger by the "Wilson plot" method. Srbislav et al. [2] studied the effect of structural parameters such as pitch and winding angle on Nussle number Nu. The following relationship has been obtained:

$$Nu = 0.5Re^{0.55}\,\mathrm{Pr}^{1/3}(\eta/\eta_w)^{0.14},\ 1000<\mathrm{Re}<9000,\ 2.6<\mathrm{PR}<6.0$$

Ma et al. [3] implemented a test for the structural parameters of the spiral wound tube shell-side fluid. The weighted order of the parameters is Re>PR>P1>P2>D. Nephron et al. [4] established a spiral-wound heat exchanger model which has six-layer coiled tubes with five-wraps. The numerical calculation was also carried out based on the experimental result. Subsequently, an experimental model of finned spiral-wound heat exchanger was established and compared with non-finned spiral-wound heat exchanger. The results show that the finned heat exchanger has higher heat transfer efficiency. Pan et al. [5] studied the relationship among the oscillating flow, heat transfer, and pressure drop in the winding pipe by computational fluid dynamics. The relationship among oscillation frequency, inlet velocity, Nussle number and drag coefficient was also analyzed according to the simulation results. Nerada's et al. [6] established an experimental model of the spiral-wound LNG heat exchanger, the heat-transfer coefficients and pressure drop for gas flow at the shell-side of heat exchanger were measured. Deng et al. [7] established a new type of corrugated tube heat exchanger, and compared the heat transfer with two different types of heat exchangers through experiments and

Content from this work may be used under the terms of the Creative Commons Attribution 3.0 licence. Any further distribution of this work must maintain attribution to the author(s) and the title of the work, journal citation and DOI.
Published under licence by IOP Publishing Ltd

simulations. Hammed et al. [8] implemented an experimental test on spiral wound tube heat exchangers and studied the effects of flow parameters, thermodynamic parameters, and geometric parameters.

2. Numerical Approach

2.1. Geometrical configuration

The main parameters of the heat exchange tube include winding radius, pitch, radius of the heat exchange tube, and the number of turns, the heat transfer inner tube is shown in Fig 1.

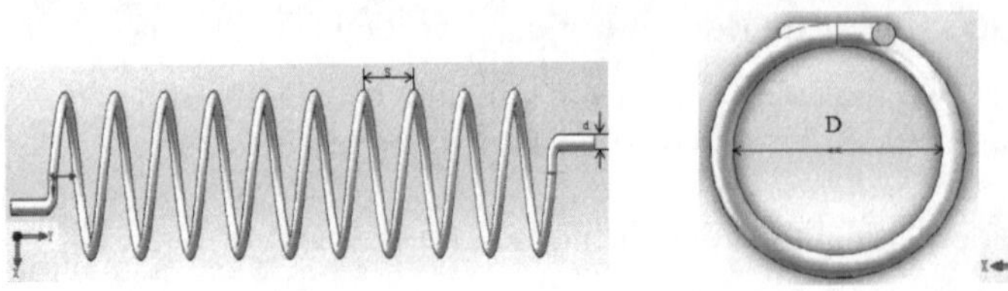

Fig 1. Heat transfer tube structure of the wound tube heat exchanger.

The calculation model of a three-dimensional spiral-wound heat exchanger was established by SolidWorks 2016. The length of the winded pipe bundle is 600mm. It is composed of two layers of six winding pipes, including three inner and outer layers.

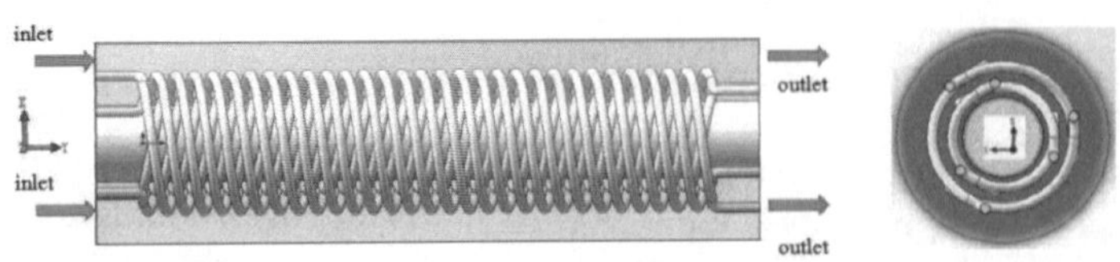

Fig 2. The schematic diagram of shell model of spiral-wound heat exchanger.

2.2. .Mathematical model

2.2.1. Basic equation. Computational fluid dynamics (CFD) is used to solve fluid mechanics based on control equations, including continuity equation, momentum equation and energy equation. The turbulent state of fluid occurred in the spiral-wound tube heat exchanger, and the control equations are shown as follow [9, 10]:

Continuity equation:

$$\frac{\partial \rho}{\partial t}+\frac{\partial(\rho u_x)}{\partial x}+\frac{\partial(\rho u_y)}{\partial y}+\frac{\partial(\rho u_z)}{\partial z}=0 \tag{1}$$

Momentum equation:

$$\rho\frac{Du_x}{D\theta}=\rho X-\frac{\partial P}{\partial x}+\mu\left(\frac{\partial^2 u_x}{\partial x^2}+\frac{\partial^2 u_x}{\partial y^2}+\frac{\partial^2 u_x}{\partial z^2}\right)+\frac{\mu}{3}\frac{\partial}{\partial x}\left(\frac{\partial u_x}{\partial x}+\frac{\partial u_y}{\partial y}+\frac{\partial u_z}{\partial z}\right) \tag{2}$$

$$\rho\frac{Du_y}{D\theta}=\rho Y-\frac{\partial P}{\partial y}+\mu\left(\frac{\partial^2 u_y}{\partial x^2}+\frac{\partial^2 u_y}{\partial y^2}+\frac{\partial^2 u_y}{\partial z^2}\right)+\frac{\mu}{3}\frac{\partial}{\partial y}\left(\frac{\partial u_x}{\partial x}+\frac{\partial u_y}{\partial y}+\frac{\partial u_z}{\partial z}\right) \tag{3}$$

Energy equation:

$$\frac{\partial(\rho E)}{\partial t} = \nabla \cdot [\vec{u}(\rho E + p)] = \nabla \cdot \left[k_{eff} \nabla T - \Sigma_j h_j J_j + (\tau_{eff} \cdot \vec{u})\right] + S_h \tag{4}$$

2.2.2. Turbulence model. Take the notational flow into consideration, the curved streamlines and son on were revised by the RNG k-ε model. Better results were obtained for complex flow phenomena such as secondary flow, swirling flow and so on. The equations of k and ε are given respectively [10]:

$$\frac{\partial}{\partial t}(\rho k) + \frac{\partial}{\partial x_i}(\rho k u_i) = \frac{\partial}{\partial x_i}\left[\left(\mu + \frac{\mu_t}{\sigma_k}\right)\frac{\partial k}{\partial x_i}\right] + G_k + G_b - \rho\varepsilon - Y_M \tag{5}$$

$$\frac{\partial}{\partial t}(\rho \varepsilon) + \frac{\partial}{\partial x_i}(\rho \varepsilon u_i) = \frac{\partial}{\partial x_i}\left[\left(\mu + \frac{\mu_t}{\sigma_\varepsilon}\right)\frac{\partial \varepsilon}{\partial x_i}\right] + C^*_{1\varepsilon}\frac{\varepsilon}{k}(G_k + C_{3\varepsilon}G_b) - C_{2\varepsilon}\rho\frac{\varepsilon^2}{k} - R \tag{6}$$

2.2.3. Wall functions. In view of the functions and models applied in this study, the scale wall function is chosen to avoid the discontinuity of the simulation results caused by the small boundary layer grid.

2.3. Grid and operating conditions

Based on the Workbench software, the geometry model was imported into Meshing function and generated meshes automatically.

Boundary conditions including the velocity inlet, the pressure outlet and the non-slip wall boundary were chosen. The SIMPLE (Semi-Implicit Method for Pressure Linked Equations) algorithm and the first-order upwind scheme were used to solve the pressure-velocity coupling problem.

3. Results and Discussions

The working medium of the heat exchanger is water with 20°C, the velocities of the inlet fluid were set at 0.1m/s, 0.3m/s, 0.5m/s and 0.7m/s, respectively.

3.1. The Velocity distribution simulation of X-Y shell section under different inlet velocity

Fig.3 shows the velocity distribution of the shell-side fluid in the heat exchanger at X-Y cross-section under different inlet velocities. It can be seen from the figure that there are low velocity regions below the tubes of the heat exchanger. The low velocity regions are not conductive to heat transfer. With the increase of the inlet flow velocity, the area of the low velocity regions gradually decreases. There are high velocity regions between the heat exchanger tubes and the outer heat exchanger wall. Because of the linear motion of the fluid flow in the high velocity regions, it is also has negative impact on heat transfer. There exists radial flow from the outer high velocity region to the winding pipes. The radial flow helps to promote the heat exchange between the outer cold fluid and the heat exchanger tube wall. With the increase of the inlet flow velocity, the radial velocity also increases. From Fig.3A, it can be seen that when the inlet velocity is 0.1m/s, the radial velocity is about 0.15m/s. From Fig.3B, it can be seen that when the inlet velocity is 0.7m/s, the radial velocity is about 1.2m/s.

3.2. The flow field simulation of X-Z cross section under different inlet velocities

3.2.1. The analysis on velocity field. Fig.5A, 5C and 5E shows the velocity distribution of X-Z section at y=120mm under different inlet flow rates. Fig.5B, 5D and 5F are the velocity distribution of X-Z section at y=600mm under different inlet flow rates. It can be seen that there are velocity boundary layers and low velocity zone (lower than 0.5m/s) near the heat exchange tube. Because of the turbulent effect, the velocity distribution in the outlet section, which has high velocity region, is more complex than that in the inlet section.

3.2.2. Temperature distribution. Fig.6A, 6C and 6E shows the temperature distribution of X-Z section at y=120mm under different inlet flow rates. Fig.6B, 6D and 6E are the temperature distribution of X-Z section at y=600mm under different inlet flow rates. The cold water was set to flow in the shell side while the hot water was set to flow in the tube side. Compared A with C and E, it can see that when the inlet velocity becomes smaller, the cold fluid stays longer in the shell side, and mixes with the hot fluid well. It can be observed that the area of low temperature region is smaller from Fig.6B. From Fig.6D and 6F, it can be seen that there are more and lager low temperature regions.

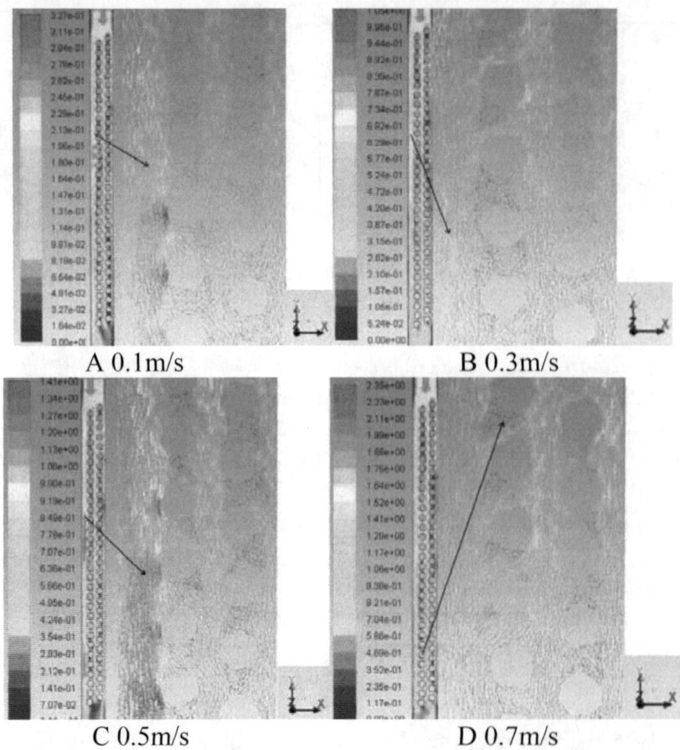

Fig 3. Velocity distribution of X-Y section under different inlet velocity.

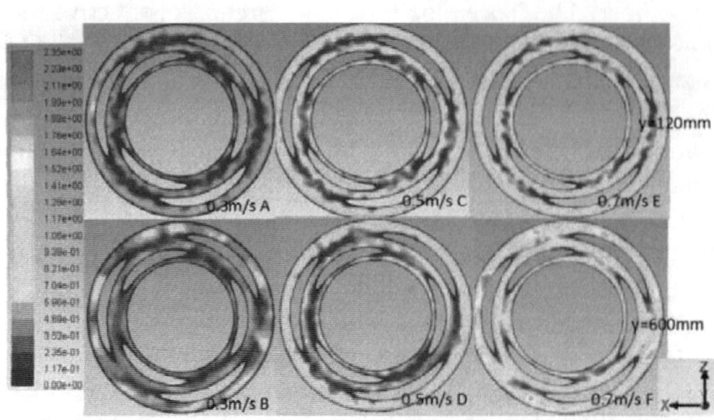

Fig 4. Velocity distribution of X-Z section at different inlet velocity.

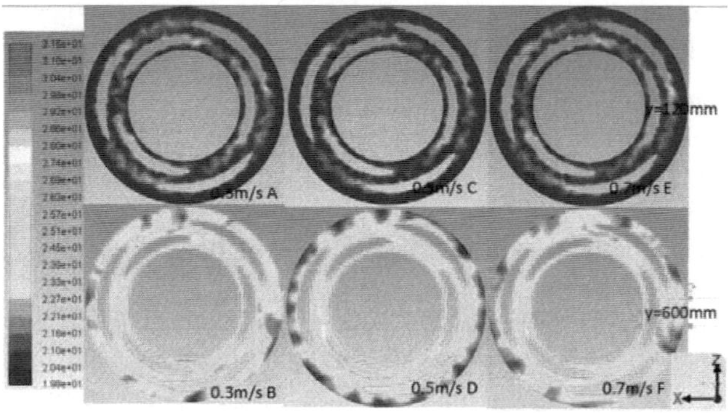

Fig 5. Temperature distribution of X-Z section at different inlet velocity.

4. Conclusions

In this paper, the flow conditions and heat transfer processes in the shell-side of a spiral-wound heat exchanger were investigated numerically. The velocity and temperature distributions of different sections were analyzed.

On the one hand, the flow field distribution of shell side X-Z section under different inlet velocity were analyzed. With the increase of the inlet flow velocity, the velocity gradient in the shell increases. The flow velocity also fluctuates automatically, and it is beneficial to heat transfer. There is a radial flow from the high speed zone of the outer layer to the winding pipes. The radial flow promotes the heat exchange between the outer layer of cold fluid and the wall of the tubes. With the increase of inlet velocity, the radial flow also increases. For the temperature distributions, the high-temperature zones and the low-temperature zones are distributed in, which promotes the heat transfer. As the inlet flow velocity increases, the shell-side temperature gradient decreases gradually.

On the other hand, the flow distribution of shell-side X-Z sections under same inlet velocity was analyzed. The results show that: With the increase of the inlet velocity, the area of the high velocity zone on the shell increases gradually. The temperature increases with the fluid flowing, through the heat transfer tube. The high and low temperature zones are distributed irregularly. The high temperature zone diffuses from the heat transfer tube to the surrounding regions. On the shell side flow field, the fluid is strongly affected by the winding tube, and the heat transfer is improved.

Acknowledgments

This work was financially supported by the transformation project of scientific and technological achievements of Qingdao (no. 16-6-2-50-nsh).

References

[1] Salimpour M R, Heat transfer coefficients of shell and coiled tube heat exchangers, Experimental Thermal & Fluid Science, 2009, 33(2):203-207.

[2] Srbislav B. Genić, Branislav M. Jaćimović, Dragan Mandić, et al. Experimental determination of fouling factor on plate heat exchangers in district heating system, Energy & Buildings, 2012, 50(7):204-211.

[3] Ma et al. Numerical simulation of heat transfer of spiral wound tube heat exchanger, Zhengzhou, 2014.

[4] Naphon P. Thermal performance and pressure drop of the helical-coil heat exchangers with and without helically crimped fins, International Communications in Heat & Mass Transfer, 2007, 34(3):321-330.

[5] C Pan, Y Zhou, J Wang. CFD study of heat transfer for oscillating flow in helically coiled tube heat-exchanger, Computers & Chemical Engineering, 2014, 69:59-65.

[6] Bengt O. Neeraas, Experimental shell-side heat transfer and pressure drop in gas flow for spiral-wound LNG heat exchanger, International Journal of Heat and Mass Transfer , 2004, 47

353–361.
[7] Deng et al.Study on the Flow and Heat Transfer Performance of Spiral Winding Heat Exchanger, Zhengzhou, 2016.
[8] Sadighi Dizaji H, Jafarmadar S, Hashemian M. The effect of flow thermodynamic and geometrical characteristics on exergy loss in shell and coiled tube heat exchangers. Energy, 2015. 91: 678-684.
[9] Tao Chen, Guoliang Zhang, Foundation of chemical engineering transfer process, Chemical Industry Press, 2002.
[10] Xiaojing Wang, Wenyan Li, Qimeng Sun, Analysis of internal flow field of the inserted link Bellows Based on FLUENT, Journal of chemical engineering in Colleges and Universities, 2016, 30(2):286-291.

IOP Conf. Series: Journal of Physics: Conf. Series **1064** (2018) 012004 doi:10.1088/1742-6596/1064/1/012004

Study on Numerical Calculation of the Shell Side Flow Field of the Space Spiral Section in Intermediate Heat Exchanger in China Experimental Fast Reactor

Guangdong Song*, Shuai Xing

China institute of Atomic Energy, Beijing, China.

*Corresponding author e-mail: songguangdong1986@163.com

Abstract. In the study of flow-induced vibration of nuclear grade heat exchanger, the distribution law of the flow field should be analyzed firstly. Taking the Intermediate Heat Exchanger(IHX) in China Experimental Fast Reactor (CEFR) as the research object, the computational fluid dynamics (CFD) method is used to establish the numerical calculation model of the shell side flow field of the elbow section of the heat exchanger in this paper. Method of the flow field distribution of the bend pipes is studied and analyzed, and the distribution characteristics of the flow field in the elbow section under different inlet velocity are analyzed. The results show that the flow angle and transverse flow velocity at the inlet of the elbow section are all larger. In the design, operation and maintenance of the heat exchanger, we should pay attention to the entrance of the elbow section.

1. Introduction

IHX is one of the most important equipment in a sodium cooled pool type fast reactor, which transports the heat from the primary sodium to the secondary sodium. Security and reliability is the most important request for development and design of nuclear grade heater exchanger, though greater demands were made for IHX [1]. Flow-induced vibration is one of the most important factors which results in damage to the tubes of IHX [2].

IHX is a vertical shell and tube type heat exchanger with primary sodium on shell side and secondary sodium on tube side. The middle part of IHX tube is a space spiral structure. The tubes of IHX are arranged in circular con-centric rows around a central down comer and the support structure of the tubes is quite special compared with other common heat exchangers [3].

The structure of IHX is special without codes and standards. So, 3D model of elbow section was build and the flow fluid distribution characteristics was researched in this paper by CFD. The distribution law of the flow field, impact of different inlet velocities on fluid distribution, fluid angle and velocity were researched by analysis of fluid distribution and fluid force change providing foundation data for flow-induced vibration of IHX.

Content from this work may be used under the terms of the Creative Commons Attribution 3.0 licence. Any further distribution of this work must maintain attribution to the author(s) and the title of the work, journal citation and DOI.
Published under licence by IOP Publishing Ltd

IOP Conf. Series: Journal of Physics: Conf. Series **1064** (2018) 012004 doi:10.1088/1742-6596/1064/1/012004

2. Model and calculation assumptions

2.1. Model
The 3D model of elbow section was shown in Fig.1. The circumferential distribution of the heat exchanger tubes was periodic at fixed angle. Considering the uniformity of fluid distribution and reducing calculated amount, only one cycle was calculated in this paper. Except the elbow section, partial straight pipe section was saved in this model in order to stabilize the flow field. The flow direction was from top to bottom and the boundary conditions of the side walls shown in Fig.1 were periodic. The tube surfaces and the drum surfaces were setted wall boundary condition.
All manuscripts must be in English, also the table and figure texts, otherwise we cannot publish your paper. Please keep a second copy of your manuscript in your office.

When receiving the paper, we assume that the corresponding authors grant us the copyright to use the paper for the book or journal in question. Should authors use tables or figures from other Publications, they must ask the corresponding publishers to grant them the right to publish this material in their paper.

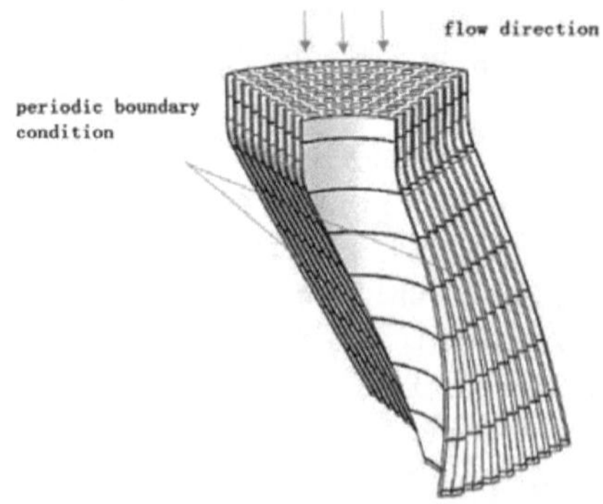

Fig.1 Model of elbow section flow field

2.2. Assumptions and monitor points
On account of the complexity of structure and flow field, the mesh quantity may be too large to calculate if considering all details. It is necessary to simplify the models. The following simplified methods were applied according to basic assumptions used in computational fluid dynamics.

(1). The fluid was treated as incompressible;

(2). Ignore the vibration control belt which impact the shell flow field little;

(3). The heat exchange tubes were assumed as rigid as a foundation for further fluid-solid coupling analysis;

(4). The axial flow was main in the inlet of elbow section based on former results and changed little, so the velocity of inlet was regarded as constant and axial.

Three boundary conditions of inlet were applied to research the flow field characters. In case 1, the parameters of inlet and outlet were from calculation on IHX whole model at the same location. In case 2 and case 3, the axial velocity was changed and pressure remained the same as case 1.

In order to monitor velocity and fluid force of transverse flow and axial flow in elbow tubes, monitor faces and points were set. Eight tubes were selected as shown in Fig.2. Two were adjacent to inside wall, two were adjacent to outside wall and two were in middle. Eight monitor surfaces with 1mm height were set along axial direction for each tube and one of them was in straight tubes section. The meaning of tubes label in Fig.2 interpreted as follow. The first number was for layer number increasing from inside to outside. The second number was for the location in the layer number. For example, the label 1.1, the first number 1 was for the first layer adjacent to inside wall, the second number was for the location adjacent to periodic wall.

IOP Conf. Series: Journal of Physics: Conf. Series **1064** (2018) 012004 doi:10.1088/1742-6596/1064/1/012004

As shown in Fig.3, 2 monitor points were set on circumferential and radial respectively. As a result, a total of 64 monitor faces, 256 monitor points and 1152 variables were obtained.

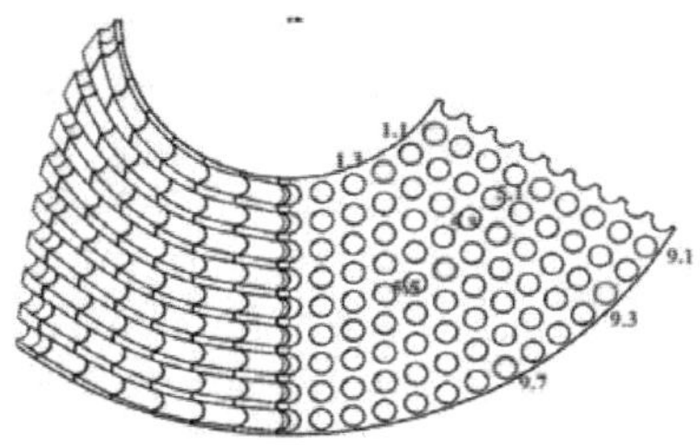

Fig.2 Monitor tubes

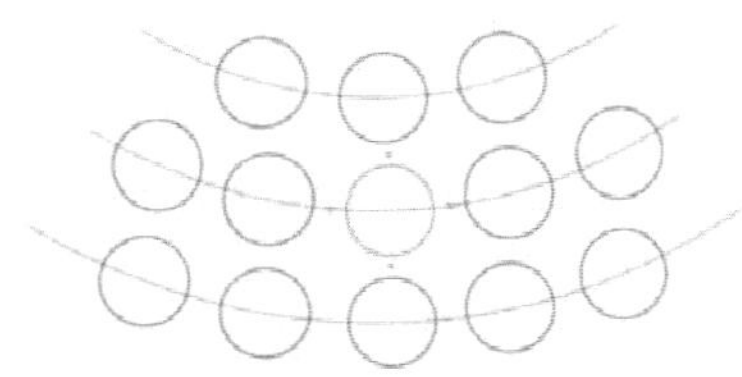

Fig.3 Monitor points

3. simulation results and analysis of flow field

3.1. simulation results

The streamlines in flow field of shell side were shown in Fig.4. The top was inlet and the bottom was outlet. Flow direction varied with the change of structure. When rotation angle of heat exchange tubes changed, the rotation angle of flow direction decreasing with flow was less than that of tubes because of inertia. As shown in Fig.4, the max angle between flow direction and tubes was in the inlet of elbow section. Due to the short length of straight tubes, the flow direction was still impacted by upper elbow tubes. On the whole, the velocity in elbow section was faster than straight section because of flow area decrease caused by the distribution of fluid was mainly in annular region between tube layers.

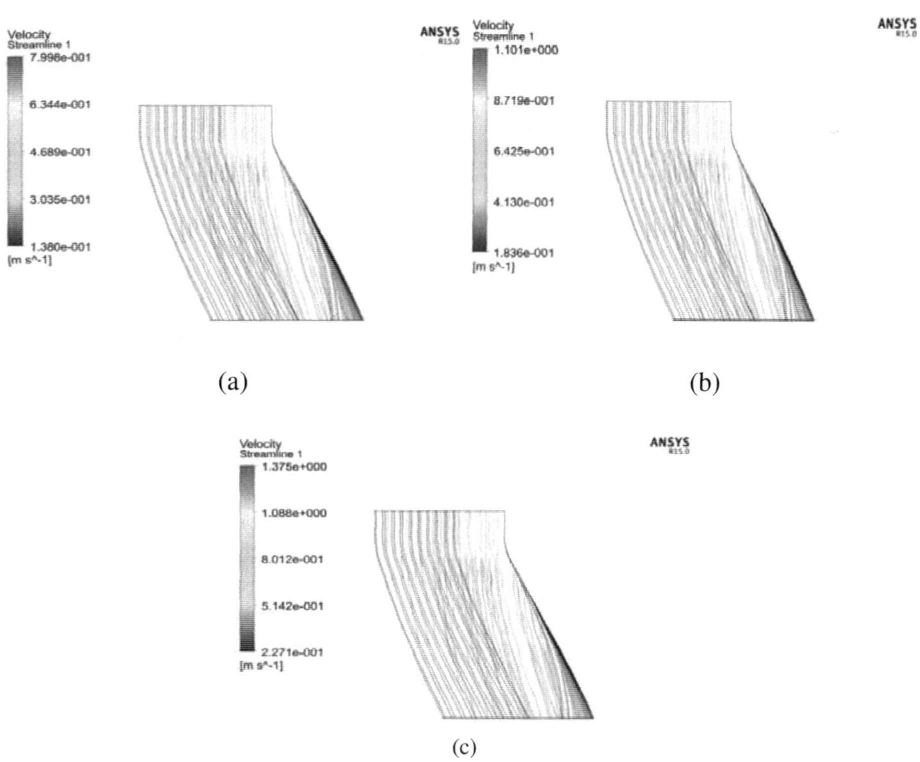

Fig.4 Streamlines in flow field of shell side

(a) case 1 (b) case 2 (c) case 3

IOP Conf. Series: Journal of Physics: Conf. Series **1064** (2018) 012004 doi:10.1088/1742-6596/1064/1/012004

3.2. change of flow angle

According the simulation results, angle between flow direction and tubes was calculated. In order to verify the accuracy of methods proposed in 1.3, take transverse flow in middle of tube 5.5 simulated by case 1 for instance. As shown in Fig.5, tube 5.5 was marked by red color, and direction vector of transverse flow marked by black line was achieved by data processing methods. As results shown in velocity vector and cloud picture, flow direction changed with elbow section structure, the velocity was faster between tube layers and transverse flow direction was vertical to radial direction. The transverse flow direction calculated by formula in 1.3 was consistent with simulation.

Change of flow angle could be calculated by formula in 1.3. As shown in Fig.6, the angle decreased in elbow section. As the fluid flowed, flow direction tended to be consistent with tubes. Moreover, the angle of inside layers was smaller than outside layers and angle increased with increase of velocity.

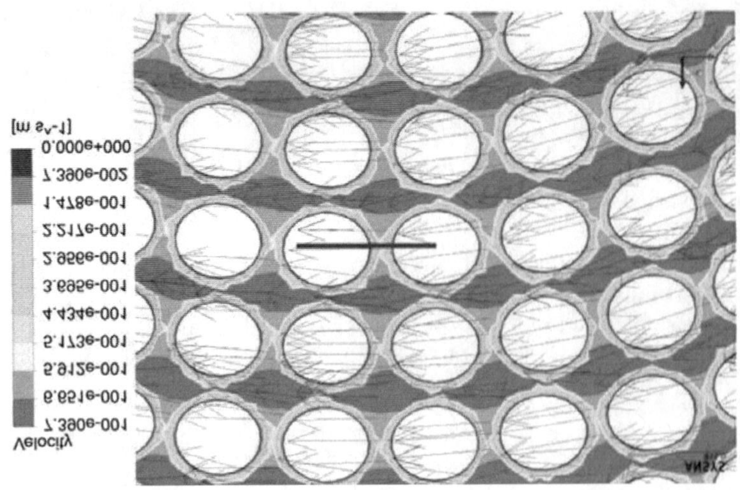

Fig.5 velocity in middle elbow section of tube5.5 in case 1

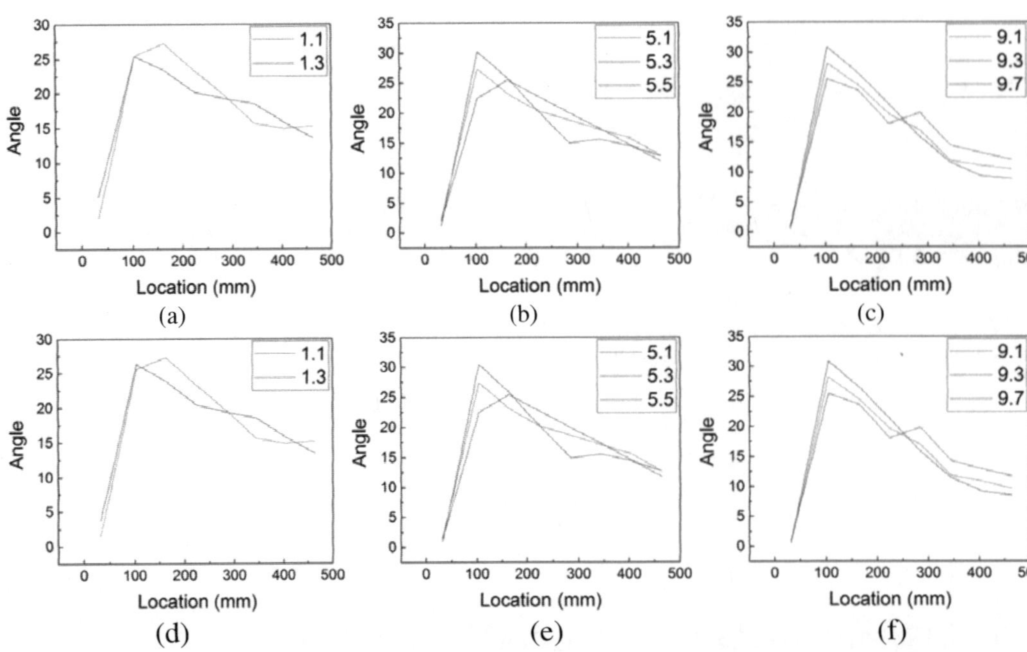

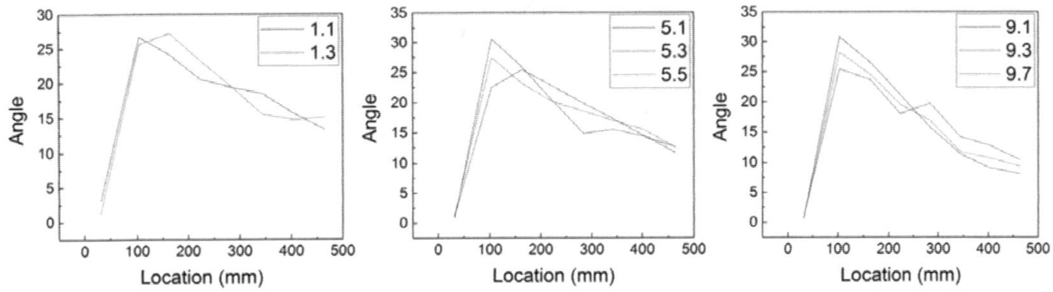

(g) (h) (i)

Fig.6 change of flow angle in elbow section

(a)-(c) inside, middle and outside tubes in case 1

(d)-(f) inside, middle and outside tubes in case 2

(g)-(i) inside, middle and outside tubes in case 3

3.3. transverse flow velocity simulation

Vibration amplitude induced by axial flow was small and impact little on equipment, in general, it could be ignored [4]. Only when velocity was much faster than normal, it will be considered. So, analysis of the transverse flow was main in this paper. Velocity of transverse flow could be achieved by data processing methods in every monitor faces and points. The results of transverse flow velocity changing along axial direction were shown in Fig7.

When the inlet velocity was the same, the change of transverse flow velocity had close relationship with model structure. Before entering the elbow section, transverse flow velocity was small but increased sharply when just flowed into elbow section because of inertia. And angle between flow direction and tubes deceased.

At the same inlet velocity, transverse flow velocity in middle layers were faster than inside and outside layers because of wall resistance.

The different inlet velocity had little impact on change trend of transverse flow velocity along axial direction, but there was a positive correlation between transverse flow velocity and inlet velocity.

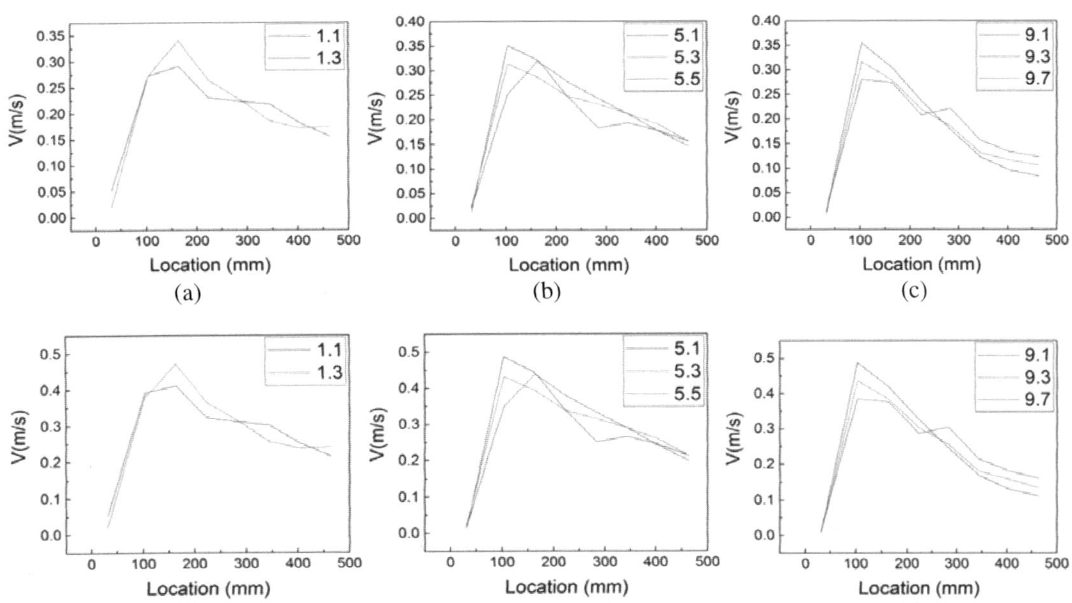

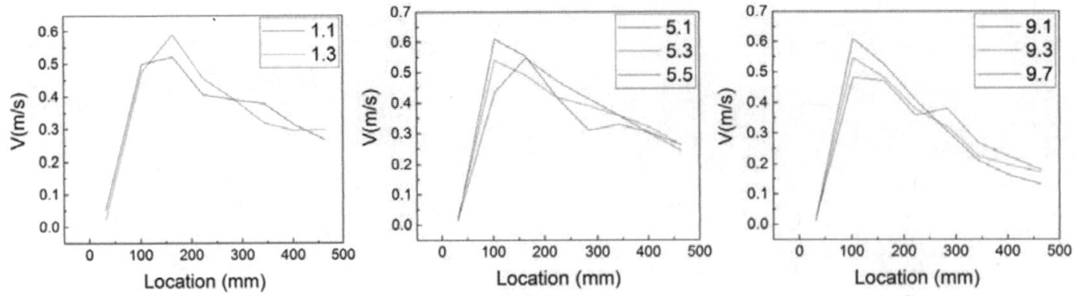

(g) (h) (i)

Fig7 change of transverse flow velocity in elbow section

(a)-(c) inside, middle and outside tubes in case 1

(d)-(f) inside, middle and outside tubes in case 2

(g)-(i) inside, middle and outside tubes in case 3

4. conclusions

The computational fluid dynamics (CFD) method was used to research the distribution characters in elbow section of IHX and the numerical calculation model was established in this paper simulating the flow field with different inlet velocity. The main conclusions were following:

The flow angle and transverse flow velocity were max in inlet of elbow section where structure changed drastically and decrease with flow. And there was a positive correlation with inlet velocity.The elbow section should be paid more attention on design, operation and maintenance of the heat exchanger.

References

[1] L Dongxiang, DONG Jinshan, JIANG Xiaozhi, elt. Calculation of Non-Conventional Heat Exchanger Tubes Natural Frequency[J]. Light Industry Machinery, 2014,(32)06,84-88.

[2] Bo Hanliang. Helical tube bundles vibration induced by low density flow[J]. Journal of Tsinghua University (Sci & Tech), 1997(5):34-37.

[3] Kakac S, Bergle A Eand Mayinger F. Heat Exchangers: Thernal-Hydraulic Fundamentals and Design[M]. McGraw-Hill, 1981.

[4] Paidoussis M P. Vibration of cylindrical structure induced by axial flow[J]. Journal of Engineering for Industry, 1974, 96: 547-552.

IOP Conf. Series: Journal of Physics: Conf. Series **1064** (2018) 012005 doi:10.1088/1742-6596/1064/1/012005

Exothermic and Insulating Riser Design of Gear Ring Seat Casting based on Anycasting

Dong Sun[1], Junhua Zhang[2], Xudong Yang[3] and Kangqi Mu[3, a]

[1]College of Mechanical Engineering, Guizhou institute of technology, Guiyang, China.
[2]Guizhou Machinery and Electronic Products Quality Supervision and Inspection Institute, Guiyang, China.
[3]College of Mechanical Engineering, Guizhou University, Guiyang, China.

[a] 1147834557@qq.com

Abstract. According to the disadvantages of design method of traditional riser, this article introduces the design method of riser which is based on the computer aided engineering technology, the casting process of exothermic and insulating riser for steel casting of gear ring seat of excavator was simulated by Anycasting software. the riser modulus was directly calculated according to the solidification time of the casting by the simulation, Meanwhile its volume and quantity could be confirmed by the feeding liquid volume method. The result shows that the feeding effect of the exothermic and insulating riser design is perfect, The method can ensure no shrinkage defect in the casting, and also improve the technological yield, thus it is validated that the technique is feasible.

1. Introduction

The riser design of steel castings appears to be more complex and difficult due to many factors. Unreasonable designed may easily lead to many defects such as shrinkage and so on which directly affect the quality of castings and productivity. Modulus and thermal pitch circle method (ratio method) are mostly used in traditional riser system design. These methods all have their own limitations, especially unable to predict the casting solidification process, cannot consider the interaction of the molding materials, chills and riser and the impact on the casting solidification process. These methods have their limitations, especially unpredictable solidification process that does not consider the shape of the material and the cold iron and other interactions with the riser and the effect on the solidification process. In recent years, due to the development of casting CAE technology, the visualization of casting filling and solidification process can be achieved by computer and the reliability and rationality of riser design can be improved [1-4]. In this paper, the riser of gear ring seat steel casting is designed by means of the South Korean Anycasting simulation software based on the actual factory condition.

Exothermic/insulating riser sleeve is made of exothermic material and insulating material which combined the advantages of insulating riser and exothermic riser. When pouring process, exothermic reaction is produced when the exothermic material in exothermic/insulating riser sleeve in contact with the molten metal, the temperature of molten metal within the riser is increased. Then the insulating material begins to sinter, insulate and the heat is not easy to be lost, the molten metal

Content from this work may be used under the terms of the Creative Commons Attribution 3.0 licence. Any further distribution of this work must maintain attribution to the author(s) and the title of the work, journal citation and DOI.
Published under licence by IOP Publishing Ltd

solidification time is extended, the riser size can be designed to be smaller compared to the insulating riser and exothermic riser. Exothermic/insulating riser sleeve can effectively improve casting quality and process yield, reducing the pressure of riser gas cutting and clearing processing workload, conform to the pursuit of the trend of near net shape casting industry, and is currently one of the most advanced technology in foundry industry.

2. Riser Design Method

There are several kinds of traditional riser design methods, modulus method, thermal pitch circle (ratio method), perimetrischen quotient method, the cubic equation method, etc. are commonly used. In recent years, the riser design method based on CAE technology appeared due to the development of casting CAE technology.

2.1 Thermal Pitch Circle Method

The cylindrical riser diameter or the circular waist width can be obtained when the pitch circle diameter of feeding position is enlarged by the a certain percentage based on the experience in thermal pitch circle method. In production, the thermal pitch circle method is widely used in the enterprise because of its convenience and is a method commonly used in riser design. Its advantage is simple calculation, but practical experience is needed due to the poor theory, a bigger safety factor is generally chosen when determining the riser proportion and the casting process yield is low.

2.2 Modulus Method

Use the square root law of modulus and solidification time, the riser modulus is a certain percentage greater than the modulus of the feeding position in order to ensure the final solidification of riser, and then make sure the riser has sufficient molten metal to feed the volume shrinkage of casting based on the feeding efficiency [5].

Modulus method is widely described in textbooks and scientific literature because of its clear physical meaning and simple derivation. However, it is often difficult to use in actual due to the complexity and the irregularity of casting parts.

2.3 Riser Design Method Based on CAE Technology

Riser design must meet the following criteria:

(1)Riser and riser neck solidification time should be greater than or equal to the casting (or feeding positions) solidification time to ensure smooth passage of feeding. This principle applies to the casting alloy with directional solidification properties such as cast steel alloy (energy principles);

(2) There must be adequate liquid metal to feed the casting (volume principle);

(3) The potential energy of riser should be higher than the fed casting (potential energy principle).

If you know the solidification time of the casting (or feeded position) and riser, and the volume need to be fed of the casting, the problem is solved. The solidification time of the casting (or feeded position) tc can be got by casting CAE simulation software. The solidification time of the riser tr can be got according to the principle of energy, and then riser modulus can be calculated out. The riser size and the number can be calculated out according to the principle of volume. Simulation and constant correction are carried out on the computer, until there are no defects in the casting.

3. Casting Process Introduction

In this paper, the test object is a steel casting of gear ring seat which is a factory production of excavator, it's made of ZG270-500, the outer contour size is 1246×1246×170(mm), the weight is 372kg, the casting process is shown in Figure 1.

Bottom gating open casting system is used in the production. The molten steel flow along the tangent line from the bottom into the cavity and then rises smoothly, the sand flushing is small and not easy to form slag. Chills are placed at the top of casting, so that the casting ring hot section is uniformly cut off and easy to place oval exothermic/insulating riser.

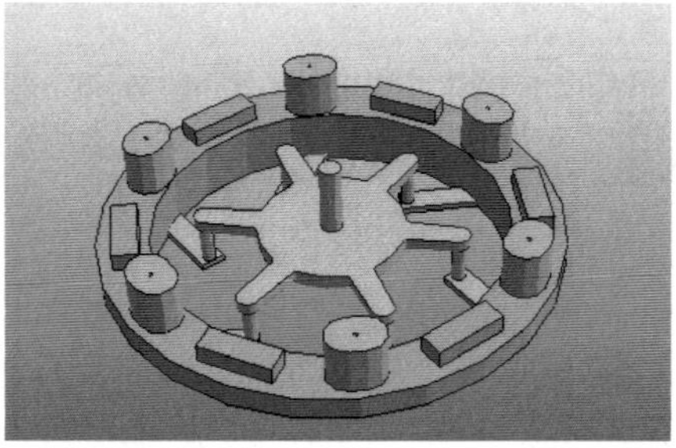

Figure 1. Casting process

4. Riser Process Design

4.1 Solidification Time

Test first for not joining in the riser of casting solidification simulation analysis, the result is shown in Fig.2: Due to the effect of chills, casting hot section in the middle of chills position and be cut off uniformly. The solidification time is 1764s. Through simulation, the final position of solidification and final solidification time can be predicted accurately which provide the basis for riser design.

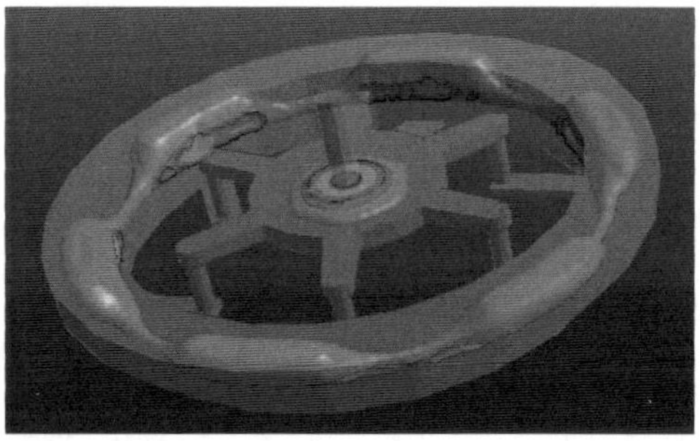

Figure 2. Casting solidification simulation

4.2 Solidification Time of Exothermic/insulating Riser

According to the Chvorinov formula, the relationship between the casting solidification time and modulus is:

$$t = kM^2$$

Where, t — Solidification Time(s); M — Modulus (mm); k — Solidification Factor (s/mm2), associated with alloy type, the same material is the same value.

Riser play the role of feeding and the energy principle is:

$$t_r > t_c$$

Where, t_r —Solidification Time of riser(s); t_c —Solidification Time of the casting (or feeded position) (s)

According to the conditions of the riser feeding effect for $t_r > t_c$, the solidification time of riser is set as 2000s temporarily. After several times simulation test of the oval exothermic/insulating riser, we know that the riser solidification coefficient is 3.65 and the effective modulus of the exothermic/insulating riser is 23.4 mm according to Chvorinov formula.

In addition to have appropriate modulus, riser design must also satisfy the principle of volume. The smallest volume of riser is decided according to the feeding area of casting volume which is called feeding liquid method, this method can be used to check the volume of riser and determine the number of riser. The maximum casting volume that a certain feeding riser volume can feed is determined by equation

$$\eta \cdot V_r > \varepsilon \cdot (V_r + V_c)$$

Where: η ——Riser feeding efficiency (%); ε ——Volume shrinkage (%); V_c——Casting volume or the volume of the feeding area (mm3); V_r——Riser volume(mm3).

The riser feeding efficiency η =25%~45%, a larger value is chosen when the heat preservation ability is more capable. At the same time, the feeding efficiency also related to the shape coefficient of the casting, and the relation are shown in table 1, q is calculated by equation

$$q = V_c / M_c^3$$

Where: V_c——Casting volume (mm3); M_c——Casting modulus (mm).

Table1. Relationship of Riser feeding efficiency and casting shape factor

Casting shape factor q	<200	200	300	400	500~1000	>1000
Riser feeding efficiency η (%)	25	30	33	35	40	45

The computed result of this casting shape coefficient is 3654, therefore the feeding efficiency is 45%, according to the Casting Process Manual[6-10], the ZG270-500 volume shrinkage rate is 4.5%, the casting weight is 372 kg, thus calculate the riser weight should be 41.33 kg. Riser FT100 - T80/120/120 produced by Jinan Shengquan group co., LTD is used in this process, type, the long axis of the ellipse is 120 mm long, short axial length is 80 mm, and the height is 120 mm, the total number is 6.The riser individual weight is 7.1 kg and the gross weight is 42.6 kg which satisfy the volume principle.

Risers are placed at the top of the casting hot section which meet the energy principle.Solidification simulation analysis of the riser is shown in figure 3 which shows the actual solidification time is 2074 s.

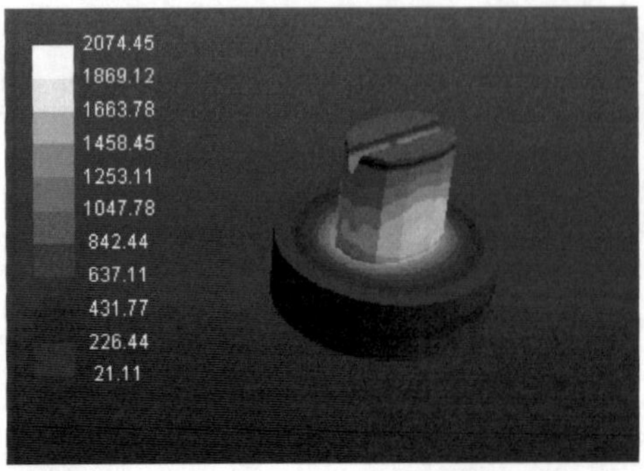

Figure 3. Solidification simulation of exothermic and insulating riser

4.3 Process validation

According to the above design and calculation, the filling and solidification simulation analysis of the casting process has been carried out. The final solidification area is located in exothermic and insulating riser, which in line with the principle of riser design [11]. A small amount of shrinkage is found in the casting, the shrinkage cavity only appear in the riser and gating system, which doesn't affect the quality of the casting. The casting anatomy section produced by this process is shown in Figure 4, the casting quality is in good condition and no shrinkage cavity defects are found in the casting, which consistent with the simulation results.

Figure 4. Dissected casting

4.4 Design Scheme Comparison

The castings modulus can be calculated by simplified method as equation

$$M_r = 1.2 M_c$$

Where: M_r ——Riser modulus (mm); M_c ——Casting modulus (mm).

The casting modulus can be directly read by Anycasting software and the value is 23.65 mm, the required riser modulus is calculated according to the equation and the value is 28.38 mm, far more than the effective modulus of 23.47 mm selected in this paper, it is proved the riser is fully able to meet the need of castings feeding, and thus we can know that the riser modulus design method based on CAE numerical simulation is more accurate and easier [12-13]. Combined with chose heat insulating riser, the process scheme theory production rate of 74%, greatly save the liquid steel raw materials, at the same time can reduce the gas cutting riser and form processing workload, improve the production efficiency.

5. Conclusion

Through production verification, the feeding effect exothermic and insulating riser designed by riser design method based on CAE is good, not only ensure there're no casting shrinkage cavity defects, but also improve the technological production rate, and consistent with the simulation results, proved the feasibility of this method. The casting process design and optimization process are completed on the computer in this method, repeated trial production and process correction are not needed and thus save time and cost, make up for the deficiency of the traditional riser design method can be made up and the level of casting process design can be improved. At the same time, under the background of energy saving and emission reduction requirements in the casting industry, the pursuit of near net shape has become a target for foundry workers, new casting technology means such as exothermic and insulating riser greatly improved the casting production rate and production efficiency. The improvement of process design method and the use of new technology means is the path to improve the level of casting technology and realize green casting.

Acknowledgments
Guizhou Province Department of Education Youth Science and Technology Talent Development Project KY [2016]231. Guizhou Science and Technology Department major project "high-speed intelligent sorting, packaging system R&D and industrialization" support [2017] 2308. Guizhou Science and Technology Plan Project [2016]2306.

References

[1] Mishra P, Ajmani S k, Kumar A, et al. Review article on physical and numerical modeling of sen and mould for continuous slab casting. Int [J]Engng Sci Tech, 2012, 4: 2234-2243.
[2] Wang L, Vitae A, Deng C, et al. Development of continuous casting technology of electrical steel and new products. [J] Iron Steel Res Int, 2012, 19: 1-6.
[3] Sindhya K, Miettinen K. New perspective to continuous casting of steel with a hybrid evolutionary multiobjective algorithm. Mater Manuf Proc, 2011, 26: 481-492.
[4] GanesanS,PoiricrDR Conservation of Mass and Moment Umfor the Flow of Interdendritic Liquid During Solidification. MetTrans,1987,18: 721-723.
[5] W. Wen. Simulation of an Automobile Wheel Casting Process'Proceedings of 61st World Foundry Congress, LiuBaicheng,ed. Beijing,1995: 5-9.
[6] ChoiJK. Application of Mold Filling and Solidification Simultion to Aluminum Wheel Castings. Proceedings of the 61 st world Foundry Congress.Beijing,China,1995: 13-27.
[7] ChiesaF Solidification Modeling of Cast Aluminum Wheels. Modem Casting 1991,(12): 26-284.
[8] NussbaumAI Semi—solid Forming of Al-Mg. Light Metal Age,1996,54: (5-6).
[9] H. Q. 10u,A. Krizman. Erosion of Materials by Alumina Slurry Partl. Wear,1989,(3): 253-269.
[10] Ruden T J, Albright D. High Ductility Magnesium Alloys in Automotive Application Advanced Mater&Processes,1994,145(6): 28-31.
[11] Clegg A. Squeeze Casting-a new Process Technology for Engineer. Foundry Trade Journal,1993,166: 484-485.
[12] J.F. Archard. Contact and Rubbing of Flat Surface.J.Applied Physics,1953,(24): 981-988.
[13] Barwell, F.T. Wear of Machine Elements,Fundamentals of Tribology. Aided by N.P.Sub,N.Saka,MIT Press,1980,(6): 401-441.

Experimental study on effect of fluid impingement location on heat transfer characteristics of a new type of elastic tube bundle

Bo Jiang [1,*], Jian Liu [2], Xianqiao Zhao[1], Kai Guo[1], Jiaxin Wang[1], Ke Liu[1], Yufeng Tang[3], Maocheng Tian[3]

[1] Shandong Electric Power Research Institute, Jinan, China.
[2] Shandong Urban and Rural Planning Design Institute, Jinan, China.
[3] School of Energy and Power Engineering, Shandong University, Jinan, China.

*Corresponding author e-mail: sdujiang@126.com

Abstract. A constant heat flux heat transfer test bed is built, and experimental study on external heat transfer characteristics of a new type of elastic tube bundle is conducted, obtaining overall and local heat transfer performance of the elastic tube bundle at different Re numbers. The results show that, the heat transfer performance for fluid entirely impacting on the free end of tube bundle is better than that of fluid completely entering from heat exchanger inlet. Besides, the greatest heat transfer performance occurs at the condition fluid entirely impinging on the small end of tube bundle. Similar heat transfer characteristics occur when fluid impinges on the big /small free end respectively, namely lower heat transfer coefficient for the fixed end and the big free end, and highest heat transfer coefficient for the small free end, the reason for which might be the differences of velocity and vibration characteristics for fluid around the tube bundle at different locations.

1. Introduction

Vibration leads to change of flow field and temperature field distribution within heat transfer range, which enhances heat transfer to a large extent [1-3]. Recently, heat transfer enhancement technology has gradually become focus of research. Gau [4] and Fu[5] conduct study on vibration and heat transfer characteristics of heat transfer tubes in air, the results of which show that vibration brings about a 40% increase for heat transfer coefficient, with Nu number increment growing less with greater Re number, and the effect of amplitude outweighs that of frequency. Bronfenbrener [6] and Lee [7] have studied the vibration and heat transfer characteristics of the heat transfer tube in water. The results show that the vibration can increase the heat transfer coefficient by 40%, the greater the number of Rev and the smaller the number of Re, the better the heat transfer. The effect of the amplitude is higher than the frequency. Cheng and Tian [8-10] put forward an elastic tube bundle heat exchanger, and the mean heat transfer coefficient of elastic tube bundle with in line arrangement reaches 3 times of that of static bundle, making significant improvement for heat transfer. However, little literature has mentioned effect of fluid impingement location on heat transfer characteristics of vibration-prone elastic element. This paper performs research on this point, and a test bed is built for

Content from this work may be used under the terms of the Creative Commons Attribution 3.0 licence. Any further distribution of this work must maintain attribution to the author(s) and the title of the work, journal citation and DOI.
Published under licence by IOP Publishing Ltd

flow and heat transfer of elastic tube bundle, studying effect of fluid impingement location on heat transfer characteristics of a new type of elastic tube bundle.

2. Experimental system and method

Electric heating constant heat flux experimental system is as shown in Figure.1 for a new type of elastic tube bundle. The test bed mainly consists of test section (the new type of elastic tube bundle heat exchanger), circulating water system, electric heating system and measurement system.

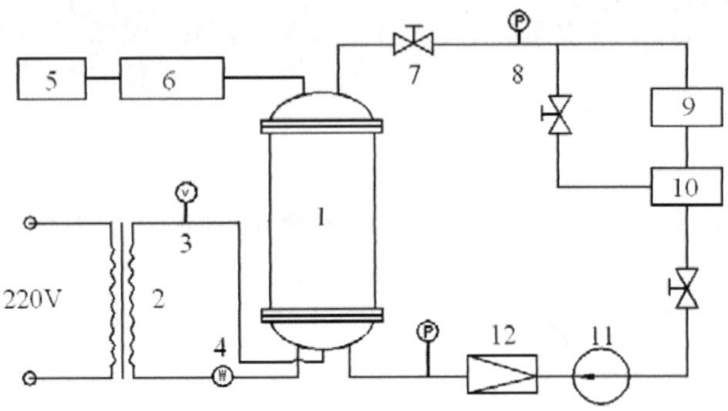

Figure. 1 Schematic diagram of electric heating constant heat-flux experiment system

1. New elastic-bundle heat exchanger, 2. Transformer, 3. Voltmeter, 4. Wattmeter, 5.computer, 6.Flunke-Netq data sampler, 7. Valve, 8. Pressure meter, 9. Cooling device, 10. Cold water tank, 11. Water pump, 12. Flow meter

Measurement system comprises temperature, pressure, flow rate and electric power measurement. Φ0.2mm copper-constantan thermal couple is adopted for temperature measurement, signals of which are all transmitted to FLUKE Net-DAQ2640A data sampler. Temperature measuring points are distributed as shown in Figure.2, with tube bundle No. from outside to inside respectively #1, #2, #3 and #4. Spring pressure gauge with accuracy class 0.4 and range 0.6MPa is adopted for pressure measurement. LWGY-50A type turbine flow meter and XSJ-30 type flow totalizator are used for flow rate measurement. PZ200E type multifunctional electric power meter is employed for power measurement.

The new type elastic tube bundle consists of 4-Φ18mm red copper tube coil connected to 2 carbon joints, the coil diameter of which are respectively 540mm, 480mm, 420mm and 360mm from outside to inside. The ideal approach for tube bundle wall temperature measurement is setting infinite measuring points, which is difficult to realize in experiment; besides, introduction of too many measuring points leads to change of external conditions for test section and damage to boundary layer, which may impact validity of experiment results to a large extent. Therefore, this experiment adopts moderate amount of measuring points evenly distributed, making the influence of wall measuring points to external heat transfer coefficient as little as possible. The measuring points layout is as shown in Figure.2.

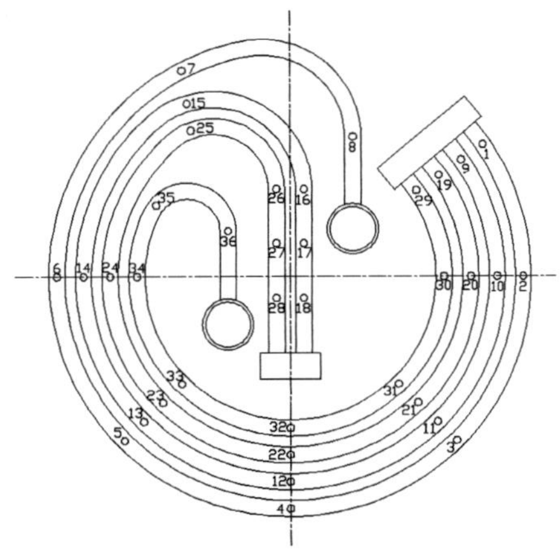

Figure. 2 Distribution of temperature measuring points of the new elastic tube bundle

3. Tube bundle arrangement mode

The experiment carries out research on heat transfer characteristics on the new type of elastic tube bundle heat exchanger with unilateral distribution-in line arrangement. Figure.3 a and b are three-dimensional overall view for the new type of elastic tube bundle heat exchanger adopted in experiment. The big free end refers to the location where four tubes join, and the small free end means the location where two tubes join.

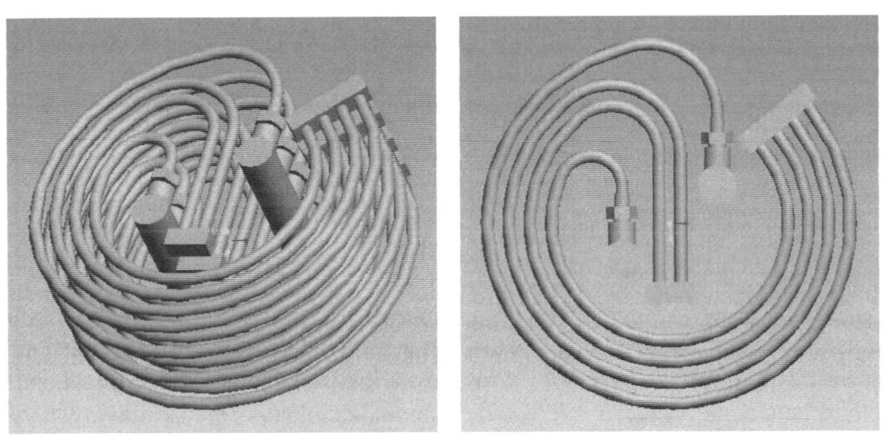

(a) (b)

Figure. 3 Three-dimensional view of unilateral distribution—in-line arrangement

4. Analysis on experimental results

4.1. Heat transfer experimental results for the new type of elastic tube bundle with different fluid impingement locations

The new elastic tube bundle is a complex elastic structure. Different fluid impingement locations lead to different vibration response characteristics, which tends to cause different heat transfer characteristics. To validate this point, the paper carries out an experimental study on heat transfer characteristics for different fluid impingement locations (big/small free end).Accordingly, the former single fluid inlet for elastic tube bundle heat exchanger is transformed to dual fluid inlets, one of which is the original inlet named A channel, and the other fluid guided by hose directly to the big/small free end, named B channel.

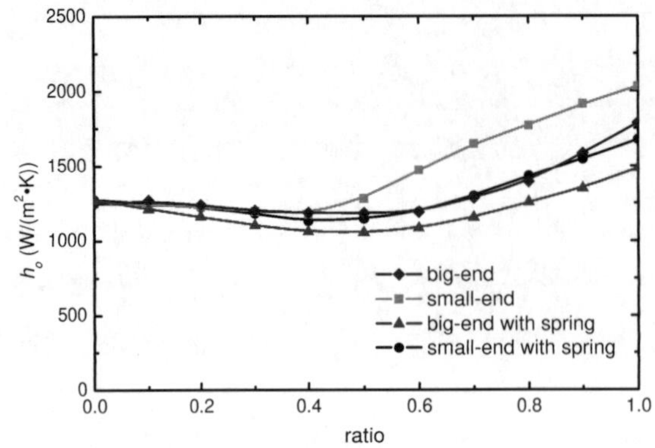

Figure. 4 Variation of external tube surface heat transfer coefficient with ratio flow impacting on the free end

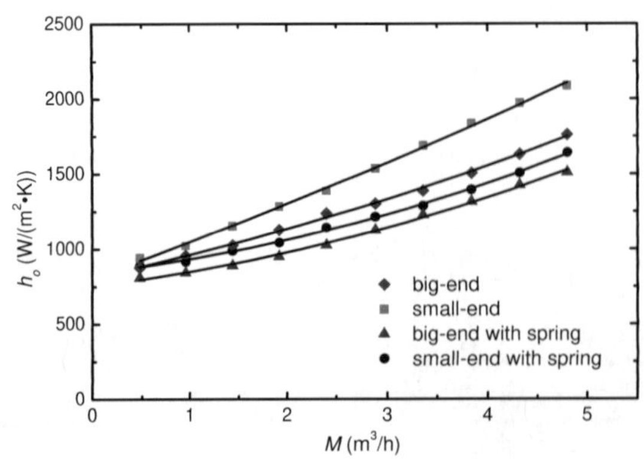

Figure. 5 Variation of external tube surface heat transfer coefficient with flow impacting on the free end

Figure.4 shows the variation of external tube surface heat transfer coefficient with ratio flow impacting on the free end. Both A and B channel are available under experiment condition, with constant total flow rate into heat exchanger. In Figure.4 the horizontal ordinate represents proportion of total flow rate impacting on the free end (B channel), and the vertical ordinate indicates external tube surface heat transfer coefficient. As is shown in the figure, the average external tube surface heat transfer coefficient presents a tendency to reduce at first and then rise with proportion of total flow rate impacting on the free end, and the valley value for heat transfer coefficient occurs at ratio flow approximately approaching 0.5(namely equal flow rate for A and B channel), showing that increase of flow rate for either channel tends to improve vibration property for the tube bundle and enhance heat transfer performance. It is found by comparison that the heat transfer performance for fluid totally impacting on the free end is better than that of fluid entering entirely through heat exchanger inlet. The best heat transfer performance occurs with fluid impinging on the small free end, and the worst occurs with fluid impacting on the big free end with spring connection; besides, the heat transfer performance for fluid impacting on the big free end and the small free end with spring connection are approximately similar. Figure.5 shows the variation of external tube surface heat transfer coefficient with flow impacting on the free end, with A channel cut off and B channel available, fluid entering heat exchanger entirely impacting on the free end. It can be seen from the figure that average heat transfer coefficient for the tube bundle grows with the increase of flow impacting on the free end.

Besides, the optimum overall heat transfer performance occurs with fluid impinging on the small free ends, which corresponds well with Figure.4.

4.2. Research on local heat transfer characteristics for elastic tube bundle under different conditions

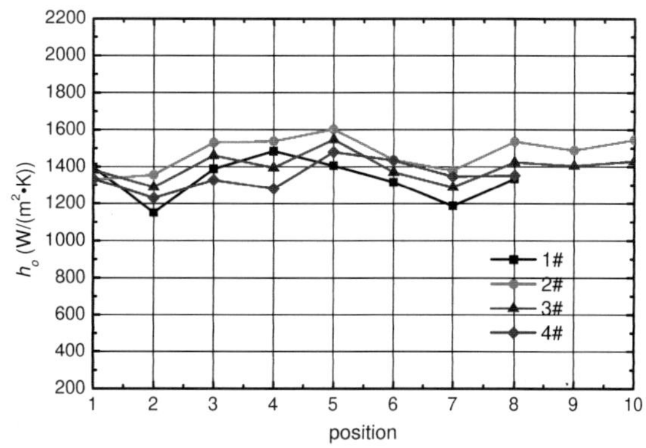

Figure. 6 Distribution of local surface heat transfer coefficient of unilateral distributed--in-line elastic tube bundle, flow impacting on the small end=2.4m^3/h

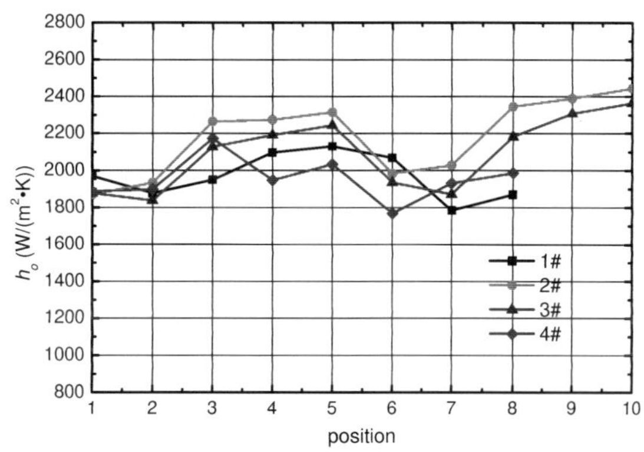

Figure. 7 Distribution of local surface heat transfer coefficient of unilateral distributed-in-line elastic tube bundle, flow impacting on the small end=4.8m^3/h

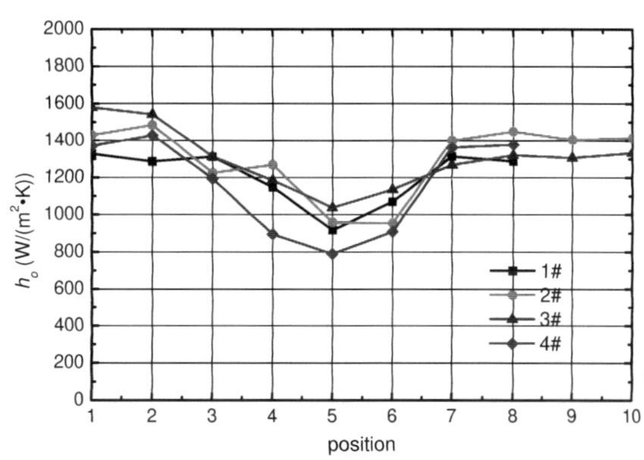

Figure. 8 Distribution of local surface heat transfer coefficient of new elastic tube bundle, flow impacting on the big end=2.4m^3/h

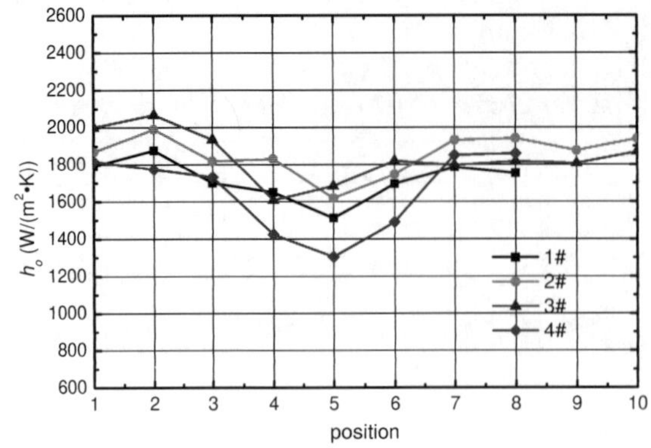

Figure. 9 Distribution of local surface heat transfer coefficient of new elastic tube bundle, flow impacting on the big end=4.8m^3/h

Figure.6 and 7 are respectively the distribution of local surface heat transfer coefficient of the elastic tube bundle, with fluid impacting on the small free end at different flow rates. It can be seen from the figures that with fluid impinging on the small end, similar heat transfer characteristics occur for the elastic tube bundle under the two conditions. Lower heat transfer coefficient occurs at the fixed end and the big free end, while higher heat transfer coefficient occurs at the small free end, the reason for which might be that fluid velocity and vibration characteristics vary with the tube bundle locations, with fluid velocity and vibration property at the small free end better than other locations. With the growth of fluid velocity, a significant increase of heat transfer coefficient for the tube bundle can be found by comparing the two figures.

Figure.8 and 9 are respectively the distribution of local surface heat transfer coefficient of the elastic tube bundle, with fluid impacting on the big free end at different flow rates. Identical to the conditions with fluid impinging on the small end, it can be seen from the figures that with fluid impinging on the big end, similar heat transfer characteristics occur for the elastic tube bundle under the two conditions. The highest heat transfer coefficient occurs at the big end, and the valley value occurs in the region apart from the free end, while little fluctuation is observed for heat transfer performance of the small free end.

5. Conclusions

1) A constant heat flux heat transfer test bed is built, and an experimental study on external tube heat transfer characteristics of a new type of elastic tube bundle is conducted, obtaining overall and local heat transfer performance of the elastic tube bundle at different Re numbers.

2) Within the experimental parameter range, better heat transfer performance is observed for the condition with fluid entirely impacting on the free end than all fluid entering through heat exchanger inlet. Besides, the best heat transfer performance occurs under the condition fluid entirely impinging on the small free end of tube bundle.

3) With fluid impacting on big/small free end, similar heat transfer characteristics are presented for the elastic tube bundle under two conditions, namely lower heat transfer coefficient for the fixed end and the big free end, and highest heat transfer coefficient for the small free end, the reason for which might be the differences of velocity and vibration characteristics for fluid around the tube bundle at different locations.

References

[1] C. H. Guo, X. G. Hu, W. Cao, D. Yu, D. W. Tang Effect of mechanical vibration on flow and heat transfer characteristics in rectangular microgrooves Original Research Article[J]. Applied Thermal Engineering, 2013, 52(2): 385-393.

[2] B. Jiang, M. C. Tian and X. L. Leng et al. Numerical simulation of flow and heat transfer characteristics outside a periodically vibrating tube[J]. Journal of Hydrodynamics, 2008, 20(5): 629-636.

[3] B. Jiang, M. C. Tian and X. L. Leng et al. Experimental research and field synergy analysis of flow and heat transfer outside a vibratory tube[J]. Journal of vibration and shock, 2009, 28(5): 102-105.

[4] C. Gau, J. M. Wu, C. Y. Liang. Heat transfer enhancement and vortex flow structure over a heated cylinder oscillating in the crossflow direction [J]. Journal of Heat Transfer, 1999, 121: 789-795.

[5] W. S. Fu, B. H. Tong. Numerical investigation of heat transfer from a heated oscillating cylinder in a cross flow [J]. International Journal of Heat Mass and Transfer, 2002, 45(11): 3033-3043.

[6] L. Bronfenbrener, L. Grinis, E. Korin. Experimental study of heat transfer intensification under vibration condition [J]. Chemical Engineering and Technology, 2001, 24(4): 367-371.

[7] Y. H. Lee, D. H. Kim, S. H. Chang. An experimental investigation on the critical heat flux enhancement by mechanical vibration in vertical round tube [J]. Nuclear Engineering and Design, 2004, 229(1): 47-58.

[8] L. Cheng, Y. Qiu. Complex heat transfer enhancement by fluid induced vibration [J]. Journal of Hydrodynamics, Ser. B, 2003, 15(1): 84 -89.

[9] M. C. Tian, L. Cheng and Y. Q. Lin et al. Heat transfer enhancement by crossflow- induced vibration [J]. Heat Transfer-Asian Research, 2004, 33 (4): 211-218.

[10] L. Cheng, T. Luan and W. Du et al. Heat transfer enhancement by flow-induced vibration in heat exchangers [J]. International Journal of Heat and Mass Transfer, 2009, 52 (3-4): 1053-1057.

IOP Conf. Series: Journal of Physics: Conf. Series **1064** (2018) 012007 doi:10.1088/1742-6596/1064/1/012007

Simulation analysis of air cooling for vehicle power battery

Liming Xiang[1, 2],*, Jingpeng Cui[1]

[1]School of Automotive and Traffic Engineering, Hubei University of Arts and Science, Xiangyang, China.
[2] Hubei Superior and Distinctive Discipline Group of "Mechatronics and Automobiles", Xiangyang, China.

*Corresponding author e-mail: 846410358@qq.com

Abstract. Because of its high voltage, strong storage and circulation ability, no memory effect, high safety, low environmental pollution, Lithium ion batteries are the most widely used as power battery in the electric cars, but its heat dissipation problem can not be ignored. When the power battery is working, the heat is accumulated. If it cannot be discharged in time, it will seriously affect the normal operation of the battery and even cause danger. As a heat dissipation medium, air can give good consideration to heat dissipation effect and cost. In this paper, CFD software was used to simulate the flow field of the power battery when working. Specific research method of this article is that the heating characteristics of lithium ion battery are studied, by studying thermal field distribution of the battery pack, and the battery pack is optimized to improve. Finally the temperature of the battery pack was kept a reasonable range.

1. Introduction

As the most important component of electric vehicle, the battery pack of electric vehicle is good enough to determine the performance of the vehicle. At the same time, heat dissipation problem is also not to be ignored. Higher capacity density will inevitably bring more heat generated, so heat dissipation problem needs to be taken seriously. Air cooling is the most widely used method in electric vehicle power system. Aiming at the parallel hybrid power connection mode, H.J. Jiao et al [1] developed the air-cooled heat dissipation device by using MH-NI battery in natural convection heat dissipation mode. They put 24 battery modules in two horizontal rows, air flowed into the battery modules in parallel way from the bottom of the battery pack. H. Liu et al [2]. Arranged 16 batteries in sequence and installed triangular baffles at the front of the battery to increase the air flow between the battery gap. The influence of the cooling structure of the battery on temperature and flow field distribution was studied by using CFD software simulation. X. Z. Li et al [3] used 10 Ni-MH batteries to form battery pack, the battery pack was tested for charge-discharge and high-temperature forced cooling. The temperature inside and on the surface of each battery was monitored, it was found that the best heat dissipation effect of the battery pack was achieved by using the insulated grating structure.

This paper mainly studied the thermal field of single battery of electric vehicle power battery and its battery pack. Firstly, by studying the structure and working principle of monomer battery, the thermogenesis of monomer battery is mastered. Then, on the basis of studying the single battery, the thermal field distribution of the battery pack was further studied. CFD software was used to simulate the flow field distribution of the battery pack.

Content from this work may be used under the terms of the Creative Commons Attribution 3.0 licence. Any further distribution of this work must maintain attribution to the author(s) and the title of the work, journal citation and DOI.
Published under licence by IOP Publishing Ltd

2. Thermal simulation analysis of power battery pack

2.1. Thermal simulation of single battery

Single battery, as the smallest battery unit in the battery pack, is the smallest heating unit in the battery pack. In order to simulate the heat dissipation problem of the battery pack, it is necessary to study the heat generation of a single battery. According to the three dimensional parameters of the battery, some software was used to build the three dimensional model. Based on the three dimensional model, it was imported into CFD software to conduct grid division and further simulation. Fig.1 (a) is the three dimensional model of the battery, and (b) is the picture after the cell was divided into grids, and materials and boundaries were set for each part respectively.

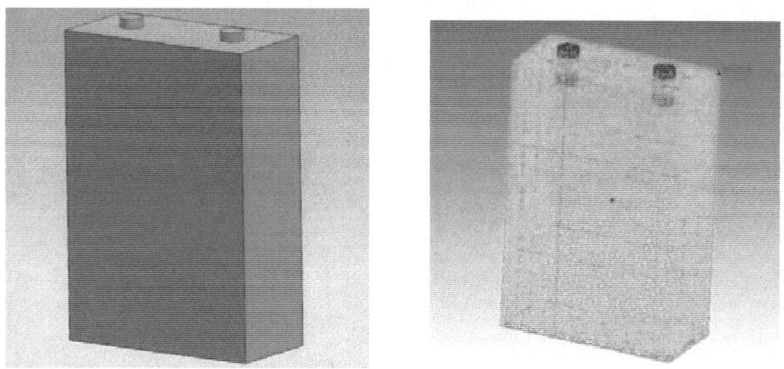

(a) (b)

Figure 1. Battery three dimensional model and grids.

The shell thickness of the battery is 3mm, the material inside the battery in the shell is regarded as uniform heating, the heat generation rate of the battery is 4749.15W/m^3 when the battery is discharged at 0.5C. The model of single battery was imported into CFD software, and the material parameters of the battery, the battery heating rate, the boundary conditions, the solver, the iteration steps, was set and calculated. The simulation results are shown in Fig. 2, and Fig.2 (a) is the overall temperature cloud diagram of the battery, andFig.2 (b) is the temperature distribution diagram of the battery surface.

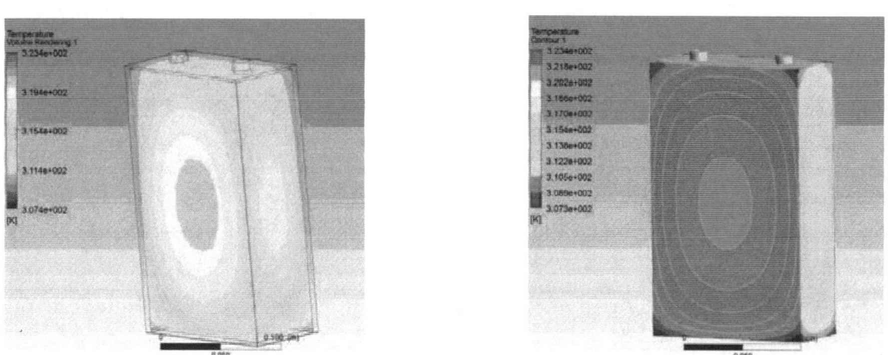

(a) (b)

Figure 2. Overall temperature cloud diagram of the battery and surface temperature cloud diagram

It could be seen from the simulation results that when the battery was discharged at 0.5C, the heat inside the battery would be generated and spread outwards, with the maximum temperature difference approaching 26K. By monomer battery simulation it could be seen that the highest temperature of the surface of the battery was 47.05 ℃, and the lowest temperature was 34.25 ℃, the difference of the temperature of was 12.8 ℃, and the temperature difference of 12.8 ℃ was more than reasonable temperature difference of 5 ℃ battery pack. Therefore, it was necessary to install the cooling system when the battery works in the battery pack.

2.2. *Heat dissipation simulation of battery pack*

On the basis of simulating the single battery, the heat distribution on the single battery was understood, and the necessity of cooling the battery was determined. Battery pack was the place where batteries were concentrated, and the 100 batteries accumulated heat in the battery pack. If the heat was not discharged in time, there would be danger, so the battery pack was simulated.

There were 100 battery cells in the battery pack. A battery module was consisted of 5 individual batteries, and a battery module was consisted of 5 individual batteries, and a total of 20 battery modules were arranged in series in the battery pack. Fig.3 was that five individual batteries were closely aligned to form a battery module.

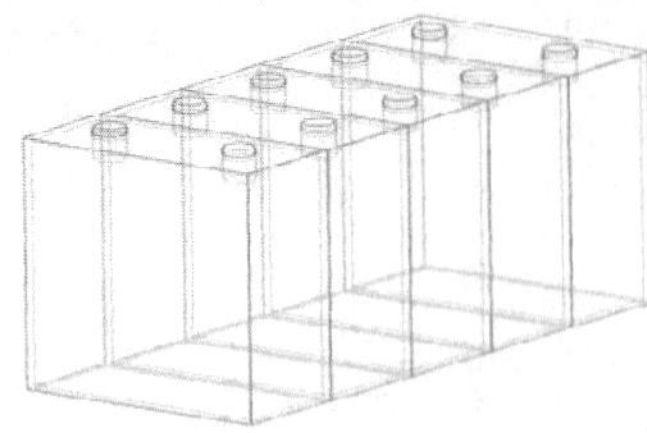

Figure 3. A battery module consisting of 5 single cells

The influence on the heat dissipation for the arrangement of the battery pack in the battery box, that is to say, the distance between the battery modules, was studied. The space was set as 10mm, 15mm, 20mm, 25mm and 30mm, respectively. The distance between the battery pack and the battery box wall was 30mm. The diameter of the inlet was 100mm, and the inlet wind speed was 2m/s. The five clearance structures were shown in Fig. 4 (a) (b) (c) (d) (e), and (f) was the battery box structure.

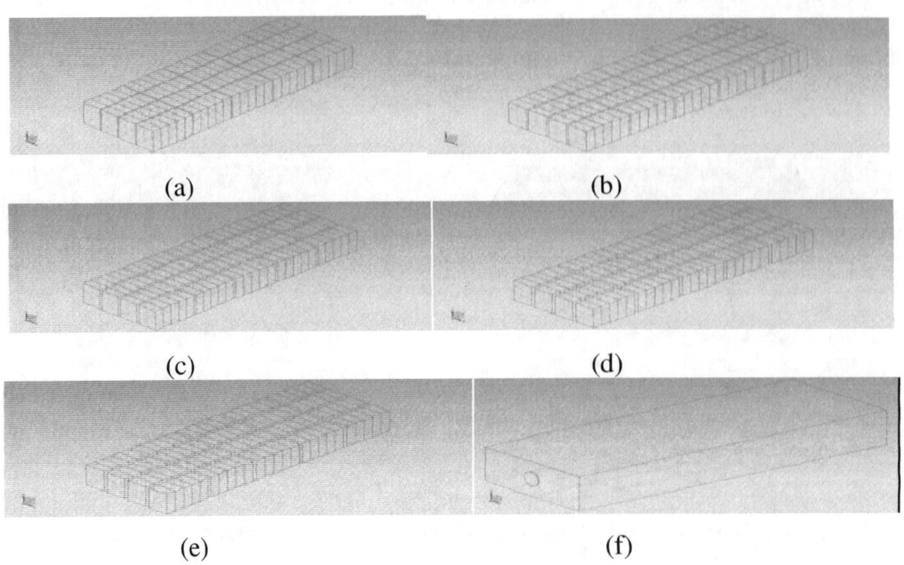

(a) (b) (c) (d) (e) (f)

Figure 4. Battery pack structure of 5 kinds of clearance and battery box

The pressure, velocity and temperature cloud maps of the five layout forms were obtained by simulation respectively. Fig. 5 showed the temperature cloud diagram of 5 groups of simulation, and (a), (b), (c), (d) and (e) were temperature clouds of simulation, and the spaces between battery modules were 10mm, 15mm, 20mm, 25mm and 30mm respectively.

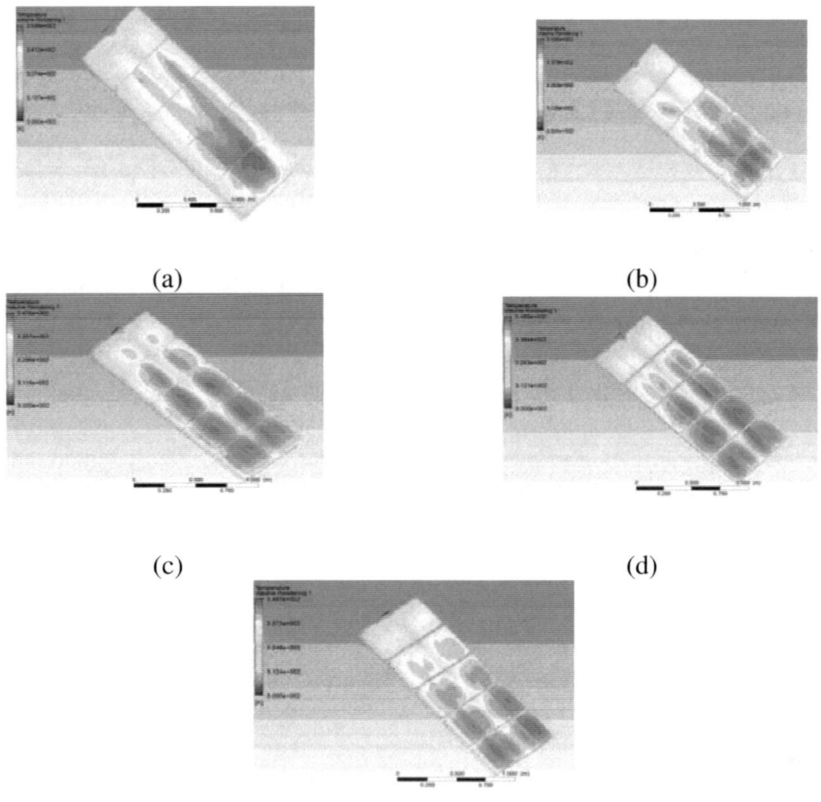

(a) (b) (c) (d) (e)

Figure 5. Temperature cloud diagram of 5 structures

As can be seen from the temperature cloud diagram, the temperature at the front of the battery box decreased significantly, and most battery packs at the back still showed high temperature. The interstitial wind speed in the middle of the battery pack was high, and the wind speed on both sides was almost zero. Due to the low inlet wind speed, the overall cooling effect of the battery box was not good.

In order to improve the heat dissipation effect of the battery pack at the back of the battery pack, the influence of wind speed on the heat dissipation of the battery pack was studied. The inlet wind speeds of 1m/s, 2m/s, 3m/s, 4m/s, 5m/s, 6m/s and 7m/s were respectively set, and the inlet diameter was 100mm. The space between the battery pack was 20mm. CFD software was used for simulation analysis and the simulation results were compared. After simulation, the temperature cloud maps of the battery pack in the battery box were shown in Fig.6 (a), (b), (c), (d), (e), (f) and (g), and the temperature cloud maps of the inlet wind speed of the battery pack were1m/s, 2m/s, 3m/s, 4m/s, 5m/s, 6m/s and 7m/s.

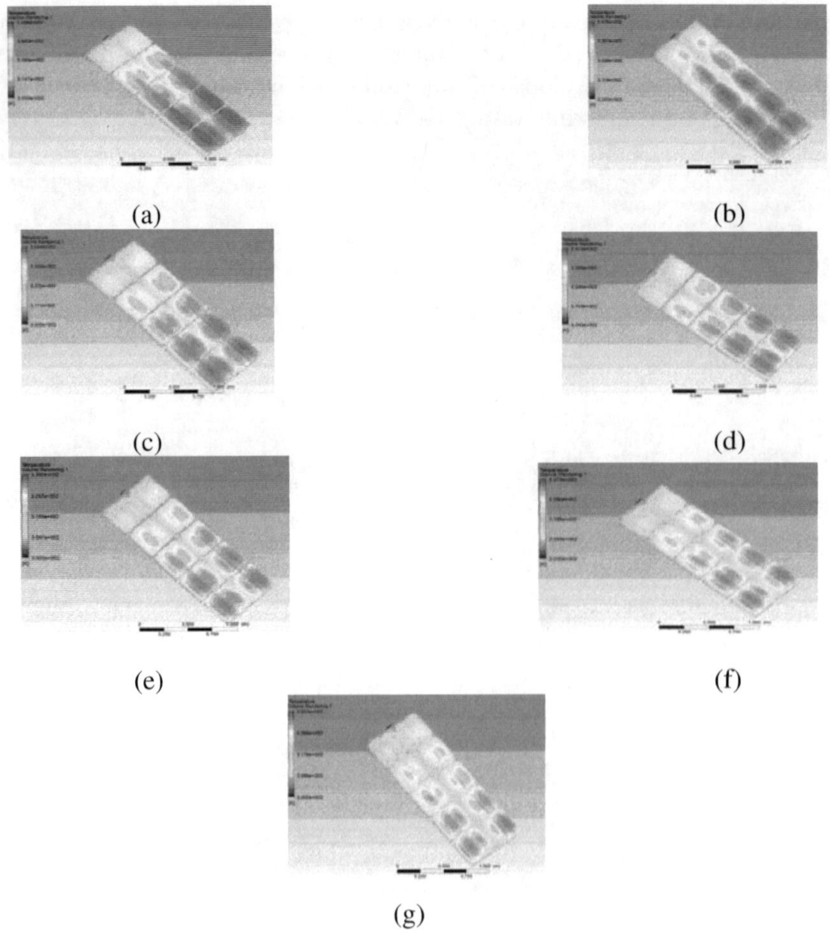

Figure 6. Temperature cloud diagram of battery pack under 7 wind speeds

With the increase of inlet wind speed, the average pressure in the battery box was increased. The increase of wind speed was beneficial to the circulation of air and enhanced the cooling effect, and the maximum temperature of the battery pack in the battery compartment was decreased gradually. As the wind speed continued to increase, the temperature was decreased steadily and the cooling effect was decreased gradually.

3. Conclusion

By setting a reasonable gap between the battery modules, the air could flow more smoothly when it entered the battery box, which could speed up the flow of air and took away the heat generated by the battery pack. The faster the wind speed, the more air flew through the battery's surface, speeding up heat exchange. In addition to accelerating heat exchange, the uniformity of battery temperature in the battery pack should be considered. By increasing the number of inlet and arranging the position properly, air could flow evenly through each battery pack.

This paper provides a theoretical basis for the design of battery pack through the simulation of wind cooling and heat dissipation of electric vehicle power battery. But the accuracy of simulation is limited, at the same time, the influence of other circuit, mechanical structure and control chip on temperature and flow field is not considered, and the simulation of the real situation is not perfect. Therefore, the research on this problem needs to be further improved.

Acknowledgments
This work was financially supported by the Educational Commission of Hubei Province of China (No.D20152603).

References

[1] H.J. Jiao, H.Y. Xiang, L.C. Ji, D. Wang, J. Fang, Development of the ventilating and cooling system of the nickel-hydrogen battery pack for hybrid electric vehicles, Automobile Technology. 1 (2003)23-25.
[2] H. Liu, K. Yang, D. Hui, H. Q. Zhang, Air cooloing structure design of Li-ion battery pack, Chinese Journal of Power Sources, 35(2011)1208-1210.
[3] X. Z. Li, F. Su, J. C. Huang, R. Rong, X. B. Xun, Effects of heat-dissipating structure on the performance of Ni-MH batteries, Chinese Battery Industry. 12(2007)27-30.

IOP Conf. Series: Journal of Physics: Conf. Series **1064** (2018) 012008 doi:10.1088/1742-6596/1064/1/012008

Research of Heat Dissipation Technologies for Electromagnetic Linear Actuator

Jianguo Dai, Shanzhen Xu*, Jianhui Zhu, and Xiaoming Hu

Huaiyin Institute of Technology, Huai'an, China.

*Corresponding author e-mail: hyitjeancool@163.com

Abstract. As the core drive component of Electromagnetic Driven Valve Train (EDVT), Electromagnetic Linear Actuator (ELA) directly determines the performance of EDVT, and thus affects the dynamic performance and fuel economy of the engine. During high-velocity motion, ELA produces heat that will negatively affect its property or even make some damages. On the basis of analysis of temperature rise and temperature distribution of ELA, three heat dissipation technologies were proposed in this study, and relevant experimental test was carried out for each of the three technologies. Through comparing data of temperature rising test and considering factors such as noise and space usage, the method of silent fan equipped with cooling fin was selected as optimal heat dissipation scheme. This study provides sound test data on heat dissipation technology of ELA and is of important engineering significant.

1. Introduction

Compared with Cam Driven Valve Train (CDVT) in conventional engine, EDVT can realize fully flexible regulation of valve movement law within the entire operating range, which makes the engine working at optimum performance and thus significantly improve the dynamics and fuel economy of the engine [1-4]. The core drive component of EDVT is ELA. The energy consumption and performance of ELA directly influence the energy consumption, high-speed adaptability and long-time work stability of EDVT, and thus affects the dynamics and fuel economy of the engine [5]. Therefore, research on ELA energy consumption and heat dissipation technologies of ELA is of theoretical significance and practical engineering value.

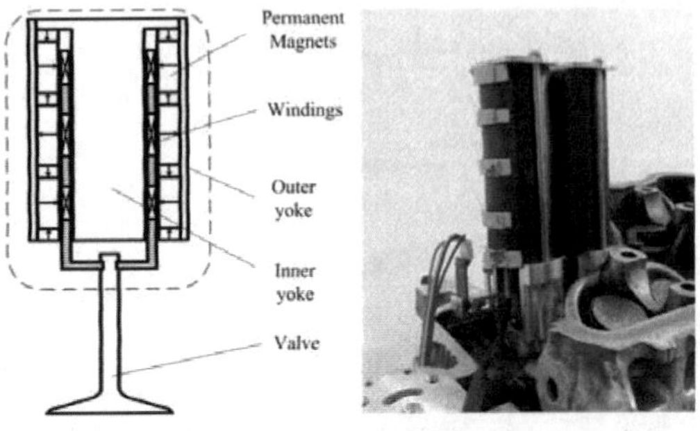

Figure 1. ELA structure and experimental prototype

Content from this work may be used under the terms of the Creative Commons Attribution 3.0 licence. Any further distribution of this work must maintain attribution to the author(s) and the title of the work, journal citation and DOI.
Published under licence by IOP Publishing Ltd

In this work, the ELA is basically a cylinder type moving coil linear motor [6], of which the structure and experimental prototype are shown in Figure 1. On the basis of previous energy consumption analysis and temperature rise test of ELA [7, 8], this paper proposes three heat dissipation technologies of ELA and carries out in-depth analysis and comparison.

2. Temperature rise of ELA

Previously, we have carried out in-depth analysis of heat transfer process of ELA and established mathematical models of radial and axial heat transfer. On this basis, the finite element model of ELA was constructed with the aid of Ansoft Maxwell and Ansys Workbench, and bidirectional coupling dynamic simulation of energy consumption and temperature rise was performed [8]. Through simulation, the temperature rise and distribution can be obtained, as shown in Figure 2 and Figure 3, respectively.

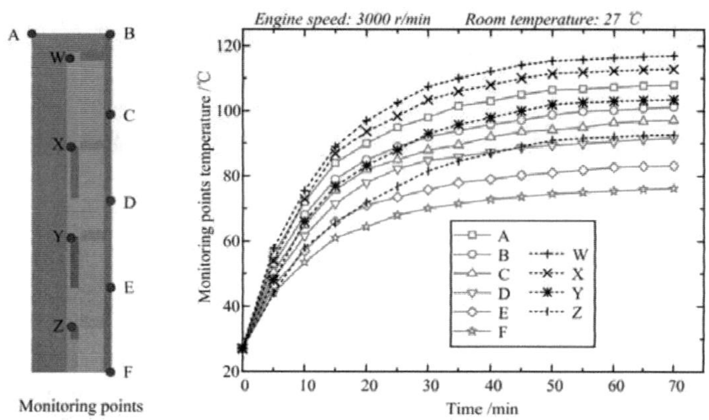

Figure 2. Temperature rise curve of ELA's internal measuring point at engine speed of 3000r/min

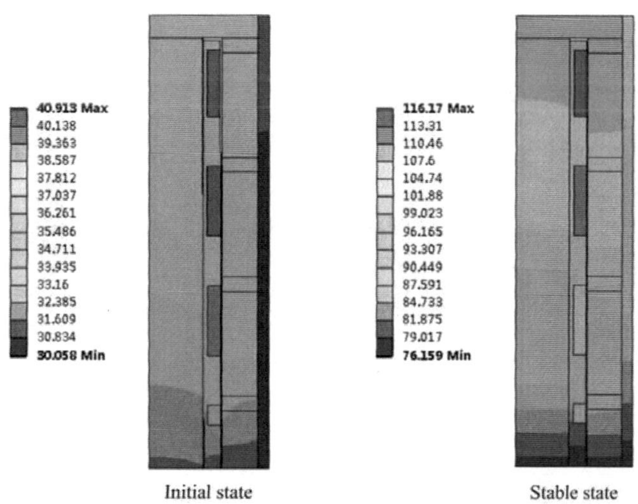

Figure 3. Temperature distribution cloud chart of ELA at initial and final stable state when engine speed is 3000 r/min

As shown in Figure 2, the temperature of ELA gradually increases with working time, and eventually reaches a stable state (where the temperature maintains at a constant value and ELA enters into stable working state). However, when the engine speed is 3000r/min, ELA enters into stable working state with maximum temperature reaching over 115℃ and minimum temperature reaching at point F reaching about 70℃. If continuing working under such high temperature, the permanent magnet material

NdFeB in ELA will be demagnetized, some plastic parts show deformation and aging, and the working performance of ELA will be reduced or even damaged. In the case when the engine speed is more than 3000r/min, the situation will be even worse. Therefore, heat dissipation of ELA is of engineering significance to guaranteeing the working stability and service life of ELA under high-intensity work load.

Figure 3 shows the comparison of temperature distributions of ELA under initial state and final stable state. It can be seen that the high-temperature zones of ELA are mainly concentrated at inner yoke and coils. In the vertical direction, the temperature gradually increases from bottom to top, and the temperature is the most concentrated especially at the top position. As for heat dissipation and cooling measures for ELA, the heat dissipation near the position of inner moving coils and at the position of upper cover should be concerned the most.

3. Heat dissipation scheme of ELA

According to the temperature rise and temperature distribution of ELA in section 2, the high-temperature zones of ELA are mainly concentrated near moving coils and at top cover, so heat dissipation at these positions are very important. Through in-depth analysis, three heat dissipation schemes were proposed in this study.

3.1. Reed valve pumping

The working principle of reed valve pumping is shown in Fig.4. The inner moving coil of ELA produces air flow, which drives the reciprocating motion of reed valve, resulting in pump suction effect. Therefore, rapid flow of air and heat convection can be realized. In such innovative design, only installation of a reed valve rather than other complex devices is needed, which is basically a minor structural alteration of ELA. Dr. Fan Xinyu has made in-depth analysis on this issue and achieved certain results [9], which will not be reiterated here.

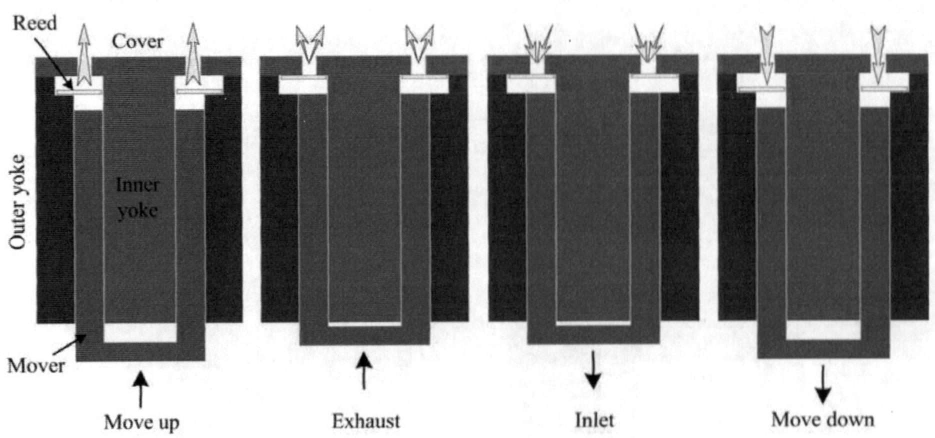

Figure 4. Working principle of reed valve pumping

Through conducting temperature rise test of ELA with reed valve pumping and comparing it with the temperature rise results of ELA without reed valve pumping, temperature variation curve of ELA at measuring point B can be obtained, as shown in Fig.5. Since point B was the point where the highest external temperature value was measured except for point A and subsequent two scheme cannot measure the temperature at point A, point B was selected as the unified comparison point in this study.

According to Figure 5, the adoption of pump suction leaf valve has certain but not significant inhibition effect on temperature rise of ELA. When the engine speed exceeds 4000r/min, it is still difficult for ELA to reach stable state at temperature of 110℃. If the engine speed is even higher, the temperature for stable state will be increased accordingly. When the temperature exceeds 120℃, the performance of ELA will get worse, and ELA will be easily damaged under such high-temperature and high-intensity working condition.

In conclusion, reed valve pumping has compact structure and innovative design, but has worse heat dissipation effect, with limited amount of heat being brought away.

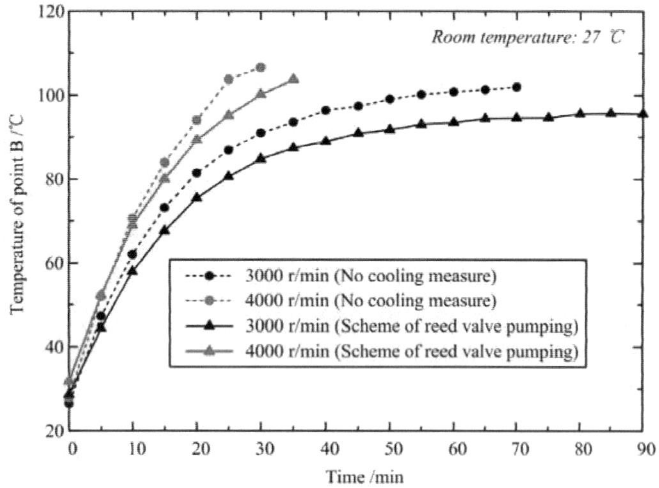

Figure 5. Comparison of temperature variation at measuring point B

3.2. Fan+ flange support

ELA is a executive device of the EDVT in engine, of which the surrounding space is compact, and the external heat abstractor cannot fully cover it. Through analysis of ELA temperature distribution, we decided to install a heat abstractor at top cover of ELA.

Considering the temperature at ELA moving coil is the highest, it is of vital significance to lower the temperature at the position of inner moving coils and air gap. However, according to the analysis result of pump suction leaf valve scheme, the air convection effect produced by ELA movement itself is not enough. So we were wondering if installation of a external forced air blast device can have sound cooling effect on ELA? In this study, we considered to employ "fan + flange support" scheme to perform forced cooling with the aid of the air vent at top of ELA, as shown in Figure 6.

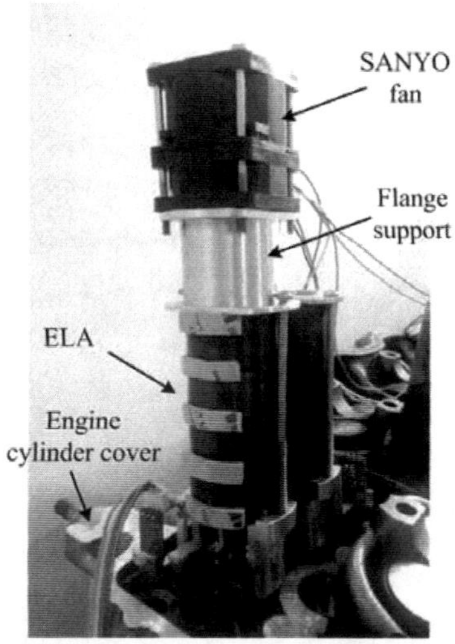

Figure 6. Layout of "fan + flange support" heat dissipation scheme

To allow air penetrate via air gap to lower end of ELA, the wind power of fan must be strong enough so as to achieve sufficient forced air cooling effect. In this scheme, the fan adopted was SANYO fan PCRA0412P4K03, with parameters listed in Table 1.

Table 1. Parameters of SANYO fan PCRA0412P4K03

Parameter / Duty cycle	Rated voltage	Rated current	Rated power input	Rated speed (Inlet)	Rated speed (Outlet)	Static pressure	Operating temperature range
100%	12V	1.6A	19.2W	17500min^{-1}	11700min^{-1}	650Pa	-10℃~70℃
0%		0.19A	2.28W	5250min-1	3510min-1	58.5Pa	

Since measuring point A was occupied, the data measured at other 5 points were obtained, as shown in Figure 7. Through performing force-air cooling for ELA with a powerful fan, the temperature rise of ELA was significantly inhibited. When the engine speed is lower than 5000r/min, temperature at all external measuring points can be finally maintained within 90℃; even when the engine speed is 6000r/min, the temperature can be finally maintained around 110℃, indicating a significant decrease of temperature of ELA.

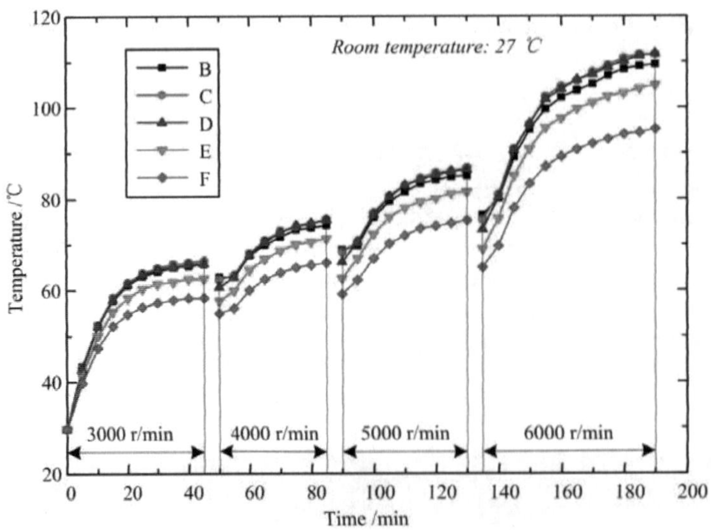

Figure 7. Temperature rise curves of ELA at 5 measuring points after adoption of "fan+ flange support" heat dissipation scheme

However, such scheme is subjected to some defects. The dimensions of fan and flange are large, which will increase the overall height of ELA after installation of fan and flange. When ELA is in high-speed movement, inertia shaking will be inevitably resulted. In actually application, some problems will be caused due to dimensional limit of engine cylinder head. In addition, there is a loud noise during the operation of the powerful SANYO fan, which is a significant noise pollution.

3.3. Silent fan + cooling fin

Regarding external heat dissipation at top cover of ELA, there is another scheme using silent fan and cooling fin, as shown in Figure 8. To reduce the noise of external installed fan, a silent fan is adopted in this scheme. Moreover, the employment of cooling fin enlarge the heat dissipation area at top cover of ELA, thus realizes effective cooling effect.

Through experimental test of such scheme, temperature rise curve can be obtained, as shown in Figure 9. When the engine speed is 6000r/min, the temperature of ELA at all external measuring points can be finally maintained within 100℃. This indicates that the scheme of silent fan + cooling fin effectively reduce the operating temperature of ELA and improve the reliability and stability of ELA during long-time operation.

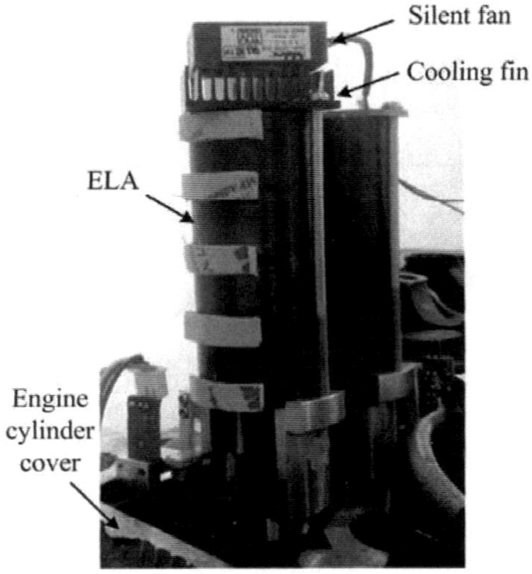

Figure 8. Layout of air cooling device using silent fan and cooling fin

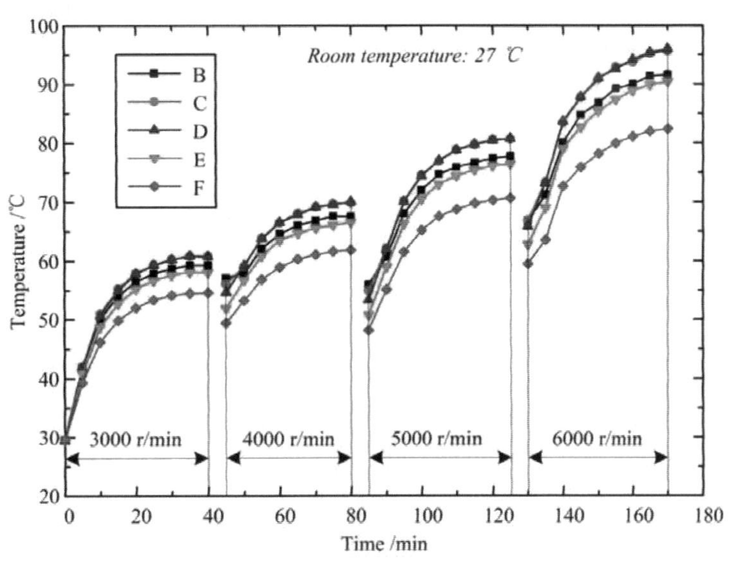

Figure 9. Temperature rise curves of ELA at 5 measuring points using silent fan and cooling fin

4. Comparison of three heat dissipation schemes

Through analysis and comparison of three different heat dissipation schemes, an optimal one was finally determined. By collecting and analyzing test data of three schemes, the comparison of temperature rise curves of ELA at measuring point B by three different schemes can be obtained, as shown in Figure 10-a.

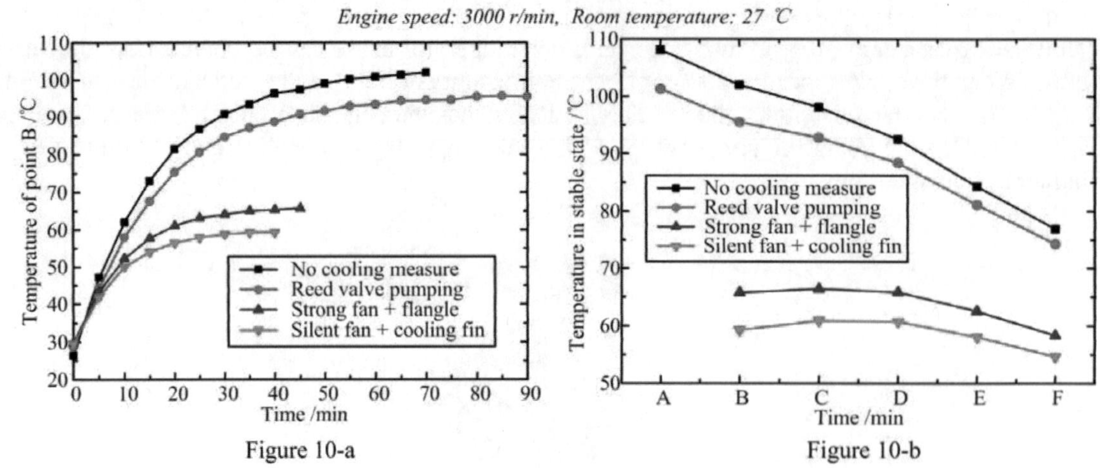

Figure 10. Comparison of data of three different heat dissipation schemes

As shown in Figure 10-a, the three schemes exert cooling effect on ELA in different degrees. Among them, the cooling effect of pump suction leaf valve scheme is relatively unsatisfactory, whereas the other two schemes using fan have significantly cooling effect as the final temperature of ELA at measuring point B can be maintained within 65℃ when the engine speed is 3000r/min. Of two schemes using fan, the scheme of silent fan + cooling fin has better cooling effect.

Figure 10-b shows the temperature values at different measuring points under different schemes when ELA reaches stable state. The results of Figure 10-b are consistent with the results in Figure 10-a, from which other laws can be obtained. All three schemes can reduce the temperature values measured at different points at stable state of ELA, wherein the cooling effect at point A, B, C is significantly better than that at point D, E, F. The two schemes using fan have better cooling effect, which make the final stable temperature at point B and C (at the two points, the temperature used to be high) lower than that at point D. The satisfactory cooling effect is associated with the installation site of fan and also the consequence of comprehensive analysis of ELA temperature distribution.

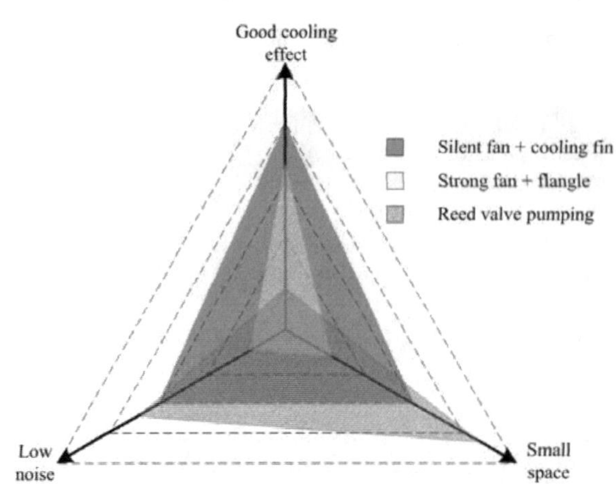

Figure 11. Diagram of comprehensive performances of three heat dissipation schemes

Through compressive analysis of three heat dissipation schemes, diagram of comprehensive performances of three schemes can be obtained, as shown in Figure 11. It can be seen that the three schemes were compared in terms of cooling effect, noise intensity and space occupation. The scheme of silent fan + cooling fin has the best comprehensive performance. The scheme of powerful fan + flange support has significant cooling effect, but has excessive noise and space occupation, which limits its comprehensive performance. In contrast, the scheme of reed valve pumping has advantages in noise and space occupation, but has unsatisfactory cooling effect. Moreover, it does not require external air cooling power supply, which is the most distinguished feature compared with the other two schemes. Pump suction leaf valve has various advantages, which is of great research value. If the cooling effect can be improved, the comprehensive performance of pump suction leaf valve can be significantly improved. Therefore, further researches on pump suction leaf valve are needed. According to comparison of three schemes at present, the scheme of silent fan + cooling fin has the best comprehensive performance.

5. Conclusion

Three heat dissipation schemes for ELA were analyzed and compared in this study, and following conclusions can be obtained:

1. High-temperature zones of ELA are mainly concentrated near inner moving coils. In vertical direction, the temperature gradually increases from bottom to top, therefore the heat dissipation at the position of moving coil and top cover should be concerned the most for cooling treatment of overall ELA.
2. Through comparing temperature rise data and evaluation from cooling effect, noise intensity and space occupation, the scheme of silent fan + cooling fin was selected as optimal heat dissipation scheme.
3. Although reed valve pumping scheme has limited cooling effect, it has various advantages including innovative design, extremely small space occupation, no requirement of external air cooling power supply, which is worth of further researches.

This study provides test data and analysis basis for further development of ELA heat dissipation technologies, which is of engineering application value.

Acknowledgments

This work was supported by the National Natural Science Foundation of China (Grant No. 51605183, 51505171), the IUR Union Prospective Program of Jiangsu Province (BY2016061-13) and the Science and Technology Support Programs of Huai'an (Grant No. HAG201603).

References

[1] Liang Liu, Siqin Chang. A moving coil electromagnetic valve actuator for camless engines. IEEE International Conference on Mechatronics and Automation, 2009, 176-180.
[2] Xinyu Fan, Siqin Chang, Liang Liu, et al. Realization and optimization of high compression ratio engine with electromagnetic valve train. Applied Thermal Engineering, 2016, 112: 371-377.
[3] S Binbin, G Song, L Bo. Research of intake valve deactivation on engine performance. IJE Transactions A: Basics, 2016, 29(04): 473-481.
[4] Jiangtao Xu, Siqin Chang, Xinyu Fan, et al. Effects of electromagnetic intake valve train on gasoline engine intake charging. Applied Thermal Engineering, 2015, 96: 708-715.
[5] Liang Liu, Siqin Chang. Improvement of valve seating performance of engine's EMV system. Mechatronics, 2011, 21:1234-1238.
[6] Siqin Chang, Liang Liu. A moving coil permanent magnet linear actuator with high power density. China Patent, CN101127474B, 2010.
[7] Jianguo Dai, Siqin Chang. Loss analysis of electromagnetic linear actuator. International Journal of Applied Electromagnetic and Mechanics, 2014, 46(3): 471-482.
[8] Jianguo Dai, Jingjing Xia, Cheng Wang, Shanzhen Xu. Thermal analysis of an electromagnetic linear actuator. Advances in Mechanical Engineering, 2017, 09(12): 1-12.
[9] Fan xinyu, Chang Siqin, Liu Liang, Dai Jianguo. Research on temperature suppression scheme of electromagnetic valve train based on pumping effect. Transactions of the Chinese Society for Agricultural Machinery, 2016, 47(01): 337-343.

IOP Conf. Series: Journal of Physics: Conf. Series **1064** (2018) 012009 doi:10.1088/1742-6596/1064/1/012009

Numerical Simulation of Tail Flow Field of Four - Nozzle Solid Rocket Engine

Bailin Zha, Zhe Hui*, Qingdong Su

Rocket Force University of Engineering, Xi'an, P. R. China

*Corresponding author e-mail: HJJhuizhe@163.com

Abstract. A solid rocket motor with four nozzles is taken as the research object. According to its actual flight environment and the simulation boundary conditions are given. Then, the integrated simulation of nozzle-flow field in exhaust plume of solid rocket motor is carried out by means of FLUENT. The velocity of flow field in exhaust plume, temperature, and pressure are obtained. The simulation results are in good agreement with the actual flow conditions and the existence of Mach disk is proved.

1. Introduction

Multi-nozzle rocket engine is widely used in high thrust rocket and variable thrust rocket. The characteristics of wake flow field of multi-nozzle rocket engine have great influence on the overall design of rocket, the selection of materials and the design of launching device. Therefore, it is necessary to study the characteristics of wake flow field. With the rapid development of CFD technology, the study cost and studying time cycle of the wake flow field characteristics can be greatly reduced, and the details of the flow field can be comprehensively analyzed. In this paper, the characteristics of wake flow field of multi-nozzle rocket engine are numerically simulated and analyzed based on FLUENT, the commercial software of computational fluid mechanics.

Hideyo et. [1] used Fluent commercial software and coupled radiative heat transfer calculation model to carry out the three-dimensional simulation calculation of the H-IIA rocket and its wake flame flow field, and the heat effect of complex wake flame flow field structure and the backflow on the tail of the rocket were obtained. Xiao et. [2] took the space multi-nozzle engine as the model, the numerical simulation of aero-engine space wake flame was carried out using computational fluid dynamics (CFD) and Monte Carlo (DSMC) method, and the characteristics of the exhaust wake flame flow field of the space multi nozzle engine were obtained. Houshang et. [3] used GIFS solver to carry out the three-dimensional numerical simulation of the wake flame flow field of the first and third-level dynamical system of the militia III, certified the simulation validity of the GIFS solver, and obtained the parameters distribution of the wake flame flow field. Hu Shengchaoet et. [4] carried out the numerical simulation study for the mechanism and radiation characteristics of multi nozzle gas jet noise, and put forward a scheme for reducing noise by using multiple nozzles instead of single nozzle, and verified the feasibility of the scheme. Sun Ping [5] carried out a three-dimensional numerical simulation of 4 nozzle engine, and got the effects of different altitude and nozzle layout on wake flame flow and radiation characteristics. Wang Yanming et. [6] obtained infrared characteristics of the 2~5μm-band spectra in the wake flame flow field of a multi nozzle engine by studying the wake flame

Content from this work may be used under the terms of the Creative Commons Attribution 3.0 licence. Any further distribution of this work must maintain attribution to the author(s) and the title of the work, journal citation and DOI.
Published under licence by IOP Publishing Ltd

flow field and infrared radiation characteristics of a multi nozzle engine at low altitude. Equipment College Nie Wansheng, et [7~9] started from the internal combustion mechanism of liquid rocket engine to build a set of internal and external integrated simulation model of engine combustion wake flame, and carried out the simulation calculation of wake flame radiation characteristic accordingly. The above documents studied the wake flame of a multi nozzle rocket engine from different angles, and provided an important computational model and simulation method for the related field work. However, with the increasing demand of carrying capacity, the number of engines in parallel work is increasing, which leads to more complicated structure of wake flame. At the same time, non-toxic non-polluting liquid propellant and gradually favored by the world, for example, the rocket engine with liquid hydrogen liquid oxygen and liquid oxygen kerosene as propellant is not only low cost, non-toxic and environmental protection, but also has the advantages of high density ratio, higher performance reliability, convenient use and maintenance, and is the best power choice for launch vehicle booster and first level. It is also the object of study in the world of various countries [10~12]. In this paper, the flow analysis of a rocket engine with 4 nozzles is carried out, and the distribution of the velocity, temperature and pressure of the wake field is studied.

2. Physical model and numerical methods

2.1. Physical model

Taking a solid rocket engine as the research object, the wake field distribution of its multi-nozzle during flight is studied. The structure diagram of the simulation object is shown in figure 1.

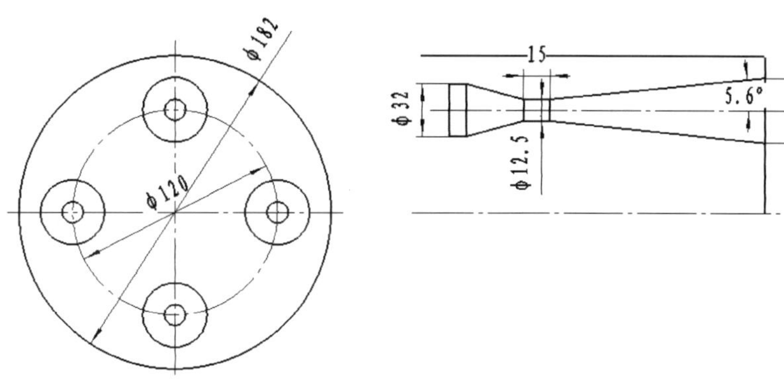

Figure 1. Multi-nozzle structure diagram

According to the actual flight environment, on condition that the following simulation boundary conditions, firebox pressure is 8.5 MPA, and the temperature of the firebox is 3200K, assuming that the gas is a single-phase ideal gas, and the specific heat at constant pressure is 1750J / kg K, and the gas law constant is 302J / kg K, the flying altitude is 5000m, and the flight Mach number is 2.

2.2. Governing equations

The mass, momentum, energy and component equations can be expressed in the following general form:

$$\frac{\partial(\rho\phi)}{\partial t}+\mathrm{div}(\rho\mathbf{U}\phi)=\mathrm{div}(\Gamma_{\phi}\,\mathrm{grad}\,\phi)+S_{\phi} \tag{1}$$

The formula is a general variable, which can represent solving variables such as u, v, T. Γ is a generalized diffusion coefficient; S_{ϕ} is a generalized source term. For a specific equation, ϕ, Γ,

S_ϕ has a specific form, table 1 shows the correspondence between the three symbols and the specific equation:

Table 1. Universal variable meaning

	ϕ	Γ	S_ϕ
Mass equation	1	0	0
Momentum equation	u_i	μ	$-\frac{\partial p}{\partial x_i}+S_i$
Energy equation	T	$\frac{\lambda}{c_p}$	S_T
Composition equation	m_i	Γ_i	0

Standard model $k-\varepsilon$ for turbulence equation selection;

$$\frac{\partial(\rho k)}{\partial t}+\mathrm{div}(\rho \mathbf{u} k)=\mathrm{div}\left[(\mu+\frac{\mu_t}{\sigma_k})\mathrm{grad}\,k\right]+G_k+G_b-\rho\varepsilon-Y_M+S_k \quad (2)$$

$$\frac{\partial(\rho\varepsilon)}{\partial t}+\mathrm{div}(\rho \mathbf{u}\varepsilon)=\mathrm{div}\left[(\mu+\frac{\mu_t}{\sigma_\varepsilon})\mathrm{grad}\,\varepsilon\right]+C_{1\varepsilon}\frac{\varepsilon}{k}(G_k+C_{3\varepsilon}G_b)-C_{2\varepsilon}\rho\frac{\varepsilon^2}{k}+S_\varepsilon \quad (3)$$

Where: G_k is turbulence kinetic energy due to laminar flow velocity gradient; G_b is turbulence kinetic energy due to buoyancy; Y_m is the wave generated by transient diffusion in compressible turbulent flow. σ_k and σ_ε are turbulent Prandtl number respectively representing k equations and ε equations; S_k and S_ε are source items respectively defined for the user.

2.3. Meshing

Use ICEM CFD to mesh the above policy geometry model. The existence of boundary layer is considered when setting the size. Considering the requirements of computer resources and calculation accuracy, the thickness of the first layer grid at the wall surface of the throat is 0.083 mm, and the boundary layer has 20 layers, with an increase ratio of 1.2. The dimensions of the remaining positions are set appropriately on this basis, and finally a grid is generated, as shown in fig. 2.

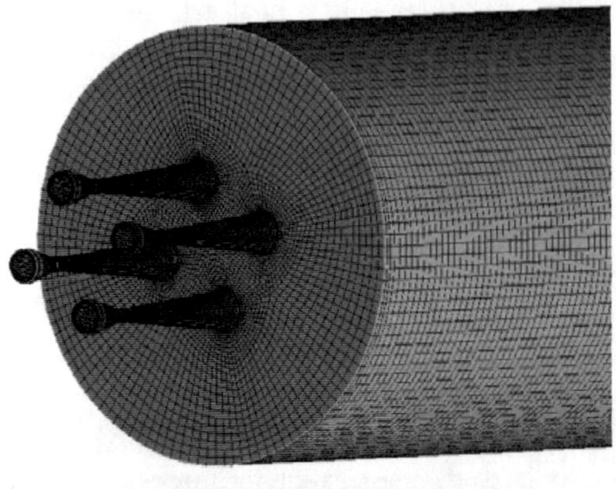

Figure 2. Meshing results

3. Calculation and analysis

Firstly, the distribution diagram of the nozzle wall y + during the simulation process is given, as shown in figure 3.

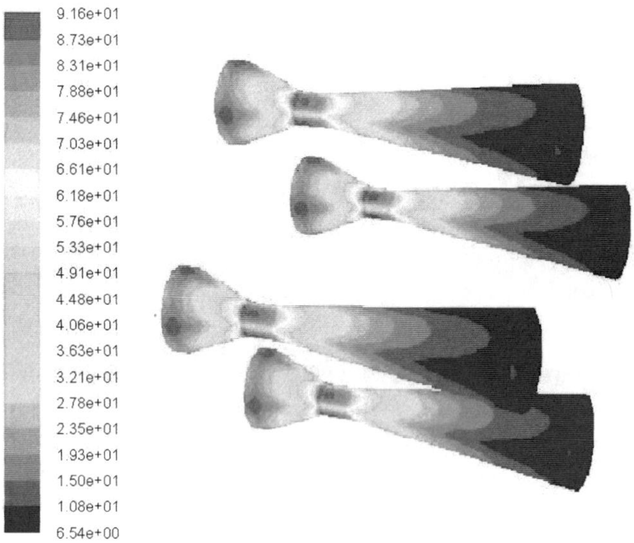

Figure 3. Nozzle wall Y + position nephogram

The above figure shows that the wall y + value is between 6.5 and 91.6 in the simulation process, and the largest y + value is located at the throat, which is basically consistent with the selected turbulence model and wall function. In addition, it also shows that the number of the first layer grids taken in the grid division process is good, and the simulation results basically reflect the flow law. In addition, it also shows that in order to obtain better flow simulation results, the first layer grid of the throat can be further encrypted and a better turbulence model can be selected.

3.1. Velocity distribution

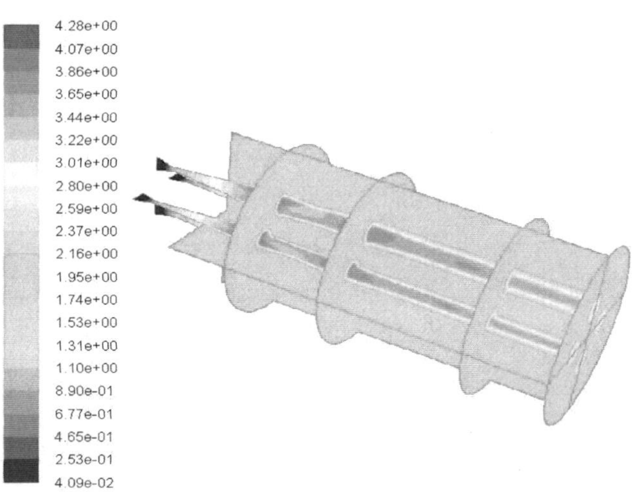

Figure 4. Velocity nephogram

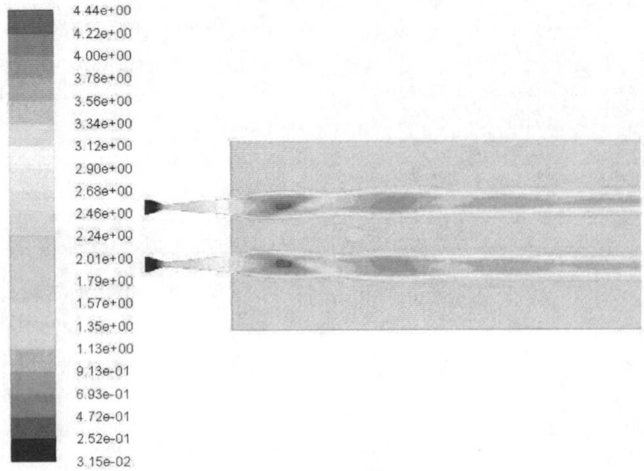

Figure 5. Velocity nephogram along flowing section

Figure 4 shows the final Mach number cloud picture. From the figure, it can be seen that the simulation results conform to the actual flow situation. The air flow enters the far field after expanding and accelerating through the laval nozzle, and forms a relatively obvious Mach disk through continuous refraction and intersection in the far field (shown in figure 5).

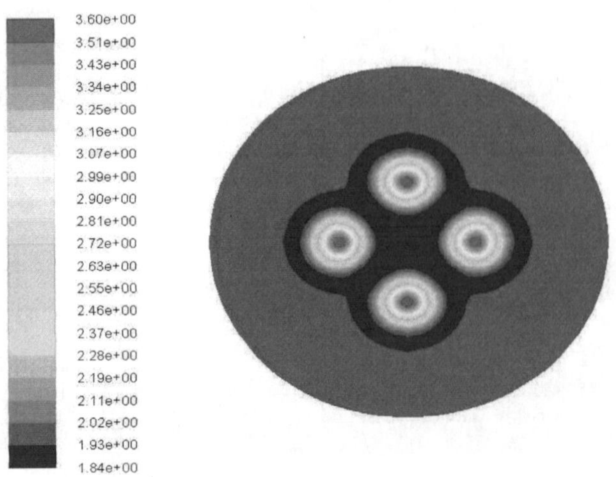

Figure 6. Velocity nephogram on far-field exit

Figure 6 shows the velocity distribution diagram of the far field exit section. from this, it can be seen that the engine wake is uniformly distributed in the radial direction and the mach number of the axial flow velocity is about 3.6 on the far field exit section.

IOP Conf. Series: Journal of Physics: Conf. Series **1064** (2018) 012009 doi:10.1088/1742-6596/1064/1/012009

3.2. Temperature distribution

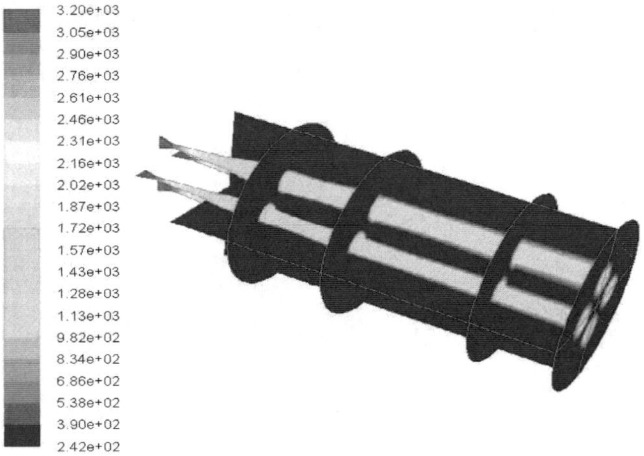

Figure 7. Temperature nephogram

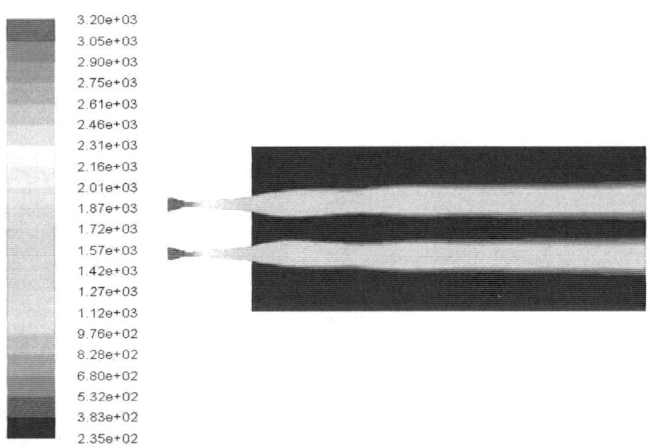

Figure 8. Temperature nephogram along flowing section

Figure 7 shows the temperature cloud picture obtained by simulation. As can be seen from the figure, the gas temperature gradually decreased from the initial 3200 k to about 2400 k after expansion of the laval nozzle. In the cloud picture, there is no obvious sign of Mach disk, but from figure 8, we can see that there is a small temperature fluctuation at the front end of the far field and there are a certain number of shock waves and expansion waves. The overall temperature is relatively uniform.

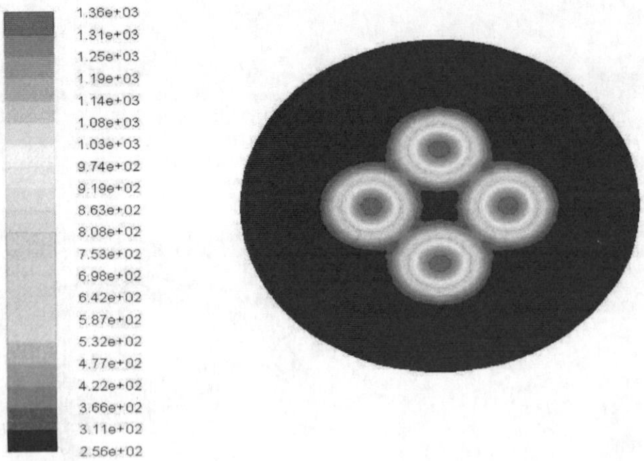

Figure 9. Temperature nephogram on far-field exit section

From the temperature distribution diagram of the far field exit section, it can be concluded that the temperature of the engine wake in the far field exit section is uniformly distributed in the radial direction, and the static temperature of the nozzle axis is about 1360 K.

3.3. Pressure distribution

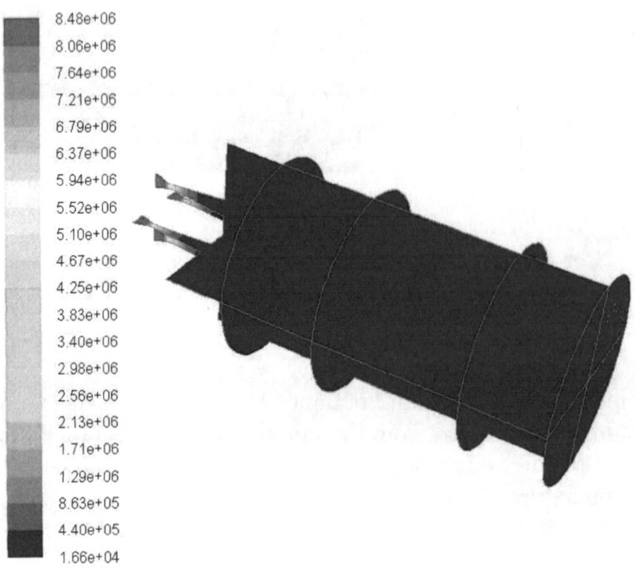

Figure 10. Pressure nephogram

Figure 10 shows the cloud picture of the simulation results of pressure. From the cloud picture, it can be seen that in the convergent section of the nozzle, the pressure is still relatively large and the decrease is not obvious. When the flow passes through the throat, the pressure of the gas begins to drop sharply, with larger amplitude. After this sharp drop, a relatively stable pressure distribution will follow. This result is consistent with the actual working condition.

IOP Conf. Series: Journal of Physics: Conf. Series **1064** (2018) 012009 doi:10.1088/1742-6596/1064/1/012009

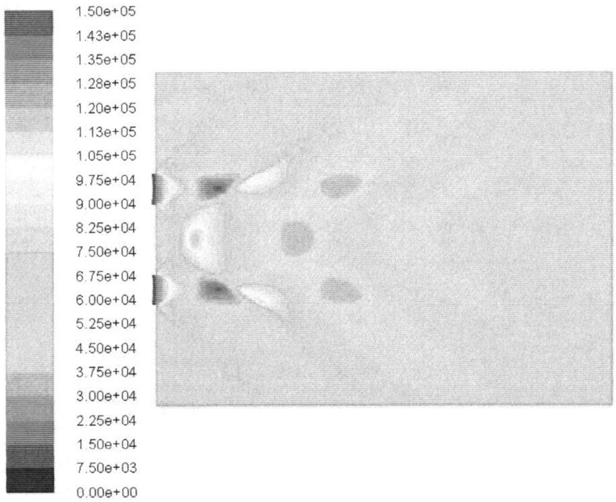

Figure 11. Pressure nephogram along flowing section

Since the pressure in the convergent section is far greater than the pressure in the far field, it is difficult to see the pressure change in the far field in the cloud picture. Figure 11 shows the pressure cloud picture in the cross section along the flow direction in the far field. The obvious pressure fluctuation can be seen in this figure, and the supersonic flow in this region indicates that there is a shock wave at the fluctuation position, which causes the pressure change.

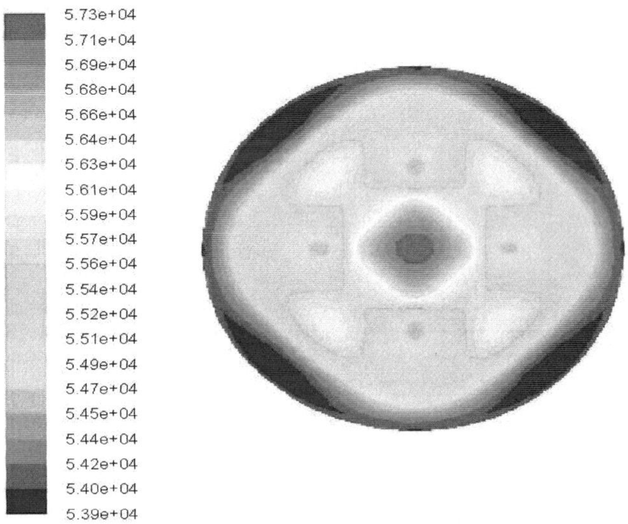

Figure 12. Pressure nephogram on far-field exit section

It can be seen from the pressure distribution diagram of the far-field exit section that the engine wake is uniformly distributed in the radial direction on the far-field exit section, and the pressure coupling between different nozzle wake is strong.

3.4. Mach disk

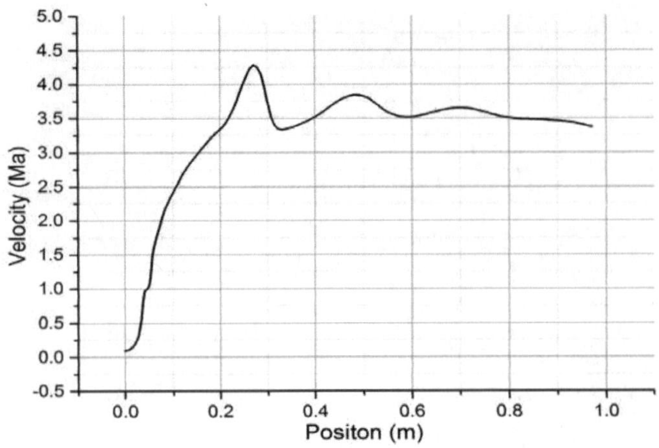

Figure 13. Velocity curve graph of the axis of a certain nozzle

From the velocity distribution map on a certain nozzle axis, it can be concluded that there is velocity fluctuation after the gas exits from the nozzle outlet. These areas prove the existence of Mach disk, but with the increase of axial length, the fluctuation tends to be gentle.

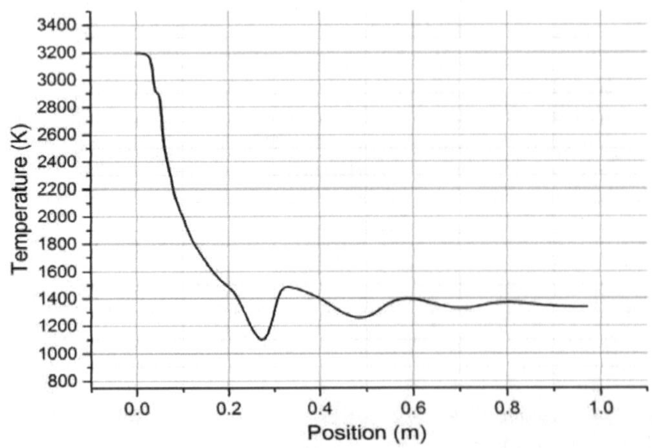

Figure 14. Temperature curve graph of the axis of a certain nozzle

From the temperature distribution map on a certain nozzle axis, it can be concluded that after the gas flows out from the nozzle outlet (the nozzle outlet cross section is x = 170 mm), it continues to expand and do work, causing the static temperature to drop.

Figure 15 shows a comparison diagram of the velocity and temperature curves on a certain nozzle axis. From the diagram, it can be seen that the change trend of the two physical quantities is completely opposite. There is a Mach disk at the location where there is fluctuation, and there is a change in the physical quantities.

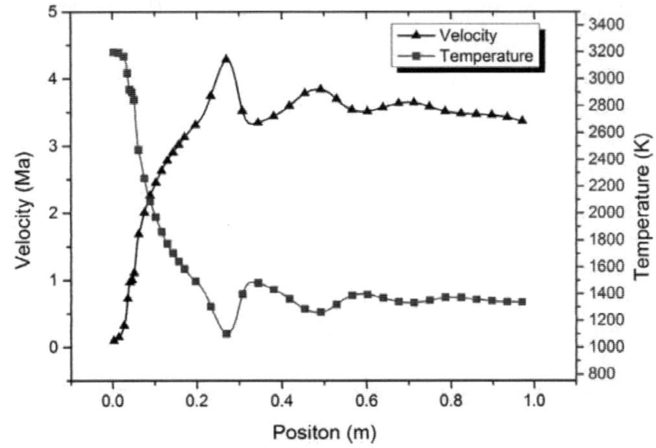

Figure 15. Contrast curve graph of velocity and temperature of a nozzle axis

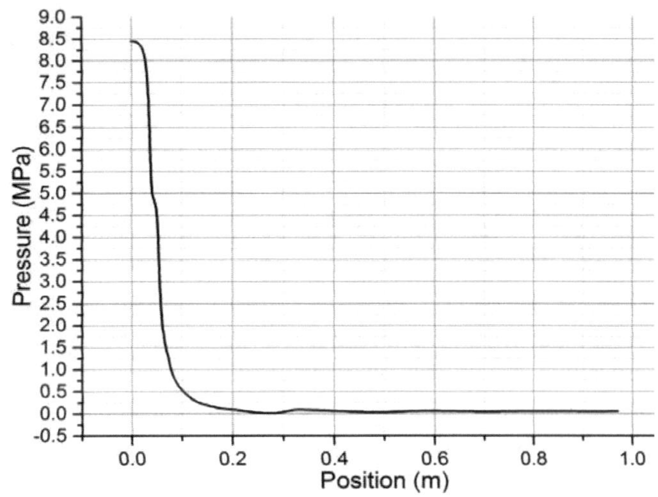

Figure 16. Curve graph of pressure on the axis of a certain nozzle

From the pressure distribution map on a certain nozzle axis, it can be concluded that there are still pressure rising areas and pressure falling areas after the gas flows out from the nozzle outlet. These areas prove the existence of Mach disk.

4. Conclusion

(1) The y + values obtained from the simulation show that the grid division is better and can better reflect the real flow situation;

(2) The simulation results are authentic, and the distribution and variation characteristics of the main physical quantities are consistent with the actual situation;

(3) Mach inventory lies in the far field, causing the fluctuation of velocity, temperature and pressure along the flow direction, and the fluctuation of velocity and temperature can correspond to each other one by one.

References

[1] Hideyo N, Nobuhiro Y, Makoto A, et al. Numerical Analysis of Plume Heating Environment for H- IIA Launch Vehicle during Powered Ascent [R]. AIAA 2007-5505.

[2] Ze-Juan Xiao, Hui-Er Cheng. Plume Interaction in Par-allel Multi- Thrusters Propulsion System and the Effect on Backflow [R]. AIAA 2006-3599.

[3] Houshang B E. Numerical Investigation of Multi-Plume Rocket Phenomenology [R]. AIAA 1997-3622.

[4] Shengchao Hu, Ang Li, Futing Bao ,etc. multi-nozzle fuel gas denoise scheme feasibility data study [J]. JTHJ, 2012, 35(2): 198-202.

[5] Ping Sun,Xinyun Fan, multi-nozzle rocket engine Wakefield flow and radiation characteristic numerical simulation [J]. Mechanical , 2008, 14(7): 41-44.

[6] Yanming Wang, Heping Tan, Shikui Dong, etc. Low –altitude multi-nozzle engine flare research on infrared characteristics [J]. Solid rocket propulsion technology, 2009, 32(6): 634-637.

[7] Wansheng Nie, Songjiang Feng. Liquid motor kinetics of combustion numerical model [M]. BJ: National Defense Industry Publishing House, 2011.

[8] Wansheng Nie, Junhui Yang,Haobo He, etc. Liquid rocket engine tail flame infrared radiation [J]. journal of Defence Science University of PRC , 2005, 27(5): 91-94.

[9] Songjing Feng, Wansheng Nie, Qingfen xie,etc. Combustor combustion model affect tail flame flow field and its radiation [J]. Rocket-propelled, 2006, 32(2): 6-10.

[10] Zhaobo Ding, Jiguo Sun, Xiaohong Lu. Foreign representative high thrust hydrogen oxygen engine thrust chamber technical schemes summary [J]. Missile and Space Carrier technology, 2012(4): 27-30, 38.

[11] Yonghua Tan, High thrust liquid rocket engine Research [J]. Acta Astronautica, 2013, 34(10): 1303-1308.

[12] Dong Li, Tangming Cheng. Prospects for the Development of China's New Generation launch vehicle [J]. China Aerospace, 2008(2): 7-10.

IOP Conf. Series: Journal of Physics: Conf. Series **1064** (2018) 012010 doi:10.1088/1742-6596/1064/1/012010

A study on combustion characteristics under enriched-oxygen condition by exergy analysis

Xiong Zhang*

School of Mechanical Engineering, Xi'an Shiyou University Xi'an China.

*Corresponding author e-mail: xlaiqinghai@163.com

Abstract. The object of this paper is to investigate the effect of operating parameters on Oxygen-enriched combustion (OEC) by exergy balance analysis during methane combustion process. In this work, the specific exergy value of fuel, feed gas and products were calculated on the basis of the second law of thermodynamics in different reaction conditions, so that availability destructions and exergy efficiencies of OEC process can be obtained based on exergy balance, and the influence of initial fuel temperature, oxygen concentration, exhaust gas recirculation ratio and equivalence ratio were addressed. It is observed that during methane OEC process, the exergy efficiency is mainly affected by products temperature, in the case of products temperature lower than 1870K, exergy efficiency of combustion increase with products temperature, and opposite changes was appeared when the products temperature higher than 1870K. Meanwhile, appropriate control of equivalence ratio can also be used to improve exergy efficiency by suppressing the dissociation of product.

1. Introduction

As a result of the rapid depletion of fossil fuel reserves and excessive emissions of greenhouse gases, it is necessary to exploit effectively natural energy resources and release the pollutants emissions at present, which are available for the science progress and the sustainable development. In 2007, almost 83.8% of total 495 quadrillion kJ energy is produced from fossil fuel. Ignoring the exhaustion of the natural resource, up to 2030, the total energy consumption will reach almost 739 quadrillion kJ, 86% of which will still be acquired from fossil fuel [1]. Hence, it is vital to increase the energy conversion efficiency through improving the combustion systems in the next several decades.

Oxygen-enriched combustion (OEC) is one of the useful energy-saving technologies for combustion systems. By concentrating the oxygen in the feed gas, the amount of inert nitrogen contained in flue gases is lowered, and many benefits may be realized: higher thermal efficiency, increased productivity, enrich the concentration of CO2 in flue gas etc. Historically, OEC has been the conventional technology used in nearly all industrial heating processes, such as glass smelting furnace, metallurgic furnace and etc[2]. In the last decade, OEC systems are becoming more common in a variety of industries because of growing concern about CO2 emissions [3, 4]. The power generation industry have done a considerable amount of work on oxy-fuel combustion (oxygen typically of greater than 95% purity), and many benefits have been demonstrated: increased power generation, lowering the energy cost for CO2 capture and capital cost [5-7]. Moreover, the comment production

Content from this work may be used under the terms of the Creative Commons Attribution 3.0 licence. Any further distribution of this work must maintain attribution to the author(s) and the title of the work, journal citation and DOI.
Published under licence by IOP Publishing Ltd

IOP Conf. Series: Journal of Physics: Conf. Series **1064** (2018) 012010 doi:10.1088/1742-6596/1064/1/012010

industry also achieved similarly benefits by using of oxy-fuel technology [8]. In industrial heat treating processes, the application of OEC can realize the use of low valuable gas (such as blast furnace gas), reducing the energy consumption and CO2 emissions [9-10]. In this paper, the fundamental research on oxygen-enriched combustion technology process is practical interest.

The concept of exergy is a direct outcome of second law of thermodynamics, which can be used to reveal the depletion of energy in energy conversion process and to evaluate the work potential of a system. Several studies have indicated that the conventional combustion process involves inherent exergy loss by 20～30%, which is attributed to both internal heat and mass transfer and chemical reactions in flame sheet.

Historically, two methods were developed to evaluate exergy efficiency in combustion process: one way of analyzing the performance of a combustor is base on the exergy balance during combustion process; the other way is to calculate the rate of entropy generation in the physical and chemical process, the rate of exergy destruction is calculated with Gouy-Stodola equation(Eloss=T0×dS. E, T and S are the exergy, reference environment temperature and entropy) [1].

In recent papers, some researchers have analyzed the rate distribution of entropy generation inside the several combustor by numerical simulation and gained the effects of operating parameters in the laminar and turbulent combustion process. It is showed that 72～77% of exergy destruction in combustion process was due to the inner heat transfer, chemical reaction (15～18%), and mass diffusion (5～10%) in CH4-air laminar flame. Then in the same way, more than 95% of the exergy destruction was due to turbulent parts in CH4-air turbulent flame.

Meanwhile, some other researcher analysis the exergy efficiency of combustion process by the exergy balance in the combustor. It is proved that reactant conditions (such as the equivalence ratio, initial temperature and pressure) may affect the exergy efficiency, and the implications of exhaust gas recirculation on exergy efficiency was also evaluated. Due to this, exergy efficiency is a important parameter to evaluate the performance of the combustion technology, but there were hardly any investigation on exergy analysis of oxygen-enriched combustion technology.

In this paper, the exergy efficiency can be obtained by exergy balance analysis after calculating the specific exergy value of fuel, feed gas and products on the basis of the second law of thermodynamics in different reaction conditions. The initial fuel temperature, oxygen concentration, exhaust gas recirculation and the equivalence ratio will affect the exergy efficiency greatly, and exergy efficiency diagram can be gained in the oxygen-enriched combustion, and a strategies that reduces the exergy loss in the combustion process were introduced, which will provide a theoretical guidance to design oxygen-enriched combustor and combustion system.

2. Theoretical basis

2.1. *Basic concept of exergy*

Thermodynamically, the exergy is defined to its"work potential" in the reference environment. The work potential implies physically the maximum theoretical obtainable work of the system, and the term exergy is sometimes replaced by availability. In general, the availability of a closed system is a composite property of the system in the reference environment, which is expressed as Eq. (1).

$$a = K + a_{tm} + a_{ch} \tag{1}$$

Where, a is the total specific availability (J/mol); K is the specific kinetic energy (J/mol); atm is the specific thermomechanical availability (J/mol); ach is the specific chemical availability (J/mol).

The specific thermomechanical availability is estimated as the sum of two components, thermal availability and pressure exergy. For open systems (subscript 0 denotes the parameters of reference environment):

$$a_{tm} = (h-h_0) - T_0(s-s_0) = C_p(T-T_0) + RT_0 \ln(\frac{P}{P_0}) - C_p T_0 \ln(T/T_0) \quad (2)$$

For closed systems:

$$a_{tm} = (u-u_0) - T_0(s-s_0) + P_0(u-u_0) = C_p(T-T_0) - C_p T_0 \ln(T/T_0) + RT_0\left[\ln(\frac{P}{P_0}) - \left(1-\frac{P}{P_0}\right)\right] \quad (3)$$

T is the temperature (K); P is the pressure (Pa); s is the specific entropy (J/ (K mol); u is the specific internal energy (J/ (K mol); Cp is the constant-pressure specific heats, (J/ (K • mol); R is the perfect gas constant (8.314 J/mol K) ;

The subscript 0 represents the reference state.

Chemical availability of species is expressed as the sum of the standard Gibbs free energy of formation and the Gibbs free energy of elements. The specific chemical availability of a species is given as:

$$a_{ch} = \Delta g_f(T_0, P_0) + \sum_i n_i g_i(T_0, P_0) \quad (4)$$

Where, Δgf(T0,P0) is the standard Gibbs free energy of formation(J/mol); gi(T0,P0) is the Gibbs free energy of elements(J/mol).

In most cases, the reactants and products of combustion systems are mixture of different species, so it is necessary to estimate the availability destruction during the mixing process. The specific availability of mixture is evaluated as:

$$a_{mix} = \sum_i^N x_i a_i + RT_0 \sum_i^N x_i \ln(x_i) \quad (5)$$

Where, amix is the specific availability of mixture (J/mol); xi is the mole fraction of i species in the mixture.

2.2. The exergy balance in the combustion process

In order to find the relations between the reactants condition and the exergy efficiency in the oxygen-enriched combustion process, a zero dimensional model is raised to simulate the combustion process, and methane is used as fuel and consider the combustion process as isobaric.

During combustion reaction, medium products are considered, generally, including the following 7 species: CO2, CO, H2O, H2, CH4, O2, and N2. Other medium products are ignored because of its lower concentration. The equilibrium products state are characterized by the equilibrium constants and minimum Gibbs free energy. The generalized combustion reaction formulas can be presented as follows:

$$CH_4 + \frac{2}{\Phi}O_2 + \frac{2}{\Phi}\left(\frac{1}{X_{O_2}} - 1\right)N_2 + \frac{F}{1-2F}\sum_i^6 n_i Y_i = \frac{1-F}{1-2F}\sum_i^6 n_i Y_i \quad (6)$$

$$\sum_i^6 n_i Y_i = n_1 CO_2 + n_2 CO + n_3 O_2 + n_4 H_2O + n_5 H_2 + n_6 N_2 + n_7 CH_4 \quad (7)$$

Where, Yi is the constituents of product i ; ni is the number of moles of elements i.

The equilibrium product temperature is obtained from the energy balance equation in the combustion process (Eqs (8)-(10)). According to Eq (6) and Eq (8), the equilibrium product temperature and constituents concentration of products have to be solved simultaneously.

$$E_r = E_p + Q \tag{8}$$

$$E_r = h_{CH_4}(T_0) + \frac{2}{\Phi} h_{O_2}(T_r) + \frac{2}{\Phi}\left(\frac{1}{X_{O_2}} - 1\right) h_{N_2}(T_r) + \frac{F}{1-2F}\sum_i^6 n_i Y_i h_{Y_i}(T_r) \tag{9}$$

$$E_p = \frac{1-F}{1-2F}\sum_i^6 n_i Y_i h_{Y_i}(T_p) \tag{10}$$

Er is the total energy of reactants(J) ; Ep is the total energy of products(J); F is the exhaust gas recirculation ratio, defined as the ratio of recirculated exhaust gas to total exhaust gas; Tr is the temperature of feed gas(K); Tp is the equilibrium temperature(K); Φ is the equivalence ratio; X_{O_2} is the oxygen mole fraction in feed gas; Q is the heat loss in the combustion process(J).

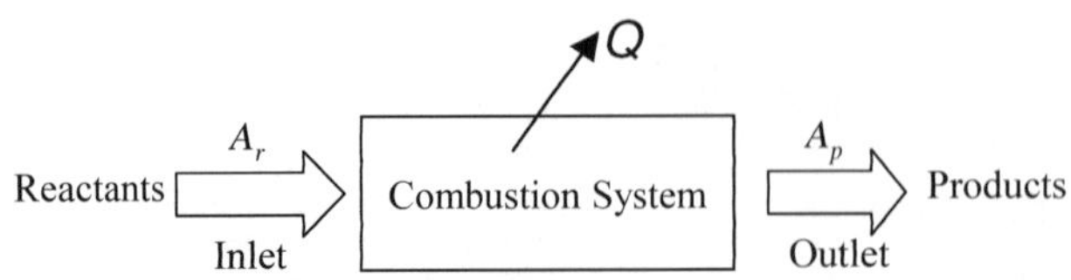

Fig. 1 Diagram of the exergy transfer in a combustion system

Fig. 1 is a diagram of the exergy transfer in a combustion system. It can be seen from Fig.1 that the availability of reactants (Ar) is divided into three parts in the combustion process, the availability of products (Ap), availability lost in the heat loss (Δ AQloss) and availability destruction in combustion process (Acomb).

$$A_r = A_p + A_{comb} + \Delta A_{Qloss} \tag{11}$$

$$A_p = A_{p\ th} + A_{p\ chem} \tag{12}$$

Exergy efficiency (η e) of the combustion process can be defined as the ratio of thermomechanical availability of products (Ap tm) to the the availability of reactants (Ar). It can be expressed as:

$$\eta_e = \frac{A_{p\ th}}{A_r} \tag{13}$$

3. Results and discussions

In the oxygen-enriched combustion process, the concentration of species in product and the equilibrium temperature are changed with the increasing of oxygen concentration. As is expected, equilibrium temperature and species concentration have a great effect on exergy efficiency of the combustion process. In this section, we are able to find the effect of oxygen concentration in feed gas, reactant temperature, exhaust gas recirculation (EGR) and equivalence ratio on exergy lost during OEC process. Finally, some methods to improve the exergy efficiency of OEC are provided.

3.1. Effect of oxygen concentration in feed gas

The effects of oxygen concentration on exergy efficiency are investigated by using oxygen mole fraction in feed gas(X_{O_2}). As previously mentioned, products temperature increases with oxygen mole fraction X_{O_2}, higher products temperature has great impact on the availability destruction (Acomb) and chemical availability lost (Ap chem) . However, the chemical availability lost increases with products temperature. The effect of X_{O_2} on exergy efficiency is shown in Fig.2 (a).

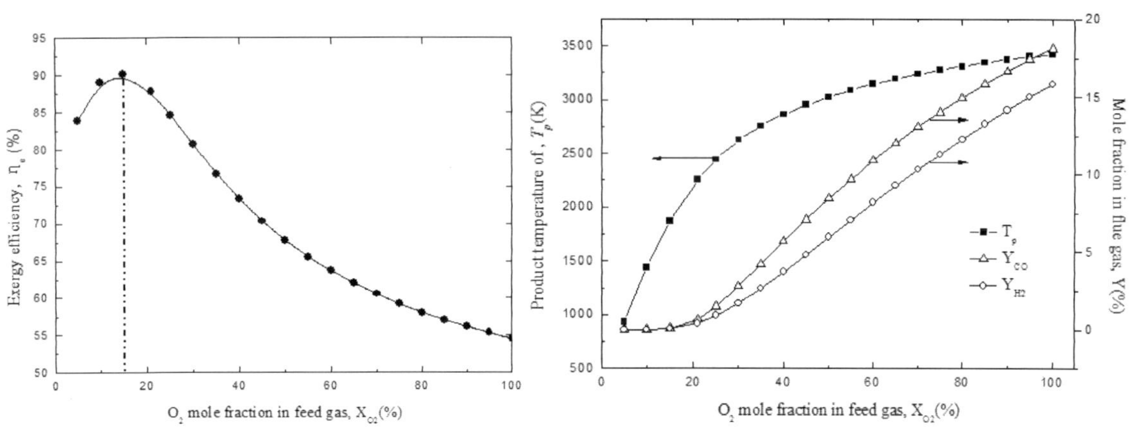

Fig. 2 The effect of oxygen enrichment on (a) exergy efficiency η e , (b) CO and H2 mole fraction in products and the products temperature Tp[Tr=298, F=0, Q=0J, Φ=1, P=P0].

As Fig.2(a) shows, exergy efficiency (η e) rises firstly, and then drops sharply with the increases of X_{O_2} and hits a maximum at X_{O_2}=15%, the products temperature is about 1868.7K at this point(shown in Fig.2(b)). The reasons mainly include: in the range of lower oxygen concentration, exergy efficiency of combustion increases due to decrease of availability destruction (Acomb); and in the range of higher oxygen concentration, exergy efficiency decreases due to the enhanced dissociation reactions at higher temperatures produce more moles of CO and H2 (shown in Fig.2(b)). In Fig.2, when X_{O_2} =40～100% , the increase of X_{O_2} only slightly increase the products temperature but largely enhance the generation of unburned species, therefore, reasonable choice of oxygen concentration in OEC process is very important.

3.2. Effect of exhaust gas recirculation ratio

Historically, exhaust gas recirculation is widely used to improve the temperature uniformity in oxygen-enriched combustion chamber, so that it is important to evaluate the influence of EGR on exergy efficiency of OEC process. It is investigated with a factor F defined in section 2.2. The conditions ranging from F=0(no flue gas recirculation) to F=0.45 are studied, and oxygen concentration X_{O_2} is considered at X_{O_2}=10%, 21% and 50%, as depicted in Fig.3.

As Fig.3(a) shows: for X_{O_2}=10%, exergy efficiency (η e) decreases continuously when raising F, due to the lower product temperature; for X_{O_2}=21%, exergy efficiency rises slowly with the increase of F in the range of 0～0.18, then drops sharply and reaches a maximum around F =0.18; for X_{O_2}=50%, exergy efficiency has a similar trend just as X_{O_2}=21%, which hits a maximum at F=0.37, and the products temperature at this point is about 1867.4K. Fig.3 (b) depicts the relations of products temperature and CO, H2 concentration with variation of the exhaust gas recirculation ratio F in the case of X_{O_2}=50%. Naturally, the products temperature is lowered by increasing F, and the low temperature is also attributed to the reduction in the amount of combustible species (mainly CO and H2).

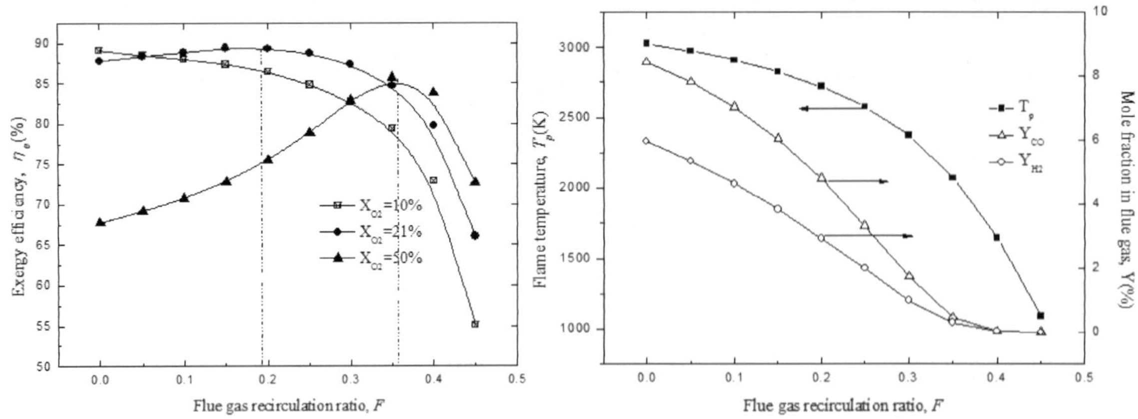

(a) (b)

Fig. 3 The effect of flue gas recirculation ratio on (a) exergy efficiency η e [Tr=298, Q=0J, Φ=1, P=P0], (b) CO and H2 mole fraction in products and the products temperature Tp [X_{O_2}=50%, Tr=298, Q=0J, Φ=1, P=P0].

Meanwhile it can be shown that the exergy efficiency difference decreases between X_{O_2}=21% and 50% with the increase of F in the range of F=0～0.45 and exergy efficiency for X_{O_2}=21% is lower than that for X_{O_2}=50% when F>0.3 from Fig.3 (a).

Combining with the previous analysis in section 3.1, the main reasons caused this changes were listed as follows: when the products temperature is below 1870K, availability destruction (Acomb) play a major role in total exergy lost in combustion process; and when the temperature is higher than 1870K, exergy efficiency is mainly affected by the chemical availability lost (Ap chem).

In conclusion, reasonable exhaust gas recirculation is necessary in the oxygen-enriched combustible process. On one hand, it can increase the temperature uniformity in combustion chamber and on the other hand, the exergy efficiency can also be improved.

3.3. *Effect of feed gas temperature*

Actually, waste gas can be recovered to preheat the feed gas via heat exchanger, and feed gas can be heated up to 1400K with the regenerative exchanger. In this paper, the relations between the exergy efficiency and the temperature of the feed gas (Tr) ranging from Tr =298K up to 1400K are studied, Fig.4(a) shows the variation of exergy efficiency with feed gas temperature in the cases of X_{O_2}=10%,

21% and 50%. The curve for X_{O_2}=10% in Fig.4(a) first increase and then decrease, and hit the maximum at Tr=870K, products temperature at this point is about 1867K (very close to 1870K that mentioned in section 3.2). In contrast, the other two curves continuously decrease with increase of the feed gas temperature.

The products temperature rises relatively as the temperature of the feed gas is raised. The temperature difference between the reactant and products diminish and cause decrease of the inner exergy loss (Acomb). The higher products temperature also result in dissociation of product species (as show in Fig.4 (b)). For X_{O_2}=21% and 50%, the products temperature is higher than 1870K at the starting point (Tr =300K). Hence, two curves continuously decrease with increase of the feed gas temperature.

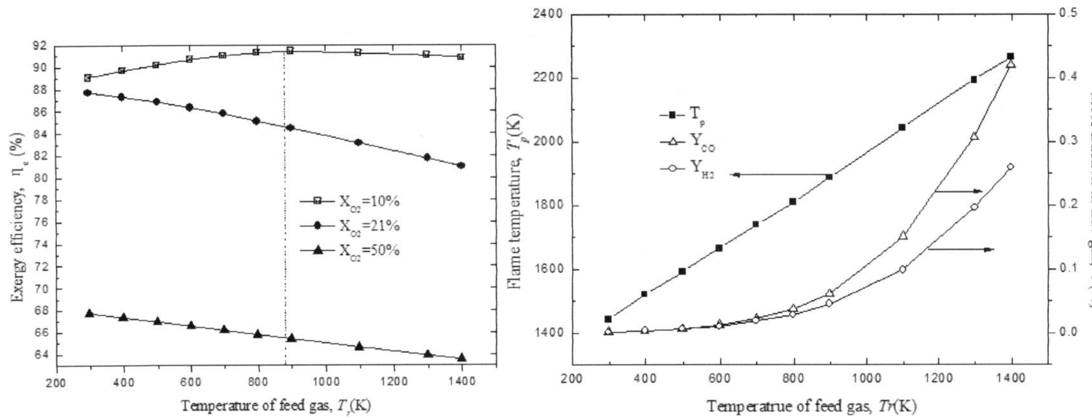

(a) (b)

Fig. 4 The effect of feed gas temperature on (a) exergy efficiency η e[F=0, Q=0J, ϕ=1, P=P0] , (b) CO and H2 mole fraction in products and the product temperature Tp [X_{O_2}=10%, F=0, Q=0J, ϕ=1, P=P0].

3.4. Effect of equivalence ratio

The equivalence ratio is an important parameter in the process of the combustion control, and has great impact on the products temperature and components. Referring to Fig.5(a), the exergy efficiency for X_{O_2}=10%, 21% and 50% first increase and then decrease with the increase of Φ , and the peak of curves shifts towards the smaller Φ as X_{O_2} increase.

It leads to a conclusions that the temperature of product and concentration of combustible product in flue gas hit their maximum around Φ=1.05 for all X_{O_2}=10%, 21% and 50% after research. In contrast, the maximum exergy efficiency point shifts towards to smaller Φ as X_{O_2} increase. The main reason is that the dissociation of product can be suppressed significantly by increasing oxygen concentration in flue gas, as shown in Fig.5 (b).

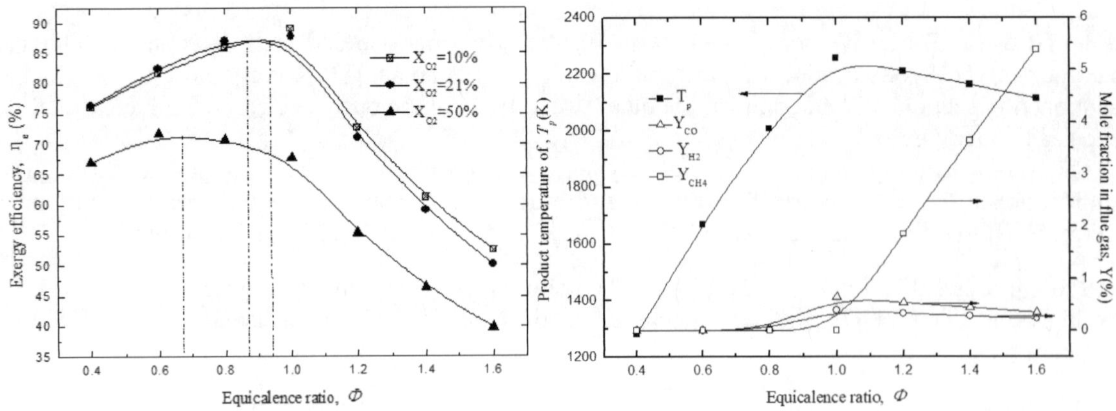

(a) (b)

Fig. 5 The effect of equivalence ratio on (a) exergy efficiency η e [Tr=298, Q=0J, ϕ=1, P=P0], (b) CO and H2 mole fraction in products and the products temperature Tp [X_{O_2}=21%Tr=298, Q=0J, ϕ =1, P=P0].

4. Conclusion

As demonstrated in this paper, the effect of important parameters on exergy efficiency of combustion is investigated to improve the efficiency of oxygen-enriched combustion system, several important conclusions drawed from the results of this work are as follows:

(1) For methane, exergy efficiency of the isobaric combustion process is mainly affected by products temperature: when products temperature is below 1870K, exergy lost in the combustion process includes inner combustion exergy lost Acomb, exergy efficiency η e rises with the increase of the products temperature; when the temperature is higher than 1870K, exergy efficiency is mainly affected by the chemical incomplete combustion exergy lost Ap chem, so that exergy efficiency η e will decrease with the increase of the products temperature .

(2) In OEC process, exhaust gas recirculation ratio F and feed gas temperature Tr have great impact on exergy efficiency. For the lower oxygen concentration X_{O_2}, exergy efficiency can be improved by increasing Tr. At the same time, the exhaust gas recirculation can significantly improve the exergy efficiency in the higher range of X_{O_2}.

(3) The reasonable equivalence ratio in the oxygen-enriched combustion process is also important to improve exergy efficiency η e by reducing the generation of the intermediate product.

Acknowledgments

This study is supported by the Foundation of Shaanxi Provincial Department of Education (Grant 17JK0604).

References

[1] Som SK, Datta A. Thermodynamic Irreversibilities and exergy balance in combustion processes [J]. Progress in energy and combustion science, 2008; 34: 351-376.
[2] Zhang X. Tong LG, Wang L. The application and analysis of oxygen-enriched combustion [J].. Energy for Metallurgical Industry 2007; 26: 41-46.
[3] Yewen T, Eric C, Mark A. Combustion characteristics of coal in a mixture of oxygen and recycled flue gas [J].. Fuel 2006; 85: 507-512.
[4] Yewen T, Mark AD, Kelly VT. CO2 capture using oxygen enhanced combustion strategies for natural gas power plants [J].. Fuel 2002; 81: 1007-1017.

[5] Buhre BJP, Elliott LK. Oxy-fule combustion technology for coal-fired power generation [J].. Progress in Energy and Combustion Science, 2005; 31: 284-305.
[6] D Singh, E Croiset, L Douglas. Techno-economic study of CO2 capture from an existing coal-fired power plant: MEA scrubbing vs. O2/CO2 recycle combustion [J].. Energy Conversion & Management, 2003;44: 3073-3091.
[7] Yang HQ, Zheng HF, Mao H. Progress in carbon dioxide separation and capture: A review [J].. Journal of Environmental Sciences 2008; 20: 14–27.
[8] Frank Z. Oxygen combustion in cement production [J].. Energy Procedia 2009; 1: 187-194.
[9] Wu KK, Chang YC, Chen CH. High-efficiency combustion of natural gas with 21-30% oxygen-enriched air [J]. . Fuel 2010; 89:2455-2462.
[10] Giacomo B, Alessandro B, Giuseppe R.. Thermodynamics applied to oxygen enrichment of combustion air [J].. Energy Conversion & Management, 2002;43:2589-2600.

IOP Conf. Series: Journal of Physics: Conf. Series **1064** (2018) 012011 doi:10.1088/1742-6596/1064/1/012011

Study on heat transfer process of a heat not burn tobacco product flow field

Zhicai Jiang, Xue Ding*, Ting Fang, Hua Huang, Wei Zhou, Qi Sun

[1] China Tobacco Zhejiang Industrial Co., Ltd. Hangzhou, China.

*Corresponding author e-mail: snowdingxue@163.com

Abstract. A new-style cigarette works by a battery-powered device heating cut-tobacco to generate an aerosol. This product heats the tobacco at low temperature and produces less harmful and potentially harmful constituents (HPHCs). In this paper, a heat not burn tobacco product has been designed, and the heat transfer process in the cigarette has been studied based on a porous medium theory. Both of viscous resistance and inertia force effect on pressure gradient have been considered using Darcy-Forchheimer model. After numerical simulation the process of gas traversing the cut-tobacco medium, the temperature field distribution has been acquired in the new-style cigarette. The result shows the heating system of the heat not burn tobacco product satisfies the demand of temperature control. The aerosol of the new style cigarette can reduce the yield of tobacco combustion produces.

1. Introduction

A few hundred million people choose to smoke despite warnings from the International Agency for Research on Cancer(IARC) [1] that adverse health effects may arise from smoking, such as human cancer [2,3]. With the continuous advancement of the anti-smoking movement, the Chinese government has been tightening its restrictions on smoking, especially in public places. It is necessary to produce a new style cigarette with less damage to satisfied the smoker. Some smokers are unwilling to sacrifice flavour for reduce tar. Therefore, this has prompted a continuing effort to develop alternative cigarette designs.

Conventional cigarette (c-cigarette) smoking, also known as tobacco smoking producing HPHCs such as carbon monoxide, carbon dioxide, hydrogen, particulate matter, nicotine and tar etc. has been causing multifarious problems and arousing concerns. Substantial evidence tended to substantiate that c-cigarette smoking is the primary etiological factor responsible for the progressing of bronchiolitis-interstitial lung disease (RB-ILD), desquamative interstitial pneumonia (DIP) and adult pulmonary Langerhans' cell histiocytosis (PLCH) [4,5]. For purpose of replacing c-cigarettes as well as minimizing the toxic effect the new style cigarettes such as heat not burn tobacco, electronic cigarette have brought out[6,7]. The traditional cigarettes produce an enormous amount of harmful substances at high temperatures of 350 ° C to 600 ° C. While the heat not burn tobacco and E-cigarettes are battery-powered devices which heat tobacco or vaporize a liquid solution into aerosol under condition of 250~550 ℃. Such products have less nicotine and other combustion products [8]. Not like traditional cigarettes, the new products heat the tobacco or liquid at under 550 °C, result in only 5% of the substances and 14% of the cytotoxicity produced by traditional cigarettes [9]. Compared with other

Content from this work may be used under the terms of the Creative Commons Attribution 3.0 licence. Any further distribution of this work must maintain attribution to the author(s) and the title of the work, journal citation and DOI.
Published under licence by IOP Publishing Ltd

novel cigarettes, their flavour can better satisfy consumers' needs and provide them with an authentic tobacco taste. Additionally, due to the condition of heating of liquid of e-cigarette in which chemical reactions could result in the formation of new compounds, chemical composition of aerosol could be different from the composition contained in liquid [10,11]. Uchiyama et al. [12] demonstrated that 70% of examined e-cigarette brands contained or generated carbonyl compounds such as formaldehyde, acetaldehyde, acrolein, crotonaldehyde and methylglyoxal. Therefore, the heat not burn tobacco product attracts people's attention recently. The heat not burn tobacco products have the following advantages: (1) they are less harmful than traditional cigarettes because they avoid combustion and, therefore, drastically reduce the production of tar and other harmful substances; (2) to a certain extent, they resolve the conflicts between smoking and the ban on smoking in public because they do not produce second-hand smoke, which damages the environment and other people's health; and (3) since they contain tobacco, they can adapt to and satisfy the consumers' physical needs, which is the most outstanding feature of the international tobacco industry in recent years [13,14]. However, as novel cigarette in the market work by heating and atomizing the liquid or gel inside a cartridge at under 200 ℃, the cut tobacco or tobacco extract cannot release the chemicals into the air at this temperature. Therefore, these products do not taste as good as traditional ones. To reach a suitable temperature for heating, studies should be conducted on the structure and heat transfer process of the invention. In 2014, Philip Morris International Inc. launched IQOS, a heat-not-burn tobacco product that uses electronics to heat specially designed tobacco units to release a nicotine-containing vapour [15]. In comparison, the new cigarette technology in China is still in its infancy. The Zhengzhou Tobacco Research Institute has made progress in studying the feasibility of research and the development of new style cigarettes in terms of heat source, heating parameters, and other factors [16]; however, since the research has been performed for only a short period of time, a lack of depth or final products as well as a low degree of industrialization has be emerged.

The difficulty in designing heat not burn tobacco products lies in controlling the suitable temperature field for the tobacco to release its chemicals at certain temperatures. Therefore, in this study, the structure of a heat not burn tobacco product has been designed, and the numerical simulation has been used to optimised for the product parameters by examining their temperature field. The heat transfer process in a porous media of cut-tobacco has been analysed. This approach will improve the products, then reduce the air pollution from smoking.

2. The structural of heat not burn tobacco product

Figure 1 illustrates the structure of the heat not burn tobacco product, which is made up of five parts: an inner shell, a heating wire, a filter mouthpiece, an outer shell, and a temperature control device. The inner shell stores tobacco units in the form of shreds, granules, or fragments. Eight pores of 1-mm diameter are evenly distributed across the bottom of the shell to ensure air permeability during

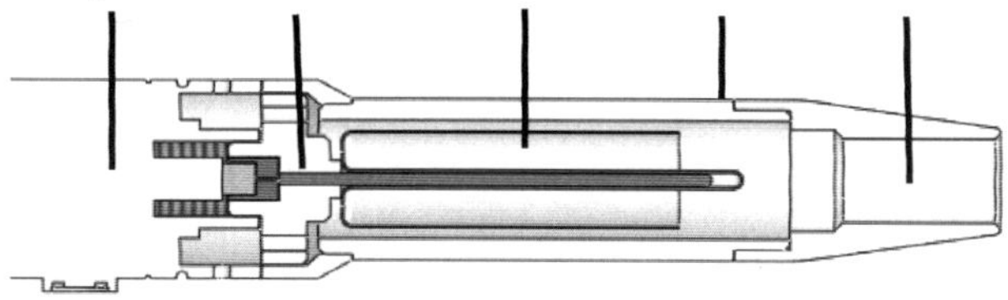

Figure 1. Schematic of a new-style cigarette

Inhalation. The heating wires are kept at a temperature of around 400, as studies show that cracking at below 500 produces fewer harmful substances [17]. The temperature is regulated by the temperature control device with an error of +2. The outer shell functions as a heat preserver, while the filter

IOP Conf. Series: Journal of Physics: Conf. Series **1064** (2018) 012011 doi:10.1088/1742-6596/1064/1/012011

mouthpiece reduces the concentration of tar and nicotine in the smoke and cools down the smoke before inhalation.

Figure 2 illustrates the flow of air during inhalation. Air enters the inner shell through the pores on the outer shell. Then, the heating needle transfers heat to the cut tobacco in the inner shell which gives off substances such as alcohols, aldehydes, and ketones as the temperature rises. These substances are mixed with air which enters the cavity through the filter mouthpiece during inhalation.

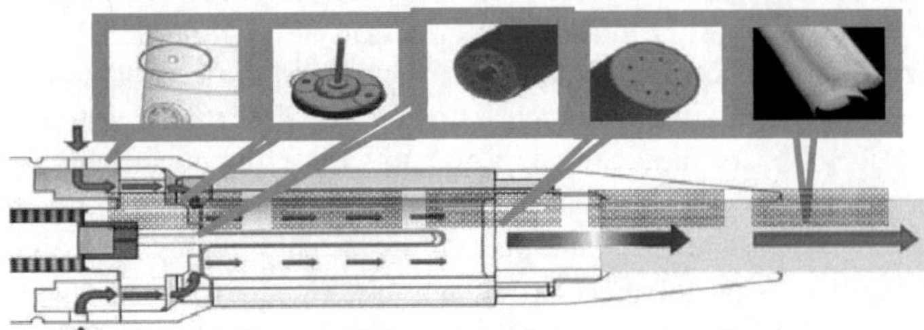

Figure 2. Schematic of fluid flow in a new-style cigarette

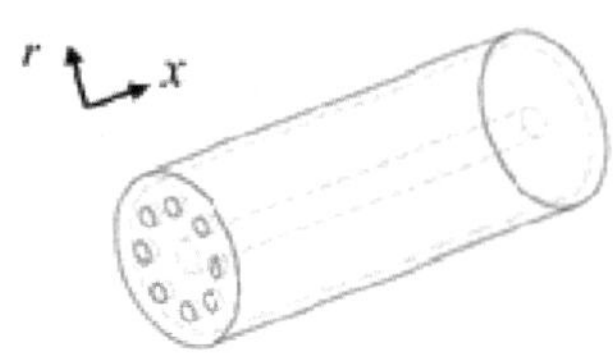

Figure 3. The geometric model

3. Physical model and mesh generation

The cylinder shown in Figure 3 is the geometric model of the flow field of a new-style cigarette. The center contains heating wires; between the wires and the interior wall of the chamber is cut-tobacco. The cylinder has a radius of 45 mm, a wall thickness of 5 mm, a length of 240mm and a wick of 8mm diameter. As the space between the interior wall and the wick is filled with cut-tobacco, air passes through from one side and takes away the heat of the wick in a laminar flow. During simulation, the following assumptions were made: 1) the porous medium was even and isotropic, 2) the porosity, specific heat capacity, density, and heat transfer coefficient were constant, and 3) natural convection and radiation were ignored.

The equations for the flow and heat transfer control of the above model are as following equations [18, 19]:

$$\frac{\partial \rho}{\partial t} + \nabla \cdot (\rho \boldsymbol{v}) = 0 \tag{1}$$

$$\frac{\partial}{\partial t}(\rho \boldsymbol{v}) + \boldsymbol{v} \cdot \nabla(\rho \boldsymbol{v}) = -\nabla p + \nu \nabla^2 \boldsymbol{v} + \rho \boldsymbol{g} + \boldsymbol{F} \tag{2}$$

$$\sigma \frac{\partial T}{\partial t} + \boldsymbol{v} \cdot \nabla T = \nabla(\alpha \nabla \mathrm{T}) \tag{3}$$

In the equations, v, p, and T denote the average velocity, pressure, and temperature of the fluid, respectively. ρ is the fluid density, ν is the kinematic viscosity, g is the gravity acceleration, σ is the heat capacity of the solid matrix and fluid in the porous medium, α is the thermal diffusivity, and F is the source term in the momentum equation of the porous medium. According to J. Pascal and H. Pascal [20, 21], for a flow at low Reynolds numbers, the pressure gradient is mainly to overcome the viscous resistance. Darcy's law is applicable. When the velocity accelerates to a certain level, the fluid inertia increases, therefore, the pressure gradient must overcome not only the viscous resistance but also the inertia; the higher the velocity, the stronger the inertia. The correlation between the pressure gradient and velocity can be expressed by the Darcy–Forchheimer law. Since the materials are isotropic, the source term can be simplified as following [22]:

$$F = -\left(\frac{\mu}{a}\boldsymbol{v} + C_2\frac{1}{2}\rho|\boldsymbol{v}|\boldsymbol{v}\right) \quad (4)$$

In this equation, μ is the kinematic viscosity of air and a is the permeability. C2 is the inertia coefficient, which means the pressure dropper unit length in the flow direction.

The governing equations were discretized by the finite volume method and solved with an unsteady implicit scheme, while the semi-implicit method for pressure-linked equations (SMPLE) algorithm was used for pressure-velocity coupling. The temperature of the wick was set at 673.15 K according to the condition of heat –not burn product, and the permeability of the cut tobacco was 0.249m/d acquired by experiment. As the wall of the cylinder is made of stainless steel, it had a heat transfer coefficient of 15 w/m2k; it was solid and 1 mm thick. The initial temperature and velocity of the air flowing into the model were 295.15 K and 17.5 ml/s, respectively. The physical parameters of cut tobacco in the wick were as follow: the density ρm = 288.98 kg/m3, the specific heat capacity C = 794.1252 J/kg·k, and the heat transfer coefficient λ= 0.08186 w/m·k. Based on the geometric and physical model, the hybrid grids of both the structured and unstructured portions were used for the mesh refinement of the central wick. The total number of meshes was 280,800.

4. Computation results and discussion

4.1. The temperature distribution in a new-style cigarette

Figure 4 displays the transient temperature distribution along the x-axis of the cigarette after 6.5 s of heat transfer. During computation, the volumetric air flow rate was fixed at the standard level of 17.5 ml/s. As shown in Figure 4a, heat from the central heating wires spreads through the cut tobacco and air; during inhalation, heat diffuses faster at the upper part of the cigarette because the air flows upwards and brings the heat with it. At the same time, air enters at a low temperature (295.15 K) and takes away the heat of the surrounding medium as it rises. The closer to the entrance, the more slowly the temperature rises, and the lower the thermal diffusivity. Hence, before the heat is transferred to the wall, the isotherm of the cut tobacco exhibits a V shape. Figure 4b shows the temperature distribution along the x- axis after 5 s of heating. It is evident that the distribution of the temperature field is radially symmetrical. As time passes, heat continues to diffuse from the centre to the surroundings. The comparison reveals that airflow drastically increases heat transfer and leads to a faster increase in the temperature of the cut tobacco; there is a huge difference in the temperature field without or without airflow.

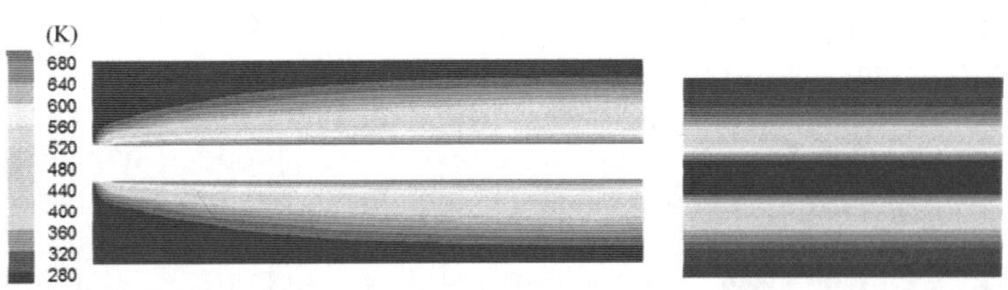

a. with airflow b. without airflow

Figure 4. The temperature distribution of a cigarette at t=6.5s along x axis

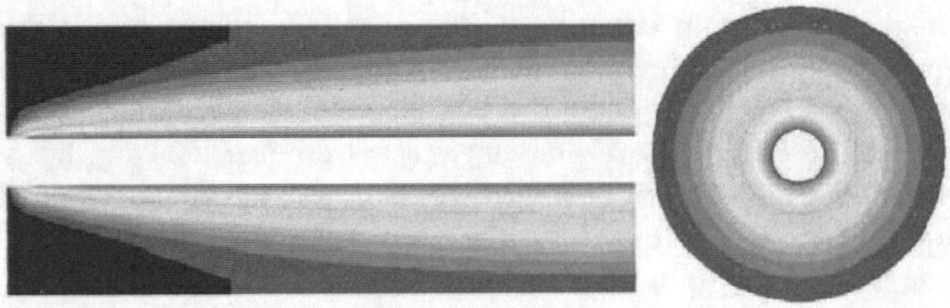

a. along x axis b. at x=55mm

Figure 5. The temperature distribution of a cigarette at t=55s

After 20 s of heating, the heat at the top diffuses to the interior wall, the temperature of which then increases significantly. It passes through the wall and spreads to the exterior. Figure 5a shows the temperature distribution along the x-axis after 55 s of heating. With increasing duration, the temperature of the wall rises gradually, first at the air exit and eventually at the entrance; there is also an increasing heat transfer to the exterior. Figure 5b shows the temperature distribution along the x-axis after 55 s of heating. At this point, the temperature of the wall exceeds the fixed ambient temperature, but the temperature field distribution is still axially symmetrical.

Figures 6 demonstrate the temperature distribution along a vertical axis 0.5 mm from the exterior

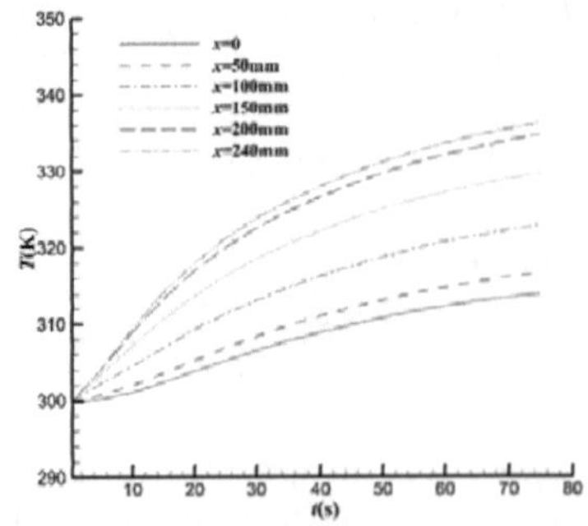

Figure 6. The temperature distribution of a cigarette at r=40mm

Wall when heating a new-style cigarettes. As shown in the figure, for this cigarette, the initial temperature of the shell equals the ambient temperature. As time passes, heat from the central wick spreads outwards and the internal temperature of the cut tobacco increases. The temperature increases while the rate of increase is on the decline with increasing of the heating time. The result clarifies that the temperature in the inner shell surface is less than 70ºC after 80 seconds heating and suction. It is satisfied the safety requirement of the product.

4.2. Chemical constituent in aerosol of a new-style cigarette

The chemical analyses have been performed on smoke generated with the standard FTC methods. The measurement shows the tar yield from the new-style cigarette being 3mg compared to 11mg from the traditional cigarette. The nicotine release from the new cigarette are 0.2mg which is a quarter of the reference traditional cigarette. The results mean the less HPHCs products of heat not burn cigarette

5. Conclusion

This study designed a structure of a new type of e-cigarette and analyzed its process of heating and internal airflow. The Darcy–Forchheimer model of porous media was used to examine the issues of convective heat transfer inside the cigarette chamber of cut tobacco. The temperature field distribution in the inner shell of the new style cigarette has been studied considering the suction process. The aerosol of this cigarette has been measured using FTC method. The research results show that: (1) the new style cigarette can acquired the temperature condition of heat but not burn tobacco; (2) the temperature of inner cut tobacco changing with suction process and heating time, finally tends to be steady; (3) the new style cigarette in this paper can reduce the yield of tobacco combustion produces and resulted in a reduction of harmful activity in smokers.

Acknowledgments
The project supported by ZJTI Science and technology project" Research on electronic cigarette device and heating characteristic (ZJZY2015A008).

References

[1] IARC, International Agency for Research on Cancer monographs on the evaluation of the carcinogenic risk of chemicals to humans. Tobacco Smoking, 38(1986)168-170.
[2] Alexandrov, L. B., Ju, Y. S., Haase, K., Van, L. P., Martincorena, I., Nik-Zainal, S., Mutational signatures associated with tobacco smoking in human cancer. Science. 354(2016) 618-622.
[3] Bilano, V., Gilmour, S., Moffiet, T., D'Espaignet, E. T., Stevens, G. A., Commar, A., Global trends and projections for tobacco use, 1990-2025: an analysis of smoking indicators from the WHO Comprehensive Information Systems for Tobacco Control, Lancet. 385(2015)966-976.
[4] Patel R.R., Ryu J.H., Vassallo R. Cigarette smoking and diffuse lung disease, Drugs. 68(2008) 1511-1527.
[5] Alexander L. E. C, Shin S., Hwang J. H.,Inflammatory Diseases of the Lung Induced by Conventional Cigarette Smoke : A Review, Chest. 148(2015) 1307-22.
[6] Besaratinia A, Tommasi S,Electronic cigarettes: the road ahead, Preventive medicine. 66(2014)65-67.
[7] Baker R.R.,Smoke generation inside a burning cigarette: Modifying combustion to develop cigarettes that may be less hazardous to health, Progress in Energy & Combustion Science. 32(2006)373-385.
[8] Hughes J.R., Keely J.P.,The effect of a novel smoking System-Accord-on ongoing smoking and toxin exposure, Nicotine &Tobacco Research. 6(2004) 1021-1027.
[9] Werley, M. S., Freelin, S. A., Wrenn, S. E., Gerstenberg, B., Roemer, E., Schramke, H., Smoke chemistry,invitro and in vivo toxicology evaluations of the electrically heated cigarette smoking system series, Regulatory Toxicology and Pharmacology. 52(2008)122-139.
[10] Flora, J. W., Wilkinson, C. T., Wilkinson, J. W., Lipowicz, P. J., Skapars, J. A., Anderson, A.,Method for the Determination of Carbonyl Compounds in E-Cigarette Aerosols, Journal of chromatographic science. 55(2017)142-148.
[11] Han, S., Chen, H., Zhang, X., Liu, T., Fu, Y.,Levels of Selected Groups of Compounds in Refill Solutions for Electronic Cigarettes,Nicotine & tobacco research.18(2016)708-714.
[12] Uchiyama, S., Ohta, K., Inaba, Y., Kunugita, N. Determination of carbonyl compounds generated from the E-cigarette using coupled silica cartridges impregnated with hydroquinone and 2,4-dinitrophenylhydrazine, followed by high-performance liquid chromatography, Analytical sciences: the international journal of the Japan Society for Analytical Chemistry. 29(2013)1219-1222.
[13] Duan, Y.X., Zhao, W., Yang, J., Han J.M., Sun Z.Y., Yang L., Research on the Particle Size Distribution of Aerosol in the Smoke of Traditional and Electronic Cigarettes, Acta Tabacaria Sinica. 21(2015) 1–5.
[14] Wang, P.J., Feng, S.A., Wu, Y.Analysis of the Volatile Ingredients of Tobacco-flavoured E-liquid, Flavour Fragrance Cosmetics. 6(2014) 27–31.
[15] Philip Morris International Inc. A aerosol producing installation with improving temperature distribution.CN201280060098.9

[16] Xu H.Chen Y.Tang J.G., Analysis on the Temperature Object of Electric-heating Cigarette Heater, Journal of Anhui Agricultural Sciences. 45(2017) 215-223
[17] Dai, L., Shi, C., Liu X.B., Xu J., Shen, K., Yuan, K., Difference Analysis on Physical Characteristics and Chemical Components of Reconstituted Tobacco Used in Traditional Cigarettes and Heat-not-burn Tobacco Products, Acta Tabacaria Sinica. 23(2017) 20–26.
[18] Yin Z.Q., Li D.S., Meng J.L., Discharge coefficient of small sonic nozzles, Thermal Science. 18(2014) 1505-1510.
[19] Wang, K.Y., Wang, D.Z., Li, P.C., Two Decoupling Methods for the Heat Transfer Model of a Plate Channel Filled with a Porous Medium, Applied Mathematics and Mechanics. 36(2015) 494–504.
[20] Pascal J.P., Pascal H., Pressure diffusion in unsteady non-Darcian flows through porous media, European journal of mechanics. B, Fluids. 14(1995) 75-90.
[21] Pascal J. P., Pascal H., Non-linear effects on some unsteady non-Darcian flows through porous media, International journal of non-linear mechanics. 32(1997) 361-376.
[22] Nield D.A., Resolution of a paradox involving viscous dissipation and nonlinear drag in a porous medium, Transport in Porous Media. 41(2000) 349-357.

Research on application of nozzles for ampoule bottle production machine based on ANSYS fluent

Tu Jun*, Pengcheng Xuan

School of Mechanical Electronic & Information Engineering, China University of Mining and Technology, Beijing 100083, China

*Corresponding author e-mail: TuJun0189@163.com

Abstract. In order to determine the suitable combustion temperature suitable for ampoule bottle production, the nozzle and combustion field were numerically simulated and analysed by using ANSYS fluent software to change the inlet diameter and inclination angle of nozzle, respectively. The influence of inlet diameter and inclination angle on the central temperature of combustion flame was determined, and the reference was provided for the condition of ampoule bottle forming. The results show that the inlet diameter has a great influence on the central temperature of the combustion flame and can obtain a more favourable processing temperature range when the inlet diameter is smaller, and the change of the inclination angle has no significant effect on the central temperature of the combustion flame. When the inlet diameter is 0.6 mm and the inclination angle is 20 °, the highest processing temperature range can be obtained. This study can provide a reference for improving the production efficiency of ampoule bottles.

Keywords: nozzle, combustion field, ampoule bottle, numerical simulation.

1. Introduction

Ampoule bottles are currently widely used in the medical field as containers for high-purity chemicals that contain injection preparations and must be insulated from the air. They are mostly made of glass [1].

The production of ampoule bottles is made of finished glass tubes for secondary processing. The key steps are heating, firing, coloring, neck marking, annealing, cooling, etc. The firing molding is the key of the bottle body forming. If the effective flame temperature is not obtained during the sintering process, it is very likely to lead to the deformation of the bottle body or the unqualified size and specification of the bottle body. In the manufacture of ampoule bottles with glass tubes, the nozzle spray flame is needed to heat the softening glass tubes. At present, with the rapid development of computational fluid dynamics (CFD) technology [2], it is possible to use CFD to simulate the dynamic working process of fluid machinery. In actual production, the internal structure of the nozzle will have a great influence on the combustion temperature. Therefore, studying the structure and key influencing factors of nozzles is a major technical problem that technicians in this field need to solve at present.

Content from this work may be used under the terms of the Creative Commons Attribution 3.0 licence. Any further distribution of this work must maintain attribution to the author(s) and the title of the work, journal citation and DOI.
Published under licence by IOP Publishing Ltd

IOP Conf. Series: Journal of Physics: Conf. Series **1064** (2018) 012012 doi:10.1088/1742-6596/1064/1/012012

2. Materials and methods

2.1. Geometric modelling

Figure. 1 is a schematic diagram of the nozzle structure. The left end of the nozzle is the intake port, and the different inlet diameters and inclination angles are respectively set. A uniform inlet pressure is set in the numerical simulation process.

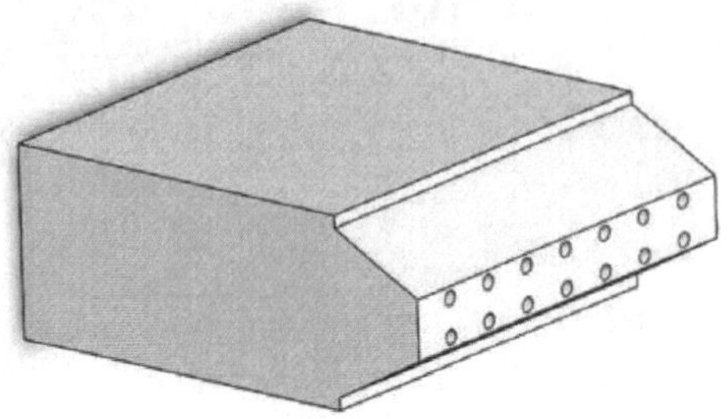

Figure 1. The three-dimensional model of the nozzle.

2.2. Grid partition

In this study, the symmetrical structure of the nozzle is studied, and the simplified model is the plane axisymmetric model. After simplification, the structure of the model is simple, and the meshing of the quadrangle grid in a structured grid can meet the simulation requirements [3], taking the symmetrical distribution of nozzle outlet wide 80mm, long 300mm as the outer combustion flow field region, as shown in Figure 2, the quadrilateral meshes are divided by using Gambit software for nozzles and external combustion flow field. In order to improve the flame effect, the inlet diameter and inclination angle are taken as follows:

$$(a1):\begin{cases} d = 0.6mm \\ \beta = 0° \end{cases}, (a2):\begin{cases} d = 0.6mm \\ \beta = 10° \end{cases}, (a3):\begin{cases} d = 0.6mm \\ \beta = 20° \end{cases};$$

$$(b1):\begin{cases} d = 0.8mm \\ \beta = 0° \end{cases}, (b2):\begin{cases} d = 0.8mm \\ \beta = 10° \end{cases}, (b3):\begin{cases} d = 0.8mm \\ \beta = 20° \end{cases};$$

$$(c1):\begin{cases} d = 1mm \\ \beta = 0° \end{cases}, (c1):\begin{cases} d = 1mm \\ \beta = 10° \end{cases}, (c1):\begin{cases} d = 1mm \\ \beta = 20° \end{cases}.$$

The combustion field temperature was studied under the above 9 conditions.

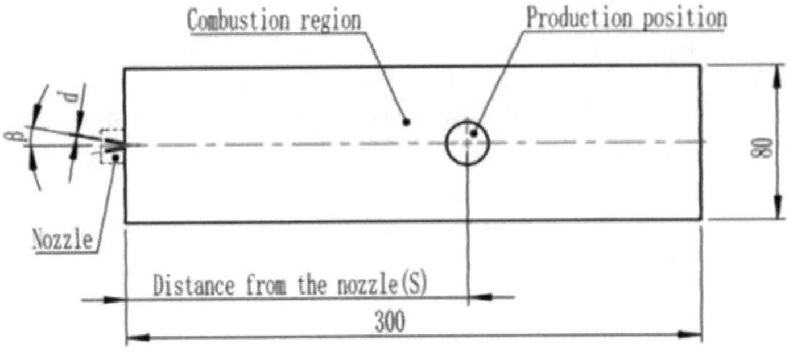

Figure 2. Nozzle and combustion plane area.

2.3. Boundary condition

In the process of numerical simulation, in order to ensure that the gas does not reflux, the pressure values of the two entrances should be set to the same, and the inlet boundary conditions should be set as pressure-inlet. According to the actual working conditions, given inlet pressure of 3KPa as the research condition, the outlet is pressure-out. in order to simulate the real working conditions, the outlet pressure is set at relative atmospheric pressure of 0 Pa, and the temperature is 300K [4].

2.4. Gas fuel pre-treatment

The fuel used in the calculation must be pretreated before the numerical simulation of combustion [5]. The fuel used in this paper is natural gas, its main component is methane, its chemical formula isCH_4, and other microelements are not considered [6].

The oxidant is O_2 in the air. When the air is selected, only N_2 and O_2 in the air are taken into account, which account for 79% and 21%, respectively, excluding other trace components.

The chemical reaction formula for methane is as follows:

$$CH_4 + 2O_2 \rightarrow 2H_2O + CO_2$$

2.5. Mathematical model

The fluid flow must follow the law of physical conservation. The basic laws of conservation include the law of conservation of mass, the law of conservation of momentum, and the law of conservation of energy. All flow phenomena in the natural world can be described by two equations: the continuous equation (that is, the mass conservation equation) and the Navies-Stokes equation (that is, the momentum conservation equation) [7].

This paper is studied under the temperature condition of 300K.The transport equation of the turbulent energy in the standard $k-\varepsilon$ model is derived from the exact equation, but the dissipation rate equation is obtained by physical reasoning and mathematical simulation of the similar original equation [9].The combustion flow is a complete turbulent flow and the molecular viscosity can be ignored, so the turbulence model is chosen as the standard $k-\varepsilon$ model [10], This model is widely used in practical engineering problems and scientific research.

The standard k-e model is a semi empirical model, mainly based on turbulent flow energy k and turbulence energy dissipation rate e. The turbulent kinetic energy equation K is an exact equation, and the turbulence energy dissipation rate e equation is an equation derived from the empirical formula. In order to ensure the closure of the model, the following assumptions are made according to the physical significance [11].

$$\text{Generate item} = C_{\varepsilon 1} \cdot \frac{\varepsilon}{k} \times Turbulent\ kinetic\ energy\ generation\ term$$

$$\text{Gradient diffusion term} = \frac{\mu_t}{\sigma_\varepsilon} \cdot \frac{\partial \varepsilon}{\partial x_i}$$

$$\text{Dissipative term} = C_{\varepsilon 2} \cdot \frac{\varepsilon}{k} \times Turbulent\ kinetic\ energy\ dissipation\ rate$$

2.6. Glass characteristic analysis

Glass is an amorphous material with no fixed melting point and softens at high temperatures. Medical glass materials are generally made of neutral borosilicate glass with better performance. The range of mass fraction of SiO_2 is about 75%, Al_2O_3 is 5%, B_2O_3 is 10%, CaO is 3%, K_2O is 1%, Na_2O is 6% and Fe_2O_3 is 0.05%. The flame temperature of secondary reprocessing of glass is 1350°C [12].Since the production of ampoule bottles is reprocessed by glass tubes, that is, the second reprocessing of glass, the temperature of ampoule processing at 1350°C is taken as the processing temperature of ampoule bottles in this paper, and the temperature distribution of combustion field is analyzed.

3. Results and discussion

3.1. Numerical simulation results of nozzle and combustion field

This study used ANSYS fluent software to perform numerical simulations on 9 experimental models. The results are shown in Figure 3. From the figure, it can be observed that as the flame gradually moves away from the exit, the fuel concentration in the combustion zone gradually becomes thinner, the combustion temperature begins to decrease, the flame propagation speed decreases, the flame thickness increases, and the flame area gradually expands. The combustion temperature of all combustion fields can reach a maximum of 2180K, but the production and processing temperature of ampoule bottle should reach 1350°C, that is 1623K.

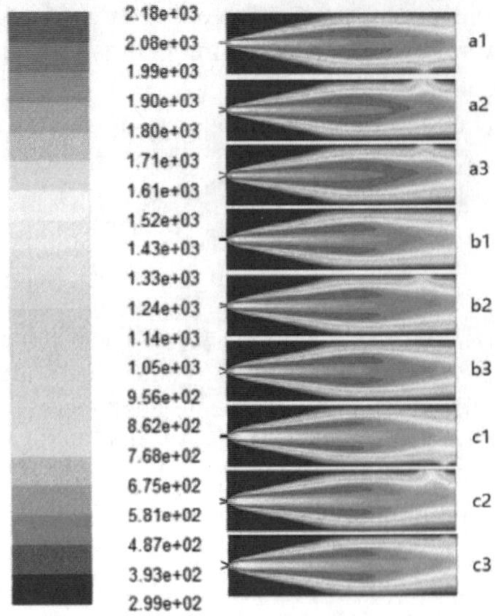

Figure 3. Temperature cloud map of combustion center.

3.2. Numerical comparison and analysis of combustion field

In order to show the effect of different adjusting position on combustion temperature field accurately and intuitively. The simulated data of the central temperature of the combustion field under the condition of the same inclination angle α and different nozzle inlet diameters d are extracted to determine the influence of the nozzle inlet diameter on the combustion temperature in this paper, as shown in figure 4. Comparing the effects of different inlet diameters, it can be found that the smaller the diameter, the easier it is to obtain a higher combustion temperature and the greater the maximum combustion temperature range. And at a position closer to the nozzle, a processing temperature of 1623K can be obtained. Its rationality is inversely related to the size of the inlet diameter.

The simulation data of the center temperature of the combustion field under the conditions of the same nozzle inlet diameter d and different inclination angles α were extracted to determine the influence of the inclination angle of this article on the combustion temperature, as shown in Figure 5. Comparing the effects of different tilt angles, it can be found that the larger the tilt angle, the higher the combustion temperature can be obtained. However, the impact is not obvious.

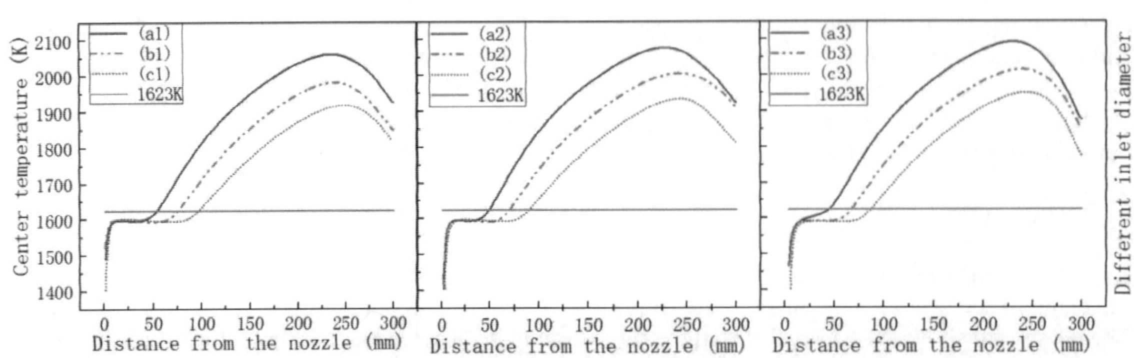

Figure 4. Temperature distribution at different inlet diameters.

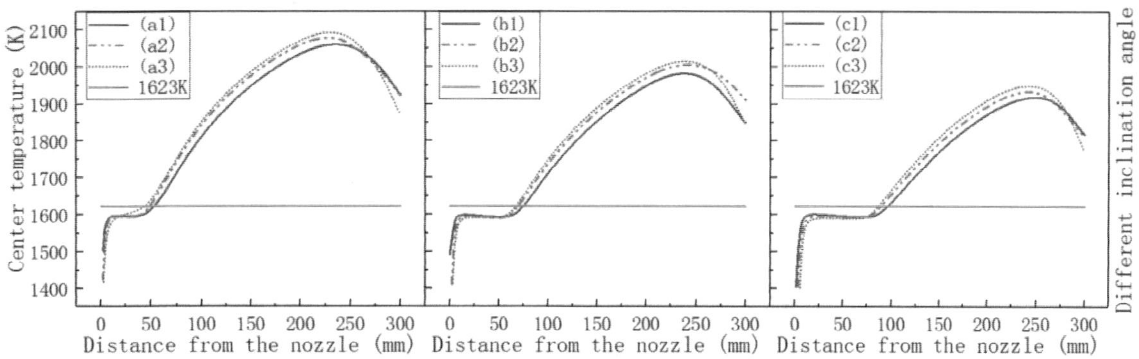

Figure 5. Temperature distribution at different inclination angles.

4. Conclusion

(1) The numerical simulation method of CFD was used to simulate the combustion temperature at different inlet inclination angles. It was found that the change of the inlet tilt angle had a certain influence on the combustion temperature. The higher the inclination angle, the higher the combustion temperature, but the effect was not obvious. It reveals the change law of the combustion temperature with the entrance inclination angle.

(2) The numerical simulation method of CFD was used to simulate the combustion temperature of different nozzle inlet diameters. It was found that the change of the inlet diameter had a great influence on the combustion temperature. With other conditions being equal, a more suitable combustion temperature can be obtained with a diameter of 0.6 mm. The maximum processing temperature range can be obtained at position (a). At the same time, the best ampule production position was determined at 155.182 mm from the outlet of the gas mixing valve.

(3) The research on the nozzle parameters of the ampoules production machine can provide reference for the follow-up engineers and technicians to improve the qualification rate of the ampoules in the nozzle parameters.

References

[1] Jiheng Ding. The development of embedded ampoule bottle liquid agent online detection system. Xi'an University Of Architecture And Technology. 2011:1.
[2] Howard H. Hu . Fluid Mechanics (Fifth Edition) [M]: Elsevier Inc, 2012: 421-425.
[3] Lijuan Wu, Mian Zheng, Caiming Zhang. The Edge Optimization Disposal in Quadrangular Meshes Generation Process[J]. Journal of engineering graphics. 2006,6:49-52.
[4] Jun Li, Qingqing Li, Yuantao Wang, Zhaoli Guo, Xueling Liu. Fundamental flame characteristics of premixed H2-air combustion in a planar porous micro-combustor[J]. Chemical Engineering Journal.2016,283:1190.
[5] Zvonimir Petranovic´, Wilfried Edelbauer, Milan Vujanovic´, Neven Duic´ . Modelling of spray and combustion processes by using the Eulerian multiphase approach and detailed chemical kinetics[J]. Fuel.2017,191:28.
[6] Mingkuan Yang. The application of natural gas combustion mode and burner[J].Glass and Enamel. 2001.29(4):31-35.
[7] Pengfei Li, Minyi Xu, Feifei Wang. Proficient in CFD engineering simulation and case practice[M]. Beijing: people post and Telecommunications Press, 2011.10:1-2.
[8] Xueguan Song, Lin Cai, Hua Zhang. ANSYS fluid-solid coupling analysis and engineering example[M]. Beijing: China Water Conservancy and Hydropower Press, 2012:3.
[9] Tianze Cao. Research on Numerical Prediction Technology of Gas Fuel Combustion Flow in Combustion Chamber[D]. Harbin Engineering University. 2012.3:24-25.
[10] Hongjun Zhu. Practical course of FLUENT fluid Analysis and Simulation[M]. Beijing: China

Post and Telecommunications Press, 2010:79-90.
[11] Alan Y.Chen and Lisa Dang,Characterization of Supercritical JP-7' Heat Transfer and Coking Properties. [C], AIAA, 2002-0005.
[12] Yingliang Tian. Study on the application of spodumene in neutral borosilicate glass[J]. Glass and Enamel. 2012-8.40(4):11-13.

Numerical study of plasma assisted combustion for a rocket combustor using GCH4/GOX as propellants

Tikai Zheng, Xueke Che, Liang Li, Chuan Chen, Wansheng Nie and Xiuqian Li*

Space Engineering University, Beijing, China

*Corresponding author e-mail: lixiuqian123@126.com

Abstract. In order to study the effect of active particles in plasma on the combustion of small rocket engines, the combustion characteristics of the combustion chamber under different ionization levels of fuel and oxidant were simulated. The results show that with the increase of the ionization degree of fuel and oxidant, the high temperature region of the combustion chamber advances and the proportion increases; the pressure of the combustion chamber increases continuously, so as to achieve the effect of increasing the engine thrust. It shows that the ionized active particles can shorten the propellant ignition delay, improve the combustion efficiency of the engine combustion chamber, making the combustion more fully; increase the combustion chamber pressure so as to increase the engine specific impulse.

1. Introduction

The spacecraft power system is of great significance to the aerospace industry. The development of aerospace projects such as manned space flight, lunar exploration and deep space exploration in China all rely on rocket power. The small-thrust rocket engine has a very wide range of applications on satellites, space shuttles, upper stage of multi-stage vehicles and other spacecraft's [1]. It is mainly used for tasks such as orbit control, attitude control and space landing et al. Most domestic and foreign spacecraft, such as rockets and satellites, currently use nitro-oxidants and thorium fuels as propellants. Although these propellants have good performance and mature technology, they are highly toxic, expensive and pollute the environment. They cannot adapt to the human space mission needs in the future.

With the rapid development of manned spaceflight, the non-toxic propellants have become an inevitable trend of development [2-3]. Methane fuel is a good choice for non-toxic propellants because of its high specific impulse, low cost, and low pollution. However, when it is used as an orbit control or attitude control engine, it is faced with problems such as difficulty of ignition under low temperature and high vacuum conditions. Spacecraft attitude and orbit control rocket engine has less injection unit so propellant mixing efficiency is low, the combustion chamber volume is small, so combustion time for propellant is short, which is not conducive to the full combustion of the propellant [4]. Spacecraft rocket engines should have a strong ability to change thrust to meet the needs of different working conditions. When the thrust is adjusted by changing the propellant flow rate and mixing ratio, the engine is prone to occur combustion instability, which will lead to lower combustion efficiency or even damage the engine. Advanced technical urgently need to be adopted to ensure the efficient and stable combustion of the engine under various working conditions [5].

Content from this work may be used under the terms of the Creative Commons Attribution 3.0 licence. Any further distribution of this work must maintain attribution to the author(s) and the title of the work, journal citation and DOI.
Published under licence by IOP Publishing Ltd

As the plasma has the potential to improve the combustion efficiency, expand the flammability, and stabilize the combustion flame under severe working conditions, the plasma assisted ignition and plasma assisted combustion have received extensive attention [6-7]. At present, researchers have summarized the mechanism of plasma combustion into the following aspects: thermal effects, chemical kinetic effects, and transport effects [8-9]. Applying plasma technology to spacecraft rocket engines, researching plasma assisted ignition and plasma assisted combustion for spacecraft rocket engine special working environment and performance requirements can promote the development of China's spacecraft rocket engine technology, supporting for China's major space projects such as manned spaceflight and deep space exploration et al.

In this paper, the combustion chamber of a gas-oxygen/gas-methane coaxial single nozzle rocket engine is constructed. The effects of different degrees of ionization of fuel and oxidant on the combustion characteristics of the rocket engine combustion chamber are studied. The relevant results can provide some reference for the future research of plasma assisted combustion for small thrust rocket engines.

2. Numerical simulation approach

2.1. Simulation object and numerical methods

The rocket engine chamber designed by Technical University of Munich, Institute for Turbo machinery and Flight Propulsion, Germany is adopted as simulation object [10]. Relevant parameters are shown in Table 1.

Table 1. Parameters of the rocket engine chamber.

Parameter	Value	Parameter	Value
Nominal chamber pressure (Bar)	20.0	Chamber diameter (mm)	12.0
Fuel mass flow (g/s)	13.5	GO_2 diameter (mm)	4.0
Oxidant mass flow (g/s)	35.1	GO_2 post thickness (mm)	0.5
Fuel inlet temperature (K)	269	GCH_4 outer diameter (mm)	6.0
Oxidant inlet temperature (K)	275	Throat diameter (mm)	7.6
Chamber length (mm)	305.0	Contraction ratio	2.5

The engine uses a gaseous propellant, so it does not need to consider evaporation and atomization during combustion, only the mixing and chemical reaction process is considered. In the simulation process, the models involved are turbulence model, chemical kinetic model, and turbulence-chemical reaction interaction model. For the simulation the Reynolds-averaged Navies-Stokes (RANS) equations were solved, turbulence is modeled using a standard k-ε model and the wall is treated using standard wall Fn. The combustion is modeled using a laminar finite rate chemistry model, i.e. turbulence chemistry interaction (TCI) is neglected. As chemical kinetic scheme a 14 species 18 reaction reduced mechanism [11] is used.

2.2. Computational zone, initial and boundary conditions

The engine is symmetrical and the computational domain contains the injector, chamber, and nozzle. In order to reduce the calculated amount, it can be simplified as a two-dimensional axisymmetric structure. The structural grid was used to divide the computational domain and the nozzles and walls were encrypted. As shown in Figure 1, the total number of grids was about 71000.

The finite volume method is used to discrete control equations. For the nonlinear mutual coupling characteristic of N-S equations, the pressure and velocity decoupling are implemented using the pressure implicit operator splitting algorithm (PISO). The convection terms are discretized using the second-order upwind scheme, and the viscous terms are discretized using the central difference scheme. At the inlet the mass flow rates are given together with injection temperature as listed in Table 1. The outlet is set to a pressure outlet.

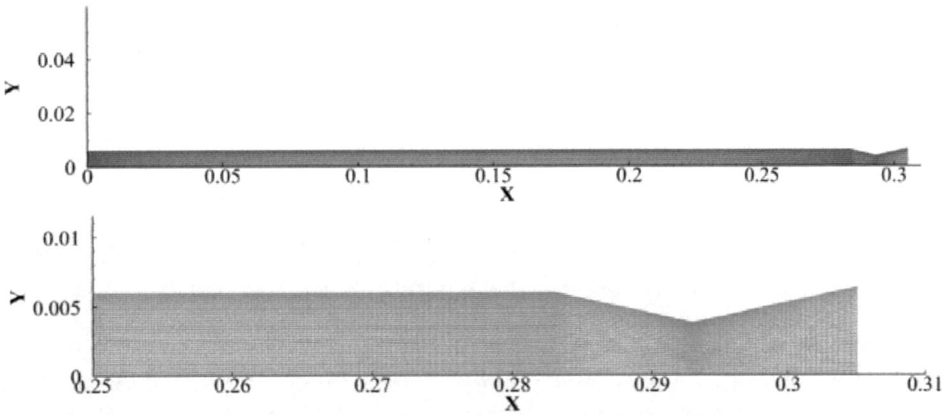

Figure 1. Computational zone and meshing

2.3. Model validation

The Technical University of Munich, Germany, has conducted experiments and numerical simulations on this engine under various conditions. The relevant data from the literature can provide verification for the simulation model in this paper. Figure 2 shows the comparison between the temperature field obtained by simulation in this paper and the temperature field in [10]. It can be seen from the figure that the temperature distributions of the two pictures are basically the same. Figure 3 shows the comparison of the wall pressure distribution of the combustion chamber simulated in this paper with the experimentally measured under the same conditions in [10]. It can be seen from the figure that the wall pressure distribution of the combustion chamber obtained by simulation agrees well with the experimental values. In summary, we can verify the correctness of this simulation model.

a) Temperature field in the literature [10] b) Temperature field calculated

Figure 2. Calculated temperature field vs temperature field in the literature

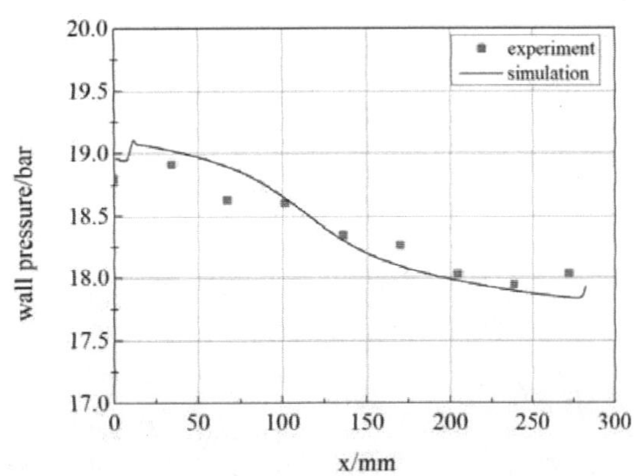

Figure 3. Wall pressure from CFD simulation vs experiment

IOP Conf. Series: Journal of Physics: Conf. Series **1064** (2018) 012013 doi:10.1088/1742-6596/1064/1/012013

3. Results and analysis

To study the combustion-supporting effect of non-equilibrium plasma in rocket engines, it is assumed that when the propellant is injected into the combustion chamber, the gas is processed by the plasma exciter. The fuel and oxidant of different ionization levels were added to the combustion flow field as the initial component, and a total of five conditions were included, as shown in Table 2.

Table 2. Ionization Degree under Different Operating Conditions

	Case 1	Case 2	Case 3	Case 4	Case 5
CH_4	0	5‰	0	5‰	10‰
O_2	0	0	5‰	5‰	10‰

3.1. Influence of plasma on the temperature distribution of the combustion chamber

Figure 4 shows the change of the center temperature of the combustion chamber along the axial direction under different working conditions. It can be seen from the figure that the temperatures in different conditions are basically the same in the front and rear sections of the combustion chamber, in the middle section of the combustion chamber, and in the reference state. Compared with Case 1, Case 2 to Case 5 all accelerated the temperature increase to different extents, that is, the distance to the same temperature was shortened, indicating that the active particles produced by ionization increased the propellant combustion efficiency. It can be seen from the figure that under the same degree of ionization, ionized CH4 has a greater effect on combustion than ionized O2.

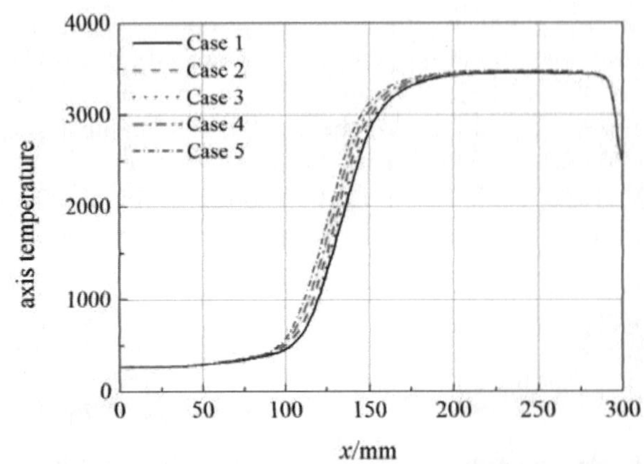

Figure 4. Temperature distribution along axis direction under different conditions.

Figure 5 compares the temperature fields of Case 1 and Case 5. It can be seen from the figure that the Case 5 high temperature zone is closer to the front of the combustion chamber and the range is larger than Case 1. It shows that the injection of propellant into the combustion chamber under the condition of Case 5 can achieve faster ignition and combustion, and the addition of active particles shortens the ignition delay, so that the propellant burns more fully under the same volume combustion chamber conditions.

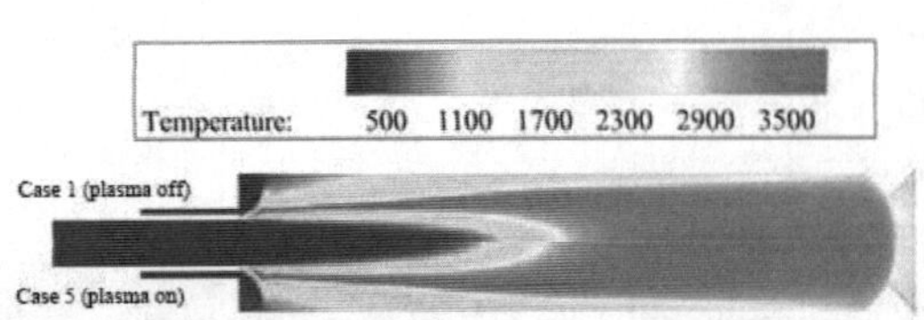

Figure 5. Comparison of temperature distribution.

3.2. Influence of plasma on combustion chamber pressure

Combustion chamber pressure is an important index to measure engine performance. The combustion chamber pressure under different operating conditions is shown in Figure 6. Figure 6 a) shows the change of the wall pressure of the combustion chamber under different working conditions. It can be seen from the figure that the addition of active particles generated by ionization increases the wall pressure of the entire combustion chamber. Under the same degree of ionization, the effect of increasing the wall pressure of the combustion chamber obtained by the ionized fuel and the oxidant is not much different. The effect of ionized O2 is slightly better than that of the ionized CH4. Figure 6 b) shows the change of the combustion chamber pressure under different working conditions. It can be seen that the addition of active particles generated by ionization can increase the pressure of the combustion chamber, and the greater the degree of ionization, the more obvious the effect.

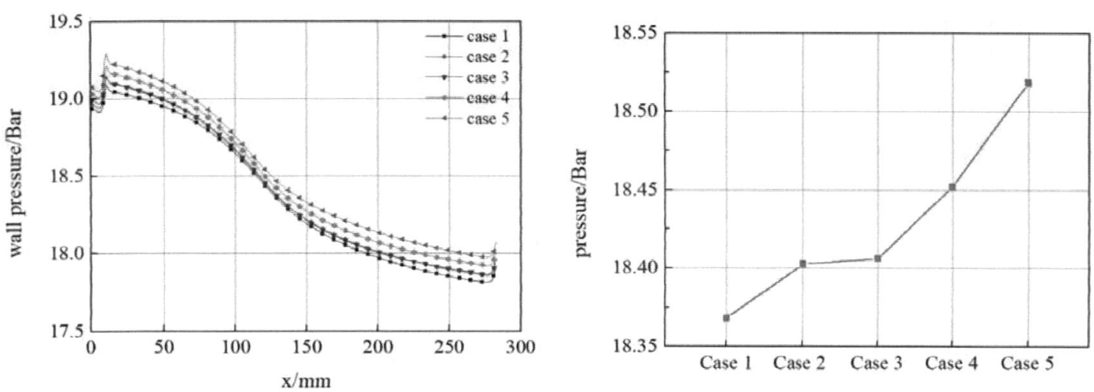

a) Wall pressure under different conditions b) Chamber average pressure under different conditions

Figure 6. Chamber pressure under different conditions.

4. Conclusion

In this paper, the impact of different ionization conditions of fuel and oxidant on engine combustion is studied for a small-thrust oxygen/gas methane rocket engine. The main conclusions are as follows:

(1) The active particles produced by ionization can shorten the ignition delay of the propellant and allow the propellant to achieve more complete combustion under the same volume of the combustion chamber.

(2) The addition of active particles produced by ionization can increase the pressure in the combustion chamber, and the greater the degree of ionization, the more obvious the effect.

(3) Under the same degree of ionization, the effect of ionized oxidant on the combustion chamber pressure is slightly better than that of ionized fuel, and the effect of ionized fuel on temperature is more pronounced than that of ionized oxidant.

Acknowledgments

This work was financially supported by the National Nature Science Foundation of China (Grant No. 51777214).

References

[1] J. Li, P. Zeng, N. Yu, Design and experiment research of a gas oxygen/propane small thrust engine, J. Journal of Aerospace Power. 10 (2011) 2352-2357.
[2] X. Fang, C, Shen, Progress of Liquid Oxygen/Methane Small Thrust Engine and Its Variable Thrust Technology, J. Winged Missiles Journal. 4 (2015) 88-95.
[3] T. Yu, Y. Li, Application Prospect of Liquid Oxygen/Methane Engine, J. Aerospace

Manufacturing Technology. 2 (2007) 4-7+13.
[4] X. Chen, X. Tian, L. Su, Theory of Spacecraft Propulsion, 1st ed, National Defense Industry Press, Beijing, 2014, pp, 152-157.
[5] W. Nie, F. Zhuang, Effect of spray characteristic on the liquid rocket combustion stability, J. Journal of Propulsion Tecnology. 3 (2001) 56-59.
[6] P. Li, H. Mu, L. Yu, Progress, key problem and prospect on low temperature plasma assisted combustion, J. High Voltage Engineering. 6 (2015) 2073-2083.
[7] Y. Wu, Y. Li, Progress in research of plasma-assisted flow control ,ignition and combustion, J. High Voltage Engineering. 7 (2014) 2024-2038.
[8] L. He, X. Liu, B, Zhao, Current investigation progress of plasma-assisted ignition and combustion, J. Journal of Aerospace Power. 7 (2016) 1537-1551.
[9] W. Nie, S. Zhou, X. Che, Review of plasma assisted combustion tecnology by nanosecond pulsed discharge, J. High Voltage Engineering. 6 (2017) 1749-1758.
[10] C. Roth, S. Silvestri, N, Perakis, Experimental and Numerical Investigation of Flow and Combustion in a Single Element Rocket Combustor using GH2/GOX and GCH4/GOX as Propellants, C. International Symposium on Space Technology and Science. 2017.
[11] H. Liu, F. Chen, H. Liu, 18-step reduced mechanism for methane/air premixed supersonic combustion, J. Journal of combustion science and tecnology. 5 (2012) 467-472.

IOP Conf. Series: Journal of Physics: Conf. Series **1064** (2018) 012014 doi:10.1088/1742-6596/1064/1/012014

PIV measurement of flow characteristic in single-head combustor with triple-stage swirler

Jun Zhang[1,*], Laijun Yan[1], Kewei Yin[1], Liujun Chen[2, 3], and Yungang Wu[1]

[1]Facility Design and Instrumentation Institute of CARDC, Mianyang, China
[2]College of Power and Energy, Northwestern Polytechnical University, Xi'an, China
[3]Science and Technology on Scramjet Laboratory, Hypervelocity Aerodynamics Institute of CARDC, Mianyang, China

*Corresponding author e-mail: zhangjun4117@163.com

Abstract. The present study investigates the flow field characteristic in a single-head combustor with triple-stage swirled, using particle image velocimetry (PIV). With pure air inlet flow, detailed mean and RMS velocities, turbulent intensity, vortices, Reynolds shear stress are obtained at several longitudinal and cross sections downstream of the swirled in the single-head combustor at flow temperature 300K and inlet combustor pressure 0.49MPa, 0.98MPa. It is seen from the investigation results that strong aerodynamic interactions among different kinds of combustor Eigen structures are existed. Multiple vortices are observed, accompanied with relevant recirculation zones. The effect of swirled is strong, and there are big and regular vortices in the cross sections in the downstream of the swirled 50mm away. The jet effect of main combustion holes and dilution holes is obvious, and the penetration of the jet flows are clearly deep with distinct effect on the flow field structure. It is shown that there are big size recirculation zones in the combustor flow field with high levels of turbulence generated due to triple-stage swirled, main combustion holes and dilution holes of the single-head combustor. It is helpful for promoting rapid mixing and stabilizing flame combustion.

1. Introduction

With the improvement of performance of aero turbine engine, the requirement for fuel atomization, mixing efficiency and flame stability is much rigorous than before, and the swirled stage develops from one to two, even three stages. Accordingly, the inner flow field of turbine engine combustor is quite complicated because of many kinds of aerodynamics structures. In order to understand the complex flow mechanism in turbine engine combustor and provide reference for design of advanced aircraft, the full field distribution of velocity and microscopic structure need to be measured accurately, especially with regard to the study of fuel injection, dilution, ignition, flame stabilization and combustion organization in combustor or afterburner[1-3]. According to the request of aero turbine engine for flame stabilization and efficient mixing fuel with air, kinds of combustor Eigen structures are necessary, such as swirled, main combustion whole, dilution hole with different parameters. As a result, it is difficult to obtain the inner flow field with wide-range velocity, strong rotational and turbulent characteristics [1, 4]. Usually, instantaneous full-field velocity distribution cannot be

Content from this work may be used under the terms of the Creative Commons Attribution 3.0 licence. Any further distribution of this work must maintain attribution to the author(s) and the title of the work, journal citation and DOI.
Published under licence by IOP Publishing Ltd

obtained through traditional measure technology such as hot wire and pressure probe. Also, only discrete velocity distribution can be measured by laser Doppler velocimetry (LDV) through single point scan method. Benefit from the instantaneous, planar and non-contact nature of PIV [5], it is possible for people to obtain the instantaneous and full-field velocity distribution of complicated inner flow field in the combustor. Up to now, there are a number of successful cases adopting PIV in turbine engine combustor.

Bahrain et al have experimentally studied the swirling flow field in a realistically complicated geometry of a gas turbine combustor [6]. M Scholl et al have successfully applied PIV in highly luminescent flames avoiding saturation of the second frame of commonly available double shutter PIV cameras. And the flow field structure of swirled is obtained with the condition of kerosene burning, 0.7MPa and 800K [7]. M Ahmed et al have studied the aerodynamic interactions between the primary air jets and the recirculation zone of a gas turbine combustor model, also the effect for pollutant emissions production such as CO, Knox is discussed [8]. N P Yadav and A Kushiro have presented an experimental investigation of non-reacting flow in a low aspect ratio dump combustor with a tapered exit, especially about effect of swirled on turbulent behavior [9].The characteristics of swirled flow field, including cold flow field and combustion flow field, in gas turbine combustor with two-stage swirled are studied using PIV by Yan Yingwen et al [10]. In the study, velocity components, fluctuation velocity, Reynolds stress and recirculation zone length are obtained. Influences of geometric parameter of primary hole, arrangement of primary hole, inlet air temperature, and first-stage swirled angle and fuel/air ratio on flow field are investigated respectively. Wang Chinju et al have designed four kinds of triple-stage twirlers, and the corresponding investigation for the combustor flow field is carried out through PIV [4, 11].

The present study investigates the flow field characteristics of a single-head combustor with triple-stage swirled, using PIV technique. With pure air flow under normal temperature at 0.49MPa and 0.98MPa pressure, detailed mean velocities and streamlines, RMS velocities and turbulent intensity(TI), vortices and Reynolds shear stresses are obtained at various cross and longitudinal sections. And the affection for flame stabilization and mixing fuel with air is analyzed and discussed.

2. Experimental details

2.1. Experimental setup

The experimental investigations are conducted at the combustion facility in CARDC. Figure 1 shows the schematic diagram of the experiment system. In order to keep the tracer particles distributing uniformly in the combustor flow, the tracer particles are injected into the main flow about 1m upstream of the inlet of combustor. For the use of camera and laser sheet in PIV measurement, there is a T-branch pipe with quartz glass window at the outlet of combustor.

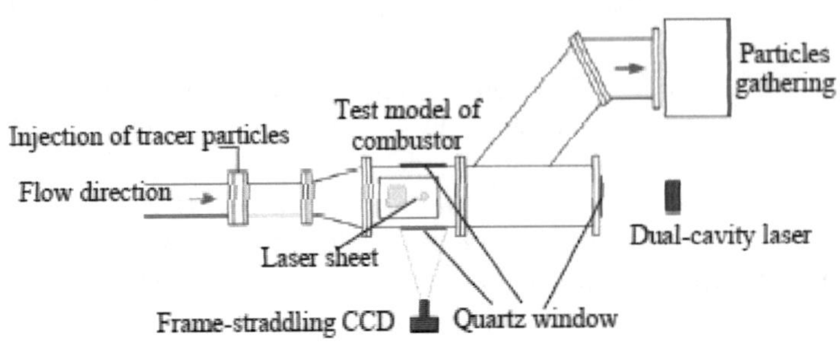

Figure 1. Schematic diagram of experiment system.

In the present work, air is fed from an air storage vessel at a maximum pressure of about 20 MP. And then, air passes through a nozzle. Downstream of the nozzle, air flows into the combustor through triple-stage swirled, main combustion holes, and dilution holes. Leaving from the combustor, air flows up at 45 degrees into the exhaust pipe. For experimenters' security, there is a special unit for particle gathering. The combustor has two side quartz windows for 116mm×86mm and one rear quartz window for 98mm×72mm.

The test model is shown in Figure 2. It is a rectangular single-head combustor with triple-stage swirled. The model is obtained from a wholly annular combustor by the way that flow velocity is constant. The inner Eigen structures are shown in Figure 3, which include fuel nozzle, triple-stage swirled, main combustion holes, dilution holes and lots of pin holes for air film cooling.

Figure 2. Test model of combustor.

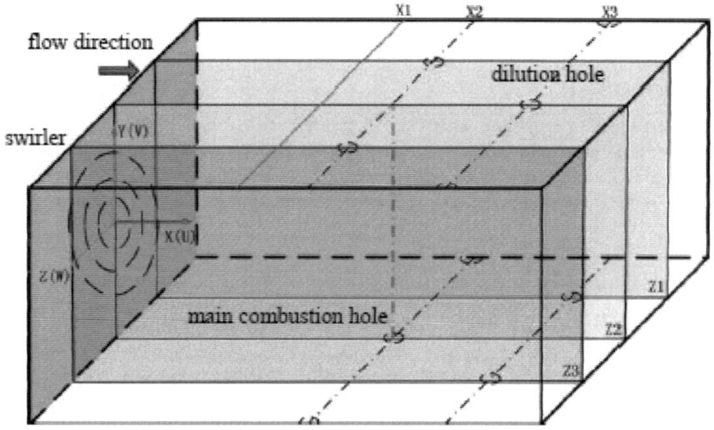

Figure 3. Schematic diagram of characteristic of combustor.

2.2. Particle image velocimetry

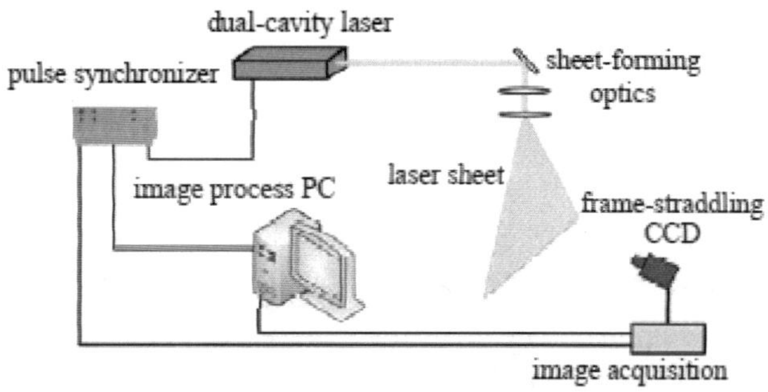

Figure 4. Schematic diagram of PIV system.

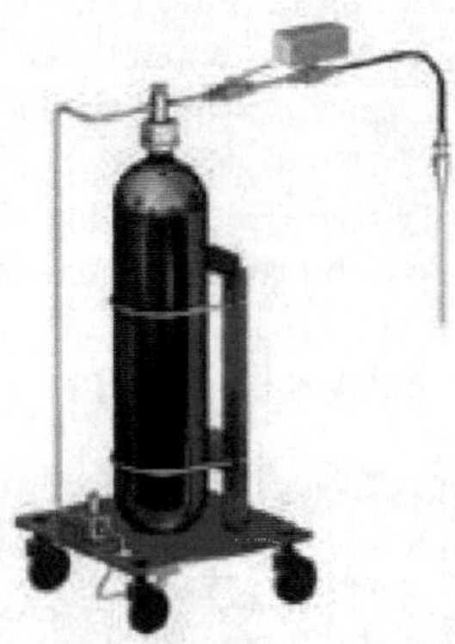

Figure 5. Schematic diagram of solid micro-particle generator for high pressure using.

Figure 4 shows the schematic diagram of PIV system used in the present work. The PIV system consisted of a frequency-doubled dual-cavity ND: YAG laser, a frame-straddling CCD camera (2048×2048 pixels) and a pulse synchronizer. The laser pulses have the duration of 10 ns and energy of 500 mg at 532 nm with 15Hz repeat frequency. After passing through sheet-forming optics, the laser beam has a height of approximately 100 mm and a thickness of 2 mm inside the combustor. The time separation between the two laser pulses is set to 4 is. Synchronization between the cameras, laser, and image acquisition is accomplished by the pulse synchronizer.

2.3. Tracer particles and seeding

It is clear from the principle of PIV that velocity measurement is based on images of tracer particles suspending in the flow. For the reasons of security and measurement accuracy, firstly the tracer particles have to trace the flow well. Secondly, they need to possess high scattering signal quality. At last, the tracer particles should provide such characteristics as non-toxin, non-cauterization, chemical stability and non-pollution. In experimental investigation, the adding particles should be small enough and have the similar density as fluid so as to trace flow well. However, if diameters of the adding particles are too small, the scattering signal will be poor, which is adverse for particles imaging. Thus, when choosing the tracer particles, all kinds of factors should be taken into account.

In general, choosing tracer particles should obey the following rules. The tracer particle density should be as similar to the fluid as possible. In the premise that the tracer particle can be imaged clearly, its diameter should be small enough. In present study, titanium dioxide (TiO2) powders are chosen for tracer particles. The powder nominal diameter is 0.3 μ m, density is 4.26g/cm3, and melting point is 1840 °C with chemical stabilization.

In fact, tracer particles have strong tendency to form agglomerates which are several times larger than the nominal size, especially at the submicron or Nan levels as a result of humidity and prolonged storage. In order to break up agglomerates and obtain uniform particles distribution, a special high pressure particles generator based on principle of cyclone separation is designed. The schematic diagram is shown in Figure 5. When used in different work conditions, the injection pressure of particle-air mixture flow should be higher than that of main flow about 2 MP to ensure particles to be seeded into the combustor successfully.

3. Results and discussions

As mentioned above, there are kinds of Eigen structures in the test combustor model such as triple-stage swirled, main combustion holes and so on. The whole combustor inner flow field has the properties with wide-range velocity, strong rotational and turbulent flow due to the interactions among the swirled, main combustion holes, dilution holes. In order to make sure that the correlation between particles image A and B is good enough for analysis, the particles displacement during frame straddling time Δt should be appropriate, and the displacement off the laser sheet must be as small as possible. Thus, it is difficult to measure the combustor inner flow with highly three-dimensional character by 2D-PIV. And there are lots of other challenges for PIV measurement, such as seeding particles uniformly, pollution of windows, system parameters setting and image post-processing. In the present work, the combustor inner flow field velocities are measured by 2C-2D PIV. The different test conditions are listed in Table 1.

Table 1. Experimental conditions of single-head combustor test.

Test conditions	Inlet pressure /MP	Inlet temperature /K	Inlet velocity /m•s^{-1}
Middle pressure	0.49	300	13
High pressure	0.98	300	13

3.1. 0.49 MP Pressure test condition

Figure 3 shows the coordinates system used in the image processing of the experimental particles image. The positive direction of X axis is same as the inlet flow direction, and the positive direction of Y axis is toward the top of test model vertically, while the positive direction of Z axis points toward side window along radial direction of the swirled. The X, Y and Z axes meet the condition of right-hand rule, and the origin of coordinate's lies at the outlet center of the swirled. In order to study the flow characteristics in the combustor, flow field velocities in several sections are measured, including longitudinal sections Z1, Z2, Z3 with Z coordinates -24.5mm, 0mm, 24.5mm respectively, and cross sections X1, X2, X3 with X coordinates 50.05mm, 67.2mm, 101.6mm. The real measurement areas are a little different to each other duo to differences of system installation and reflection of model.

The result data corresponding to the cross sectional planes X1, X2, X3 downstream of the swirled are plotted together in a 3D cascaded view as shown in Figure 6-Figure 11. The ensemble-averaged velocity vectors and streamlines in several cross-sectional planes at different locations along X axis are shown in Figure 6 and Figure 7. Due to the effect of strong swirl flow, there is a big anticlockwise vortex. The cores of vortex in different planes are off X axis somehow. Along the X axis, the flow field velocities become higher gradually, accompanied with high velocity gradient. In the X3 plane far field downstream of the swirled, the velocities are observed to be having a maximum value of 35 m/s.

In sections X1 and X2, the velocity values in vortex's left side region are higher than the right side for the reason that the flow direction of the vortex's left side is from top to bottom, just as similar as the flow direction of the top main combustion hole in longitudinal section Z3. But in the right side region, the velocity values become lower because of inverse flow direction of the top main combustion hole in longitudinal section Z1 relative to the vortex. The velocity values at the bottom are higher obviously than other region due to the bottom center main combustion whole in longitudinal section Z2.

In section X3, the velocity values in vortex's left side region are lower than the right side for the reason that the flow direction of the vortex's left side is from top to bottom, just opposite to the flow direction of the bottom dilution hole in longitudinal section Z3. But in the right side region, the velocity values become higher because of similar flow direction of vortex and the bottom dilution hole in longitudinal section Z1. The velocity values in the top region are higher obviously than other region due to the top dilution hole in longitudinal section Z2. Besides, there is a complicated flow structure of irregular vortex in the top-left region due to the effect of the strong swirl flow and the dilution holes in longitudinal section Z2 and near the left side wall.

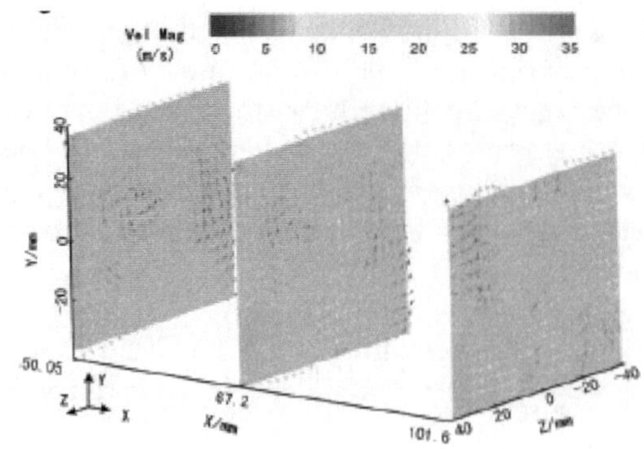

Figure 6. Mean velocity vector plots in cross sectional planes of 0.49 MP.

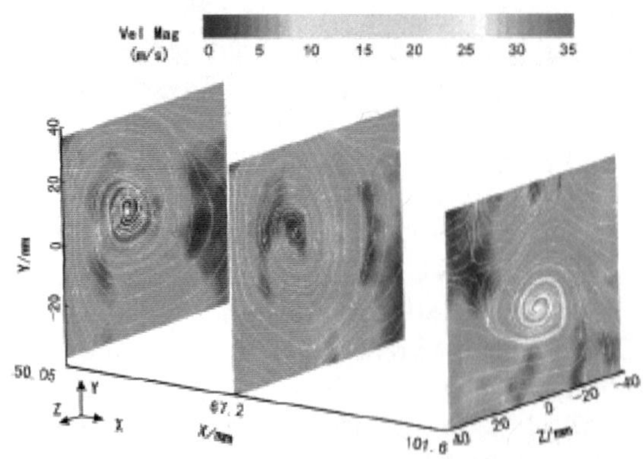

Figure 7. Mean velocity streamline plots in cross sectional planes of 0.49 MP.

The turbulent velocity fluctuations are obtained from lots of instantaneous flow field velocities. Through further computations, RMS velocity magnitude (RMS Mag) and TI can be obtained by the following definitions, where u' and v' stand for velocity fluctuations in Z and Y directions respectively, $\overline{u}$ and $\overline{v}$ stand for ensemble-averaged velocity in Z and Y directions respectively.

$$RMS\quad Mag = \sqrt{\overline{u'^2} + \overline{v'^2}} \tag{1}$$

$$TI = \sqrt{\overline{u'^2} + \overline{v'^2}} \Big/ \sqrt{\overline{u}^2 + \overline{v}^2} \tag{2}$$

The RMS and TI contours obtained are shown in Figure 8 and Figure 9. Velocity fluctuations become stronger gradually along X axis, accompanied with stronger turbulent intensity. The TI is higher than 40% in most regions in each cross section, even close to 20 somewhere. It is shown that there is high turbulent intensity in combustor inner flow, and it is helpful for efficiently mixing fuel with air.

IOP Conf. Series: Journal of Physics: Conf. Series **1064** (2018) 012014 doi:10.1088/1742-6596/1064/1/012014

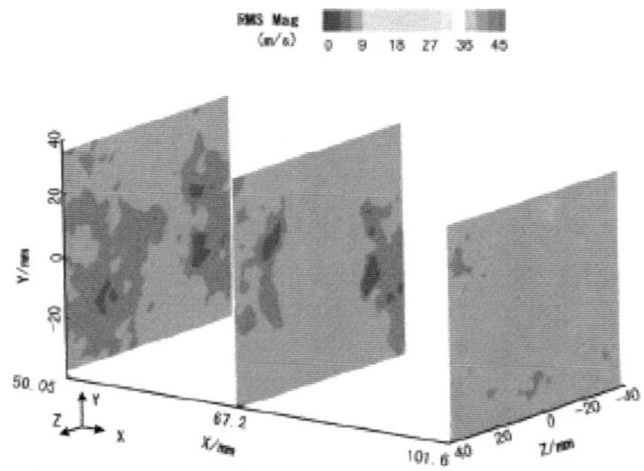

Figure 8. RMS velocity magnitude contour in cross sectional planes of 0.49 MP.

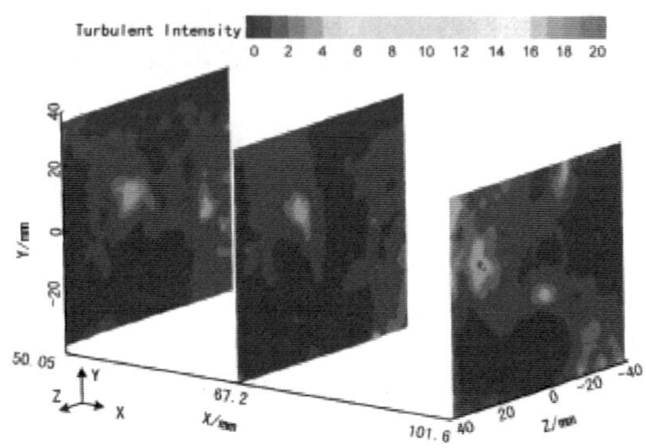

Figure 9. Turbulent intensity contour in cross sectional planes of 0.49 MP.

The mean vortices and Reynolds shear stress ($-u'v'$) contours obtained in various cross sections are shown in Figure 10 and Figure 11. The magnitude of mean vortices is observed to be varying from -2300 to 2300 s-1 depending on the direction of circulation. There is negative vortices distribution in some regions along X axis due to effect of main combustion holes and dilution holes, and the swirling flow intensity is obvious strong nearby the center-line of swirled. Also, the turbulent characteristic of swirling flow is presented by Reynolds shear stress. There are higher levels of Reynolds shear stress distribution nearby the main combustion holes and dilution holes in cross sections X2 and X3, where the flow possesses high level of velocity gradient distribution. Thus, in real combustion condition the mixing impact among main combustion whole jet, dilution whole jet and main swirling flow is strong. It is helpful for rapidly, efficiently mixing fuel with air and promoting combustion efficiency.

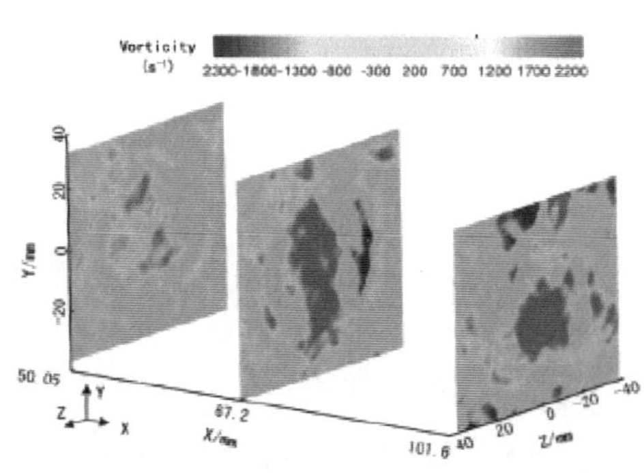

Figure 10. Mean vortices contour in cross sectional planes of 0.49 MP.

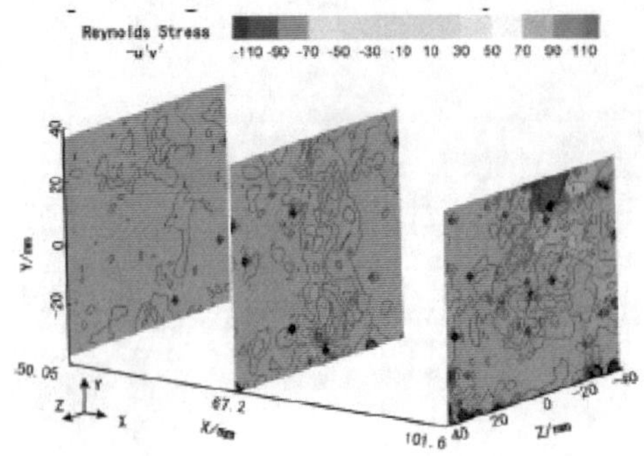

Figure 11. Reynolds shear stress contour in cross sectional planes of 0.49 MP.

The result data corresponding to the longitudinal sectional planes at Z1, Z2, Z3 are plotted together in a 3D cascaded view as shown in Figure 12-Figure 17. The ensemble-averaged velocity vector fields and streamlines in several longitudinal sectional planes at different locations along Z axis under 0.49 MP pressure test condition are shown in Figure 12 and Figure 13. The velocities are observed to be having a maximum value of 70 m/s in different longitudinal sections, and there are higher levels of velocity distribution nearby the main combustion holes and dilution holes. In the center longitudinal section Z2, outlet flow velocity distribution of swirled is symmetric about the center line of swirled? It can be seen clearly that there is tapered swirling flow due to triple-swirled, and two symmetric small vortexes are formed in the top and bottom corners after meeting the main combustion whole backflow, accompanied with two small recirculation zones. Flow velocities of main combustion hole and dilution hole are higher than other regions, and they have deep penetration. There are two big vortexes due to interaction between main combustion whole jet and main swirling flow, together with big recirculation. Located at the outlet of combustor, a vortex is formed after the bottom main combustion whole jet meeting the top dilution whole jet, then flowing out of combustor.

In section Z1, a S-shaped flow throughout flow field top and down is observed owe to effect of bottom center main combustion hole jet and swirling flow. The outlet flow velocity distribution of swirled is still symmetric about Y = 0 mm. and two symmetric small vortexes are formed in the top and bottom corners after meeting the S-shaped flow. A big vortex is formed nearby the top main combustion hole because of interaction between the main combustion whole jet and the S-shaped flow. Located at outlet of combustor, main combustion whole jet meets dilution whole jet, then flowing out of combustor downstream.

Also, in section Z3 an anti-S-shaped flow throughout flow field top and down is observed owe to effect of bottom center main combustion whole jet and swirling flow. It is a pity that effective flow field velocities nearby the swirled cannot be measured, as there is strong laser reflection of twirler's polished surface.

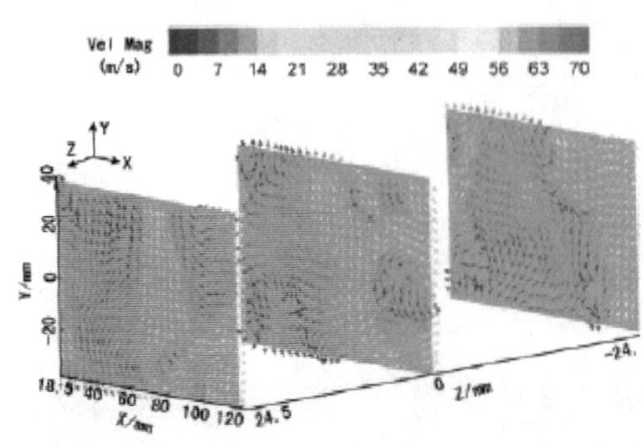

Figure 12. Mean velocity vector plots in longitudinal sectional planes of 0.49 MP.

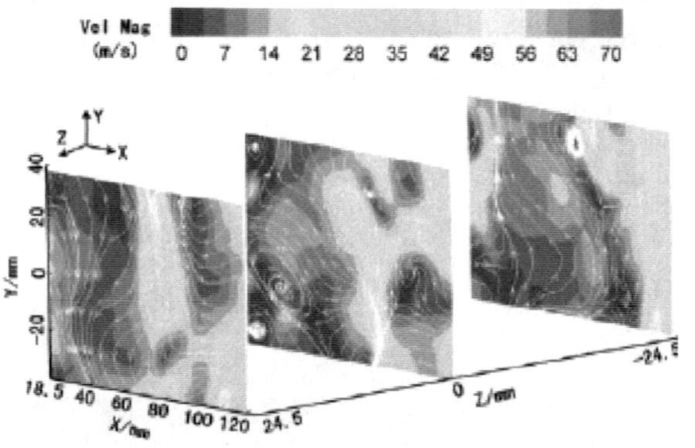

Figure 13. Mean velocity streamline plots in longitudinal sectional planes of 0.49 MP.

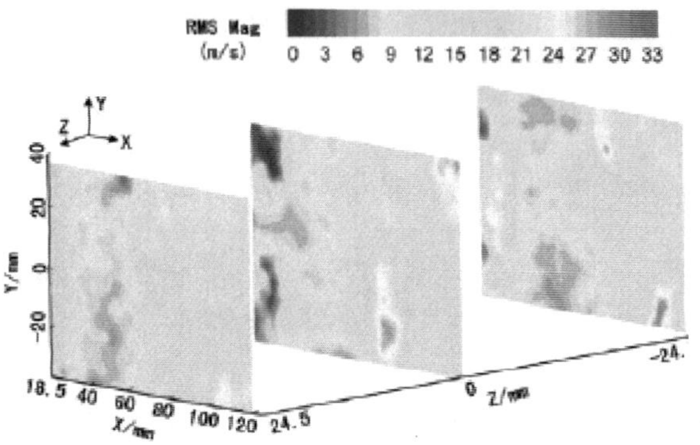

Figure 14. RMS velocity magnitude contour in longitudinal sectional planes of 0.49 MP.

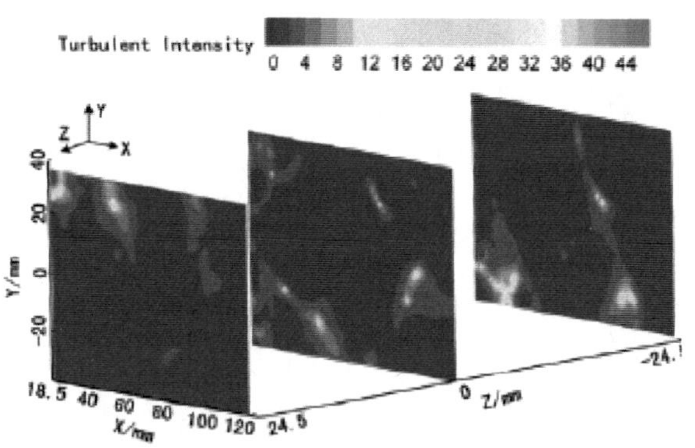

Figure 15. Turbulent intensity contour in longitudinal sectional planes of 0.49 MP.

The RMS and TI contours obtained in various longitudinal sections are shown in Figure 14 and Figure 15. It is seen that there are lower levels of velocity fluctuations nearby regions of X=40mm due to stable recirculation zone. And there are higher levels of velocity fluctuations nearby main combustion hole and dilution whole jet. The TI is about 20 in some regions in each cross section, even close to 45 somewhere.

The mean vortices and Reynolds shear stress contours obtained in various longitudinal sections are shown in Figure 16 and Figure 17. In sections Z1 and Z3, vortices nearby main combustion holes appear in pairs, which means that the jets flow forward and backward simultaneously. In section Z2, there are also positive and negative vorticities in pairs, but the positive values is much higher. In the top region of section Z2, negative vorticities are dominant due to swirling flow effect. There are higher levels of Reynolds shear stress distribution nearby the regions of the center of swirled, main combustion holes and dilution holes, which are consistent with the distribution of Reynolds shear stress in cross sections X2 and X3.

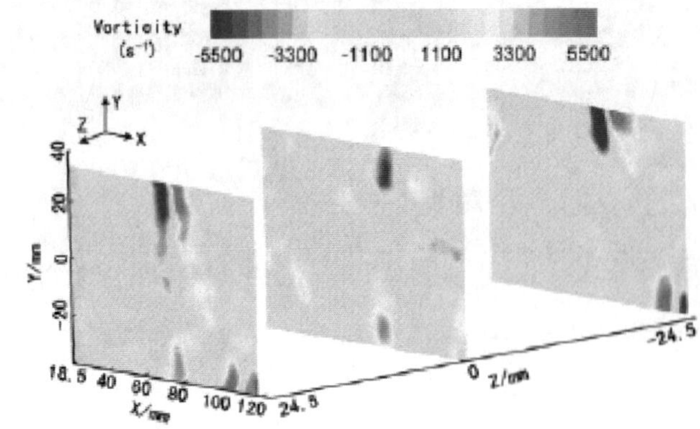

Figure 16. Mean vortices contour in cross sectional planes of 0.49 MP.

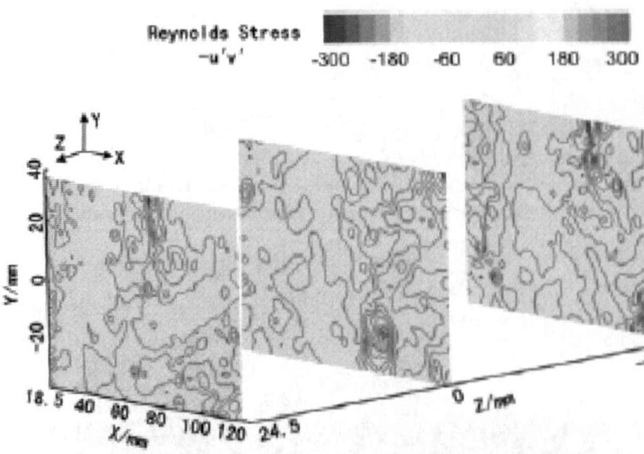

Figure 17. Reynolds shear stress contour in cross sectional planes of 0.49 MP.

3.2. 0.98MPa Pressure test condition

In order to compare characteristics of engine combustor in different test conditions, 2D-PIV measurement is carried out for longitudinal section Z2 under combustor inlet pressure of 0.98 MP. The ensemble-averaged velocity vector fields and streamlines in longitudinal section Z2 under 0.98 MP pressure test condition are shown in Figure 18 and Figure 19. Compared with the situation of 0.49MPa pressure condition, there is also a big recirculation zone between swirled and main combustion hole, which lies in the region of X<70mm. And contra-flow can be seen along the center line. The flow field structure is similar as 0.49MPa pressure condition. It is shown that the inlet pressure has little effect on the flow field structure.

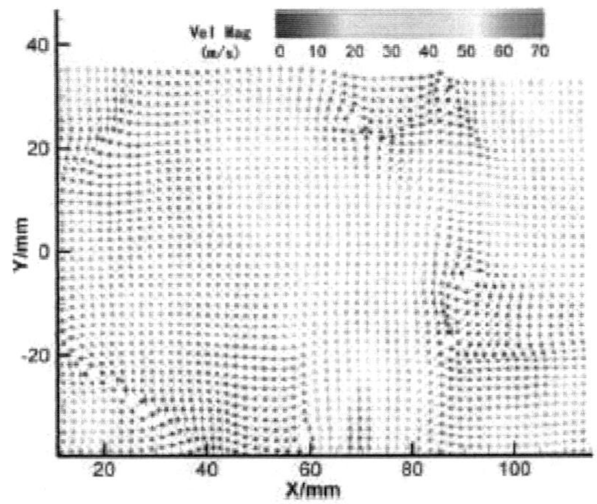

Figure 18. Mean velocity vector plots in longitudinal sectional plane Z2 of 0.98 MP.

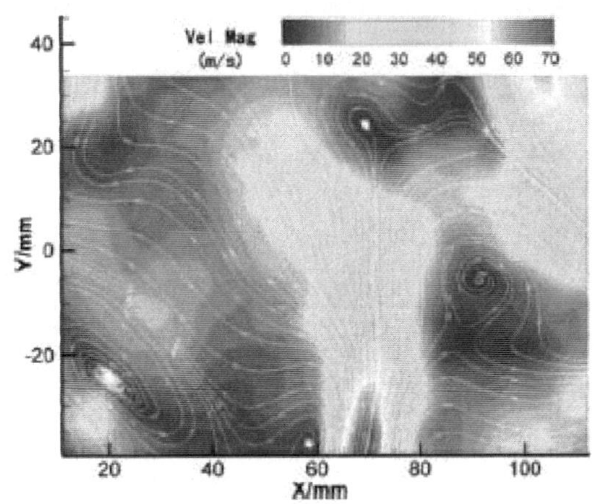

Figure 19. Mean velocity streamline plots in longitudinal sectional plane Z2 of 0.98 MP.

4. Conclusion

The present work investigates the flow field characteristics in a single-head combustor with triple-stage swirled by 2D-PIV. With pure air inlet flow, detailed mean and RMS velocities, turbulent intensity, vortices, Reynolds shear stress are obtained at several longitudinal and cross sections downstream of the swirled in the single-head combustor at flow temperature 300K. Complicated three-dimensional flow characteristics with strong swirling flow, multi-jet flow and wide velocity range are observed in the combustor. The test conditions include combustor inlet pressures of 0.49 MP and 0.98 MP under normal temperature.

A high pressure solid particle generator based on cyclone separation principle is designed, and appropriate tracer particles and seeding method are chosen. Then they are applied for PIV measurement of inner flow field of high pressure aero turbine engine combustor successfully. The measurement results show that there are multiple vortices in the inner flow field of the studied combustor, accompanied with relevant recirculation zones. The effect of swirled is strong, and there is big vortex anti-clockwise in cross section. The jet effect of main combustion holes and dilution holes is obvious; the jet flows have deep penetration, and have distinct effect on the flow field structure. Through the present research it is shown that there are big size recirculation zones in the combustor flow field with high levels of turbulence intensity generated due to triple-stage swirled, main combustion holes and dilution holes of the single-head combustor. And the strong aerodynamic interactions among different kinds of combustor Eigen structures are existed. It is helpful for promoting rapidly mixing fuel with air and stabilizing flame combustion. The flow field structure under 0.98 MP pressure condition is similar as 0.49MPa pressure condition.

References

[1] B Sekar, H J Thornburg, A M Briones, et al, Effect of trapped vortex combustion with radial vane cavity arrangements on predicted inter-turbine burner performance [R]. AIAA 2009-4603.

[2] C Fureby, F F Grinstein, G Li, et al, An experimental and computational study of a multi-swirl gas turbine combustor [J]. Proceeding of the combustion institute, 2007, 31(2):3107-3114.

[3] C Eberle, P Gerlinger, K P Geigle, et al, Soot predictions in an aero-engine model combustor at elevated pressure using URANS and finite-rate chemistry [R]. AIAA 2014-3472.

[4] C J Wang, P Jiang, X Xin, et al, Measurement of triple stage swirler cup combustor flow field based on PIV technology [J]. Journal of aerospace power, 2015, 30(5):1032-1039. (in Chinese)

[5] M Raffel, C Willert, S Wereley, et al, Particle Image Velocimetry [M]. Puducherry, India: Integra software services Pvt. Ltd, 2007.

[6] S Bharani, S N Singh, D P Agrawal, Effect of swirl on the flow characteristics in the outer annulus of a prototype reverse-flow gas turbine combustor, Exp. Therm. Fluid Sci., Vol. 25, June 2001, pp. 337-347.

[7] M Schroll, J Klinner, L Lange, et al, Particle image velocimetry of highly luminescent, pressuried combustion flows of aero engine combustors[R]. Delft, Netherlands:TU-Delft, 10th international symposium on particle image velocimetry(PIV13)-A112, 2013.

[8] M E Ahmed, J San-mou, M HuKam, The influence of primary air jets on flow and pollutant emissions characteristics within a model gas turbine combustor [R]. Reno Nevada: 44th AIAA Aerospace Sciences Meeting and Exhibit, AIAA 2006-544, 2006.

[9] N P Yadav, A Kushari, Effect of swirl on the turbulent behaviour of a dump combustor flow [J]. Proceedings of the Institution of Mechanical Engineers, Part G. Journal of aerospace engineering, Vol. 224(6), 2010, pp. 705-714.

[10] Y W Yan, J H Li, R Xu, et al, PIV measurement for swirler flow field in gas turbine combustor [J]. Transactions of nanjing university of aeronautics & astronautics, Vol. 29(4), 2012, pp. 307-317.

[11] C J Wang, Z Y Wu, D D Wang, et al, Design of triple swirler and simulation of its flow field [R]. Third International Conference on Measuring Technology and Mechatronics Automation, 2011, pp.1006-1009.

IOP Conf. Series: Journal of Physics: Conf. Series **1064** (2018) 012015 doi:10.1088/1742-6596/1064/1/012015

Numerical Simulation on Smoke Propagation and Fire Separation in Electric Power Cable Tunnel

Jiaqing Zhang[1,*], Kun Ji1, Bo Yan[1], Zhangbin Zhou[1], Minghao Fan[1] and Bosi Zhang[2]

[1]State Grid Anhui Electric Power Research Institute, Hefei, China.
[2]China University of Labor Relations, Beijing, China.

*Corresponding author e-mail: dkyzjq@163.com

Abstract. The smoke propagation process in an electric power cable tunnel was investigated numerically, and the effects of fire separation on the fire development were also explored. The numerical simulations were conducted by using the Fire Dynamics Simulation (FDS) in the present study. The electric power cable tunnel was insisted of four working wells and three tunnel sections, and the total length of this tunnel was 561 m. According to the simulation results, the smoke propagated fast in the electric power cable tunnel when there was no fire separation inside the tunnel, and the temperature of the whole tunnel increased obviously. The fire separation is an effective method to confine the fire smoke into one fire compartment. However, the flame would migrate in the whole fire compartment, mainly due to the lack of oxygen. The working well was affected by the high-temperature flame even the fire happened in the tunnel section. The temperature of the 2# working well reached almost 1000°C when the fire initially located at the 1# tunnel section in the present study.

1. Introduction

The cable tunnel is important in electric power system, since it is the main channel for long distance transmission of the electric power. There are usually large amounts of cables laying in the electric power cable tunnel. Once one of the cables is ignited due to electric fault, the other cables are likely to be ignited and finally caused the cable tunnel fires. The fires in electric power cable tunnel not only affect the normal supply of electricity, but also bring threats to the security of the entire electric grid. Therefore, studies on fire safety of the electric power cable tunnel is necessary to be conducted.

The electric fires have earned much concern. Babrauskas [1] indicated that the main reasons for electric fires included arc fault, overheating of the core and external heating. Hagimoto et al. [2], Stoliarov et al. [3] and Dunki-Jacobs [4] put attentions on the mechanism and cause of the electric fires. Niu et al [5], Wang et al. [6] and Zhang et al. [7] revealed the fire hazard of cables by using small-scale experimental instrument. However, the smoke movement in electric power cable tunnel under fire accident as well as the effects of fire separation method on smoke control have not receive enough attentions.

Content from this work may be used under the terms of the Creative Commons Attribution 3.0 licence. Any further distribution of this work must maintain attribution to the author(s) and the title of the work, journal citation and DOI.
Published under licence by IOP Publishing Ltd

2. Numerical method and Simulation model

The numerical simulations were conducted by using the Fire Dynamic Simulators (FDS), which is developed by the National Institute of Standards and Technology based upon the field modelling method. FDS is an effective way to investigate fire process and smoke propagation, and has been adopted in many fire related researches [8-10].

The sketch of electric power cable tunnel is illustrated in Figure 1. The tunnel model was consisted of four working wells (1# working well to 4# working well) and three tunnel sections (1# tunnel section to 3# tunnel section). The lengths of 1# tunnel section, 2# tunnel section and 3# tunnel section were 224 m, 68 m and 178 m, respectively. The shape of the tunnel was rounded. The diameter of the tunnel was 2.2 m, and the lower edge of the tunnel was located at 1.1 m.

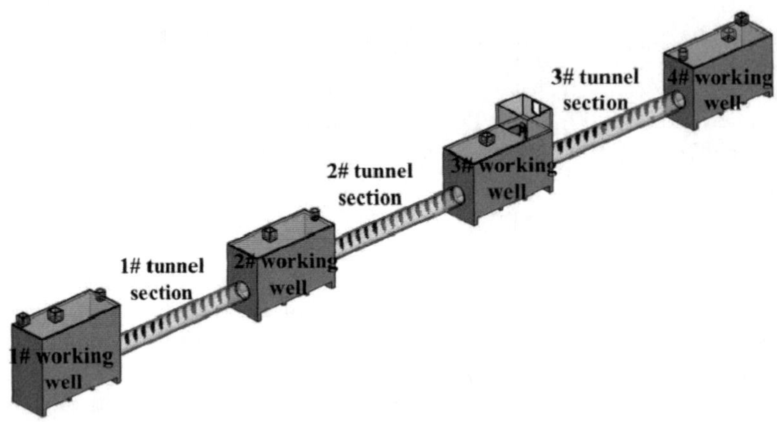

Figure 1. Sketch of the electric power cable tunnel model

Four fire scenarios were investigated in this study. In fire scenario 1 and 2, there were no fire separation installed in the electric power cable tunnel, and the smoke could propagate unrestrictedly in the whole tunnel area. The fire source was located at 2# tunnel section and 2# working well in scenario 1 and 2, respectively. The fire separation was included in scenario 3 and 4. The fire source was located at 2# tunnel section and 2# working well in scenario 3 and 4, respectively. In all of the four scenarios, the heat release rate of fire was set as 2MW in the simulations.

3. Smoke Propagation in Power Cable Tunnel

3.1. Smoke propagation under the fire source in tunnel

The smoke propagation in the electric power cable tunnel with the fire source located at 2# tunnel section are shown in Figure 2. At 35 s after fire, the 2# tunnel section, which contained the fire source, was filled of smoke and the smoke has spread into the 2# and 3# working wells. The 2# and 3# working wells were full of smoke at 120 s, and the smoke has already spread into the 1# and 3# tunnel sections. The 1# and 4# working well were contaminated with smoke at 320 s. The whole electric power cable tunnel was full of smoke after 540 s, and the concentration of the smoke keep increase with the process of fire. It is observed that the smoke spread very fast in the in the electric power cable tunnel when the fire source located at the 2# tunnel section. It is mainly due to that there is no fire separation set in tunnel in this fire scenario.

When fire happened in the electric power cable tunnel, the heat accumulated in the tunnel and the temperature increased gradually. At first, the temperature rise was only limited in the area around the fire source. With the smoke propagation in the tunnel, the temperature rise almost appeared in the whole electric power cable tunnel. The temperature of the tunnel with the fire source located at 2# tunnel section is illustrated in Figure 3. At 220 s, the temperature of the whole 2# tunnel section was almost larger than 700 °C, under which the fault of cable would happen in a short time. At 540 s, the temperatures of both 2# work well and 3# working well were above 700 °C. The temperature inside the tunnel increase continually, and the temperature of 1# and 4# working well, which were over 200 m away from the location of the fire source, was above 100 °C at 1200 s. According to the Figure 2, the increase of temperature in the tunnel was quickly.

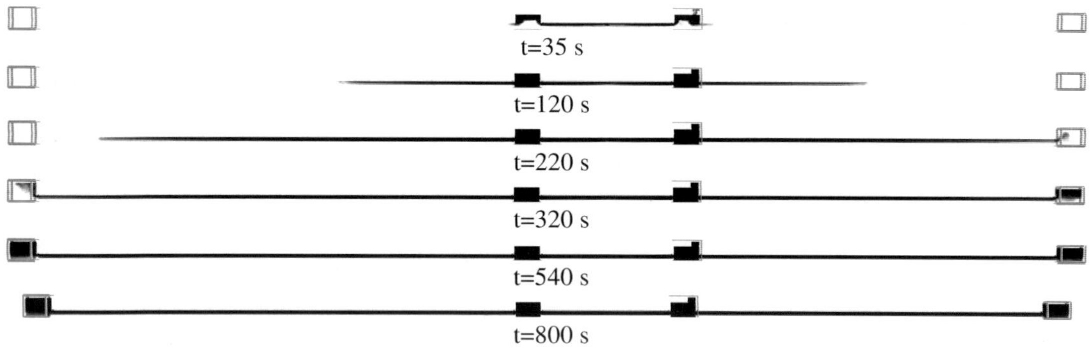

Figure 2. Smoke propagation in the electric power cable tunnel with the fire source located at 2# tunnel section

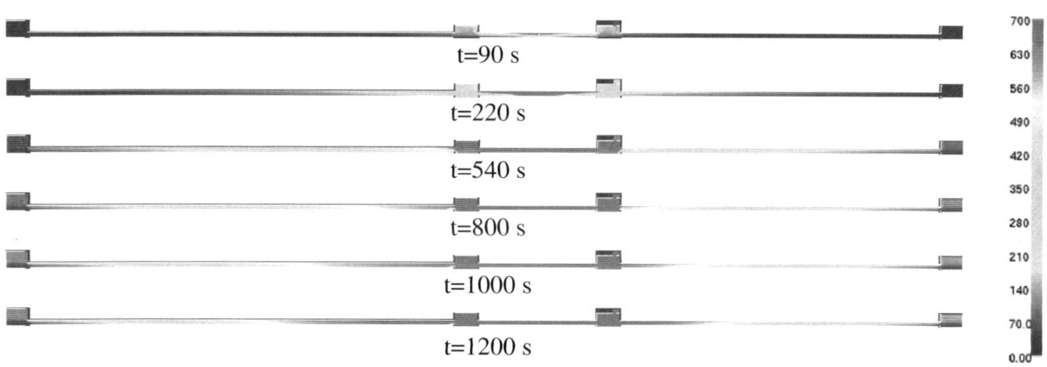

Figure 3. Temperature of the electric power cable tunnel with the fire source located at 2# tunnel section

3.2. Smoke propagation under the fire source in working well

When the fire source was located at 2# working well, the smoke propagation in the power cable tunnel is shown in Figure 3. The 2# working well was filled with smoke soon after the fire came out. The fire smoke spread into the 3# working well after 120 s, and it was filled with smoke at 180 s. At 540 s, the smoke has spread into the 1# working well, and the whole power cable tunnel was almost charged with smoke after 1200 s. The temperature rise inside the electric power cable tunnel is illustrated in Figure 4. At 30 s, the temperature of 2# working well was higher than 700°C. The temperature of the 3# working well began to increase at 120 s. At 1200 s, the temperature of 3# working well, which was about 70 m away from the fire source, reached 700°C.

It can be found that the smoke propagation quickly in the electric power cable tunnel in fire scenario 1 and 2, since there was no fire separation and the smoke flowed unrestrictedly in the electric power cable tunnel. When fires happened, the heat accumulated quickly in the tunnel because of the structure of tunnel was long and narrow. The rise of temperature provided sufficient force for smoke propagation. To confine the contaminated area of smoke in the tunnel, the effective fire separation should be set and well designed.

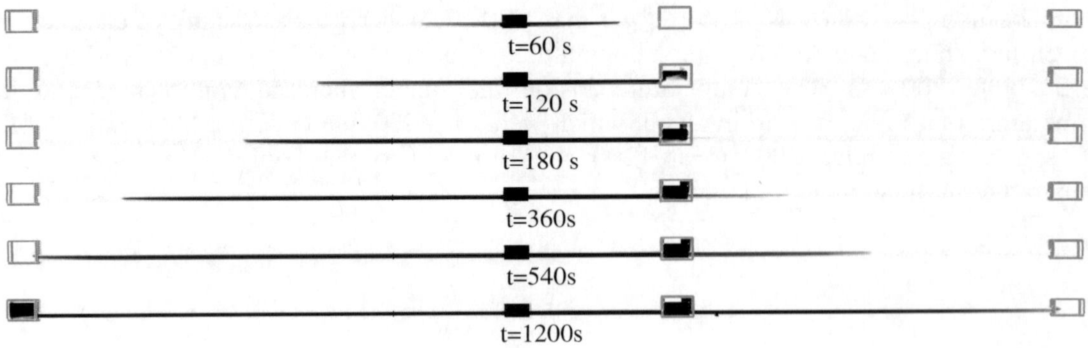

Figure 4. Smoke propagation in the power cable tunnel with the fire source located at 2# working well

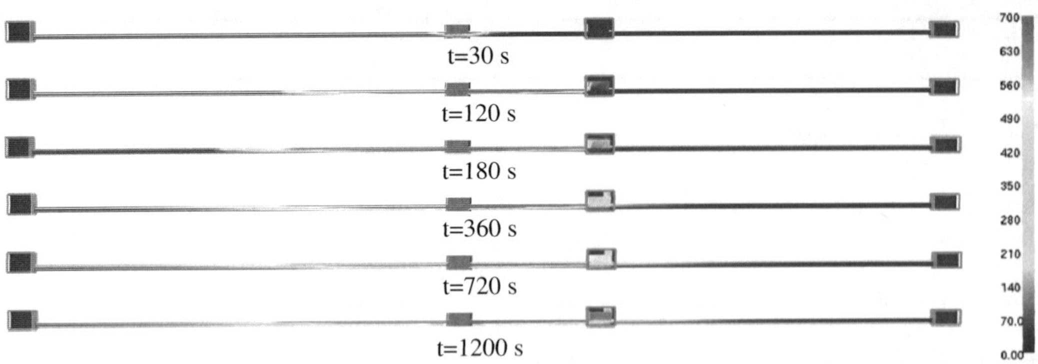

Figure 5. Temperature of the power cable tunnel with the fire source located at 2# working well

4. Fire Hazard of Tunnel with Fire Separation

4.1. Flame spread in tunnel with fire separation

The fireproof door is usually installed in the power cable tunnel as the fire separation. The distance between two adjacent fireproof doors is about 100 m. The total length of the power cable tunnel was 561m, so there were four fireproof doors installed. The 1# tunnel section, 2# working well and 2# tunnel section were included in a fire compartment that confined by two fireproof doors. The fire compartment contained both the tunnel section and the working well, therefore, the flame and smoke spread in this fire compartment should be investigated.

When the fire source located at the 1# tunnel section, the flame spread in this tunnel section is shown in Figure 6. The flame burned steady at first as shown in Figure 6(a). Then the oxygen concentration around fire decreased due to the combustion. At 60 s, the flame began to spread toward the two sides of tunnel due to the lack of oxygen in the fire area. The spread of flame became more severe at 280 s. At 420 s, the flame in the origin fire location was almost extinguished, and the flame spread into the 2# working well. It showed that the working well would be affected by the high-temperature flame even the fire source was initially located at the tunnel section.

When the fire source located at the 2# working well, the flame spread in this working well is shown in Figure 7. The flame was steady at the initial stage of fire. The steady stage of flame was longer in scenario 4 than that in scenario 3, which might due to the fact that the height of working well was higher than the height of tunnel section. The flame was stretched at 320 s, since the oxygen concentration was not sufficient around the fire. With the decrease of oxygen concentration, the flame spread toward the tunnel sections on both sides. It implies that the tunnel section would also be influenced by the flame when the initial fire location was in the working well.

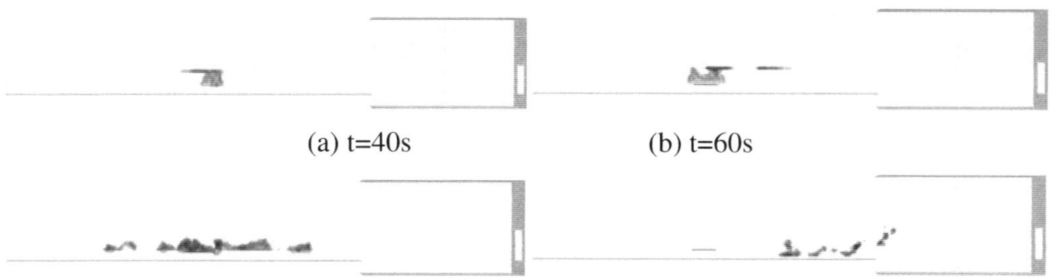

(a) t=40s (b) t=60s

(c) t=280s (b) t=420s

Figure 6. Flame spread in the 1# tunnel section in fire scenario 3

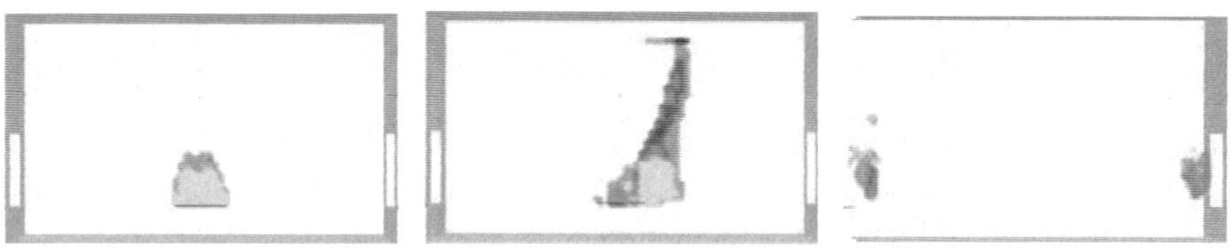

(a) t=90s (b) t=320s (c) t=380s

Figure 7. Flame spread in the 2# working well in fire scenario 4

4.2. Smoke propagate in tunnel with fire separation

The smoke propagation in tunnel with fire separation when the fire source was located at 1# tunnel section is shown in Figure 8. The smoke spread into the 2# working well within 9 s. The smoke in the 2# working well descended almost to the floor level at 100 s. After 120 s, this fire compartment was almost filled with smoke totally. The temperature of the 2# working well versus time in scenarios 3 is illustrated in Figure 9. The temperature of the upper space of 2# working well increased soon after the ignition, since the smoke spread into the 2# working well within 9 s, as shown in Figure 8(a). The temperature at the lower space of 2# working well kept low within 100 s. After 100s, the temperature of lower space of 2# working well increased obviously because of the smoke descended to the lower space. The temperature of the 2# working well trended to uniform after 300s, and it reached to 1000 °C at 600 s.

When the fire source was located at the 2# working well, the smoke propagation in tunnel with fire separation is shown in Fig. 10. The smoke could not be confined in the initial fire working well and began to spread into the tunnel sections on both side after 20 s. The smoke reached to the fireproof doors at both sides of the tunnel at about 60 s. The smoke began to flow back after it reached to the fireproof door, and the whole fire compartment was filled with smoke at 120 s. The temperature of the 2# working well versus time in fire scenario 4 is illustrated in Figure 11. The temperature of 2# working well increased quickly after the ignition. The temperatures of the upper and lower spaces the working well became almost uniformed after 150 s. The temperature of the whole working well reached 1200 °C at 300 s.

(a) t=9 s (b) t=23 s

(c) t=100 s (d) t=120 s

Figure 8. Smoke propagation in tunnel with fire separation (fire scenario 3)

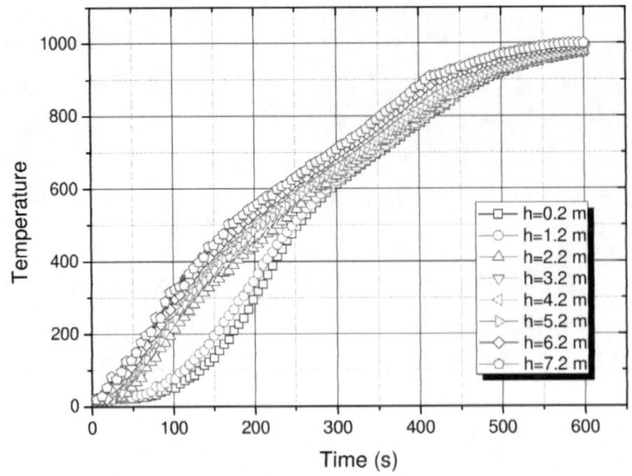

Figure 9. Temperature of the 2# working well in fire scenario 3

(a) t=20 s (b) t=40 s

(c) t=60 s (d) t=120 s

Figure 10. Smoke propagation in tunnel with fire separation (fire scenario 4)

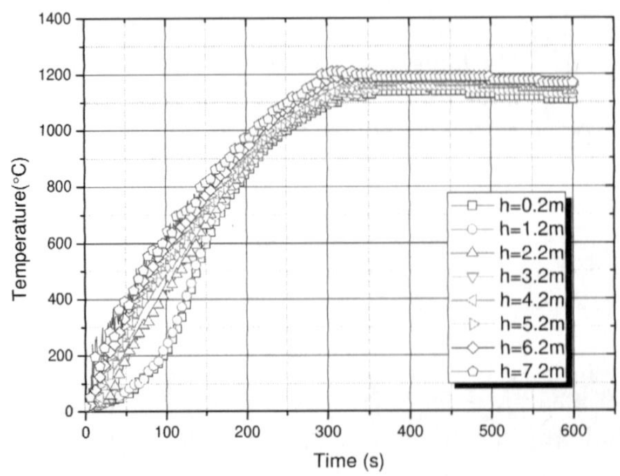

Figure 11. Temperature of the 2# working well in fire scenario 4

According to the above discussion, the fire separation is an effective way to confine the spread of smoke in electric power cable tunnel. However, the tunnel section and working well might be contained in the same fire compartment. When the fire happened in the tunnel section, the flame would spread into the working well due to the lack of oxygen, and caused extremely high temperature rise in the working well. Since the working well is very important for the electric power cable tunnel, the currently-used fire separation need further improve. When the fire source was located at the working well, the tunnel section would be affected by the flame as well.

5. Conclusions

The smoke propagation in the long electric power cable tunnel was investigated numerically. The effects of fire separation on flame behavior and burning process were also explored. The results showed that the smoke propagated quick in the power cable tunnel without fire separation, and the temperature of the whole tunnel increased obviously. The fire separation is an effective way to confine the fire smoke into the fire compartment. However, the flame would migrate in the whole fire compartment, mainly due to the lack of oxygen. The working well might be affected by the high-temperature flame even the fire happened in the tunnel section. The temperature of the 2# working well reached almost 1000°C when the fire initially located at 1# tunnel section in the present study.

Acknowledgments
This work was financially supported by the Science and Technology Project of State Grid Corporation of China.

References

[1] Babrauskas V. How do electrical wiring faults lead to structure ignitions? //Proc. Fire and Materials 2001 Conf, 2001: 39-51.
[2] Hagimoto Y, Kinoshita K, Hagiwara T. Phenomenon of glow at the electrical contacts of copper wires. Natl. Res. Inst. of Police Science Reports–Research on Forensic Science, 1988, 41: 30-37.
[3] Fisher R P, Stoliarov S I, Keller M R. A criterion for thermally-induced failure of electrical cable . Fire Safety Journal, 2015, 72: 33-39.
[4] Dunki-Jacobs J R. The escalating arcing ground-fault phenomenon. IEEE Transactions on industry applications, 1986 (6): 1156-1161.
[5] M.J. Niu, X.F.Li, J.Z. Chen et al., Study on the Flame Retardation and Smoke Emission of Flame Retarded Polyethyleneterephthalate with Cone, Polymer Materials Science and Engineering, 2002, 18(6): 202-204.
[6] W. Wang, H.P. Zhang and Y.T. Wan., A study on the burning behavior of PVC cables in the conic calorimeter test, Journal of Safety and Environment, 8(2): 117-120, 2008.
[7] J.Q. Zhang, B.S. Zhang, M.H. Fan, et al. Effects of External Heat Radiation on Combustion and Toxic Gas Release of Flame Retardant Cables. Materials Science Forum, 898: 2392-2398, 2017.
[8] Zhang, B., Zhang, J., Lu, S., & Li, C. (2015). Buoyancy-driven flow through a ceiling aperture in a corridor: a study on smoke propagation and prevention. Building Simulation, 8(6), 701-709.
[9] Liang, Q., Li, Y., Li, J., Xu, H., & Li, K. (2017). Numerical studies on the smoke control by water mist screens with transverse ventilation in tunnel fires. Tunnelling & Underground Space Technology, 64, 177-183.
[10] Ferng, Y. M., & Liu, C. H. (2011). Investigating the burning characteristics of electric cables used in the nuclear power plant by way of 3-d transient fds code. Nuclear Engineering & Design, 241(1), 88-94.

IOP Conf. Series: Journal of Physics: Conf. Series **1064** (2018) 012016 doi:10.1088/1742-6596/1064/1/012016

Study of dust diffusion at transfer point of belt conveyor based on FLUENT

Guijun Gao[1, 2,*], Jie Shen[1, 2], Qin Liang[1, 2]

[1]College of Mechanical Engineering Taiyuan University of Technology, Taiyuan 030024, China
[2]National-Local Joint Engineering Laboratory of Mining Fluid Control, Taiyuan 030024, China

*Corresponding author e-mail: Gaogj161@163.com

Abstract. The special working modes make transfer point of belt conveyor become one of the main dust sources in coal production. In this paper, the computational fluid dynamics software FLUENT was used to simulate the model of gas-solid two phases flow at transfer point. The distribution of velocity, dust particle and concentration were analysed. The results showed: the maximum value of dust concentration appeared along the tangential direction of the falling coal at the transfer point of the belt conveyor and raised the maximum height, the distribution of dust at transfer point was influenced by different airflows including belt induced flow and falling coal induced flow, and the latter was main factor, this study provided a theoretical basis for the dust treatment at transfer point of belt conveyor.

1. Introduction

With the continuous deepening of mechanization of fully mechanized mining, coal production is increasing day by day, simultaneously, the problem of dust in the process of coal mining and conveying has become more serious. The damage of coal dust to mine safety production mainly is: flammable and explosive, endangering physical and mental health of miners, and accelerating equipment wear, etc. [1-3]. Due to its special working mode, the transfer point of conveyor belt makes it one of the main dust generating sources in the production process of coal mine. According to relevant data, the concentration of dust detected at the transfer point of belt conveyor greatly exceeds the relevant national standards [4, 5]. Therefore, the study of the dust distribution at the transfer point of the belt conveyor has a positive theoretical significance for the treatment of dust.

2. The generation and motion characteristics of dust at transfer point

At the transfer point of the belt conveyor, due to the combined effects of airflows such as belt traction flow, falling coal induced flow and sheared compressed flow, a large amount of dust is generated at the transfer point [6, 7]. Coal blocks do similar parabolic motion at the transfer point, the tiny dust particles on the coal block escape from the coal block under the impacts of the air flow and air resistance in the roadway at this time, in addition, the dust particles adhering to the upper belt will enter the air due to the airflow of the roadway and the vibration of the belt [8-11]. Therefore, the characteristics of generating dust at the transfer point are: the amount of generating dust is large, and

Content from this work may be used under the terms of the Creative Commons Attribution 3.0 licence. Any further distribution of this work must maintain attribution to the author(s) and the title of the work, journal citation and DOI.
Published under licence by IOP Publishing Ltd

the dust generated is mainly at the falling of the coal block, which is mainly affected by the falling induced airflow of the coal block, most of them are small dust particle with diameters of less than 10 am, and the second fugitive dust is more serious.

At the transfer point of the belt conveyor, the motion characteristics of dust are: the dust has better turbulent follow ability after being removed from the coal blocks, at start, most of dust moves along the direction of motion of the belt and then drifts under the influence of the induced airflow and the mine wind flow. The respective movement trajectories vary with the particle size of the dust; and dust particles with diameters of less than 10 am is suspended in the air [12-14].

3. The motion of dust in space

The motion of dust in space is basically manifested as diffusion motion, based on Fick's First Diffusion Law, the number of particles passing through the unit area per unit time in the Cartesian Coordinate System can be expressed as [15]:

$$F = -D\frac{\partial C}{\partial x} \tag{1}$$

The C is the concentration of the diffusion, number/m• s, D is the diffusion coefficient m 2/s.

When the dust concentration changes so much that it must be considered, the relationship between matter quantities in body units can be analyzed by the law of conservation of mass, the basic differential equation of dust diffusion is shown in (2):

$$\frac{\partial C}{\partial t} = D\left(\frac{\partial^2 C}{\partial x^2} + \frac{\partial^2 C}{\partial y^2} + \frac{\partial^2 C}{\partial z^2}\right) \tag{2}$$

D in Equation (2) is the Stokes-Einstein diffusion coefficient; x, y, z is the direction vector in the body; t is start time.

4. Numerical simulation and the analysis of results

4.1. Geometric Modelling and meshing

According to the actual situation, the model at the transfer point of the roadway belt conveyor was simplified, a rectangular space (2mx1mx2m) containing the transfer points was intercepted as the calculation area, belt width was 500 mm, and the distance between the upper and lower belts was 500 mm.

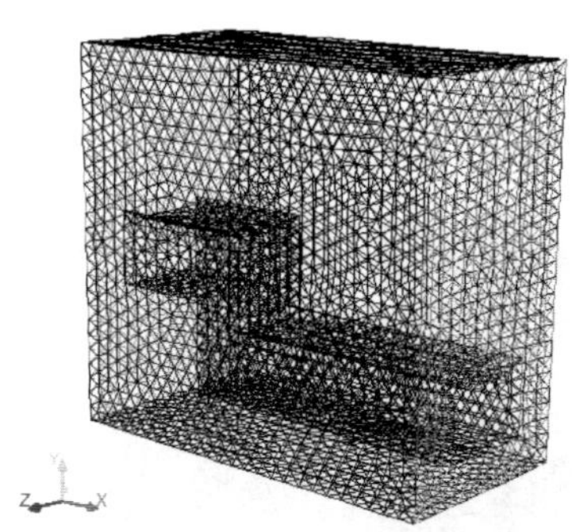

Fig 1. Grid of model.

The model was built and then meshed to unstructured grid in GAMBIT, the mesh size is 80 mm, the final mesh number is 35484 (the mesh was displayed in FLUENT as shown in Fig. 1). The types of the left end, the right end, the upper and lower belt were set to velocity-inlet, outflow, and wall. 1 and wall. 2 respectively, and others were set to wall.

4.2. The setting of boundary conditions
According to the actual situation combined with the established model, the gravity acceleration in the Y-axis direction was set to -9.8 m/s 2 in FLUENT, and the standard k-e was selected in turbulence model. The settings of discrete phase and boundary conditions were shown in Table 1, Table 2, respectively.

Table 1. Parameter setting table of discrete phase.

Parameter name	Parameter setting
Min. Diameter	1 e-06 m
Max. Diameter	0.0001 m
Mean Diameter	1 e-05 m
Velocity	0 m/s
Total Flow Rate	1.5×10^{-3} kg/s
Number of Diameters	10

Table 2. Setting table of boundary condition.

	Parameter name	Setting
	Inlet Boundary Type	Velocity-inlet
	Inlet Velocity Magnitude	0 m/s
	Hydraulic Diameter	0.93 m
	Turbulence Intensity	6.5%
	Outlet Boundary Type	Outflow
Wall. 1	Wall Motion	Moving Wall
	Motion Speed	2 m/s
	Motion Direction	x-axis positive
Wall. 2	Wall Motion	Moving Wall
	Motion Speed	2 m/s
	Motion Direction	x-axis positive

4.3. The analysis of numerical simulation results
After the above settings and computer calculations, the numerical simulation results were obtained. Furthermore, the distribution of velocity, dust particle, and dust concentration in the flow field were analyzed below.
4.3.1. Velocity distribution. Fig. 2 shows the velocity cloud at Z=0, which is the section XOY. Fig. 3 shows the velocity vector at the section Z=0.

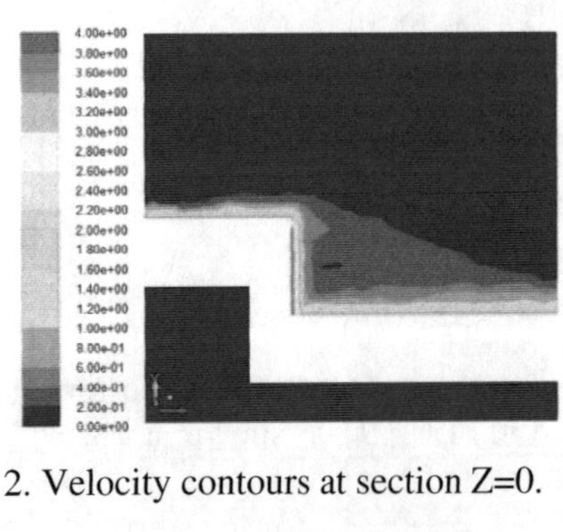

Fig 2. Velocity contours at section Z=0.

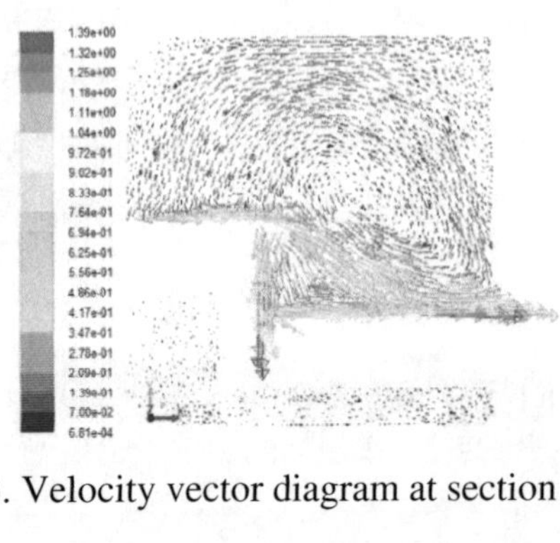

Fig. 3. Velocity vector diagram at section Z=0.

As shown from above figures, the fluid velocity around the belt and in the falling segment is high, and the large area fluid flow appears in the falling segment. This shows that the motion of the belt can drive the flow of nearby air. The drop of dust can cause turbulence in the surrounding air. As shown in Fig. 2, the velocity of air around the falling dust is significantly higher than other parts, and Fig. 3 shows that the air caused by the falling of coal dust is in turbulent gyration state. This shows that the dust generating at the transfer point of the belt conveyor is affected by various airflows such as belt induced flow and falling coal induced flow.

4.3.2. The spatial distribution of dust particles. The Fig. 4 is a particle diagram with taking particle's motion time as colouring variable in the flow field, which shows intuitively that the distribution of dust at the transfer point. Behind coal heap falling point at transfer point of belt conveyor, the amount of dust is the largest, and the height of fugitive dust is the highest.

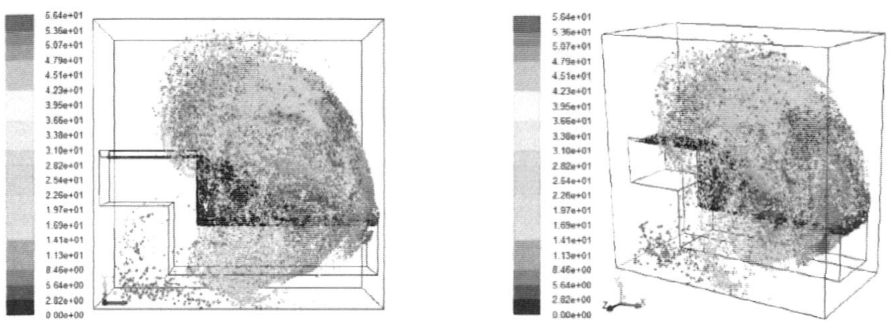

Fig 4. Particle figure by colored with particle movement time.

4.3.3. The distribution of dust concentration. Fig. 5 to Fig. 8 are dust concentration cloud diagrams at sections Z=0, 0.1, 0.2, and 0.3(m), respectively.

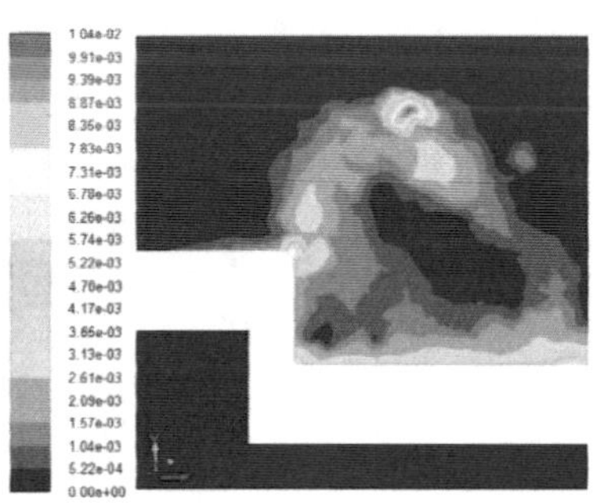

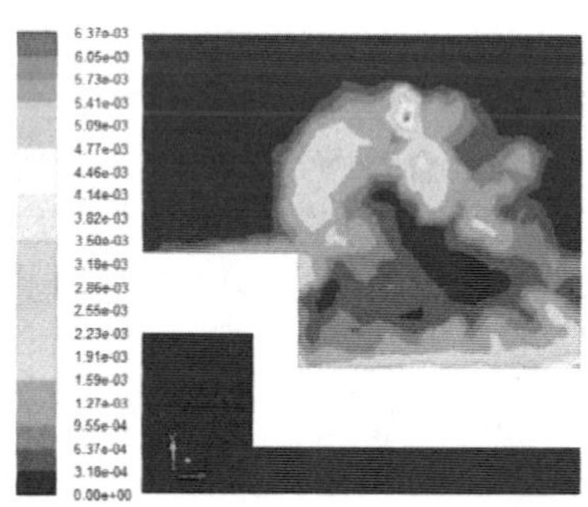

Fig 5. Concentration contours at section Z=0. Fig. 6. Concentration contours at section Z=0.1.

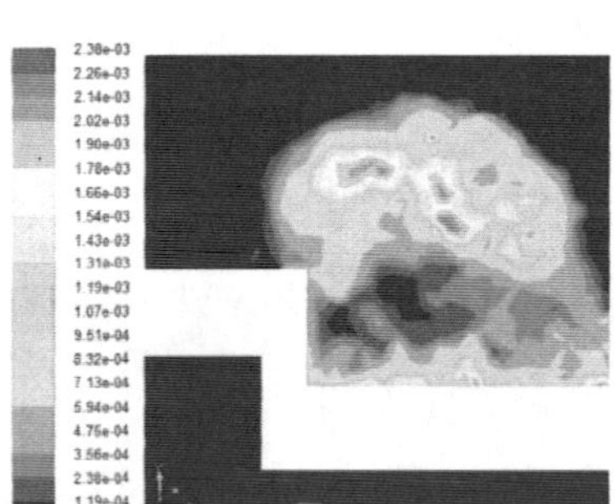

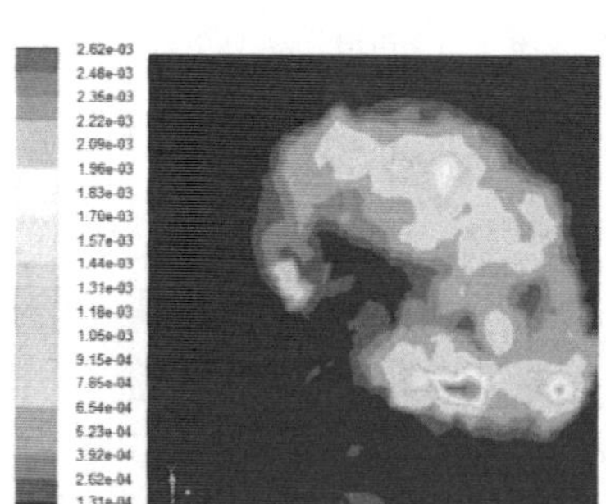

Fig 7. Concentration contours at Section Z=0.2. Fig. 8. Concentration contours at Section Z=0.3.

As shown from Fig. 5 to Fig. 8, the maximum dust concentration appears above and behind the falling coal point. This indicates that a great deal of dust is raised by falling coal induced airflow, and the dust rises rapidly after being raised and spreads to the surrounding environment.

To quantitatively analyze the concentration of dust in the flow field, the broken line charts of concentration distribution have been drawn, The Fig. 9 shows the broken line chart of dust concentration that belong to 7 straight lines drawn on XOY plane, the positions are Y=1.2, 1.3, 1.4, 1.5, 1.6, 1.7, and 1.8 (m), respectively. As shown, the dust concentration reaches the maximum between X=1.2 m and X=1.4 m, which can reach to 4200 mg/m 3, the reason why the values of dust concentration at the two straight lines, Y=1.2 and Y=1.3, reach the maximum near X=0.8 m is that it is close to the upper belt and is greatly affected by the belt traction flow of the upper belt. The dust concentration is very low at Y=1.8 (m) in the figure, in addition, several straight lines with higher dust concentrations are Y=1.4, 1.5, 1.6, and 1.7 (m), respectively.

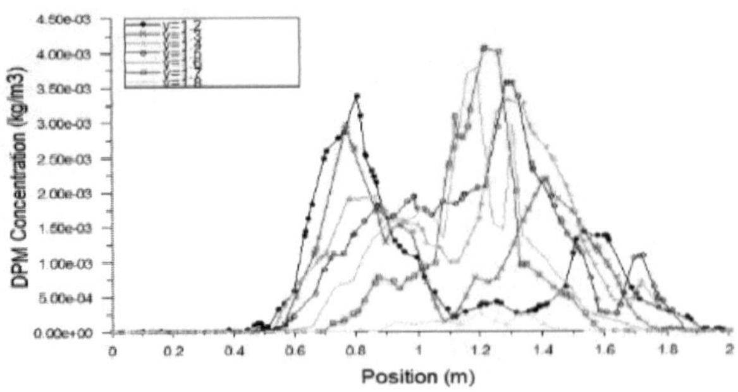

Fig 9. Line chart of concentration at different height in XOY surface

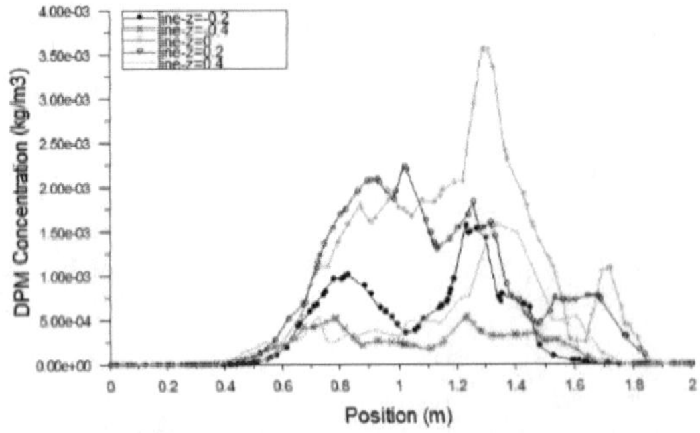

Fig 10. Line chart of concentration at different location on plane Y=1.5.

Fig. 10 shows the analysis of distribution of horizontal concentration, five straight lines at Y=1.5 m are selected, which are Z=0, ±0.2, and ±0.4 (m), respectively. As shown in figure, the maximum of dust concentration is mainly between X=1.2 m and X=1.4 m, and the maximum value of dust concentration is at the midline Z=0.

Therefore, for the layout design of the dust control device at the transfer point, according to the results of this study, when using a pressure nozzle to reduce dust, it is recommended that the spray point be located 1.2 meters above the lower belt, which can make the optimal atomization zone of the nozzle cover the range that is 0.3-0.7 m horizontally from the upper belt, and the dust generated by belt can be effectively reduced at the transfer point.

5. Conclusion

This paper uses FLUENT to simulate the dust diffusion at the transfer point of the belt conveyor, and analyzes the distribution of velocity, concentration field and dust particles, at the back of falling point of coal heap of transfer point, the dust concentration is maximum, and the height of fugitive dust is the highest, the amount of dust generated at the transfer point is affected by multiple airflows such as belt induced flow and falling coal induced flow, among them, the latter is the main factor. And this study provides a theoretical basis for the treatment of dust at the transfer point.

References

[1] Liao Xian-xin, Jiang Zhong-an, Niu Wei, et al. Numerical simulations of blasting dust migration with the use of Fluent in stope[J]. Journal of Safety and Environment,2012, 12(6): 43-46.
[2] Han Gui-bo, zhan Shui-fen, Zhang Xiao-chun, et al. Influence factors and mathematical model of coal dust particles threshold velocity[J]. Journal of China Coal Society, 2009, 34(10):1360-1 363.
[3] Zhao Zhen-bao. Experimental research on diffusion and mitigation law of dust cut generated by coal shearer[J]. College of Engineering, Peking University, 2011, 31(4): 383-386+390.
[4] Chen Ju-shi, Jiang Zhong-an, Jiang Lan. Experimental research on dust distribution and its influencing factors in belt conveyer roadway[J]. Journal of China Coal Society,2014, 39(1): 135-140.
[5] Lü Tai , Che Sheng, Guo Cheng-kun, et al. Flow field simulation and structure optimization of confined hood at transshipment point of coal conveyor belt[J]. Chinese Journal of Environmental Engineering, 2013, 7(7): 2667-2672.
[6] Chen Xi, Ge Shao-cheng. Numerical and experimental study on spray characteristics with effect of induced airflow on transshipment point[J]. Journal of China Coal Society, 2015,40(3):603-608.
[7] Niu Wei, Jiang Zhong-an, LIU Yi. Numerical simulation on dust movement regularities at fully-mechanized coal faces and its utilization. Journal of Liaoning Technical University(Natural Science), 2010, 29(3): 357-360.
[8] Zhou Gang, Cheng Wei-min, Chen Lian-jun, et al. Numerical simulation and its application of dust concentration spatial distribution regularities in fully-mechanized caving face[J]. Journal of China Coal Society, 2010, 35(12): 2094-2099.
[9] Zhang You-shi, Jing De-ji, et al. Study on dust technologies at a transfer point of a coal belt based on gas-solid two-phase flow[J]. China Coal, 2014, 40(10):122-126.
[10] Ma Yundong, Luo Genhua, Guo Shaohua. Numerical simulation on application of diffuse regulation with power dust in transshipping site [J]. Journal of Safety and Environment, 2006, 6(2): 16-18.
[11] Yao Xi-wen, Lu Guang-li, Xu Kai-li. Numerical simulation of dust generation at different procedures inclined fully-mechanized caving face[J]. Journal of China Coal Society, 2015,40(2):389-396.
[12] Wang Xiao-feng, Chen Cai-yun. Mechanism and comprehensive treatment of dust separating out at transfer site[J]. Coal Technology, 2008,27(1):124-125.
[13] Li Jie-ying. Analysis of the coal dust mechanism of separating out[J]. Shanxi Coking Coal Science& Technology, 2011,10:36-37+41.
[14] Li Yue-qing. Research on dust precipitation mechanism, diffusion regularity and control technique[J]. Journal of Capital Normal University(natural science edition), 2014,35(4):32-37.
[15] Huang Lei, Wu Ke-jia, Wang Zhen-he, et al. Research on the law of coal dust setting at transfer point and simulation analysis[J]. Coal Preparation Technology, 2009,3:10-12+77.

IOP Conf. Series: Journal of Physics: Conf. Series **1064** (2018) 012017 doi:10.1088/1742-6596/1064/1/012017

Aerodynamic Heating Research of Induction System under Different Attack Angle Condition

Zhaowei Wang*, Ran Yao and Pengju Hu

China Academy of Launch Vehicle, Beijing, China.

*Corresponding author e-mail: wangzhaowei-01@163.com

Abstract. Series aerodynamic heating problems often occurs in induction system of high mach number vehicle. Especially when accompanied by shock-shock interaction and shock-boundary layer interaction. Simulation is carried out on the strong interaction between impinging shock wave and specified induction model at Mach 5~6 under different attack angle ranging from -10° -10°. Results indicated that the attack angle largely affect the aerodynamic heating values.

1. Introduction

At high Mach number flight conditions, particularly for a long time and long range fly, the serious aerodynamic heating problems become an unavoidable key issue in the design of a induction system. This needs to take into account both aerodynamic performance and aerodynamic heating, which are actually dependent on local flow characteristics. Consequently, it is of great importance to study flow characteristics on typical parts of an induction system subjected to severe aerodynamic heating for the engineering application and development of basic subject.

The flow about the airbreathing high mach number vehicle is generally characterized by the occurrence of intense shock that form around the nose leading edge and induction leading edge, of which shock-shock wave interaction can form the complex types of interferences with large increase in pressure and temperature levels, while the shock wave-boundary layer interaction can form and separation with moderate increase in pressure and temperature levels.

In this paper, flow characteristics of the shock-shock wave interaction and shock wave-boundary interaction for two dimensional typical induction were simulated under different attack angle at Mach 6. Ten cases depending attack angle relative to incoming flow were investigated to study the aerodynamic heating rules of the specified induction system.

Content from this work may be used under the terms of the Creative Commons Attribution 3.0 licence. Any further distribution of this work must maintain attribution to the author(s) and the title of the work, journal citation and DOI.
Published under licence by IOP Publishing Ltd

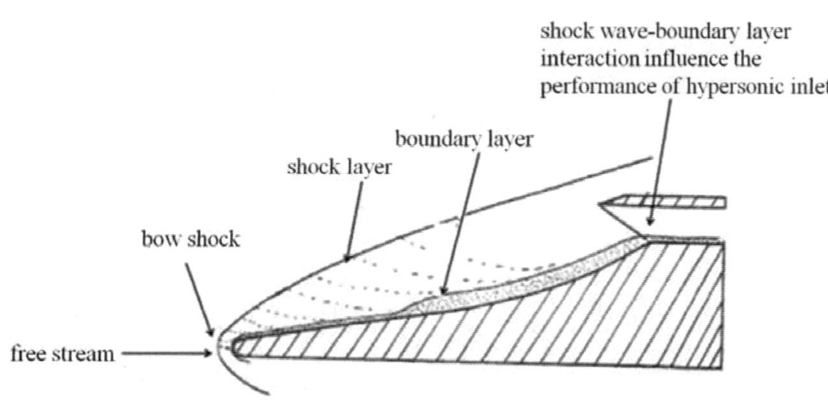

Fig.1 Characteristics of high mach number induction flow

2. Problems and modelling

It is well recognized that attack angle can largely affect the interference between the external waves and a bow shock in front of a cowl lip[1-2], thus resulting totally different heating rates distributions in high mach number induction. However the effect of incoming flow condition on aerodynamic heating of cowl lip and induction were not thoroughly studied and a clear conclusion was never drawn for engineering application[3]. It is may be due to the fact that the complicated flow conditions are difficult to simulate and results at given induction design may not easy to explain. However engineers still want to get a rule between incoming flow conditions and heating rates of high mach number induction.

In this study, research and CFD method are based on commercial software FLUNT and Navier-Stokes equations. It is a convenience tool and has better precision for complicate flow. The results were validated with experiments in reference[4-6]. The results had good agreement with the experiment results obtained at NASA scramjet investigation, indicating the CFD model and simulation methods were suitable for the induction aerodynamic heating modeling.

3. Results and Discussion

3.1. Flow characteristics of 2D high mach number induction

Flow conditions of nine different attack angles ranging from -10° to 10° have been calculated. Mach contours and static pressure of computation results (Fig.2-Fig.6) showed that the bow shock wave and compression waves converged at the cowl lip and the scramjet induction at attack angle of 0°, 6° and 10°, however the compression wave and oblique waves were out of induction with negative attack angles. With the negative angles studied, the shock wave convergence effect tend to became weaker. The static pressure of scramjet induction increased with the attack angle. The flow characteristic results indicated strong aerodynamic heating effect occurred at positive angle conditions.

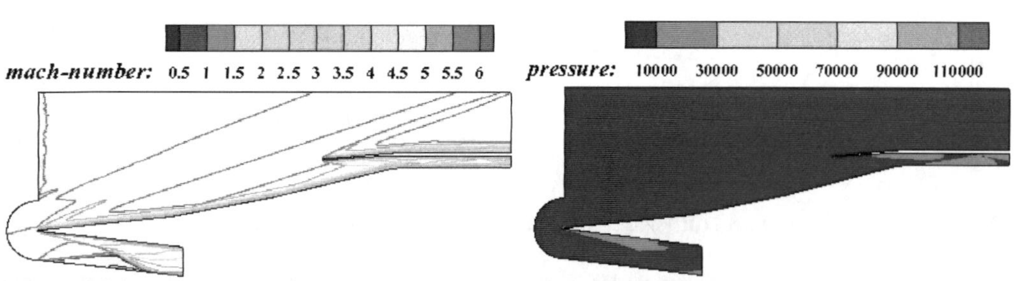

Fig.2 Numerical simulation results under Ma6, -10°

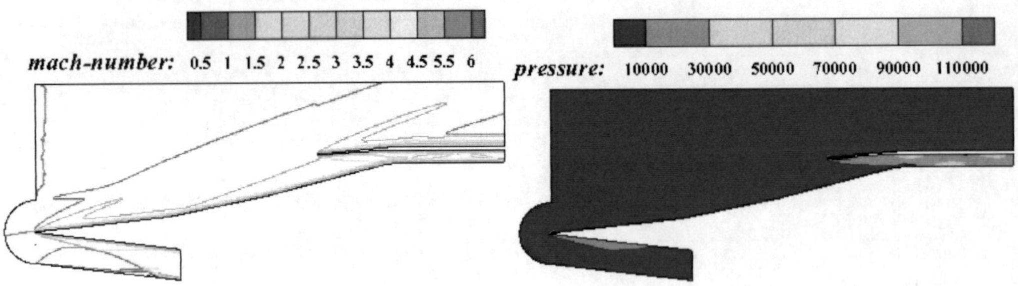

Fig.3 Numerical simulation results under Ma6, -6°

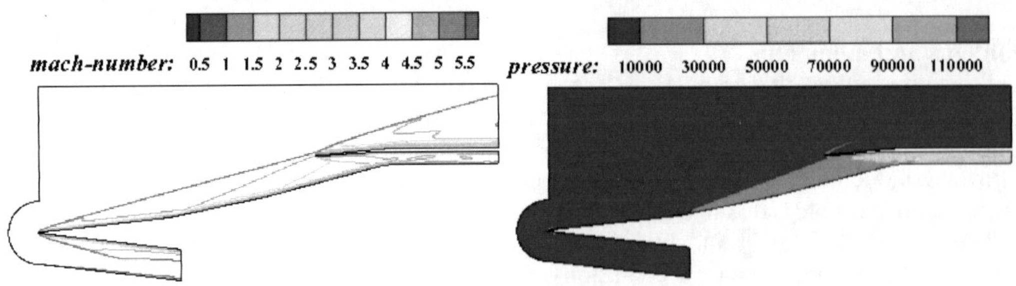

Fig.4 Numerical simulation results under Ma6, 0°

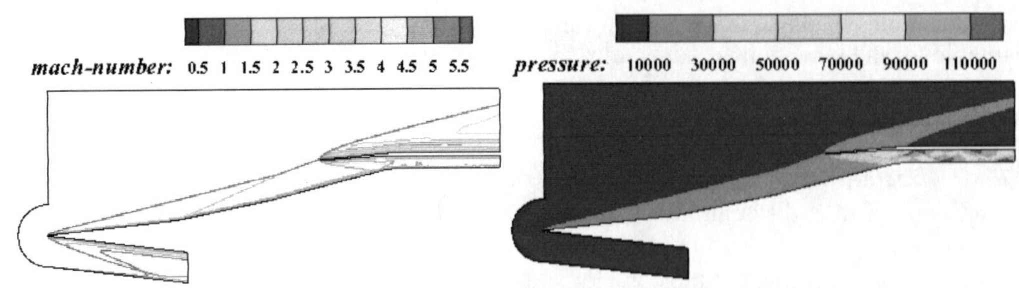

Fig.5 Numerical simulation results under Ma6, 6°

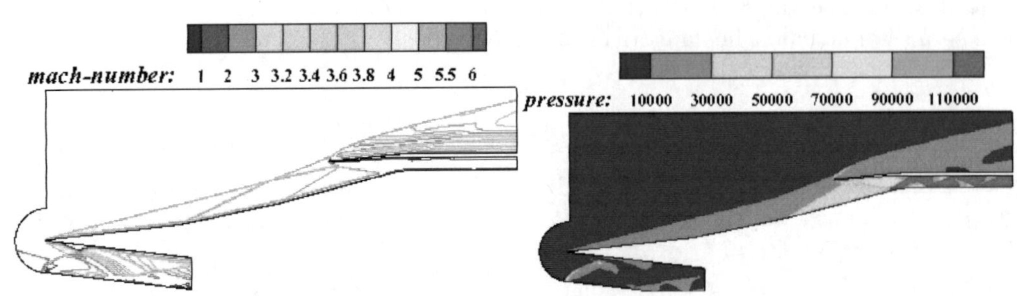

Fig.6 Numerical simulation results under Ma6, 10°

Key parameters at the exit of the isolator and the rules of variation have been given at different conditions (Tab.1). The static pressure and static temperature of high mach number induction were increased with the attack angle due to the compression effect. The total pressure recovery and mass flow coefficient showed different tendency compared with the pressure and temperature, due to the varied boundary layer thickness.

Tab.1 Key induction exit parameter under different attack angles

attack angle(α)	total pressure recovery coefficient	Mach number	static pressure rise	static temperature rise	flow per unit width(kg/s)
-10°	0.43	3.59	6.14	2.31	14.96
-6°	0.48	3.43	9.72	2.44	23.48
-4°	0.52	3.34	12.00	2.52	29.07
-2°	0.54	3.24	15.03	2.64	35.29
0°	0.55	3.10	19.13	2.81	41.84
2°	0.58	2.97	23.90	2.95	48.61
4°	0.57	2.82	29.57	3.16	54.28
6°	0.50	2.58	37.56	3.49	58.78
10°	0.37	2.18	52.17	4.14	65.34

Total pressure distribution for induction exit under negative attack angles were shown in Fig.7, indicating total pressure decreased with the attack angle. Low pressure area increased and the flow rate had great loss at negative attack angle conditions.

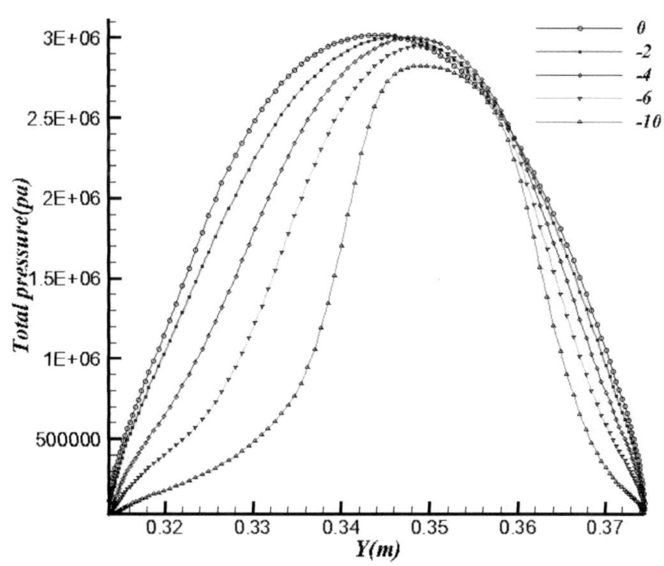

Fig.7 Total pressure distribution for induction exit under negative attack angles

3.2. Aerodynamic heating characteristics of 2D high mach number induction

Figure 8 showed heat flux distribution of induction forebody at given attack angle of -6°, 0°, 6° and 10°. In general, the heat flux increased with the attack angle, which had good agreement with the static pressure variation tendency, indicating pressure and heat flux had similar variation rules for high mach number scramjet induction. At the same time, seperation flow field may increase under strong shock/shock interaction condition, which introduced severe influence for thermal protection system.

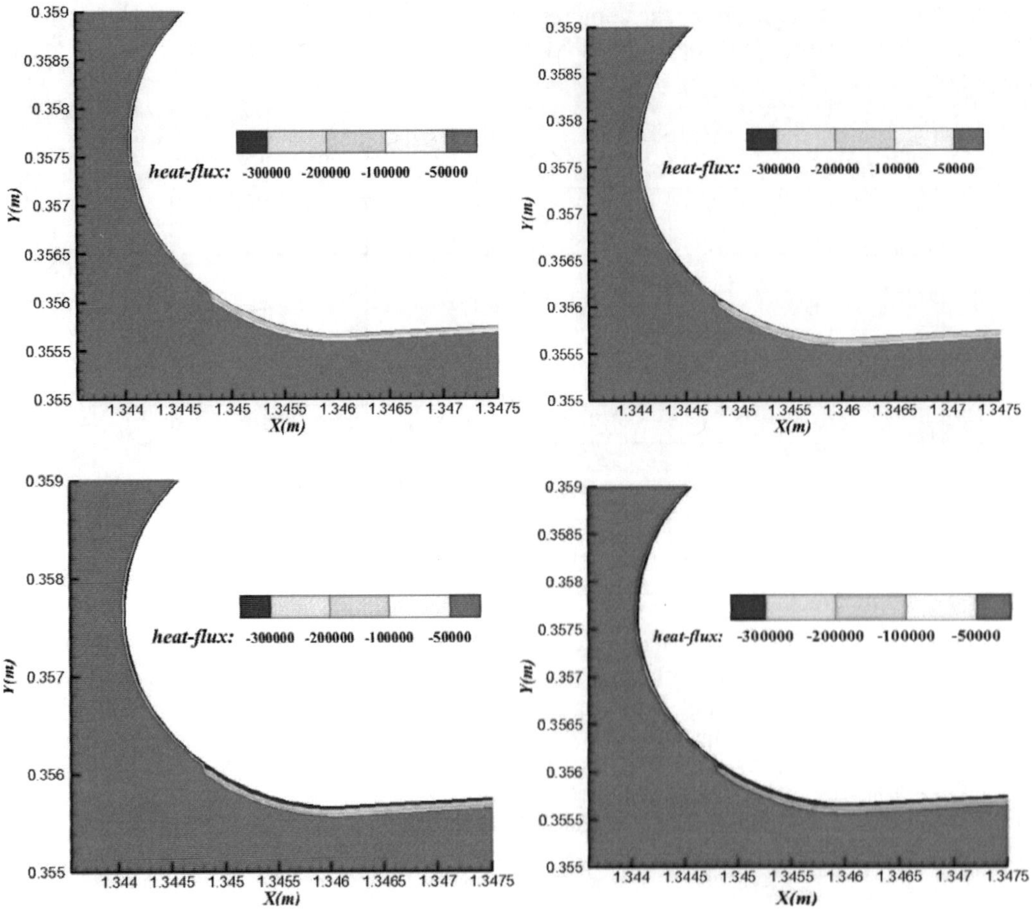

Fig.8 Aerodynamic heating results of induction under attack angle of -6°, 0°, 6° and 10°

Cowl lip static pressure distribution did not simply increase or decrease with the attack angle due to the complicated interaction between shock waves and different angle of free stream. Of all the condition studied, the static pressure of cowl lip was the largest at 10°.

4. Conclusion

The high mach number induction performance and aerodynamic heating characteristics are the main concern in engineering applications, both theoretical and experimental guidance are in great need in induction aerodynamic heating evaluation. This paper tried to obtain the rules and connections between aerodynamic heating and incoming flow conditions, such as attack angles.

(1). The static pressure of scramjet induction increased with the attack angle. The flow characteristic results indicated strong heating occurred at positive angle conditions;

(2). The negative angle is not good for forebody compression, the bow shock from leading edge and oblique shock can not be converged with the angle of less than -2°;

(3). Cowl lip static pressure distribution did not simply increase or decrease with the attack angle due to the complicated interaction between shock waves and different angle of free stream. Of all the cases studied, heat flux of cowl lip at 6°and 10°are the largest, which are almost two times larger than the 0° condition.

Reference

[1] znderson J D. High mach number and high temperature gas dynamics [M]. 2nd ed. Reston V A:

AIAA, 2006.
[2] Heppenheimer T A. Facing the Heat Barrier: A History of High mach numbers[M]. NASA History Series, NASA SP-2007-4232, 2007.
[3] Van Wie D M, Nedungadi A. Plasma Aerodynamic Flow Control for High mach number Inductions[C]. AIAA 2004-4129, 2004.
[4] Emami S., Trexler C A. Experimental investigation of induction-combustor isolators for a dual-mode scramjet at a Mach Number of 4[R].NASA Technical Paper 3502, 1995.
[5] Pan Sh, Tian ZhY, Feng DH, Ding GH, Li H. Computation and analysis of aeroheating of scramjet inlet cowl lip[J]. Journal of Aerospace Power, 2009, 24(9): 2096-2100.
[6] Holden, M.S. Studies of Scramjet Performance in the LENS Facilities[C], AIAA- 2000-2004, 2000.

IOP Conf. Series: Journal of Physics: Conf. Series **1064** (2018) 012018 doi:10.1088/1742-6596/1064/1/012018

The aerodynamic stability of a new type of Multi-tower cable-stayed bridge with rhombic stiffening cables

Xiaowei Wang [1, 2], Yingmin Li [1,*] and Weiju Song [1, 2]

[1]College of Civil Engineering, Chongqing University, Chongqing, China.
[2] City college of Science and Technology, Chongqing University, Chongqing, China.

*Corresponding author e-mail: liyingmin1968@126.com

Abstract. Aerodynamic stability is an important problem in the design and research of long-span multi-tower cable-stayed Bridges. The three-dimensional nonlinear aerodynamic stability analysis method is used to qualitatively analyze the aerodynamic stability of a new type of rhombic stiffened cable-stayed bridge in this paper, and compared with conventional cable-stayed bridges. The results show that the new rhombus stiffener can greatly increase the longitudinal stiffness of the structure and reduce the vibration period of the structure. The maximum lateral displacement, vertical displacement and torsion Angle of the structure under wind load are reduced, and the critical unstable wind speed of the structure is increased. It can be seen that the aerodynamic stability of the new type of rhombus cable-stayed bridge is better than conventional cable-stayed bridge.

1. Introduction

Long-span multi-tower cable-stayed bridge is a flexible structure [1], which is easy to vibrate and deform under wind load. In history, the damage accidents of cable-stayed Bridges due to wind vibration often occur, so the influence of wind load on cable-stayed Bridges should be paid attention to.

It is of great significance to study the strength, stiffness and carrying capacity of long-span multi-tower cable-stayed bridge under natural wind load so that it can still play the role of transportation in the process of high wind. Because of the difficulty in anchorage of the middle tower, the longitudinal stiffness of the long-span multi-tower cable-stayed bridge is low and vibration and deformation are more likely to occur under wind load. A new type of rhombic stiffener cable-stayed bridge which can increase the directional stiffness of a multi-tower cable-stayed bridge was proposed in this paper. The aerodynamic stability of the new type of rhombic stiffener cable-stayed bridge and the conventional cable-stayed bridge under the wind load was compared and analyzed by considering the wind attack Angle, vertical wind load and cross wind load in this paper. It is a reference for the application of the new cable-stayed bridge in practical engineering.

2. Dynamic characteristics of a new type of rhombus cable-stayed bridge

The stiffness control of multi-tower cable-stayed bridge is very important in design. Through reasonable layout of stiffening cables can improve the overall stiffness of multi-tower cable-stayed

Content from this work may be used under the terms of the Creative Commons Attribution 3.0 licence. Any further distribution of this work must maintain attribution to the author(s) and the title of the work, journal citation and DOI.
Published under licence by IOP Publishing Ltd

bridge [2]. But the conventional methods of stiffening cables have many deficiencies, the new stiffening cables system (rhombic Cables) was performed in this paper (as shown in Figure 1):

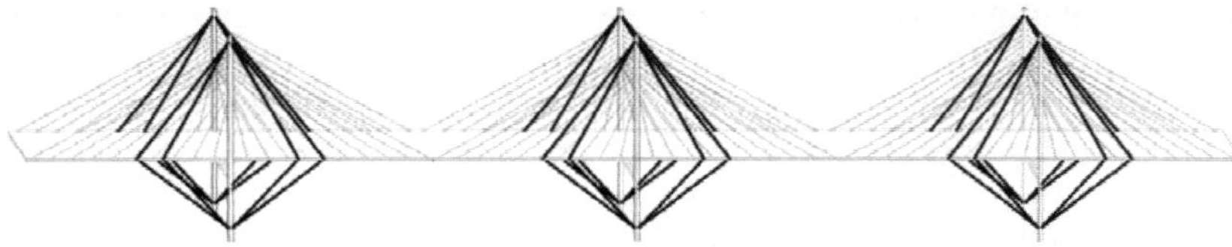

Figure 1. Multi-tower Cable bridge with rhombic cables

As shown in Figure 1, because one end of Tie-down cables was anchored on the top and root of tower and the other end is anchored on the girder, it causes that tower and girder are validly connected to become a balanced system and increase the overall stiffness of the structure.

The actual bridge design parameters were used in this paper, and diamond stiffeners were set on each tower. Compared with the conventional cable-stayed bridge, the influence of rhombic stiffening cables on the wind resistance of the bridge was analyzed.

The total length of Bridge is 1327.6m. This bridge has three towers and four spans whose midspan is 460m, side span is 203.8m and height of tower is 247.5m. The calculation model is shown in figure 2.

a. The conventional cable-stayed bridge   b. The cable-stayed bridge with rhombus cable

Figure 2. Calculation model

The above models were analysed by using the finite element software Ansys, and the first 6 order natural vibration frequencies and formations obtained were shown in table 1

Table 1. Natural vibration characteristics of cable-stayed Bridges

Num	Frequency (Hz)		The vibration shape	
	Model a	Model b	Model a	Model b
1	0.0645	0.1841	Longitudinal floating	Longitudinal floating
2	0.1851	0.1968 •	The main tower bends in the same direction	The main tower bends in the same direction
3	0.1972	0.2404 •	The side tower bends in the same direction	The middle tower bends in the opposite direction
4	0.2011	0.2736	The main beam bends in a symmetrical vertical direction	The main beam bends in a symmetrical vertical direction
5	0.2433	0.2991	The middle tower bends in the opposite direction	The main beam is opposed to vertical bending
6	0.3084	0.3188	Main girder torsion	Main girder torsion

(*Attention : Model a is calculation model of conventional cable-stayed bridge.Model b is calculation model of cable-stayed bridge with rhombus cable)

As shown in table 1, the self-vibration period of rhombus cable-stayed bridge decreases from 15 seconds to 9 seconds, and the frequency increases from 0.1045 to 0.1841, indicating that the rhombus greatly increases the stiffness of the structure along the bridge. Moreover, the vertical bending formation of the main girder is relatively late compared with the traditional cable-stayed bridge, so the vertical rigidity of the main girder can be increased.

3. The aerodynamic stability of the new rhombus stiffening cable-stayed bridge

In -3°, 0° and 3° initial wind Angle of attack, the conventional cable-stayed bridge and rhombus stiffening cable-stayed bridge had carried on the aerodynamic stability analysis by three dimensional nonlinear analysis method [3]. The bridge model was simplified to the bar model. The bridge deck main beam and bridge tower were simulated by spatial beam element. The cable-stayed cable was simulated by spatial rod element. The calculation model is shown in figure 2.The main beam takes into account the action of three components of static force. Because the main girder size of the bridge in this paper was similar to that of Taizhou Changjiang river bridge, static three-component force [4]of Taizhou Changjiang river bridge was adopted in this analysis.

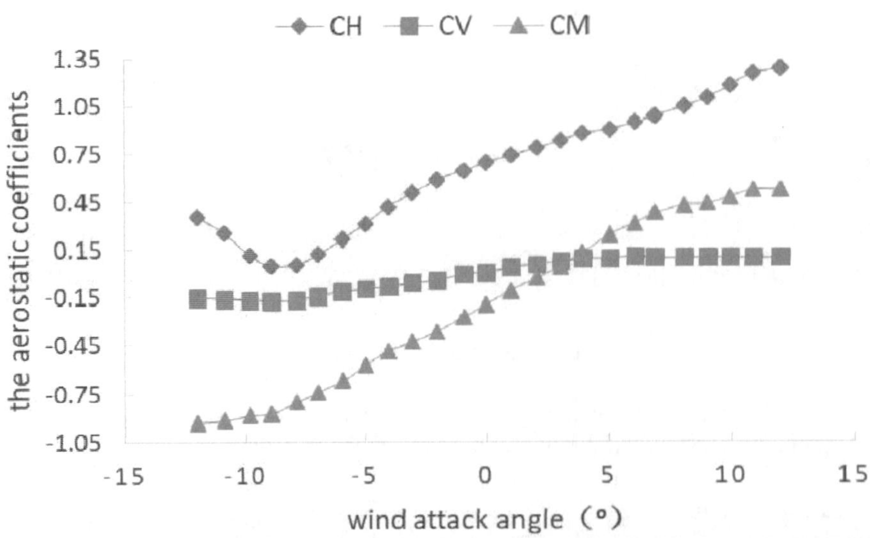

Figure 3. Static three-component force coefficient of Taizhou bridge

(*Attention :CH- Lateral component coefficient, CV- Vertical component coefficient, CM-Torque component coefficient)

The cable and the bridge tower only consider the role of the resistance component. The drag coefficient of the cable is 0.8 and the drag coefficient of the bridge tower is 2.0. The maximum horizontal, vertical and torsional displacement of the main beam under initial wind attack Angle changes with the increase of wind speed [5] ~ [8], as shown in figure 4.

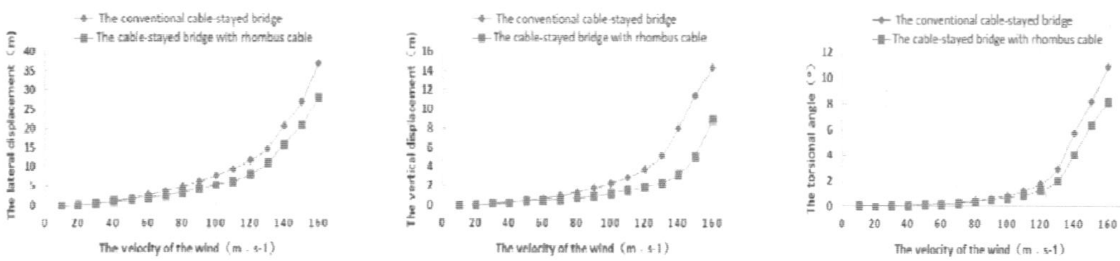

a. The wind Angle of attack in -3°

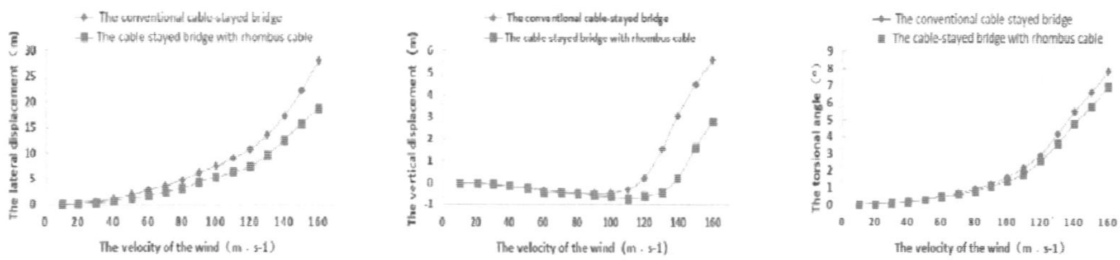

b. The wind Angle of attack in 0°

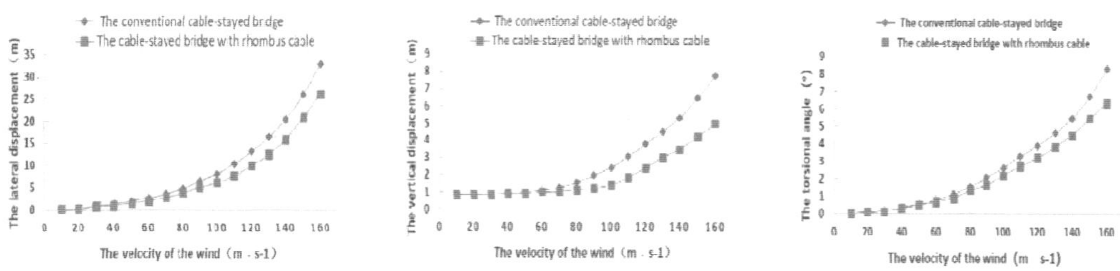

c. The wind Angle of attack in 3°

Figure 4. Evolutions of the girder's maximum displacements with increasing wind speed under different wind attack angles

As shown in figure 4, under 0 ° angle of attack wind, when wind speed is low, the conventional cable-stayed bridge and the rhombus stiffening cable-stayed bridge of displacement are close to in all directions. When the wind speed reaches 110m/s, conventional cable-stayed bridge vertical and torsional displacement sudden increase sharply, structure began to enter a state of instability, and the displacement difference between the two structures is large. The critical wind speed of rhombic stiffener cable-stayed bridge is about 130m/s. The instability of the two cable-stayed Bridges is mainly manifested by the spatial flexural and torsional coupled instability of the main girder in vertical bending and torsional deformation, and it is also associated with lateral bending deformation. It can be seen that under the same conditions, the aerodynamic stability of rhombus cable-stayed bridge is better than that of conventional cable-stayed bridge [9].

Under 3 ° and + 3 ° angle of attack wind, the displacement changing law of different wind speeds is similar. In the same wind speed, the vertical and horizontal displacement and torsion angle of the rhombic stiffening cable-stayed bridge are smaller than conventional cable-stayed bridge. As the wind speed is higher, the displacement of the rhombic stiffening cable-stayed bridge in all directions is reduced obviously. It further states that in the same case, the structure stiffness of rhombus stiffening cable-stayed bridge is larger than that of conventional cable-stayed bridge, and the aerodynamic stability of rhombus stiffening cable-stayed bridge is better than that of conventional cable-stayed bridge [10].

4. Conclusion

The three-dimensional nonlinear aerodynamic stability analysis method is used to qualitatively analyze the aerodynamic stability of a new type of rhombic stiffened cable-stayed bridge in this paper, and compared with conventional cable-stayed bridges.

The results show that the new rhombus stiffening can greatly increase the longitudinal stiffness of the structure and reduce the vibration period of the structure. The maximum lateral displacement, vertical displacement and torsion Angle of the structure under wind load are reduced, and the critical unstable wind speed of the structure is increased. It can be seen that the aerodynamic stability of the new type of rhombus cable-stayed bridge is better than conventional cable-stayed bridge.

Acknowledgments

This work was financially supported by "Static and aerodynamic stability of a new type of multi-tower cable-stayed bridge" (KJ1602303) fund.

References

[1] Han, SH, Park, JK. Practical valuations on the effect of two type uncertainties for optimum design of cable-stayed bridges. International Journal of Steel Structures, 2009, 9(2): 143-152.
[2] Man Hong-gao,Li Qiao, Zhang Yu-zhi. Design optimization of fatigue test model for the cable-girder anchorage zone of steel cable-stayed bridges. Journal of Southeast University (Natural Science Edition), 2007, 37(2): 301-305.
[3] J.B.Roberts, P.D.Spanos. Random vibration and statistical linearization. John Wiley&Sons, 1990.
[4] Zhang Xin-jun, Sun Bing-nan, Xiang Hai-fan, Nonlinear aerostatic and aerodynamic analysis of long-span cable-stayed bridges considering wind-structure interactions[J]. Journal of Wind Engineering and industrial aerodynamics, 2002,90(9):1065-1080.
[5] Chen,J.,Jiang,J.J.,Xiao,R.C.,Xiang,H.F. Nonlinear aerostatic stability analysis of Jiangyin suspension bridge[J]. Engineering Structures,2002,24:773-781.
[6] M.Nagai, X.Xie, H.Yamaguchi, Y. Fujino, Minimum girder width ensuring safety against static and dynamic instabilities of long-span cable-stayed bridges under wind load, J.Struct.Eng. JSCE.44A(1998):1191-1198.
[7] Larsen A,Walther J.H.Aeroelastic analysis of bridge girder sections based on discrete vortex simulations[J].Journal of Wind Engineering and Industrial Aerodynamics,1997,67&68:253-265.
[8] C.T.Georgalcis, C.A.Taylor. Nonlinear dynamics of cable stays.Part2: stochastic cable support excitation. Journal of Sound and Vibration, 2005, 281, 565-591.
[9] Yong Xia, Yozo Fujino. Auto-parametric vibration of a cable-stayed-beam structure under random excitation. Journal of Engineering Mechanics, 2006.3, Vo1.132, No.3: 279-286.
[10] Ho-Kyung Kim, Masanobou Shinozuka,Sung-Pil Chang. Geometrically nonlinear buffeting response of a cable-stayed bridge. Journal of Engineering Mechanics,2004.7,Vol.130, No.7:848-857.

IOP Conf. Series: Journal of Physics: Conf. Series **1064** (2018) 012019 doi:10.1088/1742-6596/1064/1/012019

The Aerodynamic Characteristics of Vehicle under Lateral Wind Action for the Process of Two Vehicles Passing Each Other

Jiguo Zhou[1,*], Yan Han[2]

[1]School of Civil Engineering, Baicheng Normal University, Jilin, Baicheng 137000, China
[2]Taishan Polytechnic, Shandong, Taishan 271000, China

*Corresponding e-mail: zhoujiguo_1986@126.com

Abstract. Research the aerodynamic characteristics of vehicle under lateral wind action for the process of two vehicles passing each other based on the CFD (Computational Fluid Dynamics) numerical simulation of the dynamic grid technology. Analysis the change rule of aerodynamic characteristics of vehicles in the process of vehicles passing each other and discuss the influence of wind speed, driving speed and the different bridge transverse driving locations upon aerodynamic characteristics of vehicles. The results show that, it has large mutation for the aerodynamic force coefficients of the windward vehicle and leeward vehicle. The overturn moment coefficient, lift force coefficient, lateral fore coefficient both reduce with increasing driving speed and both increase with increasing lateral wind speed.

1. Introduction

In recent years, the engineering network has been developing very fast. Among them, the proportion of long-span bridges are increasing. The larger spans and higher bridge above the ground, the greater influence of lateral wind load on vehicles running on bridge. The present studies of aerodynamic characteristics for vehicles running on the bridge have some valuable research. Currently, there are two methods for research on this, wind tunnel test and numerical simulation.

At present, the research situation of mutation wind load on running vehicle mainly have following several kinds, vehicle running on the bridge passing the bridge tower, vehicles running on bridge passing each other, vehicles passing out of the tunnel. The wind-vehicle-bridge system are investigated mostly with the numerical simulation method.

The study model of aerodynamic characteristics of vehicle mainly taking vehicle stationary on bridge. This article does the research on the change rule of aerodynamic characteristics using numerical method of CFD technique to simulate the process of vehicles passing each other under the lateral wind action. In the meantime, the moving grid has been applied and the three-dimensional dynamic numerical wind tunnel model has been established.

Content from this work may be used under the terms of the Creative Commons Attribution 3.0 licence. Any further distribution of this work must maintain attribution to the author(s) and the title of the work, journal citation and DOI.
Published under licence by IOP Publishing Ltd

2. Calculation model

2.1. Model of vehicle and bridge

Study the change of aerodynamic characteristics of vehicles under the lateral wind action in the process of passing each other, select the bus as research object model shown in figure 1 and select the steel box girder as research object model shown in figure 2. The influence of ancillary facilities such as railing on bridge is not considered in the numerical model and the model dimensions are modeled in 1:1 in the numerical modeling.

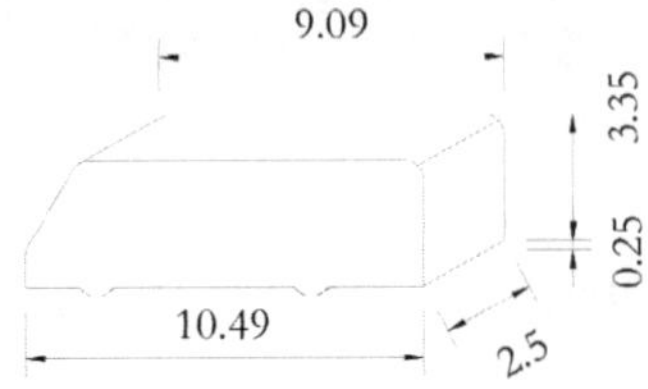

Fig.1 Vehicle model (Unit: m)

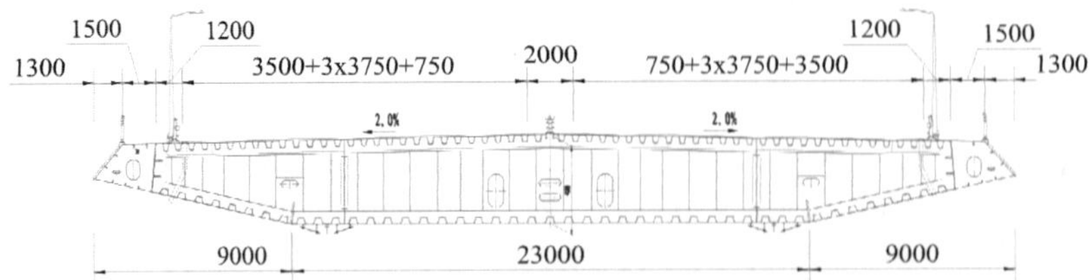

Fig.2 Cross section of bridge (Unit: mm)

Established numerical model for numerical simulating the process of vehicle passing each other. The outer surface of the two vehicles is 3.5 meters in the numerical model. In the numerical tunnel model, the windward boundary condition is velocity inlet, the leeward boundary condition is pressure outlet, both rest sides conditions are symmetry. Determine the size of the numerical model for fluid computation with the considering of clogging effect to make the windward area of model and numerical wind tunnel cross section area ratio is less than 5%. A local mesh encryption method is presented on the fluid mesh to make the wall y+ function value ($y^+ = y\sqrt{\rho\tau_w}/\mu$) between 30~300 at the vehicle and bridge boundaries. Where y is the distance from the unit center to the wall, μ is aerodynamic viscosity, ρ is air density, τw is wall shear stress. The tetrahedral prism is adapted for numerical analysis model of fluid flow and use a varied grade of cell to make the mesh could control the changes of elements size around the vehicle and bridge areas.

2.2. Aerodynamic coefficients

Vehicle under the wind load will be acted on by aerodynamic six-component force, which includes resistance of F_z, lateral force F_x, vertical lift F_y, overturning moment M_z, torque moment M_y and pitching moment Mx. If the vehicle travels speed is V_t and acted on by the lateral wind whose speed is V_w then the aerodynamic three component coefficient can be defined as Eq.1.

$$C_x = \frac{F_x}{\frac{1}{2}\rho V^2 HL} \qquad C_y = \frac{F_y}{\frac{1}{2}\rho V^2 BL} \qquad C_{Mz} = \frac{M_z}{\frac{1}{2}\rho V^2 B^2 L} \tag{1}$$

In Eq 1, C_x, C_y, and C_{Mz} respective is lateral coefficient, lift coefficient and overturning moment coefficient, besides that ρ is the air density and *V* is the synthesis of positive wind generated by vehicle traveling and the natural wind. Moreover, H, B, L respectively represents the height, width and length of vehicle.

3. Wind pressure characteristics of vehicle without lateral wind action

Firstly, study the change rule of aerodynamic characteristics of vehicle while vehicle passing each other at 80km/h speed. The change rule of aerodynamic force coefficients is shown in figure 3 while vehicle passing each other.

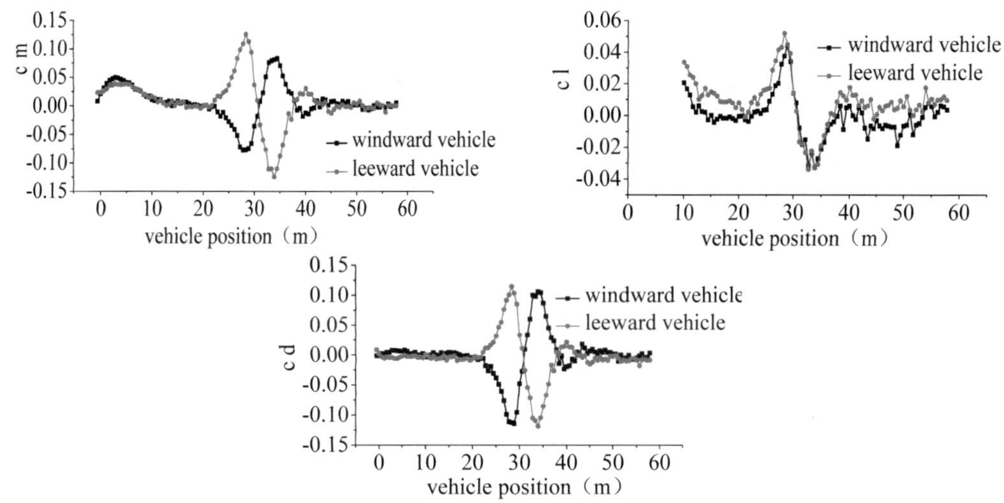

a) Overturning moment coefficients; b) lift force coefficients; c) lateral force coefficients.

Fig 3. Aerodynamic coefficients of vehicle without lateral wind action

The overturning moment coefficients and lateral force coefficients of vehicles both come to about 0 when the vehicle before and after passing each other without lateral wind effect. That is because vehicle only subjects the positive wind loads due to vehicle driving when vehicle running on bridge without lateral wind action. Then, the positive surface pressure on vehicle body is larger and the lateral surface pressure on vehicle body is smaller. So, it is accord with the actual that the overturning moment coefficients and lateral force coefficients come to about 0 and it also demonstrate the correct feasibility of numerical simulation in this paper.

The aerodynamic coefficients of windward and leeward vehicles both have a larger mutation in the process of vehicle passing each other without lateral wind action. The change rule of overturning moment coefficients and lateral force coefficients is identical for windward vehicle, which start from zero to decrease and increase to final stable. The overturning moment coefficient and lateral force coefficients of vehicle decrease to negative. That phenomenon can show that the thrust force generates on the windward vehicle deviate from the leeward vehicle because that the air surrounding vehicles met due to the vehicle passing each other. The overturning moment coefficients and lateral force coefficients of vehicle increase to positive. That phenomenon can show that pull force generates on the windward vehicle close to leeward vehicle due to the space pressure becoming smaller between vehicles. The aerodynamic force coefficients come to about zero after the vehicles passing each other. For the windward and leeward vehicles, it is in contrast for the change rule of overturning moment coefficient and lateral force coefficients but it is same for the change rule of lift force coefficients which increase to positive and decrease to negative stable in the process of vehicle passing each other.

4. The influence of wind speed and driving speed on aerodynamic characteristics

Establish the analysis model of vehicle driving respectively at 80km/h and 100km/h speed under the lateral 8m/s or 15m/s wind action to study the change rule of aerodynamic fore coefficients while vehicle passing each other. The temporal change rule of aerodynamic fore coefficients of windward and leeward vehicle are shown in figure 5~6 while vehicle passing each other.

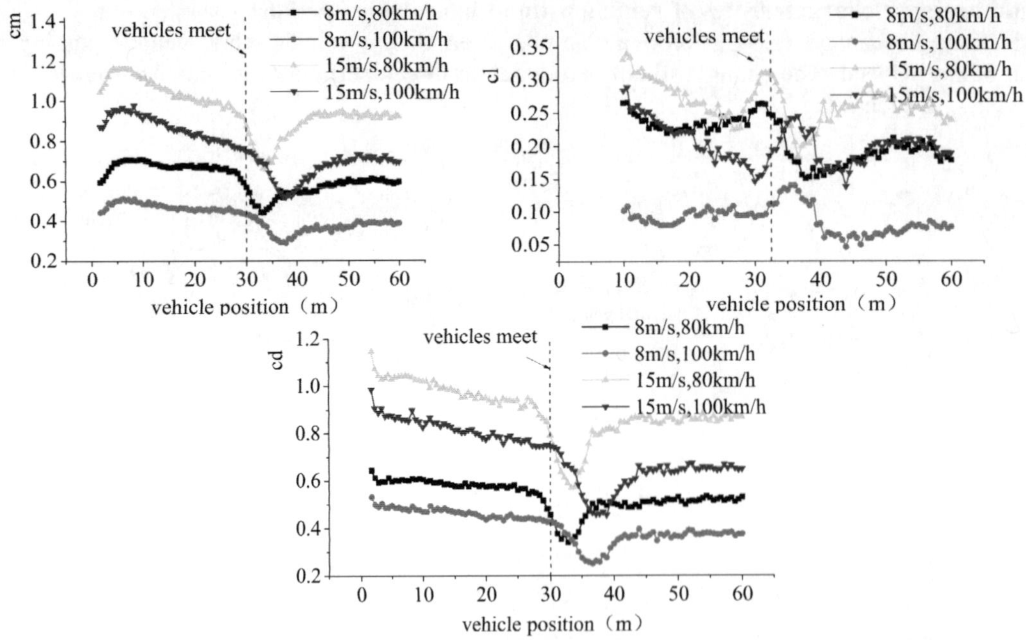

a) Overturn moment coefficient b) lift force coefficients c) lateral force coefficient

Fig 4. Aerodynamic coefficients of windward vehicle

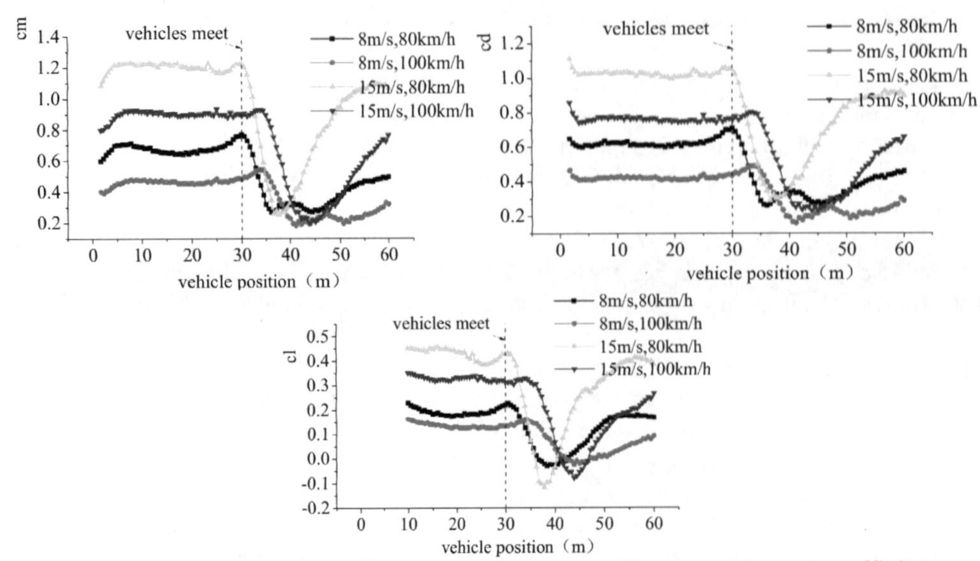

a) Overturning moment coefficients b) lift force coefficients c) lateral coefficients

Fig.5 Aerodynamic force coefficients of leeward vehicle

The leeward vehicle has certain influence on the aerodynamic coefficients of windward vehicle in the process of vehicle passing each other. The overturning moment coefficient and lateral force coefficient of windward vehicle both decrease to a certain degree in the process of vehicle passing each other, but begin to return to normal after the process of vehicle passing each other. The aerodynamic force coefficients have begun to change before the process of vehicle passing each other. The overturning moment coefficients, lateral force coefficients and lift force coefficients of windward vehicle both decrease with increasing vehicle driving speed and both increasing with increasing lateral wind speed.

The windward vehicle has certain influence on the aerodynamic coefficients of leeward vehicle in the process of vehicle passing each other. The overturning moment coefficient and lateral force coefficient of leeward vehicle both have a great change, but begin to return to normal after the process of vehicle passing each other. The overturning moment coefficients, lateral force coefficients and lift force coefficients of leeward vehicle both decrease with increasing vehicle driving speed and both increasing with increasing lateral wind speed.

5. Conclusion

It has large mutation for the aerodynamic force coefficients of the windward vehicle and leeward vehicle. The change region of aerodynamic force coefficients of leeward vehicle is larger than that of windward vehicle due to the windward vehicle blocking when the vehicles pass each other. The overturn moment coefficient, lift force coefficient, lateral fore coefficient both reduce with increasing driving speed and both increase with increasing lateral wind speed.

Acknowledgments

The authors gratefully acknowledge the financial supported by the Jilin province department of education science and technology research projects (JJKH20180305KJ), the Jilin province 13th five-year plan important project of educational science: Practical training course construction of civil engineering specialty in applied undergraduate colleges (ZD17122). This work is partially supported by the Chang'an University.

References

[1] Xu Xin-yu, Li Yong-le, CHEN Ke-yu, et al. Study of Wind-Train-bridge Coupling Vibration of Multi-Track Railway Bridge Under Meeting of Two Trains [J].Bridge Construction, 2017, 47 (1): 41-46.

[2] Hu Bao. Numerical Simulation Method and Application of Metro Vehicle-Tunnel Dynamic Nonlinear Coupled System [D]. Shanghai Jiao Tong University, 2015.

[3] Du Feng-yu. Experimental Investigation on Aerodynamic Characteristics of Train-Bridge System and Influence of Wind-Break Wall under Strong Wind [D]. Central South University, 2014.

[4] HAN Yan, Hu Jie-xuan, CAI Chun-sheng, et al. Wind tunnel measurement on aerodynamic characteristics of vehicle and bridge under cross winds [J]. Journal of Vibration Engineering, 2014, 27 (1): 67-63.

[5] Guo Weiwei, Xia He, Zhang Tian.Analysis on aerodynamic effects of bridge wind barrier and its influence on running sagety of a high-speed train [J]. Engineering Mechanics, 2015, 32(8): 112-119.

[6] CHENG Jing-wen, GAO Guang-jun, ZHU Chun-li. Detached-eddy simulation of flow around high-speed train on a bridge under cross winds [J]. Journal of Southeast University (English Edition), 2016, 23: 2735-2746.

[7] Yang Liu, Xinfing Yin, Liu Deng, et al. Ride comfort of the bridge-traffic-wind coupled system considering bridge surface deterioration [J]. Wind and Structures, 2016, 23 (1): 19-43.

[8] Se-Jin Kim, Chul-Hwan Yoo, Ho-Kyung Kim. Vulnerability assessment for the hazards of crosswinds when vehicles cross a bridge deck [J]. Journal of Wind Engineering and Industrial Aerodynamics, 2016, 156, 62-71.

IOP Conf. Series: Journal of Physics: Conf. Series **1064** (2018) 012020 doi:10.1088/1742-6596/1064/1/012020

Numerical Study on a 2-D Simplified Missile Separation from High Speed Aircrafts

Yunpeng Qin*, Xiaohui Qiao, Tianlai Gu, Geling Yin, Yang Yang and Yong Zhang

R&D Center, China Academy of Launch Vehicle Technology, Beijing, China

*Corresponding author e-mail: xuanwofenli@126.com

Abstract. In this paper, 2D numerical simulations are carried out on the symmetric plane of a full size missile separation from high speed aircrafts at the conditions of Ma=5 and AOA=0°. Aerodynamics and typical flow field characteristics of the missile at different distances away from the cavity bottom are predicted via a steady RANS solver. Results show that the flow of the missile separation from high speed aircrafts with an embedded weapon bay can be divided into four stages, which are dominated separately by the flow around the missile in the deep cavity, compressed incoming flow by the lower surface of the missile, strong interacting flow between the shear flow and the upper surface of the missile and wall effect flow of the missile. For Stage I, the flow field mode of the cavity is determined by the effective length depth ratio of the cavity, however, the effective length depth ratio is decreasing with the fall of the missile. For stage II, the incoming flow will be firstly compressed by the lower surface of the missile since it's protruding from the embedded weapon bay. After that, i.e. on the stage III, when the axis line of the missile is beyond the surface of the aircraft, the strong shear layer induced by the incoming high speed flow and the non-high speed flow in the cavity will act on the lower incoming flow, which leads to the shock wave interaction at the missile nose. Simultaneously, the strong shear layer will significantly increase the pressure on the upper surface of the missile. For stage IV, i.e. the embedded weapon bay is closed, the reflection times of the shock waves between the missile and the aircraft decrease with the increasing distance between the missile and the aircraft. In addition, the pressure on the upper surface of the missile decreases gradually. The large negative lift of the missile formed due to the extreme wall effect is beneficial to the missile separation from high speed aircrafts in some degree.

1. Introduction

With the breakthrough and maturity of key technologies of scramjet engines and combined cycle engines (RBCC, TBCC, SABRE, PATR, and ATREX), the shape and outline of future fighters have gradually seen the world. Arguably, high speed cruise will be a fundamental characteristic for future fighters. The high speed unmanned reconnaissance aircraft SR-72 proposed by Lockheed Martin may become the prototype of future fighters.

For high speed fighters, if the external weapon store of traditional subsonic fighters is adopted, the drag and Radar Cross Section (RCS) of the fighters [1] will greatly increase. More importantly, the

Content from this work may be used under the terms of the Creative Commons Attribution 3.0 licence. Any further distribution of this work must maintain attribution to the author(s) and the title of the work, journal citation and DOI.
Published under licence by IOP Publishing Ltd

thermal environment of airborne weapons, pylons and fighters will definitely be deteriorated, which may lead to the outcome like the pylon structural failure of X-15 (It is probably caused by the type IV shock wave interaction [2]). Obviously, the application of external weapons store to high speed fighter meets great challenges, and the problems to be solved are also very complicated. As for the embedded weapon bay adopted by the current 4th generation fighters, such as F-22, F-35, T50, J20 and other supersonic fighters, further thorough research and adaptable transformations are inevitably required to meet the need of high speed aircraft aerodynamic/thermal and acoustic environments, unless the embedded weapon bay under the aircraft is not adopted.

Recently, external and embedded delivery methods for airborne weapons have been widely studied all around the world. The related researches cover the aerodynamic characteristics of airborne weapons, separation safety and its control [3-6], aerodynamic acoustic and its control [7-11], suspension and delivery mode, airborne weapon type spectrum, and so on.

In this paper, the researches of airborne weapon separation from aircrafts will be summarized mainly in the point of view of aerodynamic characteristics, and emphasis will be placed on the form of embedded delivery methods. It's known that the flow characteristics of the embedded weapon bay, i.e. the cavity flow modes are influenced by the length depth ratio, the breadth depth ratio, the wedge angle of bulkheads, Mach number, AOA and so on. Among them, the length depth ratio and Mach number have been considered as the key factors [1, 3, and 12]. Q F Zhang et al. [13] studied the differences between external and embedded delivery methods at subsonic and supersonic speeds. It's found that the missile separation from the embedded weapon bay is worse than the other one, and it's recommended to increase the nose-down pitching angular velocity and make the missile pass through the shear layer of the bay with a nose-down profile, which will reduce the negative effects of the shear layer and improve the missile separation safety. S T Zhu et al. [12] studied the missile separation from aircrafts with the embedded weapon bay by six-degree of freedom simulations. It's shown that the safety of the separation decreases with the increase of Ma in the low supersonic range, and that the missile may stay with no falling down. So they suggested to increase the ejection force and pitching moment to avoid such disadvantages. C Chang and H H Ding [14] exhibits the typical flow field classifications for the separation of embedded delivery methods. Three typical stages are 1) the missile inside the cavity, 2) the missile crossing the shear layer and 3) the missile passed the shear layer, which are classified mainly by the separation process, but not classified precisely by flow characteristics. W Wang [15] pointed out Mach number influences the cavity flow modes. With Ma increasing, the transitional flow mode and the closed flow mode will turn to be the open flow mode and the transitional open flow mode. Y Liu [16] showed the flow fields around the door of the embedded weapon bay with different opening angles, and depicted the flow fields of AIM-120C released from F-22 at different times. Generally, the existed researches focus on the flow field, the aerodynamic characteristics and the trajectories of the missile during the separation. Detailed information of the flow interacting between the cavity and the missile is rare. And the researches mentioned above are all about separation at subsonic, transonic and supersonic speed, the researches on the separation at high speed are also rare.

Based on the similarity of the characteristics of supersonic and high speed flows, it can be deduced that the separation flow field for the embedded delivery methods at high speeds is similar to that at supersonic speeds. However, there also be evident differences and it's still needs to be studied in detail. For instance, the information of shock wave interactions and shock-boundary layer interactions, which have negligible influence on the missile thermal environment at subsonic, transonic and supersonic speeds, need further research at high speeds; the complicated thermal environment and acoustic environment induced by the complicated flow around the missile, cavity and aircraft (flow mentioned above and the shock/eddy interactions, etc.) at high speeds are different from those at supersonic speeds. There will be a long list of the research on high-speed separation. And these mentioned above are all key research contents in the future.

In this paper, 2D numerical simulations are carried out on the symmetric plane of a full size missile separation from high speed aircrafts at the conditions of Ma=5 and AOA=0. The main purpose of this paper is to find the main flow characteristic during the missile separation from thehigh speed aircraft with an embedded weapon bay. Besides, it's still noted that the missile separation is substantially simplified into a 2-D model and many factors are not considered in this paper. Still, the author believes that it's meaningful and valuable for such a 2-D simulation. Further 3-D simulations will be carried out later.

2. Problem Description and Numerical Method

2.1. Problem Description

In this paper, the missile separation from the high speed aircraft with an embedded weapon bay is divided into two parts, delivery before and after the bay closure. In the specific designing process, the open door are folded inward the weapon bay to reduce the severity of the thermal environment and improve the feasibility of thermal protection design, which is different from that of F - 22 , J20, etc. In addition, the door will be closed once the missile leaves the cavity. The separation safety is mainly determined by the parameters of delivery process. Accordingly, 3D effect and the viscous effect of the aircraft are neglected from the perspective of simplification, 2D numerical simulations will be carried out. According to the state of the door, the specific numerical models can be divided into the open (O) model and the closed (C) model, which are shown in Fig. 1 (a) and (b), respectively.

The length of the cavity is L_C=5.0m, the depth D=1.0m. The length of the missile is L_D=4.0m, the maximum diameter Φ =0.4m, and the nose radius R=0.01m. AOA = 0°in all simulations, no varying AOAs and attitude angles are considered in this paper. The distance between the missile and the bottom of the bay and that between the missile and the aircraft are denoted by H_O and H_C, respectively.

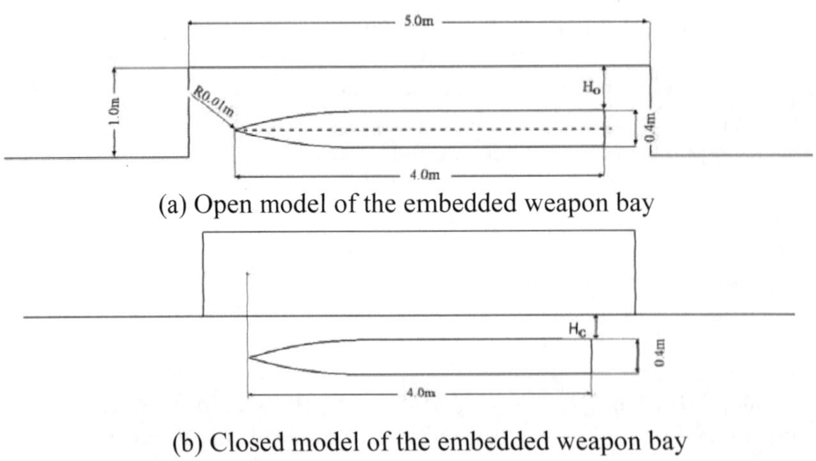

(a) Open model of the embedded weapon bay

(b) Closed model of the embedded weapon bay

Figure 1. 2-D numerical models of the missile separation from high speed aircrafts.

2.2. Computational Domain, Boundary Conditions and Grid Topology

An H-O type computational domain around the missile and an H type computational domain with multi-block structured grids are generated by the ANSYS ICEM. The grids are refined above the missile, in the missile wake, in the cavity and in the shear flow region. The non-dimensional normal distance of the first grid layer is about y^+~1. There are about 150,000 grids for the case of the open model, and 100,000 grids for the case of the closed model. The dimensions of the computational domains for the two cases are basically the same (Fig. 2 (a)), the grid topology are shown in Fig. 2 (b).

The boundary conditions are as follows: The left, right and bottom side of the domain use the pressure far-field boundary conditions. The upper side and the cavity wall adopt slipping adiabatic wall boundary conditions. (According to the principle of relative motion, the surface velocity of the aircraft should be zero. If non-slipping wall boundary condition is adopted, a very thick boundary layer will arise above the wall before the cavity, which is inconsistent with the real condition. Further study will investigate on the impact of the aircraft fore-body boundary layer on the missile separation.) The surface of the missile uses the non-slip adiabatic wall.

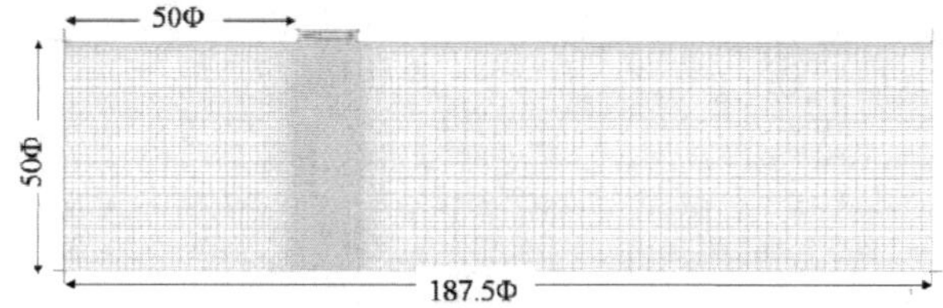

(a) Computational domain

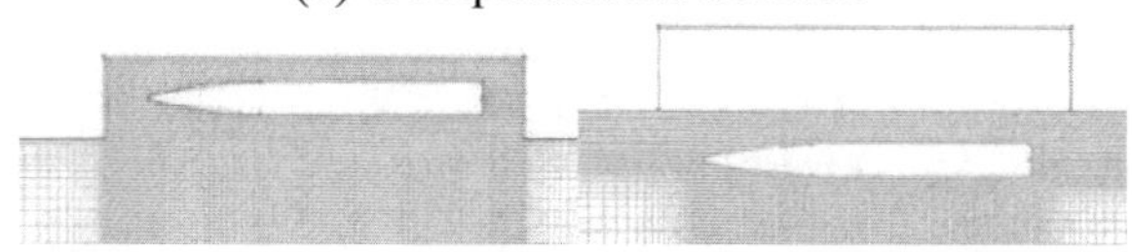

(b) Grid topologies for the two models

Figure 2. Computational domain and grid topology around the missile.

2.3. Numerical Method

ANSYS FLUENT 16.0 is used to carry out the numerical simulations with a density based solver. The steady RANS equations grounded on κ-ε turbulence model are solved under the assumption of ideal gas. The convection terms are discretized by Roe-FDS scheme, the viscosity coefficient is calculated via Sutherland Law, and the standard wall function is adopted. Specific computational conditions are as follows: Scale Ratio (SR) =1:1, Ma=5, AOA=0°, H_O=0.25Φ, 0.75Φ, 1.5Φ, 2.0Φ, 2.25Φ, 2.5Φ, 2.75Φ, H_C=0.25Φ (door closes), 0.5Φ, 1.0Φ, P_∞ =5529.31Pa, T_∞ =216.65. The dimensionless reference length is 4m.

3. Results and Discussion

3.1. Aerodynamic Characteristics of the Missile

Fig. 3 shows the lift and drag coefficients of the lower surface, upper surface and the whole missile at different H_O and H_C. (The lower surface and the upper surface are divided by the axis of the missile).

As shown in Fig. 3 (a), the total lift of the missile increases monotonously when $H_O \leq 2.0Φ$. When $H_O \leq 1.0Φ$, the positive lift of the lower surface decreases monotonously, while the negative lift (the absolute value in this paper if not mentioned specially) of the upper surface also decreases monotonously; As a result, the little negative total lift gets close to zero gradually. When $1.0Φ \leq H_O \leq 2.0Φ$, the positive lift of the lower surface increases significantly, while the positive lift of the upper surface is near zero and increases slightly; As a result, the missile sees a monotonously increasing positive lift. When $2.0Φ \leq H_O \leq 2.75Φ$, the positive lift of the lower surface increases slightly at first and then increases rapidly, while the negative lift of the upper surface increases monotonously and significantly at first and then decreases rapidly; The total lift shows a similar trend with the upper surface. It can be predicted that if the weapon bay is not closed, the lift of the upper and lower surface will soon offset each other, and the total lift will be zero. When H_O=2.75Φ (i.e. H_C=0.25Φ), the weapon bay is ordered to be closed in this paper, the negative lift of the upper surface and that of the whole missile will reach a maximum value, while the positive lift of the lower surface will decrease abruptly. With the falling of the missile, the lift of the lower surface stays almost the same, while the negative lift of the upper surface and the whole missile show a greatly monotonic decrease. It's evident that the lift increment of the whole missile is almost from the upper surface.

As shown in Fig. 3 (b), the total drag coefficient increases monotonously when $H_O \leq 2.0Φ$. When $H_O \leq 1.0Φ$, the thrust of the lower surface, the upper surface and the whole missile all decrease gradually to zero. When $1.0Φ \leq H_O \leq 2.0Φ$, the drag of the lower surface increase gradually, while that of the upper surface stays almost the same. Thus, the total drag increases gradually. When

$2.0\Phi \leq H_O \leq 2.75\Phi$, the drag of the lower and upper surface both increase at first and then decrease, and the total drag reaches the minimum at $H_O=2.75\Phi$. Similarly, it can be predicted that the drag of the lower and upper surface will continue to increase if the weapon bay is not closed. At $H_C=0.25\Phi$ with the bay closed, the drag of the lower and upper surface both increase abruptly and the total drag reaches its maximum. With the falling of the missile, the drag of the lower surface slightly increases, while the drag of the upper surface decreases significantly; finally, the total drag decreases rapidly and achieves stability.

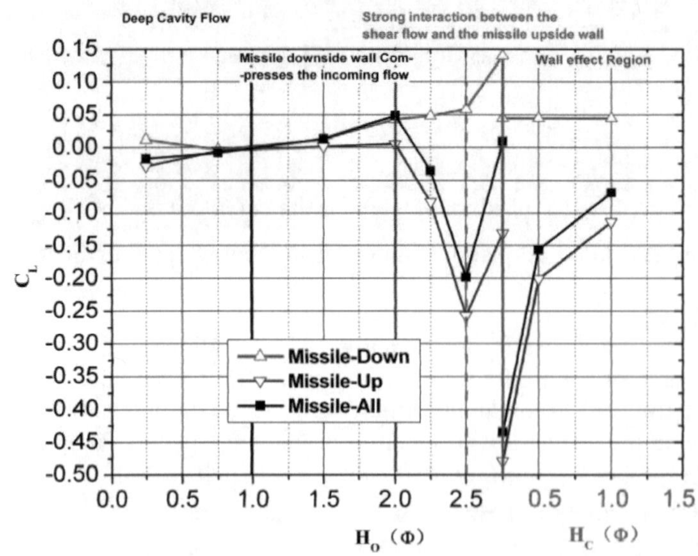

(a) Lift coefficients (Down denotes to the lower surface of missile, up the upper surface of the missile)

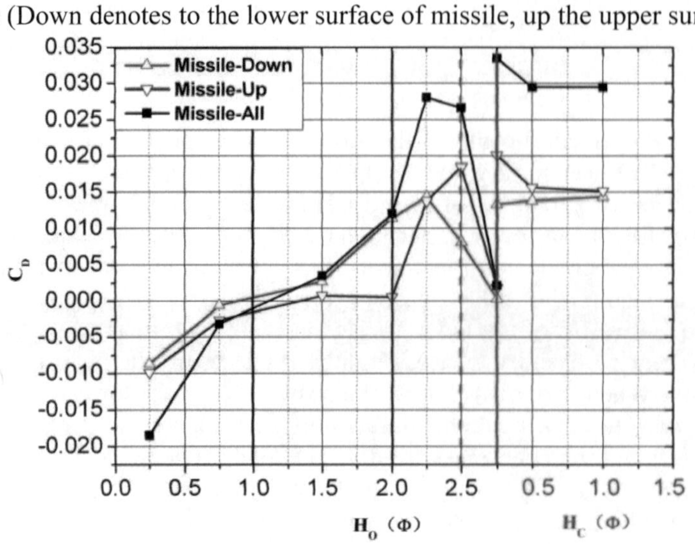

(b) Drag coefficients

Figure 3. Lift and drag coefficients of the lower surface, upper surface and the whole missile.

According to the aerodynamic characteristics of the missile and the flow fields discussed in the following sections, the flow of the missile separation from high speed aircrafts with an embedded weapon bay can be divided into four stages, which are dominated separately by the flow around the missile in the deep cavity, compressed incoming flow by the lower surface of the missile, strong

IOP Conf. Series: Journal of Physics: Conf. Series **1064** (2018) 012020 doi:10.1088/1742-6596/1064/1/012020

interacting flow between the shear flow and the upper surface of the missile and wall effect flow of the missile, as shown in Fig. 3 (a).

3.2. Flow around the Missile in the Deep Cavity

The length depth ratio of the cavity in this paper is $L_C/D=5$, the flow of the cavity without missile is of a typical open flow mode, as shown in Fig. 4. The incoming flow firstly goes through a wide weak expansion area in the middle and posterior area below the cavity, and then a local shock wave and weak expansion wake in the rear area below the cavity. In the cavity, there is a large separation circulation. Based on the flow structure mentioned above, the open flow mode of the cavity is beneficial to thermal protection structure design under high speed conditions.

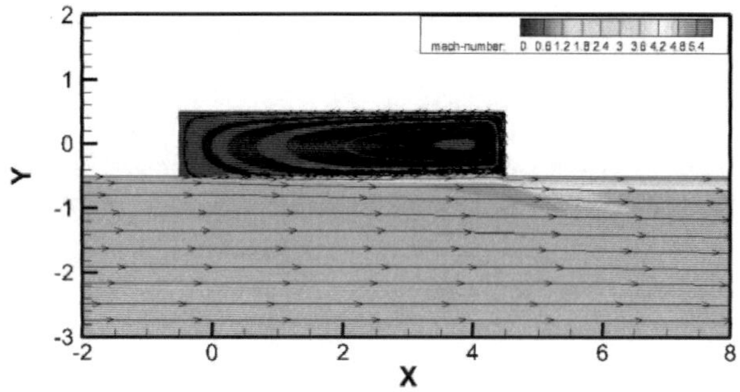

Figure 4. Flow field of the cavity at $H_O=0.25\Phi$.

Fig. 5 and Fig. 6 show the pressure coefficient contour & streamlines and Ma contour of the flow field at $H_O=0.25$ and 0.75, respectively. Fig. 7 presents the pressure coefficient curves of the upper and lower surfaces of the missile. In Fig. 5, there is a strong expansion area at the front of the cavity, while there are shock waves and expansion waves in the middle and posterior area below the cavity. There is a typical circulation area in the cavity, and the velocity magnitude is relatively small in the circulation area. An obvious separation bubble generates below the rear of the missile due to the high pressure at the rear of the cavity and the strong shear induced by the external high speed flow. Simultaneously, a smaller separation bubble is forced to generate below the missile nose. From the back to the forward, the flow between the upper surface of the missile and the wall of the cavity is similar to the flow of the nozzles. Despite the flow around the missile nose is complicated, the thermal environment around the nose is not that severe since the Ma is relatively low in this region. In general, the pressure on the upper surface is larger than that on the lower surface, thus, the missile lift is negative. Owing to the high pressure at the rear of the cavity, the missile gains thrust. Since the missile is very close to upper wall of the cavity, the effective length depth ratio of the cavity greatly increases, and the flow is characterized as the transitional open flow mode[16], which can be seen from the monotonously increasing pressure on the lower surface of the missile at $H_O=0.25\Phi$ in Fig. 7.

In Fig. 6, there are weak compression and expansion areas at the front below the cavity. The weak expansion area covers the middle area below the cavity, at the same time, there are relatively strong compression and expansion area at the rear below the cavity. The circulation area in the cavity is relatively simple, the flow between the upper surface of the missile and the wall of the cavity is similar to that at $H_O=0.25\Phi$; Below the missile, the flow velocity magnitude gradually increases front to back due to the ejection effect of the external high speed flow, and the pressure at the rear of the missile decreases significantly when compared to that at $H_O=0.25\Phi$. According to the Fig. 6 and Fig. 7, because of the relatively low pressure at the rear of the cavity, the total drag coefficient of the missile increases gradually from negative to positive. The circulation area of the in the cavity is such smooth due to the sufficient space between the missile and upper wall of the cavity, the flow mode of the cavity with the missile at $H_O=0.75\Phi$ is similar to that in Fig. 4, only slight differences exist at the front below the cavity.

Based on the flow fields, flow characteristics around the missile in the deep cavity when $H_O \leq 1.0\Phi$ can be summarized as follows: When the missile is very close to the upper wall of the cavity, the effective length depth ratio of the cavity increases, and it develops from the open flow mode to the transitional open flow mode; the missile gains thrust due to the high pressure area at the rear of the cavity. When the missile is in the middle of the cavity, the impact of the presence of the missile on the open flow mode is weak, and the missile lift and drag are close to zero.

On the aspect of thermal protection, there is no severe problem in the cavity when AOA=0°. However, when the incoming flow comes into the cavity when AOA≠0°, the middle and the rear of the missile, the rear of the cavity are all the key areas that should be considered. This problem will be studied later.

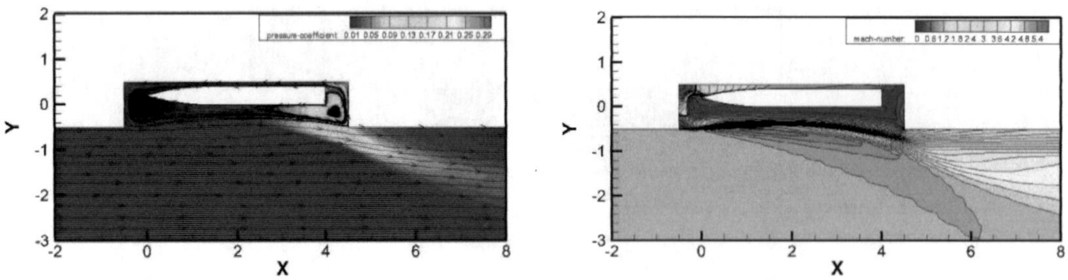

(a) Pressure coefficient contour and streamlines (b) Ma contour

Figure 5. Flow around the missile in the deep cavity at $H_O=0.25\Phi$.

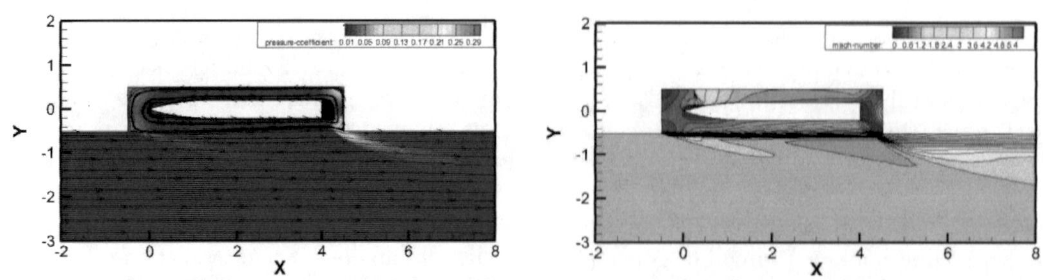

(a) Pressure coefficient contour and streamlines (b) Ma contour

Figure 6. Flow around the missile in the deep cavity at $H_O=0.75\Phi$.

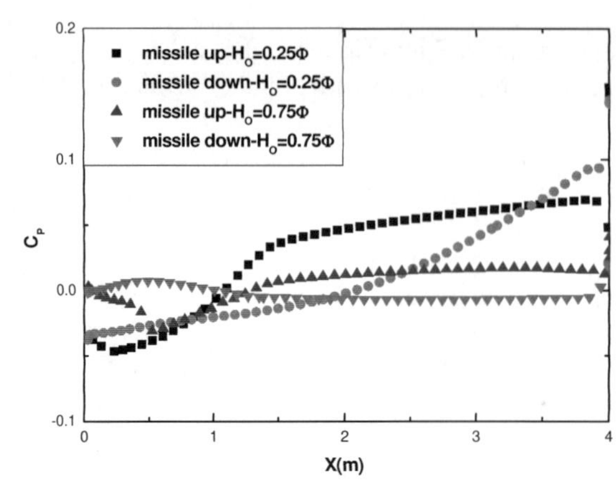

Figure 7. Pressure coefficients of the upper and lower surface of the missile at $H_O=0.25\Phi$ and $H_O=0.75\Phi$.

IOP Conf. Series: Journal of Physics: Conf. Series **1064** (2018) 012020 doi:10.1088/1742-6596/1064/1/012020

3.3. Compressed Incoming Flow by the Lower Surface of the Missile

Fig. 8 and Fig. 9 show the pressure coefficient contour & streamlines and Ma contour of the flow field at H_O=1.5 and 2.0, respectively. Fig. 10 presents the pressure coefficient curves of the upper and lower surfaces of the missile. With the falling of the missile, the interaction between the lower surface of the missile and the shear flow at the front below the cavity increases gradually. At H_O=1.5Φ, the incoming flow is slightly deflected inward the cavity region, thus there is a narrow and weak expansion area at the front edge of the cavity. But, the change of the incoming flow is negligible. With the flow developing downward, a weak shock wave generates at the shoulder of the missile nose, and a wide expansion area arises after the arc shoulder. At the rear below the cavity, the flow changes little. The compression effect on the incoming flow due to the lower surface of the missile just appears at H_O=1.5Φ. As for the flow in the cavity, since the lower surface of the missile is right parallel to the surface of the aircraft, the cavity is considered to be closed. A relatively large separation bubble generate above the rear of the missile, and the circulation flow around the missile is still preserved. The pressure increase at X=0.5m on the lower surface of the missile is due to the pressure increase behind the shock wave.

At H_O=2.0Φ, the incoming flow passing the front edge of the cavity is also slightly deflected inward and goes through an expansion area. The deflected flow passes through the lower and upper surfaces of the missile and forms a stagnation point at the nose of the missile. The flow below the lower surface of the missile experiences a strong compression due to the cambered surface at the nose and an expansion when passing the shoulder. At the same time, the kinetic energy of the flow above the upper surface of the missile decreases rapidly due to the recirculation, and the flow is expelled from the rear of the missile.

Based on the flow mentioned above, the lift increment from H_O=1.5Φ to H_O=2.0Φ is mainly due to increasing compression strength and region of the flow below the nose of the missile, which can also be reflected in the partial enlarged picture in Fig. 10.

In this region, the flow covers the missile changes from the recirculation flow in the cavity to the incoming flow. It can be speculated that the thermal environment of the missile is deteriorating rapidly, especially that of the lower cambered surface of the missile.

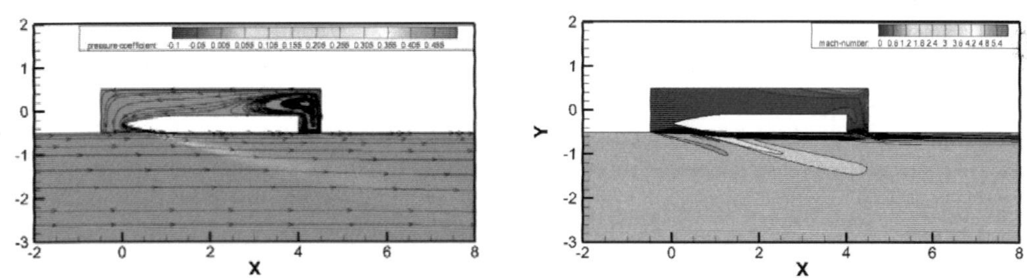

(a) Pressure coefficient contour and streamlines (b) Ma contour

Figure 8. Flow around the missile at H_O=1.5Φ.

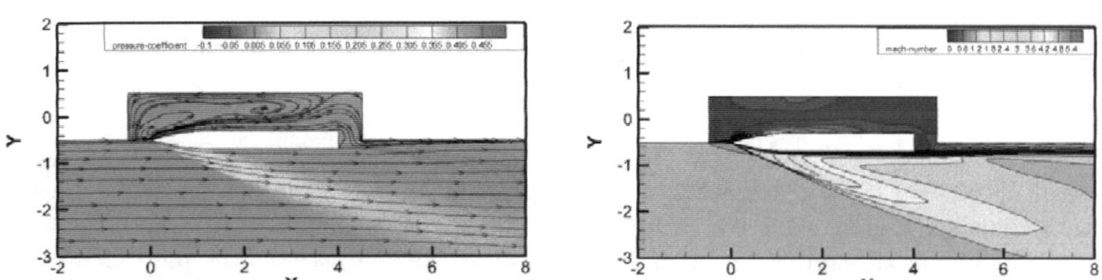

(a) Pressure coefficient contour and streamlines (b) Ma contour

Figure 9. Flow around the missile at HO=2.0Φ.

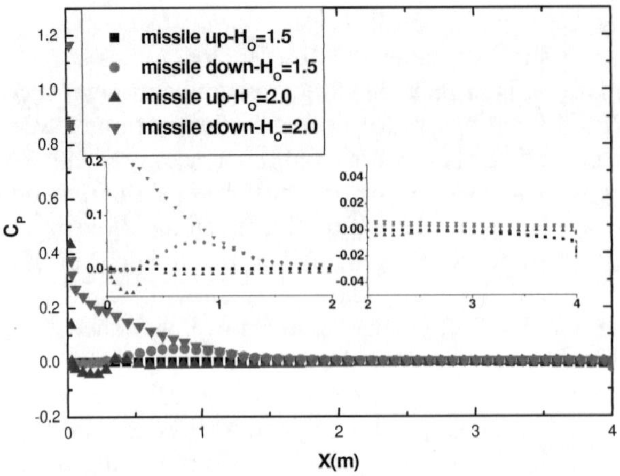

Figure 10. Pressure coefficients of the upper and bottom surface of the missile at Ho=1.5Φ and Ho=2.0Φ.

3.4. Strong Interacting Flow between the Shear Flow and the Upper Surface of the Missile

Fig. 11, Fig.12 and Fig. 13 present the pressure coefficient contour & streamlines and Ma contour of the flow field at Ho=2.25Φ, 2.5Φ and 2.75Φ, respectively. Fig. 14 presents the Ma contour & streamlines around the missile nose at Ho=2.0Φ, 2.25Φ, 2.5Φ and 2.75Φ. Fig. 15 presents the pressure coefficient curves of the upper and lower surfaces of the missile. As shown in the contours, the flow field below the missile changes little before Ho=2.5Φ, While the flow field above the missile changes significantly. At Ho=2.75Φ, both of them change greatly. It's noted that the change of the flow field above the upper surface is closely related to the shear layer formed at the front edge of the cavity. Besides, the complicated flow phenomena such as shock wave interaction at the nose of the missile should also be concerned.

As shown in Fig.14, with the falling of the missile from Ho=2.0Φ to Ho=2.75Φ (before the weapon bay closed), the shear layer formed at the front edge of the cavity gradually changes from a slight inward deflection to an outward deflection. The shear layer is like a protruding wedge. The oblique shock wave hits on the bow shock wave right before the missile nose, as a result, the shock wave interaction happens. The type of the shock wave interaction is related to the positon of the missile. It's deduced from Fig. 14 that there should be six types of shock wave interaction around the missile nose from Ho=2.25Φ to Ho=2.75Φ (The position interval in this paper is relatively large). Since the shock wave interaction has a great impact on the thermal environment of the missile nose, it needs further study and the type IV shock wave interaction requires special attention.

During the falling of the missile, the strong shear flow formed at the front edge of the cavity right passes over the upper surface of the missile. The complicated circulation in the cavity will interact with the strong shear flow, leading to a more complicated flow channel between the upper surface of the missile and the circulation. In Fig. 15, the pressure on the upper surface of the missile increases gradually from Ho=2.0Φ to Ho=2.5Φ, which is consistent with the increase of the negative lift on the upper surface. At Ho=2.75Φ, the strong shear flow over the missile cannot withstand the adverse pressure gradient at the shoulder of the missile, which leads to a relatively large separation region after the shoulder of the missile. The separation shock wave generates at the front of the separation, thus the high pressure region appears at the middle and rear of the missile. However, the pressure is much lower than that at Ho=2.5Φ. Meanwhile, the complicated flow field downstream the missile also induces a separation region at the rear below the missile, and, the shock wave before the separation and the resulting high pressure region after the shock wave generate. Therefore, the lift of the missile is just slightly larger than zero.

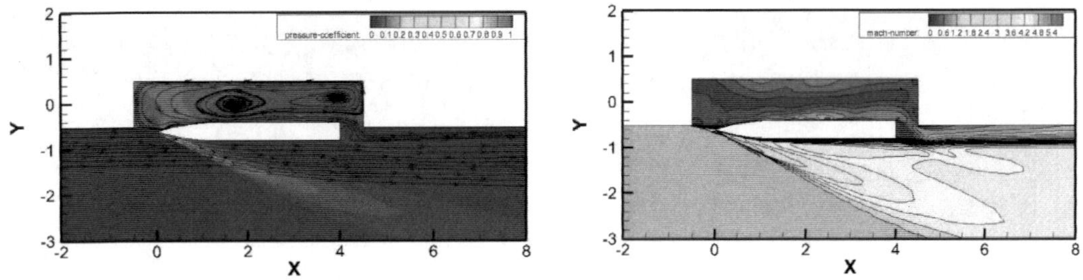

(a) Pressure coefficient contour and streamlines (b) Ma contour

Figure 11. Flow around the missile at H_O=2.25Φ.

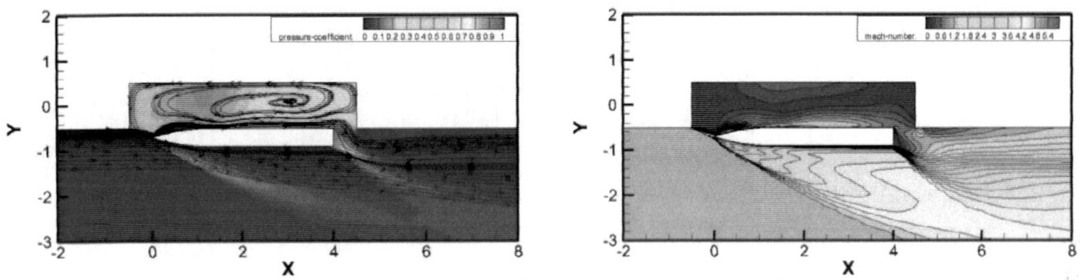

(a) Pressure coefficient contour and streamlines (b) Ma contour

Figure 12. Flow around the missile at H_O=2.5Φ.

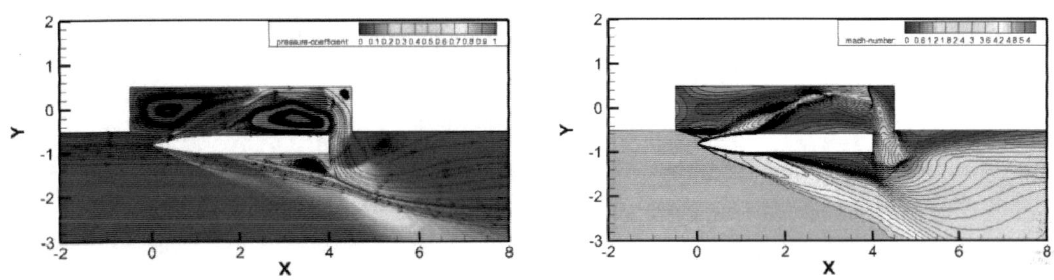

(a) Pressure coefficient contour and streamlines (b) Ma contour

Figure 13. Flow around the missile at H_O=2.75Φ.

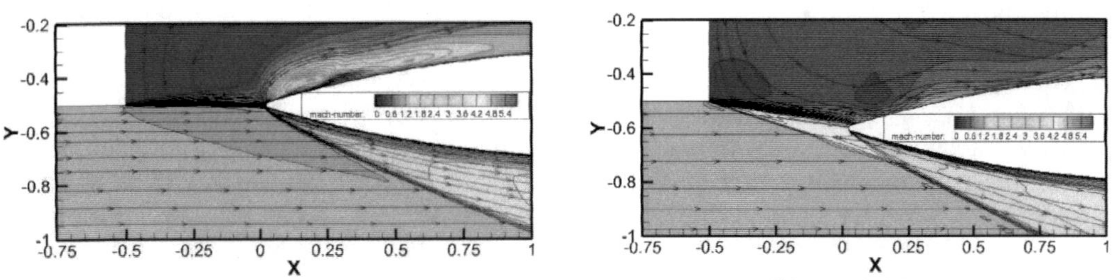

(a) H_O=2.0Φ (b) H_O=2.25Φ

IOP Conf. Series: Journal of Physics: Conf. Series **1064** (2018) 012020 doi:10.1088/1742-6596/1064/1/012020

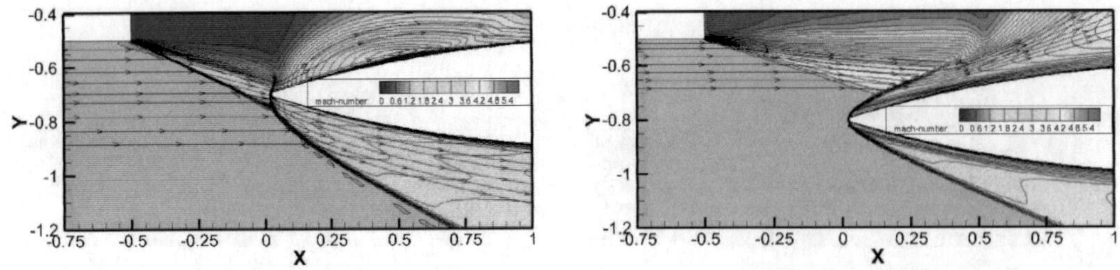

(c) H_O=2.5Φ (d) H_O=2.75Φ

Figure 14. Flow field around the missile nose at H_O=2.0Φ, 2.25Φ, 2.5Φ and 2.75Φ.

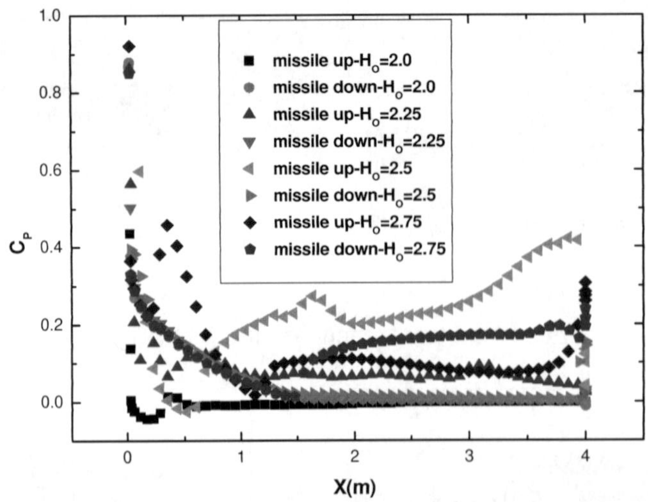

Figure 15. Pressure coefficients of the upper and lower surface of the missile at H_O=2.0Φ, 2.25Φ, 2.5Φ and 2.75Φ.

3.5. Wall Effect Flow of the Missile

After the missile separates from the cavity, whether the embedded weapon bay should be closed quickly may be a problem to be discussed. From the perspective of safety, there are disadvantages for closing the bay quickly. However, it can reduce the thermal and acoustic environment severity and increase the negative lift (In Fig. 3 (a)) if the bay closes quickly. In this paper, the weapon bay is ordered to be closed quickly when the missile is at H_C=0.25Φ.

Fig. 16, Fig. 17 and Fig. 18 present the pressure coefficient contour & streamlines and Ma contour of the flow field at H_C=0.25Φ, 0.5Φ and 1.0Φ, respectively. Fig. 19 and Fig. 20 present the stream-wise density gradient contour and the pressure coefficient curves of the upper and lower surfaces of the missile, respectively. According to these figures, the pressure on the lower surface and the flow field below the missile vary slightly. The variation of the aerodynamic characteristics of the missile is mainly determined by the flow between the upper surface of the missile and the door of the bay. The flow mentioned above is denoted as the wall effect flow.

In the wall effect region, the main flow characteristics are shock wave reflections and shock-boundary layer interactions (Because the shear on the aircraft is neglected, the viscosity effect only embodies on the surface of the missile). When the missile is very close to the wall (H_C=0.25Φ), the incident shock wave from the missile nose is reflected by the aircraft wall, the reflected shock wave then strongly interacts with the boundary layer at the shoulder of the missile. Consequently, a local small separation is forced by the strong adverse pressure gradient induced by the high pressure after the shock wave. The small separation makes the posterior flow channel invariant, and the pressure gains its balance gradually along the flow through weak shock waves and expansion waves. Later, a weak adverse pressure gradient in the flow channel keeps and develops to the rear of the missile. With H_C increasing, the incident point of the oblique shock wave on the bay door moves downstream, the interaction strength between the reflected shock wave and the boundary layer on the missile declines gradually, and the reflection times also decrease gradually. Finally, the influence of the wall completely disappears. The pressure on the upper surface is dominated by the reflection times. With the reflection times decreasing, the pressure on the upper surface decreases gradually (as shown in Fig. 20), and the negative lift decreases, too.

In Fig. 20, the high pressure on the rear of the missile is likely to induce a nose-up pitching moment, which is harmful to the separation safety. Accordingly, it is necessary to balance the influence of the negative lift and the nose-up pitching moment on the separation safety. If we can increase the negative lift and decrease the nose-up pitching moment simultaneously, it will be more favorable for the separation safety.

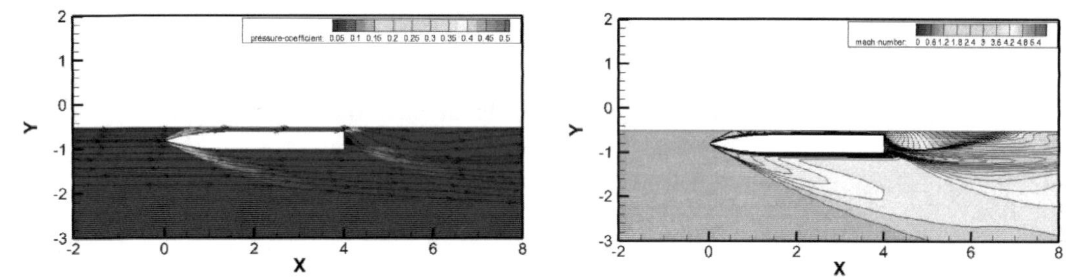

(a) Pressure coefficient contour and streamlines (b) Ma Contour

Figure 16. Flow around the missile in wall effect at H_C=0.25Φ.

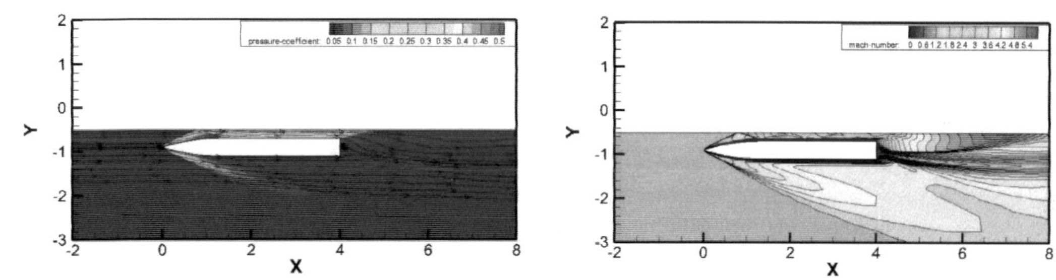

(a) Pressure coefficient contour and streamlines (b) Ma Contour

Figure 17. Flow around the missile in wall effect at H_C=0.5Φ.

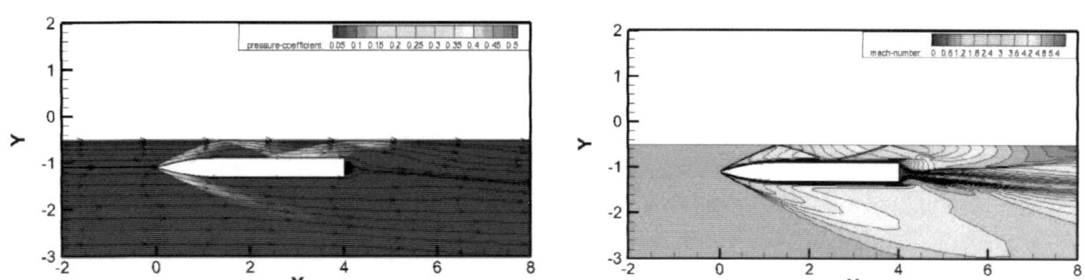

(a) Pressure coefficient contour and streamlines (b) Ma Contour

Figure 18. Flow around the missile in wall effect at H_C=1.0Φ.

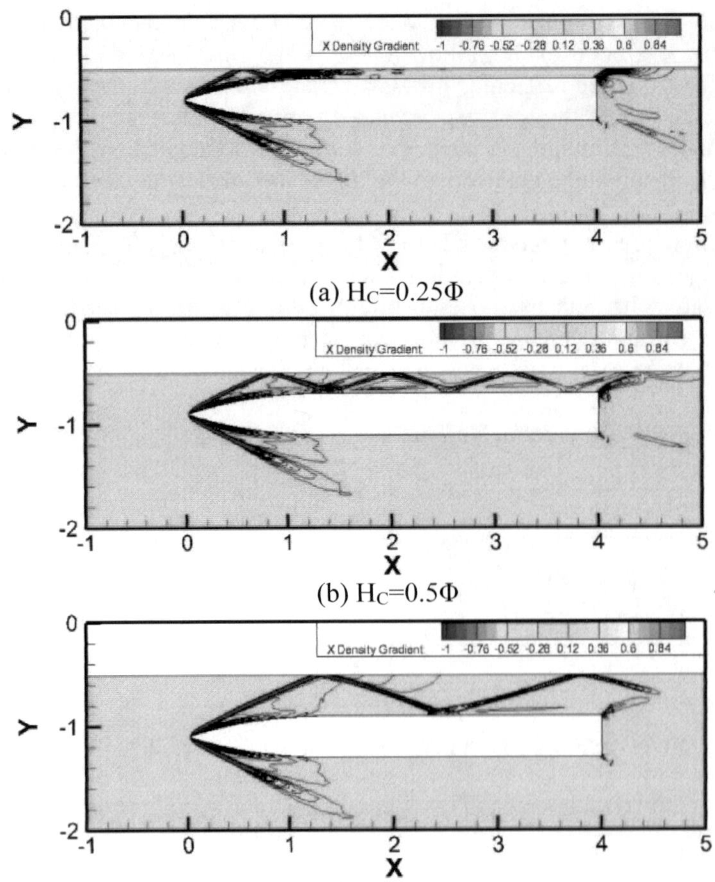

(a) $H_C=0.25\Phi$

(b) $H_C=0.5\Phi$

(c) $H_C=1.0\Phi$

Figure 19. Stream-wise density gradient contour at $H_C=0.25\Phi$, 0.5Φ and 1.0Φ.

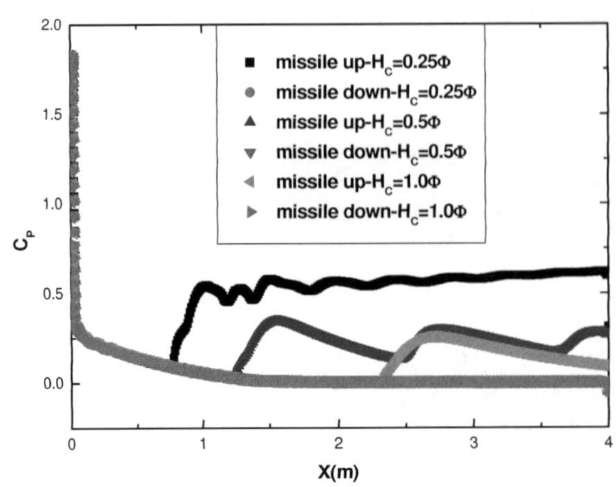

Figure 20. Pressure coefficients of the upper and lower surface of the missile at $H_C=0.25\Phi$, 0.5Φ and 1.0Φ.

4. Conclusion

In this paper, 2D numerical simulations are carried out on the symmetric plane of a full size missile separation from high speed aircrafts at the conditions of Ma=5 and AOA=0°. Aerodynamics and typical flow field characteristics of the missile at different distances away from the cavity bottom are predicted via a steady RANS solver. The main conclusions are summarized as follows:

1) The flow of the missile separation from high speed aircrafts with an embedded weapon bay can be divided into four stages, which are dominated separately by the flow around the missile in the deep cavity, compressed incoming flow by the lower surface of the missile, strong interacting flow between the shear flow and the upper surface of the missile and wall effect flow of the missile.

2) For stage I, when the missile is very close to the cavity wall, the effective length depth ratio of the cavity increases and the flow around the cavity develops from the open-type flow to the transitional open-type flow. When the missile is in the middle of the cavity, the impact due to the presence of the missile on the open-type flow is weak.

3) For stage II, the incoming flow will be firstly compressed by the lower surface of the missile since it's protruding from the embedded weapon bay. After that, i.e. on the stage III, when the axis line of the missile is beyond the surface of the aircraft, the strong shear layer induced by the incoming high speed flow and the non-high speed flow in the cavity will act on the lower incoming flow, which leads to the shock wave interaction at the missile nose. Simultaneously, the strong shear layer will significantly increase the pressure on the upper surface of the missile.

4) For stage IV, i.e. the embedded weapon bay is closed, the reflection times of the shock waves between the missile and the aircraft decrease with the increasing distance between the missile and the aircraft. In addition, the pressure on the upper surface of the missile decreases gradually. The large negative lift of the missile formed due to the extreme wall effect is beneficial to the missile separation from high speed aircrafts in some degree. In addition, the closed weapon bay is also favorable to improve the thermal and acoustic environments.

The missile separation from high speed aircrafts is a key problem for airborne weapons in the future. It involves the aerodynamic characteristics, thermal environment, thermal protection structure design, acoustic environment, instruments, materials, and many other aspects. Whether the traditional embedded weapon bay of supersonic aircrafts can be adapted to the high speed one should be considered with caution.

The traditional embedded weapon bay with great simplification has been studied in this paper. There are many aspects that need to be improved based on the preliminary results. Besides, considering the predicting accuracy of complex flows such as cavity flow, shock-boundary layer interaction, and so on, there are lots of issues to be further studied.

References

[1] J F Feng, S T Yang, W J Liu. Review of key technologies for embedded weapons of fighters. Aerodynamic missile Journal. 2010 (10).
[2] D. A. John, Y Yang, D Y Li. Hypersonic and high temperature gas dynamics. Aviation Industry Press. 2013.
[3] J F Wu, X F Luo, L W Xu, et al. Investigation on internal weapon separation characteristics and flow control methods. ACTA Aerodynamics Sinica, 2014, 32 (6).
[4] L Guo, C Wang, B Ye, et al. Investigation on characteristics of store release from internal bay in supersonic flow under flow control. Acta Aeronautica et Astronotica Sinica, 2015. 06. 36 (6): 1752-1761.
[5] J F Wu, Z L Fan, X F Luo. Analysis of the effects of mass-injection at the leading edge on cavity flow characteristics. Modern Physics Letters B. 2009, 22 (1): 413-416.
[6] E Lazar, G Elliott, N Glumac. Control of the shear layer above a supersonic cavity using energy deposition. AIAA. 2007-1229.
[7] J F Wu, L W Xu, Z L Fan, et al. Aeroacoustic characteristics and flow control method of open cavity flow.Acta Aeronautica et Astronotica Sinica, 2015. 07. 36 (7). 2155-2165.
[8] Y D Wang, L Guo, M B Tong, et al. Active control and nonlinear numerical simulation for oscillating pressure of high-speed aircraft cavity. Acta Aeronautica et Astronotica Sinica, 2015, 36 (1): 213-222.

[9] S F Mcgrath, L L Shaw. Active control of shallow cavity acoustic resonance, AIAA-1996-1949. Reston: AIAA, 1996.
[10] V Sarohia, P F Massier. Control of cavity noise. Journal of Aircraft, 1977, 14 (9): 833-837.
[11] B R Smith, T J Welterlen, B H Maines, et al. Weapons bay acoustic suppression from rod spoilers, AIAA-2002-0662. Reston: AIAA, 2002.
[12] S T Zhu, L P Cao, P W Feng, et al. Simulation of missile separation from internal weapon bay. Electronics Optics & Control, 2012.09. 19 (9).
[13] Q F Zhang, P P Yan, J Li. Comparison between external store separation and buried store separation of fighter. Journal of Beijing University of Aeronautics and Astronautics. 2017.06. 43 (6).
[14] C Chang, H H Ding. Review on missile store safety separation technology of embedded ejection weapons. Modern Defense Technology. 2012.10. 40 (5).
[15] W Wang, Simulaiton of aerodynamic characteristics of embedded missiles. Harbin Engineering University. 2011.03.
[16] Liu Yu. Investigation on weapon separation process from internal weapon bay. Nanjing University of Aeronautics and Astronautics. 2010.10.

IOP Conf. Series: Journal of Physics: Conf. Series **1064** (2018) 012021 doi:10.1088/1742-6596/1064/1/012021

CFD Simulation of a Finned Smart Bullet with Micro-actuator

Chengqing Zhang[1], Huiyuan Wang[1, *], Pengjun Zhang[1], Shuhua Gao[2], Hao Xiong[1]

[1] College of Mechatronic Engineering, North University of China, Taiyuan 030051, China
[2] Chongqing Jialing special equipment Co., Ltd., Chongqing 400032, China

*Corresponding author e-mail: wanghuiyuan@nuc.edu.cn

Abstract. The geometric configuration of the micro-actuators is known to have a large impact on control authority. This study describes a finned smart bullet (FSB) to support the requirements of various missions. The flow control system consists of micro-actuator located at the rear body of FSB, alter the flow field in the rear region of FSB to create asymmetric pressure distributions and thus produce aerodynamic control forces and moments. This work seeks to derive an optimal mass centre (MC) for FSB and explore the FSB configuration with rear micro-actuator. Computational fluid dynamics simulations are performed to study the MC position for FSB stable flight and review the aerodynamic characteristics FSB configuration with actuator. For control trajectory of the FSB, the further work will explore the FSB best configuration to finish the projectile dynamic model for standard 6DOF projectile flight dynamic model in order achieve flight control.

1. Introduction

Direct-fire projectile are fired by line-of-sight aiming from ground-based platforms, helicopters, and fixed-wing aircraft, especially individual soldier. A number of conditions can cause projectiles to miss an intended target. These conditions include manufacturing inaccuracies of the gun tube, propellant, and projectile, along with variable atmospheric conditions, firing platform motion, and aiming errors. With rapid development of modern high-tech ammunition, it tends to have high-efficiency and high-reliability of combat capability. The digital age has brought about profound changes in long established areas of weapons technology, and the artillery is no exception with digital technology penetrating into guided artillery rounds and navigation systems. Precision guidance technology has been mature in big artillery and missile application, but it is still in the stage of research in small caliber projectile or bullet, such as 0.50-inch caliber EXACTO round and DARPA's self-guided bullet [1-5].

The FSB additionally relates to non-spinning small caliber having forward viewing optical sensor, fixed strakes and electrically actuated control fins for steering a bullet toward the target [6]. Therefore, a variety of novel control effecter concepts designed specifically for gun-launched precision munitions have been proposed. These include three main classifications control mechanisms, such as aerodynamic load mechanisms, jet thrust mechanisms, and internal load mechanisms. Example of aerodynamic control mechanisms gives canards, ram air deflectors, and moveable noses, which count on aerodynamic

Content from this work may be used under the terms of the Creative Commons Attribution 3.0 licence. Any further distribution of this work must maintain attribution to the author(s) and the title of the work, journal citation and DOI.
Published under licence by IOP Publishing Ltd

effect on control force and moment. Others, such as gas jet thrusters and explosive thrusters, rely on on-board thrust mechanisms. While the third class of actuator devise, examples of inertial control mechanisms include internal translation of a control mass and internal rotation of an unbalanced part [6-12].

In general, the theoretical calculation method of projectile flight stability is to calculate the stable margin in several positions at different work situation. There are different methods to modify its position of projectile or bullet [13], one way is to employ applicable material, and others is to change bullet mass distribution or nose structure. It is required 12%-20% stability margin to achieve a better flight stability for the fin-stabilized projectile or missile. Here, if static margin is too large, not only lead to wind disturbance, but also increase vibration frequency. In order to flexibly operate the guided projectile or missile, it requires static margin ranging from 2% to 5% and even static instability design [14].

Therefore, the aerodynamic characteristic of FSB is investigated in this work. The rest of the paper is organized as follows: A detailed description about static margin analysis on FSB is discussed in Section 2. The geometries model and computational girds of FSB is provided in Section 3. The experimental of range flight is shown in Section 4. The numerical study for MC position optimization and this FSB configuration is analyzed in Section 5. The conclusions are summarized in Section 6.

2. Static margin analysis

The aerodynamic force and moment diagram acting on a rigid FSB is shown in Fig 1. To describe projectile or bullet aerodynamic force and moment, a ground-based frame OXYZ is used as the absolute coordinate frame fixed on the earth ground and body fixed reference frame O_1ABC. Vector of O_1A is direction of bullet velocity and total angle of attack (AOA) α .OA is axis of the projectile or bullet, bullet velocity is measured vertically upward and is simplified to MC, O_1 is the position of MC and P is the pressure center (PC) position. This rigid-body model is similar to use in next [15]:

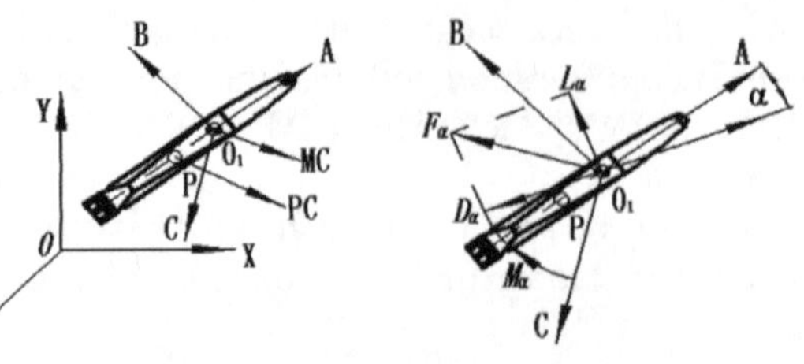

Fig. 1 Schematics of aerodynamic force and moment acting on FSB

There is a need to state that α is equivalent to total AOA in the numerical calculation. The aerodynamic force and moment coefficient are related to bullet shape, Mach number and attack angle [16]. The static aerodynamic relationship can be achieved from Fig.1 M_α , L_α and D_α have a certain steady-state relationship for a symmetric bullet, that is:

$$M_\alpha = D_\alpha l_d \sin\alpha + L_\alpha l_d \cos\alpha \tag{1}$$

Where M_α , L_α , D_α are the drag force, lift force and pitching moment at AOA, the l_d is the distance between MC and PC.

When attack angle α is small, lift force coefficient and pitching moment coefficient can be expressed with its differential coefficient to AOA, just as shown in the next:

$$C_{L\alpha} = C'_{L\alpha}\alpha \tag{2}$$

$$C_{M\alpha} = C'_{M\alpha}\alpha \tag{3}$$

$$sin\alpha = \alpha \tag{4}$$

$$cos\alpha = 1 \tag{5}$$

Where $C'_{L\alpha}$, $C'_{M\alpha}$ are the coefficients of derivative for lift force and pitching moment.
If equations are substituted in, the following of equations can be provided:

$$C'_{M\alpha} = \frac{l_d}{l}\left(C'_{L\alpha} + C_{D\alpha}\right) \tag{6}$$

As usual for fin stabilized projectile or bullet as expressed below: $C'_{M\alpha} >> C_\alpha$,

$$R = \frac{l_d}{l} \approx \frac{C'_{M\alpha}}{C'_{L\alpha}} = \frac{C_{M\alpha}}{C_{L\alpha}} \tag{7}$$

The non-dimensional ratio R called static margin mainly reflects the static stability of the bullet, and it is important that R be efficient to keep the bullet flight performance. Formula (7) shows that the distance between PC and MC is a decisive factor for its static margin when the bullet reference length is a given parameter [17].

3. FSB Geometries model and Computational Girds

3.1. FSB Geometries model Setup

This paper is part of a larger research effort to characterize the aerodynamic performance of FSB and design them in order to achieve the best performance to support the requirements of various missions. The FSB structure adapted is to be fired from smooth bore gun barrels, when light weapon is desired to increase the accuracy of placing the bullet on a target from long range (e.g. 2000 meters and beyond) [6]. It is self-guided to a target illuminated by a laser target designator. To compute the control effects of micro-actuators, a series of CFD simulations was performed in order to finish the study of the pre-program feasibility for the FSB aerodynamic configuration which can provides the control authority.
The rare micro-actuator for FSB is chosen as the main subject used for all aerodynamic prediction and simulation studies throughout this paper. The bullet modeled in this study is triangle fins of main body, a cone-cylinder-finned configuration. A schematic diagram of this FSB provides basic geometry normalized by the bullet reference caliber is shown in Fig.2 and two micro-actuators mechanism which is composed of two set of orthogonal and same size of micro-actuator located at the rear body. The length of the bullet is 11 cal and the diameter is 12.7mm, and the nose is 3 cal, the connecting piece is 1.1 cal and is followed by 5.8 cal fixed fins, the rear body length is 0.78 cal. For the case in this paper, the micro-actuator width is set to 6mm and the depth is 1.5mm. Unless otherwise noted, the spoiler full extension from the surface of the projectile is assumed to be 0.2mm.

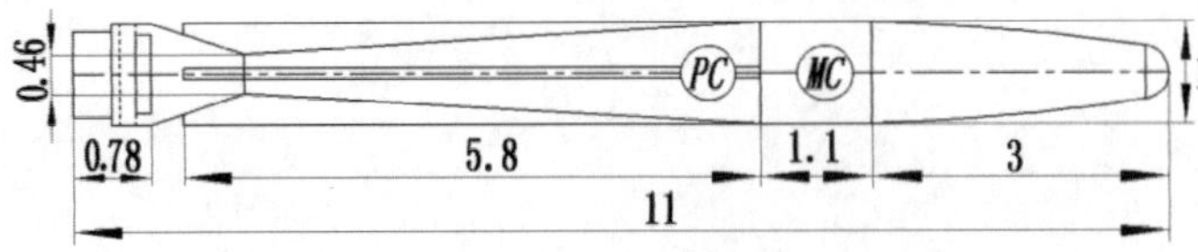

Fig. 2 Schematics of the finned smart bullet

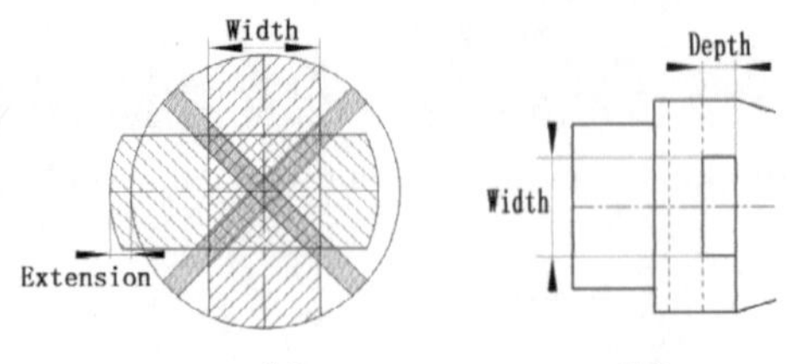

(a) (b)

Fig. 3 Micro-actuator geometry parameterization

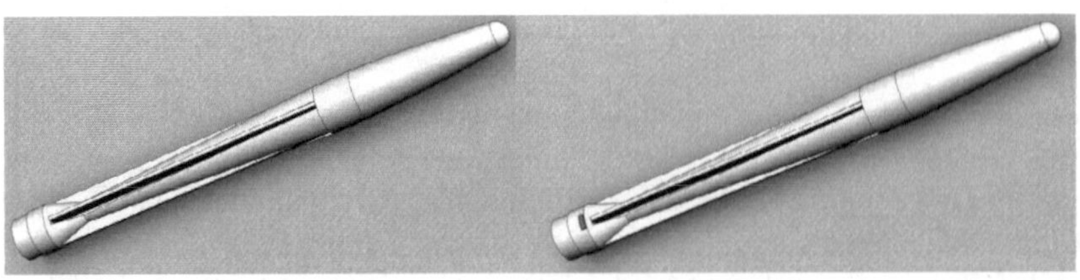

(a) (b)

Fig. 4 Finned body geometry (a) without and (b) with rear micro-actuator mechanisms

Fig 4 shows the 3D computational models of FSB both with and without micro-actuator. Fig 4.(b) and presents single micro-actuator control mechanisms at the rear body, and they are intended to create asymmetric pressure distributions and flow field in the rear of FSB and provide control force and moments needed for bullet control maneuver.[18]

Different nose material and various nose structure meets the electronic components, optical system and other parts in assembly, and there are MC position range located at between 86mm and 100mm from the rear body. Five types of MC location are accomplished in order to explore the logical position for bullet static-stability. As shown in Fig.4 (a), the different kinds of MC location the bullet without micro-actuator are 86mm, 88.5mm, 93mm, 96.5mm and 100mm for initial study of bullet. Steady-state computation have been performed for FSB without of different MC position at Mach number from 0.3 to 3Ma and at AOA from 0° to 8° . For the body without micro-actuator, the different kinds of MC location produce essentially aerodynamic force and moment.

With the micro-actuator installed in the rear body of FSB, the drag force perturbation δ_D and the lift moment perturbation δ_M actually generate control in opposite directions. However, the moment perturbation typically dominates the effects of the drag force δ_D because it directly generates the AOA [19]. Furthermore, although the micro-actuator increases the resulting perturbation moment, it also increases drag. When evaluating the performance of the rear micro-actuator configuration, it is important to settle terminal error correction for a notional FSB equipped with micro-actuator. The FSB mission for the correction only and the trajectory error is not known until the bullet-target distance is small and thus control is only exerted over a short period of time near the end of the trajectory. The following reward function may be defined for comparing geometric configuration for the correction scenario:

$$K = \frac{\delta_M}{D\delta_D} \tag{8}$$

Note that, in Eq. (8) the reference diameter D is included so that K is non-dimensional. In order to fulfill the mission, it is important that the micro-actuator be efficient in producing control authority by generating as large a control moment as possible for the smallest sacrifice in drag force. Additionally, the drag is of less importance, and the overall magnitude of the control moment δ_M may be used directly to compare the effectiveness of various configurations.

3.2. Computational Girds

In order to analyze external flow field of FSB, it can be considered cylindrical slender body approximately, so the 400mm×800mm computational region around of FSB was created by modeling software with Boolean. It was a full three-dimensional unstructured hexahedral mesh for the simulation of FSB. Fig 5 reveals the computational mesh for the bullet configuration without micro-actuator. The domain to be solved is divided into two parts, the inner domain was the bullet surface in purple, and out domain was in green. Bullet surface of mesh is shown in Fig 5.b).The boundary conditions in out domain were set as pressure far-field, the bullet surface was set to no-slip adiabatic and stationary wall condition. Because gun of firing bullet is a direct fire weapon, the firing gun is similar to a flat-fire ballistic trajectory, and the firing process of the bullet is analogous to the weather feature [19-23]. Let us suppose that the flow is ideal, the flight simulation condition is shown in Table 1.

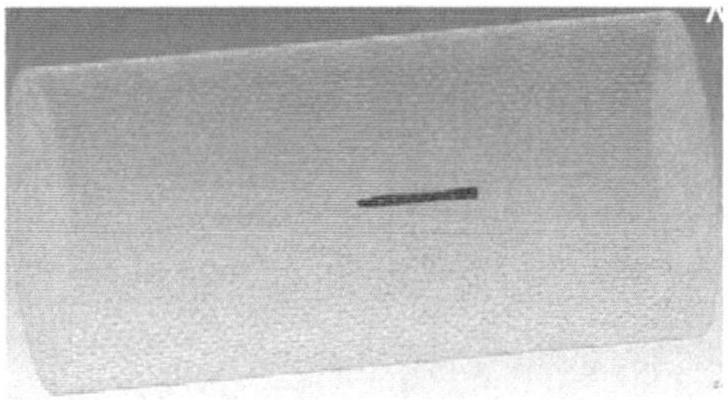

a). Computational and bullet surface of mesh

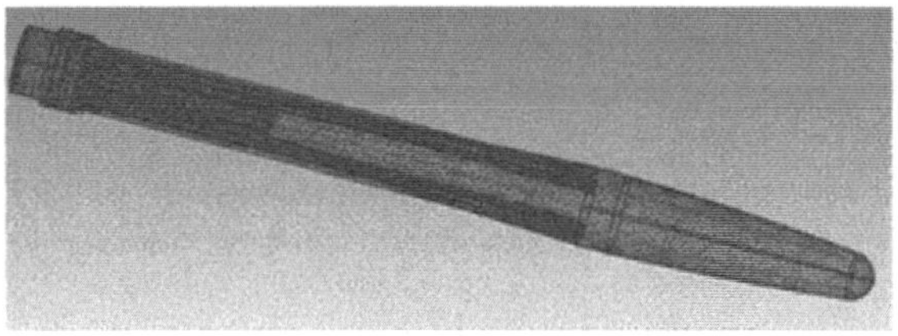

b). Bullet surface of mesh

Fig. 5 Schematics of computational and bullet surface of mesh

Table 1. Flight simulation condition

Parameter	Symbol, unit	Value
Pressure	P[Pa]	101325
Temperature	Ta[K]	288.13
Far field velocity	v [Ma]	0.3-3
AOA	α [degree]	0-8

IOP Conf. Series: Journal of Physics: Conf. Series **1064** (2018) 012021 doi:10.1088/1742-6596/1064/1/012021

4. Experimental of Range Flight

Bullet range flight without micro-actuators was obtained experimentally at Chongqing Jialing special equipment Co., Ltd. During such tests, a set of prototype ammunition were fired along a flat trajectory through an enclosed building and high-speed photography at 2m from gun muzzle was used to catch the flight attitude of FSB. Because of flight stabilization of FSB, a 12.7mm smooth bore gun tube was selected for this experimental in order to eliminate rotation effects. For ease of manufacturing, as shown in Fig 6, the entire test bullets were combination of different components and nose body and main body were connected by M 0.5 screw. The final test FSB can be seen in Fig 6, (a) exhibited the FSB with sabot and there was no sabot for FSB in (b). In total, 15 shots were recorded on film, and it was without micro actuator to finish new aerodynamic configuration verification. The final design parameters for range test FSB are listed in Table 2. Total FSB weight was about 38.2g, with its MC located at about 86mm of the overall FSB length measured from the rear base.

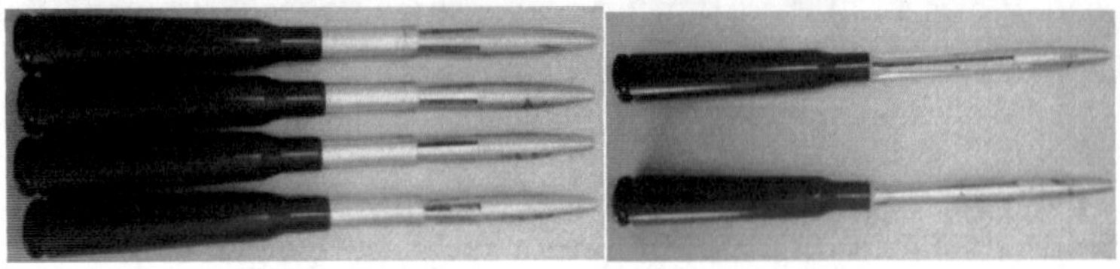

a) FSB with sabot b) FSB without sabot

Fig. 6 Prototype bullet and FSB ammunition

Table 2. Parameter of assemblage FSB

Number	1#-a	1#-b	2#-a	2#-b	3#-a	3#-b
Powder(g)	12.5	13	12.5	13	12.5	13
Bullet weight(g)	38.5	38.1	38.2	38.2	38.15	38.05
X_{MC}(mm)	84.88	86.78	85.95	86.54	86.84	87.05

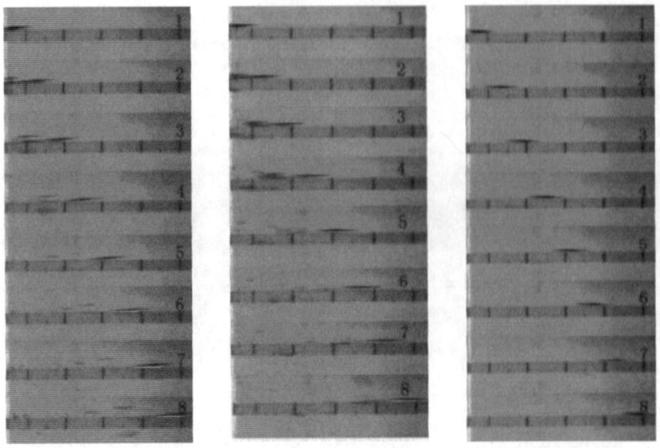

a) 1#-a FSB b) 2#-a FSB c) 3#-a FSB

Fig. 7 High-speed photography frames of the gun muzzle

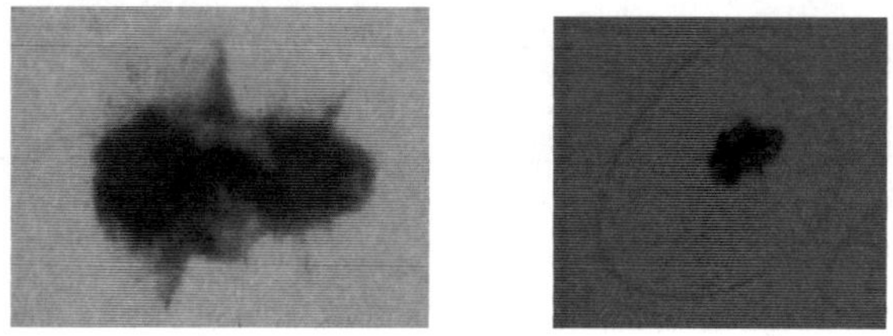

Target sheet 35m Target sheet at 70m

a) Target sheet holes of 1#-a FSB

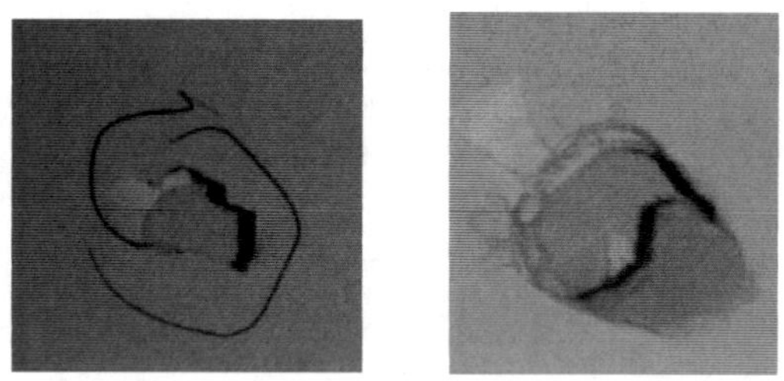

Target sheet 35m Target sheet at 70m

b) Target sheet holes of 2#-a FSB

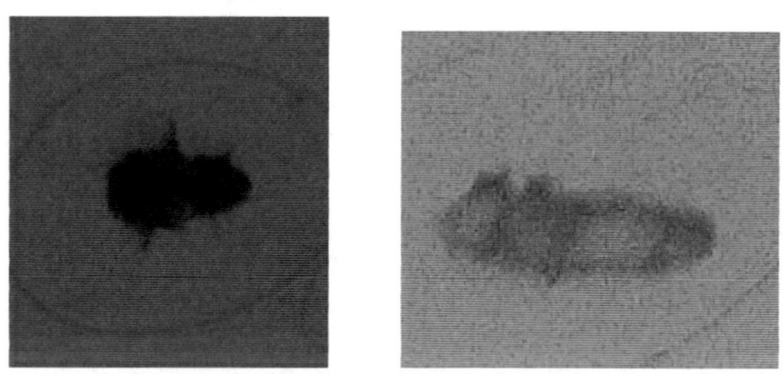

Target sheet 35m Target sheet at 70m

b) Target sheet holes of 3#-a FSB

Fig. 8 Two target sheet holes of different FSB

Fig 7 provides a series of picture of FSB in supersonic flight when FSB 1#-a, 2#-a, 3#-a FSB

left the gun muzzle and Fig 8 shows target sheet holes of different FSB, the mark of 35m target sheet had slightly deformed round hole and 70m distance sheet gave circular hole for FSB 1#-a. The experimental results consist of Fig 8 a), b), c) indicate that test FSBs play an important role in critical stability flight performance and that the MC position needs to be further optimized to meet the FSB without micro-actuator flight performance firstly. In next section, the influence of MC location on FSB flight performance is simulated and analyzed concretely.

5. Numerical Study

5.1. MC position optimization

The Fig 13 and Table 3 present that static margin can be achieved by the parameters of aerodynamic characteristics and 86mm MC position, and the static margin range of 86mm MC position is from 1.05% to 9.52%, maximum R is 9.25% at 1Ma and 2 AOA and there is approximate 1.5% static margin at higher than 1.5Ma supersonic flight, and Fig 9 shows the velocity contours of numerical calculation for different Mach number. However, the flight experiment happened to critical flight performance of FSB and was not deemed acceptable.

Table 3. Static margin with 86mm MC position

Ma	α			
	2	4	6	8
0.3	0.01658	0.01059	0.01286	0.03092
0.5	0.02068	0.01708	0.02883	0.03721
1	0.19522	0.1469	0.11286	0.10647
1.3	0.07701	0.03063	0.04961	0.04967
1.5	0.0467	0.0159	0.03106	0.03798
2	0.01979	0.01394	0.02138	0.02279
2.5	0.01594	0.01422	0.02753	0.02543
3	0.01268	0.01559	0.01122	0.02647

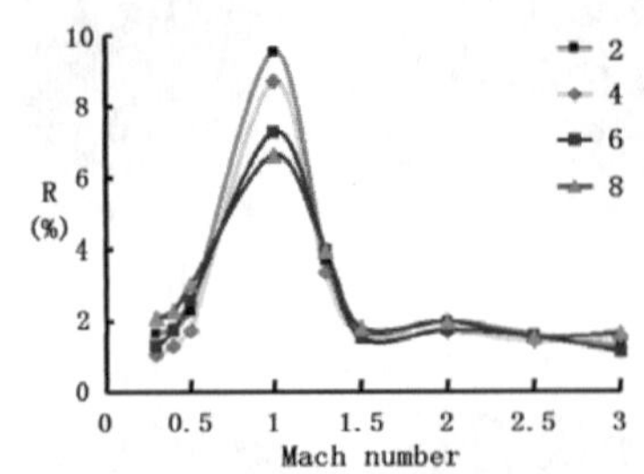

Fig. 9 Static margin vs. Mach number with 86mm MC position

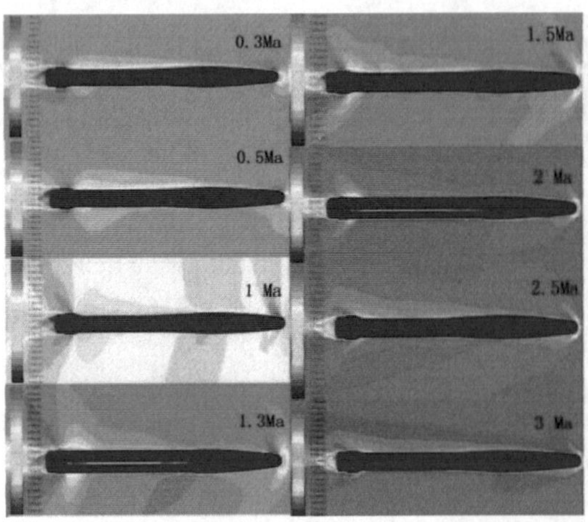

Fig. 10 Velocity contours of numerical calculation for different Mach number

How to judge the static margin is reasonable in this case, it is important that the different MC position be efficient in variety of static margin by interfering sizable aerodynamic characteristics. The following reward function may be defined for comparing geometric MC position for the flight performance:

$$J = \frac{R_\alpha}{C_{D\alpha}} \tag{9}$$

Note that, in the aerodynamic coefficients Eq. (9) are included so that J is non-dimensional, and value J can be a symbolic date for the flight performance of different MC position FSB. Especially, as shown in Table 4, which reveals FSB configuration static margin, drag coefficient and value J of 2° AOA at different MC position.

Table 4. FSB configuration of 2° AOA at different MC position

MC	86	88.5	93	96.5	100
R	0.0925	0.1645	0.1883	0.2321	0.2401
C_D	0.7891	0.7425	0.7215	0.7712	0.8834
J	0.1172	0.2215	0.2276	0.3048	0.2717

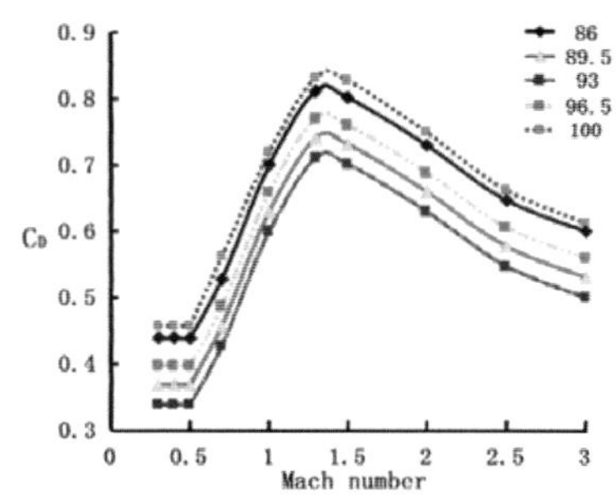

Fig. 11 Drag coefficients vs. Mach number of 0 AOA at different MC location

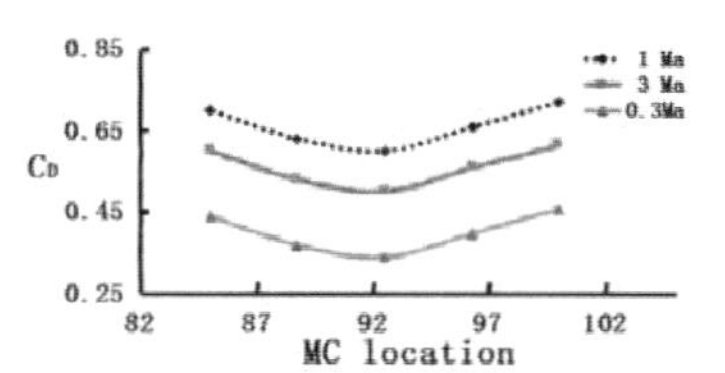

Fig. 12 Drag coefficient vs. MC location of 0 AOA at different Mach number

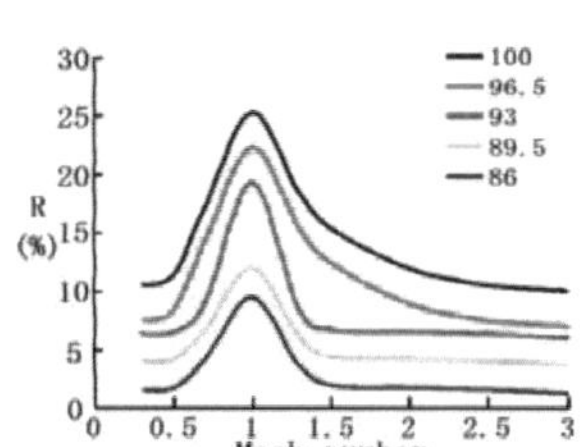

Fig. 13 Static margin vs. Mach number at AOA 0°

With the premise of ensuring FSB flight performance, a reasonable location with lower drag is chosen to determine the optimal MC location. Increase of static margin is the important value for MC position, but the value of drag coefficient is the main parameter determining the FSB effective firing range. The geometrical parameters defining each of different MC position cases and the results of reward value J are shown in Table 4. Starting with the based MC 86mm used in the previous simulation and range test, this MC position is not the ideal configuration for the flight performance. As expected, this is noted that by comparing the position minimum and maximum, although the percentage increase in static margin

IOP Conf. Series: Journal of Physics: Conf. Series **1064** (2018) 012021 doi:10.1088/1742-6596/1064/1/012021

was greater when the MC max position 100mm than the previous prototype position 86mm, and also clearly produced the highest value 0.88 in drag coefficient. Overall, examining the specific value function shown in Table 4, 93mm position configuration yields an impressive 100% increase in static margin with a better 8.6% decrease in drag. The 96.5mm position configuration had a value K of 0.3048 and an outstanding value in static margin and still the value only 2.2% decrease in drag. There is drag coefficient variation trend of different MC position in Fig 11 and Fig 12, and the maximum drag is as high as 0.88 at the 100mm MC and these coefficients are also bigger than another MC position at various AOA. The cave of different MC shown in Fig 13, there are something noted that the maximum margin always appeared at 1Ma at different MC position and the R almost changed rarely when the relative airflow velocity of FSB is exceeding 1.5Ma range. The percentage increase in the flight static margin is greater four times at the supersonic flight when the MC position moves from the 86mm to 100mm. Considering the three factors of stability margin, drag coefficient and ratio J , the 93mm MC position is the relatively superior result. This 93mm MC location is used in further investigations throughout this study.

5.2. Geometric with micro-actuator study

This purpose of study is focused on the investigation of aerodynamic characteristics and control authority problem caused by the given structure of FSB with 6mm×0.2mm micro-actuator. This micro-actuator in this configuration is 6mm width and approximately 0.2mm extension, the deflection angel between the micro-actuator and the bullet body is 90. For this case with rear micro-actuator, computations were performed to quickly provide the extent of control force that could be generated using the micro-actuator. The results of in the previous section were restricted to Mach 1 only for FSB configurations. Additional CFD simulations were performed for configuration F at the range of Mach number between 0.3 and 3, again at a zero AOA. Fig 14 shows the results of these studies for Mach number-dependent, there are no any significant change for control force and moments at below 1 Mach number and all these are linearly variation in the case of supersonic flight. Note the general trend that control forces and moments, as well as drag, increase as a function of Mach number as expected due to the higher dynamic pressure on the upwind side of the micro-actuator surface.

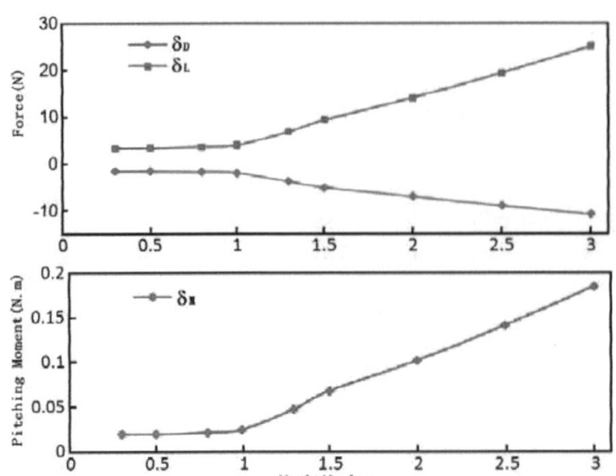

Fig .14 Forces and moments vs. Mach number (AOA=0°)

All the results present here for different AOA are again at the same Mach number 1Ma. Again the effect of AOA is studied use the effective candidate configuration with 6mm×0.3mm micro-actuator. Computed results have been obtained for this configuration at diverse AOA from 0 to 8. In this paper, because the rear micro-actuator special structure of FSB can up-down and left-right retractable movement, so the small range positive AOA indicate the control force and moment variation tendency. The results aerodynamic delta parameters for AOA-dependent are shown in Table.5. These results seem to indicate that control force and moment, and even $_J$ value are generated across the range of AOA is not as significant for AOA, and the AOA 4° has better $_J$ value which is no qualitative change.

Table 5. Variation of delta force and moments due to micro-actuator with AOA at 1Ma.

α (°)	δ_D, N	δ_L, N	δ_M ,N·m
0	-1.98	0.93	0.024
2	-1.75	0.89	0.022
4	-1.63	0.94	0.021
6	-1.85	0.99	0.023
8	-1.67	0.91	0.021

6. Conclusion

The study reported here details the FSB overall configuration and the aerodynamic characterization with and without micro-actuator for FSB. Through a serious of steady computational fluid simulations, the main two research aspects concentrate on the MC position and the given micro-actuator aerodynamic study. A combination of steady computational fluid simulations and range test of flight can be adopted to analyze the FSB without rear micro-actuator aerodynamic characteristics for the MC position optimization, steady simulation for five different kinds of MC position give the conclusion that 93mm MC location can be called the optimal results relative to four others, when comprehensive study of stability margin, drag force and ratio value K is conducted to comparatively analyze at 1 Mach number and 0° AOA. In this case, 93mm position configuration yielded an impressive 100% increase in static margin with a better 8.6% decrease in drag. When MC position moved forward to the nose from 86mm to 93mm, the static margin maintains 6.5% increases three times at the supersonic flight exceed 1.5Ma range. In the previous study, steady-state solutions were firstly gained for FSB without rear micro-actuator at a 1Ma. This configuration of FSB with 6mm×0.3mm rear show that some control authority of micro-actuator exists even at flight trajectory. For control trajectory of the FSB, the further work will explore the FSB best configuration to finish the projectile dynamic model for standard 6DOF projectile flight dynamic model in order achieve flight control.

References

[1] Warnash, A., and Killen, A., Low Cost, High G, Micro ElectroMechanical (MEMS), Inertial Measurements Unit (IMU) Program,2002 IEEE Position Location and Navigation Symposium, PalmSprings, CA, April 2002.
[2] Silton, S. Navier-Stokes Computations for a spinning projectile from subsonic to supersonic speeds. J. Spacecr. Rockets, 42(2005), 223–231.
[3] M.D Mikhailov, A.P. Silva Freire, The drag coefficient of a sphere: An approximation using Shanks transform, J. Powder Technology, 237(2013) 432-435.
[4] Robert L. CCMoy, Modern Exterior Ballistics (Revise 2nd Edition), M. Schiffer, Attlen, PA 19310 USA, 2012,PP.33-36.
[5] Lei Junmian, Engineering Prediction Methods of Aerodynamic Characteristics for Guided Weapon, Beijing Institute of Technology Press, 2015. 2(in Chinese).
[6] James F. Jones, Brian A. Kast, Marc W.Kniskern, et al US Patent 7,781,709 B1(24 Aug 2010).
[7] Massey, K., McMichael, J., Warnock, T., and Hay, F., Mechanical Actuators for Guidance of a Supersonic Projectile, Journal of Spacecraft and Rockets, Vol. 45, No. 4, 2008, pp. 802–812.doi:10.2514/1.31709.

[8] Massey, K. C., and Guthrie,K. B., Optimized Guidance of a Supersonic Projectile Using Pin Based Actuators, 23rd AIAA Applied Aerodynamics Conference, AIAA Paper 2005-4966, June 2005.

[9] Zipeng Han, et al, Exterior Ballistics of Projectlies and Rockets, M. Beijing: Beijing institute of technology Press, 2014.2(in Chinese).

[10] M. Zeidler, E. Garnier, R. Cayzac, and A. Merlen, Fluidic control of a 155 millimeter spin-stabilized projectile using Coanda effect, AIAA Journal, vol. 53, no. 5, pp. 1146–1158, 2015.

[11] J. McMichael, A. Lovas, P. Plostins, J. Sahu, G. Brown, and A. Glezer, Microadaptive flow control applied to a spinning projectile,Tech. Rep. ARL-TR-3589, 2005.

[12] M. Samimy, I. Adamovich, B. Webb et al., Development and characterization of plasma actuators for high-speed jet control, Experiments in Fluids, vol. 37, no. 4, pp. 577–588, 2004.

[13] T. Corke and M. Post, Overview of plasma flow control: concepts, optimization, and applications, in Proceedings of the 43rd AIAA Aerospace Sciences Meeting and Exhibit, AIAA 2005-563, Reno, Nev, USA, January 2005.

[14] J. C. Lin, Review of research on low-profile vortex generators to control boundary-layer separation, Progress in Aerospace Sciences, vol. 38, no. 4-5, pp. 389–420, 2002.

[15] J. Lopera, T. Ng, M. Patel, and S. Vasudevan, Coning motion control of a blunt-nose projectile using small deployable strakes, in Proceedings of the 25th AIAA Applied Aerodynamics Conference, AIAA 2007-3812, Miami, Fla, USA, 2007.

[16] Zhang Wen-dian, Wang Yi-bai, Liu Yu, Aerodynamic study of theater ballistic missile target, J. Aerospace Science and Technology 24 (2013) 221-225.

[17] Bradley T. Burchett, Aerodynamic parameter identification for symmetric projectile: An improved gradient based method, J. Aerospace Science and Technology 24 (2013) 119-127.

[18] HF Nerson, MV Talpallikar, Fin-planform effect on lift and center pressure for supersonic missiles, J. Journal of Spacecraft &Rockets,30(2015)707-714

[19] B.Burchett, Aerodynamic parameter identification for symmetric projectiles:Comparing gradient based and evolutionary algorithms, in:2011 AIAA Atmospheric Flight Mechanics Conference, Portland, Oregon, August 8–11, 2011.

[20] Dykes, J., Montalvo, C., Costello, M., and Sahu, J., Use of Microspoilers for Control of Finned Projectiles, Journal of Space craft and Rockets, Vol. 49, No. 6, 2012, pp. 1131–1140.doi:10.2514/1.A32274.

[21] Scheuermann, E., Costello, M., Silton, S., and Sahu, J., Aerodynamic Characterization of a Microspoiler System for Supersonic Finned Projectiles, Journal of Spacecraft and Rockets, Vol. 52, No. 1, 2015, pp. 253–263.

[22] T. Birch, I. E. Wrisdale, and S. Prince, CFD predictions of missile flowfields, in Proceedings of the 18th Applied Aerodynamics Meeting, AIAA Paper No. 2000-4211, Denver, Colo, USA, August 2000.

[23] J. Sivasubramanian, R. D. Sandberg, Numerical investigation of transitional supersonic base flows with flow control, AIAA Journal of Spacecraft and Rockets,vol. 44, no. 5, pp. 1021–1028, 2007.

IOP Conf. Series: Journal of Physics: Conf. Series **1064** (2018) 012022 doi:10.1088/1742-6596/1064/1/012022

Automatic Simulation Method for Performance Analysis of Kerosene-Based Turbo-Pump

Xuan Jin[1,2,*], Chibing Shen[1,2]

[1]College of Aerospace Science and Engineering, National University of Defence Technology, Changsha, China.
[2]Science and Technology on Scramjet Laboratory, National University of Defence Technology, Changsha, China.

*Corresponding author e-mail: 1540445629@qq.com

Abstract. Kerosene-based turbo-pump is crucial to pump-fed expansion cycle scheme of the scramjet system in aerospace applications. An automatic simulation method was developed to conduct performance analysis of kerosene-based turbo-pump at low costs. Turbulence simulation combined with a multi-species surrogate model of kerosene was employed to study objective characteristics by simulating the flow-field inside the turbo-pump at 100000 rpm. Automatic simulation process, based on Design of Experiments technique, response surface model and high-fidelity numerical simulation, was applied for similarity analysis and Pareto front analysis, respectively. Results indicate that, all the characteristic curves of pump developed head (ΔH_p) and turbine shaft power (P_t) with respect to mass flow rate at various rotating speeds, agree well with Similarity Criterion; ΔH_p is negatively related to the efficiency of pump (η_p) in general, but the contrary is the case for P_t and the efficiency of turbine (η_t).

1. Introduction

The pump-fed expansion cycle scheme of scramjet system has become a potential in solving the problems such as high combustion performance, fuel supply efficiency and long duration thermal protection, and having advantages of simple configuration and light weight [1, 2]. Through exchanging heat with the walls of cooling channels, supercritical/cracking kerosene is formed to drive the turbo-pump to pressurize the fuel (kerosene at room temperature and atmosphere pressure). Then, all the fuel flows into the scramjet combustor, and reacts with air to produce heat release and thrust.

Turbo-pump pressurization by supercritical/cracking kerosene is the key part of the expansion cycle scheme[3]. As the rotating machinery, many structural details have significant effects on turbo-pump performance[4]. Thus, the study on kerosene-based turbo-pump operating characteristics is the indispensable part for building the expansion cycle scheme of scramjet system.

The core of performance analysis in fluids mechanics lies in the database containing the results of all the computational fluids dynamic (CFD) computations. For each sample, the inputs are the geometrical parameters, fluid properties and the flow-field boundary conditions used by the 3D flow solver. The outputs are the developed head, shaft power and efficiency, which characterize the hydrodynamic performance. Nowadays, the exponential increase of computation power has allowed the development of automatic simulation methods by coupling commercial CFD softwares[5]. These

Content from this work may be used under the terms of the Creative Commons Attribution 3.0 licence. Any further distribution of this work must maintain attribution to the author(s) and the title of the work, journal citation and DOI.
Published under licence by IOP Publishing Ltd

methods lead to an analysis process that relies more on a systematic methodology than on experience. Compared with experimental study and traditional approaches, they consume less time and resource.

Many efforts have been directed to corresponding researches. Shah S.R.[6] and Jaiswal N.P.[7] reviewed key technologies regarding to the centrifugal pump design and numerical simulation. Pei et al.[8] proposed an optimization method based on ANSYS 14.5-CFX, Latin Hypercube method and Response Surface Model for centrifugal pump to improve the hydraulic performance. Hu et al.[9] presented a collaborative framework in combination with a variable-complexity modeling technique for the multidisciplinary coupling analysis and design of a shrouded turbine blade. Ahmed et al.[10] carried out numerical simulation of turbine flow-field using 3D Reynolds-Averaged Navier-Stokes models with Shear Stress Transport turbulence model in ANSYS 15-CFX.

Based on the previous experimental work[11], an automatic simulation method for performance analysis of kerosene-based turbo-pump was developed. It was desired that the method should be easily modified and extended to handle more complex problems. In this paper, initial design scheme of turbo-pump was acquired according to typical design method; then, the simulation model for studying turbo-pump performance was developed and validated. Based on automatic simulation process organized by iSIGHT software, similarity analysis and Pareto front analysis of turbo-pump performance parameters were carried out.

2. Model Calculation and Analysis

2.1. *Kerosene-Based Turbo-Pump Design Scheme*

According to the operating Mach number (4~7) and altitude (17km~30km) of scramjet system, the operation parameters of kerosene turbo-pump at design state were designed as Table 1.

Table 1. Operation parameters of the kerosene turbo-pump at design state.

Parameters	Nomenclature	Value
ω	Rotating speed of turbo-pump (rpm)	100000
q_{mp}	Fuel mass flow rate of pump (kg/s)	1
p_{p1}	Fuel pressure at pump inlet (MPa)	≤0.5
p_{p2}	Fuel pressure at pump outlet (MPa)	6.2
T_p	Fuel temperature at pump inlet (K)	300
q_{mt}	Fuel mass flow rate of turbine (kg/s)	0.95
p_{t1}	Fuel pressure at turbine inlet (MPa)	5.5
p_{t2}	Fuel pressure at turbine outlet (MPa)	2.5
T_t	Supercritical/cracking fuel temperature at turbine inlet (K)	850

The physical model in present paper is the structure schematic drawing of kerosene-based turbo-pump shown in Fig. 1, mainly consisting of turbo-pump shell, inducer, centrifugal pump impeller, turbine disk, shaft, and so on. The main structure parameters of turbo-pump are shown in Table 2.

2.2. *Thermophysical Property Evaluation of Supercritical/Cracking Kerosene*

Huang et al.[12] showed that the cracked product distribution of n-alkanes is similar to that of hydrocarbon fuel; and Xu et al.[13] verified that the thermophysical properties of n-decane are similar to kerosene. Therefore, n-decane was chosen as the surrogate of aviation kerosene for the present study. According to the experimental result of Ward et al.[14], the total reaction model for mild conversion of n-decane under pressure 3.45MPa ~ 11.38MPa was employed:

$$C_{10}H_{22} \rightarrow 0.151H_2 + 0.143CH_4 + 0.256C_2H_4 + 0.126C_2H_6 + 0.230C_3H_6 + 0.180C_3H_8 +$$
$$0.196C_4H_8 + 0.102C_4H_{10} + 0.171C_5H_{10} + 0.124C_5H_{12} + 0.195C_6H_{12} + 0.089C_6H_{14} +$$
$$0.169C_7H_{14} + 0.072C_7H_{16} + 0.152C_8H_{16} + 0.012C_8H_{18} + 0.053C_9H_{18} + 0.003C_9H_{20}$$

Table 2. Structure parameters of the kerosene turbo-pump at design state.

Parameters	Nomenclature	Baseline	Range
r_2	The radius of impeller outlet (mm)	42.8	[38, 50]
r_u	The radius of short blade inlet (mm)	19.2	[17, 22]
θ	The offset angle of short blade (°)	45	[40, 54]
B_1	The inlet installation angle of blade (°)	20	[20, 28]
B_2	The outlet installation angle of blade (°)	30	[20, 40]
b_1	The width of impeller inlet (mm)	6	[5, 5.6]
h_{en}	The outlet height of turbine nozzle (mm)	5.3	[4.5, 6]
d_{av}	The intermediate diameter of turbine blade (mm)	26	[24, 28]
b_b	The width of turbine blade (mm)	8	[7, 9]
n	The number of turbine nozzles	3	/

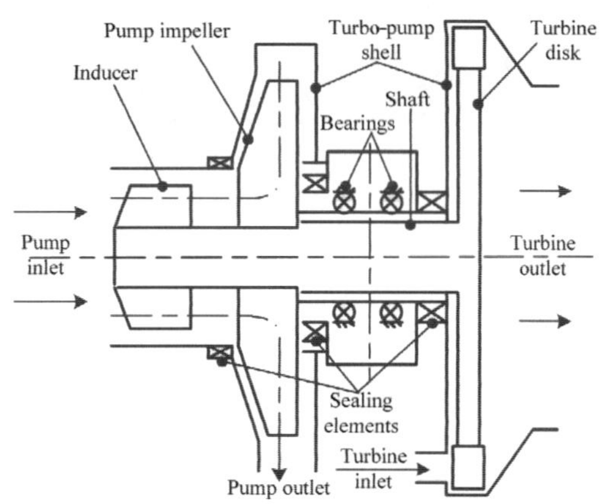

Fig. 1 Structure schematic drawing of kerosene-based turbo-pump.

The molar reaction rate of n-decane is:

$$-\frac{dC_R}{dt} = A \cdot e^{-\frac{E_a}{RT}} \cdot C_R \tag{1}$$

where A refers to the pre-exponential factor, 1.6×10^{15} s^{-1}; E_a refers to the activation energy, 263.7 kJ/mol; R refers to the molar gas constant, 8.314 J/(mol·K).

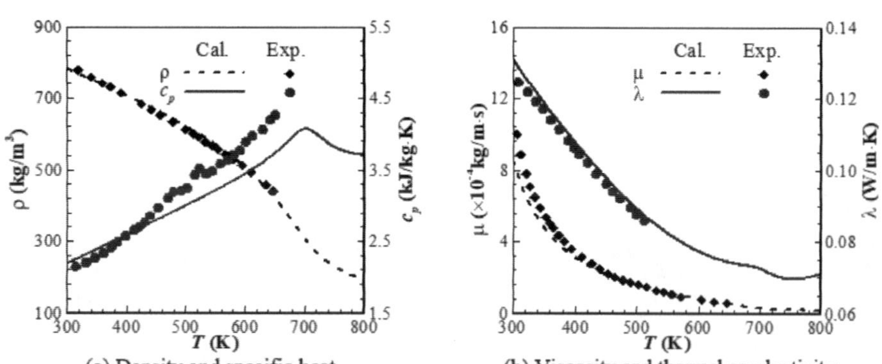

Fig. 2 Comparison between calculated thermophysical properties and experimental data at 5 MPa.

Based on the above pyrolysis mechanism, NIST Supertrapp software[15] was applied for approximately predicting the thermophysical properties of the cracked products. This property-evaluation method briefly described in this section have been used to calculate the kerosene thermophysical properties at 5 MPa. As shown in Fig. 2, comparisons were made to the published experimental data up to 650 K[16-18] except for the thermal conductivity, for which the available experimental data is limited to 513 K[19]. All the average errors between the calculated data and experimental one is within 13%. Good agreement has been reached between the two sets of data.

2.3. Numerical methods and mesh generation

In order to simulate the kerosene flows inside turbo-pump, the conservation equations of mass, momentum, energy, and species were numerically solved through finite volume method provided by the pressure-based solver in the CFD commercial package, ANSYS® 15- FLUENT. RNG k-ε turbulence model was adopted for treating turbulent flow, and standard wall functions were used to determine the flow variables in the near-wall region. The SIMPLEC algorithm was used to resolve the coupling between velocity and pressure, and the discretization scheme of second-order (upwind) was chosen to handle all terms in governing equations. Additionally, efficient thermodynamical properties of supercritical/cracking kerosene were computed as a function of temperature, pressure and n-decane mass fraction; then, User-defined-function (UDF) about the pyrolytic kinetics model and property-evaluation method was coded and implemented into FLUENT software.

Herein, the Solidworks software was used to generate 3D model of centrifugal pump and axial-flow turbine, which can be seen in Fig. 1. The solution domains of turbo-pump were divided into three sections, respectively; and 3D unstructured mesh with different grid density was applied to different sections of turbine and pump. The grid density was clustered at the vicinity of the wall and rotors of turbo-pump, for the required resolution precision in the boundary layer. Then, all sections of turbine and pump was jointed on the basis of the continuous joint grid technology, respectively; and no slippery wall was specified here. Non-inertial coordinate system was applied for the flow-field around pump impeller and turbine rotor-cascade[20, 21]. The boundary conditions of mass flow inlet and pressure outlet were employed for pump, and the boundary conditions of pressure inlet and pressure outlet for turbine; other variables at boundary regions were extrapolated from the interior.

Prior to detailed numerical studies, the independence of turbo-pump's performance parameters from grid number was checked to ensure the numerical accuracy. The meshes shown in Fig. 3 and Fig. 4 were adopted as the grid-independent solution for steady-state calculations hereinafter; the grid sums of different sections of pump and turbine were about 3170000 and 2630000, respectively.

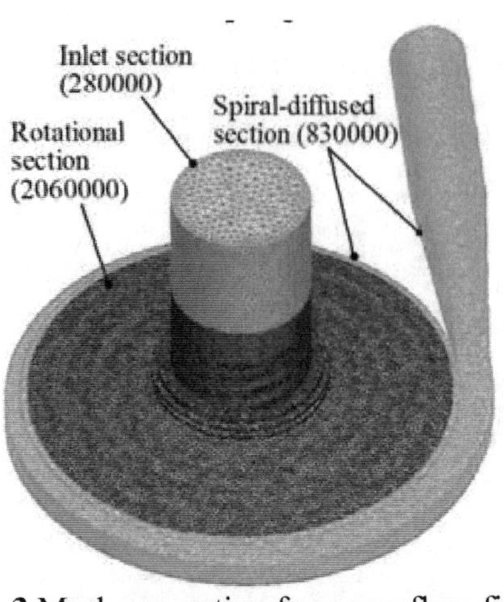

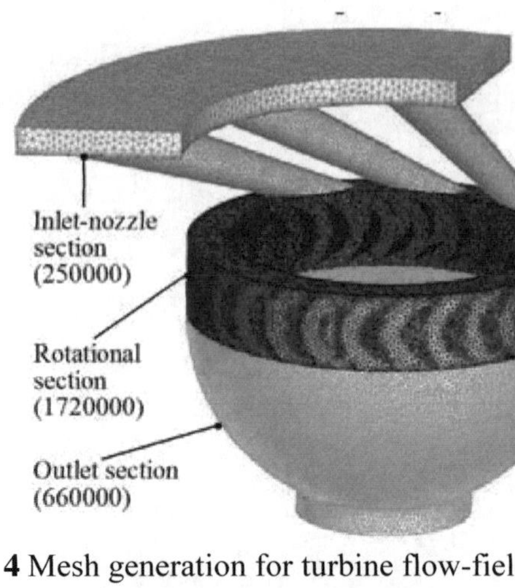

Fig. 3 Mesh generation for pump flow-field. **Fig. 4** Mesh generation for turbine flow-field.

2.4. Preliminary analysis on internal flow field of turbo-pump

The relevant boundary conditions of the fundamental calculation case for pump and turbine were specified according to the operating parameters in Table 1. The cross-sectional flow-fields of the centrifugal pump and axial-flow turbine at design state were predicted as shown in Fig. 5.

From Fig. 5 (a), it could be observed that no vortex exists in the rotational section of pump, and the streamline is smooth around the blades; the fuel pressure at the inlet of the impeller is lowest, then increases rapidly as the blades rotate to work. When the liquid flows into the diffused channel, the effect of viscous converts a small amount of kinetic energy into pressure energy; moreover, a vortex appears at the outlet of the pump. Fig. 5 (b) displays that the static pressure of the subsonic working medium undergoes a process of abrupt decrease when flowing inside the nozzle of turbine. This means that large pressure energy is translated into kinetic energy for driving turbine blades, just as the streamline direction shows. Then the kinetic energy of working medium decreases notably after doing work, which results in a vortex at the exit of turbine blades. In general, the simulation results are consistent with the basic flow-field characteristics of the turbo-pump.

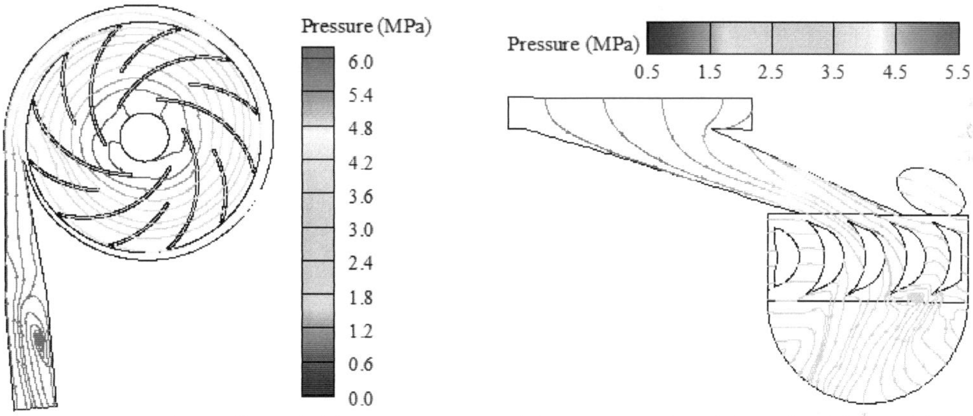

(a) Cross-sectional flow field for pump. (b) Cross-sectional flow field for turbine.

Fig. 5 Detailed distributions of streamline and static pressure within turbo-pump at design state.

In this paper, shaft power (P_t) and efficiency (η_t) are the performance parameters for turbine, while developed head (ΔH_p) and efficiency (η_p) for pump. According to the above numerical result and Eqs. (5) ~ (8), the values of ΔH_p, η_p, P_t and η_t are 5.78 MPa, 50.2 %, 25.95 kW and 53.5 %, respectively.

$$\Delta H_p = p_{p2} - p_{p1} \tag{2}$$

$$P_t = M \cdot \frac{2\pi \cdot \omega}{60} \tag{3}$$

$$\eta_p = \frac{q_{mp} \cdot \Delta H_p}{\rho \cdot P_p} \tag{4}$$

$$\eta_t = \frac{P}{q_{mt} \cdot L_{ad}} = \frac{P}{q_{mt}} \cdot \frac{k-1}{k \cdot R \cdot T} \left[1 - \left(\frac{p_{t2}}{p_{t1}} \right)^{\frac{k-1}{k}} \right]^{-1} \tag{5}$$

where M is the torque of pump; L_{ad} is the isentropic expansion work; k is the specific heat ratio.

3. Results and Discussion

3.1. Automatic simulation method for kerosene-based turbo-pump

The comprehensive performance analysis of turbo-pump requires the accurate data from large amount of the experiment or high-fidelity numerical simulation, which needs huge experimental fees or calculation cost. Hereinafter, automatic simulation method was adopted to alleviate such difficulty by balancing the requirements between the calculation cost and the accuracy.

Automatic simulation method synthetically adopts Design of Experiments (DOE) technique, surrogate model (such as polynomial-based response surface (PRS) model, Kriging model, and so on), and high-fidelity numerical simulation method in Section 2. The desired results of performance analysis could be acquired efficiently through the following process in Fig. 6.

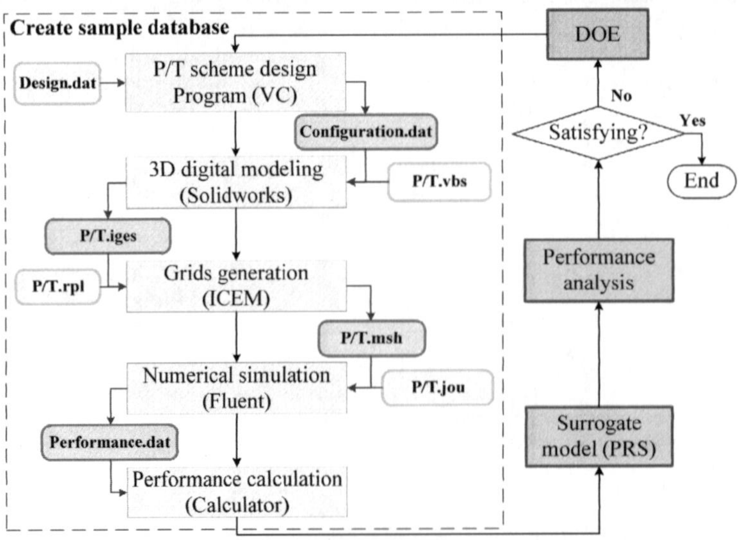

Fig. 6 Automatic simulation process organized by iSIGHT platform.

Herein, DOE technique is employed to explore the given design space and obtain enough sampling points. In order to train the sampling points efficiently, iSIGHT software is used to integrate VC (Microsoft Visual C++) compiler, Solidworks software, ICEM software and Fluent software by calling corresponding batch files (Design.dat, P/T.vbs, P/T.rpl, P/T.jou). The steps (scheme design, 3D modeling, mesh generation, numerical simulation and performance calculation) were implemented subsequently, and the information exchanging between them includes parameterized data files (Configuration.dat, P/T.iges, P/T.msh and Performance.dat) and well-defined variable mapping relationship. Based on the trained sampling database, surrogate models (PRS) were obtained for corresponding performance analysis of turbo-pump.

As the most frequently applied surrogate model, PRS model of first-order or second-order (Eq. (9)) can map complex design space effectively and filter some data noise easily. The coefficients that best fits the trained database was acquired by regression process such as linear Least-Squares Analysis[22].

$$y = c_0 + \sum_{i=0}^{n} c_i x_i + \sum_{i=0}^{n} c_{ii} x_i^2 + \sum_{i=0}^{n-1} \sum_{j=i+1}^{n} c_{ij} x_i x_j \tag{6}$$

where y refers to the response of PRS; x refers to n-dimensions design variables; c refers to the quadratic polynomial coefficients.

3.2. Similarity analysis

Similarity Criterion[23] for viscous fluid flow contains geometric similarity, kinematic similarity and dynamic similarity. Thus, for the obtained design scheme with baseline structure parameters in Table 2, the following relationships exist among ΔH_p, P_t and ω:

$$\frac{\Delta H_p}{\Delta H_{rated}} = \left(\frac{q_{mp}}{q_{mrated}}\right)^2 = \left(\frac{\omega}{\omega_{rated}}\right)^2 \quad (7)$$

$$\frac{P_t}{P_{rated}} = \left(\frac{q_{mp}}{q_{mrated}}\right)^3 = \left(\frac{\omega}{\omega_{rated}}\right)^3 \quad (8)$$

Based on a serious of flow-field simulations at rated rotating speed of 100000 rpm, the characteristic curves of ΔH_{rated} (second order) and P_{rated} (first order) with respect to mass flow rate (q_{mrated}) were obtained through polynomial fitting (Eq. (9)). Then, the characteristic curves of ΔH_p and P_t with respect to mass flow rate, at different rotating speed (80000, 90000, 110000, and 120000 rpm), were derived from Eqs. (12) and (13).

$$\Delta H_p = \Delta H_{rated}\frac{\omega^2}{\omega^2_{rated}} = \left(c_{2p}q^2_{mrated} + c_{1p}q_{mrated} + c_{0p}\right)\frac{\omega^2}{\omega^2_{rated}} = c_{2p}q^2_{mp} + c_{1p}q_{mp}\frac{\omega}{\omega_{rated}} + c_{0p}\frac{\omega^2}{\omega^2_{rated}} \quad (9)$$

$$P_t = P_{rated}\frac{\omega^3}{\omega^3_{rated}} = \left(c_{1t}q_{mrated} + c_{0t}\right)\frac{\omega^3}{\omega^3_{rated}} = c_{1t}q_{mt}\frac{\omega^2}{\omega^2_{rated}} + c_{0t}\frac{\omega^3}{\omega^3_{rated}} \quad (10)$$

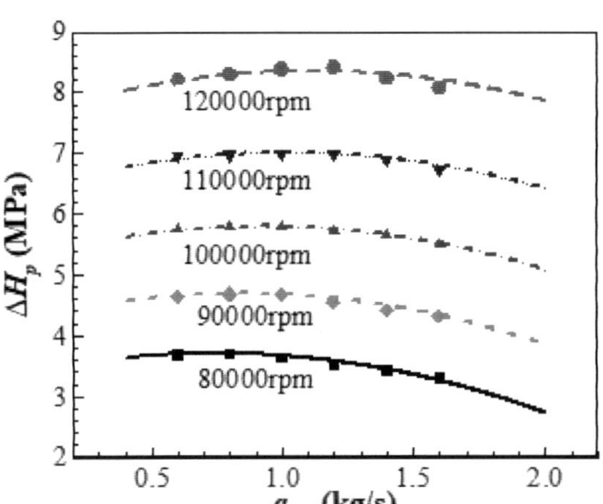

Fig. 7 Similarity analysis of ΔH_p.

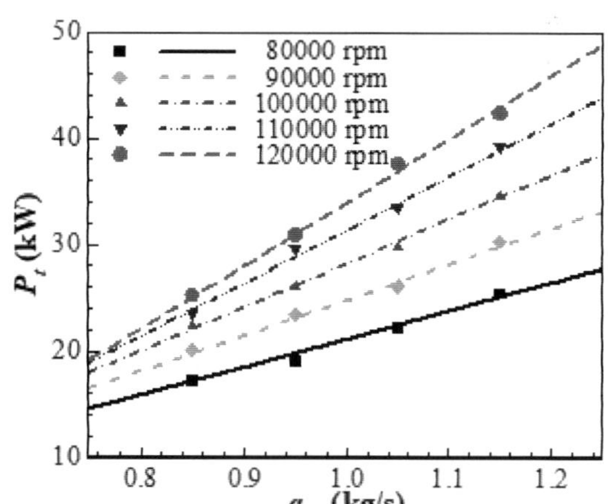

Fig. 8 Similarity analysis of P_t.

Fig 7 and 8 display the five characteristic curves of ΔH_p and P_t with respect to mass flow rate at various rotating speeds of 80000, 90000, 100000,110000, and 120000 rpm; and the spot data from automatic simulation process was used to verify corresponding characteristic curves. The maximum discrepancy between characteristic curves and spot data is below 5%; thus ΔH_p and P_t agree well with

Similarity Criterion[23] in general, proving the accuracy and reasonability of the turbo-pump design scheme and simulation method.

3.3. Pareto front analysis

Due to the complex structure of turbo-pump, it will be difficult to find out the variation range of performance parameters within the limits of structure parameters. According to the automatic simulation method, the given ranges of main structure parameters in Table 2 were taken as the design space; then 256 sampling points were obtained by means of Optimal Latin Hypercube method in DOE, which spread evenly and afford to capture higher-order effects. At last, second-order PRS models (Eq. (9)) of four performance parameters (ΔH_p, η_p, P_t and η_t) with respect to main structure parameters in Table 2 were obtained. Based on error analysis of R^2, the confidence levels of four second-order PRS models are all above 90%, which meets the precision requirement for prediction.

In order to comprehensively understand the relationship between the performance parameters, Pareto front depicted in Fig. 9 was also obtained on the basis of the four second-order PRS models. Fig. 9 presents 6000 sampling points from the design space shown in Table 2. These sampling points were trained by the established second-order PRS models; this method is less time-consuming than high-fidelity numerical simulation while precision guaranteed.

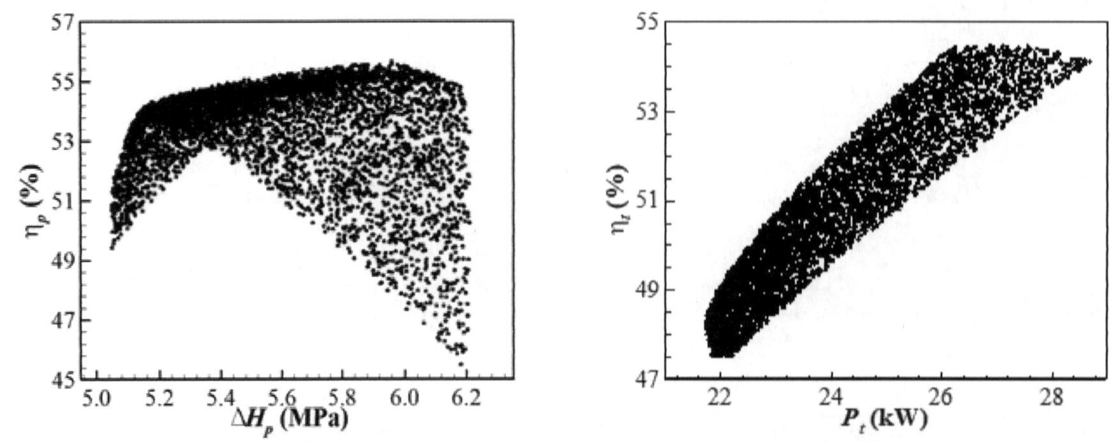

(a) ΔH_p versus η_p. (b) P_t versus η_t.

Fig. 9 Pareto front for turbo-pump performance parameters.

From Fig. 9 (a), η_p rises first and falls later as ΔH_p increases. This can be concluded that ΔH_p and η_p have their own optimal value; although the two optimal values are close to each other, they cannot reach the optimal solution at the same time. However, from Fig. 9 (b), it is clearly observed that η_t decreases with the decrease of P_t. The objective value of P_t should be ascertained according to the required shaft power of pump (P_p), while the optimal value of η_t is fixed in the design space of turbine.

4. Conclusion

In this paper, an original investigation was carried out on the automatic simulation method for kerosene-based turbo-pump with small flow rate, which is mainly used in the fuel supply system of kerosene-based scramjet with expansion cycle.

By means of the automatic simulation method coupled with high-fidelity numerical simulation, the performance analysis process of kerosene-based turbo-pump saves time and effort compared to traditional approaches. The similarity analysis indicates that ΔH_p and P_t agree well with Similarity Criterion, proving the accuracy and reasonability of the turbo-pump design scheme and simulation method; the Pareto front analysis shows that ΔH_p is negatively related to η_p in general, but the contrary is the case for P_t and η_t . As for other aspects of applications, further investigation could be carried out to exploit the potential of the automatic simulation method.

Acknowledgments
This research work was financially supported by National Natural Science Foundation of China (Grant No. 11272344 and 11572346).

References

[1] H. Zhang, C.B. Shen, X.Y. Wu. Design and optimization of hydrocarbon-fueled scramjet start-up scheme with expansion cycle. 62st International Astronautical Congress. Cape Town, 2011.
[2] C.B. Shen, X.Y. Wu, X.F. Chen. Study on the thermodynamic cycle characteristics of the scramjet fuel feed system. 65st International Astronautical Congress. Toronto, 2014.
[3] X.Y. Wu, J. Yang, H. Zhang. System design and analysis of hydrocarbon scramjet with regeneration cooling and expansion cycle. Journal of Thermal Science, 2015, 24 (4): 350-355.
[4] J.H. Kim, H.C. Lee, J.H. Kim. Design techniques to improve the performance of a centrifugal pump using CFD. Journal of Mechanical Science and Technology, 2015, 29 (1): 215-225.
[5] M.P. Errera, S. Chemin. Optimal solutions of numerical interface conditions in fluid-structure thermal analysis. Journal of Computational Physics, 2013, 245 (14): 431-455.
[6] S.R. Shah, S.V. Jain, R.N. Petal. CFD for centrifugal pump:a review of the state-of-the-art. Procedia Engineering, 2013, 51 (5): 715-720.
[7] N.P. Jaiswal. CFD analysis of centrifugal pump: a review. International Journal of Engineering Research and Applications, 2014, 4 (5): 175-178.
[8] J. Pei, W.J. Wang, S.Q. Yuan. Optimization on the impeller of a low-specific-speed centrifugal pump for hydraulic performance improvement. Chinese Journal Mechanical Engineering, 2016, 29 (5): 992-1002.
[9] D.Y. Hu, J.X. Mao, R.Q. Wang, et al. Optimization strategy for a shrouded turbine blade using variable-complexity modeling methodology. AIAA Journal, 2016, 54 (9): 2808-2818.
[10] M.D. Ahmed, A.J. Ayad, M. Saad. Development of three-dimensional optimization of a small-scale radial turbine for solar powered brayton cycle application. Applied Thermal Engineering, 2017, 111: 718-733.
[11] X.Y. Wu, J. Yang, X. Jin. Experimental study of operation performance for hydrocarbon fuel pump with low specific speed. Jouranl of Thermal Science, 2017, 26 (5): 440-447.
[12] H. Huang, D.R. Sobel, L.J. Spadaccini. Endothermic heat-sink of hydrocarbon fuels for scramjet cooling. AIAA 2002-387.
[13] K.K. Xu, H. Meng. Analyses of surrogate models for calculating thermophysical properties of aviation kerosene RP-3 at supercritical pressures. Science China Technological Sciences, 2015, 58 (11): 510-518.
[14] T.A. Ward, J.S. Ervin, S. Zabarnick. Pressure effect on flowing mildly-cracked n-decane. Journal of Propulsion and Power, 2005, 21 (2): 344-355.
[15] J.F. Ely, M.L. Huber. NIST standard reference database 4-NIST thermophysical properties of hydrocarbon mixtures. National Institute of Standards, Gaithersburg, MD, 1990.
[16] H.W. Deng, C.B. Zhhang, G.Q. Xu, et al. Viscosity measurements of endothermic hydrocarbon fuel from (298 to 788) K under supercritical pressure conditions. Journal of Chemical and Engineering Data, 2012, 57 (2): 358-365.
[17] H.W. Deng, K. Zhu, G.Q. Xu, et al. Isobaric specific heat capacity measurement for kerosene RP-3 in the near-critical and supercritical regions. Journal of Chemical and Engineering Data, 2012, 57 (2): 263-268.
[18] H.W. Deng, C.B. Zhang, G.Q. Xu, et al. Density measurements of endothermic hydrocarbon fuel at sub- and supercritical conditions. Journal of Chemical and Engineering Data, 2011,

56 (6): 2980-2986.

[19] G.Q. Xu, Z.X. Jia, J. Wen, et al. Thermal-conductivity measurements of aviation kerosene RP-3 from (285 to 513) K at sub- and supercritical pressures. International Journal of Thermophysics, 2015, 36 (4): 620-632.

[20] S.B. Riffat, X. Zhao, P.S. Doherty. Investigation of the flow in an impulse micro turbine using numerical CFD predictions and laboratory testing. International Journal of Ambient Energy, 2002, 23 (4): 171-178.

[21] J.H. Kim, B.M. Cho, S. Kim. Design technique to improve the energy efficiency of a counter-rotating type pump-turbine. Renewable Energy, 2016, 101 (7): 647-659.

[22] X.Y. Wu, S.B. Luo, X.Q. Chen, et al. Global design optimization for hypersonic scramjet propulsive flowpath. AIAA 2006-1911.

[23] N.C. Zhu, G.Q. Liu. Principle of liquid-propellant rocket engine. Beijing: China space press,1994.

IOP Conf. Series: Journal of Physics: Conf. Series **1064** (2018) 012023 doi:10.1088/1742-6596/1064/1/012023

Numerical Simulation and Analysis for the Decay Heat Exchanger of China Prototype Fast Reactor

Mengmeng Liu*, Guangdong Song and Kangwei Peng

China Institute of Atomic Energy, Beijing, China.

*Corresponding author e-mail: lmm_hit@163.com

Abstract. Decay heat exchanger (DHX) is an essential piece of equipment of the decay heat removal system (DHRS) in a pool-type sodium-cooled fast reactor. The main function of DHX is to transfer heat from primary circuit fluid (liquid sodium) to secondary circuit fluid (liquid sodium) of DHRS, which can reduce the high temperature caused by decay heat in the reactor core. In this paper, CFD method was used to simulate the statistic three-dimensional flow field of DHX which is used in China Prototype Fast Reactor under several different operating conditions. The numerical results under rated operating conditions were compared with the design values and the comparison was fit in with expectation, which demonstrated the validity of the numerical model. The study of the flow field of DHX under designed operating conditions was conducted to obtain the heat transfer features and the flow characteristics. The results show that the design of DHX in China Prototype Fast Reactor is sufficient to the purpose to timely bring the decay heat out of the reactor core and DHX meets the safety design requirements from the perspective of heat transfer. The work in this paper provided references for further design improvement of the DHX in a pool-type sodium-cooled fast reactor.

1. Introduction

China Prototype Fast Reactor is a kind of pool-type sodium cooled fast reactor, reactor core and most pieces of the essential equipment are immersed in the liquid sodium pool. The decay heat removal system (DHRS) which consists of 4 independent loops is an important safety system. There are one DHX and one Sodium-Air Heat Exchanger in each loop. The whole system is driven by natural circulation and delivers the heat to the air finally through the Sodium-Air Heat Exchanger. The primary fluid flow driving force of DHX is the temperature difference and the potential difference between the reactor core and the DHX, while the secondary fluid flow driving force of DHX is the temperature difference and the potential difference between the Sodium-Air Heat Exchanger and the DHX.

Up to now, several scholars have done research about the decay heat removal system and have applied for the patent of the fully passive decay heat removal system which utilizes one kind of decay heat exchanger [1, 2].

Research about the DHX of China Experimental Fast Reactor (CEFR) have been done by Chinese scholars [3, 4]. However, the installed capacity of China Prototype Fast Reactor is much higher than that of CEFR, which means that DHX designed thermal power of China Prototype Fast Reactor is larger than that of CEFR. The design of DHX has already been conducted to obtain the structure parameters which meet the demand of the higher thermal power from DHRS and other design requests.

Content from this work may be used under the terms of the Creative Commons Attribution 3.0 licence. Any further distribution of this work must maintain attribution to the author(s) and the title of the work, journal citation and DOI.
Published under licence by IOP Publishing Ltd

The DHX designed for China Prototype Fast Reactor has been simulated by using CFD method with ANSYS CFX. The temperature and thermal power have been calculated to make sure that the design of DHX is reasonable. Heat transfer and flow characteristics have been analyzed according to the results of numerical simulation to further understand how DHX functions and what the disadvantages are for heat transfer, which can help to improve the design of DHX in the future.

2. Design parameters and geometric model

Decay heat exchanger (DHX) consist of Central pipe Shell, Outlet pipe, Discharge chamber, Inlet window, Outlet window, Tube bundle and Bottom chamber. Inner and outer Shell form the flow pass for the primary fluid, while Central pipe, Bottom chamber and tube bundle form the pass for the secondary fluid.

The heat exchanging area is the area between the upper tube plate and the bottom tube plate.

When reactor accident happens, the temperature of the sodium pool is 535℃, which is the inlet temperature of DHX. The designed thermal power of one single DHX is 9MW. The outlet temperature is requested to be 450℃. Other design parameters are listed in Table 1.

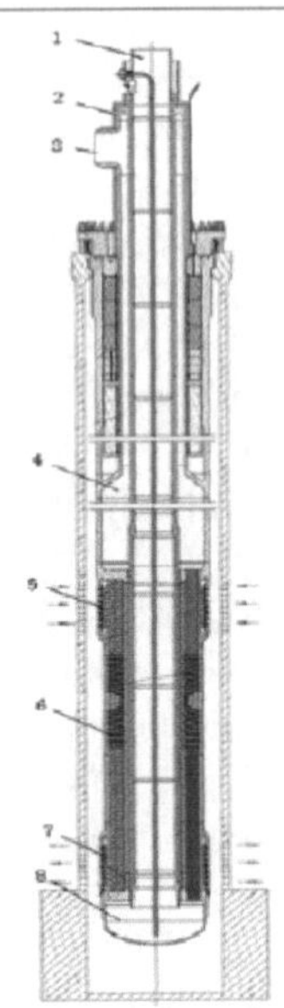

No.	Name of component
1	Central pipe
2	Shell
3	Outlet pipe
4	Discharge chamber
5	Inlet window
6	Tube bundle
7	Outlet window
8	Bottom chamber

Figure 1. The schematic of DHX

Table 1. Design parameters of DHX

No.	Parameters	Units	Values
1	Designed thermal power	MW	9
2	Primary inlet temperature	℃	535
3	Primary outlet temperature	℃	450
4	Secondary inlet temperature	℃	380
5	Secondary outlet temperature	℃	520
6	Primary mass flow	kg/s	84
7	Secondary mass flow	kg/s	50
8	Designed temperature	℃	580
9	Secondary designed pressure	MPa	1.0
10	Secondary operating pressure	MPa	0.8

Physics properties of liquid sodium for calculation is obtained from reference [5]. The calculating formulas are as follows:

$$\rho = 16.0185 \times [59.566 - 7.9504 \times 10^{-3}(1.8t + 32) \\ -0.2872 \times 10^{-6} \times (1.8t + 32)^2 + 0.0603 \times (1.8t + 32)^3] \tag{1}$$

$$Cp = 4186.8 \times [0.389352 - 1.10599 \times 10^{-4}(1.8T) \\ +3.41178 \times 10^{-8} \times (1.8T)^2] \tag{2}$$

$$\lambda = 1.72958 \times [54.306 - 1.878 \times 10^{-2} \times (1.8t + 32) \\ +2.0914 \times 10^{-6} \times (1.8t + 32)^2] \tag{3}$$

$$\eta = 4.134 \times 10^{-4} \exp\{2.302585 \times [1.0203 \\ +397.17 / (1.8T) - 0.4925 \times \lg(1.8T)]\} \tag{4}$$

ρ is the density of the liquid sodium (kg/m3), Cp is the constant-pressure specific heat (J/(kg • K)), λ is the thermal conductivity (W/(m • K)), and η is the viscosity (Pa • s). These properties are all functions of temperature T (K) or t(℃).

The structure of the outlet window and the inlet window are similar which is a annular plate with orifices showed in Fig 2. There are 15 rows of orifices with a diameter of 50mm and each row contents 24 orifices.

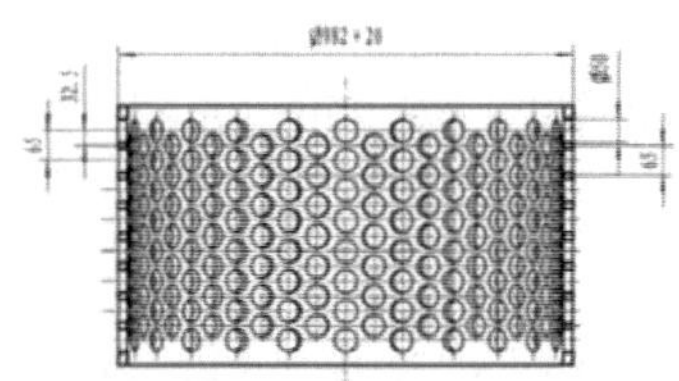

Figure 2. Structure of the inlet window

Figure 3 shows the structure of the tube bundle. Heat transfer tube bundle is formed by 285 and is divided into 5 layers. There are 69 tubes evenly distributed in the innermost layer. The number of each layer increases by 6. Figure 4 shows that the tubes are periodically distributed in a period of 120°. To reduce the computational complexity, 1/3 of the total model is selected for calculation, such as Fig 3 shows.

Figure 3. Tube bundle geometric model

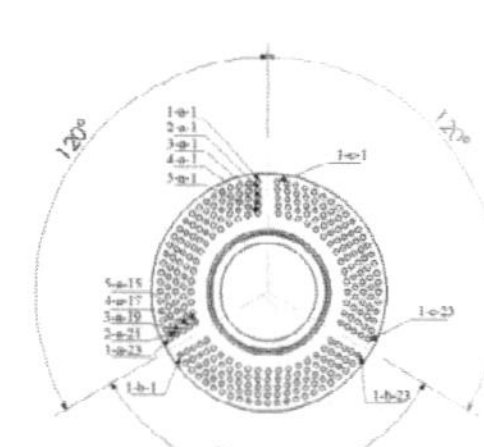

Figure 4. Tube bundle geometric model

The calculation model consist of the tube bundle area, primary fluid flow field and the secondary fluid field.

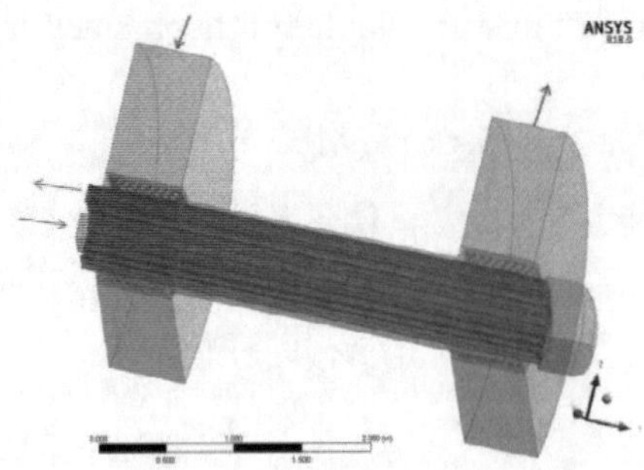

Figure 5. 1/3 part of the DHX model for calculation

3. Mathematic Models

3.1. Basic Assumptions

According to the physic model of the DHX described above, several assumptions have been made as follow:

1) The heat transfer system is at a steady state and adiabatic;

2) Because the fluid flow driving force of DHX is the temperature difference and the potential difference, it is hard to set the boundary conditions and to give an exact inlet flow direction. In order to make the flow condition be as close as possible to the real situation, an annular region was added surrounding the inlet and outlet window and the mass flow of the outer wall of the annular region was set as a constant to make sure that the fluid flows from a far place and flows into the inlet window with a low velocity. In a similar way, there is also another annular region surrounding the outlet window to make sure that the fluid flow to different directions after it comes out of the outlet window.

3) In early analyzing stage, the study focus on the characteristics of the entire flow field. It has been proved that using straight tubes instead of regional spiral tubes have little effect on the heat transfer features near inlet/outlet area and most of the straight tubes area. Thus it is assumed that the tubes in the DHX are all straight tubes.

3.2. Basic Governing Equations

The flow and heat transfer of fluid can be described by the mass conservation law, the momentum conservation law and the energy conservation law. The governing equations are the mathematical descriptions whose universal form of Euler method is as follows:

$$\frac{\partial(\rho\phi)}{\partial t} + div(\rho U\phi) = div(\Gamma_{\phi} grad\phi) + S_{\phi}) \tag{5}$$

ϕ is an universal variable which can represent variables such as mass-flow rate, velocity, temperature and enthalpy of one particular fluid control volume. ρ is the density and U is the velocity of the fluid in the control volume. Γ_{ϕ} is the Generalized diffusion coefficient. S_{ϕ} is the source term.

4. Boundary Conditions and Mesh

Table 2 shows the boundary settings for numerical simulation. Primary and secondary inlets are all set as mass-flow-inlet.

IOP Conf. Series: Journal of Physics: Conf. Series **1064** (2018) 012023 doi:10.1088/1742-6596/1064/1/012023

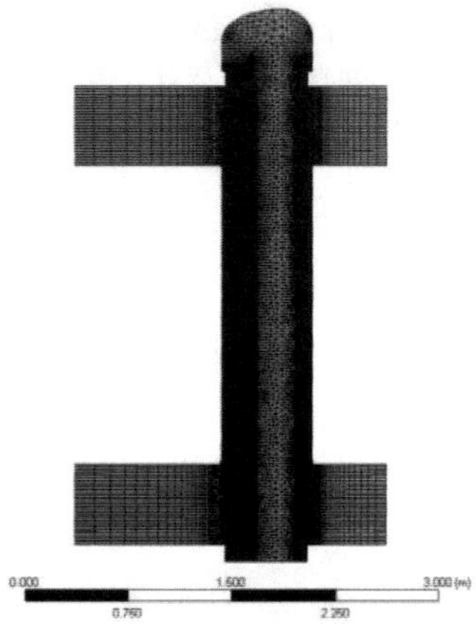

Figure 6.(a) Mesh of the calculation area

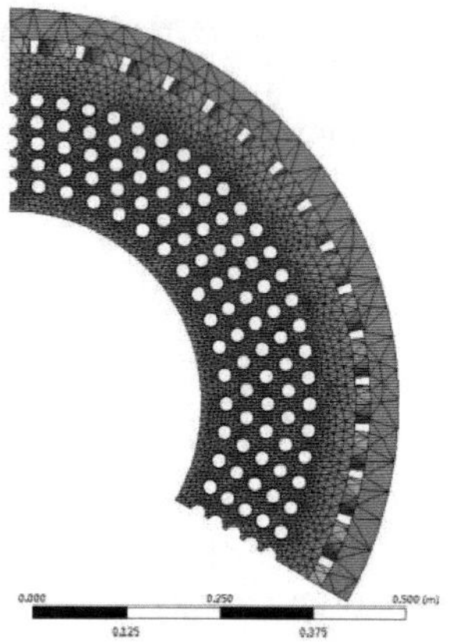

Figure 6.(b) Mesh of primary inlet area

Figure 6. The mesh of DHX for numerical simulation

Table 2. Boundary settings for numerical simulation

Boundary name	Boundary type	Values
Primary inlet	mass-flow-inlet	28kg/s
Secondary inlet	mass-flow-inlet	16.67kg/s
Primary outlet	opening	-
Secondary outlet	outlet	-

Figure 6 shows the mesh of DHX for numerical simulation. The grid quantity is about 30 million.

5. Organization of the Text Results and Analysis

5.1. The Outlet Temperature of The Primary and Secondary Fluid

Primary/Secondary outlet thermal parameters (calculated values) are listed in Table 3. It can be found that the outlet temperature of the primary and secondary fluid have a tiny deviation from the design value.

Table 3. Primary/Secondary outlet thermal parameters

Section	Design value	Calculated value	Deviation from design value
Primary outlet cross section mean temperature	723.15K/ 450℃	719.46K/ 446.31℃	-0.01%
Secondary outlet cross section mean temperature	793.15K/520℃	803.73K/ 530.58℃	2.03%
Primary thermal power	9MW	9.39MW	Deviations between primary and secondary thermal power 2.56%
Secondary thermal power	9MW	9.63MW	

The deviation may be caused by the simplification of the calculation model (such as the assumption that the tubes are all straight). The deviation between primary and secondary thermal power is 2.56%, which is acceptable and meets the calculation requirements.

IOP Conf. Series: Journal of Physics: Conf. Series **1064** (2018) 012023 doi:10.1088/1742-6596/1064/1/012023

5.2. Temperature Distribution Features of The Primary/Secondary Flow Field

The temperature distribution of the tube bundle outlet plane is shown in Fig 7. It can be found that the temperature of the innermost layer is the lowest and the temperature in tubes increases from inside to outside, the highest temperature is at the outermost layer.

Figure 7. Tube bundles outlet plane temperature field (secondary flow field)

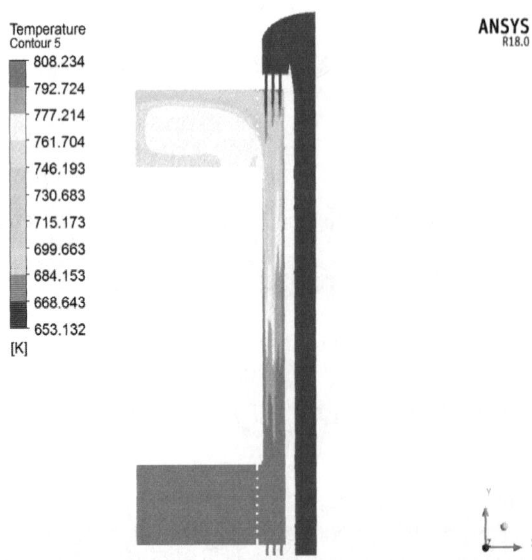

Figure 8. Temperature distribution at the axial section

The heat transfer performance of a shell-and-tube exchanger depends on the coupling effect between fluid in the tube pass and the shell pass. In the regions near the inlet window, the primary fluid scours the tube bundle with cross flow which is perpendicular to the axis of the tube bundle. Along the primary fluid flow direction, the flow is gradually changing into axial flow until the fluid comes to the outlet window and flows cross the tube bundle.

To study further about how the flow characteristics influent the heat transfer, the temperature at the axial section has been extracted as showed in Fig 8.

It can be found in Fig 8 that the temperature distribution of both the primary fluid and the secondary fluid is relatively even in the region near the inlet windows, while in the region near the outlet window the temperature distribution is relatively large.

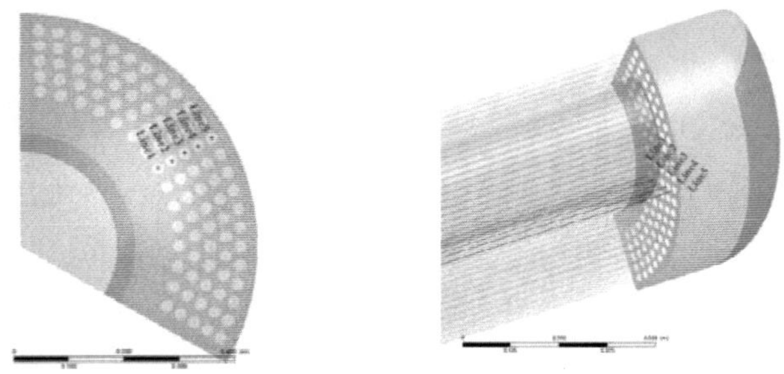

Figure 9. The positions of the lines inside the tubes

To get a more precise regulation of the distribution along the tube, the temperature data is extracted from five central lines in five different tubes from each tube layer.

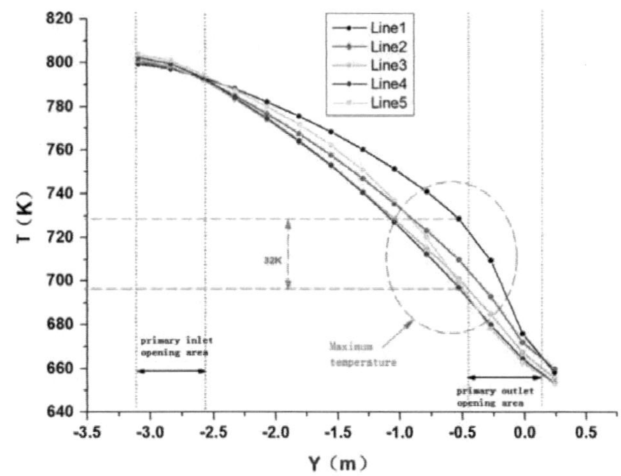

Figure 10. Temperature distribution along the five lines inside the tubes(secondary fluid flow region)

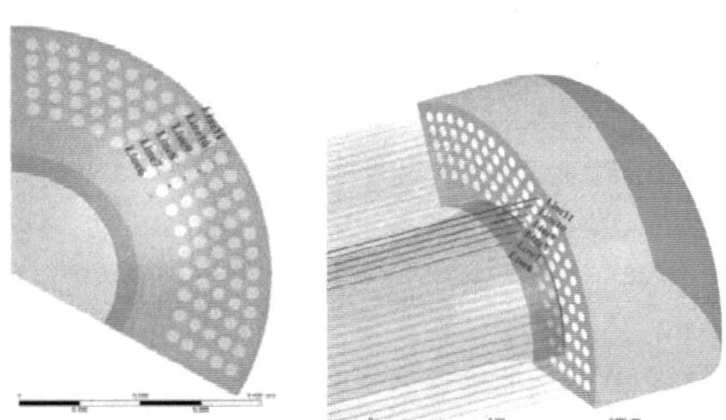

Figure 11. The positions of the lines outside the tubes

It can be found in Fig 10 that, the secondary fluid is heated by the primary fluid along the secondary flow direction (-Y direction). The maximum temperature difference happens in the region near the primary outlet window with a difference of 32K.

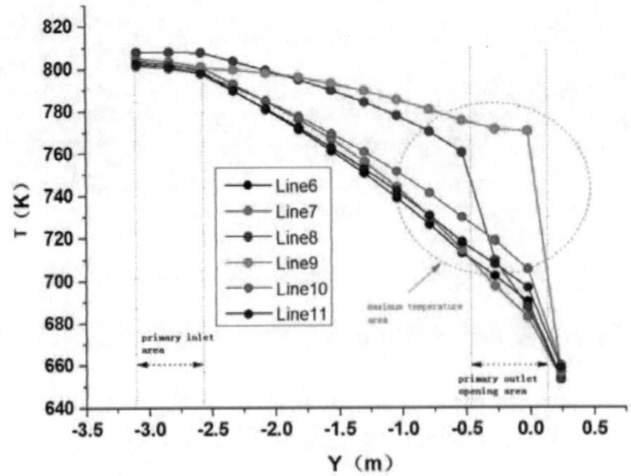

Figure 12. Temperature distribution along the five lines outside the tubes(primary fluid flow region)

The temperature data near the outer wall of the tubs is extracted from five lines outside five different tubes from each tube layer, as showed in Fig 11.

It is shown in Fig 12 that the primary fluid is cooled by the secondary fluid along the primary flow direction (+Y direction). The maximum temperature difference happens in the region near the primary outlet window with a difference of 90K.

6. Conclusions

The numerical simulation of the DHX designed for China Prototype Fast Reactor has been conducted to obtain the heat transfer and the flow characteristics in this paper. The conclusions are as follows:

The primary outlet average temperature of DHX is 446.31℃ while the secondary outlet average temperature is 530.58℃. The deviations between primary and secondary thermal power 2.56% which is acceptable and meet the calculation requirements.

According to the analysis about the temperature distribution inside and outside of the heat transfer tubes in different layer, the maximum temperature difference happens in the region near the outlet window both for the primary fluid and the secondary fluid. The maximum temperature difference in primary fluid is about 90K, while that in the secondary fluid is about 32K.

The uneven distribution in both primary and secondary fluid has negative effect on the heat transfer efficiency and may cause thermal stress on the device. Thus the future work is to focus on the improvement of the flow field to lower the unevenness of the temperature distribution based on the study in this paper in order to improve the design of the DHX.

References

[1] Eoh J H, Lee T H, Han J W, et al. Fully Passive Decay Heat Removal System for Sodium-cooled Fast Reactors that Utilizes Partially Immersed Decay Heat Exchanger[J]. 2014.

[2] Eoh J H, Lee T H, Han J W, et al. Fully Passive Decay Heat Removal System for Sodium-cooled Fast Reactors that Utilizes Partially Immersed Decay Heat Exchanger: US, US8670518[P]. 2014.

[3] Zuguo Chen, Yijun Xu. Thermal Hydraulics Analysis for Modified Sodium-to-Sodium Decay Heat Exchanger of CEFR [J]. Nuclear Science and Engineering, 2016, 36(4):441-448. (in Chinese)

[4] Yuheng Feng, Xuedong Qiao. Numerical Simulation of Immersion Cooler For China Experimental Fast Reactor.[J].Atomic Energy Science and Technology, 2007. (in Chinese)

[5] Huaiming Ju. Thermal Transfer Medium Properties Calculation Program and Data Manual.[M]. Atomic Energy Press, 1990. (in Chinese)

Numerical calculation of rising behavior and mass transfer for single CO_2 bubble in water

Xin Li[1], Jianlong Li[1], Pan Zhang[2] and Guanghui Chen[1,*]

[1]College of Chemical Engineering, Qingdao University of Science and Technology, Qingdao, China.
[2]College of Electromechanical Engineering, Qingdao University of Science and Technology, Qingdao, China.

*Corresponding author e-mail: guanghui@qust.edu.cn

Abstract. The rising behavior and mass transfer process of CO_2 bubble in the water were simulated and analyzed. The VOF model was applied to track the interface boundary. The CO_2 was added through an orifice with a diameter of 2.4 mm at the bottom. The bubble shapes through the rising process were shown and the aspect ratio was used to reflect the bubble deformation. The flow patterns around the bubble were also analyzed and its influence on mass transfer was discussed. The results show that the bubble equivalent diameter keeps constant though the bubble shapes change dramatically. Besides, the mass transfer process in the bubble rise section is different from that in the bubble formation and detachment sections.

1. Introduction

Bubble rising phenomenon often happens in the practical production processes, especially for distillation and reaction processes. The behaviors are complex when the conditions change. Bubble trajectories are quite different in various liquids and some parameters of bubbles, such as dimension, rising velocity, and flow patterns inside or outside the bubbles are also different. When the gas substances can be absorbed into the liquids, the mass transfer processes always occur with rising.

Mass transfer processes have been investigated by experiments in different devices and the results deepen the understandings of relevant fields. Hosoda et al. [1] studied the CO_2 bubble rose through water in a vertical pipe. Three kinds of pipes with different diameters were used and different initial bubbles behaviors were investigated. The ratio of pipe diameters and bubble diameters are applied as an important variable. In general, the rising velocity increases with bubble diameter increasing. However, the velocity stops increasing under some ratios in the vertical pipes. Besides, a diameter correlation was presented and its predictive accuracy was verified to be good. Aoki et al. [2] investigated the CO_2 bubble rising in the contaminated water based on the works of Hosoda et al. Only the pipe with a diameter of 12.5 mm was applied and the liquid was contaminated by surfactant. The mass transfer coefficient of contaminated water is lower compared with that of purified water and it decreases with the content of surfactant increasing. Then, Aoki et al. [3] studied the bubbles behaviors in alcohol aqueous solutions. Four different solutions, which are 1-pentanol, 1-heptanaol, 1-octanol, and 1-decanol, are applied as the liquids. The critical bubble diameters are reported under different

Content from this work may be used under the terms of the Creative Commons Attribution 3.0 licence. Any further distribution of this work must maintain attribution to the author(s) and the title of the work, journal citation and DOI.
Published under licence by IOP Publishing Ltd

pipe diameters. As for mass transfer, the coefficient of bubbles decreases with content of alcohol solutions increasing.

The CO_2 bubbles are more or less soluble in most liquids so that their diameters reduced continuously. Relevant experiment methods are quite important to improve the study accuracy. Nock et al. [4] analyzed the relationship between the mass transfer and properties of single bubbles. The image analysis sequence was applied to calculate the bubble diameters. And the camera was set on a moving platform to be able to record the whole bubble rising process. A mobile bubble surface transits to an immobile surface can be reflected by the transition from high mass transfer rate to the low one. Legendre and Zevenhoven [5] used the similar image analysis sequence to calculate the single bubbles and individual bubbles in small bubble swarms. The bubble behaviors in a tall tower were investigated. Though the theoretical studies about mass transfer were implemented, the experiments or numerical calculations of how the mass transfer influence the rising behavior or vice versa were not enough.

Saito's research team has focused on the experimental observation of CO_2 mass transfer for several years. Saito and Toriu [6] firstly used the LIF/HPTS method to visual the CO_2 dissolution. Three sections of rising single bubble were visualized: the linear-ascent, the first-inversion, and the second-inversion of the zigzag motion. The results show that the instantaneous mass-transfer coefficients increased in the linear-ascent section and second-inversion section. Huang and Saito [7] then studied the influence of bubble-surface contamination on instantaneous mass transfer. Compared with the CO_2 mass flux in the purified water, that in the contaminated water was two or three times less. On the photos of the bubble, there is a dark region below the bubble and the area of the region in the purified water is larger than that in the contaminated water. Huang and Saito [8] indirectly considered the action of the Marangoni convection and analyzed the forces on the bubbles. Three-dimensional wake structure was obtained from LIF visualization and this makes the mass transfer mechanism clearer [9].

Numerical calculations about CO_2 mass transfer can overcome the shortcomings of data acquisition in the experiments. The data can be calculated and obtained directly. In this paper, the CO_2 bubble rising with mass transfer in the water was calculated by the computational fluid dynamics software Fluent. VOF method was applied to track the gas-liquid interface. The mass transfer mechanism on the liquid side was implemented by User Defined Function. The bubble dimensions and shapes were calculated and shown. The external flow patterns around the bubble and their influences on the mass transfer process were analyzed.

2. Methodology

2.1. Governing equations

Both the gas and liquid are assumed as incompressible and obeyed the continuity and momentum equations which are shown as Eq.1 and Eq.2:

$$\nabla \cdot \vec{u} = 0 \tag{1}$$

$$\frac{\partial(\rho\vec{u})}{\partial t} + \nabla \cdot (\rho\vec{u}\vec{u}) = -\nabla p + \nabla \cdot [\mu(\nabla\vec{u} + \nabla\vec{u}^{T})] + \rho\vec{g} + F_S \tag{2}$$

Where Fs stands for the surface tension source term which is implemented by the continuum surface force (CSF) model. The CSF model correlation is written as Eq.3:

$$F_S = \sigma \frac{\rho\kappa\nabla\alpha}{0.5(\rho_1 + \rho_2)} \tag{3}$$

Where σ is surface tension, $\kappa = \nabla \cdot \hat{n}$, $\hat{n} = \frac{n}{|n|}$, $n = \nabla\alpha_q$.

VOF method introduced the volume fraction to describe the phases and interface. Take the Fv as the gas phase volume fraction for example, gas phase is determined when the Fv is 1. Liquid phase is determined when the Fv is 0. When Fv is equal to positive values, interphase is determined.

A User Defined Scalar Equation was used in the simulation. The scalar was set as CO_2 mass fraction w in the water. The equation was shown as follow:

$$\frac{\partial(\alpha_l \rho_l w)}{\partial t} + \nabla \cdot (\alpha_l \rho_l \vec{u} w - \alpha_l D_l \nabla w) = S_l \tag{4}$$

Where Dl is the diffusion coefficient and Sl is the mass transfer source term. Take the Henry's law into consideration, the source term can be described as follow [10]:

$$S_l = k_l \alpha (M_g \frac{P}{H} - \rho_l w) \tag{5}$$

Where H is the henry constant, kl is the mass transfer coefficient.

2.2. Mesh geometry and solution methods

The calculation domain in 2D geometry is 50 mm*100 mm, which is large enough to study the CO_2 bubble rising with mass transfer. The mesh number is 500000 and it is enough to give satisfactory simulation results. The calculation domain is filled with water initially and the gas is added through an orifice with a diameter of 2.4 mm at the bottom. The gas flux is 10 mL·min-1 and the corresponding velocity is 0.037 m·s-1. The wall boundary condition is set as no-slip wall and the initial gas velocity is zero.

The pressure-velocity coupling method is PISO scheme and the pressure discretization method is PRESTO! method. The momentum discretization method is QUICK and the volume fraction is discretized by Geo-Reconstruct method. The scalar equation is discretized by First Order Upwind scheme.

3. Results and discussion

3.1. Bubble dimension and shapes

The rising process of a CO_2 bubble in the water was recorded in Fig.1. The Phase 2 indicated the gas phase. The bubble deforms through the rising process and it rises in a straight-up path initially but then rises along a zigzag path. The rising path is related to the gas and liquid properties, especially for the liquid viscosity. The properties are shown in Table 1.

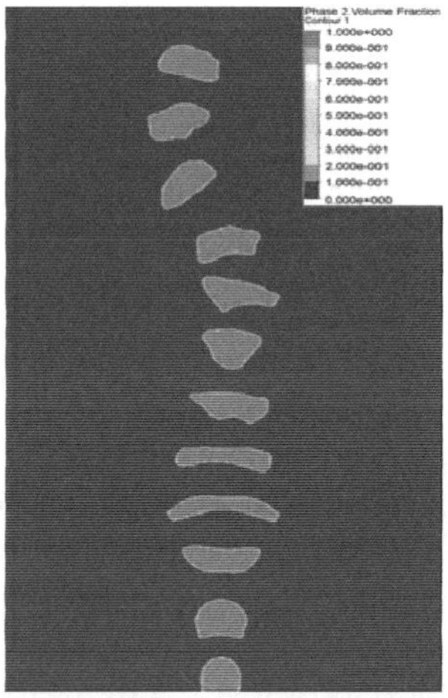

Figure 1. The rising process of CO_2 bubble.

Table 1. Material properties.

	Molecular Weight (g/mol)	Density (kg/m^3)	Viscosity (kg/m-s)	Surface tension (N/m)
CO_2	44	1.7878	1.37e-5	0.07196
water	18	998.2	1.003e-3	

The viscosity of water is only 1.003e-3 kg/m-s so that the bubble shape tends to change when the diameter is large enough. To reflect the bubble deformation, aspect ratio is usually used. It is calculated by the ratio of vertical length and horizontal length of the bubble. The calculation equation is shown as follow:

$$A_R = \frac{L_y}{L_x} \tag{6}$$

where A_R is the aspect ratio, L_y and L_x are the vertical length and horizontal length, respectively.

Figure 2 shows the results of aspect ratio for the single bubble. It should be noticed that the bubble detaches from the orifice after 0.3s so that the aspect ratio can be obtained since then. The aspect ratio fluctuates between 0.27 and 1.07, which agrees with the bubble deformation shown in Figure 1. All the bubble shapes are irregular but the equivalent diameters almost constant at 5.7 mm.

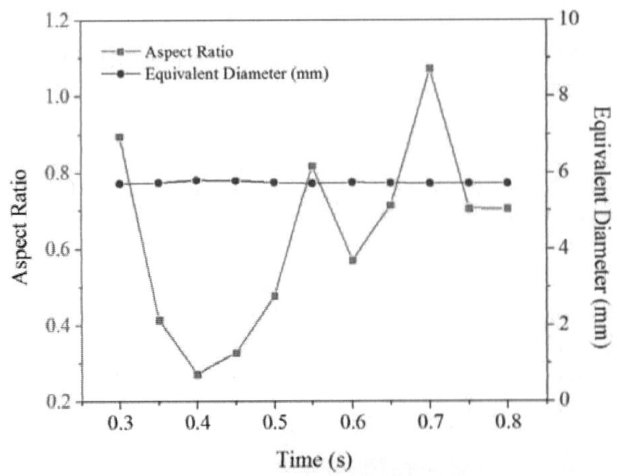

Figure 2. The aspect ratio and equivalent diameter of CO_2 bubble.

3.2. Liquid side mass transfer of CO_2 bubble

The bubble behavior can be divided as three sections, which are bubble formation, bubble detachment, and bubble rise.

In the bubble formation section, the bubble was forming at the orifice and the mass transfer process happened. Figure 3a shows the CO_2 mass fraction around the bubble at 0.1s. On the liquid side, the region of CO_2 distributed at the top of the bubble is larger compared with other parts. Figure 3b shows the velocity distribution near the bubble. At the top of the bubble, the velocity is higher, so the large CO_2 region may result from the high velocity. Same phenomenon can be seen in bubble detachment section from Figure 3c and Figure 3d which show the bubble at 0.25s. When the bubble is forming or detaching, the gas from the orifice enters the liquid along the longitudinal and then make the water above the bubble flow quickly. The faster flow velocity makes more fresh water contact with the bubble boundary and then increases the CO_2 mass fraction gradient which intensifies mass transfer.

Therefore, at the top of the bubble, the CO_2 mass fraction is higher and the mass transfer of CO_2 is better.

In the bubble rise section, the mass transfer conditions are different. Take the bubble at 0.5s for instance, the CO_2 transfers mainly from the bottom of the bubble instead of the top (see Figure 3e). There are several CO_2 mass fraction wakes occurred below the bubble. Through analyzing the velocity distribution in Figure 3f, it is clear that the wakes are caused by the liquid flow patterns around the bubble.

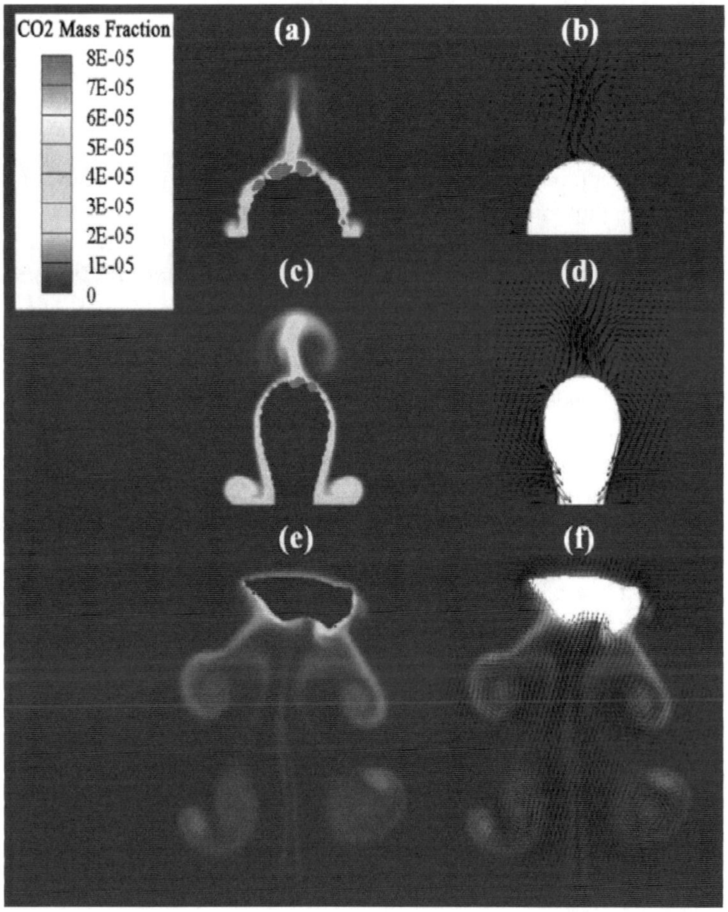

Figure 3. CO_2 mass fraction and velocity vectors around the bubble:
(a)-(b) at 0.1s; (c)-(d) at 0.25s; (e)-(f) at 0.5s

4. Conclusion

The rising behavior and the mass transfer process of CO_2 bubble in the water were simulated in this paper. Through analyzing the bubble dimension and deformation, the flow patterns around the bubble, and the CO_2 mass fraction distribution, the main results are concluded as follow:

(1) Though the shapes of CO_2 bubble change dramatically, the bubble equivalent diameter keeps constant through the rising process.

(2) In the bubble formation and detachment sections, the CO_2 transfers better at the top of the bubble because the liquid flows quickly and more fresh water in this region can contact with the bubble to intensify the mass transfer.

(3) In the bubble rise section, the main mass transfer region is below the bubble. There are wake structures occurred and the flow patterns influence the CO_2 mass fraction distribution.

Acknowledgments
This work was financially supported by the transformation project of scientific and technological achievements of Qingdao (no. 16-6-2-50-nsh).

References

[1] S. Hosoda, S. Abe, S. Hosokawa and A. Tomiyama, Mass transfer from a bubble in a vertical pipe, Int. J. Heat. Mass. Tran. 69 (2014) 215–222.
[2] J. Aoki, K. Hayashi, A. Tomiyama, Mass transfer from single carbon dioxide bubbles in contaminated water in a vertical pipe, Int. J. Heat. Mass. Tran. 83 (2015) 652–658.
[3] J. Aoki, Y. Hori, K. Hayashi, S. Hosokawa and Akio Tomiyama, Mass transfer from single carbon dioxide bubbles in alcohol aqueous solutions in vertical pipes, Int. J. Heat. Mass. Tran. 108 (2017) 1991–2001.
[4] W.J. Nock, S. Heaven, C.J. Banks, Mass transfer and gas–liquid interface properties of single CO_2 bubbles rising in tap water, Chem. Eng. Sci. 140 (2016) 171–178.
[5] D. Legendre, R. Zevenhoven, Detailed experimental study on the dissolution of CO_2 and air bubbles rising in water, Chem. Eng. Sci. 158 (2017) 552–560.
[6] T. Saito, M. Toriu, Effects of a bubble and the surrounding liquid motions on the instantaneous mass transfer across the gas–liquid interface, Chem. Eng. J. 265 (2015) 164–175.
[7] J. Huang, T. Saito, Influence of Bubble-Surface Contamination on Instantaneous Mass Transfer, Chem. Eng. Technol. 38 (2015) 1947–1954.
[8] J. Huang, T. Saito, Influences of gas–liquid interface contamination on bubble motions, bubble wakes, and instantaneous mass transfer, Chem. Eng. Sci. 157 (2017) 182–199.
[9] J. Huang, T. Saito, Discussion about the differences in mass transfer, bubble motion and surrounding liquid motion between a contaminated system and a clean system based on consideration of three-dimensional wake structure obtained from LIF visualization, Chem. Eng. Sci. 170 (2017) 105–115.
[10] D. Bao, X. Zhang, H. Dong, Z. Ouyang, X. Zhang and S. Zhang, Numerical simulations of bubble behavior and mass transfer in CO_2 capture system with ionic liquids, Chem. Eng. Sci. 135 (2015) 76–88.

IOP Conf. Series: Journal of Physics: Conf. Series **1064** (2018) 012025 doi:10.1088/1742-6596/1064/1/012025

Numerical analysis of flows in main nozzle with convex point in local needle and acceleration tube

Sihu Li and Min Shen*

School of Mechanical Engineering and Automation, Wuhan Textile University, Wuhan, 430073, China

* Corresponding author: min_shen18@163.com

Abstract. When the thin weft yarn through the slightness acceleration tube, the weft may drift in the cavity due to the turbulence flow in the main nozzle. In order to analyse the turbulence influence during weft insertion, a kind of new main nozzle is designed by adding convex point in local needle and inside the acceleration tube. This paper focus on a numerical analysis of transonic/supersonic flows in the new main nozzle with a three-dimensional model. A compressible Reynolds Average Navier-Strokes method is used. The pressure and velocity along the center line of the main nozzle are obtained by changing the air tank pressures and number of air diversion groove.

1. Introduction

An air-jet loom inserts the weft into the shed by using the high pressure thrust force and skin friction force along the yarn. The main nozzle is a key part of an air-jet room, which is composed of body, needle, conical sleeve and acceleration tube. Currently, the aperture of acceleration tube of main nozzle is the same size in principle. The different fabrics required in the weft insertion system, the weft yarn is divided into coarse and thin weft yarn normally. If the inner diameter of the tube is too small, the coarse weft yarn is difficult to pass through the tube. However, the diameter of tube is too large, the weft yarn will drift along the axis of tube with the turbulence flow in the main nozzle.

Air flow inside the air-jet loom main nozzle shows subsonic, transonic and supersonic flow. There are some difficulties to measure the flow inside nozzle throat region by experiment because of its small cross-sectional area and composite shape. However, most important flow phenomena, such as shock waves and flow separations occur inside this nozzle throat area. Hence, a simulation approach may be necessary.

Fundamental research in flow field of main nozzle is especially necessary to design optimum shapes of the main nozzles. Mohamed and Salama [1-2] studied the effects of diameter and length of acceleration tubes on flow velocity. Okajima [3] measured velocity and pressure weft insertion channel with notched and un-notched cylindrical tube. Oh et al. [4] studied transonic flow in the main nozzle of an air-jet loom. The effects of air tank pressure, acceleration tube length and nozzle shape are investigated. sang et al. [5] studied unsteady intermittent flows in main nozzle. The pressure and Mach number behaviors along the center line of the main nozzle were presented by changing the inlet condition. Song and Shen [6] established a synthetic flow field model to compute the flow field of the air-jet loom. Liu et al. [7] analyzed the distribution of airflow velocity and static pressure along axial and radial direction in normal ZA type nozzle. Feng et al. [8] studied the internal flow fields of a new kind of main

Content from this work may be used under the terms of the Creative Commons Attribution 3.0 licence. Any further distribution of this work must maintain attribution to the author(s) and the title of the work, journal citation and DOI.
Published under licence by IOP Publishing Ltd

nozzle composed of two nozzle needles connected in series. In this paper, the velocity distribution of the airflow along the axial direction is obtained and the drag force on the weft yarn is calculated. The simulation results show that the drag force can be improved. In this study the compressible Navier-Stokes equations are solved to study turbulence flows inside a new type of main nozzle. We also analyze the effect of air diversion groove on the flow field.

2. Structure and flow field model of main nozzle

The structure of a new type of main nozzle is composed of 4 parts, a nozzle body, a needle, a conical sleeve and an acceleration tube. The difference between the new and normal nozzle is that the acceleration tube and needle constructed a series of regular circular holes.

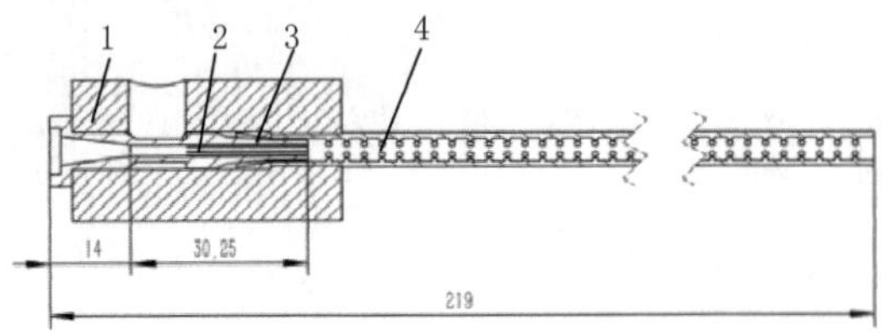

1.nozzle body 2. needle 3. conical sleeve 4. acceleration tube

Figure 1. The Schematic of new main nozzle

Figure 2. The Schematic of needle and acceleration tube

Figure 1 shows the geometry configuration of main nozzle. The inner flow channel of the main nozzle is divided into 3 parts, the compressed air acceleration channel, the weft ejection zone and the weft acceleration zone. This paper used Solid works for geometry modeling based on the ZA type main nozzle and adding the convex point inside the local needle and acceleration tube. As shown in Figure 2, the axis along with the direction of the nozzle is x axis and nozzle needle entrance are the origin of the xoy coordinate system. The main parameters of this new type of main nozzle structure model are shown in Table 1.

Table 1. Structural parameters of main nozzle

Parameter	Value(units)
Length of acceleration tube	181mm
Inter diameter of acceleration tube	4mm
Inter diameter of needle tip	6mm
Inter diameter of weft inlet	8mm
Inter diameter of air tank inlet	10mm
Throat area	5.8mm^2
Slot number	4,6,8

Before the calculation, the 3D model will be meshed by professional software ICEM, because the structure is symmetrical, only the 1/2 area can be reduced by half the number of grids, to reduce the time iteration model. The number of meshes is about 800000, the average distortion of the grid is less than

IOP Conf. Series: Journal of Physics: Conf. Series **1064** (2018) 012025 doi:10.1088/1742-6596/1064/1/012025

0.9, the quality is better, and the meshes are well divided as shown in Figure 3. Then the boundary conditions are set up in the ICEM software. The boundary conditions include: pressure inlet, pressure outlet, pressure far-field, symmetrical surface and wall condition.

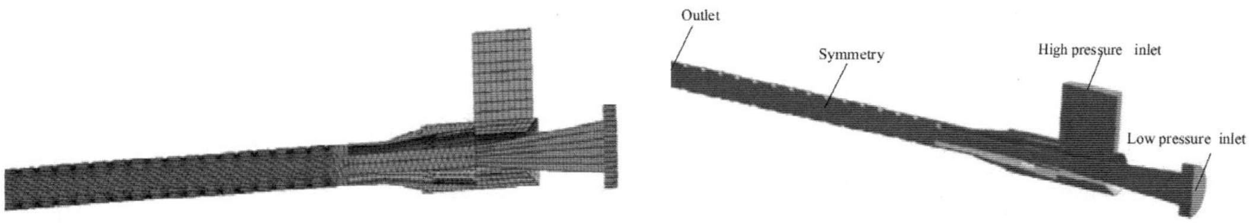

(a)The grid model of main nozzle needle (b)The boundary condition setting

Figure 3. Grid and boundary condition of main nozzle

3. Numerical Analysis

The outlet velocity of main nozzle of air-jet loom may 150~300m/s, and the air flow in main nozzle is turbulence flow. The motion control equations of compressible Newtonian fluid in Euler coordinates are as follows:

Continuity equation:

$$\frac{\partial \rho}{\partial t}+\frac{\partial(\rho u_i)}{\partial x_i}=0 \tag{1}$$

The energy conservation equations:

$$\frac{\partial(\rho T)}{\partial t}+\frac{\partial(\rho u_i T)}{\partial x_i}=\frac{\partial}{\partial x_i}\left(\frac{\lambda}{c_p}\frac{\partial T}{\partial x_i}\right)+S_T \tag{2}$$

The Navier-Stokes equation

$$\rho\left(\frac{\partial u}{\partial t}+u\frac{\partial u}{\partial x}+v\frac{\partial u}{\partial y}+w\frac{\partial u}{\partial z}\right)=-\frac{\partial p}{\partial x}+\mu\left(\frac{\partial^2 u}{\partial x^2}+\frac{\partial^2 u}{\partial y^2}+\frac{\partial^2 u}{\partial z^2}\right)+\rho g_x \tag{3}$$

$$\rho\left(\frac{\partial v}{\partial t}+u\frac{\partial v}{\partial x}+v\frac{\partial v}{\partial y}+w\frac{\partial v}{\partial z}\right)=-\frac{\partial p}{\partial y}+\mu\left(\frac{\partial^2 v}{\partial x^2}+\frac{\partial^2 v}{\partial y^2}+\frac{\partial^2 v}{\partial z^2}\right)+\rho g_y \tag{4}$$

$$\rho\left(\frac{\partial w}{\partial t}+u\frac{\partial w}{\partial x}+v\frac{\partial w}{\partial y}+w\frac{\partial w}{\partial z}\right)=-\frac{\partial p}{\partial z}+\mu\left(\frac{\partial^2 w}{\partial x^2}+\frac{\partial^2 w}{\partial y^2}+\frac{\partial^2 w}{\partial z^2}\right)+\rho g_z \tag{5}$$

The turbulence kinetic energy equation

$$\frac{\partial(\rho k)}{\partial t}+\frac{\partial(\rho k u_i)}{\partial x_i}=\frac{\partial}{\partial x_j}\left[\left(\mu+\frac{\mu_t}{c_p}\right)\frac{\partial k}{\partial x_j}\right]+G_k-\rho\varepsilon \tag{6}$$

The rate of dissipation equation

$$\frac{\partial(\rho\varepsilon)}{\partial t}+\frac{\partial(\rho\varepsilon u_i)}{\partial x_i}=\frac{\partial}{\partial x_j}\left[\left(\mu+\frac{\mu_t}{c_p}\right)\frac{\partial k}{\partial x_j}\right]+\rho C_1 S\varepsilon-\rho C_2\frac{\varepsilon^2}{k+\sqrt{v\varepsilon}} \quad (7)$$

where $\mu_t = \rho C_\mu k^2 / \varepsilon$, $\eta = Sk / \varepsilon$, $S = 2\sqrt{S_{ij}S_{ij}}$ and the parameter C_μ are constant.

The numerical simulation is carried on Fluent. To implement the simulation, some assumptions are made as given below:

The air is assumed as ideal gas

The air viscosity is $\mu = \mu_0 \cdot (T / T_0)^{2/3}$

The out static pressure is atmosphere pressure

This paper adopts RNG $k-\varepsilon$ two equation models. The RNG $k-\varepsilon$ model considers the turbulent vortex and provides an analytical formula for low Reynolds number viscous flow. The formulas can deal with the near wall region towards the correct.

4. Results and Discussion

Air supply pressure is a key parameter influence on the energy consumption and weft flight performance. Figure 4. shows the static pressure along the centerline under air supply pressure varies from 0.2Mpa, 0.3Mpa, 0.4Mpa and 0.5Mpa. The air diversion groove of needle is number 6. It is seen from Figure 4. that there is a negative pressure zone near axis position 50mm and the negative pressure value increases with the increase of gas supply pressure. Oh, and Kim5 had shown the pressure plot of normal main nozzle and the length of negative pressure zone is about 20mm. In this paper, we can see the length of negative pressure zone is 50mm. The extension of the negative pressure area can increase the external air flow into the main nozzle and increase the rates of weft insertion.

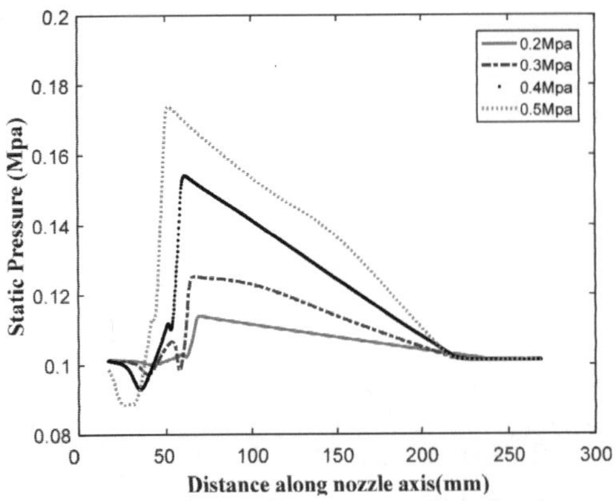

Figure 4. The static pressure distribution along axis at various air supply pressures

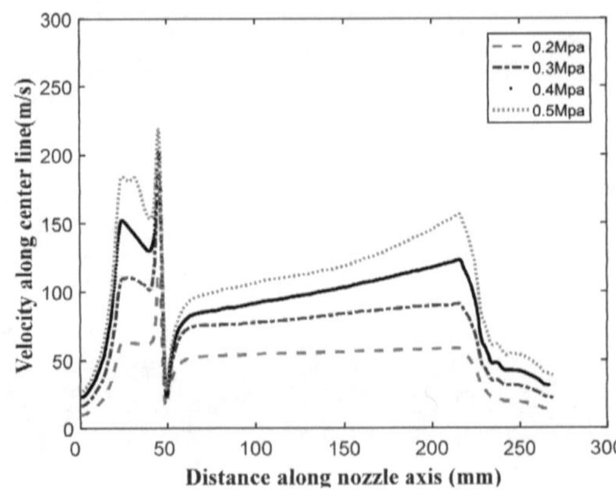

Figure 5. Airflow velocity distribution along axis at various air supply pressures

Figure 5 shows the airflow velocity along the nozzle center under air supply pressure varies from 0.2,0.3, 0.4 and 0.5Mpa. As shown in Figure 5, the airflow velocity at the acceleration tube exit is raised with increasing air tank pressure. It is beneficial for the weft insertion. When the air supply pressure is increased over 0.5Mpa, the improvement of the airflow velocity outside the acceleration tube is not obvious.

Figure 6 plots the static pressures along centerline with various number of air groove. The air supply pressure is 0.2Mpa. The air diversion groove of the needle is divided into number 6, 8 and 10, respectively. As seen from Figure 6, all air pressures near the outlet of the needle are less than the atmospheric pressure and form a negative pressure zone. Then, the weft yarn could be sucked into the acceleration tube.

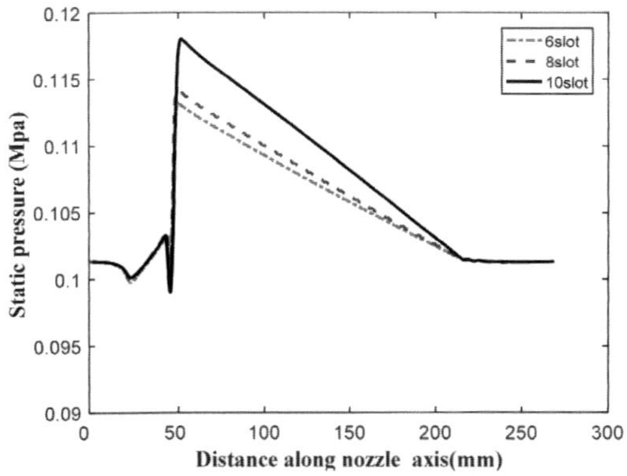

Figure 6. The static pressure along centerline at air supply pressure 0.2Mpa

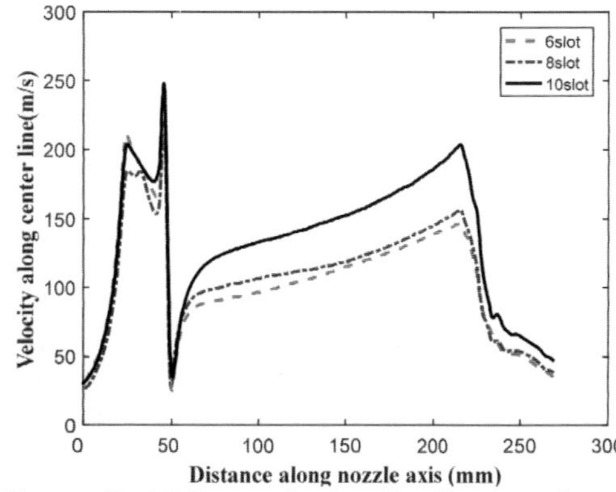

Figure 7. Airflow velocity distribution along centerline at air supply pressure 0.5Mpa

Figure 7 shows the air velocity along the centerline with various number of air groove. The air supply pressure is 0.5Mpa. It is seen from Figure 7 that the velocity is increase with the number of air diversion groove increasing. Hence, it is an effect way increasing the number of air diversion groove for improvement the rate of weft insertion.

Figure 8 shows the velocity contours and streamline of the symmetric surface from axis position 0.025m to 0.08m. The air supply pressure is 0.5Mpa. As can be seen from Figure 8 that there is obviously vertex flow at the end of the nozzle needle. Because the sharp change at the throat of needle. This vertex causes the velocity of the weft yarn fluctuation.

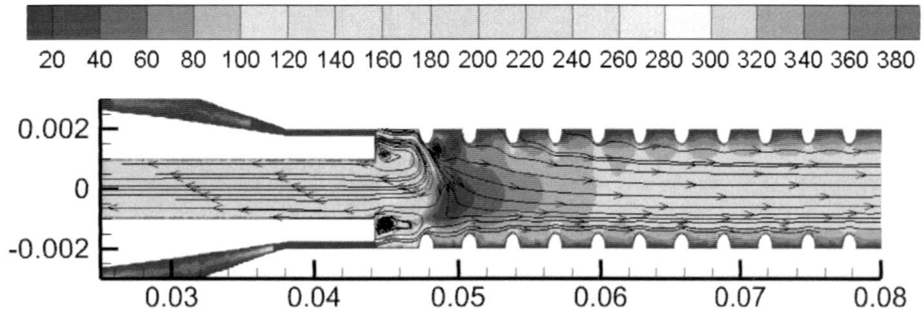

Figure 8. Airflow velocity contours and streamline of needle with 8 air diversion groove at symmetry surfaces

5. Conclusion

(1) Adding the convex point inside the needle and acceleration tube, the length of negative pressure zone is extended to 50mm more than the normal ZA type main nozzle. It helps increase the rates of weft insertion.

(2) It is an effect way increasing the number of air diversion groove to improve the rate of weft insertion.

(3) There is obvious vertex flow due to the sharp change at the outlet of the nozzle needle.

Acknowledgements
This work was financially supported by the Natural Science Foundation of Hubei Province of China (No.2014CFB766), China National Science Foundation (No.51505344), Scientific Research Foundation for Returned Scholaors, Ministry of Education of China and Hubei Provincial Education department outstanding young and middle-aged scientific and technological innovation team project (No. T201607).

References

[1] M.H. Mohamed and M. Salama, Mechanics of a single Nozzle Air-jet Filling Insertion System Part.I, Nozzle design and performance. Textile Research Journal, 1986, 56(11):683-690.
[2] M.H. Mohamed and M. Salama, Mechanics of a single Nozzle Air-jet Filling Insertion System:part II: Velocity distribution and design of the air guide system. Textile Research Journal, 1986, 56(12):721-726.
[3] M. Ishida, A. Okajima, Flow Characteristics of the Main Nozzle in an Air-Jet Loom: Part I: Measuring Flow in the Main Nozzle. Textile Research Journal, 1994, 64(1): 10–20.
[4] T.H. Oh, S.D. Kim and D.J. Song, Flows in the Axisymmetric main nozzle of an air-jet loom, Textile Research Journal, 2001, 71(9):783-790.
[5] D.K.Song, A computational analysis of unsteady Transonic/Supersonic flows over backward facing step in Air jet Nozzle, Journal of Mechanical Science and Technology, 2007, 21(2):336-347.
[6] S.J. Song, D.F. Shen, Synthetic flow-field analysis of main and auxiliary nozzles in air-jet looms[J] Advanced Materials Research, 2011, 411:16-20.
[7] D.D.Liu, Z.H.Feng, B.H.Tan, Y.P. Tang Numerical simulation and analysis for the flow field of the main nozzle in an air-jet loom based on fluent, Applied Mechanics and Materials,2012, (105-107):172-175.
[8] L. Chen, Z.H.Feng, Z.H.Teng, W.H.Wang, S.Liu, Numerical simulation of the internal flow field of a new main nozzle in an air jet-loom based on fluent, Textile Research Journal, 2015, 85(15):1590-1601.

IOP Conf. Series: Journal of Physics: Conf. Series **1064** (2018) 012026 doi:10.1088/1742-6596/1064/1/012026

Three dimensional numerical simulation of ice crystal melting under different influencing factors

Jiachen Zhang[1,*], Weifang Chen[1] and Lifen Zhang[2]

[1]School of Aeronautics and Astronautics, Zhejiang University, Hangzhou 310027, China.
[2]School of Power and Energy, Northwestern Polytechnical University, Xian 710072, China.

*Corresponding author e-mail: 18291447884@163.com

Abstract. The icing of the aeroengines caused by ice crystals has gradually become a research hotspot in recent years. In this study, the influence laws of different factors of ice melting on melting process of ice crystal were studied. Performed unsteady simulation of ice crystal melting under forced convection of air based on VOF method coupled with solidification and melting model in CFD simulation analysis. The influence laws of different factors of ice melting on melting process of ice crystal were revealed.

1. Introduction

The icing of the aeroengines caused by ice crystals has been confirmed and gradually become a research hotspot in recent years. The melting state of ice crystals directly affects icing of the inside of engines, and different factors of ice melting have an influence on melting of ice crystals [1]. The most common such as particle diameter, air temperature, air speed, particle shape and particle attitude.

The melting of ice crystals under different influencing factors creates different flow fields. The flow fields in melting of ice have been studied with experimental method and computational fluid dynamics (CFD) simulation [2-8]. In recent years, CFD analysis has been developed to be a useful and essential method on researching the flow field. Researchers [2] studied the mechanism of melting and attachment of ice crystals inside aeroengines. However, there are no relevant results yet on the influence laws of different factors on melting of ice crystals outside. In this paper, the flow fields in melting of ice crystals were numerical simulated by approach of CFD, three-dimensional model was built and the simulated flow field such as stream line and temperature distribution was correspondingly analyzed. Meanwhile, this paper showed the influence laws of different factors on melting of ice crystals.

2. The melting of ice crystals

2.1. Numerical Simulation

We used a calculation domain of a quarter or one-half of the total flow field due to the symmetrical characteristic of the field. The computational domain and the corresponding grids were created by ICEMCFD. The flow field was obtained by solving the N-S equations through the commercial

Content from this work may be used under the terms of the Creative Commons Attribution 3.0 licence. Any further distribution of this work must maintain attribution to the author(s) and the title of the work, journal citation and DOI.
Published under licence by IOP Publishing Ltd

software FLUENT 16.0. The volume of fluid (VOF) method and solidification & melting model (SMM) [3-8] were applied to run the simulations. Laminar model was used due to simulation conditions that flow states were laminar. Pressure-velocity coupling used semi-implicit method pressure linked equation (SIMPLE) algorithm. Figure 1 shows the three-dimensional model of the ice crystals with different shapes and equal quality. Figure 3 shows the corresponding mesh dividing, hexahedral mesh and double O-grid blocks were applied in this model.

For the study on different attitudes of ice crystals with ellipsoid-shaped, the rotation angles 0°,30°,60° and 90° were only chosen due to the symmetrical characteristic of the angle, as shown in Figure 2. The calculation domain is large enough to not affect the melting of ice crystals.

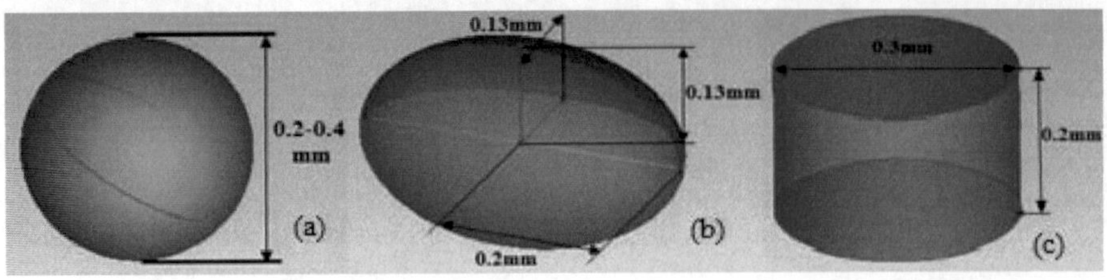

Figure 1. Numerical simulated model of ice crystals. (a) is the ice crystal with sphere-shaped, (b) is the ice crystal with ellipsoid-shaped, (c) is the ice crystal with cylinder-shaped.

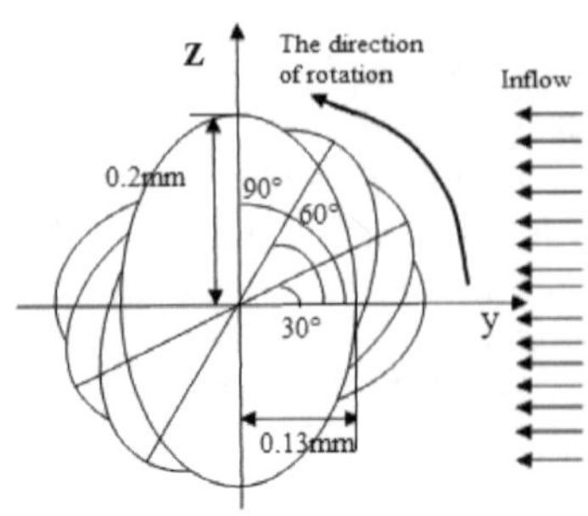

Figure 2. Rotation diagram of the ice crystal with ellipsoid-shaped.

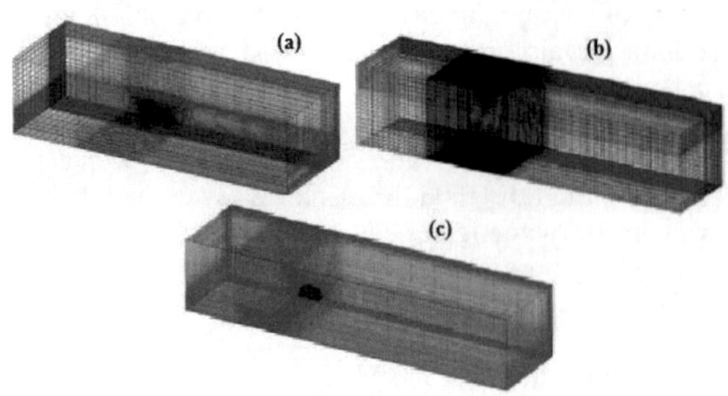

Figure 3. The mesh dividing of the model for the whole flow field.(a) contains the ice crystal with sphere-shaped, (b) contains the ice crystal with ellipsoid-shaped, (c) contains the ice crystal with cylinder-shaped.

2.2. *Method Validation*

In order to study the influence of the number of grids on the calculation results, the ice crystal with sphere-shaped was taken as an example to study the melting of ice crystals under four grids. Figure 4 shows the results, which verify the grid independence [9]. In addition, the experiment of melting of ice crystals [10] with sphere-shaped was numerically verified in order to verify the reliability of the simulation method. Table 1 shows the results, which indicates that the simulation method is feasible.

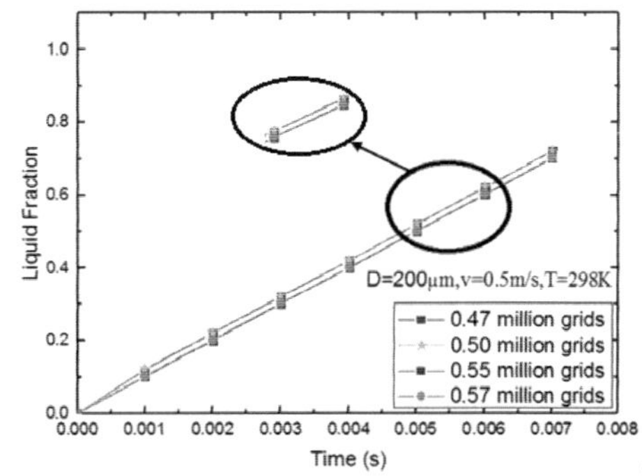

Figure 4. Comparison of liquid fraction under four grids.

Table 1. Comparison of numerical value.

Comparative object	Time/s				
	0.4	1.2	2.0	2.8	3.3
Experiment/%	16	50	68	80	96
Numerical simulation/%	17	39	62	84	99
Error value/%	6.25	22	8.82	5	3.13

3. Result and discussion

The ice crystals with sphere-shaped was taken as an example (diameter is 0.3mm, air speed is 0.5m/s and temperature is 298K). Figure 5(a) shows the time evolution of temperature distribution in flow fields. It's obvious that the inflow surface of ice crystals melts faster and the backflow surface melts slower. Figure 5(b) shows the velocity vector distribution cloud diagram of the flow field close to the backflow surface at a certain moment. There is a vortex formed close to the backflow surface which leaves the surface at some location, where it's in the range of 120° to 150° that's the point melting slowest.

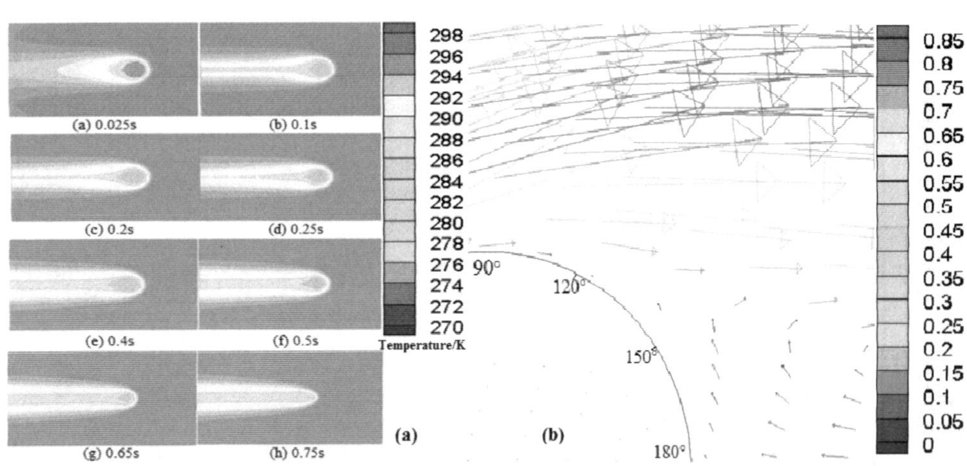

Figure 5. Cloud diagram of flow fields. (a) is the temperature distribution, (b) is the velocity vector distribution.

Figure 6(a) shows the time evolution of liquid fraction of ice crystals with sphere-shaped under different diameters. Figure 6(b) shows the time evolution of liquid fraction of ice crystals with sphere-shaped under different air speeds. Figure 6(c) shows the time evolution of liquid fraction of ice crystals with sphere-shaped under different air temperatures. Figure 6(d) shows the time evolution of liquid fraction of ice crystals under different particle shapes. Figure 6(e) shows the time evolution of liquid fraction of ice crystals with ellipsoid-shaped under different attitudes.

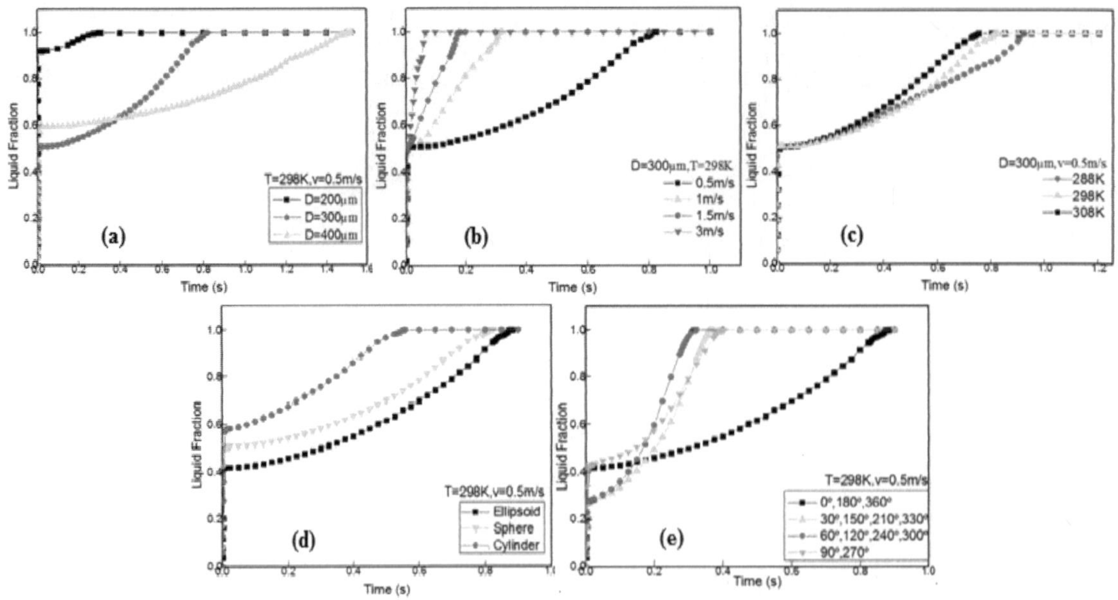

Figure 6. The time evolution of liquid fraction. (a) is under different particle diameters, (b) is under different air speeds, (c) is under different air temperatures, (d) is under different particle shapes, (e) is under different particle attitudes.

4. Conclusion

Three dimensional numerical simulation models were built for simulating the flow field in melting of ice crystals. Meanwhile, the grid independence and the reliability of simulation methods were verified. The result shows that the larger the particle diameter, the longer the ice crystal melts completely. The air temperature has little effect on the melting of ice crystals, but the air speed has a great influence. In addition, the particle shape and the particle attitude are also important factors affecting the melting of ice crystals.

This study illustrated that the laws of different external factors affecting the melting of ice crystals are also different and the melting of ice crystals can be controlled by adjusting external factors.

Acknowledgments

This research was supported by the National Natural Science Foundation of China (Grant NO.11502232, 11572284 and 61627901) and the National Basic Research Program of China (Grant NO.2014CB340201).

References

[1] Sutton, M., and Hammond, D, "The Development of the Facilities for the Simulation of Ice Crystal Conditions", Proceedings of the International Icing Symposium, Montreal, Canada.1995.

[2] Mason, J. G., "Current Perspectives on Jet Engine Power Loss in Ice Crystal Conditions: Engine Icing", Seventh AIRA Research Implementation Forum Boeing.2008.

[3] L. Fan, "A Numerical Simulation on the Solidification and Melting of ice", Publisher of Nanjing University of Aeronautics and Astronautics.2005.
[4] Y. J. Qiu, "Numerical Research of Solidification and Heat Transfer Characteristics for an Air-Cooling Molten Blast Furnace Slag Droplet", Publisher of Chongqing University.2014.
[5] N. Qin, "The Experimental Study and Numerical Simulation of Ice Water Changing Between Solid and Liquid", Publisher of Qingdao University of Science and Technology.2011.
[6] T. T. Guo, "The Theoretical Resrarch on Transformation Bestween Solid and Liquid of Water anf Ice", Publisher of Qingdao University of Science and Technology.2011.
[7] X. H. Geng, "The experimental Study and Numerical Simulation on the Effect of Humidity and Temperature Between Solid and Liquid of Ice Water", Publisher of Qingdao University of Science and Technology.2014.
[8] Y. Qin, "A Numerical Simulation on the Interaction of the Interface and the Convection Heat Transfer", Publisher of Dalian University of Technology.2005.
[9] W. Q. Tao, "Numerical Heat Transfer", Publisher of Xi'an Jiaotong University.1998.
[10] Sihong Yan, Jose Palacios, "Experimental quantification of partial melting in a single frozen droplet", AIAA 2016-3741.

Simulation of High-Speed Pantograph Dynamic Performance Based on Finite Element Model and Aerodynamic Pantograph Model

Lei Li[1], Ge Chen[2]

[1]College of Information Science, Ocean University of China, Shandong, Qingdao, 266100, China.

[2] Sinopec Qingdao Institute of safety engineering, Shandong, Qingdao, 266100, China.

Abstract. A coupled finite element model for the pantograph-catenary system and an aerodynamic pantograph model were built. Based on the models, the dynamic current collection and aerodynamic performances of the pantograph were studied using MSC-Marc and STAR-CD simulations of software. In the finite element model, the pantograph and catenary subsystems are integrated into a whole system through contact coupling between them to simulate the dynamic behavior of the pantograph-catenary system. Then, the aerodynamic performance of the high-speed pantograph was simulated on the three-Dimensional aerodynamic pantograph model. The simulation results show that the dynamic uplift and average contact force of simple suspension are smaller than those of stitched suspension.

Keywords: current collection; pantograph; finite element model; aero dynamic performance

1. Introduction

The current-receiving quality of the pantograph is the key to the safety of high-speed trains, such as the stability of high-speed bow-nets, high-speed aerodynamics, etc. Therefore, it is necessary to study the dynamic characteristics of high-speed current reception and reduce the range of such dynamic changes.

In the study of the theory of the flow-receiving of the pantograph system, the literature [1] proposed to improve the flow of the pantograph system by improving the uniform distribution of the contact network elasticity. The literature [2] studied the linear model of pantograph and the nonlinear model together, and discussed the process of linearization of the pantograph nonlinear models. The literature [3, 4] studied the theory of flow acceptance from the aspects of the dynamic characteristics of the pantograph system, the velocity of contact line fluctuations, and the off-line rate. The literature [5, 6] analyzed the development of the stable flow-receiving of high-speed electrified railway pantograph system, and studied the aerodynamics of high-speed trains. The literature [7] proposed some measures to improve the flow-receiving of the pantograph system by improving the uniform distribution of the contact network. The literature [8-11] studied the high speed. The suspension type and related parameters of the electrified railway catenary have theoretically explored and analyzed the characteristics of the dynamic flow-receiving characteristics. The literatures [12-14] gave the influence of the pre-sag of the catenary on the flow-receiving of the high-speed bow-net.

Content from this work may be used under the terms of the Creative Commons Attribution 3.0 licence. Any further distribution of this work must maintain attribution to the author(s) and the title of the work, journal citation and DOI.
Published under licence by IOP Publishing Ltd

In this paper, the finite element method is used on the basis of the above research on the theory of flow-receiving. And numerical simulation of fluid mechanics to study the aerodynamic performance of pantograph receiving and pantograph at high speed, and extracted the pantograph and catenary from calculations. The main parameters and related mathematical models are given, and comparative analysis of the simulation results are discussed, respectively.

2. The proposed model for contact net pantograph system

The contact net is formed by bearing cables, contact wires, and suspension chains. The height, slackness, etc. need to be finely adjusted. The power receiving is composed of the upper frame, the lower frame, the bow head, the skateboard and other structures. The analysis for the contact net pantograph system is difficult to carry out, so a suitable model is desired. In the finite element numerical simulation calculations, the contact coupling is an interaction system.

In the numerical simulation, the Euler-Bernoulli beam is used in the finite element model of the catenary. The displacement of the clues under tension is simulated by the movement of the struts. The positioner on the strut causes the contact line to produce pull values within 1 span. The connection with the sliding board and the upper and lower frame is connected by a hinge and a non-linear spring. The hinge allows a relative rotation in the horizontal direction through the lifting bow spring. The lifting force is exerted on the boundary conditions, and friction exists between the pantographs. These are the main factors that cause the nonlinear interaction of the pantograph.

3. Pantograph Aerodynamic Model

High-speed trains operate in an air-medium space, and the research object is simplified to the problem of interaction between air fluids and rigid bodies. In the kinetics study, the molecular structure of the flow system is neglected and the fluid is considered as a continuous medium. There is no vacuum, no gaps between molecules and molecular transport. Because of the continuity of the fluid, the basic quantities (speed, pressure, etc.) of the kinematics of the fluid are regarded as certain physical quantities (time and space, the continuous function of the standard) solves the aerodynamic problem with a mathematical model of the continuous function. Fig. 1 shows the aerodynamic simulation model of the pantograph. The length of the field in the negative direction of the incoming flow is 5 times the height of the pantograph; in the positive direction of the incoming flow, ie, the length of the wake, it is 6 times the height of the pantograph; The width of the plane in both directions is 5 times the height of the pantograph; the height is 5 times the height of the pantograph.

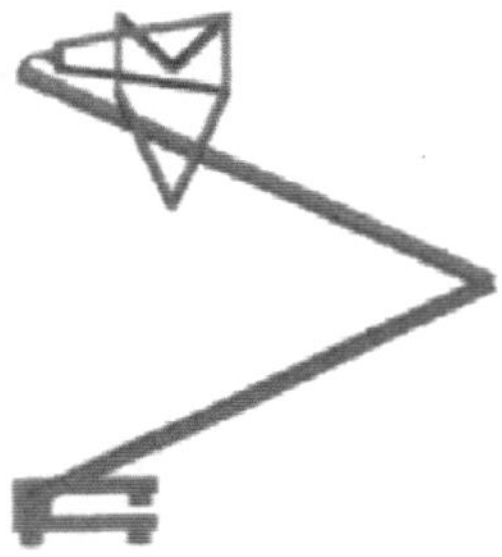

(a) Three-dimensional geometrical model

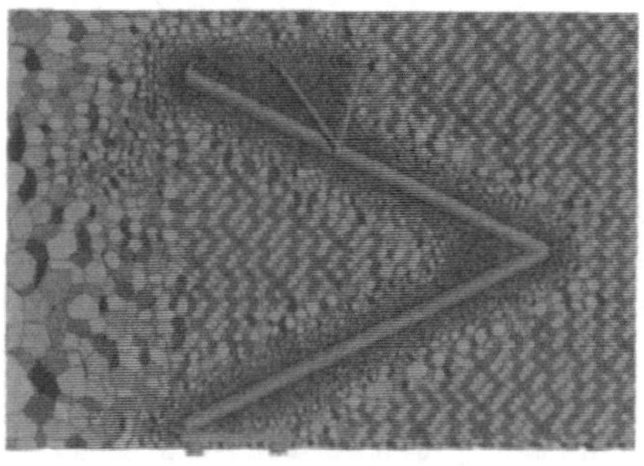

(b) Fluid simulation model

Fig. 1 Aerodynamic analysis model of pantograph

4. Results from Pantograph Dynamic Performance Simulation

Simulation and analysis of the interaction between different contact suspension modes and high-speed pantographs are demonstrated. The calculation condition is 250m/h, and pantograph calculation parameters are shown in Table 1 and 2, respectively.

Tab. 1 Parameters of catenary

Suspension type	Simple suspension	Elastic suspension
Span /m	50	50
Tension /kN	27+21	25+25
Cross-sectional area /mm^2	120	120
Hanging string material	Cu	Cu
Cross-sectional area /mm^2	0	0

Tab. 2 Parameter of pantograph

Pantograph components	Calculating condition
Skate density / g cm^{-2}	7.85
Upper frame member cross-section /mm^2	7850
Material / g cm^{-3}	1256
Down frame member cross-section /mm^2	7850
Pusher cross-sectional area / mm^2	1256
Material / g cm^{-3}	7.85

4.1. Bownet dynamic simulation

In the bownet dynamic simulation, the MSC-Marc and STAR-CD software are used to simulate the pantograph system and elastic chain type. The calculation result of the dynamic lifting amount of the contact line of the suspension (the contact point of the positioning point and the contact line) are shown in Fig. 2.

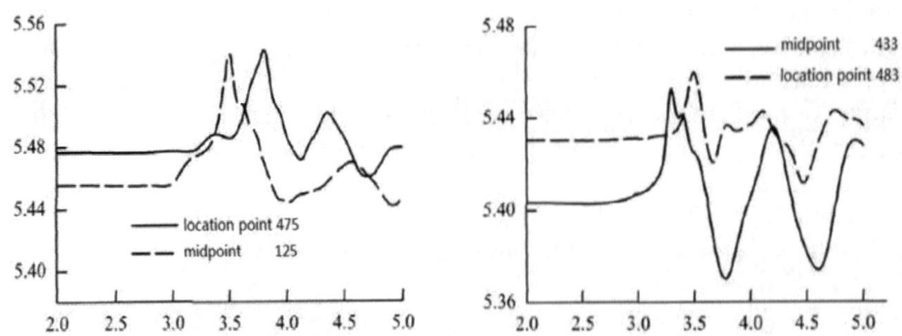

(a) Dynamic uplift of elastic suspension (b) Dynamic uplift of simple suspension

Fig. 2 Dynamic uplift of the contact wire with different suspension

From Fig. 2(a), it can be seen that: When the bow is not passing, due to the existence of clue gravity and sag, the static position in the cross is lower than the positioning point of 23mm; when the pantograph passes, the positioning point and cross. The dynamic lift at the midpoint is 65mm and 86mm, respectively. It can be seen from Fig. 2(b) that due to the absence of the pantograph, the sag and line are reserved. Cable gravity effect, the static position in the cross is lower than the positioning point of 27mm; when the pantograph passes, the dynamic lifting points of the positioning point and the midpoint are 30mm and 50mm, respectively. Through the simulation, it can be found that the simple chain suspension and the elastic chain type Hanging positioning points and mid-lift lift are small, but lifted Significant changes in volume, indicating that the simple chain suspension is more elastic than the elastic chain suspension.

The simulation results from the bownet contact pressure under the elastic suspension are shown in Fig. 3. It can be seen from Fig. 3 that the elasticity of the contact pressure of the chain suspension varies drastically with the span, the frequency is increasing, and the average contact pressure is large. With obvious cycle variation in the span, the average contact pressure is small, there will be a phenomenon that the contact pressure is 0, if the contact pressure is reduced to 0, it will make as a result of off-line and spark. Therefore, the uniform pressure change of the bownet contact pressure is an important factor to ensure reliable flow performance.

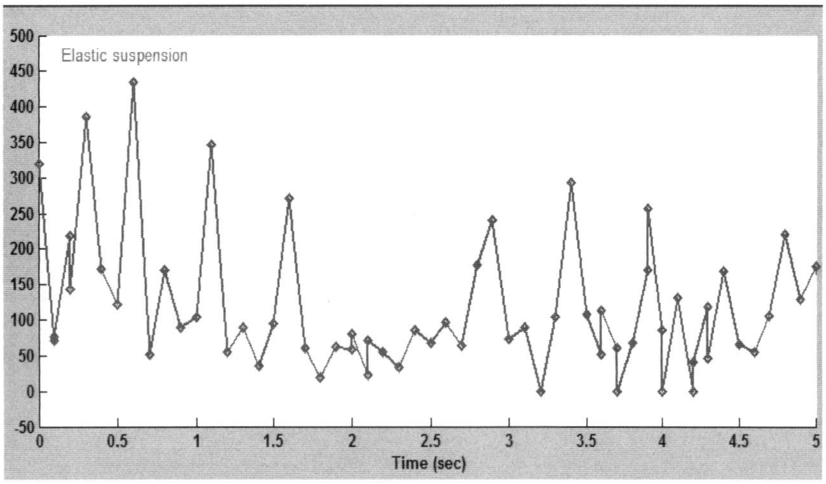

Fig. 3 Pantograph catenary contact pressure for elastic suspensions

5. Conclusion

The finite element model coupled with the pantograph and contact net established in this paper can correctly reflect the dynamic flow-receiving performance of the pantograph. Compared with the elastic chain suspension, the type of suspension reduces the amount of dynamic uplift, and the average bow contact pressure decreases, also the frequency of change is slower.

References

[1] Chen biao, Zha Wan ming. Performances of catenary of high-speed railways[J]. Journal of Southwest Jiaotong University[J]. 1997, 32(5): 497-501.

[2] MEI Gui ming, ZHANG Wei hua. Linear model of vertical dynamics of pantograph and study on characteristics[J].Journal of the China Rail way Society. 2003, 24(5): 16-20.

[3] LIU Yi, ZHANG Weihua, MEI Gui ming. Study of dynamics tress of the catenary in the pantograph /catenary vertical coupling movement[J].JournaloftheChinaRailwaySociety, 2003, 25(4): 23-26.

[4] YU D,LO K L. Analysis on catenary current-carrying the ory of high-speed railway[J].J.Chin.RailwaySoc.,1998, 20(5): 58-64.

[5] MITSUO A, ISSEIN, HIROKA ZU K. Current collection characteristic and improvement methods of high-tension overhead catenary systems [J].Electrical Engineering in Japan, 1998, 123(4): 66-76.

[6] YU Zheng ping. Aerodynamic exterior design for the high speed electriclo comotives [J]. Electrical Drive for Locomotive,1998, 10(5-6): 78-80..

[7] YU Zheng ping. A study on high-speed catenary-pantograph system [J].China Railway Science, 1999, 19(1): 59-72.

[8] ZHANG Jun jie. Current collection quality caused by the elastic of the catenary [J]. Electrical Railway, 2001(3): 24-25.

[9] GAJDAR T,BURLACU C,SUDA Y, et al. Simulation of the electric drive system of electricrol lingstocks[C] // IEEE International Conference. Budapest: Intelligent Engineering System, 1997: 365-370.

IOP Conf. Series: Journal of Physics: Conf. Series **1064** (2018) 012027 doi:10.1088/1742-6596/1064/1/012027

[10] YU Di. Analysis on catenary current-carrying the ory of high-speed railway[J]. Journal of the China Railway Society, 1998,20(5): 58-64.
[11] YU Wan ju. Analysis of stable current-collection of high-speed catenary[J]. Xi ’an Railway Technology, 1996(2,3): 38-44.
[12] FUJIIY, MANABEK. Computer simulation for dynamics of overhead catenary-pantograph system[C] // Proceedings of the International Conference on Computer Aided Design. Manufacture and Operation in Railway and other Advanced Mass Transit computers in Railways, Technology. WashingtaonDC: Computational Mechanics Publ., 1992.
[13] ANDREA C, STEFANO B. Numerical simulation of pantograph-overhead equipment ineraction [J]. Vehicle System Dynamics, 2002, 38(4): 261-291.
[14] ANDREASK, UWER. Numerical simulation of the pantograh/catenary system [J]. Eisenbahning, 1995, 46(3): 162.

IOP Conf. Series: Journal of Physics: Conf. Series **1064** (2018) 012028 doi:10.1088/1742-6596/1064/1/012028

Curved Measurement Theory of Honing Pneumatic Measurement System and Optimization of Measurement Parameters

WanJun Zhang[1, 2, 3, a], Feng Zhang[1, b], Jingxuan Zhang[1, c] Jingyi Zhang[2, d], Jingyan Zhang[2]

[1]Quanzhou Institute of Information Engineering, Fujian362000, China
[2]School of Mechanical Engineering, Xian Jiao tong University, Shanxi710049, Chin
[3]Lanzhou Industry and Equipment Co, Ltd, Lanzhou 730050, China

[a]gszwj_40@163.com, [b]zhangwanjun40@163.com, [c]116543048@qq.com
[d]zhangwj_40@163.com

Abstract. In the previous research on dynamic process of honing pneumatic measurement system are studied by measuring plane measurement instead of the surface, but to achieve high-precision measurement of aerodynamic honing, instead of surface measurement by plane measurement is not scientific. Based on the study of the theoretical basis of pneumatic measurement and air dynamics, through the dynamic study of nozzle baffle mechanism in the measuring system of gas honing, studies have shown that the measurement accuracy is higher than the baffle surface characteristic equation characteristic equation of plane baffle. Using this characteristic equation, the design of gas momentum meter and measurement system is more accurate, and it provides a reliable reference for improving the measurement accuracy of honing pneumatic measurement. For the equation of curved baffle property verification and optimization of the parameters of the nozzle, based on the entity model, gas path structure and nozzle of gas measuring device of dynamic curved baffle FLUENT software simulated. The simulation results show that the air flow changes during the operation of the pneumatic measuring device, and the pressure values are obtained. The pressure clearance curve of the curved baffle is plotted by drawing software, and the correctness of the characteristic equation of the surface baffle is verified by comparing with the theoretical curve. By adjusting the model parameters, the pressure clearance curves of different main and measuring nozzle apertures are obtained by repeated simulation. Then the law is found according to the results and the parameter design is optimized.

1. Introduction

In recent years, honing has become an indispensable process technology for precise coupling parts [1-3] such as engine cylinder sleeve, cylinder bore, and important hydraulic cylinders in construction machinery. Honing is finishing and finishing grinding method on the surface of the work piece, the advantages of high precision honing is safe, economic, reliable, durable and efficient, is widely used in machining of cylinder, cylinder, piston rod, pin hole, hole pump cylinder hole, hydraulic valve hole,

Content from this work may be used under the terms of the Creative Commons Attribution 3.0 licence. Any further distribution of this work must maintain attribution to the author(s) and the title of the work, journal citation and DOI.
Published under licence by IOP Publishing Ltd

bearing hole and bearing etc. [1]. The traditional honing process adopts the repeated process of "processing, cutting back, measuring and reprocessing", and its size measurement is usually measured manually by off-line manual measurement. The measurement accuracy and efficiency are relatively low. Applying on-line pneumatic measurement technology to bore size measurement of bore hole, it can improve production efficiency and realize automation of inner whole honing process.

The online measurement of the non-contact type pneumatic measuring technology [4-7] is introduced into the hole honing aperture size, the gas electric conversion module in real time will be in transition pore size change for the aperture signal, then the signal is processed to display and control system for subsequent analysis and control [8-9], which is helpful to the numerical control loop hole honing machine of high precision processing control.

With the further research of engine operation mechanism in engineering field, higher requirement of honing accuracy is put forward. The accuracy of honing machine determines the machining accuracy of honing machine. At present, pneumatic measurement is widely used in honing system. Lanzhou University of technology, Lanzhou Industry and Equipment Co. Ltd ZhangWanjun .et, at researchers [15-21] have researched interpolation algorithms for NURBS curves honing pneumatic measurement system. Here has been attempts to change the feed rate in NURBS interpolation honing pneumatic measurement system.

In the past research process of honing pneumatic measurement system, plane measurement was used instead of surface measurement. The design of measurement parameters mainly adopts empirical estimation and experimental comparison method. However, it is not scientific to use the plane measurement instead of the surface measurement to achieve the high precision of the honing pneumatic measurement. Therefore, it is necessary to study the theory of surface measurement and the optimization of measurement parameters. So, the subject of honing pneumatic measurement system in the surface measurement theory and optimization of measurement parameters, to achieve high-precision honing pneumatic measurement, in order to meet the requirements of high-precision honing processing.

The parameter design of the traditional gas momentum instrument mainly adopts the method of empirical estimation and experimental comparison. Based on the theoretical research of aerodynamics and fluid mechanics, a theoretical model for parameter design of honing gas momentum meter is established. By computer simulation, the design parameters of the gas momentum meter are further optimized. The design of the pneumatic measurement system is guided in theory, so that the pneumatic measurement can be better applied to machinery and various industries.

2. Aerodynamic measurement model

The basic principle of pneumatic measurement is introduced with pneumatic back pressure pneumatic measuring system, and the schematic diagram is shown in Fig.1.

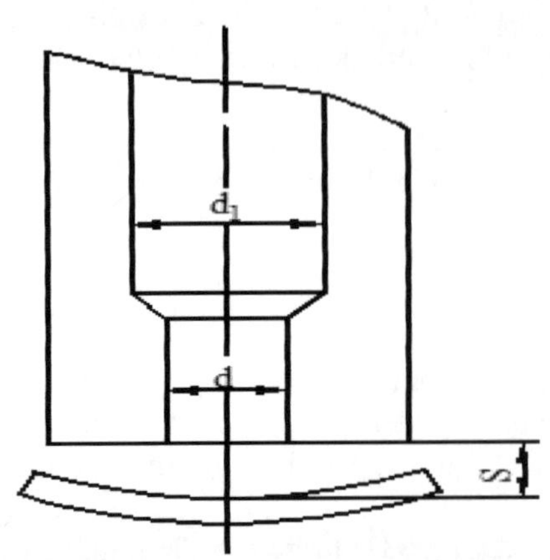

Figure 1. Schematic diagram of gas path of back pressure pneumatic meter

Compressed air from the air filter 1 by 2 by 3 after the filter inlet valve into the regulator regulator 4, subsequent gas path pressure stabilized at p_c . Compressed air of pressure p_c flows through the orifice 5 toward the measuring head (usually the nozzle flapper mechanism) and flows from the probe into the atmosphere. The pressure p_c in the pipe section between the orifice 5 and the pneumatic measuring head is called back pressure.

3. Calculation of equivalent clearance of curved nozzle flapper mechanism

When the back pressure p_x gradually drops from p_c to $p_x \geq 0.0894MPa$; the main nozzle is subcritical under the condition III, and the gas flow velocity at the baffle is in a critical state. According to fluid dynamics knowledge, the flow rate g_1 through the main nozzle and flow through the baffle are g_2 , respectively:

$$Q = cF\sqrt{2g\frac{n}{n-1}p_1\gamma_1\left[\left(\frac{P_2}{P_1}\right)^{\frac{2}{n}} - \left(\frac{P_2}{P_1}\right)^{\frac{n+1}{n}}\right]} \tag{1}$$

Type: Q - weight flow through the cross section, Kg/s;
c—discharge coefficient;
F—Flow area of nozzle, m^2;
n—Constant entropy coefficient, For air, n=1.4;
g—Gravitational acceleration, g=9.81m/S2;
P_1—Absolute pressure at the front of the nozzle, kpa;
P_2—Absolute pressure behind the nozzle, palmer;
γ_1—P_1 Gas gravity under.

When the flow state of the gas passing through the nozzle is in the subcritical state, i.e. $p_2 / p_1 \geq$ 1.894, the flow rate is expressed as:

$$Q = cF\left(\frac{2}{n+1}\right)^{\frac{1}{n-1}}\sqrt{2g\frac{n}{n+1}P_1\gamma_1} \tag{2}$$

Supposed

$$Q = K_1F \tag{3}$$

The flow of gas through the nozzle is in a subcritical state,

$$K_1 = c\sqrt{2g\frac{n}{n-1}P_1\gamma_1\left[\left(\frac{P_2}{P_1}\right)^{\frac{2}{n}} - \left(\frac{P_2}{P_1}\right)^{\frac{n+1}{n}}\right]} \tag{4}$$

When the flow of gas through the nozzle is in a critical state,

$$K_1 = c\left(\frac{2}{n+1}\right)^{\frac{1}{n-1}} \sqrt{2g\frac{n}{n+1}P_1\gamma_1} \quad (5)$$

The formula 4 or formula 5 is substituted formula 6, and the order K is obtained

$$G = KdS \quad (6)$$

As shown in Fig.2, the nozzle bore diameter is d, the outer diameter of the nozzle is d_2, and the end face is flat. The bezel is a circular cylinder with a radius of R. The relative position of the nozzle and the baffle is the axis of the nozzle hole intersecting the axis of the round cylinder, and the axis of the cylinder is parallel to the end face of the nozzle. The maximum distance from the end face of the nozzle to the inner surface of the cylinder is a gap S. It is obvious that when the clearance S is determined, the flow rate Q is not equal to KdS.

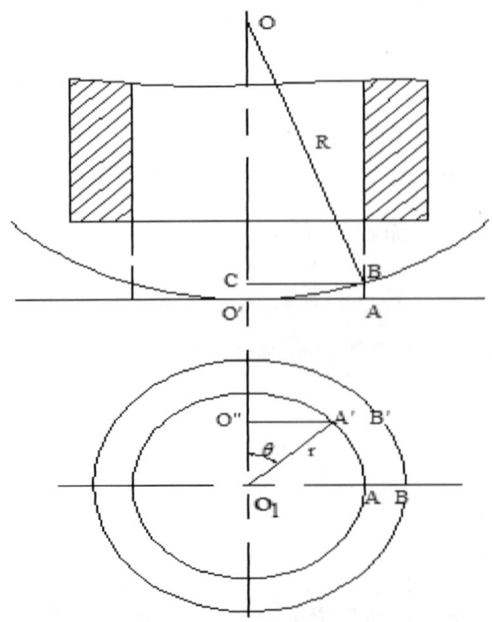

Figure 2. Surface nozzle baffle diagram

Supposed, relationship between Q and S for the gap flow

$$Q = KdS - G_R \quad (7)$$

Then, the clearance S_R of the parallel plane nozzle flapper mechanism with a nozzle diameter of D is called the clearance clearance of the curved baffle mechanism when the flow rate is G_R

$$G_R = KdS_R \quad (8)$$

So,

$$Q = KdS - KdS_R = Kd(S - S_R) \tag{9}$$

The equivalent clearance S_R is an important parameter for designing an air momentum meter. When the S is determined, the difference between the surface area of the flat baffle and the curved baffle is F_R, as shown in Fig. 5,

$$\overline{AB} = \overline{OO'} - \overline{OC} = R - \sqrt{R^2 - \overline{BC}^2} = R - \sqrt{R^2 - \overline{AO'}^2} \tag{10}$$

Because the surface baffle is a cylinder block, the $\overline{AO'}$ is $\overline{A'O''}$ on different cross sections. If the nozzle radius is r, then

$$\overline{A'O''} = \overline{O_1A'}\sin\theta = r\sin\theta \tag{11}$$

Outflow area of F_R for any projection in the plane of the circular baffle nozzle diameter on a A' to $\overline{A'B'}$ curved baffle distance $rd\theta$ and length of product in θ from0~ 2π points, i.e.

$$\overline{A'B'} = R - \sqrt{R^2 - r^2\sin^2\theta} \tag{12}$$

The formula (12) is substituted (13)
Instant,

$$F_R = \int_0^{2\pi}\left(R - \sqrt{R^2 - r^2\sin^2\theta}\right)rd\theta \tag{13}$$

In accordance with the definition of equivalent clearance, $S_R = \frac{F_R}{\pi d}$, so

$$S_R = \frac{r^2}{4R} = \frac{d^2}{8D} \tag{14}$$

Therefore, when the flow state of the gas passing through the nozzle is in the subcritical state, that is p_2 / p_1 <1.894, the flow rate of the surface baffle is:

$$Q = c\pi d\sqrt{2g\frac{n}{n-1}P_1 r_1\left[\left(\frac{P_2}{P_1}\right)^{\frac{2}{n}} - \left(\frac{P_2}{P_1}\right)^{\frac{n+1}{n}}\right]}\left(S - \frac{d^2}{8D}\right) \tag{15}$$

The characteristic equations of the plane baffle measurement system under four operating conditions are [13]:

Condition Ⅰ:

$$P_x = P_c - \frac{P_c\left[\left(\frac{P_o}{P_c}\right)^{\frac{2}{n}} - \left(\frac{P_o}{P_c}\right)^{\frac{n+1}{n}}\right]\left(\frac{c_2 d}{c_1 \frac{d_1^2}{4}}\right)^2 S'^2}{\frac{n-1}{n} + \left[\frac{2n-2}{n}\left(\frac{P_o}{P_c}\right)^{\frac{2}{n}} - \frac{n-1}{n}\left(\frac{P_o}{P_c}\right)^{\frac{n+1}{n}}\right]\left(\frac{c_2 d}{c_1 \frac{d_1^2}{4}}\right)^2 S'^2} \tag{16}$$

Condition II:

$$P_x = \left[\frac{1}{2}P_0^{\frac{n-1}{n}} + \frac{1}{2}\left(P_0^{\frac{2n-2}{n}} + 4P_0^{-\frac{2}{n}}\frac{B^2 P_c^2}{A^2 S'^2}\right)^{\frac{1}{2}}\right]^{\frac{n}{n-1}} \tag{17}$$

Condition III:

$$P_x = P_c\left[0.5 + 0.5\left(1 + 4\frac{A^2}{B_1^2}S'^2\right)^{0.5}\right]^{\frac{n}{1-n}} \tag{18}$$

Condition IV:

$$P_x = \frac{c_1 \frac{d_1^2}{4}}{c_2 d S'} P_c \tag{19}$$

Where, S' is the measuring clearance.

C_1 — The flow coefficient at the orifice (main nozzle);

d_1 — Orifice diameter;

P_c — work pressure (absolute pressure);

P_x—back pressure (absolute pressure);

n - The constant entropy index, for air;

γ_c — p_c under the severe gas.

By comparing with the equation of curved baffle measurement system characteristics, we can draw in the measuring chamber pressure p_x and other parameters of the nozzle under the same measuring clearance error measuring gap curved baffle and plane baffle is $\frac{d^2}{8D}$, the measurement accuracy is calculated according to the curved baffle characteristic equation characteristic equation of $\frac{d^2}{8D}$ is higher than the plane baffle.

According to the requirements of this project, the accuracy of honing aerodynamic measurement is 1 mu m. When the $D \geq 125d^2$ is available, it is necessary to study the characteristic equation of the surface baffle, instead of the characteristic equation of the plane baffle. When $D \geq 125d^2$ is used, the characteristic equation of the plane baffle can be used to approximate the characteristic equation of the curved baffle.

According to whether the flow velocity of the main nozzle and the measuring nozzle is in a critical state, the working conditions of the gas momentum instrument can be divided into four working conditions types, as shown in Fig. 3:

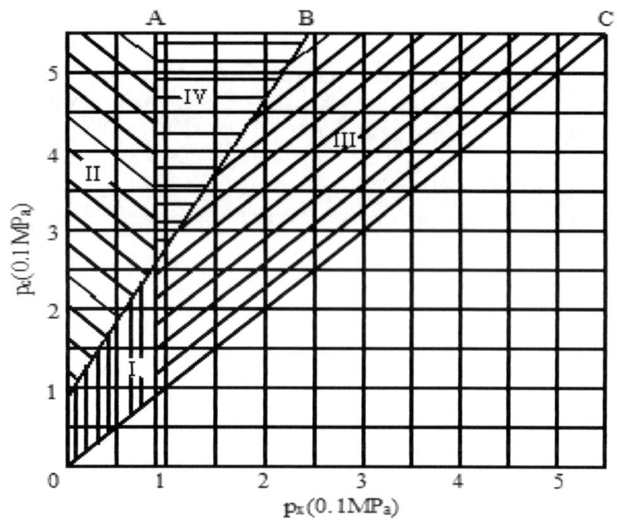

Figure 3. Working area diagram of air momentum instrument

The lines A, B and C intersect into four zones, i. e., II, III and IV respectively, which are the four conditions of back pressure meter, corresponding to the four working conditions listed in table 3.1.

The influence of the characteristic equation of the curved baffle and the characteristic equation of the baffle plate on the measurement accuracy is analyzed and compared. It is shown that the measurement accuracy of the characteristic equation of the curved baffle is higher than that of the planar baffle, and the accuracy is $\frac{d^2}{8D}$. When the honing accuracy is m 1 m and $D < 125d^2$, the influence of the diameter of the curved baffle on the measurement accuracy must be considered. The characteristic equation of the baffle cannot be used to study the characteristic equation of the curved baffle.

4. Feeding method of honing fluid

The honing fluid flows downward along the axis of the honing head in the annular narrow slot between the inner wall of the workpiece and the grinding head, which can be approximately viewed as a concentric cylindrical annular gap flow. As the hydraulic diameter of the gap (relative to the diameter of the circular section) is very small, as the working fluid honing fluid has a certain viscosity, honing fluid flow in the gap when the Reynolds number is small, generally belong to laminar flow range. The mechanism of honing liquid annular gap in motion has two: one is the flow due to the pressure difference, this flow is called flow differential pressure; the other is due to the composition of gap wall with relative motion (i.e. honing head movement) and the liquid flow in the gap, known as shear flow, the superposition of the two is called the pressure difference shear flow. The differential pressure flow is the dominant flow mode.

As shown in Fig. 4, the gap height is:

IOP Conf. Series: Journal of Physics: Conf. Series **1064** (2018) 012028 doi:10.1088/1742-6596/1064/1/012028

$$\delta = (D_2 - D_1)/2 \tag{20}$$

D_2 - workpiece bore diameter; D_1 - grinding head; δ -symmetrical baffle spacing.

As the gap size δ is small, the concentric cylindrical annular slot flow can be approximately regarded as a parallel plate gap flow with a width of $b = \pi D_1$, as shown in Fig. 5.

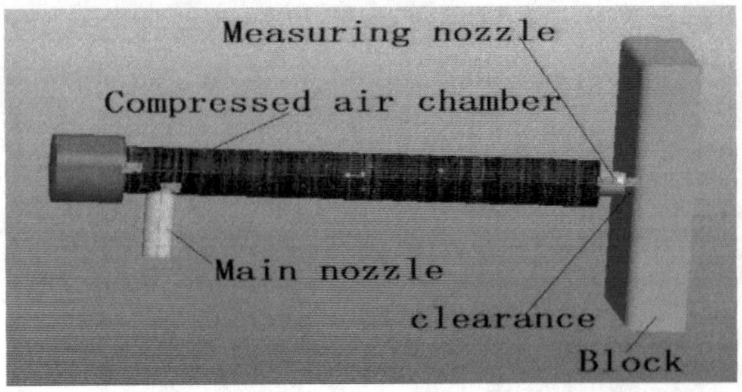

Figure 4. Gap flow between parallel plates

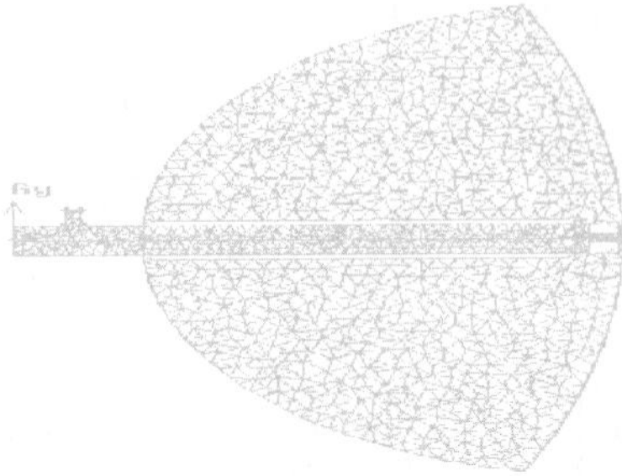

Figure 5. Mesh diagram of a cross-section of a rotating shaft

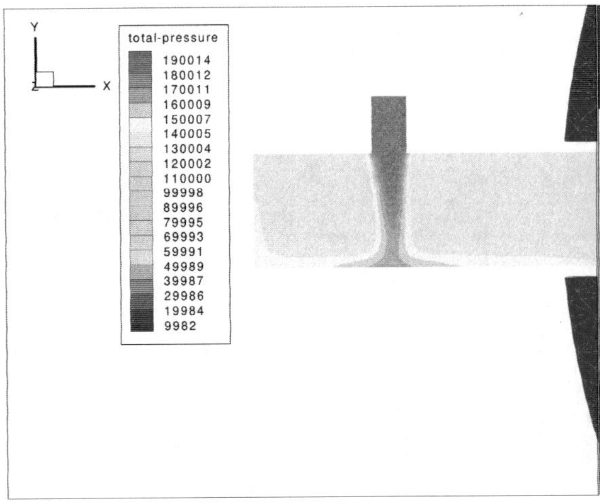

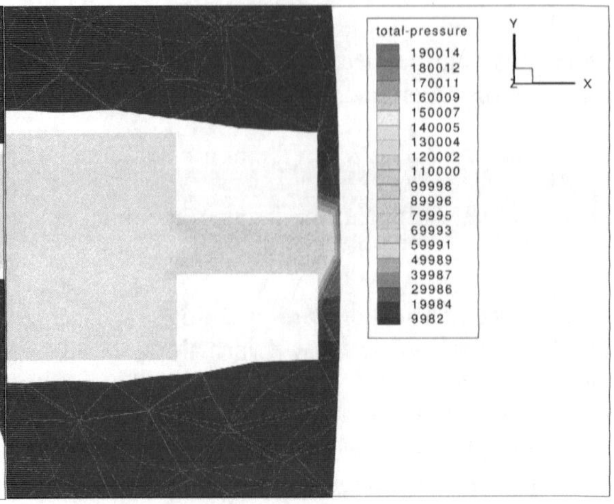

Figure 6. Total pressure inlet diagram **Figure 7.** Total pressure exit diagram

IOP Conf. Series: Journal of Physics: Conf. Series **1064** (2018) 012028 doi:10.1088/1742-6596/1064/1/012028

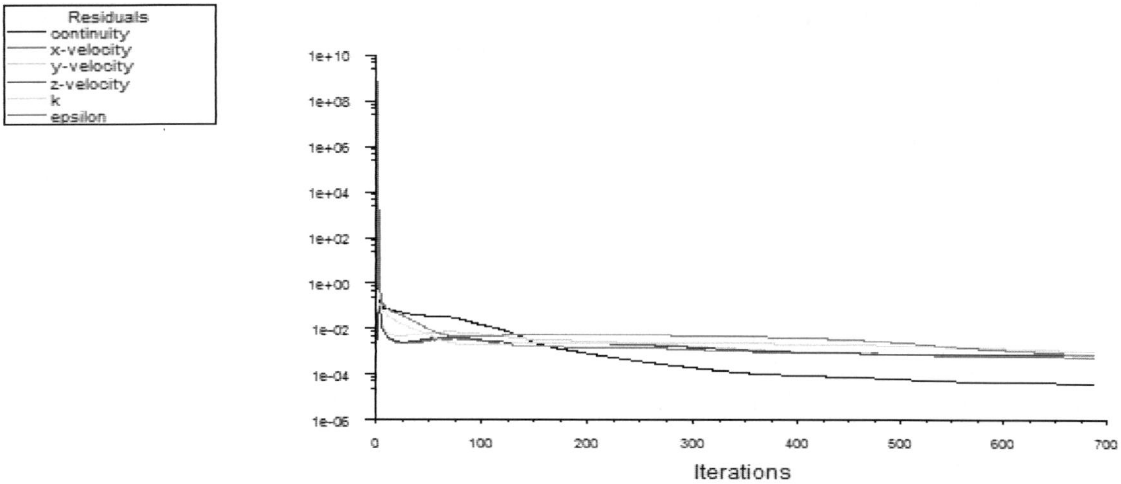

Figure 8. Residual curve

We use Flunt to make Fig.6, Fig.7 and Fig.8, the correctness of the characteristic equation of the surface baffle is validated by the fluid simulation software FLUENT. The simulation results of FLUENT software are used to draw the Px-S (back pressure clearance) curve, and compare with the theoretical equation curve. It proves that the characteristic equation of the surface baffle is correct. The simulation results show that the air flow changes during the operation of the pneumatic measuring device.

5. Experiment simulation and analysis

In our laboratory, the main nozzle, measuring nozzle experimental data diagram, Simulation, curve, of, NC, honing, machine diagram, as shown in Fig.9, the nozzle and measuring nozzle experimental data are shown in Fig.10 and Fig.11, respectively:

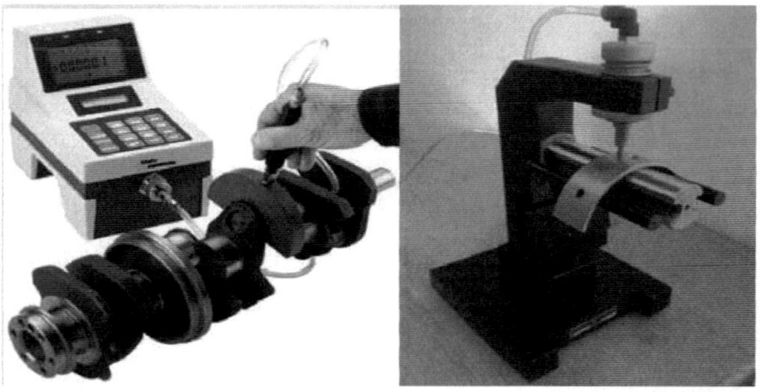

Figure 9. Simulation curve of NC honing machine

The overall design of the project adopts high-pressure differential pressure pneumatic measurement method, and the signal pickup and conversion process is as follows:

(1) The clearance S between the probe and the curved baffle is converted to the back pressure of the air chamber Px;

(2) Use a piezoresistive pressure converter to convert the back pressure Px into current value I output;

(3) The regulating circuit converts the current parameter I, which is output by the pressure converter, to voltage parameter U;

(4) The 16 bit conversion precision A/D converter quantifies analog voltage values into binary digital values; the binary voltage value acquired by the computer is converted to the actual voltage value;

(5) After measuring the gap voltage to the gap value, the gap value is compared with the measuring inductance of the inductance micrometer used as calibration, and the corresponding gap value is obtained;

(6) Display measurement results.

Source provided by air compressor, pressure is 104~106Pa, with links to the high pressure gas source to remove moisture, oil droplets and solid particles, the use of precision gas regulator to link voltage stable pressure 0.2MPa in the pressure of work. The gas after the pressure adjustment goes to the measuring nozzle and picks up the measuring signal. A/D conversion data acquisition card will collect the binary voltage value to the computer, the computer processing to obtain surface measurement parameters and display output.

(1).Main nozzle

We will measure the nozzle diameter was 0.4mm and the entrance pressure of 0.2MPa (g), the main nozzle simulation in different aperture of the pressure values are listed in Tab.1.

Table 1. Pressure (P_x/P_a) comparison tables for three main nozzle models

clearanceS/mm	Inside diameter of main nozzle d_1 /mm		
	0.2	0.4	0.6
0	199998	199989	199998
0.005	50522	183478	196597
0.100	31	142474	186745
0.150	—	94012	171588
0.200	—	54555	156012
0.250	—	34670	134071
0.300	—	23884	129498
0.350	—	17463	93040
0.400	—	13330	70182
0.450	—	10508	54880
0.500	—	8498	44126
0.550	—	7015	36263
0.600	—	5892	30342

Draw the pressure value shown in Tab.1 as the corresponding Px-S curve, as shown in Fig.10.

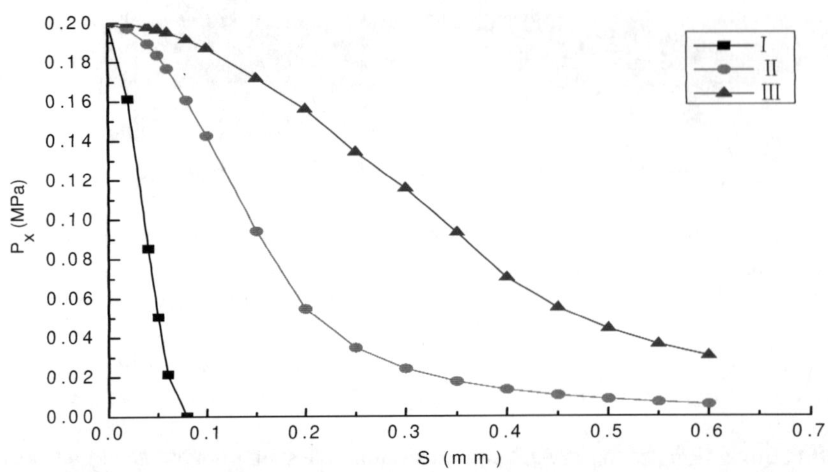

Figure 10. Performance comparison of three models of main nozzle

Ⅰ: d_1=0.2mm II: d_1=0.4mm III: d_1=0.6mm

Figure 10. Is a diagram of the clearance - back pressure gas path drawn to change the bore diameter of the main nozzle. You can see from Fig.10:

1. With the increase of the bore diameter of the main nozzle, the amplification factor (the slope of the straight line) of the pneumatic conversion section decreases;
2. As the diameter of the main nozzle increases, the viscous part of the curve increases;
3. As the diameter of the main nozzle increases, the linear part in the middle of the curve increases.

These results suggest that measuring gap initial and maximum measuring gap size can be increased due to increase of main nozzle pneumatic instrument, fixed in the case of measuring nozzle can also by fine-tuning the main nozzle aperture to change the magnification of gas dynamic link with the need to transform the pneumatic measuring instrument overall magnification.

6. Measuring nozzle

The main nozzle bore diameter is 0.4mm, the inlet pressure is 0.2MPa. The pressure value of the nozzle under different aperture is measured in Tab.2.

Table 2. Comparison of pressure measurements (Px/Pa) of three measuring nozzle models

clearanceS/mm	measuring nozzle bore d /mm		
	0.2	0.4	0.6
0	199999	199999	199999
0.05	195671	183479	165461
0.1	183354	142471	94569
0.15	164780	94019	31453
0.2	145011	54556	23995
0.25	148819	34670	15241
0.3	100423	23885	10533
0.35	72469	17465	7721
0.4	54829	13328	5908
0.45	42969	10508	4661
0.5	34611	8499	3772
0.55	28477	7016	3135
0.6	23849	5891	261

The pressure values shown in Tab. 2 are plotted as corresponding Px-S curves, as shown in Fig.11.

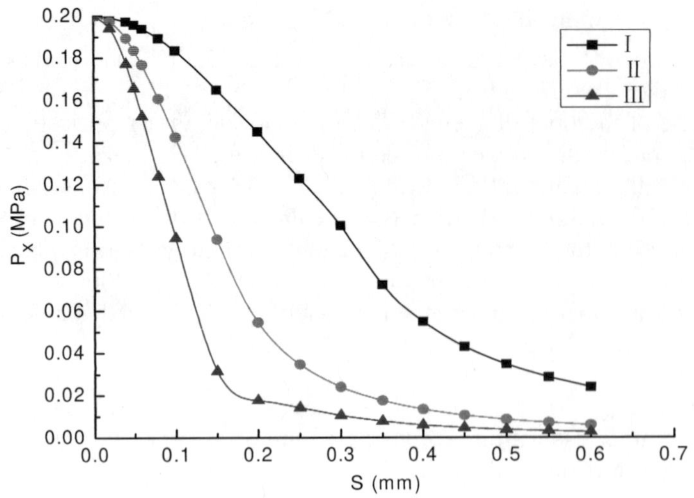

Figure 11. Performance comparison of three models for measuring nozzles

Ⅰ: d =0.22mm II: d =0.42mm III: d =0.62mm

Fig.11 is a change in the diameter of the measuring nozzle bore clearance - back pressure gas path characteristics diagram. You can see from Fig.11:

1. As the diameter of the nozzle increases, the amplification (slope of the straight line) of the pneumatic conversion section increases;

2. As the diameter of the nozzle increases, the viscosity of the curve decreases;

3. As the diameter of the nozzle increases, the linear part in the middle of the curve decreases relatively. These conclusions show that increasing the magnification of the gas meter can increase the magnification of the air momentum meter, but also reduce the initial measurement gap and increase the gas consumption of the gas source.

By combining the above two points, we can conclude that when the diameter of the measuring nozzle is equal to the diameter of the main nozzle, the optimum value of amplification and linear range can be obtained.

7. Summary

(1) Through theoretical research, the expressions of the back pressure and clearance function and the change rate of the middle and high pressure back pressure pneumatic measuring system under four working conditions are obtained when the nozzle baffle is curved surface. It provides a powerful theoretical basis for the research of dynamic honing, pneumatic measurement and measurement accuracy under experimental conditions.

(2) The relationship between the diameter of the curved baffle plate and the accuracy of the measuring system is studied. Back pressure chamber pressure p_x is not only related to measuring system working pressure p_c, main nozzle inner diameter d, measuring nozzle inner diameter d 1 and measuring gap S, but also related to the diameter of surface baffle D. And the larger the diameter of the curved baffle is, the smaller the pressure D of the back pressure air chamber is p_x. The larger the magnification of the pneumatic shift is. The slope of the line

(3) The influence of the characteristic equation of the curved baffle and the characteristic equation of the baffle plate on the measurement accuracy is analyzed and compared. It is shown that the measurement accuracy of the characteristic equation of the curved baffle is higher than that of the planar baffle, and the accuracy is $\frac{d^2}{8D}$. When the honing accuracy is m 1 m and $D < 125d^2$, the influence of

the diameter of the curved baffle on the measurement accuracy must be considered. The characteristic equation of the baffle cannot be used to study the characteristic equation of the curved baffle.

(4) The air path structure and nozzle parameters of the curved nozzle flapper mechanism in the honing pneumatic measuring system are optimized. Under the condition that the baffle is curved surface, the pressure clearance curve of different main and measuring nozzles is obtained by adjusting the model parameters, and then the law is found according to the result, and the design parameters are optimized.

Acknowledgements

The authors thank the financial supports from National Natural Science Foundation of China(Grant no. 51165024) and Science and Technology Major Project of “High-grade NC Machine Tools and Basic Manufacturing Equipment” (2010ZX040001-181).

References

[1] Roy J, Xu Y.X. Form and orientation tolerance analysis for cylindrical surfaces incomputer-aided inspection [J].Computers in Industry, 1995, 26: 127 - 134.

[2] Hodgson T. J, Kay M. G, Mittal R, et al. Evaluation of cylindricity using combinatorics [J].IIE Transactions, 1999, 31: 39 - 47.

[3] Chou S.Y, Sun C.W. Assessing cylindricity for oblique cylindrical features. International Journal of Machine Tools & Manufacture, 2000, 40: 327 - 341.

[4] Lai H, Jywe W, Chen C. K, et al. Precision modelingerrors for cylindricity evaluation using genetic algorithms [J]. JournalInternational Societies for Precision Engineering and Nanotechnology,2000, 24: 310 - 319.

[5] Zhu L.M, Ding H. Application of kinematic geometry to computational metrology: distance function based hierarchical algorithms for cylindricity evaluation [J]. International Journal of Machine Tools Manufacture, 2003, 43: 203 – 215.

[6] Lao Y. Z, Leong H.W, Preparata F.P, et al. Accurate cylindricity evaluation with axis-estimation preprocessing [J].Precision Engineering, 2003, 27: 429 – 437.

[7] Choi H, Seo J, Lee S. Machining error compensation of extern cylindrical grinding using a thermally actuated rest [J].Journal of Materials Processing Technology, 2002, 127: 280 – 285.

[8] Kirsten C, Placid F. Verification of form Tolerances Part: Cylindricity and straightness of a median line. Computers & Industrial Engineering, 2002, 1: 43.

[9] Shao Z.S, Ma G.W. Thermo-mechanical stresses in functionally graded circular hollow cylinder with linearly increasing boundary temperature [J].Composite Structures, 2008, 83: 259 – 265.

[10] Bugra K, Juan A. Aguirre C, et al. Inspection of the cylindrical surface feature after turning using coordinate metrology [J]. International Journal of Machine Tools & Manufacture, 2007, 47: 1893 – 1903.

[11] Venkaiah N, Shunmugam M.S. Evaluation of form data using computational geometric techniques—Part II: Cylindricity error [J]. International Journal of achine Tools & Manufacture, 2007, 47: 1237 – 1245.

[12] Cheraghi S. H, Jiang G.H, Jamal S. A. Evaluating the geometric characteristics of cylindrical features [J]. Precision Engineering, 2003, 27: 195 – 204.

[13] Liu X.M, Shi Z.Y. Development and application of convex hull in the assessment of roundness error [J]. International Journal of Machine Tools & Manufacture, 2008, 48: 135 – 139.

[14] Cho N, Tu J.F. Quantitative circularity tolerance analysis and design for 2D precision assemblies [J]. International Journal of Machine Tools & Manufacture, 2002, 42: 1391 – 1401.

[15] Zhang Wanjun, Zhang Feng , Zhang Guohua. Research on a algorithm of adaptive interpolation for NURBS curve. [J].Applied Mechanics and Materials, Vol. 687 - 691, pp.1600-1603, December 2014.

[16] Zhang Wan-Jun, Zhang Feng, Zhang Guohua. Research on modification algorithm of Cubic B-spline curve interpolation technology. [J].Applied Mechanics and Materials, Vol. 687 - 691, pp. 1596 - 1599, December 2014.

[17] Zhang wanjun. A device for controlling workpiece pneumatic measurement [P]. China: 201420094751. X, 2014.
[18] Zhang wanjun. A control system for a workpiece pneumatic measurement [P]. China: 201420094752. 4, 2014.
[19] Zhang wanjun. A device for workpiece pneumatic measurement [P]. China : 201420094730. 8. 0, 2016.
[20] Zhang wanjun. A pneumatic measuring instrument control system based on PLC [P]. China: 201420173572. 5, 2014.
[21] Zhang wanjun. An integrated detection system for PLC control of workpieces [P]. China: 201520309360X, 2015.

IOP Conf. Series: Journal of Physics: Conf. Series **1064** (2018) 012029 doi:10.1088/1742-6596/1064/1/012029

Flow field analysis and parameter optimization of main and measured nozzles of differential pressure type gas momentum instrument based on CFD

Wanjun Zhang[1, 2, 3, *], Feng Zhang[2], Jingxuan Zhang[1], Jingyi Zhang[1], Jingyan Zhang [1]

[1]Quanzhou Institute of Information Engineering, 362000, China
[2]School of Mechanical Engineering, Xian Jiao tong University, 710049, China
[3]Lanzhou Industry and Equipment Co., Ltd., Lanzhou 730050, China

*Corresponding author e-mail: gszwj_40@163.com

Abstract. In honing on-line pneumatic measurement, the actual working conditions of honing have not been taken into account. While honing and online measurement of aerodynamic working environment is relatively small semi closed annular narrow space, various parameters will influence the movement characteristics of the measurement of gas, gas does not consider the previous online honing the actual working conditions of the dynamic measurement accuracy is hard to guarantee, does not meet the requirement of automatic processing of high precision. Through the combination of theoretical research and simulation experiment, through the study of honing, pneumatic measurement and aerodynamics, the characteristic equations of three working conditions under honing actual working conditions are obtained. Using fluid simulation software FLUENT to the actual working parameters of honing pneumatic measuring mechanism model is simulated, simulation results show that when the working temperature is less than the optimum value, to achieve amplification and linear range, complex parameter measurement, to meet the honing high dimensional accuracy and high shape precision and high automation requirements.

1. Introduction

With the development of automobile industry, higher requirements have been put forward for cylinder block and cylinder sleeve [1-3]. At present, high precision NC honing machine has been monopolized by foreign countries. Imported machine tools are expensive and become the bottleneck of promoting the development of the automotive industry. Therefore, it is imperative to develop domestic CNC honing machines. Among them, the research on the accuracy analysis of honing on-line pneumatic measurement technology has important practical significance.

Pneumatic measurement [4-7] is non-contact measurement, has good flexibility, non-destructive and stability, high sensitivity, strong anti-interference ability, easy realization of composite parameter measurement and self-cleaning function and other unique characteristics, to meet the honing high dimensional accuracy and high shape precision and high automation requirements.

With the further research of engine operation mechanism in engineering field, higher requirement of honing accuracy is put forward. The accuracy of honing machine determines the machining accuracy of

Content from this work may be used under the terms of the Creative Commons Attribution 3.0 licence. Any further distribution of this work must maintain attribution to the author(s) and the title of the work, journal citation and DOI.
Published under licence by IOP Publishing Ltd

honing machine. At present, pneumatic measuring system is widely used in honing process. In the previous research on dynamic process of honing pneumatic measurement system is only for the static characteristics, dynamic characteristics of the few analysis on the actual conditions of honing gas dynamic characteristics analysis of measurement system is almost not involved, the design of pneumatic measuring parameters mainly adopts empirical estimation method of experiment. But to achieve the high precision of honing pneumatic measurement, it is not scientific to use static measurement instead of the actual working condition, and there is great error. Therefore, the dynamic analysis of pneumatic measurement and the actual conditions of the impact of measurement accuracy is essential. Therefore, this subject through the honing pneumatic measurement system dynamics theory and actual honing conditions of the measurement accuracy, to achieve high-precision honing pneumatic measurement, in order to meet the requirements of high-precision honing processing.

With the further research of engine operation mechanism in engineering field, higher requirement of honing accuracy is put forward. The accuracy of honing machine determines the machining accuracy of honing machine. At present, pneumatic measuring system is widely used in honing process. Lanzhou University of technology, Lanzhou Industry and Equipment Co. Ltd ZhangWanjun. ET, at researchers [15-21] have researched interpolation algorithms for NURBS curves honing pneumatic measurement system. Here has been attempts to change the feed rate in NURBS interpolation honing pneumatic measurement system.

From the domestic and international research status and related literature review, we can see that there are still some blind spots in the pneumatic measurement, especially for the pneumatic measurement of honing, and the key problems that need to be solved are as follows:

1) For pneumatic measurement, the domestic scholars have only studied the static measurement mechanism of pneumatic measurement, but the dynamic mechanism has not been studied. In real-time and on-line measurement, the measured object measurement system or one side will work in a particular state of motion, and pneumatic working medium - Measurement of compressed gas (due to the fluid characteristics) the dynamic characteristics and static characteristics are very different from the inevitable. Pneumatic dynamic mechanism is to achieve real-time online measurement of pneumatic measurement is essential, the past simple use of static measuring mechanism instead of dynamic mechanism of measurement approach has considerable error, cannot satisfy the demand of high accuracy.

2) In honing on-line pneumatic measurement, the actual working conditions (pressure change, While honing and online measurement of aerodynamic working environment is relatively small semi closed annular narrow space, various parameters will influence the movement characteristics of the measurement of gas, gas does not consider the previous online honing the actual working conditions of the dynamic measurement accuracy is hard to guarantee, does not meet the requirement of automatic processing of high precision.

This paper through theoretical research on aerodynamics and fluid mechanics, the establishment of honing pneumatic instrument model, and further optimize the working parameters of pneumatic instrument through computer simulation, dynamic debugging and measuring system on gas in theory and use give guidance to pneumatic better applied in machinery and other industries can measure.

In the previous research on dynamic process of honing pneumatic measurement system is only for the static characteristics, dynamic characteristics of the few analysis on the actual conditions of honing gas dynamic characteristics analysis of measurement system is almost not involved, the design of pneumatic measuring parameters mainly adopts empirical estimation method of experiment. But to achieve the high precision of honing pneumatic measurement, it is not scientific to use static measurement instead of the actual working condition, and there is great error. Therefore, the dynamic analysis of pneumatic measurement and the actual conditions of the impact of measurement accuracy is essential. This subject through the honing pneumatic measurement system dynamics theory and actual honing conditions of the measurement accuracy, to achieve high-precision honing pneumatic measurement, in order to meet the requirements of high-precision honing processing.

Pneumatic measurement [1] [2] is a non-contact measurement. It has high precision, small measuring force and self-cleaning function to the work piece under test. It is not affected by the surface material of the work piece. This method uses compressed air as the medium, the use of air in the pipe flow or pressure with the nozzle and the measured work piece gap between different change characteristics, size or the amount of displacement transformation flow changes or pressure signal, so as to achieve the measurement. However, due to the small size of the nozzle and the small size of the air jet, the selection of the parameters such as the main nozzle and the nozzle aperture in the aerodynamic measurement has been a difficult problem. Have a huge impact, and the main measuring nozzle aperture ratio and the measurement accuracy and sensitivity, so the analysis of differential pressure type gas flow path optimization, pneumatic measurement of the main measuring nozzle aperture parameters in pneumatic measurement is of great significance. By establishing the theoretical model parameters of pneumatic instrument design, analysis, flow field calculation and flow field prediction function flow using powerful CFD software can analyze and display the phenomenon of the flow field, in a relatively short period of time can predict object performance, and optimize various parameters.

2. The basic principle of differential pressure pneumatic measurement

2.1. Differential pressure gas path system

At present, two kinds of measuring methods are commonly used in the pneumatic on-line measuring system, backpressure and differential pressure. Simple pressure has its disadvantage of gas measurement, too high, or when the distance between the nozzle and the work piece is changed, once the air pressure fluctuations, work pressure will change, thus causing the measurement error, stability of [3] instrument failure. The differential pressure measurement can solve the problem and improve the performance of the meter. So the differential pressure measurement technique is used in this system.

The differential pressure pneumatic measuring principle diagram is shown in figure 1. The compressed air source through the filter 1 to 2, 3 and 4 after the intake valve pressure regulator P_c, with constant PC, divided into two flows. Through a throttle hole 6 through the nozzle flapper 9 into the atmosphere. The other path through a throttle hole 5 by regulating valve 8 into the atmosphere [1]. Obviously, measuring two back pressure pneumatic gas route (zero chamber and the measuring chamber, two) when measuring the backpressure gas path in the same ambient pressure and temperature, the external environment (pressure, temperature and humidity) to measure the impact of the minimum, high accuracy and stability of [4]. Two cavity pressure, according to fluid mechanics knowledge, through a cross-section of the subcritical state of the flow formula, force difference is: formula $\Delta P = P_t - P_x$; $\Delta P = f(S)$: P_t - Zero Zero cavity pressure; P_x - measuring chamber pressure. Generally adjustable flow valve in the measurement before adjusting the back pressure in the measurement process is the same, that is P_t, P_t is a constant. The measuring chamber pressure P_x changes with the change of measurement gap s. Using silicon piezoresistive differential pressure sensor 10 to detect ΔP , as the measured signal, amplified and output, you can measure the size of the parameter S. Schematic diagram of gas path of back pressure pneumatic meter, as shown in Fig. 1.

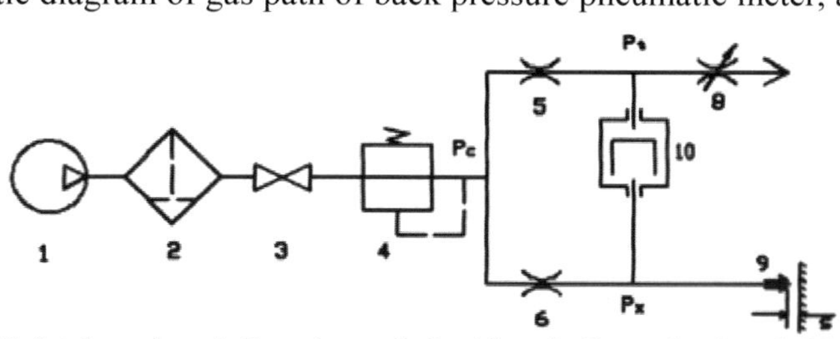

1. air source, 2. filter, 3. intake valve, 4. Regulator, 5, 6 orifice, 8. Control valve, 9. Nozzle baffle mechanism, 10. Silicon piezoresistive differential pressure sensor.

Figure 1. Schematic diagram of gas path of back pressure pneumatic meter.

3. Aerodynamic measurement model

The differential pressure pneumatic measuring principle is shown in figure 1. Compressed air from the gas source 1, after filter 2, intake valve 3 and pressure reducing valve 4, has a constant pressure of PC, divided into two flows. Through a main nozzle 5 and measuring nozzle 6 into the atmosphere, high pressure input of the other way directly connected to the differential pressure sensor. Apparently, the gas path is the difference between the constant pressure input PC and the measured chamber pressure P_x:

$$\Delta p = p_c - p_x \tag{1}$$

Where: p_c - constant pressure at the input; p_x - measuring chamber pressure.

The basic principle of pneumatic measurement is introduced with pneumatic back pressure pneumatic measuring system, and the schematic diagram is shown as follows:

The back-pressure indicator of p_x in different ways, its function is the variation of the pressure will be converted into Δp_x instructions (such as the original pointer, buoy, liquid column etc.) the amount of the mobile ΔL.

The back pressure type pneumatic meter is composed of at least two conversion links (or amplification links).

$$\Delta S \rightarrow \Delta px \rightarrow \Delta L \tag{2}$$

Among them, the change of ΔS is to measure the gap, ΔS→Δpx is called pneumatic converter, the PX and the L amplification step instructions. The ratio of back pressure type pneumatic momentum instrument Δpx→ΔL is also determined by the transfer function of these two links, that means:

$$K = \frac{\Delta L}{\Delta S} = \frac{\Delta L}{\Delta p_x}\frac{\Delta p_x}{\Delta S} \tag{3}$$

The $\frac{\Delta p_x}{\Delta S}$ is called the conversion ratio of the pneumatic system, and $\frac{\Delta L}{\Delta p_x}$ is called the indicating conversion ratio. Some of the back pressure gas momentum instrument, and its amplification instructions link is also composed of a number of links, but all back pressure type pneumatic momentum instrument, its pneumatic change link working principle is the same.

When the measuring head is a nozzle baffle, the back pressure and the clearance S of the nozzle baffle are in one-to-one correspondence function relationship. The magnitude of the measured parameter S can be obtained as long as the back pressure value is read from the pressure indicator section 6. Between the pressure and the measurement of the gas chamber pressure nozzle flapper gap S function: p_x=f(S) is shown in Fig.2.

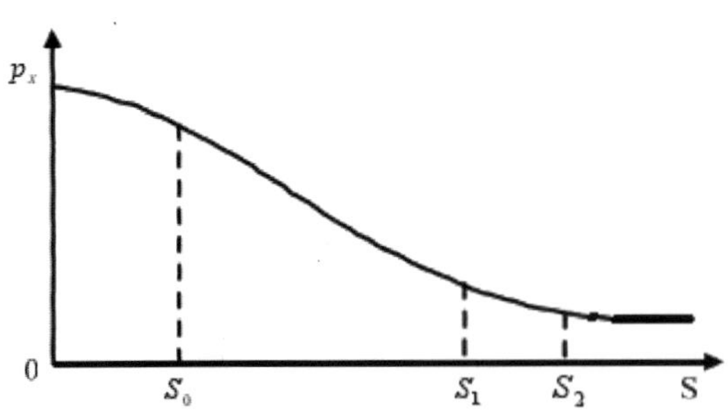

0- S_0: viscous part; S_0 - S_1: linear part; S_1 - S_2: transition part; after S_2: horizontal part

Figure 2. Back pressure clearance curve (characteristic curve).

4. Characteristic equation of baffle

The characteristic equations of the plane baffle measurement system under four operating conditions are [13]:

Condition I:

$$P_x = P_c - \frac{P_c\left[\left(\frac{P_o}{P_c}\right)^{\frac{2}{n}} - \left(\frac{P_o}{P_c}\right)^{\frac{n+1}{n}}\right]\left(\frac{c_2 d}{c_1 \frac{d_1^2}{4}}\right)^2 S'^2}{\frac{n-1}{n} + \left[\frac{2n-2}{n}\left(\frac{P_o}{P_c}\right)^{\frac{2}{n}} - \frac{n-1}{n}\left(\frac{P_o}{P_c}\right)^{\frac{n+1}{n}}\right]\left(\frac{c_2 d}{c_1 \frac{d_1^2}{4}}\right)^2 S'^2} \tag{4}$$

Condition II:

$$P_x = \left[\frac{1}{2}P_0^{\frac{n-1}{n}} + \frac{1}{2}\left(P_0^{\frac{2n-2}{n}} + 4P_0^{-\frac{2}{n}}\frac{B^2 P_c^2}{A^2 S'^2}\right)^{\frac{1}{2}}\right]^{\frac{n}{n-1}} \tag{5}$$

Condition III:

$$P_x = P_c\left[0.5 + 0.5\left(1 + 4\frac{A^2}{B_1^2}S'^2\right)^{0.5}\right]^{\frac{n}{1-n}} \tag{6}$$

Where, S' is the measuring clearance. C_1 - The flow coefficient at the orifice (main nozzle); d_1 - orifice diameter; P_c - work pressure (absolute pressure); P_x -back pressure (absolute pressure); n - the constant entropy index, for air; γ_c - p_c under the severe gas.

Working area diagram of air momentum instrument, as shown in Fig. 3. By comparing with the equation of curved baffle measurement system characteristics, we can draw in the measuring chamber pressure p_x and other parameters of the nozzle under the same measuring clearance error measuring gap curved baffle and plane baffle is $\frac{d^2}{8D}$, the measurement accuracy is calculated according to the curved baffle characteristic equation characteristic equation of $\frac{d^2}{8D}$ is higher than the plane baffle.

The lines A, B and C intersect into four zones, i. e., II, III and IV respectively, which are the four conditions of back pressure meter, corresponding to the four working conditions listed in table 3.1.

The influence of the characteristic equation of the curved baffle and the characteristic equation of the baffle plate on the measurement accuracy is analyzed and compared. It is shown that the measurement accuracy of the characteristic equation of the curved baffle is higher than that of the planar baffle, and the accuracy is $\frac{d^2}{8D}$. When the honing accuracy is m 1 m and $D < 125d^2$, the influence of the diameter of the curved baffle on the measurement accuracy must be considered. The characteristic equation of the baffle cannot be used to study the characteristic equation of the curved baffle.

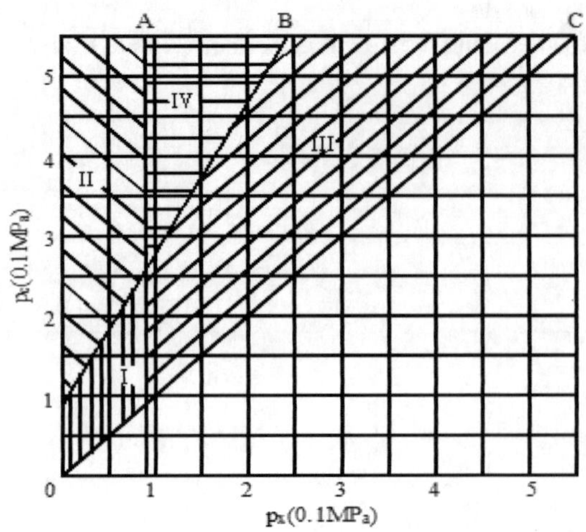

Figure 3. Working area diagram of air momentum instrument.

Through simulation experiments, we obtained the pressure of back pressure of gas chamber under different clearance of nozzle baffle p_x. Entrance pressure constant is 0.2MPa (g), $p_x - S$ function as the back pressure condition into the value of simulation and under the same conditions the value of the $p_x - S$ curve were drawn from 0 different gap corresponding to the start value, as shown in Fig.4.

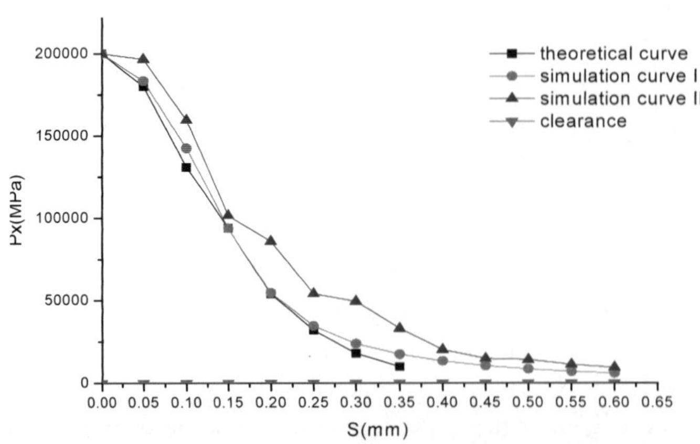

Figure 4. Comparison of theory and simulation

As can be seen from Fig.4, the theoretical curve and the simulation curve are similar in shape. They are composed of four sections, namely, the viscous part, the straight line part, the transition part and the horizontal part, and the slope is approximately the same.

5. Analysis of honing fluid flow field based on FLUENT

Compared with the back pressure type gas path, the advantages of differential pressure type gas path is to reduce the measurement error caused by the output pressure relief valve P_c changes, so the differential pressure type pneumatic measuring instrument, measuring the main nozzle as the research object were emphatically analyzed. According to the requirements of the project, the assembly diagram of the double feed grinding head of the river and the pneumatic transformation principle of differential pressure pneumatic measurement are drawn. The plane sketch map of the pneumatic transformation section of the differential pressure pneumatic instrument is drawn. In order to facilitate the research, the main nozzle and the measuring nozzle are built in the same model. Compared with the back pressure type gas path, as shown in Fig.5.

Fig. 5 is the pressure nephogram of the air path. From the static pressure diagram, it can be seen that the gas pressure distribution in each part is very uniform. From the dynamic pressure diagram, it can be seen that the pressure of the gas varies very sharply at the throttle orifice and the nozzle baffle. The pressure in the pressure chamber and the back pressure chamber is relatively stable.

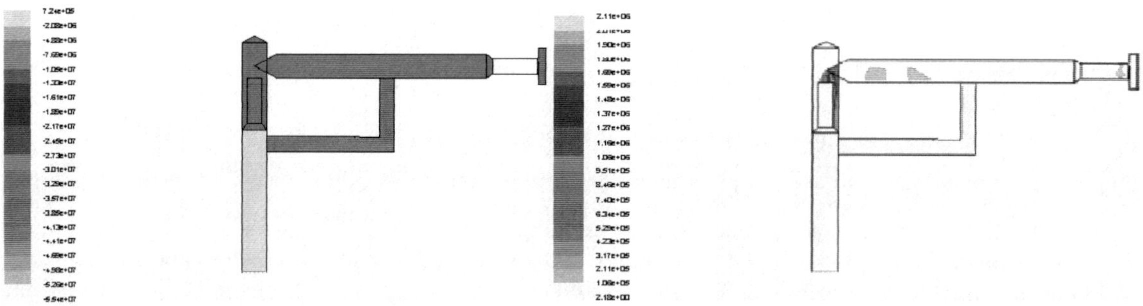

A) Static pressure diagram B) Dynamic pressure diagram

Figure 5. Pressure nephogram

6. Solution calculation and analysis

Using FLUENT for simulation solution, the first step is to initialize the flow field. Initialization means giving the initial values of each flow parameter so that the iterative computation can have a starting point. Initial flow field in two ways, one is the all cell initialization of the field with the same value of field variables, the other is a flow variable selection in selected cell regions in the coverage of a value or function [12-14].

After doing initialization work, the simulation iterative calculation [22-29] can be carried out, and the monitoring residual error of the solution process is also set up. The convergence condition of this paper is as follows: the residual energy is less than 10-6, and the convergence requirement is achieved. The exit velocity monitoring, shown in Fig. 6, Fig.7 and Fig.8.Eventually tends to remain constant, further monitoring the good convergence performance of the iterative computations.

After calculation, you can see two results of data and graphics. Figure 4 is the speed of cloud of gas path, in the figure can be seen in the velocity of fluid in each position, in the entrance orifice and nozzle baffle, flow rate of gas is maximum, supersonic phenomenon occurs even in local small range. At the same time, the airflow emitted by the nozzle is blocked by the baffle plate, and the speed is lowered instantly, and the throttling of gas is formed at the outlet.

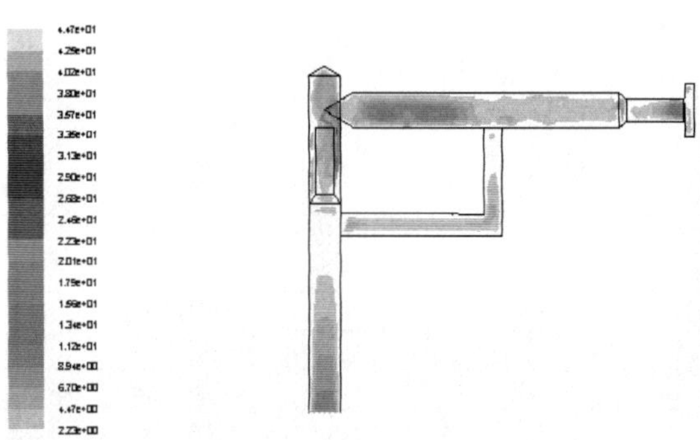

Figure 6. Static speed diagram.

Figure 7. Tangential velocity diagram.

Figure 8. Measuring the total pressure of the nozzle

7. Experiment simulation and analysis

In our laboratory, the main nozzle, measuring nozzle experimental data diagram, Simulation, curve, of, NC, honing, machine diagram, as shown in Fig.9, the nozzle and measuring nozzle experimental data are shown in Fig.10.

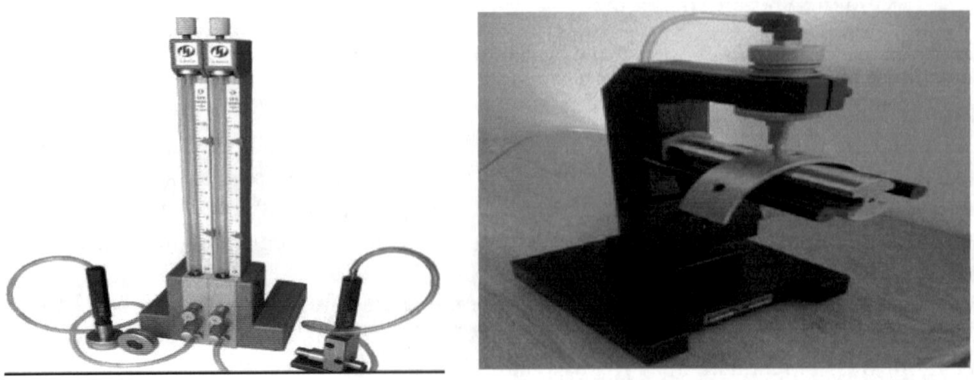

Figure 9. Simulation curve of NC honing machine.

The inlet pressure is 0.2MPa, and the pressure values under different working conditions are listed in Tab.1.

Table 1. Comparison of model pressure (p_x/p_a) at three operating temperatures

Clearance S/mm	Kinematic viscosity of honing fluid ($cst(40°C)$)		
	3	5	8
0	19999	199999	199998
0.05	180101	183475	196591
0.1	130902	142402	186742
0.15	94004	94018	171587
0.2	54002	54552	156011
0.25	32082	34670	134078
0.3	17948	23882	129490
0.35	9902	17462	93049
0.4		13338	70186
0.45		10502	54880
0.5		8496	44122
0.55		7014	36263
0.6		5892	30342

Draw the values shown in Tab.1 to the corresponding curves, as shown in Fig.10.

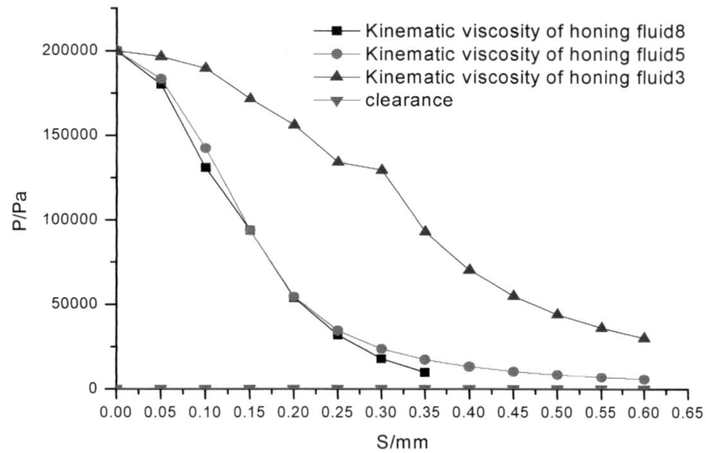

Figure 10. Comparison of characteristic curves of the model under three operating conditions.

As you can see from Fig. 8 that we can draw the following conclusions:

1). With the increase of honing fluid viscosity, the amplification (slope of the straight line) of the pneumatic conversion section decreases;

2). As the honing liquid viscosity increases, the viscosity of the curve increases with the increase;

3). As the honing fluid viscosity increases, the linear part of the curve increases with the increase.

These indicate that the honing liquid viscosity increases required to reduce the initial clearance measurement of pneumatic measuring instrument and the maximum measuring gap, fixed on the measuring device parameters, the viscosity of the oil film is not blowing through or not completely blown open, should avoid such a situation, a kinematic viscosity control in liquid honing the following.

Table 2. Comparison of model pressure (p_x/p_a) at three operating temperatures

Clearance S/mm	Working condition temperature ($^{\circ}C$)		
	15	25	45
0	199999	199999	199999
0.05	195671	183475	196524
0.1	183352	142476	189675
0.15	164787	94016	185988
0.2	145014	54554	180012
0.25	148812	34671	176436
0.3	100428	23882	174673
0.35	72460	17468	173437
0.4	54829	13323	170309
0.45	42960	10504	168943
0.5	34616	8496	163560
0.55	28475	7012	162645
0.6	23842	5897	160783

Draw the values shown in Tab.2 to the corresponding $p_x - S$ curve, as shown in Fig.11.

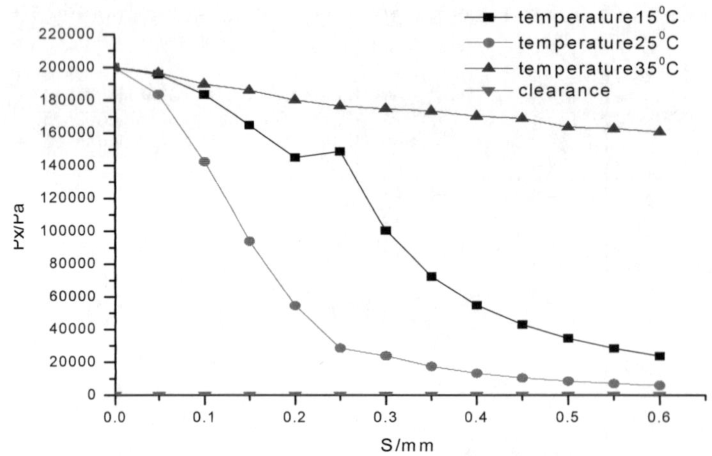

Figure 11. Comparison of characteristic curves of the model under three operating conditions.

As you can see from Fig.11, the temperature has a great influence on the characteristic curve. With the increase of working condition temperature:

1). Amplification (slope of the straight line) of the pneumatic conversion section decreases;

2). Viscosity of the curve will increase to a certain extent.

The linear part in the middle of the 3. Curve is also considerably reduced.

This shows that the increase in temperature conditions of honing pneumatic measurement have a negative impact, will make the measurement of the magnification is reduced, the linear part of the available decreases and the measurement accuracy is reduced, so the working temperature should be strictly controlled. Operating conditions shall not exceed the temperature.

8. Summary

This paper establishes a theoretical model parameters of pneumatic instrument design of differential pressure, flow field analysis using CFD software, the curve of differential pressure ΔP with flapper gap S change background structure parameters and the optimization of the main nozzle, test, provide a reliable reference for designers. These are important to design and study differential pressure pneumatic meter.

Acknowledgements

The authors thank the financial supports from National Natural Science Foundation of China(Grant no. 51165024) and Science and Technology Major Project of "High-grade NC Machine Tools and Basic Manufacturing Equipment" (2010ZX040001-181).

References

[1] Roy J, Xu Y.X. Form and orientation tolerance analysis for cylindrical surfaces incomputer-aided inspection. Computers in Industry,1995,26:127-134.

[2] Hodgson T. J, Kay M. G, Mittal R, et al. Evaluation of cylindricity using combinatorics. IIE Transactions,1999,31:39-47.

[3] Chou S.Y, Sun C.W. Assessing cylindricity for oblique cylindrical features. International Journal of Machine Tools&Manufacture,2000,40:327-341.

[4] Lai H, Jywe W, Chen C.K, et al. Precision modelingerrors for cylindricity evaluation using genetic algorithms. JournalInternational Societies for Precision Engineering and Nanotechnology, 2000,24:310-319.

[5] Zhu L.M, Ding H. Application of kinematic geometry to computational metrology: distance

function based hierarchical algorithms for cylindricity evaluation. International Journal of Machine Tools Manufacture,2003, 43:203–215 .

[6] Lao Y.Z,Leong H.W, Preparata F.P, et al. Accurate cylindricity evaluation with axis-estimation preprocessing. Precision Engineering,2003,27:429–437.

[7] Choi H,Seo J, Lee S. Machining error compensation of extern cylindrical grinding using a thermally actuated rest. Journal of Materials Processing Technology,2002,127:280–285.

[8] Kirsten C, Placid F. Verification of form Tolerances Part: Cylindricity and straightness of a median line. Computers & Industrial Engineering, 2002,1:43.

[9] Shao Z.S, Ma G.W. Thermo-mechanical stresses in functionally graded circular hollow cylinder with linearly increasing boundary temperature. Composite Structures,2008,83:259–265.

[10] Bugra K, Juan A. Aguirre C, et al. Inspection of the cylindrical surface feature after turning using coordinate metrology. International Journal of Machine Tools & Manufacture, 2007, 47: 1893–1903.

[11] Venkaiah N, Shunmugam M.S. Evaluation of form data using computational geometric techniques-Part II: Cylindricity error[J]. International Journal of achine Tools & Manufacture, 2007, 47:1237–1245.

[12] Cheraghi S. H, Jiang G.H, Jamal S. A. Evaluating the geometric characteristics of cylindrical features[J]. Precision Engineering, 2003, 27:195–204.

[13] Liu X.M, Shi Z.Y. Development and application of convex hull in the assessment of roundness error[J]. International Journal of Machine Tools & Manufacture, 2008,48:135–139.

[14] Cho N, Tu J.F. Quantitative circularity tolerance analysis and design for 2D precision assemblies[J]. International Journal of Machine Tools & Manufacture, 2002, 42:1391–1401.

[15] Zhang Wanjun, Zhang Feng , Zhang Guohua. Research on a algorithm of adaptive interpolation for NURBS curve. [J].Applied Mechanics and Materials , Vol. 687-691, pp.1600-1603, December 2014.

[16] Zhang Wan-Jun, Zhang Feng, Zhang Guohua. Research on modification algorithm of Cubic B-spline curve interpolation technology. [J].Applied Mechanics and Materials, Vol. 687-691, pp.1596-1599, December 2014.

[17] Zhang wanjun. A device for controlling workpiece pneumatic measurement[P]. China: 201420094751. X,2014.

[18] Zhang wanjun. A control system for a workpiece pneumatic measurement[P]. China: 201420094752.4,2014.

[19] Zhang wanjun. A device for workpiece pneumatic measurement [P].China: 201420094730.8. 0,2016.

[20] Zhang wanjun. A pneumatic measuring instrument control system based on PLC [P].China: 201420173572.5,2014.

[21] Zhang wanjun. An integrated detection system for PLC control of workpieces[P].China: 201520309360X,2015.

[22] Zhang Wanjun, Zhang, Gao Shanping, Zhang Sujia. Innovation research on Taylor's iteration algorithm of NURBS curve and simulation. [J]. Advances in Materials, materials, Machinery, Electronics I, 2017, 2,Vol. 1820. 080014-1- 080014-8.

[23] Zhang Wanjun,Zhang,Gao Shanping,Zhang Sujia. M NURBS curve method Taylor's launch type of interpolation arithmetic. [J]. Advances in Engineering Research, 2016,12,Vol. 118. 43-52.

[24] Zhang Wanjun,Zhang,Gao Shanping,Zhang Sujia. A Novel of Improved algorithm adaptive of NURBS curve. [J]. Advances in Engineering Research, 2016,12,Vol. 118. 53-60.

[25] Rahalnan M,Seethaler R,YeUowley I,et a1.A new approach to contour error control in high speea machi-lling 『J].International Journal of Machine Tools& Manufacture,2015,88: 42-50.

[26] Zhang Wanjun,Zhang,Gao Shanping,Zhang Sujia. A novel on high-grade CNC machines tools for B-Spline curve method of High-speed interpolation arithmetic. [J]. 2016 International Conference on Automotive Engineering, Mechanical and Electrical Engineering, 2017,3,Vol. 118. 53-60.

IOP Conf. Series: Journal of Physics: Conf. Series **1064** (2018) 012029 doi:10.1088/1742-6596/1064/1/012029

[27] Zhang Wanjun,Zhang,Gao Shanping,Zhang Sujia. Study on Embedded CNC system for NURBS curves method of interpolation arithmetic. [J]. advances in Engineering Research, 2017,3,Vol. 118. 53-60.

[28] Zhang Wan-Jun, Zhang Feng, Zhang Guo-hua. Research on modification algorithm of Cubic B-spline curve interpolation technology.[J].Applied Mechanics and Materials, Vol. 687-691, pp.1596-1599, December 2014.

[29] Zhang Wan-Jun, Zhang Feng, Zhang Wan-Liang. Research on a NURBS curve of timing / interrupt interpolation algorithm for CNC system [J].Chinese Journal of Manufacturing Technology & Machine Tool , 4(4), pp.183-187,April 2015.

IOP Conf. Series: Journal of Physics: Conf. Series **1064** (2018) 012030 doi:10.1088/1742-6596/1064/1/012030

Research on the Fluid Excitation Force of Seal and Its Influencing Factors Based on CFD

Hao Cao*, Weihua Zhou, Xiaowen Wu, Sheng Hu, Ling Lu, Jiwen Peng

State Grid Hunan Electric Power Corporation Limited Research Institute, Changsha, 410007, China

*Corresponding author e-mail:caohao45@126.com

Abstract. The flow field of seal has been modeled by CFD method in this paper, and the effects of rotor rotational speed and eccentricity on the fluid excitation force are analyzed. In order to study the mechanism of induced vibration force in seal, the velocity field and pressure field of fluid in seal are analyzed in detail in this paper, the calculation result shows that the rotor rotation will form a spiral flow when the fluid flows into the seal, it causes the uneven pressure distribution in the seal cavity and the high pressure point is not match to the minimum clearance, thus producing a tangential exciting force perpendicular to the eccentric direction of the rotor. It's the ultimate reason of rotor instability.

1. Introduction

The effect of fluid excitation force on rotor vibration is more and more obvious, which is directly related to the safe and reliable of turbo machineries. In recent years, many theoretical and experimental studies have shown that the tangential excitation force in the seal is the main cause of the instability of the seal rotor system [1-3]. An advantage of using CFD is its capacity to analyze a large number of complex design configurations and parameters, and the use of CFD analysis in many rotary machines has been increasing rapidly in recent years with the development of commercial software [4-6]. Because of the restriction of field test for practical annual seal, it is a valuable work to study the influence of seal fluid excitation force on rotor stability by CFD. So, in this paper considering the above experiences, we also built a CFD model of seal to study the fluid excitation force of seal and its influencing factors.

2. Calculation Model

The parameters of seal calculation model in this paper is built according to the size of actual seal on our test rig. The diagram of seal model is shown in Figure 1. There are 6 rings of seal installed in the test cylinder, the compressed gas enters from the middle of the cylinder, and then flows out through 3 rings seal on both sides, and each seal have 4 teeth. For the convenience of calculation, the model seal has been simplified to one seal ring with 12 teeth as the Figure 2 show. The specific parameters of seal calculation model are shown in Table 1. Figure 2 also shows the computational grid diagram of the sealing model in this paper, the density of the mesh increases as it gets closer to the wall and tooth tip. A grid independence test has been carried out in this paper to ensure the accuracy of the final flow field calculation results.

Content from this work may be used under the terms of the Creative Commons Attribution 3.0 licence. Any further distribution of this work must maintain attribution to the author(s) and the title of the work, journal citation and DOI.
Published under licence by IOP Publishing Ltd

IOP Conf. Series: Journal of Physics: Conf. Series **1064** (2018) 012030 doi:10.1088/1742-6596/1064/1/012030

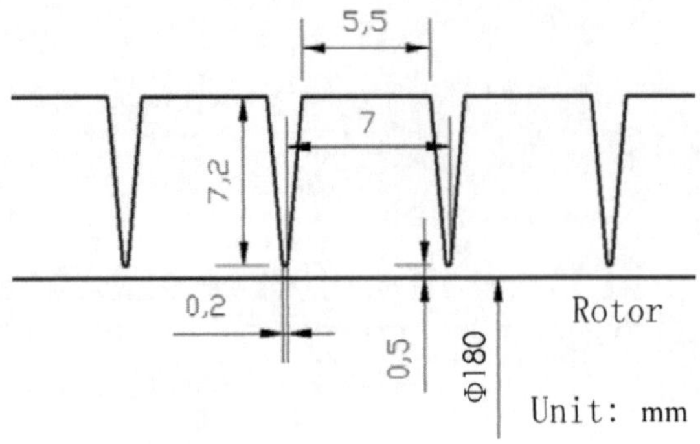

Figure 1. Diagram of Seal Size

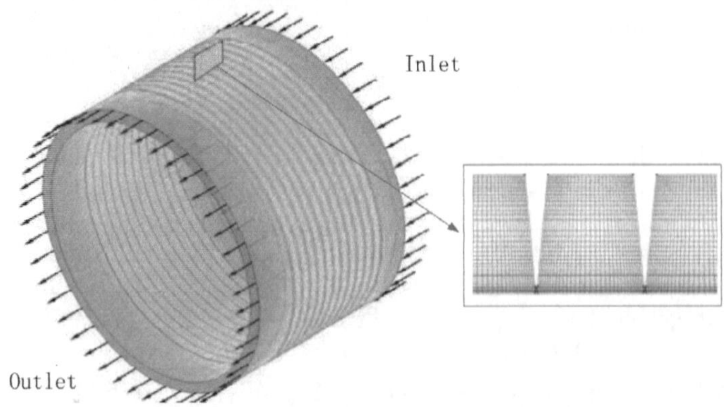

Figure 2. Seal Calculation Grid

Table 1. Parameters of Seal Calculation Model

Parameter Name	Value	Parameter Name	Value
Teeth Number	12	Rotation Speed/rpm	1000~6000
Inlet Pressure/Mpa	0.2~0.8	Inlet Temperature/K	366.7
Outlet Pressure/Mpa	0.1	Pre-rotation Speed/m.s^{-1}	0
Fluid Type	Ideal Gas	Turbulence Model	k-ε Model
Eccentricity	0~0.7	Average Gap/mm	0.5

3. Calculation Results Analysis

3.1. Velocity Field Analysis

Figure 3 shows the circumferential velocity cloud map of the axial seal chamber when the rotor speed increase from 2000 r/min to 6000 r/min under the condition of no inlet prerotation. It can be clearly seen from the figure that the circumferential flow is slowly produced because the fluid is driven by the viscous friction of the rotor. When the fluid enters the seal, the circumferential velocity is almost zero. As the axial flow increases chamber by chamber, the fluid has obvious circumferential velocity in the last chamber. Also, with the increase of the rotational speed, the increasing trend of the circumferential velocity is becoming more and more obvious. The generation of circumferential flow will cause the fluid to perform complex 3D spiral flow in the sealing clearance.

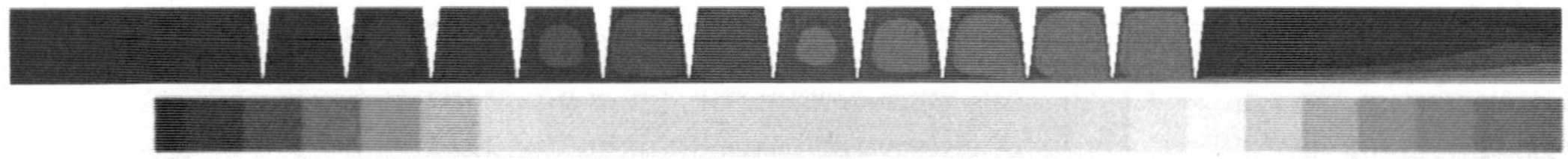

(a) 2000 rpm

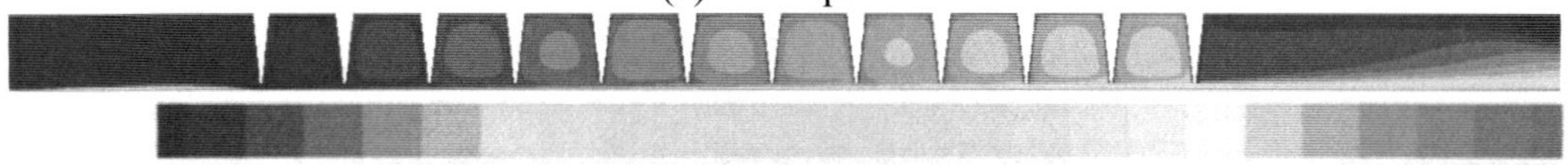

(b) 4000 rpm

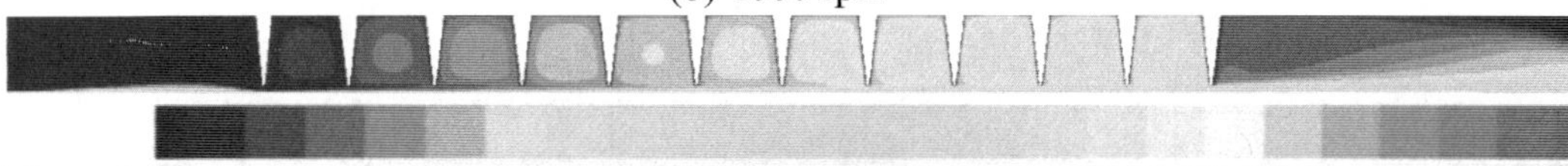

(c) 6000 rpm

Figure 3. Circumferential Velocity Cloud Map of Axial Seal Chamber with Rotational Speed

3.2. Pressure Field Analysis

Figure 4 shows the circumferential pressure cloud map of one seal chamber when eccentricity ε=0.5. It can be seen obviously that the pressure distribution along the circumferential direction is uneven, and the pressure value at the small seal clearance is higher than the value at the big clearance.

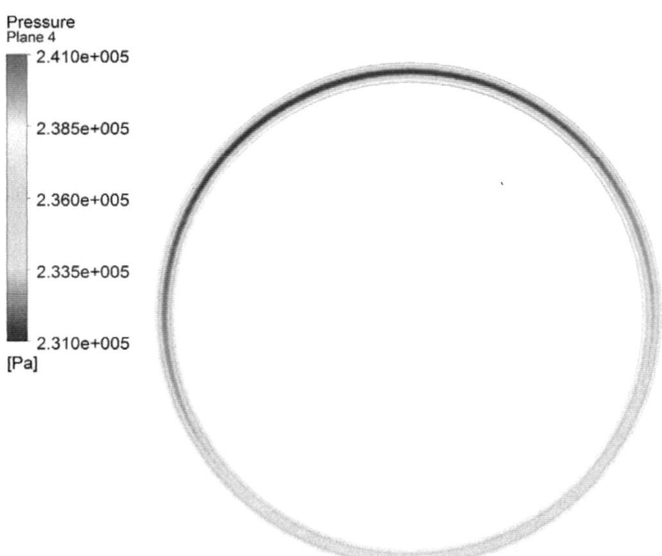

Figure 4. Circumferential Pressure Cloud Map

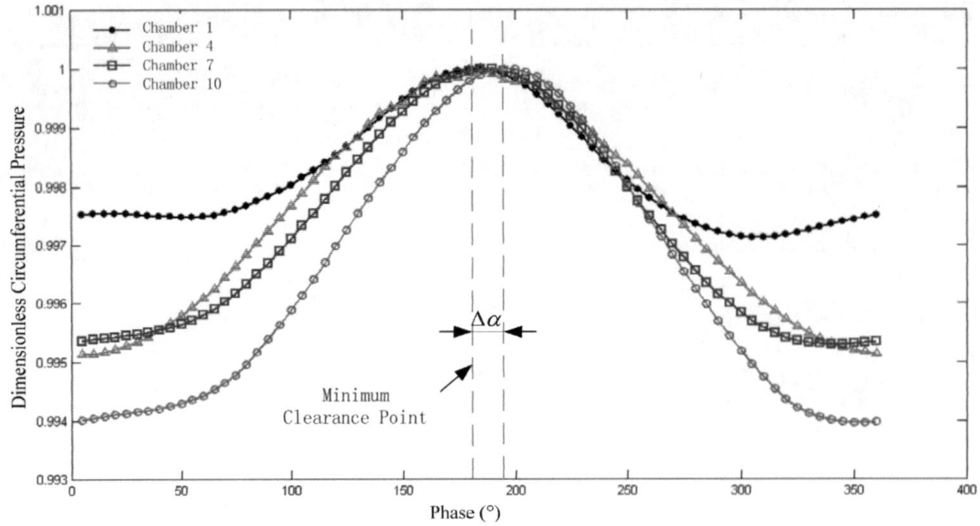

a (ω=2000rpm, ε=0.5)

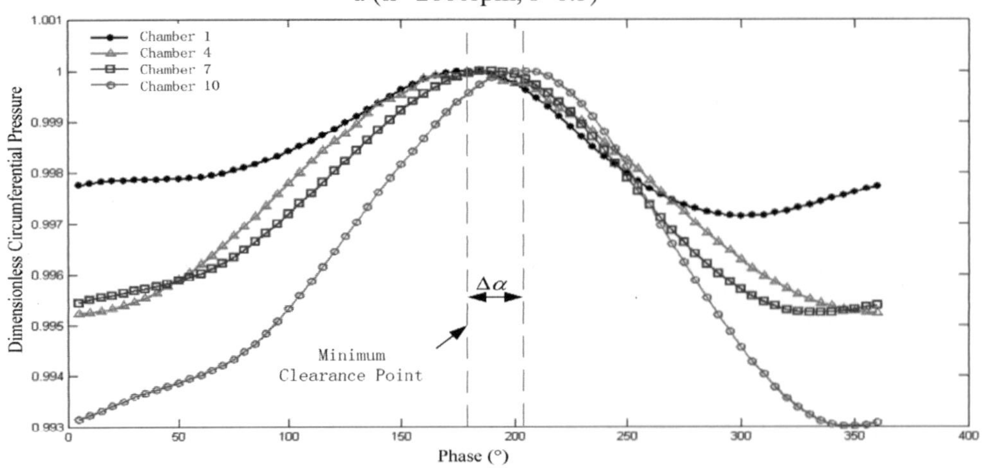

b (ω=4000rpm, ε=0.5)

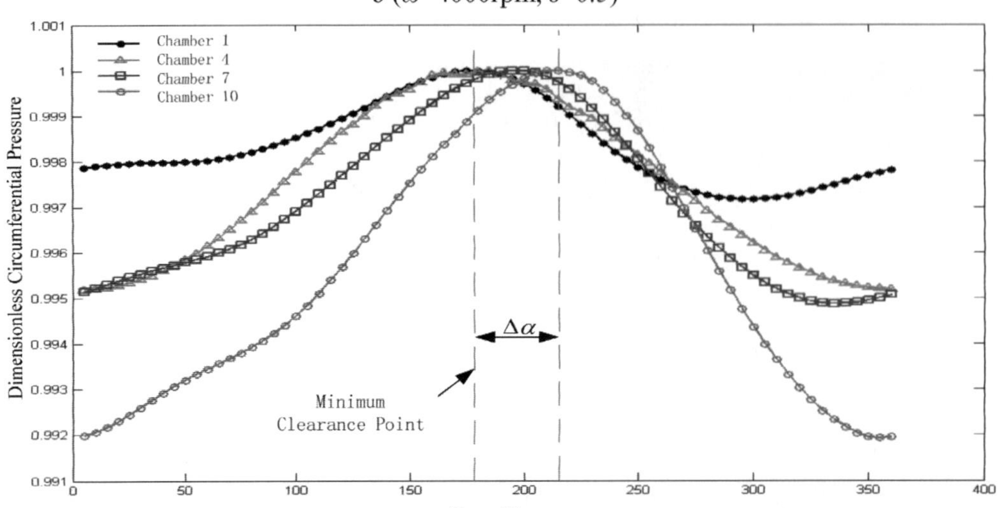

C (ω=6000rpm, ε=0.5)

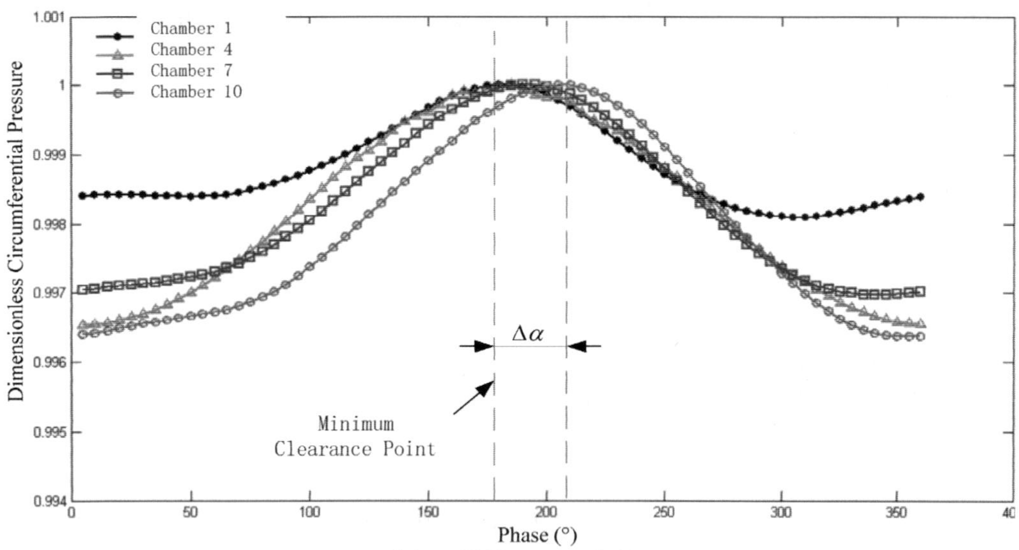

d (ω=6000rpm, ε=0.1)

Figure 5. Curve of Dimensionless Circumferential Pressure in No 1, 4, 7, 10 Seal Chamber

As shown in Figure 5, the dimensionless circumferential pressure distribution curves of No1, 4, 7, 10 seal chamber along the axial flow direction at rotational speed ω is 2000 r/min, 4000 r/min and 6000 r/min. The eccentricities in Figure 5a, 5b, 5c are 0.5, and the eccentricity of Figure 5d is 0.1. The dimensionless pressure can be derived from the following equation:

$$\bar{P}(i) = \tilde{P}(i) / P_{max}$$

In the above equation, $\tilde{P}(i)$ is the calculated actual pressure distribution along the circumference of each chamber, and P_{max} is the maximum of $\tilde{P}(i)$.

Each curve has been set evenly 72 points along the circumferential direction, and the minimum clearance at the phase angle of 180°.

In the Figure 5, $\Delta\alpha$ represents the deviation of the highest pressure in the first and tenth seal chamber. The pressure distribution of each seal chamber along the circumference is approximately cosine. With the seal fluid flows from one chamber to the next, the pressure gradient of each chamber along the circumferential direction increases, also the angle of the pressure high point deviating from the minimum clearance of each chamber is increasing.

It can be seen from the Figure 5a, 5b, 5c, the pressure high points of each seal chamber under the conditions of 2000 r/min, 4000 r/min, 6000r/min are changed, as the rotating speed increases, the $\Delta\alpha$ is gradually increased. It shows that with the increase of rotational speed, the spiral flow is becoming more and more intense, the action point of seal force deviates more and more from the minimum clearance, which leads to the increase of tangential excitation force finally.

Compared with Figure 5d and Figure 5c, under the same rotational speed, the pressure curve with small eccentricity is more flat than the big one. It means the circumferential pressure gradient decreases when the eccentricity becomes smaller. Also the $\Delta\alpha$ in Figure 5d is smaller than that in Figure 5c, it shows that with the eccentricity becoming smaller, the resulting tangential excitation force becomes smaller.

3.3. Influence Factor Analysis for Excitation Seal Force

The seal force can be obtained by integrating the rotor surface pressure along circumference direction. Figure 6 shows the variation of radial and tangential forces with rotational speed, when the rotor

IOP Conf. Series: Journal of Physics: Conf. Series **1064** (2018) 012030 doi:10.1088/1742-6596/1064/1/012030

eccentricity ε=0.6, the inlet pressure is 0.6 Mpa, the outlet pressure is 0.1 Mpa and the rotational speed from 1000 r/min to 6000 r/min. The result shows that the amplitude of radial force and tangential force both increase with the increase of rotational speed. The increased amplitude of tangential force is bigger than that of radial force.

Figure 7 shows the variation of radial and tangential forces with eccentricity, when the rotor rotational speed is 6000 r/min, the inlet pressure is 0.6 Mpa, the outlet pressure is 0.1 Mpa and the eccentricity from 0 to 0.7. The result shows that the fluid excitation force generated in the seal is very small when the relative position of the rotor and the stator in an ideal concentric state. When the eccentricity increases gradually, the radial and tangential excitation forces increase obviously.

4. Conclusion

In this paper, the flow field of seal has been modeled by CFD method, the velocity field and pressure field of fluid in seal are analyzed in detail to study the mechanism of flow induced vibration force in seal. From the above analysis, it can be seen that because the rotor eccentricity produces wedge clearance, the fluid pressure near the minimum clearance in the direction of rotation increases intensely. An apparent fluid flow excitation force opposite to the eccentric direction of the rotor has been generated. Due to the spiral flow of the seal flow driven by the rotor, the high point of the fluid pressure along the flow direction deviates gradually from the minimum clearance, which forms the tangential excitation force perpendicular to the eccentric direction of the rotor, and this force is the important cause of rotor instability.

Finally, the effects of rotor rotational speed and eccentricity on the fluid excitation force are analyzed. The result shows that the amplitude of radial force and tangential force both increase with the increase of rotational speed. The increased amplitude of tangential force is bigger than that of radial force. The result also shows that the fluid excitation force in the seal is very small when the relative position of the rotor and the stator in an ideal concentric state. But when the eccentricity increases gradually, the radial and tangential excitation forces increase obviously.

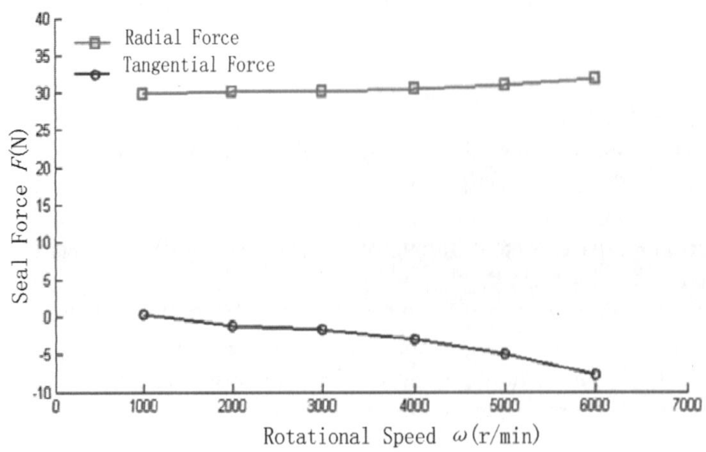

Figure 6. Seal Forces with Rotational Speed

IOP Conf. Series: Journal of Physics: Conf. Series **1064** (2018) 012030 doi:10.1088/1742-6596/1064/1/012030

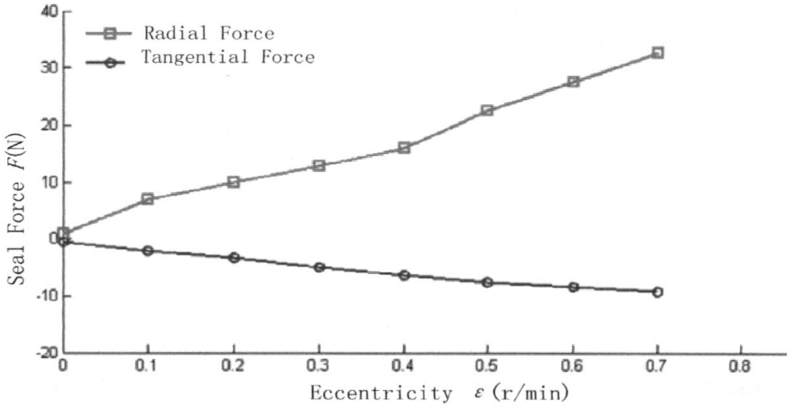

Figure 7. Seal Forces with Eccentricity

References

[1] Wanfu Zhang, Jiangang Yang. Experimental identification of fluid-induced force in labyrinth seals. Journal of Mechanical Science and Technology 25 (7), 2011, 1 - 10.

[2] Soto E. A., Childs D. W. Experimental rotordynamic coefficient results for (a) a labyrinth seal with and without shunt injection and (b) a honeycomb seal. Journal of Engineering for Gas Turbines and Power Transactions of The ASME, Vol. 121, Issue 1, 1999, p. 153 - 159.

[3] Kim N., Park S. Y., Rhode D. L. Predicted effects of shunt injection on the rotordynamics of labyrinth seals. Journal of Engineering for Gas Turbines and Power Transactions of The ASME, Vol. 125, Issue 1, 2003, p. 167 - 174.

[4] Rajakumar C., Sisto F. Experimental investigations of rotor whirl excitation forces induced by labyrinth seal flow. Journal of Vibration and Acoustics, Vol. 112, 1990, p. 515 - 522.

[5] Sreedharan S. S., Vannini G., Mistry H. CFD assessment of rotordynamic coefficients in labyrinth seals. ASME Turbo Expo: Turbine Technical Conference and Exposition, Düsseldorf, Germany: 2014, p. V7B - V32B.

[6] Untaroiu A., Untaroiu C. D., Wood H. G., et al. Numerical modeling of fluid-induced rotordynamic forces in seals with large aspect ratios. Journal of Engineering for Gas Turbines and Power Transactions of the ASME, Vol. 135, 2013, p. 0125011.

A Quantitative Study on the Cycle Length of Refracted Rays Affected By Ocean Eddies

Shuang Zhang[1, a], Yinquan Zhang[2, *], Chunjian Sun[2, b], Guofu Li[1, c], Siyu Gao[2, d], Lianxin Zhang[2, e], Tao Hu[3, f]

[1]National Ocean Technology Center, Tianjin, 300112, China
[2]National Marine Date and Information Service, Tianjin, 300171, China
[3]Key Laboratory of Underwater Acoustic Environment, Institute of Acoustics, Beijing 100190, China

*Corresponding author e-mail: zhyq_nmdis@126.com, [a]maymayed2007@126.com, [b]sczero@live.cn, [c]zhenglimuyun@sina.com, [d]gaosiyu198979@126.com, [e]lxz_nmdis@163.com, [f]hutao76@aliyun.com

Abstract. A local linear layered (LLL) model is proposed in this paper and the underwater sound channel is described by a segmental linear function. Taking advantage of the LLL model, cycle length of refracted rays can be derived as an explicit formula and the factors that could affect cycle length can be reduced to several parameters with intuitive physical meanings. Numerical simulations have shown that the results of the formula agree well with Bellhop.

1. Introduction

Ocean eddies are common phenomenons in the ocean, which have attracted the attention of both oceanographers and underwater acousticians for decades. A primary environmental characteristic of ocean eddies is a large distortion of the normally horizontal isotherms. Studies have shown that cold eddies are responsible for isothermal uplifts of 500m or more, with similar depressions for the warm case [1, 2, 3]. Due to the large temperature variations, ocean eddies cause substantial perturbations in sound speed, resulting in conspicuous fluctuation of underwater acoustic field.

The impacts of ocean eddies on acoustic propagation have been intensely studied and modeled because of their importance to underwater engineering and sonar detection. Vastano [4] and Gemmill [5] used ray-theoretical models to investigate the effects of eddies on transmission loss and ray geometry. Baer [6] utilized a primary 3D version of the parabolic equation to study propagation through an eddy. Lu [7] took advantage of experimental results to illustrate the influence of a cold eddy on acoustic transmission. Jian [8] investigated the effects of sound speed perturbations induced by three cold eddies on long-range sound propagation. Li [9] studied the influence of eddies on the shifting of convergence zones (CZ). All investigators indicated that major physical properties of the sound speed profile (SSP), such as sound speed, sound speed gradient, depth of SOFAR axes, would be influenced by ocean eddies, resulting in the shifting of convergence zones, altered multipath arrival sequences and horizontal refraction. Then questions arise: Among those physical properties, which one plays the dominate role in the fluctuations of acoustic field? Is there any way to quantitatively

Content from this work may be used under the terms of the Creative Commons Attribution 3.0 licence. Any further distribution of this work must maintain attribution to the author(s) and the title of the work, journal citation and DOI.
Published under licence by IOP Publishing Ltd

describe the impacts of each property on the fluctuations? Unfortunately, previous researches about acoustic propagating through ocean eddies were mostly studied qualitatively.

This article is devoted to solving the above questions. Taking into account of the fact that ray trace is an intuitive way to reflect the acoustic field and horizontal cycle length is a representative characteristic of the ray trace, the horizontal cycle length seems to be an effective parameter to intuitively reflect the fluctuations of the acoustic field. Therefore, this paper will take the horizontal cycle length of ray trace as the research object and quantitatively study the impacts of the physical properties of SSP on the cycle length in an eddy environment.

First and most important, one needs to explicitly formulate the relationship between the cycle length and the SSP. To achieve this goal, a local linear layered (LLL) model is proposed in this paper. The LLL model can be considered as an approximation of the real sound speed profile in the ocean. Numerical studies demonstrate that the cycle length in the LLL model is in good agreement with that in the real SSP. The LLL model consists of a sequence of parallel layers with constant sound speed gradients and the sound speed profile of the LLL model can be described by a segmental linear function. The LLL model has two distinct advantages: (1) The cycle lengths of rays can be explicitly formulated in the LLL model; (2) Parameters in the formula has intuitive physical meanings and correspond specific physical properties of SSP. Taking advantage of the LLL model, the formula of cycle lengths is derived and impacts of the physical properties of SSP on the cycle lengths of rays is quantitatively studied.

In this paper, numerical simulations are carried out to verify the cycle length formula. The results in an eddy environment are compared with those obtained in the absence of eddies to demonstrate the effects of eddies on the cycle lengths of ray traces. Besides, the impacts of the parameters in the formula on the cycle lengths have also been numerically studied and the physical mechanism is explained briefly. As it will be shown later, results of the cycle length formula in the LLL model are in good agreement with those calculated by Bellhop program in the real SSP.

This paper is organized as follows. In Sec. II, the local linear layeredmodel is introduced. In Sec. III, the cycle length of rays in the LLL model is formulated using Snell's law. In Sec. IV, numerical simulations are carried out in an cold eddy environment. The performance of the cycle length formula is compared intuitively with the results obtained by Bellhop program. The conclusions follow in Sec. V.

2. The Local Linear Layered Model

In this section, the local linear layered (LLL) model is introduced.

Typical North Pacific winter profile of sound speed versus depth is shown in Fig.1, which is representative of those encountered in many tropical and sub-tropical deep-ocean areas. There have already existed several analytical models about the profile that have been widely used, such as Munk profile [10], GDEM model [10] and LSSPM [12]. However, the horizontal cycle lengths of the ray traces cannot be formulated explicitly in the above models. To overcome the problem, the LLL model is proposed in this paper.

Since the horizontal gradient of the sound velocity is high-order smaller than the vertical one in most circumstance, the LLL model regards the local SSP as an linear layered medium. According to the characteristics of the sound speed gradients, the local linear layered model further divides the local SSP into five layers by five characteristic depths.

Just below the sea surface is the mixed layer where the sound speed increases with depth. The base of this layer is termed the mixed layer depth (MLD) denoted by z_1, which is associated with the near-surface maximum in sound speed. In the LLL model, sound speed in the mix layer is approximated as a linear function.

$$C(z) = c_1 + a_0(z - z_1), 0 \le z \le z_1 \tag{1}$$

Where a_0 is used to denote the average gradient of sound speed in the mix layer and c_1 represents the sound speed at the MLD z_1.

Below the mixed layer lies the transition layer, a region where the sound speed decreases slowly with increasing depth. Sound speed in the transition layer can be expressed as

$$C(z) = c_1 + a_1(z - z_1), z_1 \le z \le z_2 \quad (2)$$

Where a_1 is used to denote the average gradient of sound speed in the transition layer.

Below the transition layer is the thermocline layer, a region of the water column where the temperature and sound speed decrease rapidly with depth. This region is characterized by a negative sound-speed gradient. Sound speed in the thermocline layer is approximated by a linear function.

$$C(z) = c_2 + a_2(z - z_2), z_2 \le z \le z_3 \quad (3)$$

Where a_2 represents average gradient of sound speed in this layer. z_2 And z_3 are used to denote the depths of the top and bottom boundary of the thermocline layer respectively, and corresponding sound speeds are termed c_2 and c_3.

Between the conjugate depths of c_3 is the axis layer. The conjugate depths refer to a pair of depths that have the same value of sound speed but which lie on opposite sides of the sound channel axis. Here the pair of conjugate depths of c_3 is denoted by z_3 and z_4. In the axis layer, sound speed decreases with depth first from z_3 to the sound channel axis z_a, then increases from the axis to z_4. Sound speed in this layer can be approximated by a segmental linear function,

$$C(z) = \begin{cases} c_a + a_3^+(z - z_a), z_3 < z \le z_a \\ c_a + a_3^-(z - z_a), z_a < z \le z_4 \end{cases} \quad (4)$$

Where a_3^+ represents the average gradient of sound speed between z_3 and z_a, and a_3^- represents the average vertical gradient of sound speed between z_a and z_4. c_a Is the sound speed of the sound channel axis?

Below the axis layer, and extending to the sea floor, is the deep isothermal layer. This layer has a nearly constant temperature in which the speed of sound increases with depth due to the effects of pressure. Sound speed can be expressed as,

$$C(z) = c_3 + a_4(z - z_4), z_4 < z \le z_5 \quad (5)$$

Where a_4 is the average gradient of sound speed in this layer.

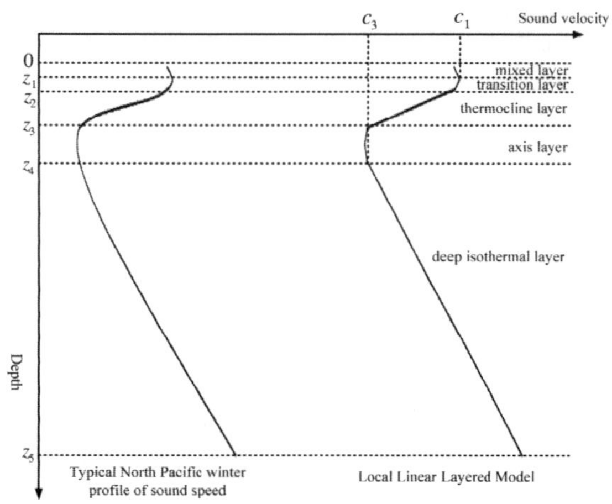

Figure 1. The local linear layered model.

3. Horizontal cycle lengths of ray traces in the LLL model

In this section, the basic equations of the horizontal cycle lengths of ray traces in the LLL model is derived by using of Snell's law.

3.1. Horizontal distance of a ray in a horizontally stratified medium

In a horizontally stratified medium, at every z, a ray must satisfy the Snell's law,

$$\frac{c(z)}{\cos\theta(z)} = \frac{c(z_0)}{\cos\theta_0} = const \tag{6}$$

Where θ_0 is the grazing angle which a ray makes with planes z=const. The constant in Eq. (6) is determined by the grazing angle θ_0 of a ray at the source z_0.

For an arbitrary small element of a ray, the total horizontal distance covered by the ray is.

$$D = \left| \int_{Z_0}^{Z} \frac{dz}{\tan\theta} \right| \tag{7}$$

Here D is assumed to be a single valued function of z. If this is not the case (for instance, in the underwater sound channel where a ray repeatedly returns to some fixed depth), Eq. (7) is applied to those parts of the ray for which it is single valued.

Considering the simplicity and wide application of the constant gradient approximation, the medium is divided into a set of layers where the dependence of velocity on depth is linear. For an arbitrary layer, we denote layer boundaries by z_{i-1}, z_i. Let c_i and θ_i be the sound velocity and the grazing angle at the lower boundary of the layer, respectively. The horizontal distance D_i covered by the ray in the layer can be expressed as,

$$D_i = \frac{c(z_{i-1})}{\cos\theta_{i-1}} \left| \frac{\sin\theta_i - \sin\theta_{i-1}}{a_i} \right| \tag{8}$$

Where $a_i = \dfrac{c(z_i) - c(z_{i-1})}{z_i - z_{i-1}}$ is the sound velocity gradient in the layer. Substitute Eq. (6) into Eq. (8), one gets.

$$D_i = \frac{c(z_0)}{\cos\theta_0}\left|\frac{\sin\theta_i - \sin\theta_{i-1}}{a_i}\right| \tag{9}$$

For the total horizontal distance covered by the ray, we have.

$$D = \sum_i D_i = \frac{c(z_0)}{\cos\theta_0}\left|\frac{\sin\theta_{b_2} - \sin\theta_{b_1}}{\overline{a}}\right| \tag{10}$$

Where

$$\overline{a} = \frac{1}{\sum_i \frac{1(\sin\theta_i - \sin\theta_{i-1})}{a_i(\sin\theta_{b_2} - \sin\theta_{b_1})}} \tag{11}$$

Here θ_{b_1} and θ_{b_2} are the grazing angles at the boundaries of the medium, respectively. The effective average sound speed gradient of the medium $\overline{a}$ can be regarded as the weighted mean of the sound speed gradients in these layers. It is worth noting that Eq. (11) is suitable in the situation where D is a single valued function of z. If this is not the case, Eq. (11) is applied to those parts of the ray for which it is single valued.

3.2. *Horizontal distance of channel rays in the LLL model*

In the deep ocean, studies have shown that channel rays trapped bellow the transition layer play a dominate role in the convergence zone. In this section, we derive the horizontal distances of a channel ray in the thermocline layer, the axis layer and the deep isothermal layer, which are denoted by D_2, D_3 and D_4 respectively in Fig.2.

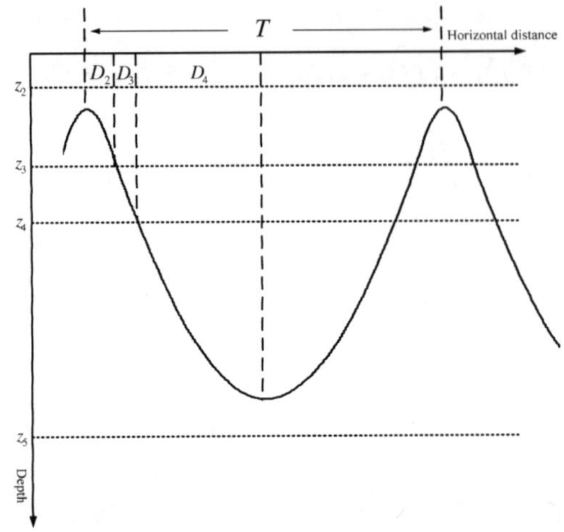

Figure 2. The diagram of a channel ray.

Combining Eq. (3) and Eq. (10), the horizontal distance of a channel ray from the up turning depth (corresponding grazing angle is 0) in the thermocline layer to z_3 can be express as

$$D_2(r) = \frac{c_0}{\cos\theta_0}\left|\frac{\sin\theta_3(r)}{a_2(r)}\right|. \tag{12}$$

Here θ_3 is the grazing angle at z_3. Since the sound speed varies with range in real ocean environment, the parameters in eq. (12) varies with horizontal range r.

Similarly, horizontal distance of a ray from z_4 to the down turning depth in the deep isothermal layer can be described as

$$D_4(r) = \frac{c_0}{\cos\theta_0}\left|\frac{\sin\theta_3(r)}{a_4(r)}\right| \tag{13}$$

In the axis layer, the sound speed profile is approximated by a segmental linear function Eq. (4), the horizontal length of a ray in this layer consists of two parts.

$$D_3(r) = D_3^+(r) + D_3^-(r) \tag{14}$$

Where the horizontal distance from z_3 to z_a is.

$$D_3^+(r) = \frac{c_0}{\cos\theta_0}\left|\frac{\sin\theta_3 - \sin\theta_a(r)}{a_3^+(r)}\right| \tag{15}$$

And the horizontal distance from z_a to z_4 is.

$$D_3^-(r) = \frac{c_0}{\cos\theta_0}\left|\frac{\sin\theta_3 - \sin\theta_a(r)}{a_3^-(r)}\right| \tag{16}$$

Here θ_a is the grazing angle at the sound channel axis z_a.

From the above derivation, it should be noted that the sound speed gradients a_2, a_3^+, a_3^-, a_4 in the LLL model actually refer to the effective average sound speed gradients of corresponding layers.

3.3. Cycle length of channel rays in the LLL model

In the context, horizontal distance between two successive up turning points of a channel ray, is named cycle length of the ray. As shown in Fig.2, cycle length of channel rays can be expressed as a summation of horizontal distances in the thermocline layer, the axis layer and the deep isothermal layer.

$$T(r) = 2\left[D_2(r) + D_3(r) + D_4(r)\right] \tag{17}$$

Substituting Eqs. (12)- (16) Into Eq. (17) and considering the grazing angle at the turning depth is zero, one obtains.

$$T = 2\frac{c_0}{\cos\theta_0}\left\{\left|\frac{\sin\theta_3(r)}{a_2(r)}\right| + \left|\frac{\sin\theta_3 - \sin\theta_a(r)}{a_3^+(r)}\right| + \left|\frac{\sin\theta_3 - \sin\theta_a(r)}{a_3^-(r)}\right| + \left|\frac{\sin\theta_3(r)}{a_4(r)}\right|\right\} \quad (18)$$

Here the range parameter is omitted for simplicity. Eq. (18) will be simplified in the rest of this section.

According to Eq. (6), one obtains.

$$\sin\theta = \sqrt{1 - \frac{c^2}{p^2}} \quad (19)$$

Where $p = \frac{c_0}{\cos\theta_0}$ is the sound speed at the turning depth.

Then divide the sound speed into two parts,

$$c(z) = p - \Delta c(z), 0 \le \Delta c(z) << p \quad (20)$$

Where $\Delta c(z)$ is the deviation of sound speed at z with respect to p .

Substituting Eq. (20) into Eq. (19) and adopt the first order approximation, one gets.

$$\sin\theta \approx \sqrt{\frac{2\Delta c}{p}} \quad (21)$$

Finally, combining Eq. (18) and Eq. (21), one gets.

$$T = 2\sqrt{p}\left\{\left|\frac{\sqrt{2\Delta c(z_3)}}{a_2}\right| + \left|\frac{\sqrt{2\Delta c(z_3)}}{a_4}\right| + \frac{2c_3 - 2c_a}{\sqrt{2\Delta c(z_a)} + \sqrt{2\Delta c(z_3)}}\left(\left|\frac{1}{a_3^+}\right| + \left|\frac{1}{a_3^-}\right|\right)\right\} \quad (22)$$

As mentioned in Section III.B, the physical meanings of a_3^+ and a_3^- are the effective average sound speed gradients of the axis layers. To simplify the above formula, we adopt the following approximation.

$$a_3^+ \approx \frac{c_3 - c_a}{z_3 - z_a}, a_3^- \approx \frac{c_3 - c_a}{z_4 - z_a} \quad (23)$$

Combining Eq. (22) and Eq. (23), the cycle length of channel rays can be expressed as.

$$T = 2\sqrt{p}\left\{\left|\frac{\sqrt{2\Delta c(z_3)}}{a_2}\right| + \left|\frac{\sqrt{2\Delta c(z_3)}}{a_4}\right| + \frac{H}{B}\right\} \quad (24)$$

Where

$$B = \frac{\sqrt{2\Delta c(z_a)} + \sqrt{2\Delta c(z_3)}}{2} \tag{25}$$

$$H = z_4 - z_3 \tag{26}$$

Eq. (24) is the explicit formula for the cycle length of channel rays. Obviously, the formula is derived briefly and easy to be understood by taking advantage of the LLL model. What's more, each parameter in Eq. (24) has an intuitive physical meaning and reflect a specific physical property of the sound speed profile. In the following, the physical meanings of these parameters are discussed in detail.

The parameter p corresponds to the sound speed at the turning depth of a channel ray. Given the grazing angle of a ray omitted from a source is unchanged, the parameter p could be influenced by the sound speed disturbance at the source. Parameters a_2 and a_4 are the effective average sound speed gradients of the thermocline layer and the deep isothermal layer respectively, which have already been illustrated in Section III.B. According to the definition of Eq. (20), the parameter $\Delta c(z_3)$ represents the deviation of sound speed at the boundary of the axis layer with respect to the parameter p. Similarly, the parameter B reflects the average deviation of sound speed in the axis layer with respect to the parameter p. At last, the parameter H is actually the thickness of the axis layer, as shown in Eq. (26). From a certain point of view, the above six parameters could be regarded as six physical properties of the SSP. And the sound speed profile just affects the cycle length of channel rays through the six physical properties. Taking advantage of Eq. (24), the impacts of the physical properties on the cycle length can be quantitatively studied in an eddy environment.

It should be emphasized that the parameters in Eq. (24) is actually range dependent and that the values of these parameters should corresponding to the local SSP where the ray propagates. Besides, Eq. (24) is suitable for the situation where the horizontal gradient of sound speed is high-order smaller than the vertical one. If this is not the case, the incline of the sound velocity isoclines due to the horizontal gradient would not be neglect, and the LLL model would be inapplicable. In the environment of ocean eddies, Eq. (24) is shown to be valid by numerical simulations, which will be illustrated in detail in the next section.

4. Numerical simulations

In this section, numerical simulations in an cold eddy environment are conducted to verify the validation of Eq. (24). Taking advantage of the equation, the impacts of the physical properties of SSP on the cycle length of channel rays are quantitatively studied.

4.1. Sound speed environment in a cold eddy

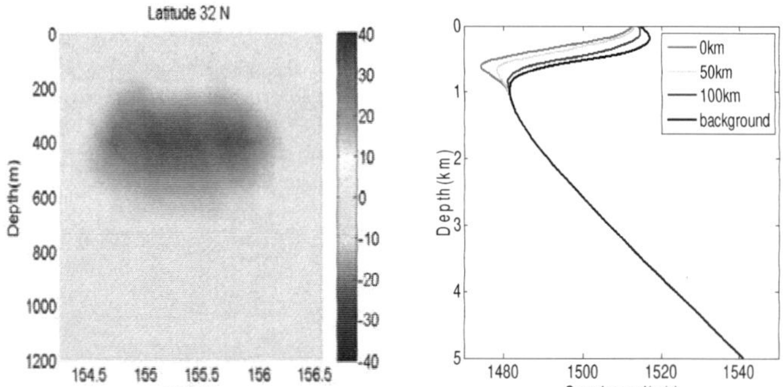

Figure 3. Sound speed disturbance (unit: m/s) in the cross section of sound propagation (left). Sound speed profile at several distances with respect to the eddy core (right).

The sound speed structure of ocean eddies is obtained by ocean reanalysis. Ocean reanalysis is a method of combing historical ocean observations with a general ocean model (typically a computational model) driven by historical estimates of surface winds, heat, and freshwater, by way of a data assimilation algorithm to reconstruct historical changes in the state of the ocean. In this paper, hycom global ocean reanalysis data (downloaded from the website https://hycom.org) is adopted, which can provide 3-dimensional fields of temperature, salinity and density. The hycom data has a uniform 0.08 degree lat/lon grid between 80.48S and 80.48N and is interpolated to 40 standard z-levels in depth, which is sufficient to extract the sound speed structure for ocean eddies.

Then a typical cold eddy in Northwest Pacific is extracted from the sound speed field. Sound speed disturbance in the cross section of sound propagation and several sound speed profiles are shown in Fig.3.

4.2. The cycle length of a channel ray

In this section, the cycle length formula is tested in the background sound speed profile and the cold eddy environment, respectively. The validity of the cycle length formula is tested by Bellhop model in the following way:

(1) Ray traces are calculated by the Bellhop model in a real sound speed field and the cycle lengths of channel rays are extracted.

(2) The real sound speed field is approximated by a LLL model and the cycle lengths of channel rays are estimated by Eq. (24).

(3) The two kinds of cycle length are compared to test the validity of Eq. (24).

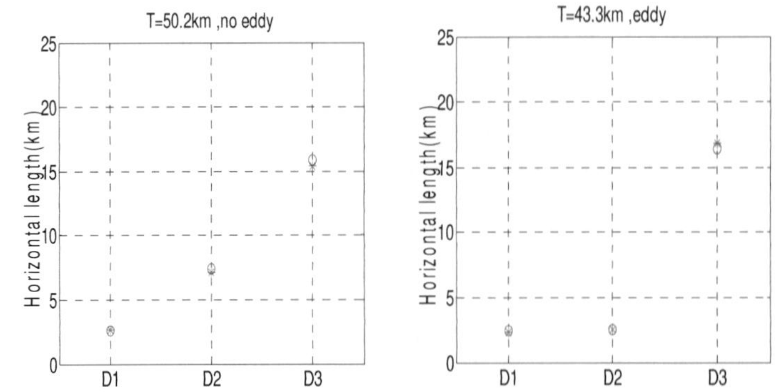

Figure 4. The values D1, D2, D3 in the background environment (left) and in the eddy core (right). Red stars denote Bellhop results, blue circles denote the values of eq. (24).

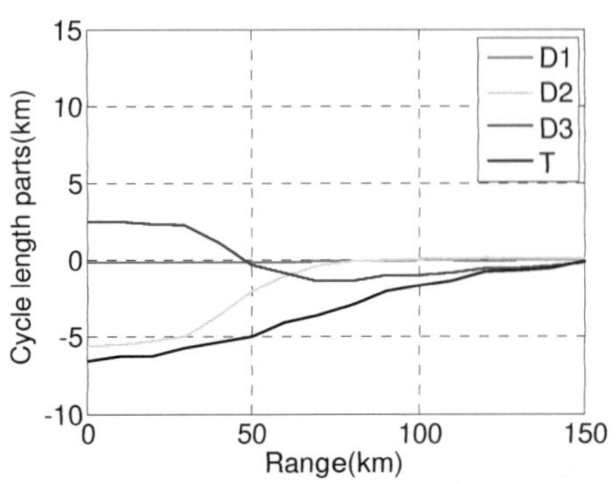

Figure 5. The variations of D1, D2, D3 in the cold eddy.

Sound source is posed at 600m. Fig.4 demonstrates the values D1, D2, D3 for a channel ray with grazing angle -9 degree. Relative deviation of cycle length T is less than 3%, which demonstrates that the estimated values of eq. (24) agree well with the true values both in eddy environment and background circumstance.

Fig.5 shows the variations of D1, D2, D3 with respect to the background environment when the ray propagates through the cold eddy. It can be found obviously that D2 and D3 play dominate role. But they counteract with each other since their signs are not the same.

5. Conclusion

A local linear layered (LLL) model is proposed in this paper and the underwater sound channel is described by a segmental linear function. Taking advantage of the LLL model, cycle length of refracted rays can be derived as an explicit formula and the factors that could affect cycle length can be reduced to several parameters with intuitive physical meanings. Numerical simulations have shown that the results of the formula agree well with Bellhop. In the simulation circumstance, the ray traces in the axis layer and the deep isothermal layer dominate the variations of cycle length when a ray propagates through a cold eddy.

Acknowledgments

This work was financially supported by Opening Fund from Key Laboratory of Underwater Acoustic Environment, Institute of Acoustics, Chinese Academy of Sciences, Tianjin Natural Science Foundation (18JCQNJC01200), National Key Research and Development Program of China (2016YFC1401806), National Program on Global Change and Air-Sea Interaction of China (GASI-IPOVAI-04), National Natural Science Foundation of China (41606039, 41506039).

References

[1] Andrews J Cand Scully-Power P D. (1976). "The structure of an East Australian current anticyclonic eddy". J Phys Oceanogr, 6: 756–65.
[2] Parker C E. (1971). "Gulf stream rings in the Sargasso Sea". Deep-Sea Res, 18: 981–93.
[3] Henrick RF, Siegmann WL and Jacobson M J. (1977). "General analysis of ocean eddy effects for sound transmission applications". J Acoust Soc Am, 62: 860–70.
[4] Vastano C and Owens G E. (1973). "On the acoustic characteristics of Gulf Stream cyclonicrings". J Phys Oceanogr, 3: 470–8.
[5] Gemmill W and Khedouri E. (1974). "A note on sound ray tracing through a Gulf Stream Eddy in the Sargosso Sea". US Naval Oceanogr Office, Washington DC, 6150: 21–74.
[6] Baer R N. (1980). "Calculations of sound propagation through an eddy". J Acoust Soc Am, 67: 1180–5.
[7] Lu L G, F Qiao, H X Chen, and Y L Yuan (2006). "Acoustic transmission in the cold eddy in the southern East China Sea". J. Geophys. Res., 111, C11S11.
[8] Jian Y J, Zhang J, et al. (2009). "Effect of mesoscale eddies on underwater sound propagation". Applied Acoustics, 70: 432–440.
[9] Li J X, Zhang R, et al. (2011). "Modeling of ocean mesoscale eddy and its application in the underwater acoustic propagation". Marine Science Bulletin, 30(1): 37-46.
[10] Munk W H. (1974). "Sound channel in an exponentially stratified ocean, with application to SOFAR". J Acoust Soc Am, 55(2): 220-226.
[11] Davis T M, Countryman K A and Carron M J. (1986). "Tailored acoustic products utilizing the NAVOCEANO GDEM (a generalized digital environmental model)". In proceedings , 36th Naval Symposium on Underwater Acoustics. San Diego: Naval Ocean Systems Center.
[12] Zhang X, Zhang Y G, et al. (2011). "A new model for calculating sound speed profile structure". ACTA OCEANOLOGICA SINICA, 33(5): 54-60.

IOP Conf. Series: Journal of Physics: Conf. Series **1064** (2018) 012032 doi:10.1088/1742-6596/1064/1/012032

Experimental Study on the Pressure-Difference Fluctuation Characteristics of Slug Flow in Horizontal T-Junction Tube

Zhian Deng[a], Xiao Sun*

College of Petroleum Engineering, Xi'an Shiyou University, Xi'an 710065, China

*Corresponding author e-mail: 370927228@qq.com, [a]1064514507@qq.com

Abstract. In practice, two-phase fluids may be often mixed or separated in T-junction tube, and alternating flow of liquid plug and gas plug may cause its working condition to be complex and changeable, so the study on it needs to be improved at present. Based on air and water, the characteristics of response to differential pressure of gas-liquid two-phase slug flow in two kinds of transparent horizontal T-junction PMMA tubes with different diameters are studied in depth and systematically, and data collection and processing are conducted by LabVIEW platform to obtain the probability density function (PDF) and the power spectral density function (PSD) characteristics of slug flow in different periods and forms. All characteristics provide reliable slug flow identification for flow pattern recognition.

1. Introduction

The gas-liquid two-phase slug flow is a kind of flow phenomenon that the liquid slug and large bubble formed in the pipe flow forward alternately along the pipeline. It will give rise to the wild fluctuation of pressure in the pipe due to the intermittency and instability of its flow and the pipe is subjected to intermittent stress shock [1]. Therefore, it is of great important to identify the slug flow and analyze its characteristics, thus ensuring the safe operation of the pipeline. Most of the relevant researchers use the signal characteristics of pressure difference to distinguish the flow patterns of multiphase flow. Weisman [2] et al concluded the fluctuation characteristic of slug flow through comprehensive analysis based on pressure-difference fluctuation amplitude, liquid slug frequency and form of sequence signal while researching the gas-liquid two-phase flow pattern in horizontal pipe. Chinese scholars Xiao Quansheng [3] and Zhu Ruian [4] studied the phase separation in equal-diameter T-junction tube and believed that the phase separation were mainly influenced by the flow pattern of two-phase flow at inlet, mass quantity of inlet, total mass inhalation ratio of bypath, inlet flow rate and geometrical parameter of bypath, etc. CHARRONY and WHALLEYPB [5] studied the phase separation in T-junction tube by using the air and water as the medium. All above scholars does not study the flow distribution of the mixture when gas-liquid two-phase slug flow goes through fork of horizontal T-junction tube and the response characteristics of pressure difference under such circumstances, so this paper conducts relevant research on this aspect.

Content from this work may be used under the terms of the Creative Commons Attribution 3.0 licence. Any further distribution of this work must maintain attribution to the author(s) and the title of the work, journal citation and DOI.
Published under licence by IOP Publishing Ltd

2. Test equipment and process

2.1. Physical test system

The physical system for this test is mainly composed of gas path metering system, water path metering system, gas-liquid two-phase mixing system, visual test section and gas-liquid mixture loop system. The gas path metering system mainly includes imported spiral lobe compressor, gas storage tank with volume of $1.5m^3$, gas dryer, gas mass flow meter, vortex street flowmeter and air flow regulating valve. The water path metering system has water storage tank with volume of $1.5m^3$, multiple-stage centrifugal pump, liquid mass flowmeter, vortex street flowmeter and water regulating valve. The test process is shown in the **Figure 1**.

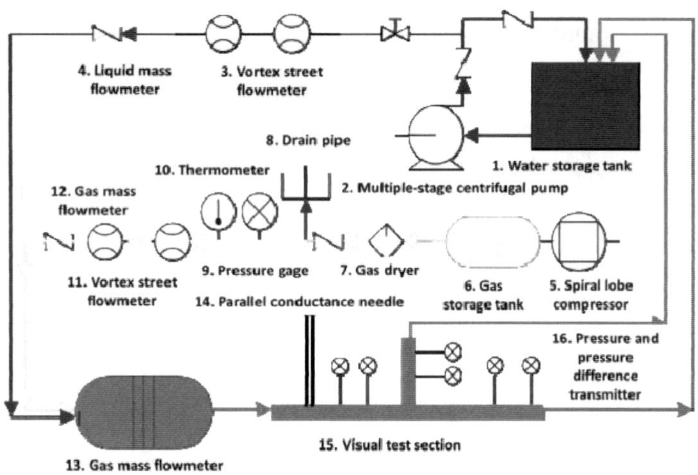

Figure 1. The test process

First, the multiple-stage centrifugal pump takes the water from the water storage tank to the water path circulation system and the water inside the circulation system entries into the gas-liquid mixing stage after it is measured by vortex street flowmeter. Next, the spiral lobe compressor compresses the air in the external environment to the gas storage tank and then flows into gas-liquid mixing stage after the buffer of gas storage tank, drying of dryer, record of thermometer temperature, pressuremeasurement of pressure gauge, and measurement of vortex street flowmeter.

Then, the water supplied by water storage tank and gas supplied by spiral lobe compressor are fully mixed in the gas-liquid mixing section and then flow into visual test section, in which the fluid mixing flows into open water tank via two loops after the data collection. Finally, the liquid can be reused for circulation system and the gas is emptied directly.The visual test section is composed of transparent PMMA pipe with inner diameter of 25mm and wall thickness of 10mm, 9 ring pressure measuring kits and 1 tee joint.

2.2. Data acquisition and analysis system and analysis theory of test data

2.2.1. Analysis theory of test data. The dynamic parameters, such as pressure and pressure difference, are fluctuating with the flow of two-phase flow mixture inside the pipe, and if multiphase flow mixture is different, the fluctuation characteristics of the parameters are different. The fluctuation process of these parameters contains rich information about the flow structure and so on. The flow pattern can be identified by the fluctuation analysis of the parameters.

Definition 2.1: The probability of defining the value of the random process x (t) in the range of Δx near the x value is as follows:

$$P(x) = P_{rob}\left[x < x(t) \le x + \Delta x\right] = \lim_{T\to\infty}\frac{T_x}{T} \tag{1}$$

It is called PDF of x (t). The probability expression for defining the value of X (t) in the interval [x, x +Δx] is as follows:

$$PDF\ (x) = \lim_{\Delta x\to 0}\frac{1}{\Delta x}P_{rob}\left[x \le x(t) < x + \Delta x\right] = \lim_{\Delta x\to 0}\frac{1}{\Delta x}\left[\lim_{T\to\infty}\frac{T_x}{T}\right] \tag{2}$$

It is called PDF of x (t), where Tx means the total time of x (t) in the interval [x, x +Δx] within T time, of which the probability of its amplitude in the interval [x, x +Δx] can be reflected by Tx/T. The statistical characteristics of a random process are characterized by the power spectral density function (PSD) of the process in the frequency domain.

Definition2.2: The discrete Fourier transform of random sequence x (n) is conducted by defining N observed value of random sequence x (n) as a set of data sequences with limited energy, to obtain X (e^{jw}).

$$X\left(e^{jw}\right) = \sum_{n=-\infty}^{+\infty} x_n e^{-jnw} = \sum_{n=0}^{N-1} x_n e^{-jnw} \tag{3}$$

It is divided by N after taking the square of $X\left(e^{jw}\right)$ to obtain:

$$\hat{S}_x\left(e^{jw}\right) = \frac{1}{N}\left|X\left(e^{jw}\right)\right|^2 \tag{4}$$

It is defined as power (PSD) of sequence x (n).

3. Measurement results and analysis

The literature shows that the frequency fluctuation characteristic of the slug flow can be more clearly characterized by pressure-difference fluctuation signal than the pressure fluctuation signal [6]. The pressure difference signal in this paper is measured by a differential pressure transmitter. According to the hydrodynamic model of slug flow, the pressure drop in the slug element is mainly concentrated in the liquid slug area, and the pressure gradient in the large bubble area is very small, which can be ignored.

For pressure difference signals, its characteristics in the time domain are mainly characterized by the probability density distribution function (PDF). Therefore, the probability density distribution function (PDF) and the power spectral density function (PSD) in the time domain are analyzed in order to further get the detailed rule of pressure difference fluctuation of main pipe, radial pipe and axial pipe after flow distribution when gas-liquid two-phase slug flow through horizontal T-junction equal diameter pipe and variable diameter pipe. The PDF of a gas-liquid two-phase slug flow through a horizontal T-junction tube after flow distribution is shown in the Figure 2, and the power spectral density functions of pressure-difference fluctuation curve for each pipe under constant superficial gas velocity and superficial liquid velocity are respectively shown in Figure 3 and Figure 4.

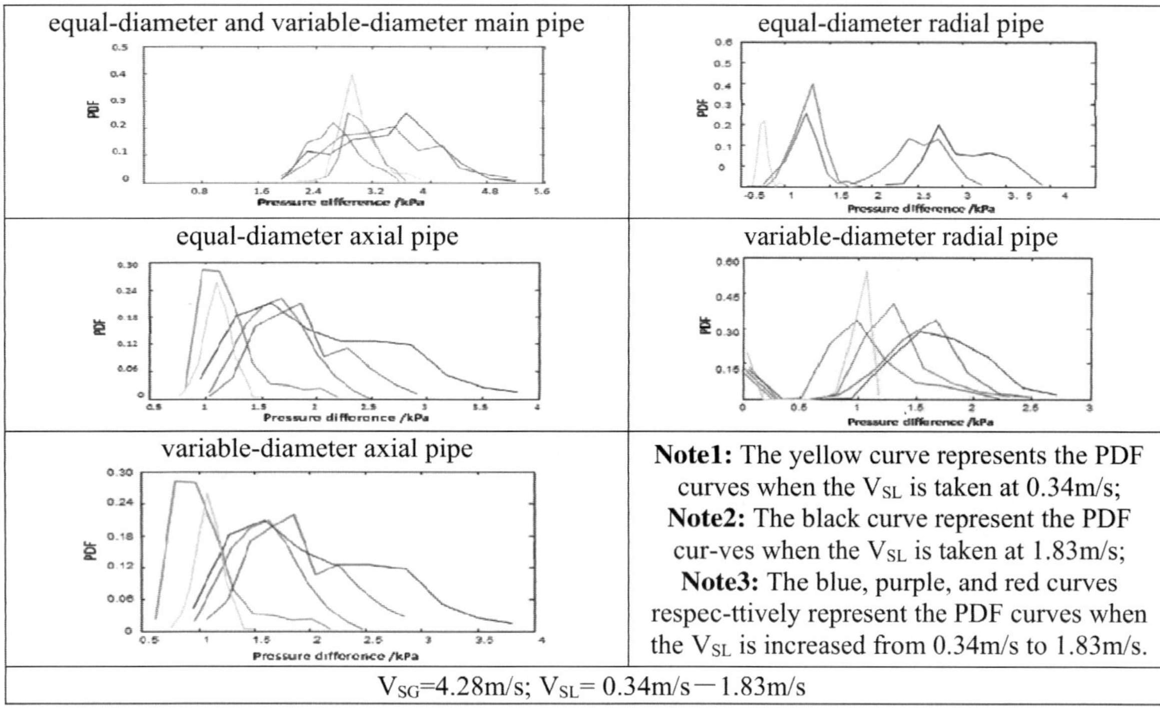

Figure 2. (a)PDF of Pressure-Difference Fluctuation Curve for Each Pipe

According to the observation and analysis of Figure 2, with regard to the PDF for pressure-difference fluctuation curve of equal-diameter and variable-diameter main pipe, radial branch pipe and axial branch pipe, there is a common characteristic that the dispersity becomes stronger as the superficial gas-liquid velocity increases. The difference is that the PDFs of equal-diameter main pipe, equal-diameter axial branch pipe, variable-diameter radial branch pipe and variable-diameter axial branch pipe are distributed in single peak. Combined with other research, the probability density of pressure-difference signal for slug flow is distributed in single peak, which can provide theoretical basis for identifying flow pattern of the slug flow. In particular, only the equal-diameter radial pipe is transformed from single peak to multi-peak to single peak, caused by the instability of the flow pattern of the equal-diameter radial branch pipe. The observation results show that the axial branch pipe is smaller than radial branch pipe due to the influence of flow distribution, and variable-diameter pipe is smaller than equal-diameter pipe, which can provide theoretical basis for flow distribution of the gas-liquid two-phase slug flow.

equal-diameter and variable-diameter main pipe

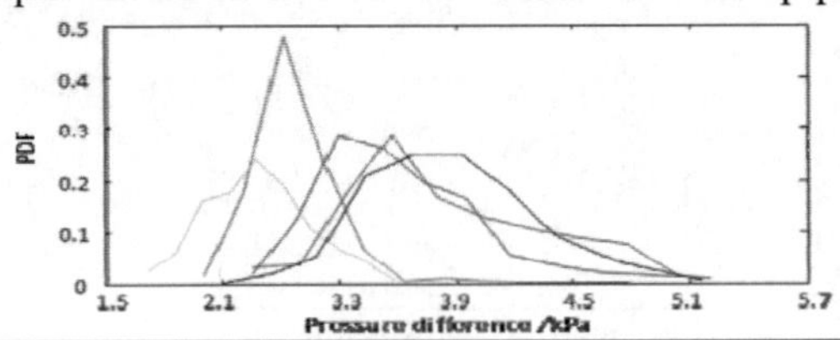

equal-diameter radial pipe

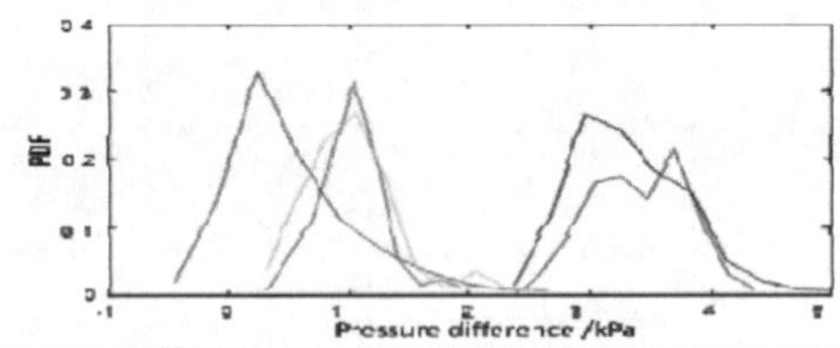

equal-diameter axial pipe

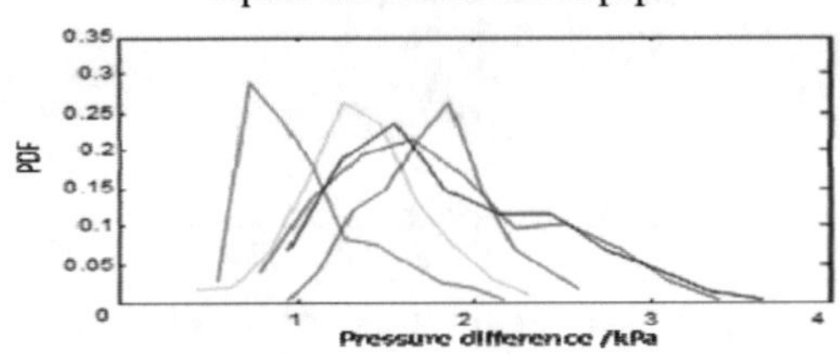

variable-diameter radial pipe

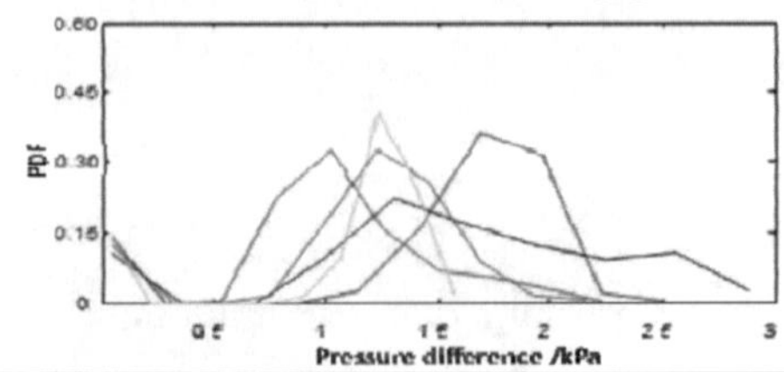

variable-diameter axial pipe

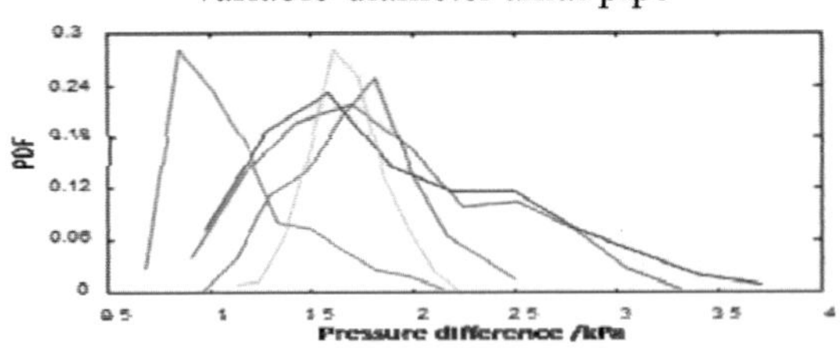

Note1: The yellow curve represents the PDF curves when the V_{SG} is taken at 4.32m/s;
Note2: The black curve represent the PDF cu-rves when the V_{SG} is taken at 11.03m/s;
Note3: The blue, purple, and red curves resp-ec-ttively represent the PDF curves when the V_{SG} is increased from 4.32m/s to 11.03m/s.

VSL=1.03m/s; VSG=4.32m/s－11.03m/s

Figure 2. (b)PDF of Pressure-Difference Fluctuation Curve for Each Pipe

Based on the above analysis, the fluctuation range of power spectral function of equal-diameter main pipe, equal-diameter axial branch pipe and variable-diameter axial branch pipe increases as another-phase superficial velocity increases, and the fluctuation range of pressure-difference fluctuation power spectral function of equal-diameter radial branch pipe is the largest if the superficial velocity of one phase is constant. The fluctuation range increases because there is a certain periodicity in gas-liquid two-phase flow, but very irregular, and there are some smaller fluctuations between each fluctuation cycle, showing that a series of two-phase flow with different local structure flows through measurement section. After the superficial gas-liquid velocity is increased, the flow pattern is more irregular, and the steam block and bubble with different shapes and sizes are unevenly distributed in the liquid phase. Therefore, the pressure-difference fluctuation process shows the characteristic of dispersive spectrum. The amplitude changing of equal-diameter radial branch pipe and variable-diameter radial branch pipe does not change obviously. This is because the gas will take precedence over the liquid into the radial branch pipe, and the pressure difference fluctuates not obviously relative to that of the main pipe and the axial pipe. Combined with other research [16], there is larger the amplitude of the power spectrum of slug flow is larger, which can provide theoretical basis for identifying flow pattern of the slug flow.

	equal-diameter main pipe	equal-diameter radial pipe	equal-diameter axial pipe
1	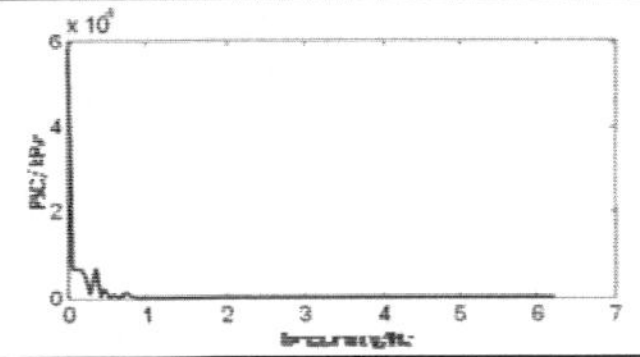	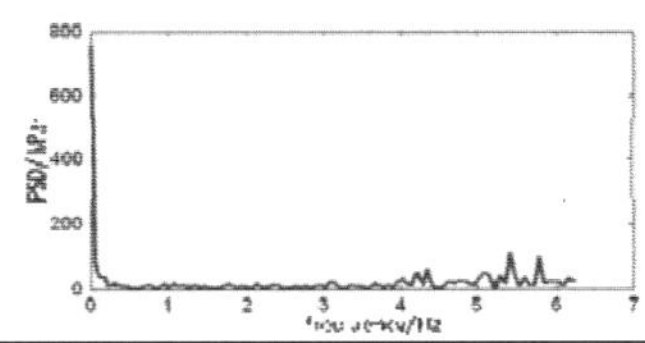	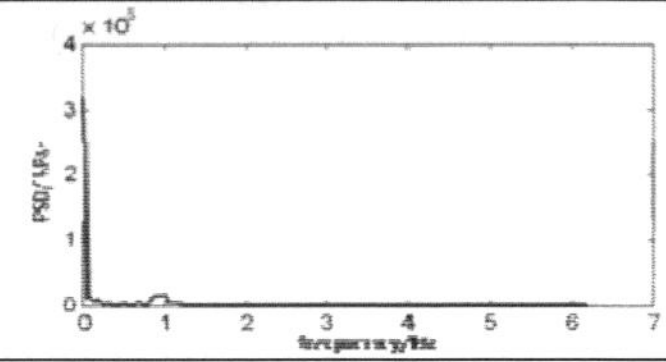
2	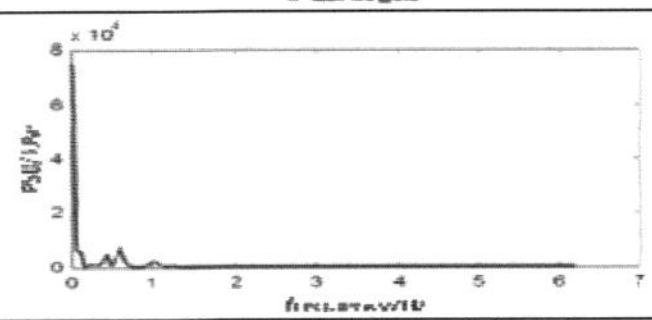	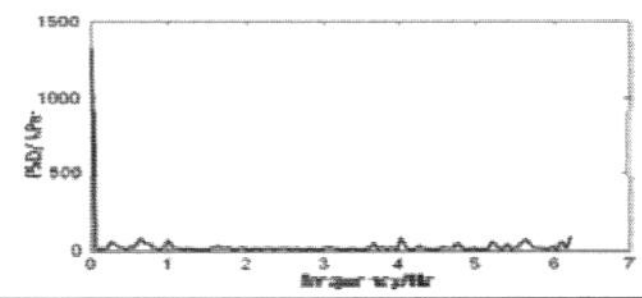	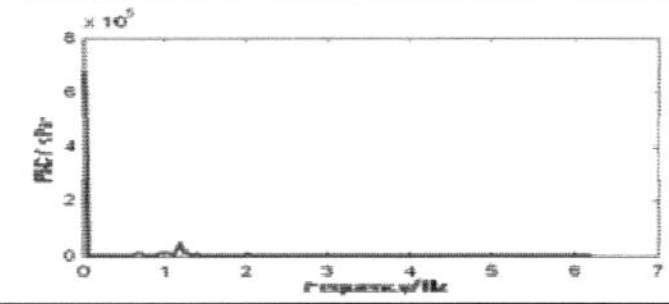
3	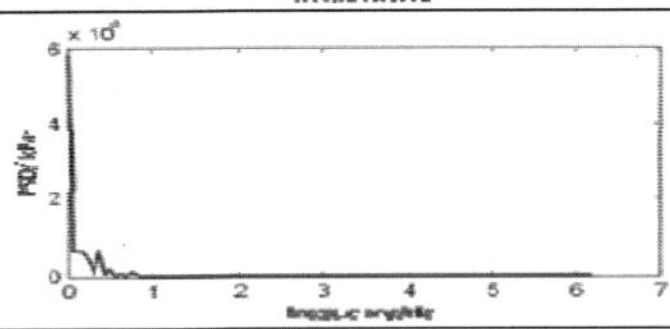	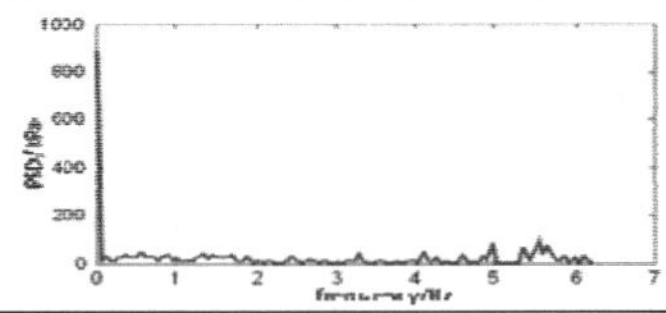	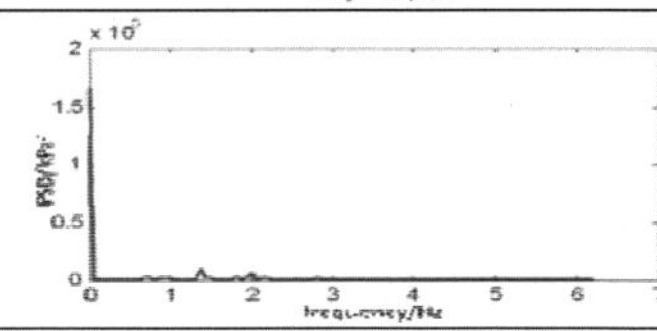
4	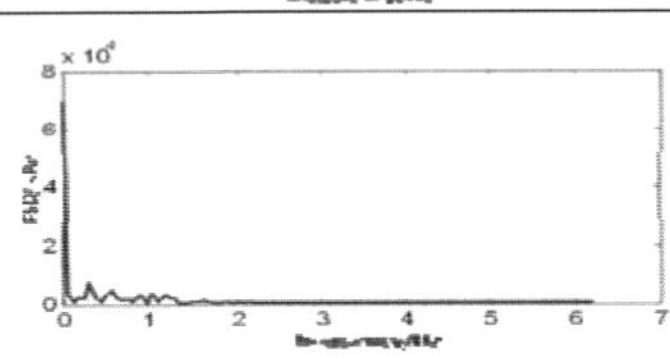	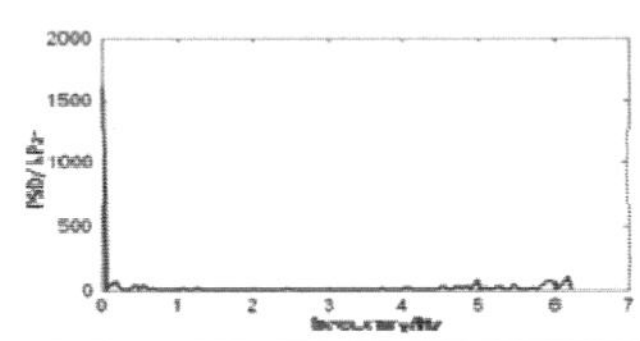	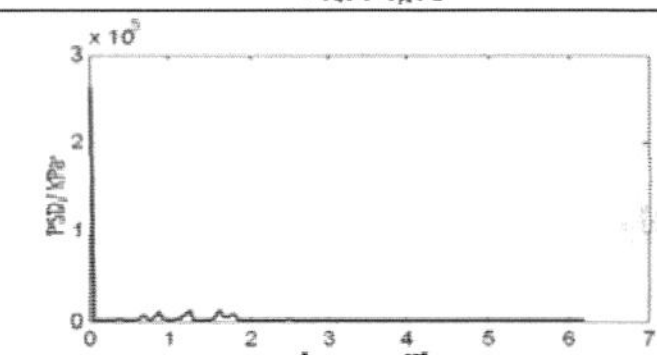
5	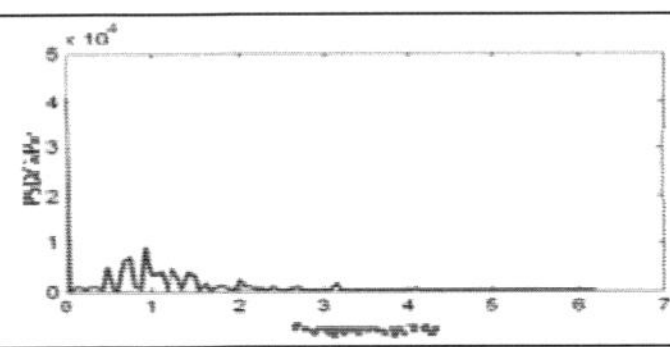	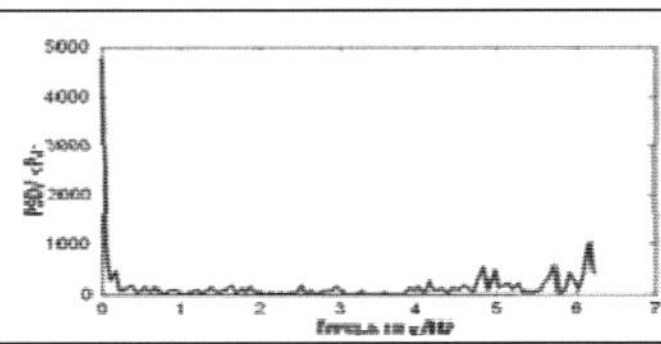	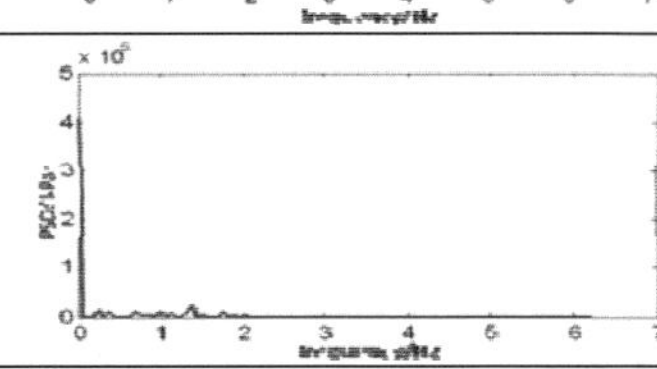
	Note1: The above curves are all obtained by using equal diameter T-tube, and the V_{SG} are 4.28m/s; **Note2:** The line 1 curves represent the PSD curves when the V_{SL} is taken at 0.34m/s; **Note3:** The line 5 curves represent the PSD curves when the V_{SL} is taken at 1.83m/s; **Note4:** The line 2, 3, and 4 curves represent three PSD curves when the V_{SL} increases from 0.34m/s to 1.83m/s.		

(a) PSD for Pressure-Difference Fluctuation Curve of the T-junction equal diameter pipe under Constant Superficial Gas Velocity.

IOP Conf. Series: Journal of Physics: Conf. Series **1064** (2018) 012032 doi:10.1088/1742-6596/1064/1/012032

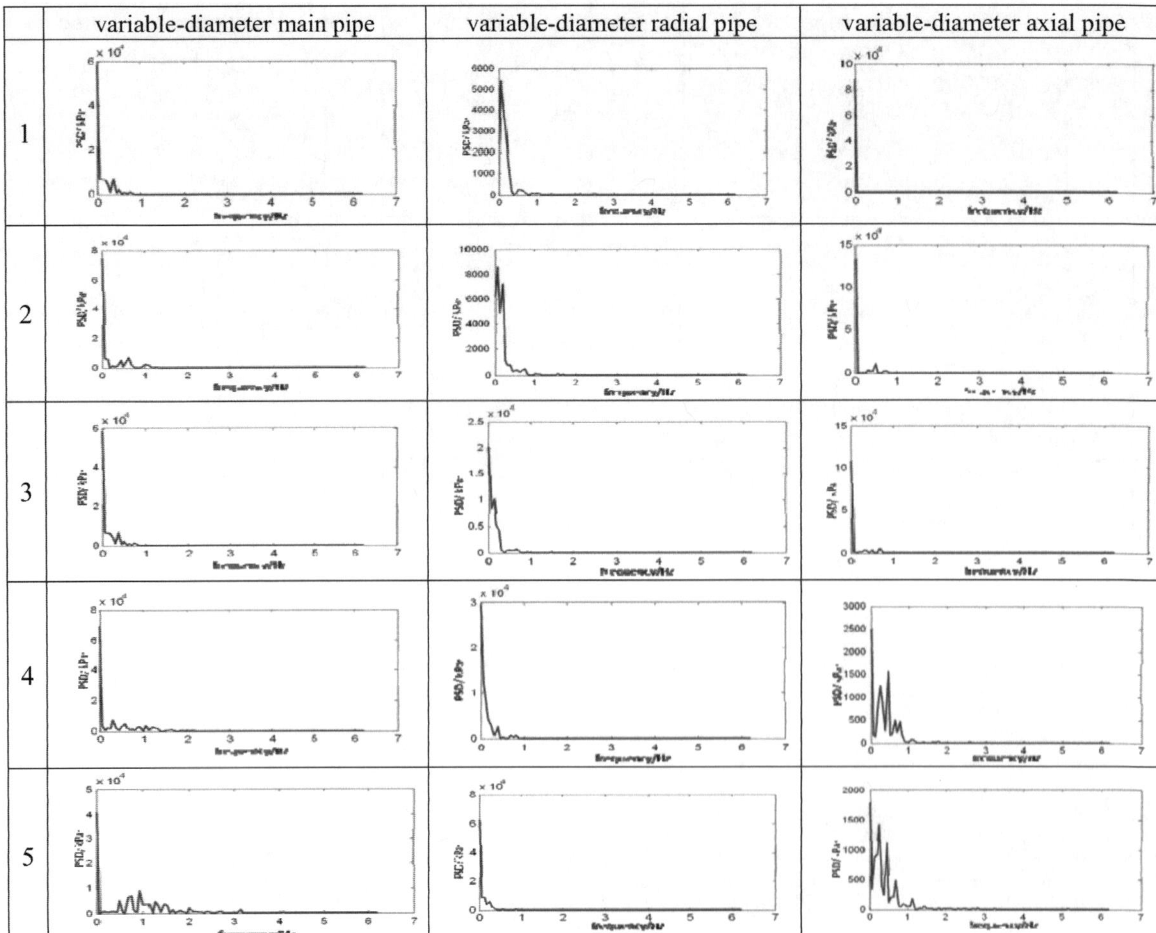

Note1: The above curves are all obtained by using variable diameter T-tube and the V_{SG} are 4.28m/s;
Note2: The line 1 curves represent the PSD curves when the V_{SL} is taken at 0.34m/s;
Note3: The line 5 curves represent the PSD curves when the V_{SL} is taken at 1.83m/s;
Note4: The line 2, 3, and 4 curves represent three PSD curves when the V_{SL} increases from 0.34m/s to 1.83m/s.

(b) PSD for Pressure-Difference Fluctuation Curve of T-junction variable diameter pipe under Constant Superficial Gas Velocity

Figure 3. PSD for Pressure-Difference Fluctuation Curve of Each Pipe under Constant Superficial Gas Velocity

IOP Conf. Series: Journal of Physics: Conf. Series **1064** (2018) 012032 doi:10.1088/1742-6596/1064/1/012032

	equal-diameter main pipe	equal-diameter radial pipe	equal-diameter axial pipe
1	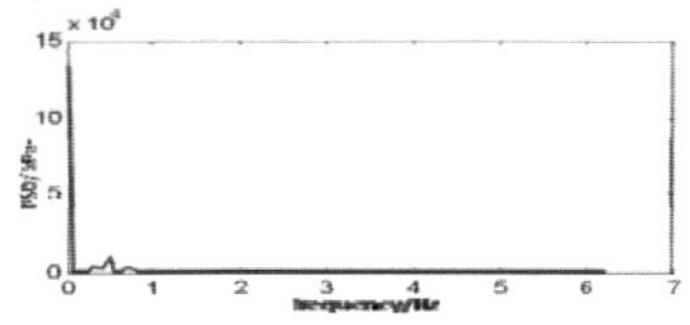	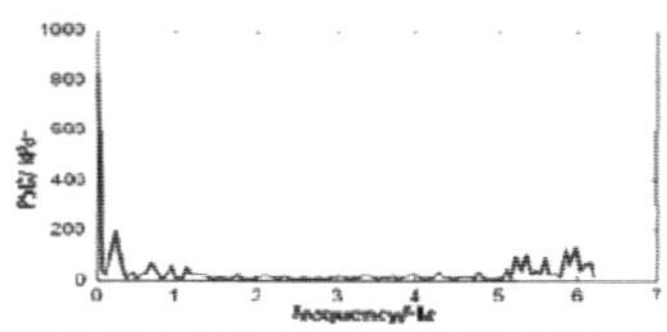	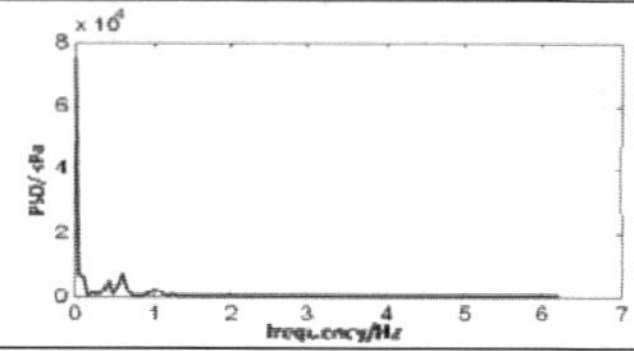
2	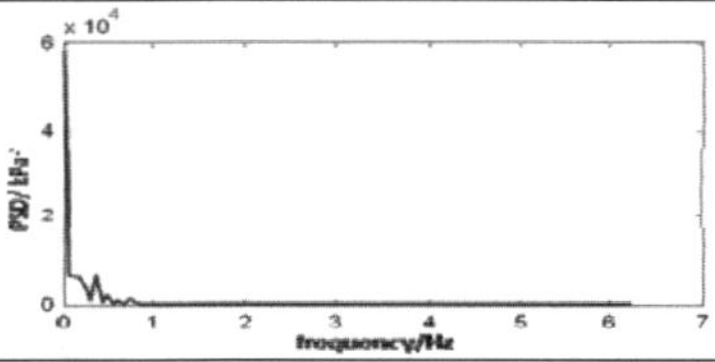	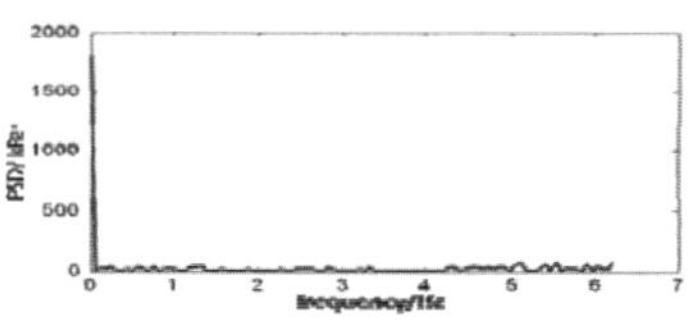	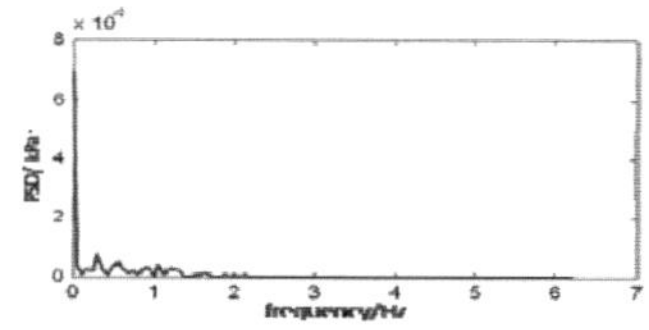
3	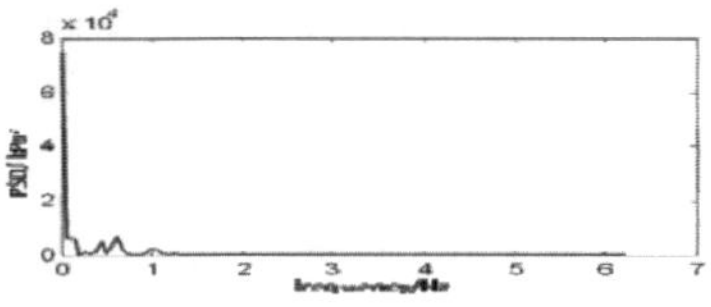	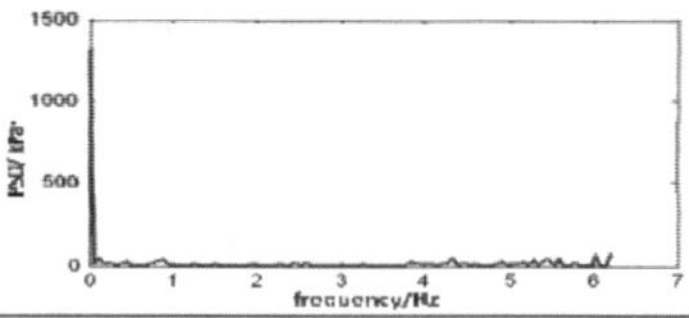	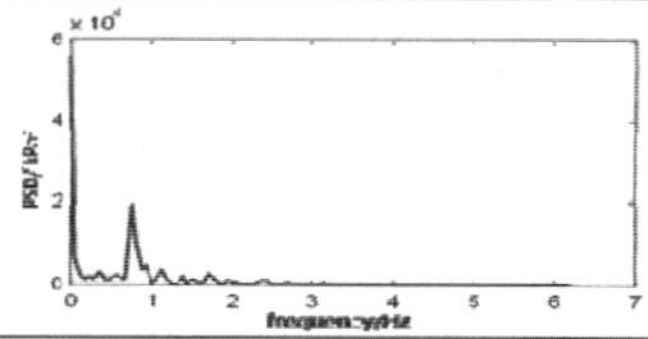
4	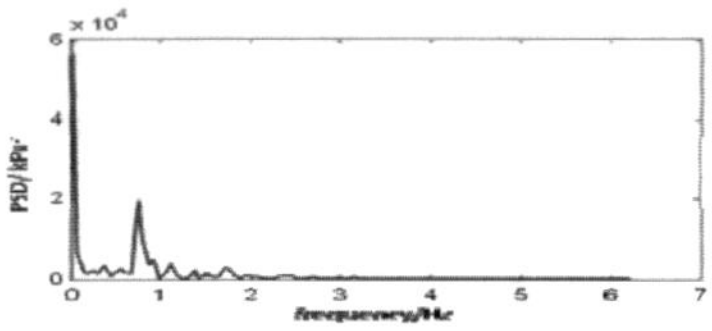	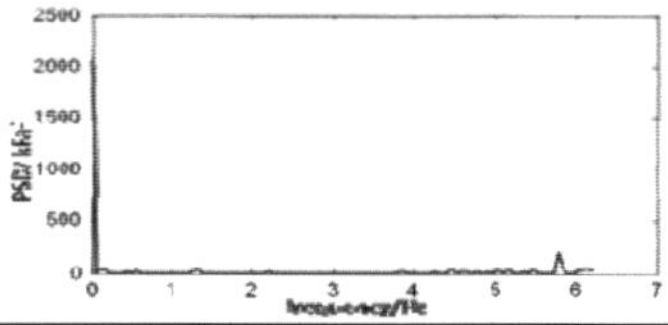	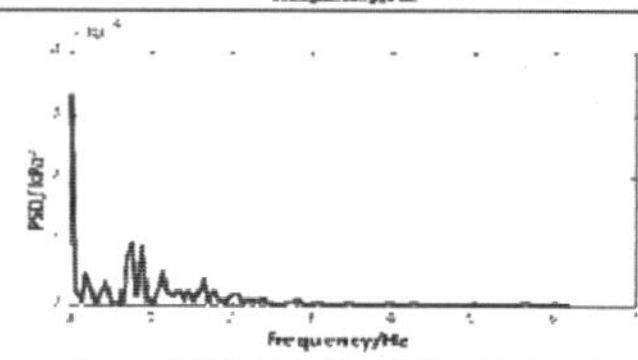
5	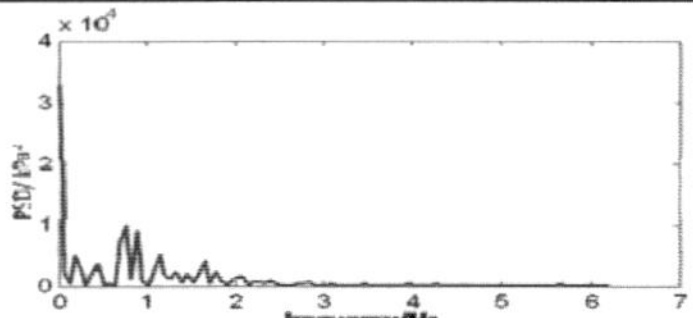	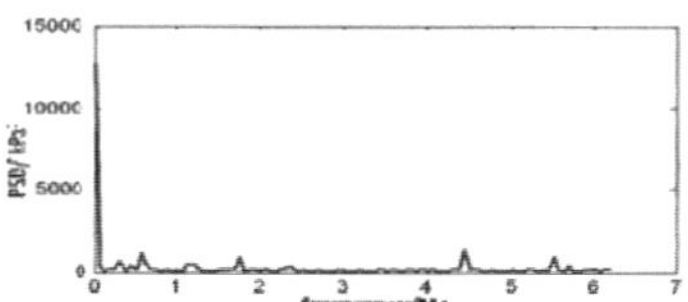	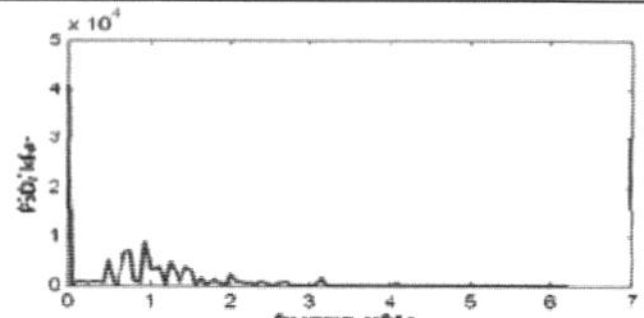
	Note1: The above curves are all obtained by using equal diameter T-tube, and the V_{SL} are 1.03m/s; **Note2:** The line 1 curves represent the PSD curves when the V_{SG} is taken at 4.32m/s; **Note3:** The line 5 curves represent the PSD curves when the V_{SG} is taken at 11.03m/s; **Note4:** The line 2, 3, and 4 curves represent three PSD curves when the V_{SG} increases from 4.32m/s to 11.03m/s.		

(a) PSD for Pressure-Difference Fluctuation Curve of T-junction equal diameter pipe under Constant Superficial Liquid Velocity

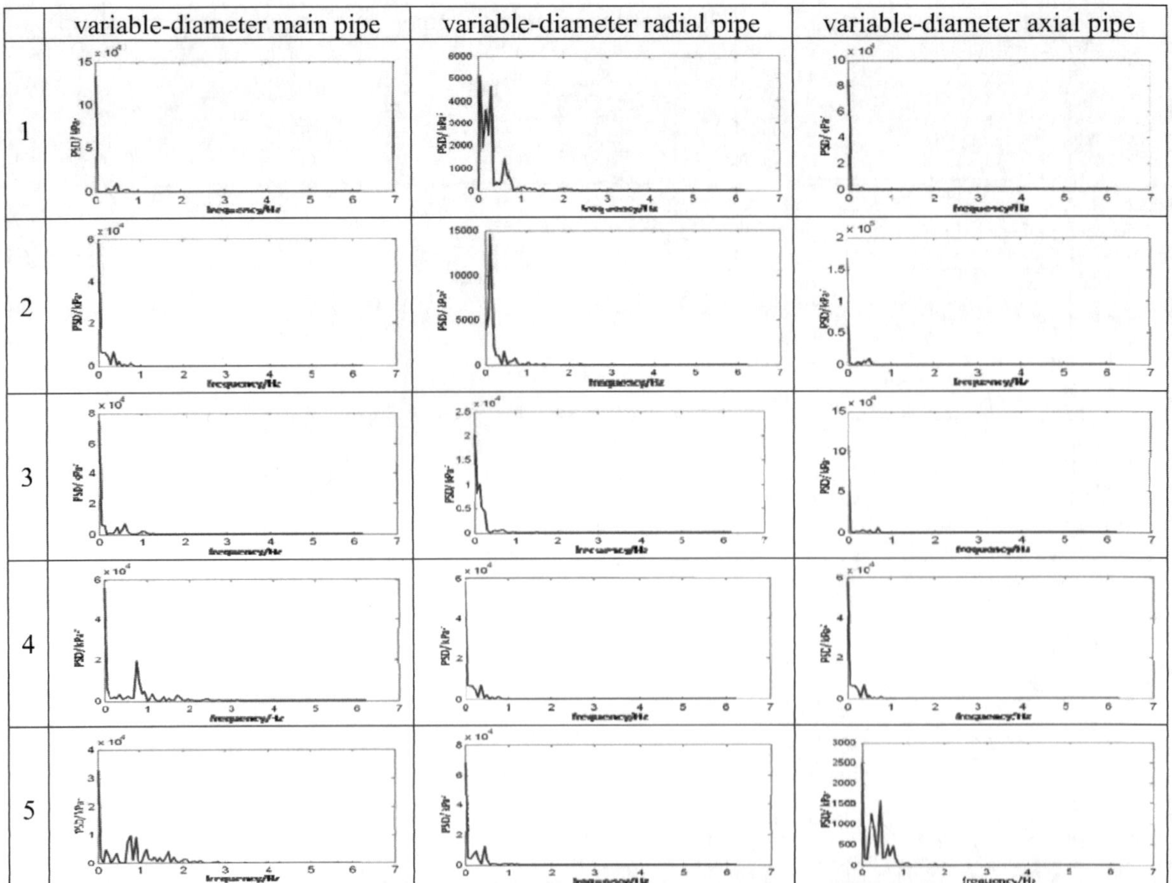

Note1: The above curves are all obtained by using variable diameter T-tube, and the V_{SL} are 1.03m/s;

Note2: The line 1 curves represent the PSD curves when the V_{SG} is taken at 4.32m/s;

Note3: The line 5 curves represent the PSD curves when the V_{SG} is taken at 11.03m/s;

Note4: The line 2, 3, and 4 curves represent three PSD curves when the V_{SG} increases from 4.32m/s to 11.03m/s.

(b) PSD for Pressure-Difference Fluctuation Curve of T-junction variable diameter pipe under Constant Superficial Liquid Velocity

Figure 4.PSD for Pressure-Difference Fluctuation Curve of Each Pipe under Constant Superficial Liquid Velocity

4. Conclusions

This paper studies the response characteristics of response to gas-liquid two-phase slug flow in equal-diameter and variable-diameter horizontal T-junction tube, radial branch pipe and axial branch pipe under different working conditions in a transparent horizontal T-junction PMMA tube by taking multiphase flow test room of Xi'an Shiyou University as the platform and using the water and air under normal temperature as the medium. The pressure-difference fluctuation curve and PDF of slug flow under different working conditions for each section are collected, drawn and analyzed respectively, of which the conclusions are as follows:

1. Flow pattern change: The flow pattern of mixtures in equal-diameter main pipe and axial branch pipe is the slug flow in the case of low superficial gas-liquid velocity, and that of radial branch pipe is not the slug flow because of more gas in it. As the superficial gas-liquid velocity increases, the flow pattern of mixture in equal-diameter horizontal T-junction tube main pipe, radial branch pipe and axial

branch pipe is the slug flow. Similarly, since there is a lot of air in the radial branch pipe, the flow pattern of the mixture in the whole pipeline system is not the slug flow as the superficial gas-liquid velocity continues to increase. For the variable-diameter pipe, the flow pattern of gas-liquid two-phase mixture in main pipe, radial branch pipe and axial branch pipe is the slug flow in the case of low superficial gas-liquid velocity, and the slug flow bubble in radial branch pipe is more obvious than that of axial branch pipe. The flow pattern of slug flow in radial branch pipe disappears first, and then the flow pattern of the whole pipeline system is not the slug flow as the superficial gas-liquid velocity continues to increase. These conclusions can provide theoretical basis for identifying flow pattern of the gas-liquid two-phase slug flow.

2. Characteristics of pressure-difference fluctuation: there is greater difference in pressure-difference fluctuation between main pipe, radial branch pipe and axial branch pipe when the gas-liquid two-phase slug flow flows through the fork of horizontal T-junction tube: ① The PDF of equal-diameter main pipe, equal-diameter axial branch pipe, variable-diameter radial branch pipe and variable-diameter axial branch pipe has a single peak distribution, and that of equal-diameter radial branch pipe is distributed in multi-peak. As the superficial gas-liquid velocity increases, the dispersity is stronger. ② The fluctuation range of power spectral function of equal-diameter main pipe, equal-diameter axial branch pipe and variable-diameter axial branch pipe increases with the increase of another phase superficial velocity, and that of the equal-diameter radial branch pipe is the largest if the superficial velocity of one phase is constant, without obvious amplitude changing. The fluctuation range and amplitude of pressure-difference fluctuation power spectrum function of variable-diameter radial branch pipe are not obvious. These conclusions can provide theoretical basis for increasing service life of downstream equipment and long-distance pipework and identifying the flow pattern of the slug flow.

References

[1] Liu Lei, Zhou Fangde, Li Huixiong. Progress in the research of density wave phenomena of two-phase flow, ADV ANCES IN MECHANICS, 1998, 28 (1).

[2] Wambasganss M W, Jendrzejczyk J A,France D M,Determination and Characteristics of the Transition to Two-Phase Chug Flow in Small Horizontal Charnels.ASME J. of Fluid Enginering,1994, 116 (1): 140-146.

[3] Weisman J,et al. Effects of Flow Properties and Pipe Diameter on Two-Phase Flow Patterns in Horizontal Lines.Int.J.Multiphase Flow, 1979, 5 (6): 437-462.

[4] Annunziato M,Girardi G. Horizontal Two-Phase Flow:A statistical Method for Flow Pattern Recognition. In:The 3rd International Conference on Multiphase Flow. Hague, 1987.169-185.

[5] CHARRON Y, WHALLEY P B. Gas-liquid annular flow at a vertical tee junction-Part I.Flow separation [J].International journal of multiphase flow, 1995, 21 (4): 569-589.

[6] LUO, Xiaoming, HE Limin, Lv Yuling. Fluctuation characteristics of gas-liquid two-phase slug flow in horizontal pipeline [J] .Journal of Chemical Industry and Engineering(China), 2008 (11): 2781-2786.

IOP Conf. Series: Journal of Physics: Conf. Series **1064** (2018) 012033 doi:10.1088/1742-6596/1064/1/012033

Study on the Wave Loads and Structural Stresses of Container Ship

Qian Sun[1,*], Hong Zhou[2], Qinping Gu[2] and Hongli Yuan[1]

[1]School of Marine Engineering, Jimei University, Xiamen, China.
[2]School of Ocean and Civil Engineering, Jiangsu University of Science and Technology, Zhenjiang, China.

*Corresponding author e-mail: sunqian1226@163.com

Abstract. In this article ship motions and wave loads are analyzed by using three dimensional motion of the ship program. As an example of a 5500TEU container ship, the heading angles of maximum motions and loads, amplitude of movement under the action of unit regular wave, load conversion functions, conditions and RAO of middle section at different angles have been gotten. The maximum dominant load parameters are also obtained. Then the long-term wave loads are calculated through the spectrum theory, and the results are compared with the values of classification rules. At the end forces and stresses distribution of the ship hull structure under dominant wave loads are also calculated by the dynamic load method with ANSYS software. The conclusions can be used to analyze ultimate loads of the container ships, and can also be the reference of ship structural design.

1. Introduction

Container ship is a kind of ship with large openings. In addition to the vertical bending moment in the design process, we must also consider the torque and horizontal bending moments produced by the oblique waves on the hull. At present, the structure of the general merchant ship is designed according to the standard of the classification societies. Because of the data and the formula of the classification rules are based on the statistical data, existing practice experiences and basic theory, which did not fully take into account the specific hull lines, size, speed, sea conditions and other factors, so the calculation according to standard out of the hull bending moment and shear force and torque is not very accurate. If the calculated value is too large, the hull structure designed according to this will cause material waste and dead weight reduction. On the other hand, if the calculated value is too small, the strength of the hull structure designed based on this will not be enough. Therefore, it is necessary to calculate the wave load of the specific ship first, then estimate the limit value according to the long term analysis, and finally use the dynamic load method to analyze the structure. This method can be more accurate than the empirical formula, and can also reduce the cost of hull structure design while improve the design efficiency.

There are two main methods for the study of three-dimensional hydrodynamic theory: the free surface Green function method and the Rankine source method [1]. In Green function method, velocity potential of flow field is determined by distributing sources and sinks in the wet hull surface, the Green function can satisfy all boundary conditions except the surface conditions. By using the

Content from this work may be used under the terms of the Creative Commons Attribution 3.0 licence. Any further distribution of this work must maintain attribution to the author(s) and the title of the work, journal citation and DOI.
Published under licence by IOP Publishing Ltd

Green second formula, the velocity potential is expressed as the sum of the area integration of the distribution source along the wet surface of the hull and the line integration of the water surface with the hull line. The Rankine source method is a calculation method, which distributes singular points on both the body surface and the free surface. The method is simple in calculating the distribution singularity, and the ship's speed effect can be considered. These two methods have made some progress in recent years. The performance of Bessho form translating-pulsating source Green's function is discussed and a numerical integration method was proposed by Yao Chaobang(et al., 2014) [2]. A 3D time domain program based on transient Green function method was developed by Sun Wei(et al., 2014) to analyze hydrodynamic problems involving ship moving at a constant forward speed [3]. The precise integration method and some mathematical derivations were employed to evaluate the memory part of the transient Green function and its derivatives. Program used to solve the initial boundary value problem based on hybrid Green function method was proposed by Tang Kai(et al., 2014,2015) to effectively solve the floating numerical divergence problem of ship with flare [4,5]. Method of Rankine source theory combined with three dimensional hydroelastic theory to solve medium and high speed condition was developed by Li Hui(2013) [6]. Aiming to predict ship performance such as wave drag, sinkage and trim, the steady ship wave problem was solved based on the high-order Rankine source method by Chen Xi(et al., 2015) [7]. Linearized free-surface condition with respect to double body potential was adopted in the calculation. Parametric roll prediction for a large containership in head waves was performed by Zhao Chunhui(et al., 2016) based on 3D Rankine source method, taking into account the square terms of the velocity potential in the Bernoulli equation [8]. The commonly used software for prediction of ship's three-dimensional motion and wave load are SESAM, SCADIS and HSC etc.

The main work of this paper is to use ship three-dimensional motion program (HSC) to predict short-term and long-term wave loads on a 5500TEU container ship, and compare them with the calculated values of current classification rules. The program uses the free surface Green function method to calculate the motions in 6 degrees of freedom and wave loads of the ship.

2. Navigational Conditions Setting

The analysis object in this paper is a 5500 TEU container ship, it's main parameters are presented in Table 1.

Table 1. Principal parameters of the container ship.

L_{OA}/m	L_{WL}/m	L_{BP}/m	B/m	D/m	Draft(Designed)/m	C_b	C_w
277.350	268.657	266.000	40.000	24.000	12.000	0.6136	0.853

In order to get the maximum wave loads of the ship, main load parameters under different sea conditions and different heading angle combinations should be set up to analyze the structural response. As container ship with large openings, the most important load parameters are the vertical wave shear, vertical wave bending moment and wave torsional moment.

In order to get the maximum value of the main load parameters, the frequency response function of ship motion can be started with. It is assumed that the sine wave frequency range of unit wave height is 0.05 ~2.00 rad/s and meets the hull in different directions. The speed is assumed to be 75% design speed, then frequency response functions along the different angle of attack can be calculated. In this example, the navigation conditions are as follows: sailing speed is 9.6 m/s; heading angles are 0°, 30°, 60°, 90°, 120°, 150° and 180°; range of phase angles is -180°~180° and wave frequency is from 0.05 to 2.00 rad/s.

Distribution diagram of small shell of ship under water used by HSC program in this paper is shown in Figure 1.

IOP Conf. Series: Journal of Physics: Conf. Series **1064** (2018) 012033 doi:10.1088/1742-6596/1064/1/012033

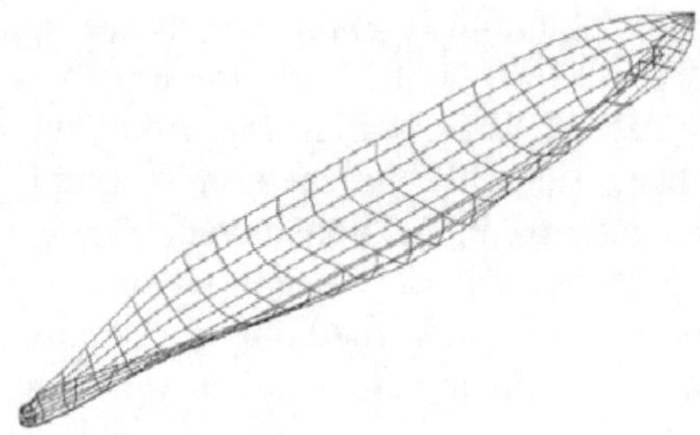

Figure 1. Distribution diagram of small shell of ship under water.

3. Motion Amplitudes of Hull under the Action of Regular Wave in Unit Wave Amplitude(ξ_0)

Surge(ξ_1), sway(ξ_2), heave(ξ_3), roll(ξ_4), pitch(ξ_5) and yaw(ξ_6) amplitudes of the ship at different heading angles are presented in Figure 2 to Figure 7.

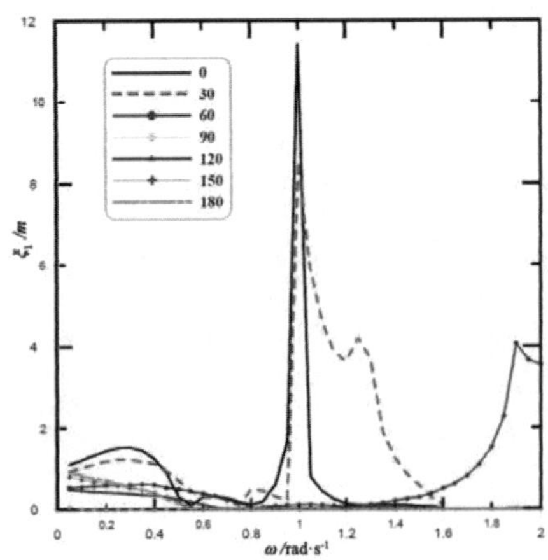

Figure 2. Surge amplitude of the ship at different heading angles.

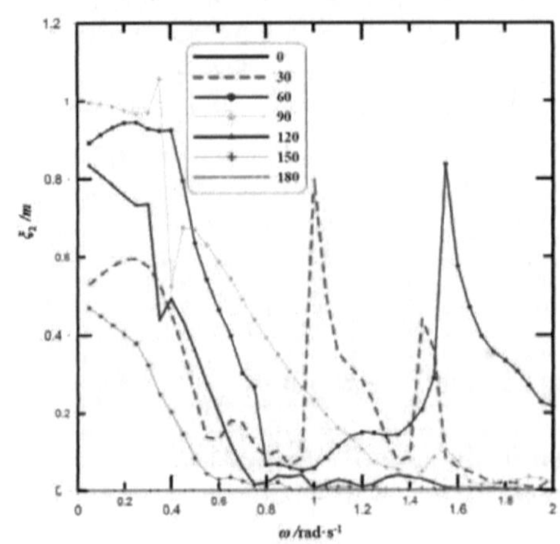

Figure 3. Sway amplitude of the ship at different heading angles.

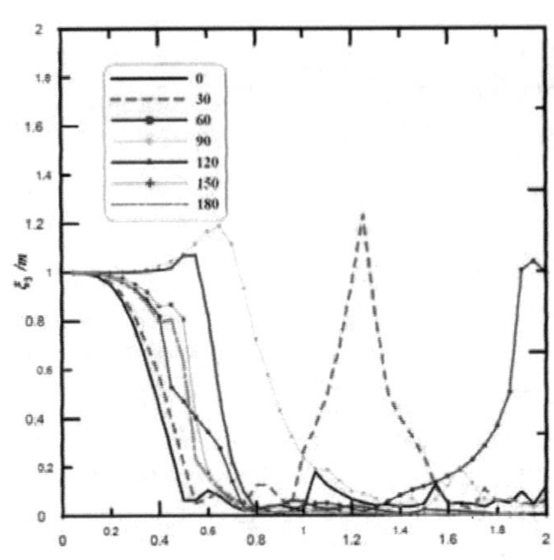

Figure 4. Heave amplitude of the ship at different heading angles.

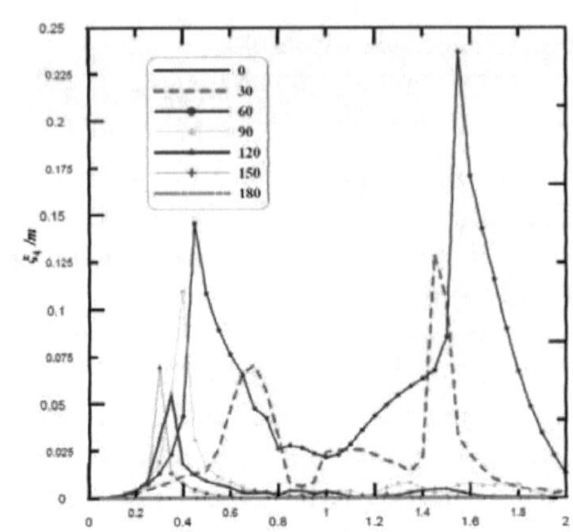

Figure 5. Roll amplitude of the ship at different heading angles.

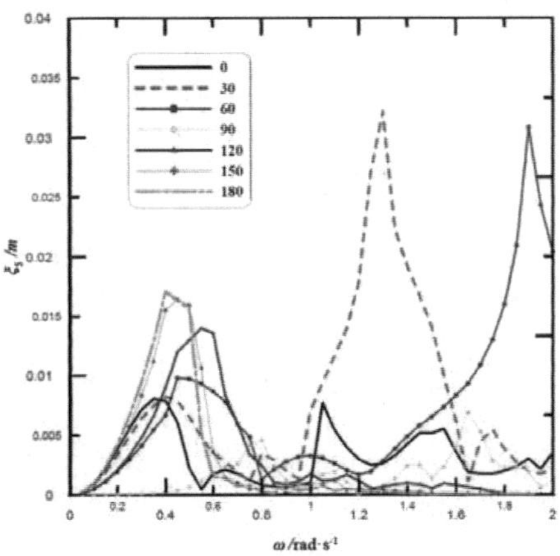

Figure 6. Pitch amplitude of the ship at different heading angles.

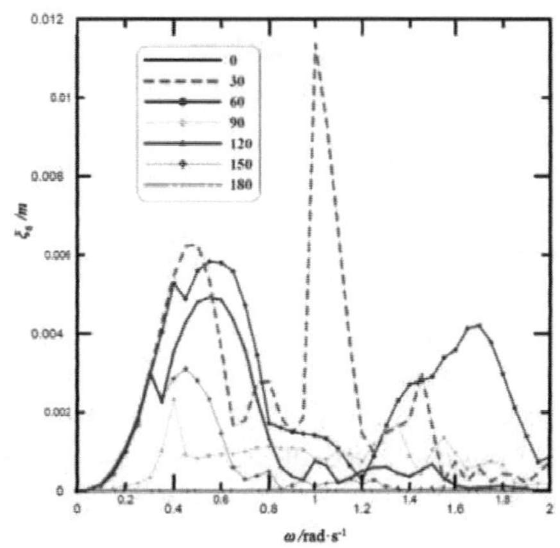

Figure 7. Yaw amplitude of the ship at different heading angles.

It can be seen from Figure 2 to Figure 7 that the heading angles corresponding to the maximum amplitudes of different motions are 0°, 90°, 30°, 60°, 30° and 30°.

4. Force Conversion Function and Response Amplitude Operator (RAO)

Phase spectrum of different force conversion functions such as longitudinal shear force(F1), transverse shear force(F2), vertical shear force(F3), torsional moment(F4), vertical bending moment(F5) and horizontal bending moment(F6) can be calculated by using ship three-dimensional motion program. Based on the results, it can be found that heading angles corresponding to each kind of maximum force are 60°, 60°, 180°, 60°, 180° and 120°.

After calculating the spectrum of ship motion in all directions and the force spectrum of the hull, it is necessary to select the Response Amplitude Operator (RAO) of the main load parameters. The main loads are vertical shear force(F3), torsional moment(F4) and vertical bending moment(F5). Because the RAO of ship's three-dimensional motion program is divided into multiple segments along the L_{bp}, so the output data must be analyzed, and the location with the maximum amplitude of RAO can be selected. Through the analysis of data, the maximum vertical shear force(F3) occurs at 165m from after perpendicular (A.P.), the maximum torsional moment (F4) occurs at A.P.66m and the maximum vertical bending moment (F5) occurs at A.P.133m (amidship). In addition, the maximum amplitude, heading angle, frequency and phase angle of the maximum amplitude should be selected from the spectrum diagrams also.

Some important results are presented in Table 2.

Table 2. Calculating results of main loads RAO.

	F3	F4	F5
Section	165m	66m	133m
Head	180°	60°	180°
w(freq)	0.55rad/s	0.5rad/s	0.45rad/s
RAO(max)	8.11×10^{6} N	2.05×10^{8} N·m	5.71×10^{8} N·m
Phase	-0.627°	80.5°	-179.87°

5. Long Term Wave Load Analysis

5.1. Long Term Wave Load Analysis

Ships may encounter various sea conditions during the whole service life (usually 20 years). Under rough sea conditions, although the operation time is short, the load response value can reach the extreme point. According to the statistics, the operating time of the ship in the rough sea condition accounts for about 5% of the total voyage [9]. Therefore, it is necessary for ship designers to pay attention to the maximum wave load that may occur during the whole life period of the ship.Wave spectrum and frequency should be selected during the analysis of the long-term wave load. In this paper P-M(Pearson-Moszkowicz) wave spectrum is chosen because a set of empirical formula are derived from the observation of the Atlantic open sea area, where is also the sailing route of 5500 TEU container ship. The wave frequency table recommended by IACS WP/S committee is selected to be used in the program calculation. After calculating, the loads of different directions along the length of ship can be obtained, as shown in Table 3.

Table 3. Loads in different directions along the length of ship by long term wave load analyses.

Distance from A.P./m	Longitudinal Shear Force/N	Transverse Shear Force/N	Vertical Shear Force/N	Torsional Moment/N·m	Vertical Bending Moment/N·m	Horizontal Bending Moment/N·m
-5	0.00	0.00	0.00	0.00	0.00	0.00
10.65	6.22×10^{6}	3.17×10^{6}	2.72×10^{7}	9.83×10^{7}	1.95×10^{8}	2.63×10^{7}
24.35	1.92×10^{7}	1.18×10^{7}	4.87×10^{7}	2.40×10^{8}	6.77×10^{8}	1.22×10^{8}
38.72	3.10×10^{7}	1.77×10^{7}	5.78×10^{7}	3.34×10^{8}	1.39×10^{9}	3.21×10^{8}
56.7	4.20×10^{7}	2.34×10^{7}	6.12×10^{7}	6.40×10^{8}	2.43×10^{9}	6.96×10^{8}
66	4.70×10^{7}	2.42×10^{7}	6.22×10^{7}	8.69×10^{8}	2.94×10^{9}	8.99×10^{8}
74.1	5.00×10^{7}	2.38×10^{7}	5.88×10^{7}	1.03×10^{9}	3.47×10^{9}	1.09×10^{9}
88.94	5.36×10^{7}	2.39×10^{7}	4.95×10^{7}	1.09×10^{9}	4.23×10^{9}	1.40×10^{9}
99	5.57×10^{7}	2.28×10^{7}	4.16×10^{7}	1.12×10^{9}	4.65×10^{9}	1.60×10^{9}
103	5.67×10^{7}	2.25×10^{7}	3.87×10^{7}	1.12×10^{9}	4.77×10^{9}	1.66×10^{9}
118	5.74×10^{7}	2.21×10^{7}	3.04×10^{7}	1.04×10^{9}	5.03×10^{9}	1.81×10^{9}
133.34	5.78×10^{7}	2.05×10^{7}	3.04×10^{7}	1.02×10^{9}	5.07×10^{9}	1.86×10^{9}
148.14	5.57×10^{7}	2.02×10^{7}	4.35×10^{7}	1.10×10^{9}	4.76×10^{9}	1.78×10^{9}
162.9	5.23×10^{7}	2.11×10^{7}	6.21×10^{7}	1.21×10^{9}	4.10×10^{9}	1.60×10^{9}
165	5.14×10^{7}	2.12×10^{7}	6.44×10^{7}	1.22×10^{9}	3.98×10^{9}	1.57×10^{9}
177	4.62×10^{7}	2.24×10^{7}	6.22×10^{7}	1.31×10^{9}	3.11×10^{9}	1.34×10^{9}
192.54	3.87×10^{7}	2.24×10^{7}	6.13×10^{7}	1.40×10^{9}	1.98×10^{9}	1.05×10^{9}
198	3.53×10^{7}	2.26×10^{7}	5.97×10^{7}	1.37×10^{9}	1.47×10^{9}	9.23×10^{8}
207.34	2.94×10^{7}	2.16×10^{7}	5.23×10^{7}	1.27×10^{9}	9.62×10^{8}	7.37×10^{8}
222.14	2.05×10^{7}	1.82×10^{7}	3.99×10^{7}	9.51×10^{8}	4.27×10^{8}	4.58×10^{8}
236.94	1.29×10^{7}	1.36×10^{7}	1.59×10^{7}	4.91×10^{8}	4.80×10^{8}	2.50×10^{8}
250.7	7.93×10^{6}	7.14×10^{6}	3.45×10^{6}	8.74×10^{7}	3.89×10^{8}	1.64×10^{8}
271.4	7.64×10^{5}	1.18×10^{6}	2.63×10^{6}	5.89×10^{7}	3.76×10^{8}	1.70×10^{8}

5.2. Comparison between the Results of Long Term Wave Load Analysis and Classification Rules Formula

The vertical shear force values in the long-term wave load analysis are similar to those calculated by the formula in the ABS and CCS [10, 11]. The middle part of long-term load calculation by PM spectrum value (3.04×107 N) and the standard value (3.06×107 N) is very close, but at length range A.P.40~60 m and A.P.150~210 m long term load vertical shear force is significantly higher than the standard value (A.P.56.7 m for example, the long-term load value is 6.12×107 N, standard value is

IOP Conf. Series: Journal of Physics: Conf. Series **1064** (2018) 012033 doi:10.1088/1742-6596/1064/1/012033

3.25×107 N). From A.P.220 m to the bow, the standard value is higher than the vertical shear force predicted by long term load. Taking A.P.222.14 m as an example, the long-term load value is 3.99×107 N, the standard value is 4.38×107 N; in A.P.250.7 m, for example, the long-term load value is 3.45×106 N, and the standard value is 1.47×107 N.

The vertical wave bending moment in the long-term wave load analysis results can be compared with the calculated value of the standard formula of the classification society, and it can be seen that the vertical wave bending moment calculated by the two methods is basically the same distributing trend along the length. The difference at sagging condition results is 16.31% (long term load value is 5.07×10^9 N·m, standard value is 4.243×10^9 N·m), hogging condition difference is 32.46% (standard value of hogging condition is 3.424×10^9 N·m). From A.P.40 to A.P.160m part of the ship at sagging, and from A.P.25 to A.P.175m at hogging condition, vertical wave bending moment standard value of these ship sections is smaller than the long-term wave load prediction value.

6. Analysis of Container Ship Hull Structure under Wave Load

6.1. Establishment of Dynamic Finite Element Model

The hull profile and rib lines plan are drawn by Autocad firstly, then the main node coordinates are input to ANSYS software. ANSYS is used to do the pre-processing work and to finish the establishment of hull dynamic finite element model, as shown in Figure 8.

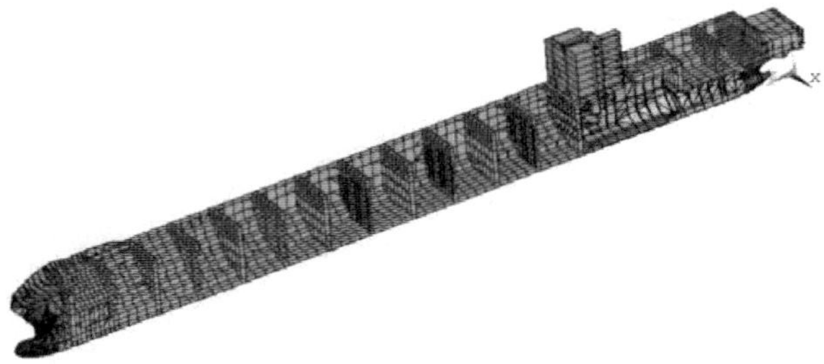

Figure 8. Dynamic finite element model (right half).

After the establishment of the whole ship dynamic load model, it need to adjust the static state of weight distribution and buoyancy distribution according to the design of the original ship plan, and confirm that the model is consistent with the original ship.

The pressure load of ships in waves can be obtained by using HSC program, as shown in Table 3. Water pressure distribution and movement amplitude can be converted to nodal force of the dynamic load model through compiling external program, further more the movement of the ship caused by unbalanced force is converted to acceleration, exerted on each node as the form of inertial force in order to achieve the balance of force.

When the nodal forces are completely exerted, the appropriate boundary conditions are added to the model before the analysis is carried out according to the requirements of the relevant rules, including the rigid body displacement constraint, as shown in Figure 9.

Figure 9. The whole ship model with dynamic loads and boundary constraints.

6.2. Analysis of Container Ship Hull Structure under Wave Load

6.2.1. Analysis of Vertical Shear Force (F3). It can be seen from Table 3 that the maximum vertical shear force lies on A.P.38 m~A.P.88 m and A.P.162 m~A.P.198 m. Figure10 and Figure 11 show the stress distribution diagram of the hull structure of A.P. 38 m~A.P.88 m. Figure 12 and Figure 13 show the stress distribution diagram of the hull structure of A.P.162 m~A.P.198 m.

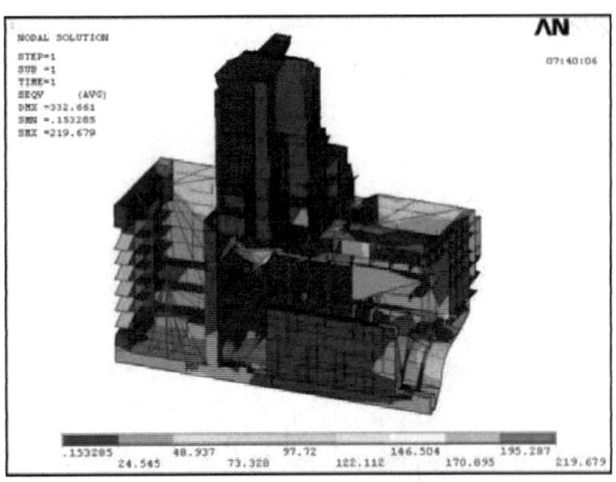

Figure 10. Stress distribution diagram at A.P. 38m~88m under vertical wave shear (Hogging).

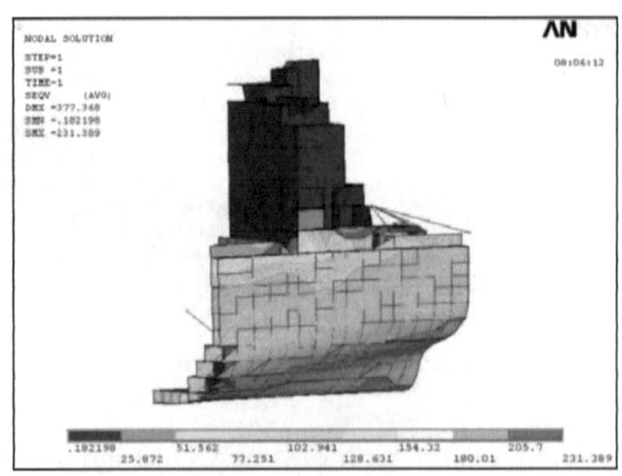

Figure 11. Stress distribution diagram at A.P. 38m~88 m under vertical wave shear (Sagging).

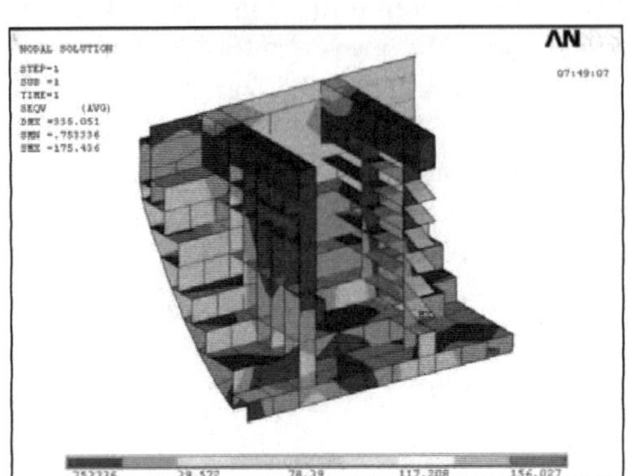

Figure 12. Stress distribution diagram at A.P. 162m~198m under vertical wave shear (Hogging).

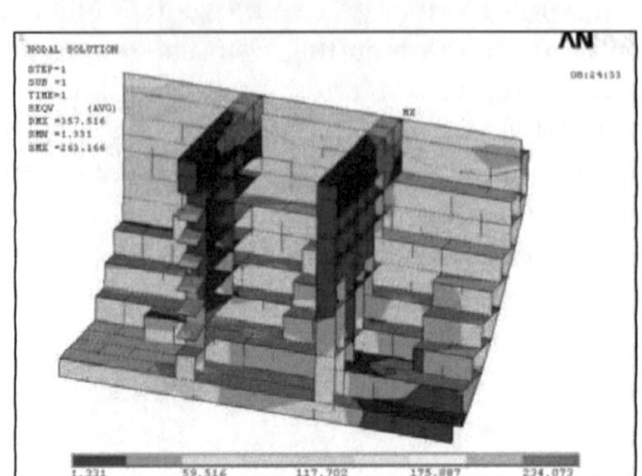

Figure 13. Stress distribution diagram at A.P. 162m~198m under vertical wave shear (Sagging).

6.2.2. Analysis of Wave Torsional Moment (F4). The maximum wave torsional moment lies on A.P.66m~A.P.103m and A.P.177m~A.P.222m by referring to Table3. Figure 14 shows the stress distribution diagram of the hull structure of A.P.66m~A.P.103m. Figure15 shows the stress distribution diagram of the hull structure of A.P.177m~A.P.222m.

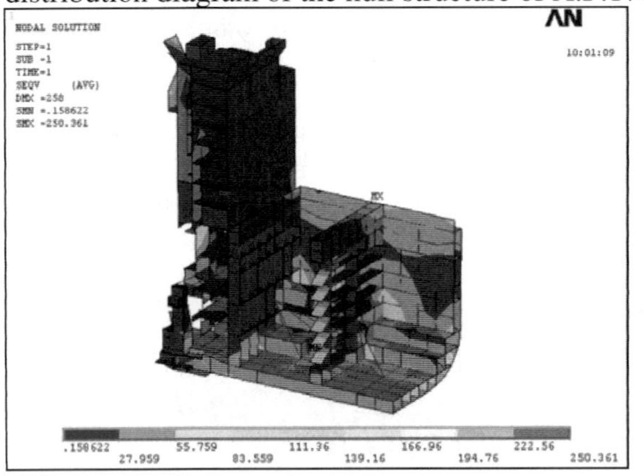

Figure 14. Stress distribution diagram at A.P. 66m~103m under wave torsional moment.

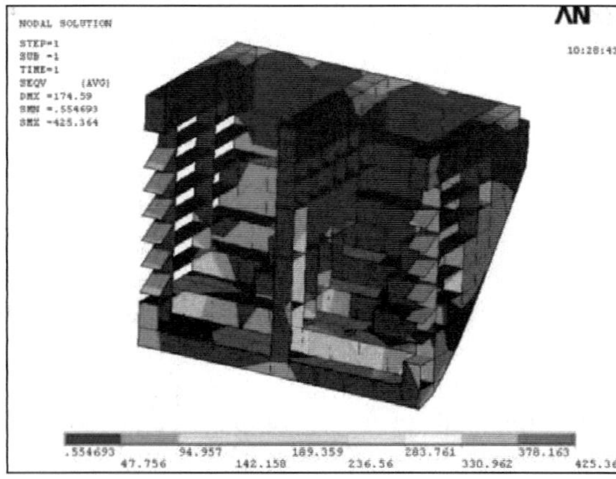

Figure 15. Stress distribution diagram at A.P. 177m~222m under wave torsional moment.

6.2.3. Analysis of Vertical Bending Moment (F5). From table 3 it can be found that the maximum vertical bending moment lies on A.P.103m~A.P.162m. Figure 16 and Figure 17 show the stress distribution diagram of the hull structure of A.P.103m~A.P.162m.

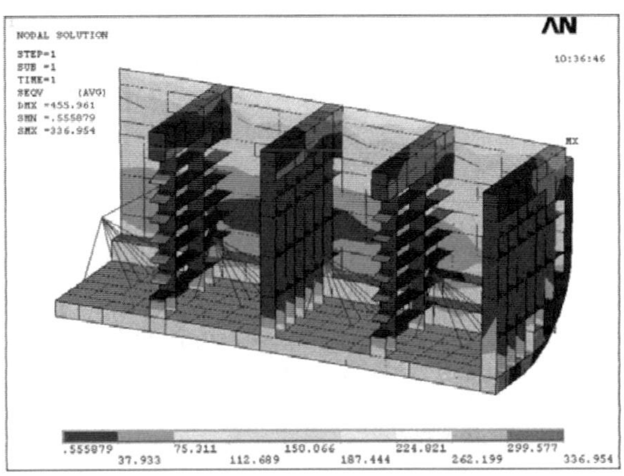

Figure 16. Stress distribution diagram at A.P. 103m~162m under vertical bending moment (Hogging).

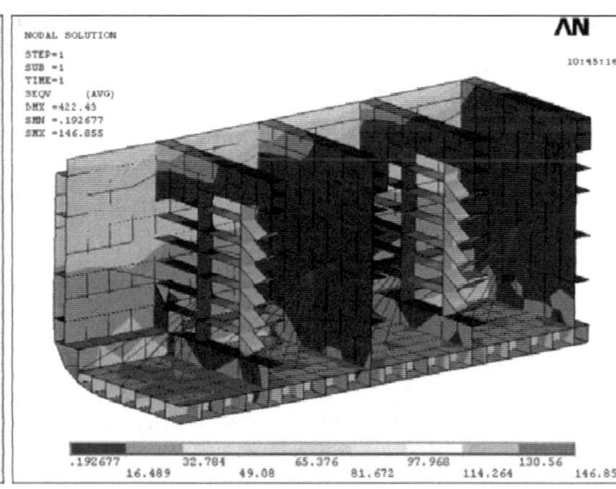

Figure 17. Stress distribution diagram at A.P. 103m~162m under vertical bending moment (Sagging).

7. Conclusion

The results of the above calculation and the analysis of the data can be concluded as follows:

(1) The position and maximum value of the main load parameters and the corresponding navigation conditions are obtained through the analysis of ship motions and wave loads.

(2) The results of long term wave load analysis and classification rules are compared, the vertical shear force of classification rules in the middle and the bow has reached 20 years of long-term load standard, but shear force near the stern is lower than 20 years standard. The vertical wave bending

moment of classification rules is less than 20 years standard, but in sagging condition the bending moment is close to 20 years standard.

(3) It can be seen from Von Mises stress distribution diagrams that when the hull structure is subjected to the vertical wave bending moment of twenty year long term wave loads, the stress will exceed the allowable stress of 22.2kg/mm^2 under hogging condition. The other ship's stress distribution state are still within the allowable stress range.

(4) According to the wave spectrum and sea state statistics recommended by the IACS WP/S Committee, the long-term load values are greater than the loads calculated by the current classification society formula, and it is conservative to predict the wave load according to that sea state. From the perspective of comparative strength of hull, in order to coordinate with the IACS standard, if the long-term prediction is made by using wave spectrum and sea state recommended by IACS to calculate the wave load value, the allowable stress may be appropriate to increase or the probability level used to forecast the wave load values can be adjusted from 10^{-8} to $10^{-5.5}$.

Acknowledgments

This work was financially supported by the Natural Science Founds of Fujian Province (No. 2018J01493).

References

[1] Y. Xu, W.C. Dong, A Fast Integration Method and Characteristics Research for 3D Translating-pulsating Green's Function of Havelock Form in Deep Water, China Ocean Engineering. 25 (2011) 365-380.

[2] C.B. Yao, W.C. Dong, A Fast Integration Method for Translating-pulsating Source Green's Function in Bessho Form, Journal of Zhejiang University. 15 (2014) 108-119.

[3] W. Sun, H.L. Ren, H. Li, Numerical Solution for Ship with Forward Speed Based on Transient Green Function Method, Journal of Ship Mechanics. 18 (2014) 1444-1452.

[4] K. Tang, R.C. Zhu, G.P. Miao, Analysis of Motions of Floating Body in Waves by Hybrid Green Function Method and Discretization of Parameters, Shipbuilding of China. 56 (2015) 102-113.

[5] K. Tang, R.C. Zhu, G.P. Miao, Hybrid Green Function Method in Time Domain Analysis of Motions of Floating Body in Waves, Journal of Shanghai Jiaotong University. 48 (2014) 508-514.

[6] H. Li, Y.H. Zhang, H.L. Ren, Long/short Term Prediction of Ship Wave Loads and Its Characteristics, Journal of Huazhong University of Science and Technology. 41 (2013) 112-116.

[7] X. Chen, R.C. Zhu, G.P. Miao, Calculation of Ship Sinkage, Trim and Wave Drag Using High-Order Rankine Source Method, Shipbuilding of China. 56 (2015) 1-12.

[8] C.H. Zhao, Y.H. Xie, H.M. Wang, Nonlinear Parametric Roll Containership in Time Domain Based on Rankine Panel Method, Naval Architecture and Ocean Engineering. 32 (2016) 22-25,58.

[9] Zakaria, Effect of Ship Size, Forward Speed and Wave Direction on Relative Wave Height of Container Ships in Rough Seas, The Journal of the Institution of Engineers. 72 (2009) 21-34.

[10] China Classification Society, Rules for the Classification of Sea-going Steel Ships, China Communications Press, Beijing, 2015.

[11] American Bureau of Shipping, Rules for Building and Classing Steel Vessels. Dover Publications, New York, 2010.

Influence of the receiver aircraft's bow wave on the motion characteristics of the drogue in hose-drogue refueling system

Xiaofei Zhang, Mengquan Hu*, Dong Zhang, Pengyun Tian and Ning Wang

Aeronautics Engineering College, Air Force Engineering University, Xi'an, China.

*Corresponding author e-mail: jubeiweihao2@sina.com

Abstract. In the docking process of hose-drogue aerial refueling system, the bow wave effect of an approaching receiver aircraft will produce a strong aerodynamic effect on the drogue which intensify the deviation of drogue and increase the difficulty of the docking. This paper presents a receiver aircraft's bow wave effect modeling method based on potential flow theory and considers the major shape characteristics of the receiver aircraft. This method is used in a probe-drogue aerial refueling dynamic simulation system, which combined with the hose-drogue dynamic model and tanker wake dynamic model. The motion characteristics of the drogue under different flight altitude, flight velocity and docking velocity are calculated and analyzed. The results show that when the other conditions are certain, when the flight altitude increases, the bow wave effect decreases the vertical and longitudinal influence on the drogue, increases the lateral influence. When the flight velocity increases, the bow wave effect increases the vertical and longitudinal influence on the drogue, decreases the lateral influence. When the docking velocity increases, the drogue's swing displacement and the swing velocity are all increased.

1. Introduction

The aerial refueling is a power multiplier in modern naval and air force. The hose-drogue aerial refueling system is simple and widely used [1, 2]. But the hose-drogue system of flexible structure which is highly sensitive to external factors such as atmospheric turbulence, tanker wake and the receiver-aircraft's bow wave. In the actual docking process, there is the problem of the hose-drogue swing, which increased the difficulty of the docking and security risks.

In the docking process, when the receiver-aircraft is close to the drogue, the drogue will deviate from the balance position and then quickly return back. This phenomenon is called as the bow wave effect or the forebody effect [3, 4]. The drogue movement caused by the bow wave brings great uncertainty to the success of the docking. The rise of unmanned aircraft vehicle (UAV) in recent years has opened up new areas for aerial refueling [5]. Therefore, it is required to conduct more in-depth studies on the hose-drogue aerial refueling to ensure the safety of the docking process, whether manned or unmanned.

So far, the research on atmospheric turbulence and tanker wake is relatively mature. However, there is less research on the bow wave effect modeling. Most of people use half Rankine body [6] to simulate the bow wave effect. This method has a small amount of calculation, but also has the problem of accuracy is poor. A CFD simulation was carried out by Wang and Dong [7]. They used the

Content from this work may be used under the terms of the Creative Commons Attribution 3.0 licence. Any further distribution of this work must maintain attribution to the author(s) and the title of the work, journal citation and DOI.
Published under licence by IOP Publishing Ltd

FLUENT software to analyze the variation of bow wave with Mach number and Angle of attack. Khan and Masud [8] used CFD to simulate the aerial refueling of the "Thunder" fighter and analyze the docking strategy of reducing bow wave effect. The CFD simulation results are more reliable. However, due to different receiver aircrafts have different complex configuration structure, the CFD simulation requires a large amount of computing resources and time. Moreover, the CFD results are inconvenient for the controller design. NASA Dryden Flight Research Center tested the bow wave effect by F/A-18 receiver-aircraft and estimated the range of the bow wave effect [5, 9].

This paper propose a fast and efficient receiver aircraft's bow wave effect modeling method based on potential flow theory and considers the major shape characteristics of the receiver aircraft. This method is used in aerial refueling docking simulation system. The motion characteristics of the drogue with different flight altitude, flight velocity and docking velocity are calculated and analyzed.

2. Frame definitions and modeling assumptions.

2.1. Reference frames

As shown in Fig. 1. This paper defines the ground coordinate system OgXgYgZg. The body frame of the receiver aircraft ObXbYbZb and the body frame of the tanker aircraft OaXaYaZa. Ob is fixed on the top of the receiver aircraft. Oa is fixed on the refueling pod.

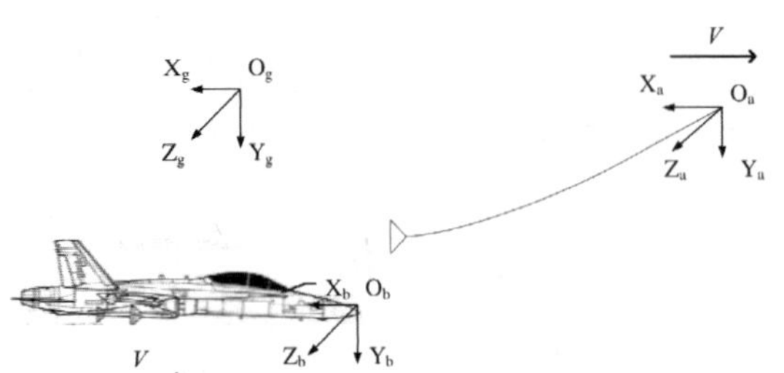

Figure 1. Overview of the reference frames

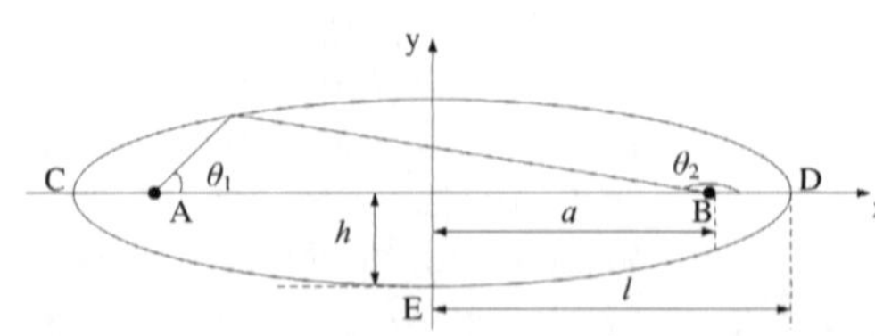

Figure 2. Overview of the Rankine body.

2.2. Modeling assumption

To simplify the modeling of the head wave effect, the following assumptions are made.

(1) In the shape structure of the receiver aircraft, there is a large influence on the flow field near the nose, fuselage, cockpit, main wing and strake wing. Suppose the aerodynamic effects of each structure are independent to each other. According to the superposition principle, the induced velocity of the head wave effect can be expressed as:

$$\boldsymbol{v}_{bow} = \boldsymbol{v}_{body} + \boldsymbol{v}_{cockpit} + \boldsymbol{v}_{maiwing} + \boldsymbol{v}_{strakewing} \quad (1)$$

Where $\boldsymbol{v}_{bow}$ denotes the induced velocity produced by the bow wave, $\boldsymbol{v}_{body}$, $\boldsymbol{v}_{cockpit}$, $\boldsymbol{v}_{mainwing}$ and $\boldsymbol{v}_{strakewing}$ denotes the induced velocity produced by the fuselage, cockpit, main wing and strake wing.

(2) Considering the angle of attack of the receiver aircraft is small. So it can be assumed that the fuselage is axisymmetric.

3. Bow wave effect modeling

3.1. Receiver aircraft's fuselage and cockpit modelling method

To simplify the receiver aircraft model establishing, seem the fuselage as Rankine body [10], whose flow field is superposition by the three dimensional point source, the three dimensional point sink and

IOP Conf. Series: Journal of Physics: Conf. Series **1064** (2018) 012034 doi:10.1088/1742-6596/1064/1/012034

three-dimensional straight uniform stream Finally, the fuselage flow field model is obtained by elliptical transformation. The cockpit is modeled in the same way as the fuselage.

Assumption that the three-dimensional point source strength of Rankin body is m and was placed in A (-a, 0) as shown in Fig. 2. The three-dimensional point sink strength of Rankin body is m and was placed in B (a, 0). Straight uniform stream U is going along the x axis. The velocity potential and stream function can be expressed as:

$$\varphi = -Ux + \frac{m}{\sqrt{(x+a)^2 + y^2}} - \frac{m}{\sqrt{(x-a)^2 + y^2}} \tag{2}$$

$$\psi = -\frac{1}{2}Uy^2 + \frac{m(x+a)}{\sqrt{(x+a)^2 + y^2}} - \frac{m(x-a)}{\sqrt{(x-a)^2 + y^2}} \tag{3}$$

Eq. (3) can be transformed to the form as:

$$\psi = -\frac{1}{2}Ur^2 \sin^2\theta + m(\cos\theta_2 - \cos\theta_1) \tag{4}$$

Cause the boundary condition rules that $\psi = 0$. The function can be rewritten as

$$y^2 + b^2(\cos\theta_1 - \cos\theta_2) = 0 \tag{5}$$

$$b^2 = \frac{2m}{U} \tag{6}$$

Assuming that OC=OD=l, OE=h. Because of point C is the stationary point. The velocity of the free stream is offset by the velocity induced by source and sink.

$$-\frac{m}{(l+a)^2} + \frac{m}{(l-a)^2} = U \tag{7}$$

Cause the geometric relationship shown as in Fig. 2

$$\left(l^2 - a^2\right)^2 = 2ab^2 l \tag{8}$$

Where $2b^2 \cos\alpha = h^2$, Eq (8) can be rewritten as

$$\frac{2a}{\sqrt{h^2 + a^2}} = \frac{h^2}{b^2} \tag{9}$$

$$h^2 l\sqrt{h^2 + a^2} = 2ab^2 l \tag{10}$$

Combine the two equations

$$h^2 l\sqrt{h^2 + a^2} = \left(l^2 - a^2\right)^2 \tag{11}$$

Based on the receiver aircraft's parameters. Setting the length of the fuselage is 2l, and the maximum radial width is h. The three dimensional coordinates and strength m of point source and point sink are calculated. The induced velocity $\boldsymbol{v}_{body}$ at any point in space can be calculated as

$$\boldsymbol{v}_{body} = \frac{m}{\boldsymbol{r}_s} - \frac{m}{\boldsymbol{r}_{-s}} \tag{12}$$

Where $\boldsymbol{r}_s$ is the vector from the source to the point. $\boldsymbol{r}_{-s}$ Is the vector from the sink to the point.

During refueling process, the aircraft keeps the subsonic flight. Based on the fuselage model in the incompressible flow. The Gothert Rule is used for compressibility modification. The corresponding induced velocity is obtained in the compressible flow.

While for some receiver aircraft like F/A-18, the cross section of the nose is like an ellipse. In order to obtain more accurate simulation results. The ellipse transformation is necessary.

As is shown in Fig. 3, the scale transformation is presented as

$$x' = x, \quad y' = \frac{a}{b} y, \quad z' = z \tag{13}$$

Where a and b are the short axis and long axis of the ellipse. The coordinates of the desired point are converted by Eq. (13) and obtain the $\boldsymbol{v}_{body}' = \left[v_x', v_y', v_z'\right]^T$ by Eq. (12). After that, the ellipse transformation is applied again to transform $\boldsymbol{v}_{body}'$ back to the original scale as

$$v_x = v_x', \quad v_y = \frac{b}{a} v_y', \quad v_z = v_z' \tag{14}$$

Modified fuselage induced velocity shown as

$$\boldsymbol{v}_{body} = \left[v_x, v_y, v_z\right]^T \tag{15}$$

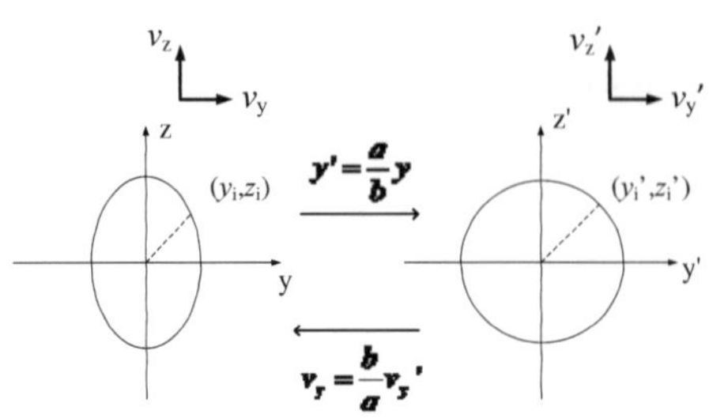

Figure 3. Illustration of the ellipse transformation

3.2. Receiver aircraft's aerofoil modelling method

In the process of aerial refueling, the flight velocity is low. In this case, compared with the general CFD method, the vortex lattice method (VLM) has a small precision and high computational efficiency [11]. Under the assumption of the inviscid flow, the VLM can be used to find the flow field of the aerofoil easily and efficiently.

The VLM is a practical numerical method based on lifting surface. The VLM considers the wing as a network with a horseshoe vortex superimposed on the aerofoil. First, the aerofoil is divided into

M×N grids, and each grid is equipped with a horseshoe vortex, and the induced velocity generated by each horseshoe vortex is obtained by the Biot-Savart Law. The ith control point's velocity induced by M×N horseshoe vortexes is obtained. According to the boundary conditions of control points, a set of equations of M×N control points can be obtained to solve the vortex strength of each horseshoe vortex, so as to obtain the induced velocity of the finite discrete horseshoe vortex on any point in space. The Gothert Rule is used for compressibility modification too. The corresponding induced velocity is obtained in the compressible flow.

4. The simulation process

In the aerial refueling, the receiver aircraft and tanker in the same flight altitude and velocity. The receiver aircraft is close to the drogue with a stable docking velocity behind the tanker, until the connection is successful, and the relative position of the drogue is no more changes. In the process of receiver aircraft slowly close to the drogue, the bow wave effect caused by the receiver aircraft destroys the original force balance state of the drogue and produces swing. The relative position and velocity between the receiver aircraft and drogue is constantly changing. To simulate this process accurately. The digital simulation process is as follows.

(1) Calculate the equilibrium position of the hose under the no turbulence condition [12].

(2) Calculate the tension of each hose element, relative airflow velocity, external force and equivalent hose restoring force.

(3) Calculate the fuselage and cockpit's strength and coordinates of the source and sink. After that, calculate the induced velocity at the point in space.

(4) Calculate the velocity induced by the main wing on the drogue by VLM.

(5) Calculate the bow wave induced velocity at the point of the drogue by Eq. (1).

(6) The acceleration of the drogue is calculated by the induced velocity, and the acceleration of the drogue is obtained by using the fourth-order Runge-Kutta method [13].

Repeat the above steps to obtain the real-time motion of the drogue.The receiver aircraft is subject to the F/A-18. The simulation is conducted under the condition without turbulence. The simulation parameters used in this paper are illustrated in Table 1. The parameters of the tanker and hose-drogue system are with the same parameters in Ref. 13

Table 1. Parameters of the simulation.

Parameter	Value
Tanker	
The length of the fuselage	17.07m
The maximum radius of the fuselage	0.808m
span	11.43m
root chord	4.04m
wing sweep	25.82°
Drogue	
Weight	200N
Diameter	0.6m

5. Simulation results and analysis of the motion characteristics of the hose-drogue.

5.1. Influence of the flight altitude on the drogue.

The flying velocity is 120m/s, the docking velocity is 1.0m/s. The flight altitude is 3000m, 5000m and 7000m respectively. The simulation results are as follows:

In the process of docking, the deviation of the offset in the vertical and lateral of the drogue is shown in the Fig. 4. The drogue begins to float to the right and upward. And then swings back to the connected position from the maximum offset position until the drogue is connected with the probe, after that the displacement is no longer changed. The other conditions remain unchanged, when the flight altitude increases, the vertical maximum deviation of the drogue from the equilibrium position will decrease, and the lateral maximum of the drogue from the equilibrium position deviation will increase. The drogue begins to deviate from the equilibrium position at a later time point, which means that the influence area of the bow wave effect on the longitudinal axis (X-axis) decreases with the increase of altitude.

When the flight velocity is constant, the circulation of the tanker and receiver aircraft's wing attached vortex and free vortex increase with the increase of altitude. The downwash and sidewash velocity of the tanker put on the hose-drogue system is increased, and the upwash and sidewash velocity of the receiver aircraft is increased. Combined with the bow wave effect, the flight altitude increases, the vertical deviation of the drogue decreases, the lateral deviation increases, and the longitudinal influence range decreases.

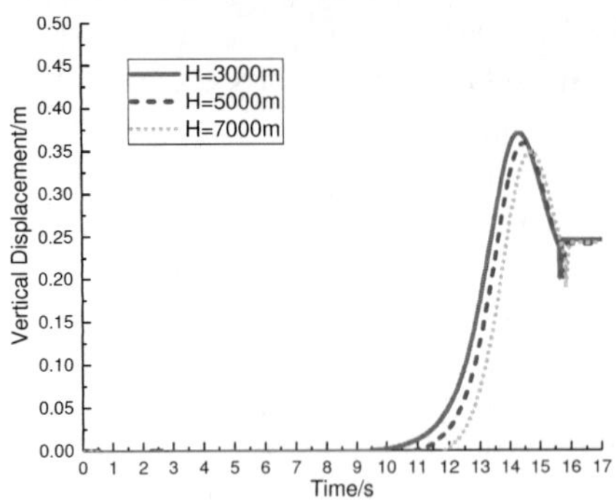

(a) Y direction

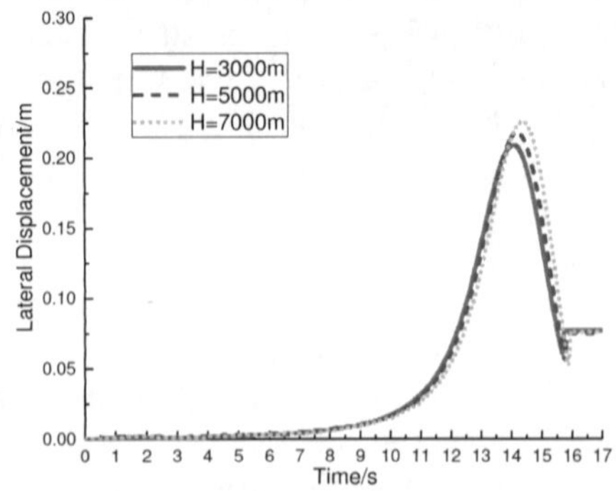

(b) Z direction

Figure 4. Deviation of drogue in Y and X direction with different flight altitude.

5.2. Influence of the flight velocity on the drogue.

The flying altitude is 3000m, the docking velocity is 1.0m/s. The flight velocity is 120m/s, 150m/s and 180m/s respectively. The simulation results are as follows:

In the process of docking, the deviation of the offset in the vertical and lateral of the drogue is shown in the Fig. 5. The other conditions remain unchanged, when the flight velocity increases, the vertical maximum deviation of the drogue from the equilibrium position will decrease, and the lateral maximum of the drogue from the equilibrium position deviation will increase. The drogue begins to deviate from the equilibrium position at a later time point, which means that the influence area of the bow wave effect on the longitudinal axis (X-axis) decreases with the increase of velocity.

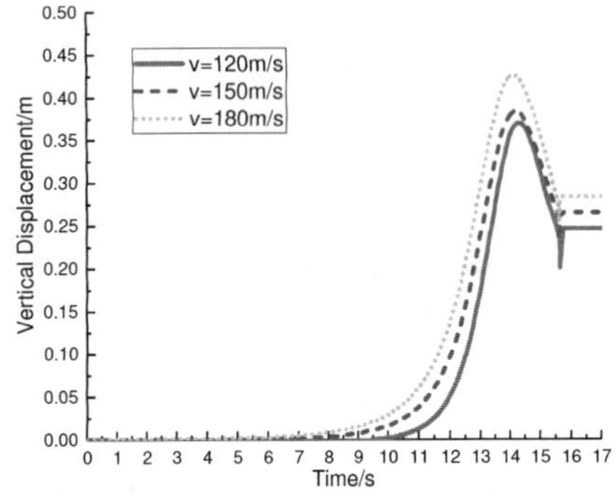

(a) Y direction

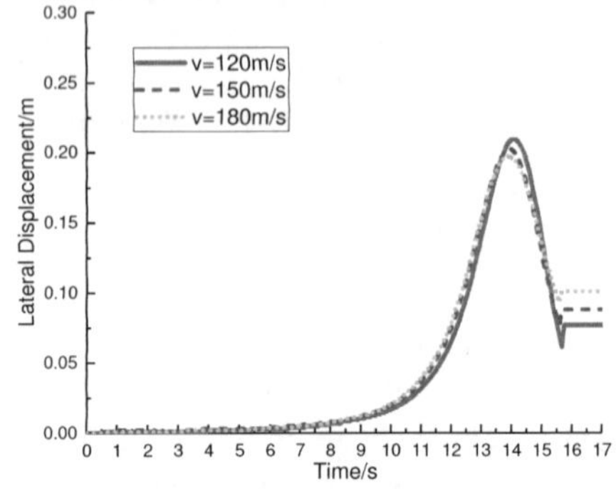

(b) Z direction

Figure 5. Deviation of drogue in Y and X direction with different flight velocity.

When the flight altitude is constant, the circulation of the tanker and receiver aircraft's wing attached vortex and free vortex decrease with the increase of velocity. The downwash and sidewash velocity of the tanker put on the hose-drogue system is decreased, and the upwash and sidewash velocity of the receiver aircraft is decreased. Combined with the bow wave effect, the flight velocity increases, the vertical deviation of the drogue increases. The lateral deviation decreases and the longitudinal influence range increases.

5.3. Influence of the docking velocity on the drogue.

The flying altitude is 3000m, the flight velocity is 150m/s, the docking velocity is 0.5m/s, 1.0m/s and 1.5m/s respectively. The simulation results are as follows:

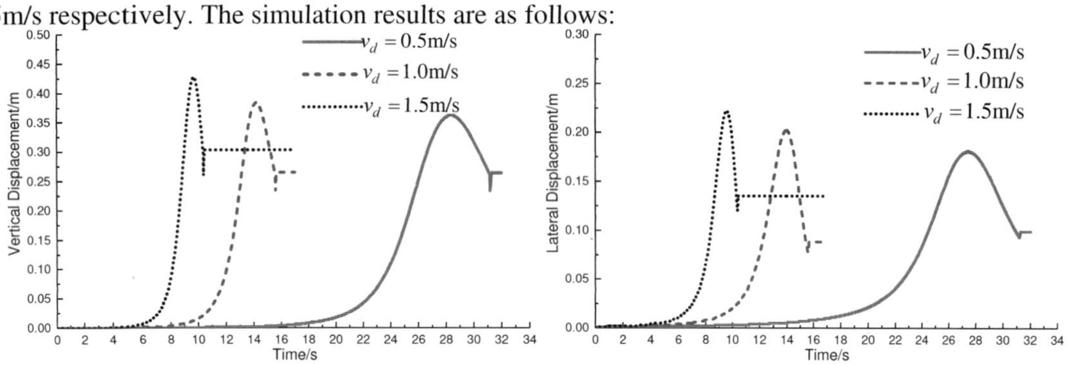

(a) Y direction (b) Z direction

Figure 6. Deviation of drogue in Y and X direction with different docking velocity.

The deviation of the lateral and vertical deviation of the drogue during the docking process is shown as Fig. 6. The larger the docking velocity is, the larger the deviation of the drogue will be, and the larger the slope of the offset curve will be, which means the larger the swing rate is. It is indicated that the larger the docking velocity will increase the swing displacement and the swing velocity of the drogue which affected by the bow wave effect and increase the difficulty of the docking.

6. Conclusion

(1). In this paper, a simple and efficient modeling method is proposed based on potential flow theory, and a mathematical model of fuselage, cockpit and aerofoil is established by means of source, sink and finite discrete horseshoe vortex. Through the compressibility modification and elliptical modification, the model is as accurate as possible to simulate the aerodynamic characteristics of the bow wave effect.

(2). Based on the shape features and parameters of F/A-18, a model of bow wave effect was established. The numerical simulation of the aerial refueling docking process was carried out. The motion characteristics of drogue under different docking state are analyzed. The results show that when the other conditions are certain, when the flight altitude increases, the bow wave effect decreases the vertical and longitudinal influence on the drogue, increases the lateral influence. When the flight velocity increases, the bow wave effect increases the vertical and longitudinal influence on the drogue, decreases the lateral influence. When the docking velocity increases, the drogue's swing displacement and the swing velocity are all increased.

Acknowledgments
This work was financially supported by National Natural Science Foundation of China (61473307).

References

[1] QUAN Q, WEI Z B, GAO J, et al. A survey on modeling and control problems for probe and drogue autonomous aerial refueling at docking stage [J].Acta Aeronautica et Astronautica

Sinica. 2014, 35(9): 2390-2410(in Chinese).
[2] LU Y P, YANG C X, LIU Y Y, A survey of modeling and control technologies for aerial refueling system[J].Acta Aeronautica et Astronautica Sinica. 2014, 35(9): 2375-2389(in Chinese).
[3] Atilla Dogan, William Blake. Modeling of Bow Wave Effect in Aerial Refueling[C]. AIAA Atmospheric Flight Mechanics Conference, Toronto, Ontario Canada: AIAA, 2010.
[4] Ujjar Bhandari, Peter R Thomasy, Steve Bullockz, Thomas S Richardson. Bow Wave Effect in Probe and Drogue Aerial Refuelling[C]. AIAA Guidance, Navigation, and Control (GNC) Conference, Boston, MA: AIAA, 2013.
[5] Ryan P Dibley, Michael J Allen. Autonomous Airborne Refueling Demonstration Phase I Flight-Test Results[C]. AIAA Atmospheric Flight Mechanics Conference and Exhibit, Hilton Head, South Carolina: AIAA, 2007.
[6] ZHANG J. Dynamic modeling and control of the motion of the hose-drogue [D]. Nanjing: Nanjing University of Aeronautics and Astronautics, 2016:36-40(in Chinese).
[7] WANG J, DONG X M, XU Y J, et al. Simulation and analysis of the bow wave effect of the receiver in hose-drogue aerial refueling [J]. Flight Dynamics, 2016, 34(1): 54-58(in Chinese).
[8] Khan, J Masud. Trajectory Analysis of Basket Engagement during Aerial Refueling[C]. AIAA Atmospheric Flight Mechanics Conference, National Harbor, Maryland: AIAA, 2014.
[9] Jennifer L Hansen, James E Murray, Norma V Campos. The NASA Dryden Flight Test Approach to an Aerial Refueling System[R]. NASA/TM-2005-212859, Edwards, California: NASA Dryden Flight Research Center, 2005.
[10] L.M. Milne Thomson. Theoretical hydrodynamics [M]. LI Y L, YAN M W, translated. Beijing: China Machine Press, 1984.9. 545-551(in Chinese).
[11] QIAN Y J. Aerodynamics [M]. Beijing: Beihang University Press, 2004.9:153-165(in Chinese).
[12] HU M Q, LIU P, NIE X, et al. Influence of air turbulence on the movement of the hose-drogue [J]. Flight Dynamics, 2010, 28(5): 20-23(in Chinese).
[13] HU M Q, NIE X, WANG L M. Determination of hose static catenary shape in “probe-drogue” in-flight refueling system [J]. Journal of Air Force Engineering University: Natural Science Edition, 2009, 10(5): 22-26(in Chinese).

Analysis of Gas-Solid-Thermal Coupling on Micro-Vibration of Aerostatic Bearing at Micron-Scale

Hui Chai[1,*], Caiqi Li[2] and Wei Long[1]

[1]School of Mechanical and Electrical Engineering, Kunming University of Science and Technology, Kunming, China.
[2] School of Materials Science and Engineering, Kunming University of Science and Technology, Kunming, China.

*Corresponding author e-mail: Henry0806@126.com

Abstract. Based on gas lubrication theory and rarefied gas dynamics, combined with boundary slip and gas-solid-thermal coupling analysis technology, the local deformation of aerostatic bearing gas film caused by temperature change and the effect on gas-vortex and micro-vibration are studied. The results show that the maximum deformation of the gas-film increases linearly with the increase of temperature. The move speed of gas-vortex motion considering thermal coupling is much less than that obtained by traditional numerical analysis method. Considering the gas-solid-thermal coupling condition, the influence of different working temperatures on the micro-vibration of aerostatic bearing is also different. According to the experiments available, with the increase of temperature, the micro-vibration intensity decreased gradually, but the temperature does not change its natural frequency.

1. Introduction

Aerostatic bearings with high precision, low friction power consumption and non-polluting advantages, are widely used in high-precision testing equipment, ultra-precision machining machine tools and aircraft flight testing and other fields [1]. Along with the changing working environment, all kinds of precision instruments are more and more demanding on the reliability and working stability of bearing, not only to ensure the normal work in the range of (-250~315) ℃, but also to ensure the stability of the spindle during the rotation and the bearing seat, and to relieve the impact and vibration between them. On the macroscopic scale, based on the isothermal assumption, ignoring the temperature change can satisfy the work requirement. However, with the decreasing of the gas film thickness and the continuous improvement of the precision, the flow form of the gas film in the micro-scale will present the transition from the layer to the slip flow [2].

Domestic and foreign scholars have analyzed the dynamic and static characteristics of aerostatic bearing. CHEN, et al [3] and KONG, et al [4] analyzed the micro-vibration characteristics of the aerostatic bearing, observed the gas-vortex phenomenon and put forward a numerical analysis method of gas-solid coupling. LI, et al [5] based on the large eddy simulation of aerostatic bearing self-excited micro-vibration characteristics of the analysis, observed the generation of gas-vortex. Although the existing mechanical models have made many assumptions about the flow characteristics of gases in

Content from this work may be used under the terms of the Creative Commons Attribution 3.0 licence. Any further distribution of this work must maintain attribution to the author(s) and the title of the work, journal citation and DOI.
Published under licence by IOP Publishing Ltd

tiny gaps and throttles, the influence of temperature on their internal microscopic properties is ignored. This will not only affect the shape and thickness of the lubricating gas film, but also further change the flow characteristics and amplitude frequency characteristics of the gas film through the way of thermal coupling, which will lead to a larger deviation between the numerical simulation results and the experimental data. Therefore, it is necessary to study the effect of the gas-solid-thermal coupling on the temperature sensitivity and the micro-vibration on the micro-scale gas film. The design calculation method of the aerostatic bearing is effectively supplemented, which provides technical support for the development of the ultra - precision shaft system and related engineering application.

2. Working principle and mathematical model

In this paper, the disc type central gas supply small hole throttle static thrust bearing with a diameter of D=50 mm is selected as the research object. Bearing structure schematic diagram as shown in Fig. 1 solid line outline, gas hole diameter d0=0.2 mm, gas cavity depth δ=0.2 mm, gas cavity diameter d1=3 mm.

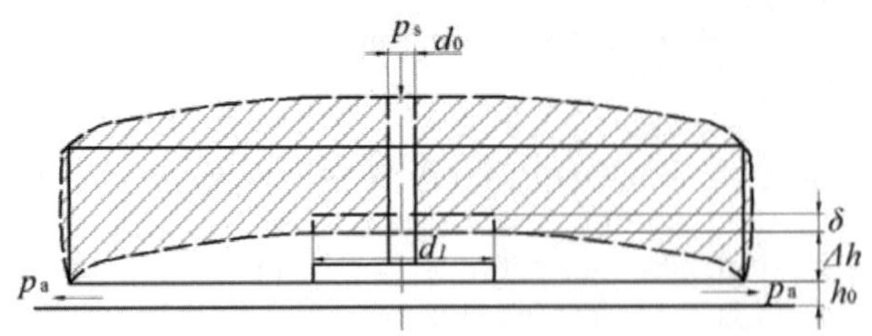

Ps - the gas pressure of the throttle hole, Δh - the maximum deformation of the gas film, h0 - the average thickness of gas film, Pa - atmospheric pressure

Fig. 1 Gas-solid-thermal coupling model

Under the condition of normal temperature, the effect of thermal coupling and velocity slip is not considered. Based on the principle of aerostatic lubrication, the motion law of the flow field in the gas film can be expressed as Navier-Stokes (N-S) equation.

The momentum equation in the X direction:

$$\frac{\partial p}{\partial x} = \mu \frac{\partial^2 u}{\partial y^2} \tag{1}$$

The momentum equation in the Z direction:

$$\frac{\partial p}{\partial z} = 0 \tag{2}$$

The momentum equation in the Y direction:

$$\frac{\partial p}{\partial y} = \mu \frac{\partial^2 w}{\partial y^2} \tag{3}$$

Gas State equation:

$$p = \rho RT \tag{4}$$

Continuity equation:

$$\frac{\partial(pu)}{\partial x} + \frac{\partial(pv)}{\partial y} + \frac{\partial(pw)}{\partial z} + \frac{\partial \rho}{\partial t} = 0 \tag{5}$$

In the form: ρ - gas density, μ - dynamic viscosity of gas, p - gas pressure, u, v, w - velocity component in x, y, z direction, T - temperature.

However, when the gas film thickness of the aerostatic bearing is in the micron-scale or even below, due to the small scale of the fluid characteristic, the Kn number will increase, which makes the gas at the micro scale have the same characteristics as the rarefied gas. The flow pattern of supporting gas film is in the transition stage of slip flow or laminar flow slip flow. The traditional N-S equation will not well reflect the characteristics of boundary slip flow and micro vibration. No slip boundary conditions will no longer apply to micro-scale flow analysis. Based on the influence of the boundary slip on the [6] and the gas stratification theory [7], the velocity slip boundary conditions need to be supplemented at this time:

$$z = 0\text{时}, \quad u = U + l'\frac{\partial u}{\partial z} \quad v = l'\frac{\partial w}{\partial z} \quad w = 0 \tag{6}$$

$$z = h\text{时}, \quad u = -l'\frac{\partial u}{\partial z} \quad v = -l'\frac{\partial w}{\partial z} \quad w = u\frac{\partial h}{\partial x} + v\frac{\partial h}{\partial y} \tag{7}$$

The above formula (1), (2), (3) is brought into the formula (6), (7) to solve the u, v, w and into the formula (4), (5) the micro-scale of the Reynolds equation can be simplified as:

$$\frac{\partial}{\partial x}\left(ph^3\frac{\partial p}{\partial x}\right) + \frac{\partial}{\partial y}\left(ph^3\frac{\partial p}{\partial y}\right) = \frac{6\mu U}{1+6k_n'}\frac{\partial(ph)}{\partial x} + \frac{12\mu}{1+6k_n'}\frac{\partial(ph)}{\partial t}, \quad k_n' = \frac{2-\sigma_v}{\sigma_v} \tag{8}$$

In the form: σ_v - molecular tangential momentum adjustment coefficient.

In actual work, with the increase of ambient temperature and bearing local temperature, the uneven thermal deformation of the bearing in height direction and the expansion of gas inside the gas film lead to the change of the flow field inside the flat supporting gas film, as shown in Fig. 1, the outline of the dashed line. At this time, the gas film is deformed in the vertical height direction, and the deformation amount is Δh. The deformation of the edge of the gas cavity is the largest. It is recorded as Δhmax and is nonlinear decreasing to the radius direction. The change of gas film thickness can be expressed by the formula as follows.

$$h = \begin{Bmatrix} h_0 + \Delta h + \delta_i \delta, r = \left(0, \frac{d_1}{2}\right) \\ h_0 + \Delta h, r > d_1 \end{Bmatrix}, \quad \delta_i = \begin{Bmatrix} 1, \sqrt{x^2+y^2} < \frac{d_1}{2} \\ 0, \frac{d_1}{2} < \sqrt{x^2+y^2} < \frac{D}{2} \end{Bmatrix} \tag{9}$$

In the form: P - the pressure of gas supply, h - the effective gas film thickness, h0 - the average thickness of gas film, δi - Kronecker delta, η - the viscosity of gas after pressurization, u, v - the velocity of x, y in the gas film, respectively, r- distance along the flow direction.

The formula (9) is brought into the formula (8), and the formula of Reynolds equation after the gas film deformation is obtained.

$$\frac{\partial}{\partial x}\left(p(h_0+\Delta h)^3\frac{\partial p}{\partial x}\right) + \frac{\partial}{\partial y}\left(p(h_0+\Delta h)^3\frac{\partial p}{\partial y}\right) = \frac{6\mu U}{1+6k_n'}\frac{\partial\left(p(h_0+\Delta h)\right)}{\partial x} + \frac{12\mu}{1+6k_n'}\frac{\partial\left(p(h_0+\Delta h)\right)}{\partial t} \tag{10}$$

3. Temperature sensitivity and gas-vortex phenomenon of gas film

In this paper, a two-way gas-solid-thermal coupling model of aerostatic bearing is established. The Fluent and Static Structural modules in Ansys Workbench are used to initialize the fluid domain and solid domain respectively. The setting mainly includes the boundary condition, the gas-solid-thermal coupling surface, the solution model and the monitoring setting, and the convective temperature should be set in the boundary condition. It needs special description that in the Static Structural module, the three-dimensional flow field in the micro gas film gap of the Fluent module can be coupled to the aerostatic bearing in the solid domain, and then the maximum deformation of the aerostatic bearing under certain working conditions can be obtained. The large eddy simulation (LES) should be used in the analysis of the fluid domain model, because it can better calculate the flow characteristics of gas in the gas film. The grid distortion is 0.05, which satisfies the calculation precision. The bearing material is aluminum alloy.

The maximum deformation Δhmax of the gas film height direction is obtained, and the variation rule shown in Fig. 2 is obtained.

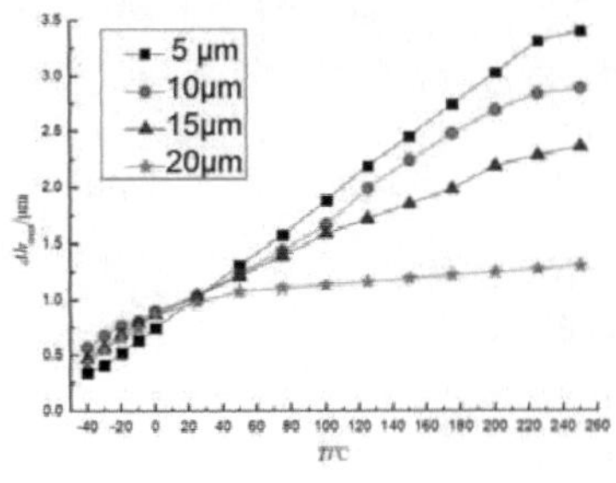

Fig. 2 The maximum deformation Δh_{max} of different gas film thickness under ps=0.4MPa

As shown in Fig. 2, as the temperature increases, the gas film maximum deformation overall increases, but the increasing amplitude decreases with the increase of the gas film thickness. Before 25℃, the maximum deformation of 5μm gas film is the smallest, and the maximum deformation of (10~20) μm gas film is very similarity. After 25℃, with the increase of gas film thickness, the maximum deformation of 5μm gas film rapidly increased, and the maximum deformation of (10~20)μm gas film began to widen. The bigger the gas film thickness, the smaller the maximum deformation is causes. Especially at 25℃, when the gas film thickness is (5~20) μm, the maximum deformation Δhmax of the gas film is maintained at about 1μm.

It is necessary to point out that when the gas film thickness is 20μm, the deformation amount is between (1~1.5) μm with the increase of temperature. This is mainly because the gas film is in a fully developed laminar state when the gas film thickness is 20μm or above. The sensitivity of the gas film to temperature is reduced. The internal flow can be described by the traditional N-S equation. At that time, the sensitivity of the gas film to temperature is very low. Therefore, the maximum deformation Δhmax is lower than that of other gas film thickness. In contrast, when the gas film thickness is reduced to 5μm or below, by the gas stratification theory, slip zone proportion increases, resulting in the N-S equation and the laminar flow hypothesis is not the traditional well reflect the internal flow characteristics. So it is necessary to use the thin layer flow velocity slip conditions, the gas film in large deformation in the range of temperature increase linearly. The flow state and pressure distribution of the gas film shows the work inside will have a temperature sensitive characteristic strongly, the flow characteristics and pressure distribution in the gas film will change.

From the above analysis, it can be seen that for the aerostatic bearing with micro scale gas film, the change of temperature will cause the gas film to deform in the direction of height, resulting in the change of the gas flow characteristics. Thus, the change of the position and size of the gas-vortex in the gas cavity will be affected, and then the micro-vibration will be affected. On this basis, the gas flow characteristics of the gas film are studied.

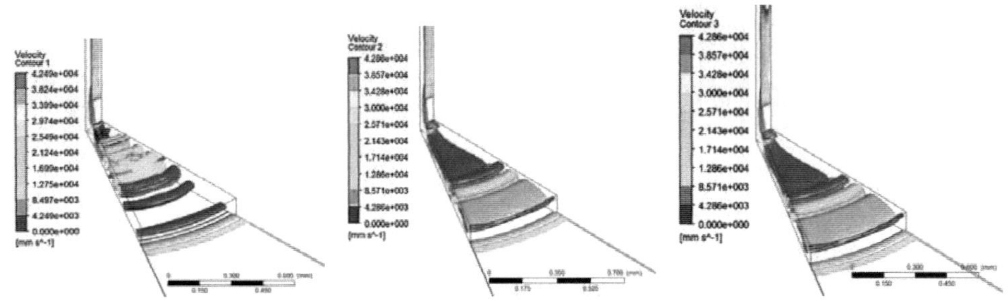

a) 16 ms b) 32 ms c) 48 ms

Fig. 3 Vortex and gas-vortex location at different time

Under ps=0.4MPa,h=10μm, T=-40 ℃ conditions, the turbulent vortex structure near the orifice exit presents complex three-dimensional characteristics, the scale of the vortex structure and the process of motion development become irregular, as shown in Fig. 3-a), Fig. 3-b), Fig. 3-c). The location, size and intensity of the gas-vortex are not fixed, but with the passage of time, the position of the gas-vortex gradually moves to the edge of the gas cavity. This is mainly due to the impact of high velocity gas flow into the gas cavity. The gas streamline is sharply bent. Due to the limited space and the large gas kinetic energy, the gas-vortex energy is very high and the momentum is large. Gradually forming an annular large scale gas-vortex. Then, in the wall of the gas-vortex appeared under the influence of continuous fluctuation, the gas-vortex scale increases gradually and moves in the radial direction. After a certain stage of development, because the flow cross-sectional area of the larger, quickly rupture under the action of convection, gas-vortex is divided into a large number of irregular small scale vortex. Then the velocity of gas-vortex decreases gradually, and finally dissipates gradually under the action of gas viscosity.

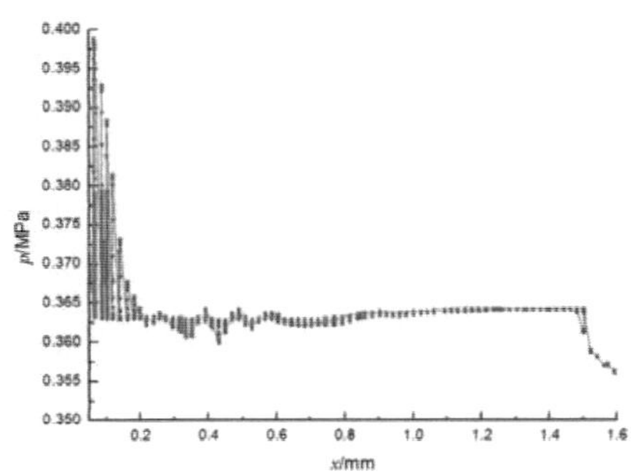

Fig. 4 Pressure distribution in gas cavity under temperature coupling condition

As shown in Fig. 4, under ps=0.4MPa, h=10μm, T=-40℃ conditions, the gas film pressure of the aerostatic bearing in the gas cavity and near the throttle hole shows an alternating upward and downward trend, resulting in the alternating variation of wave peaks and troughs. This results in micro-vibration of the aerostatic bearing. In addition, when observing the location of gas-vortex from the center of the gas cavity to the radius of 0.64 mm, the intensity of gas-vortex decreases gradually and the amplitude fluctuation is small, which proves that the location, size and intensity of the gas-vortex are not fixed. Based on this, the comparison between ideal condition and gas-solid-thermal coupling condition is made at different times, and the intensity of micro-vibration is analyzed by comparing the position of gas-vortex and the size of pressure drop in gas cavity. It will provide theoretical support and technical help for effectively suppressing or reducing micro-vibration.

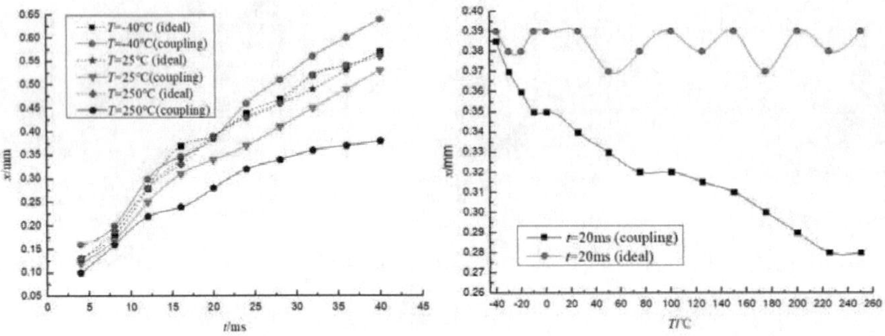

a) The gas-vortex position at different time b) The gas-vortex position at different temperatures

Fig. 5 The gas-vortex position in gas cavity under ideal and gas-solid-thermal coupling conditions

As shown in Fig. 5-a), under ps=0.4MPa, h=10μm conditions, T=-40℃, 25℃ and 250℃ were chosen as the research objects according to the different operating temperature of the aerostatic bearing. Under the ideal condition, the position of the gas-vortex is gradually spread to the periphery along the radius direction of the gas cavity with time. Under different temperature conditions, the position of the gas-vortex is approximately the same. This is mainly due to the ideal condition of no damping and viscidity. Without taking into account the error factors, the gas flow will become unhindered, and the velocity of gas-vortex will be similar. However, when the gas-solid-thermal coupling is applied, the position of the gas-vortex still moves outward along the radius of the gas cavity, but with the increase of temperature, the movement speed of the gas-vortex gradually decreases. Combined with Fig. 5-b), under the ideal conditions of t=20 ms, the position of the gas-vortex in the gas cavity of the aerostatic bearing is (0.37~0.39) mm. This may be due to the fact that, with the increase of temperature, the thermal motion of the gas molecules intensifies, the collisions between molecules become more frequent, and the gas in the gas film tends to move irregularly. Gradually counteracting with the direction of gas-vortex rotation, the gas viscosity gradually increases, which results in the decrease of gas-vortex velocity.

From the above analysis, it is known that the increase of temperature will cause the reduction of the velocity of the gas-vortex. Based on this, the size of the pressure drop in the gas cavity at different temperatures is studied, as shown in Fig. 6.

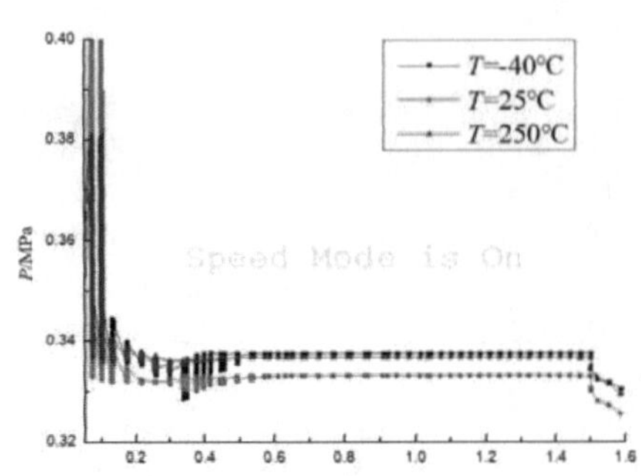

Fig. 6 Pressure distribution in gas cavity under different temperature conditions

As shown in Fig. 6, under the condition of ps=0.4MPa, h=10μm, t=16ms, There is a phenomenon of pressure drop in the gas film inside the gas cavity. When T =-40℃, the moving velocity of the gas-vortex in the gas cavity is fast , and the pressure drop is large. However, at T=25℃, the moving velocity of the gas-vortex in the gas cavity changed little, but the pressure drop in the pressure distribution was lower than that at T=-40℃. At T=250℃, the moving velocity and the pressure drop of

IOP Conf. Series: Journal of Physics: Conf. Series **1064** (2018) 012035 doi:10.1088/1742-6596/1064/1/012035

the gas-vortex in the gas cavity are decreased greatly. This is mainly due to the increase of temperature leading to the increase of the distance between the gas molecules and the gradual decrease of the interaction force. But the energy of the gas increases, the gas molecules are more inclined to do irregular motion and gradually replace the trajectories of the gas-vortex, thus reducing the velocity and the micro-vibration of the gas-vortex.

4. Design and Test Scheme

The phenomenon of gas-vortex in the gas film causes micro-vibration. An experimental set-up is constructed which makes it possible to measure the micro-vibration, the schematic drawing and test set-up are shown in Fig. 7 respectively.

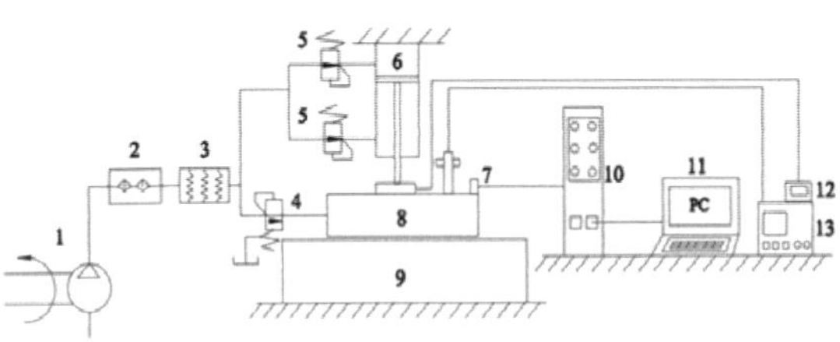

a) Micro-vibration test schematic diagram b) Micro-vibration test equipment

1. Air source 2. Air source filtrating equipment 3. Space heater 4. SMC precision pressure regulating valve 5. Pressure regulating valve 6. Cylinder 7. PCB acceleration sensor 8. Aerostatic bearing 9. Marble platform 10. LMS dynamic monitor 11.Personal computer 12. Pressure transducer 13. Micro-displacement measuring instrument

Fig. 7 Micro-vibration test

The temperature is measured by the standard glass mercury thermometer in the experiment, the measuring range is (-30~300)℃, with the zero scale. Using dry ice to make low temperature environment. All the measurement procedures are instantaneous values.

5. Results and Analysis

Based on the above experiments, the effect of gas-solid-heat coupling on the micro-vibration of aerostatic bearing is analyzed. As is shown in Fig. 8 respectively.

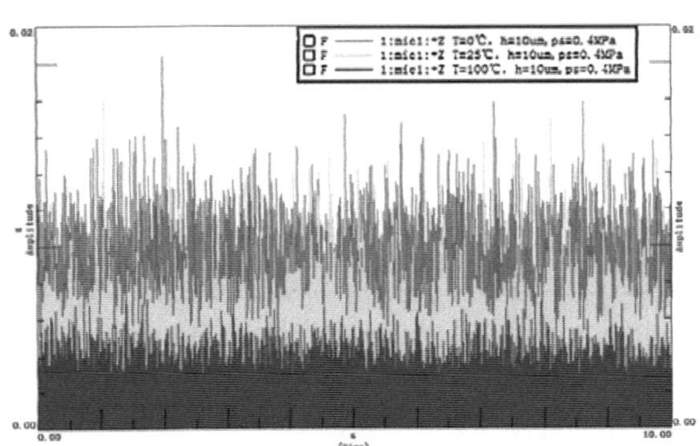

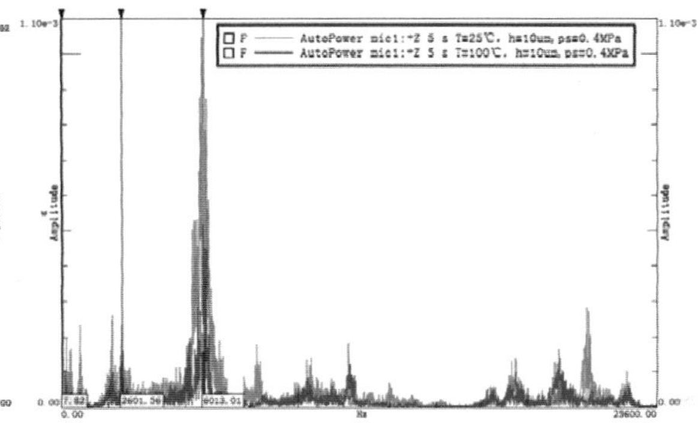

a) Time-domain signals b) Frequency-domain signals

Fig. 8 Time-domain signals of aerostatic bearings at different temperatures

As shown in Fig. 8-a), under the condition of h=10μm, ps=0.4MPa, the influence of temperature on the micro - vibration of the aerostatic bearing is analyzed when the working temperature is T=0℃, T=25℃ and T=100℃ respectively. It is obvious that with the increase of temperature, the micro-vibration amplitude of the aerostatic bearing is the highest at T=0℃, which is about (1.1~2) times the thickness of the initial gas film. As the temperature rises to T=100℃, the micro-vibration amplitude of the aerostatic bearing is the lowest, the amplitude range is (4~7)μm fluctuation, it is (0.4~0.7) times of the initial gas film thickness. As a result, the micro-vibration amplitude of the aerostatic bearing decreases with increasing temperature, so that the micro-vibration intensity can be reduced as the temperature increases. As shown in Fig. 8-b), the temperature increase does not alter its natural frequency. In addition, it can be seen from the figure that there is a large peak at frequencies of 2601Hz and 6015Hz. This is mainly caused by the system resonance due to the load of the aerostatic bearing and the influence of the environmental factors.

6. Conclusion

By comparing the above numerical analysis and experimental results, we can draw the conclusion that when the working temperature of aerostatic bearing increases gradually, the gas-vortex will gradually weaken, and the flow characteristics of gas film will be close to laminar flow. However, when the operating temperature is below a certain critical value, the flow characteristics of the gas film are closer to the turbulent, and the intensity of the gas-vortex increases gradually, which leads to the increase of the micro-vibration intensity. Mainly reflected in the gas vortex volume and time domain signal fluctuation amplitude value. Therefore, it is necessary to select suitable working parameters according to different working conditions in order to weaken the micro-vibration, and to avoid the occurrence of resonance phenomenon according to the frequency-domain signal graph.

References

[1] ZHU Jincheng. Dynamic Characteristics and Nano vibration of Aerostatic Bearings [D]. Wu Han: Huazhong University of Science and Technology, 2014.
[2] ZHANG Haijun, ZHU Changsheng, Yangqin. Analysis of Rarefaction Effect on Steady Characteristic of Micro-gas Lubrication Journal Bearings [J]. Journal of Vibration and Shock, 2008, 27(S): 19-22.
[3] CHEN Xuedong, Bao Xiulan, HE Xueming. Vibration analysis of ultra-precision stage with air-bearing [J]. Journal of Huazhong University of Science and Technology (Nature Science Edition), 2007, 35(11): 10-13.
[4] KONG Zhongke, TAO Jizhong. Study on internal cyclone phenomenon in pressure chamber of aerostatic bearing[C] National Symposium on Gas lubrication and dry gas seals. 2012.
[5] LI Yuntang, LIN Yingxiao, ZHU Hongxia et al. Analysis of the Micro Self-vibration of Aerostatic Thrust Bearing Based on Large Eddy Simulation[J]。 Journal of mechanical engineering, 2013, 49(13): 56-62.
[6] FAN Jing. Rarefied Gas Dynamics: Progress and Application [J]. Advances In Mechanics, 2013, 43(2):185-201.
[7] NING Fangwei, LONG Wei, LIU Yan. Analysis of Gas Flow Mechanism and Pressure Characteristics between Plates in Micro Scale [J]. China Mechanical Engineering, 2016, 27(14):1862-1865.

IOP Conf. Series: Journal of Physics: Conf. Series **1064** (2018) 012036 doi:10.1088/1742-6596/1064/1/012036

Evolution of the driving forces during convective drying of carrot slices

Xianxi Liu*, Xingbo Wang, Zhiyu Xu and Tao Zou

Faculty of Mechanical and Electrical Engineering, Kunming University of Science and Technology, Kunming 650500, China.

*Corresponding author e-mail: xxiliu@tom.com

Abstract. To explore the drying characteristics, a model combining the Fick's second law of diffusion with a convective boundary condition on the material surface is solved numerically. And the evolutions of the moisture gradient within the sample (internal driving force) and the relative humidity difference between the bulk air and the material surface (external driving force) with drying time are obtained. These obtained results show that the relative humidity difference remains its maximum for some period at the beginning of drying, and then falls progressively to zero for samples with high initial moisture content. However, it falls from its maximum to zero without constant period for samples with low initial moisture content. And the moisture gradient in the outer is always greater than that in the inner within the sample. Correspondingly, the moisture migration flux in the outer is always greater than that in the inner within the sample. These observations indicate that both the external and the internal driving force simultaneously control the whole drying process, and the transferring rate of internal moisture cannot be equal to or higher than that of surface moisture. Thus, the magnitude of drying rate is directly dominated by the external driving force.

1. Introduction

Moisture transfer in porous materials during the drying process is usually considered to take place in two successive steps: (1) from the interior to the surface and (2) evaporation from the surface to the air. It is said that the drying process occurred at a constant drying rate period if the transferring rate of internal moisture was equal to or higher than that of surface moisture. And the drying process was at falling drying rate period if the internal moisture transferring rate was lower than that of surface moisture [1]. As proposed by Debase et al. [2], a two-stage drying process model proposed was based on two hypotheses: (1) the resistance to the solid-gas mass transfer of water was assumed to be located near the external surface of the solid particles, in the gas phase and the solid water content was assumed uniform during the constant rate period; (2) in the falling rate period, an internal resistance to mass transfer, corresponding to the moisture diffusion in the solid particle, must be taken into account. Up to now, there have been many researches on both mathematical modeling and experimental studies about the drying behavior of various materials based on the view of moisture diffusion resistances of the external and internal samples, such as wheat [3], cheese [4], and fruit pectin gels [5], mate leaves [6] and paper [7]. However, the studies on the hot-air drying behaviors of materials from the point of

Content from this work may be used under the terms of the Creative Commons Attribution 3.0 licence. Any further distribution of this work must maintain attribution to the author(s) and the title of the work, journal citation and DOI.
Published under licence by IOP Publishing Ltd

driving forces are scarce in the literature. In this work, such an effort was made: the evolution of the driving forces of moisture diffusion, i.e. the internal moisture gradient and the humidity difference between the surface and the bulk air during the drying process were estimated and the drying characteristics of high initial moisture content and low initial moisture content product in view of the drying driving forces was also further analyzed.

2. Mass transfer model

One-dimensional transient mass transfer for an infinite slab and for a homogeneous product with constant moisture diffusivity, and a boundary condition considering convective mass transfer, can be described by the following equations:

$$\frac{\partial m}{\partial t} = D\frac{\partial^2 m}{\partial x^2} \text{ For } 0 < x < L;\ t > 0 \quad (1)$$

Where m is the moisture content at time t at point x, and D stands for the diffusion coefficient of water through the carrot samples.

The following initial condition is accepted:

$$m(x,0) = m_0 \text{ For } 0 \le x \le L;\ t = 0 \quad (2)$$

Which means that the moisture content in the samples before drying is uniform and equal to m0.

The mass flux of water from the carrot slice surface [8], J, is proportional to the external driving force, i.e., the difference between the humidity in the bulk air, yam, and the gas humidity just above the surface of the carrot slice, yin.

$$J = h_m(y_{in} - y_a)\rho_a \text{ For } x = 0; t > 0 \quad (3)$$

The humidity of bulk air, yam, can be calculated from the relation between absolute and relative humidity [9]. In addition, the surface humidity, yin, is a function of the local moisture content at the surface of carrot slice, mms, proposed by Aversa et al [10].

Food bottom:

$$\frac{\partial m}{\partial x} = 0 \text{ For } x = L; t > 0 \quad (4)$$

The local moisture content change obeys the conventional diffusion equation (1) with initial condition (2) and boundary conditions (3) and (4). This initial boundary problem (2), (3), (4) and (1) are solved numerically by the finite difference method with an explicit algorithm. The sample is divided into 10 subintervals by points' xi= (i-1/2) Ax (i=1, 2, 3, ⋯, 9) with Ax=L/10 along the thickness; introduce discrete time TN=knit; and replace equations (1), (2), (3), and (4) by their finite difference approximation:

$$m(t_{n+1}, x_i) = m(t_n, x_i) + D\frac{\Delta t}{\Delta x^2}\left[m(t_n, x_{i+1}) - 2m(t_n, x_i) + m(t_n, x_{i-1})\right] \quad \text{i=2, 10} \quad (5)$$

$$m(0, x_i) = m_0 \quad \text{i =1, 11} \quad (6)$$

$$D\rho_s \frac{m(t_n, x_{i+1}) - m(t_n, x_i)}{\Delta x} = h_m(y_{in}(t_n) - y_a(t_n))\rho_a \qquad i = 1 \tag{7}$$

$$\frac{m(t_{n+1}, x_{10}) - m(t_n, x_i)}{\Delta t} = 2D\frac{m(t_n, x_{i-1}) - m(t_n, x_i)}{\Delta x} \qquad i = 11 \tag{8}$$

3. Results and discussion

Figure 1 compared the experimental data of moisture content vs. drying time with that predicted results by the proposed model with an initial moisture content of 9.21 kg water/kg dry matter [8]. Obviously, the theoretical values well superimposed on the experimental values, indicating that the proposed model based on a convective boundary condition on the material surface was adequate for predicting the drying process of carrot slice.

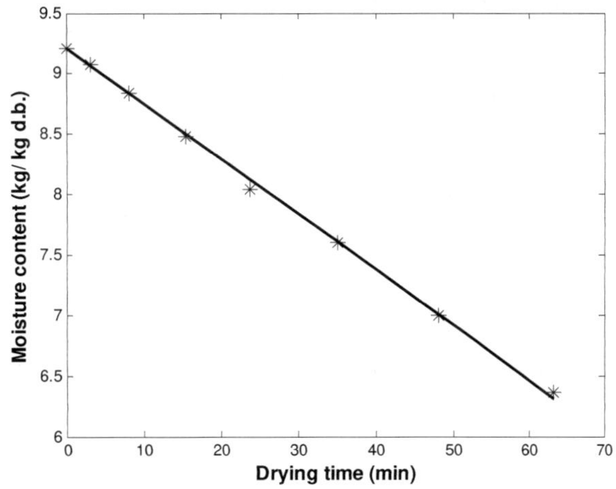

Figure 1. The plot of experimental (star) and calculated (line) moisture content as function of time for carrot slice natural convective drying with air temperature 60°C and relative humidity 15%.

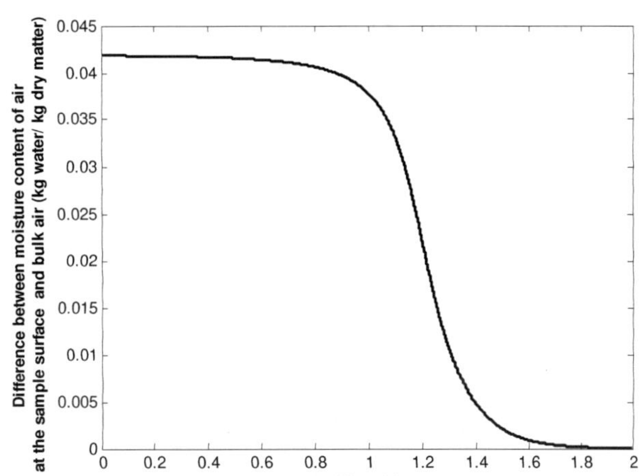

Figure 2. The evolution external driving forces of moisture diffusion for the carrot slices drying process simulation with m0=9.21 kg water/ kg dry matter, L=0.004m, r=0.007m, Ta=60°C, VA=0.269m/s, ψ=15%.

Figures 2 and 3 show the external and internal driving forces of moisture diffusion for the carrot slices drying process simulation with m0=9.21 kg water/ kg dry matter, L=0.004m, r=0.007m, Ta=60°C, van=0.269m/s, ψ=15%. As expected in the Figure 2 the external driving force i.e. difference between moisture content of air at the sample surface and bulk air was the maximum and almost constant for the material with high moisture content in the initial phase of drying. The reason is under a certain drying condition, the external driving force is only associated with the water activity of the sample surface. And if the local moisture content of the sample surface is equal to or greater than its critical moisture content, the water activity of the sample surface will remain constant and is equal to 1. With the water vapor removed away from the sample surface, when the local moisture content reached its critical moisture content, the water activity would decrease with moisture content of sample surface. Correspondingly, the external driving force decreased with increase in drying time as shown in Figure 2. These observation are in agreement with convective drying of saturated porous media (clay samples) [12, 13].

From Figure 3, the initial internal driving force of moisture diffusion was null in the sample. At the beginning of drying, the evaporation of liquid water at the surface resulted in a moisture gradient between the surface and its neighbour. This moisture gradient drove the moisture diffusing from the neighbour to the surface. In turn, the neighbour local moisture decreasing formed the gradient between the neighbour and the next neighbour, and the moisture gradient drove the next neighbour local moisture diffusing to the neighbour. In this manner, the internal driving forces were formed progressively in the whole sample, the moisture diffused from the inner to the surface and with these driving forces. It is worth noting that the driving forces increased from zero to their maximum, and remained a period, then decreased and approached to zero with increase in drying time. And the inner driving force was less than the outer during the whole drying process. These observations gave the dynamic nature of internal driving forces during mass transport in porous materials with initially high moisture content. In addition, these observations provided validation to driving force rather than resistance on the change of drying rate.

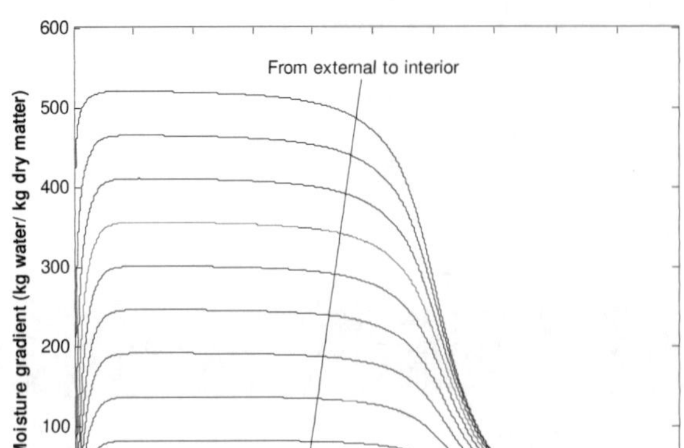

Figure 3. The evolution of the internal driving forces of moisture diffusion for the carrot slices drying process simulation with m0=9.21 kg water/ kg dry matter, L=0.004m, r=0.007m, Ta=60°C, van=0.269m/s, ψ=15%.

The external and internal driving forces as function of drying time for carrot slice with m0=2.21 kg water/ kg dry matter, L=0.004m, r=0.007m, Ta=60°C, van=0.269m/s, ψ=15%, were shown in Figures 4 and 5. The driving forces appeared significantly different from the driving force curve with highly initial moisture content (Figures 2 and 3). Upon examining the results shown in Figures 4 and 5, it was immediately apparent that the external driving force decreased with drying time. The internal driving force increased with drying time firstly, and then it fell progressively. Unlike the highly initial moisture content, the transition phase of increasing to falling for the internal driving force was sharp.

For all initial moisture contents, the flux of water migrating from the inner to the outer neighbour were less than the flux of water migrating from the outer to the outer adjacent, and the rate of moisture migration to the surface of an carrot slice was less than that of moisture diffusion from the particle surface to the bulk air. Contrary to the view of Dev let et al. [1] and Perez-Alonso et al. [14], Drying rate only depended on the relative humidity difference between the bulk air and the sample surface. If the local moisture content of the sample surface was equal or greater than its critical moisture content, the water activity of sample surface will remained constant and was equal to 1. Corresponding to the drying characteristic diagram, this phase was the constant drying rate period. When the local moisture content of the sample surface was less than the critical moisture content, the water activity of the sample surface will decrease with the local moisture content of the sample surface. And the drying process was in the falling rate period. With the analysis of the driving force of the moisture diffusion, it was apparent that the whole drying processes were simultaneously controlled by the external and internal driving forces. And the magnitude of the drying rate was dominated by the external drying force.

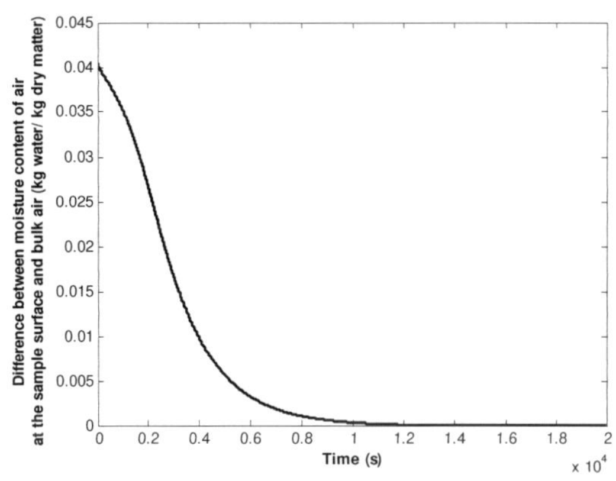

Figure 4. The variations of external driving force as function of drying time for carrot slice with m0=2.21 kg water/ kg dry matter, L=0.004m, r=0.007m, Ta=60°C, van=0.269m/s.

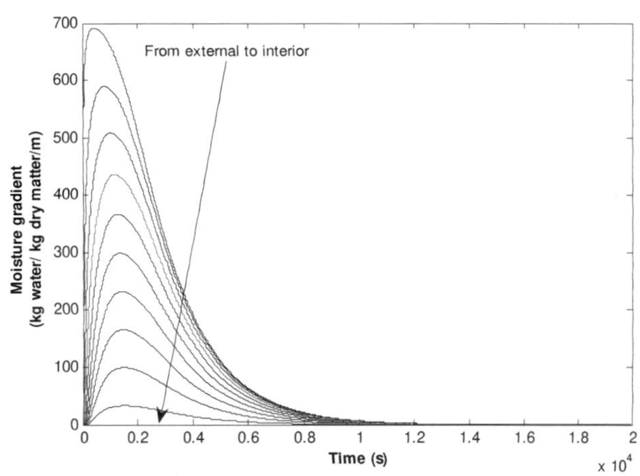

Figure 5. The variations of internal driving force as function of drying time for carrot slice with m0=2.21 kg water/ kg dry matter, L=0.004m, r=0.007m, Ta=60°C, van=0.269m/s.

4. Conclusion

A model combining the Fick's second law of diffusion with a convective boundary condition on the material surface was solved numerically and compared with experimental data. The obtained evolutions of the moisture gradient within the sample (internal driving force) and the relative humidity difference between the bulk air and the material surface (external driving force) indicated that both the external and the internal driving force simultaneously controlled the whole drying process. And the magnitude of drying rate was directly dominated by the external driving force.

Acknowledgments

This work was financially supported by the fund of National Science Foundation of China (No.51566006).

References

[1] Devlet Demirel; Mahir Turhan. Air-drying behavior of Dwarf Cavendish and Gros Michel banana slices. Journal of Food Engineering 2003, 59(1), 1-11.
[2] Debaste F.; Halloin V.; Bossart L.; Haut B. A new modeling approach for the prediction of yeast drying rates in fluidized beds. Journal of Food Engineering 2008, 84(2), 335-347.
[3] Kuts P. S.; Lyuboshits A. I.; Nikolaichik L. V.; Korateev I. G. Relative diffusion resistance in drying wheat. Inzhenerno-Fizicheskii Zhurnal 1976, 31(5), 839-843.
[4] Simal S.; Sánchez E. S.; Bon J.; Femenia A.; Rosselló C. Water and salt diffusion drying cheese ripening: effect of the external and internal resistances to mass transfer. Journal of Food Engineering 2001, 48(3), 269-275.
[5] Sergio A. Giner. Influence of internal and external resistances of mass transfer on the constant drying rate period in high-moisture foods. Biosystem Engineering 2009, 102(1), 90-94.
[6] Everton Fernando Zanoelo. A theoretical and experimental study of simultaneous heat and mass transport resistances in a shallow fluidized bed dryer of mate leaves. Chemical Engineering and Processing: Process Intensification 2007, 46(12), 1365-1375.
[7] Lars Nillson. Heat and mass transfer in multicylinder drying: partⅡ. analysis of internal and external transport resistances. Chemical Engineering and Processing: Process Intensification 2004, 43(12), 1555-1560.
[8] Cabardo S. J.; Langrish T. A. G.; Dickson R.; Joe B. Variability in transport properties regarding drying behavior for blackbutt timber in new south wales. Drying Technology 2006, 24(2), 211-224.
[9] Mujumdar A. S. Handbook of Industrial Drying; Taylor and Francis Group, LLC: New York, 2006, 7.
[10] Aversa M.; Curcio S.; Calabrò V.; Iorio G. An analysis of the transport phenomena occurring during food drying process. Journal of Food Engineering 2007, 78(3), 922-932.
[11] Liu X.; Chen J.; Hou H. Theoretical analysis of water diffusivity estimated by Crank's equation. Chemical Engineering and Processing: Process Intensification 2012, 55(5), 24-28.
[12] Mihoubi D.; Bellagi A. Drying-induced stresses during convective and combined microwave and convective drying of saturated porous media. Drying Technology 2009, 27(7/8), 851-856.
[13] Kowalski S. J.; Rajewska K. Convective drying enhanced with microwave and infrared radiation. Drying Technology 2009, 27(7/8), 878-887.
[14] Pérez-Alonso C.; Cruz-Olivares J.; Ramírez A.; Román-Guerrero A.; Vernon-Carter E. J. Moisture diffusion in allspice (Pimenta dioica L. merril) fruits during fluidized bed drying. Journal of Food Processing and Preservation 2011, 35(3), 308-312.

The physicochemical mechanism and mathematical model of the pseudo-four-phase flow in the movable gel system

Jun Wang[1], Wenchen Liu[1], Jian Zhou[2, 3], Jia Wang[2, 3], Tongjing Liu[2, *] and Rui Liu[2]

[1]SINOPEC Research Institute of Petroleum Engineering, Beijing, China.
[2]Enhanced Oil Recovery Institute, China University of Petroleum, Beijing, China.
[3]Beijing Jinshiliyuan Technology Co. Ltd., Beijing, China.

*Corresponding author e-mail: ltjcup@cup.edu.cn

Abstract. The movable gel system (MGS) has the dual function of "regulation" and "flooding", which has been carried out in a large number of pilot test and industrial application. Laboratory experiments and field test results show that approximate solid particles exist in MGS which is different from the aqueous phase and have the seepage characteristics of "blocking but never shutting off ", the micro reason is the presence of movable gel transform into gel particles. Therefore, this paper divided MGS into two parts of movable gel and gel particles, and we established mathematical model of the pseudo-four-phase flow for MGS, built relatively comprehensive characterization methods for MGS; the gel particles are regarded as separate phase which differ from the oil, gas and water phase, we proposed the pseudo-four-phase flow model and the method for the treatment of the relative permeability of the pseudo-four-phase flow. The new pseudo-four-phase seepage mathematical models is closer to the actual field seepage characteristics and improve the accuracy of the results for MGS simulation, this is an important work of scientific guidance for the industrial promotion of MGS.

1. Introduction

Movable gel profile control system includes the advantages of near well profile control and polymer flooding [1, 2]. The system is composed of polymer and crosslinking agent in a certain ratio, also known as CDG, weak gel, crosslinking polymer solution, deep profile control agent and so on [3]. It performs both highly viscous of the gel and the fluidity of polymer solution, has a very good deformation ability and pore plugging effect [4].

Major oil fields in China have conducted pilot tests and industrial application study of MGS and have accumulated lots of laboratory data and field implementation data. Laboratory studies have proved the phenomenon that movable gel is able to transform into gel particles. Consequently, it contains approximate solid particles, of which the percolation characteristics are different from water phase. But this significant distinction has not yet been applied to the characteristic equation or numerical simulation method. Therefore, current "oil-gas-water" three phase flow theory is unable to reflect the characteristic of MGS [5-10]. The simulation mechanisms and results of the system based on the old theory will not fit the field application.

Content from this work may be used under the terms of the Creative Commons Attribution 3.0 licence. Any further distribution of this work must maintain attribution to the author(s) and the title of the work, journal citation and DOI.
Published under licence by IOP Publishing Ltd

To fill the research gap mentioned in last paragraph, this paper considers MGS as two parts of movable gel and gel particles based on microscopic experiment results in Section 2. In Section 3, this paper also gives out a computation method for relative permeability of pseudo-four-phase flow. In Section 3, a numerical model is established to fit the core displacement experimental data to validate the whole method.

2. Microscopic characteristic of MGS

The field application has verified the characteristic of "blocking but never shutting off" of MGS. Chinese scholars figure out by studying the micro mechanism from experimental data and field data, that movable gel system has the displacement mechanism of "migration-blocking- remigration" and the phase transition characteristic of "Amoeba" during flowing through porous media.

This paper verified former achievements by sand-packed experiment, shown in Figure 1. At the beginning of the displacement, movable gel system move through a short distance, the concentration of system is high, contains large gel particles; in the middle of the term, the concentration decreases, and the number of particles gets lower, also the size of particles become smaller; in the later period, the concentration continuously gets down, and there are fewer particles; in the end of the displacement, particles no longer exist, the concentration of the system is at a particular low level.

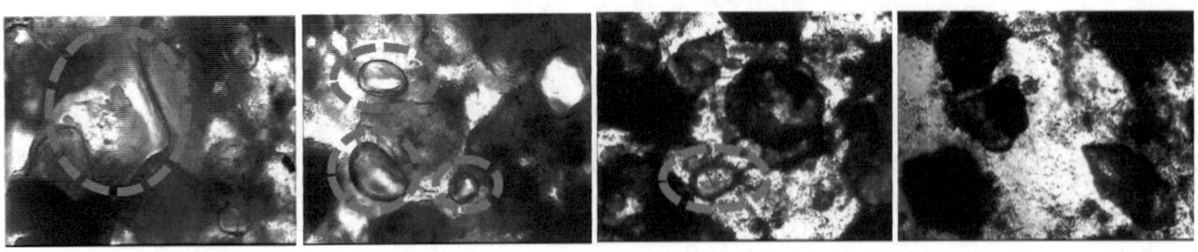

(a) (b) (c) (d)

(a) Sampling from the entry point of 0.2 m. (b) Sampling from then entry point of 0.8 m.
(c) Sampling from then entry point of 1.6 m. (d) Sampling from then entry point of 2.0 m.

Figure 1. Picture of gel particles from different section in the sand-packed pipe.

During the process of flowing through the porous media, the movable gel system deforms and smashes into pieces naturally. It is substantially the process of movable gel transform into gel particles, and this process shows the characteristic of "Amoeba". The displacement characteristic of "migration-blocking- remigration" is that movable gel and particles transform into each other due to the change of concentration. When concentration is high, it will transform into gel particles, the particles play the role of plugging the pores. When concentration is low, it will transform into movable gel and contribute to profile control. Therefore, in next section, we establish the corresponding equation to describe this phenomenon by chemical reaction and relative permeability.

3. Characterization methods of pseudo-four-phase flow

3.1. Chemical reactions

Types of polymer crosslinking reaction system are numerous, sums up into three categories as: a) transition metal crosslinking system under the effects of carboxylic acid groups of the polymer molecules and transition metal ions, such as chromium, zirconium and aluminium, b) aldehyde crosslinking system under the effects of amide groups of polymer molecules and aldehyde material (mainly formaldehyde) , and another one kind, c) composite crosslinking system which combines both advantages of the former systems.

At present, only the crosslinking reaction kinetics model of transition metal crosslinking system is partly presented, models of the other two kinds of crosslinking system are still in the experiment stage.

Take transition metal ion chromium as an example, its chemical reaction equations describe as follows:

$$\begin{cases} nCr^{3+} + polymer \xrightarrow{K_1} mg & \left(C_{mg} \le C_{mg\max}\right) \\ nCr^{3+} + polymer \xrightarrow{K_2} Gel & \left(C_{mg} > C_{mg\max}\right) \\ mg \xrightarrow{K_3} Gel & \left(C_{mg} > C_{mg\max}\right) \end{cases} \tag{1}$$

According to the equation, we can conclude that after the crosslinking reaction, the MGS can be divided into movable gel and gel particles, which agreed with the result of microscopic displacement experiment in the porous media for MGS.

Thus the reaction kinetics characteristics equation of the generative process can be described as follows:

$$\begin{cases} R_{mg}^{+} = \dfrac{d[Gel]}{dt} = -\dfrac{1}{n}\dfrac{d\left[Cr^{3+}\right]}{dt} & \left(C_{mg} \le C_{mg\max}\right) \\ R_{mg}^{-} = \dfrac{d[mg]}{dt} = -K_3[mg]^{l} & \left(C_{mg} > C_{mg\max}\right) \\ R_{gel}^{+} = \dfrac{d[Gel]}{dt} = -\dfrac{1}{n}\dfrac{d\left[Cr^{3+}\right]}{dt} & \left(C_{mg} > C_{mg\max}\right) \end{cases} \tag{2}$$

Temperature, pH value, shear and ionic concentration can all lead to the degradation of movable gel. In addition, the process of reaction varies with time, makes the reduction of gel particles strength. The characterization equation for this process is:

$$R_{gel}^{-} = \frac{dS_{gel}}{dt} = -K_d \cdot S_{gel} \tag{3}$$

3.2. Relative permeability

Taking the relative flow of gel particles into account, we established the description of four phase relative permeability. The relative permeability curve of the oil phase and gaseous phase were still considered as three phase permeability, additionally added one curve of relative permeability of gel particles and amended the relative permeability curve of the water phase.

On the basis of relative permeability curves of oil-water and oil-gas by experiment data, we assumed that the permeability of wetting phase and non-wetting phase are only functions of saturation, then the relative permeability of water and gas in the four-phase seepage model can be directly read from the relative permeability curves of oil-water and oil-gas or be calculated of interpolation from the curves separately. And the relative permeability of the middle wetting phase oil phase can be calculated from the Stone formula:

$$K_{ro} = K_{rocw}\left[\left(\frac{K_{row}}{K_{rocw}} + K_{rw}\right)\left(\frac{K_{rog}}{K_{row}} + K_{rg}\right) - K_{rg} - K_{rw}\right] \tag{4}$$

The average pore-throat radius decreases when the wetting saturation increases. As the average pore-throat radius turns to be lower than the hydrodynamic radius of the movable gel (corresponds to the saturation S'gel), the gel particles will lose the ability of flowing. The relative permeability curve of gel particles can be expressed as:

$$K_{rgel} = \begin{cases} K_{rw0} & \left(S_w \le S'_{gel}\right) \\ 0 & \left(S_w > S'_{gel}\right) \end{cases} \tag{5}$$

The correction equation of water phase relative permeability curve is:

$$K_{rw} = \begin{cases} K_{rgel} & \left(S_w \le S'_{gel}\right) \\ K_{rw0} - K_{rgel}\left(S'_{gel}\right) & \left(S_w > S'_{gel}\right) \end{cases} \tag{6}$$

Fig.2. shows the diagram of the relative permeability in the four-phase flow model.

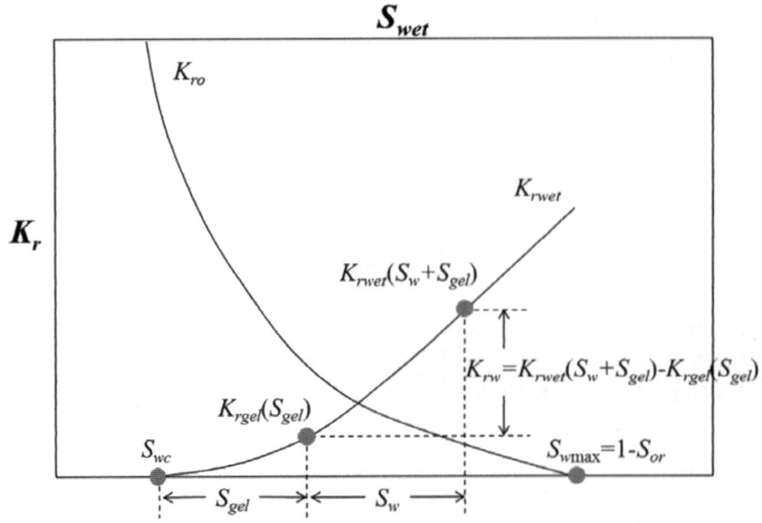

Figure 2. The relative permeability in the four-phase percolation model.

4. Result analysis

According to the parameters of injection and production adopted in the experiment, the water-cut, the injection pressure and the oil recovery curve (Figure 3) are calculated by the newly established mathematical model in section 3. During the calculation, the residual resistance coefficient of copolymer produced in the core is unknown and is obtained by fitting the subsequent water-flooding pressure.

Compared with the experimental results, the results of numerical simulation can better reflect the process of pressurization and water shut-off of sealing channeling system. The subsequent water-flooding pressure is about 1500 kPa and the water-cut funnel is obviously. Table 1 shows the comparison of experimental result and calculated result. When the water-cut rises to 91.1%, the corresponding oil recovery is 23.7%. After the injection of modified starch gel, the water-cut decreases to 63.4%, the corresponding oil recovery is 26.4%, the corresponding staged oil recovery is 2.5% and the injection pressure increases to 221 KPa. The water-cut reaches to 96.9% when injected modified starch gel volume is 1.88PV, the corresponding oil recovery is 43.0% and the corresponding staged oil recovery is 17.3%. The residual resistance coefficient of copolymer obtained by fitting is 350, much larger than the conventional gel.

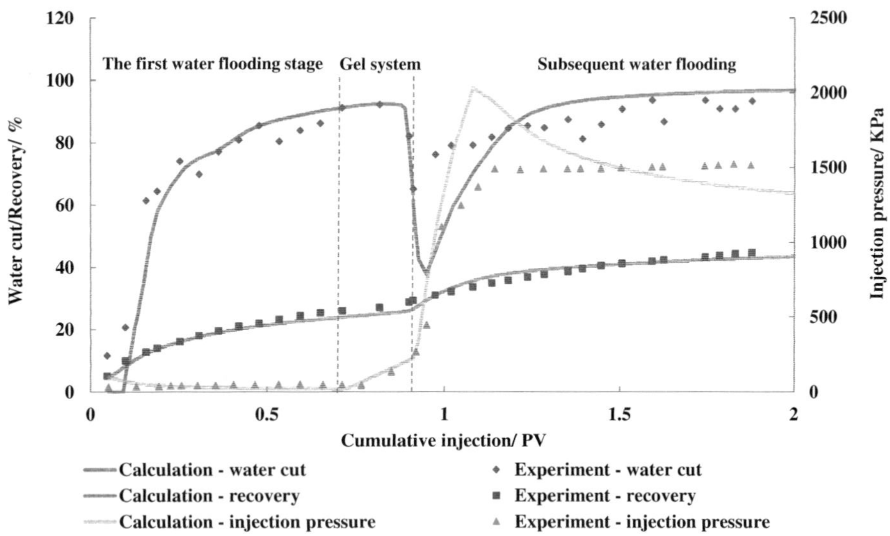

Figure 3. Curves comparison between experiment results and calculation results.

Table 1. Parameters comparison between experiment results and calculation results.

Results	The first water flooding stage	Injection stage of modified starch gel system				The secondary water flooding stage		
	Recovery /%	Recovery /%	Level recovery /%	Water cut at the end of stage /%	Injection pressure /KPa	Recovery /%	Level recovery /%	Injection pressure /KPa
Experiment	25.9	27.7	1.8	65.0	267	44.6	16.9	About 1,500
Calculation	23.7	26.4	2.7	63.4	221	43.0	16.6	1,600~ 1,300

5. Conclusion

The main understandings are as follows.

(1) Microscopic experiments proved that movable gel system in porous media can be divided into two different fluids: movable gel and gel particles, with clear interface, different viscosities and percolation properties, and with the change of concentration, the two fluids can be transformed into each other. The transformation process of gel particles intuitively reveals the change of movable gel system in field test and the percolation mechanism of "rise and then fall" of injection pressure.

(2) On the basic law of microcosmic distribution and migration of oil and water in porous media, we established a relative permeability model with a comparatively clear interface, but no strict phase interface, and gave out the handling method and characterization method of pseudo-four-phase relative permeability, built a mathematical representation method for the percolation characterization of "approximate solid particles", which is a reference for the study of the porous media transport theory about grain plugging agent.

(3) According to the parameters of injection and production adopted in the experiment, the water-cut, the injection pressure and the oil recovery curve are calculated by the newly established mathematical model. Compared with the experimental results, the results of numerical simulation can better reflect the process of pressurization and water shut-off of sealing channeling system.

IOP Conf. Series: Journal of Physics: Conf. Series **1064** (2018) 012037 doi:10.1088/1742-6596/1064/1/012037

6. Nomenclature

Table 2. Nomenclature

Cr^{3+}	The crosslinking agent.
mg	The movable gel produced in the reaction.
Gel	The gel particles.
C_{mg}	The concentration of movable gel, μg/cm^3.
C_{mgmax}	The critical concentration of movable gel, μg/cm^3.
K_1, K_2, K_3	The equivalent ratio between crosslinking agent and polymer, measured by experiment.
l	The reaction order, determined by the experiment.
-	The consumption of quality.
K_d	The degradation constant, measured by experiment.
S_{gel}	The saturation of gel particles.
K_{ro}	The relative permeability of oil phase.
K_{rocw}	The relative oil phase permeability of bound water.
K_{row}	The corresponding relative oil phase permeability on the oil-water relative permeability curve.
K_{rw}	The water phase relative permeability.
K_{rog}	The corresponding relative oil phase permeability on the oil-gas relative permeability curve.
K_{rg}	The gas phase relative permeability.
K_{rw}	The water phase relative permeability.
K_{rgel}	The relative permeability of gel particles.
K_{rw0}	The water phase relative permeability in the three-phase percolation model.
S_w	The saturation of water phase.
S'_{gel}	The critical saturation when the gel particles become unable to flow.
K_{rw}	The water phase relative permeability in the four-phase percolation model.

Acknowledgments
This work was supported by National Science and Technology Major Projects (2017ZX05009004), and Beijing Natural Science Foundation (2173061).

References

[1] Yuan Shiyi, Han Dong, Miao Kun, et al. Application of Flowing Gel Profile Control Technique to Complex Block Reservoir. Acta Petrolei Sinica, 2004, Vol.25 No.4: 50-57.
[2] Wu Xingcai, Wang Hongguang, Li fengxia, et al. EOR Mechanisms and Field Practice of Flowing Gel Profiling/Flooding. Oilfield Chemistry, 2009, Vol.26 No.1: 79-83.
[3] Ye Yinzhu, Wu Xingcai, Wang Zhengbo. Development of Moveable Gel System and Mechanism Analysis. Petroleum Geology and Engineering, 2010, Vol.24 No.3: 73-75.
[4] Yao Tongyu, Li Jishan. Plugging property of mobile gel and its effect on porous flow. Oil Drilling & Production Technology, 2009, Vol.31 No.4: 102-104.
[5] Feng Qihong, Yuan Shiyi, Han Dong, Yang shan, et al. A new method of performance prediction of flowing gel for profile control. Acta Petrolei Sinica, 2006, Vol.27 No.4: 76-80.
[6] Feng Qihong, Yuan Shiyi, Han Dong, Yang shan. A New 3D Stream line Smulation Model for Weak Gel Driving. Journal of Basic Science and Engineering, 2005, Vol.13 No.2: 146-153.
[7] Ma Qingkun, Wu Xingcai, Wang Yuhai, et al. Study Situation and Development Orientation of Moveable Gel Flow through Porous Medium. Oil Drilling & Production Technology, 2006,

Vol.28 No.3: 41-44.

[8] Cui Yinghuai, Zhu Weiyao, Sun Yukai, et al. Mathematical models of nonlinear porous flow for weak gel flooding system. Journal of Liaoning Technical University (Natural Science), 2009, Vol.28 Suppl: 283-285.

[9] Wu Xingcai, Zhu Weiyao, Ma Qingkun, et al. Study on Nonlinearity Seepage Characteristic and Mathematical Model of Movable Gel. Oil Drilling & Production Technology, 2006, Vol.28 No.5: 42-45.

[10] Ma Qingkun, Zhu Weiyao, Gao Min, et al. Characterization and Rheometry of Porous Flow for Weak Gel System. Acta Petrolei Sinica, 2007, Vol.28 No.5: 85-88.

Experimental study on atomization characteristic of tripropellant gas-liquid coaxial injector

Fang Zhao[1,*], Zebin Ren[1,2] and Chibing Shen[3]

[1]Facility Design and instrumentation institute, China Aerodynamics Research & Development Center, Mianyang, China.
[2]State Key Laboratory of Aerodynamics, China Aerodynamics Research & Development Center, Mianyang, China.
[3]Science and Technology on Scramjet Laboratory, National University of Defense Technology, Changsha, China.

*Corresponding author e-mail: 1968683223@qq.com

Abstract. Tripropellant gas-liquid coaxial straight flow injector was designed and manufactured to meet the practical needs of combustion air heater (CAH). Water and air are used as the simulant, and Malvern optical instrument and Single Lens Reflex (SLR) camera are applied to investigate the effect of recess size and the circular seam size of air on the discharge and spray characteristics. The results show that the discharge characteristic remains basically the same in different recess sizes; the gaseous phase significantly improves the atomization performance, as well as increases the spray cone angle; the recess size of injector and the circular seam size of air basically have no impact on the spray cone angle; the Saunter Mean Diameter (SMD) of particle is influenced by the gas pressure drop, the recess size of injector and the circular seam size of air.

1. Introduction

Combustion air heater heats air directly with a fuel-oxidizer reaction. Fuel is burned in the air stream in a combustor similar to a ramjet combustor. The exhaust gases are used for ramjet testing after the consumed oxygen is replenished. It is widely used in aerospace engineering, such as scramjet ground test heating system [1-3].

As one of the key components of combustion air heater [4], its function lies in mixing the fuel and oxidant thoroughly by atomization, thus producing highly efficient and stable combustion. At present, various kinds of injectors including pressure-swirl injector and straight flow injector, etc. have been applied to the combustion air heater [5-6], and the coaxial straight flow injector is widely used due to its broad turndown ratio. Liquid fuel and oxidizer into the combustion chamber through the injector spay, then atomization, evaporation and mixing the uniformity of mixture directly affects the combustion stability and combustion efficiency, thereby affecting the quality of nozzle outlet flow field of air heater [7].

With the wide application of coaxial straight flow injector in the field of combustion air heater and liquid rocket engine, the experimental research of coaxial injector is quite common[8-11]. Zhang et al. [8] presented the atomization characteristic is better with the increase of recess size of injector. Sun et

Content from this work may be used under the terms of the Creative Commons Attribution 3.0 licence. Any further distribution of this work must maintain attribution to the author(s) and the title of the work, journal citation and DOI.
Published under licence by IOP Publishing Ltd

al. [9] concluded that the coexistence of gas liquids has little influence on the discharge characteristic of gas-liquid coaxial straight flow injector. Tien et al. [10] found that a suitable recess size will help achieve good atomization characteristic and proper discharge performance. With the help of high pixel camera, Malvern optical instrument, etc. related researches of tripropellant (alcohol, oxygen and air) coaxial straight flow injector are carried out in this paper, and preliminary results are obtained.

2. Experimental methods

2.1. Experimental facilities

As shown in Fig.1, the experimental apparatus are composed of a propellants feed system, a tripropellant coaxial injector, a spray collector, a SLR camera and a Malvern optical instrument system. Considering the safety factors, the experiment was simplified as follows: water is used as the simulant medium for fuel; air is used as the simulant medium for oxygen.

Spray experiments were conducted at atmospheric pressure, with filtered water and dry air supplied through a pressurized feed system. Pressure sensors with an accuracy of 0.5% FS were used to measure the pressure in liquid and gas manifold. Liquid mass flow rate was measured by a turbine flow meter with an accuracy of 0.5% FS. The volume flow rate of gaseous propellant was measured by a turbine flow meter with an accuracy of 0.5% FS. To calculate the mass flow rate of gaseous propellant, the pressure and temperature of gas at the inlet of the gas flow meter were also measured.

Schematic of the tripropellant coaxial injector is featured in Fig.2 (named Case00). The liquid alcohol is injected through the center hole; the dried oxygen and air are injected through the middle circular seam and the outside circular seam, respectively.

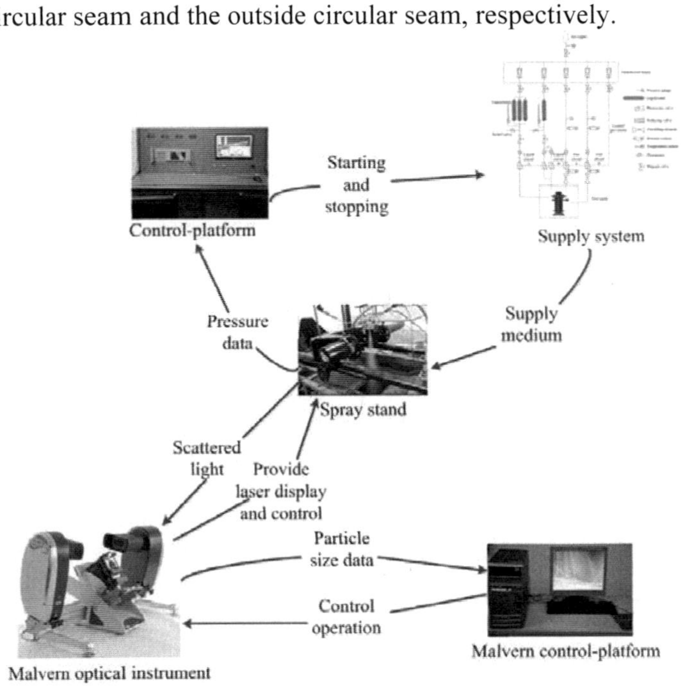

Figure 1. Flow diagram of spray test system

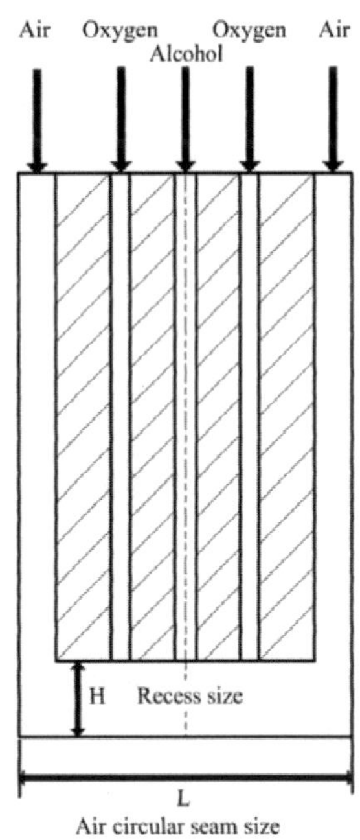

Figure 2. Schematic of injector

2.2. Experimental conditions

The injection pressure drops and the corresponding mass flow rates of gas and liquid in these tests were listed in Table 1. Moreover, the effect of structural parameters of injector on atomization characteristics were studied in this paper, the comparison cases with various parameters are shown in Table 2.

IOP Conf. Series: Journal of Physics: Conf. Series **1064** (2018) 012038 doi:10.1088/1742-6596/1064/1/012038

2.3. Experimental techniques

A SLR camera was used to shoot the injector spray field and get the spray cone angle after post-proposing. Moreover, A Malvern optical instrument was employed for measuring the size distribution of spray in real time [12].

Table 1. Experiment parameters of injector

Component	Mass flow rate(kg/s)	Pressure drop(MP)
Air	0.022~0.088	0.5~1.95
Oxygen	0.0065~0.016	0.75~1.95
Alcohol	0.0096~0.0145	0.68~1.55

Table 2. Cases with different sizes

Variable name	Case No.	Size(nondimensionalize)
Reference case	Case00	H; L
Recess size	Case01	(H-3.5); L
	Case02	(H+1.5); L
Air circular seam size	Case03	H; (L+0.85)
	Case04	H; (L+1.7)

3. Results and discussions

3.1. Discharge characteristic

Discharge coefficient is an important parameter to evaluate the discharge characteristic. The liquid discharge coefficient refers to the ratio between actual mass flow rate and theoretical mass flow rate. The actual flow rate is obtained by experiments, and the theoretical flow rate can be calculated from continuity equation and Bernoulli equation. Calculation formula for liquid flow rate as shown below:

$$\dot{m} = C_A A \sqrt{2\rho \Delta P} \tag{1}$$

Where $\dot{m}$ refers to liquid flow rate; C_A is discharge coefficient; A is flow size; ρ refers to density of working fluid; ΔP is injection pressure drop. Therefore, discharge coefficient of liquid can be calculated from the following formula:

$$C_A = \dot{m} / (A\sqrt{2\rho \Delta P}) \tag{2}$$

Sonic phenomenon existed because of the gas compressibility, thus calculation for gas mass flow rate should be divided kind of cases, namely, sonic condition and subsonic condition. Calculation formula as shown below:

$$\dot{m} = \begin{cases} C_A AP\sqrt{\dfrac{2k}{(k-1)RT}\left((\dfrac{P_e}{P})^{\frac{2}{k}} - (\dfrac{P_e}{P})^{\frac{k+1}{k}} \right)} & , \sigma_{cr} < \dfrac{P_e}{P} \le 1; \\ C_A AP\sqrt{\dfrac{k}{RT}\left(\dfrac{2}{k+1}\right)^{\frac{k+1}{k-1}}} & , \dfrac{P_e}{P} \le \sigma_{cr}. \\ \sigma_{cr} = \left(\dfrac{2}{k+1}\right)^{\frac{k}{k-1}} & \end{cases} \tag{3}$$

Where P refers to the gas total pressure; σ_{cr} is critical pressure ratio; P_e refers to the outlet pressure; k is the specific heat ratio; R refers to the gas constant; T is the total temperature.

The ratio of pressure before injecting and combustor pressure is obviously greater than the critical value, thus the gas in the injector is not sonic under the condition of hot test. However, the downstream pressure of the injector is atmospheric pressure under the condition of cold test. Therefore, the gas discharge coefficient can be calculated from the following formula:

$$C_A = \dot{m} / (AP\sqrt{\frac{k}{RT}\left(\frac{2}{k+1}\right)^{\frac{k+1}{k-1}}}) \tag{4}$$

Fig. 3 shows the curve of discharge coefficient of alcohol hole with various injection pressure drops. Discharge coefficient remains at around 0.74 within the pressure drop range in test. Due to the error caused by injector processing and flow rate measurement, discharge coefficient fluctuates within a narrow range. The overall trend shows that the discharge coefficient is less affected by injection pressure drop. With the increase of pressure drop, the friction force between the spray air and recess section increases, resulting in the loss of discharge characteristic, thus the discharge coefficient of air circular seam decreases with the increase of pressure drop, as shown in Fig. 4.

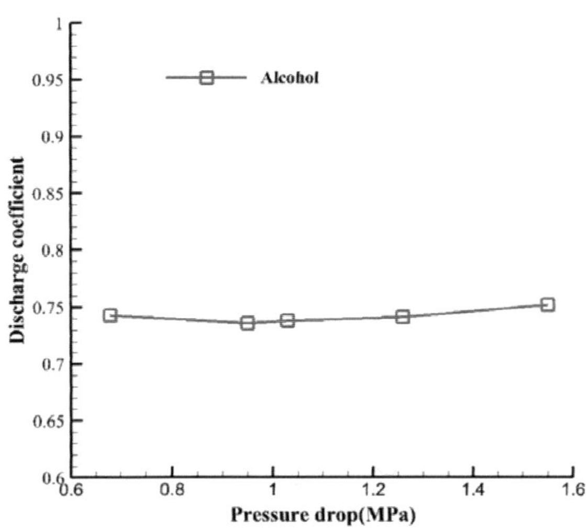

Figure 3. Discharge coefficient of alcohol hole with pressure drop

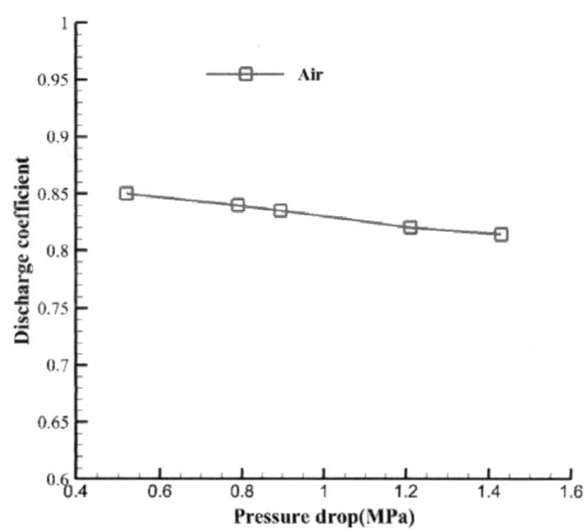

Figure 4. Discharge coefficient of air circular seam with pressure drop

Discharge characteristics of cases with different recess sizes under various air pressure drops are shown in Fig. 5 and 6, keeping the injection pressure drop of alcohol and oxygen constant. As shown from the curves, discharge characteristics of cases behave very similarly.

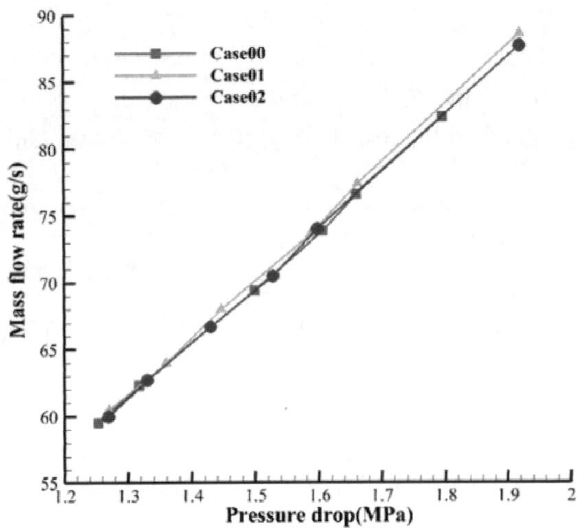

Figure 5. Air mass flow of different cases with injection pressure drop

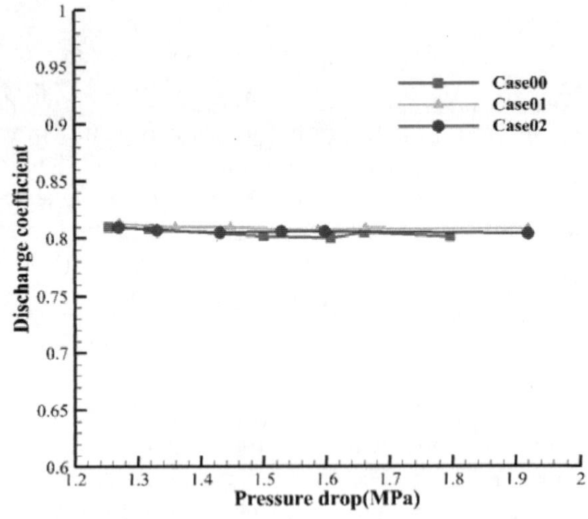

Figure 6. Air circular seam discharge coefficient of different cases with pressure drop

3.2. Spray cone angle

For detailed research, spray cone angle of monopropellant, bipropellant and tripropellant were photographed and analyzed, and the specific results are as follows.

Fig. 7 shows the spray fields of Case00 under different oxygen pressure drops on the condition of bipropellant, leaving the alcohol pressure drop unchanged. It can be seen that the addition of gas component obviously expands the atomization angle, thus improving the jet atomization. In addition, the spray cone angle increases from 17.4 degree to 21 degree when the oxygen pressure drop changed from 0.74 MPa to 1.3 MPa.

(a) 0.74Mpa (b) 1.3Mpa

Figure 7. Spray field of bipropellant with different oxygen pressure drop

On the condition of constant alcohol pressure drop, the spray fields of Case00 under different air pressure drops are shown in Fig. 8. With the increase of pressure drop, the action of the outermost air on the jet is enhanced, thus the spray cone angle decreases. Experimental results show agreeably to the theoretical analysis, and the spray cone angle decreases from 19.7 degree to 11.5 degree when the air pressure drop increased from 0.63 MP to 1.55 MP.

Three cases of different recess sizes under the same working condition in the premise of tripropellant were investigated. The injection pressure drops of air, oxygen and alcohol are about 0.96MPa, 1.14MPa and 1.05MPa, respectively. Fig. 9 shows the spray field of the three cases. As shown in the Figure, the injector recess size affects the atomization performance. The longer the recess size, the larger the spray cone angle. However, the overall trend is not obvious.

(a) 0.63Mpa (b) 1.55Mpa

Figure 8. Spray field of bipropellant with different air pressure drop

In order to study the effect of air circular seam size on spray cone angle, test on three groups of injector models were carried out. In the contrast experiment, air flow rate keeps constant, and the injection pressure of oxygen and alcohol are about 1.14MPa and 1.05MPa, respectively. To have a thorough analysis, the spray field of three cases are photographed in Fig. 10, which shows that on the premise of constant air flow rate, the spray cone angle gets larger with the increase of air circular seam size. Because in case of constant air flow rate, the increase of air circular seam size means the decrease of air injection pressure drop, which enlarge the spray cone angle, and the result is consistent with the conclusion above.

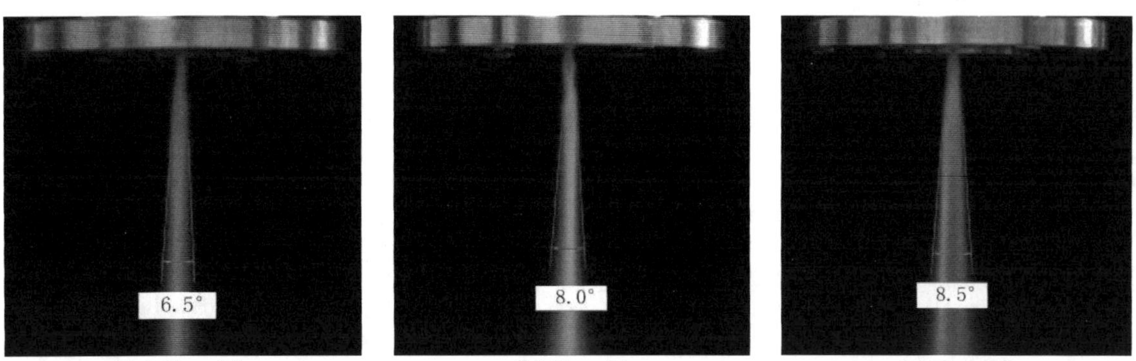

(a) Case00 (b) Case01 (c) Case02

Figure 9. Spray field of cases with different recess sizes

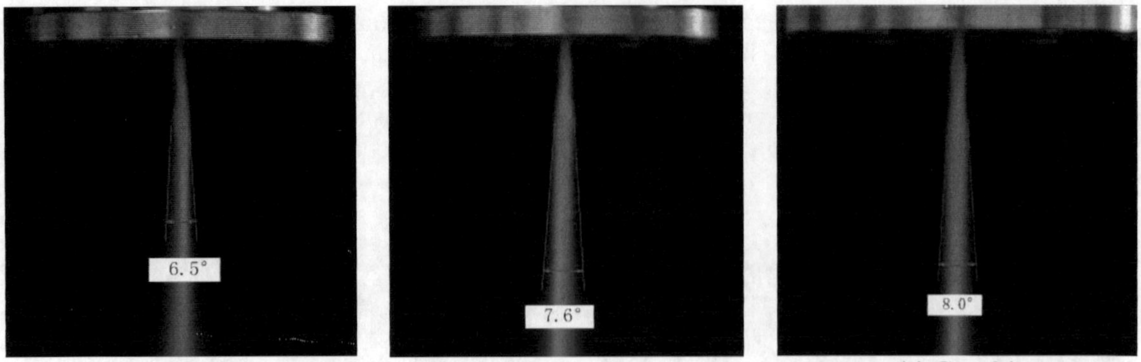

(a) Case00 (b) Case03 (c) Case04

Figure 10. Spray field of cases with different air circular seam sizes

3.3. Saunter mean diameter (SMD)

Injector particle diameter usually refers to the SMD (Saunter mean diameter) for CAH. Its physical meaning is the same size with the diameter of droplets instead of actual droplets, keeping the surface size to volume ratio the same as the real droplet group. SMD is an important index to evaluate the atomizing performance of injector, and the smaller the SMD value is, the better the atomizing performance of injector is. Meanwhile SMD research supplies reference for the design and improvement of injector, on the other hand, provides numerical simulation of CAH with a certain initial conditions.

Figure 11 shows the distribution of SMD in different distance cross-sections (measuring distance refers to the height from injector exit to the laser axis center) of Case00 under different injection pressure on the condition of bipropellant, in which the alcohol pressure drop was kept at about 0.99 MP, while the air injection pressure were set at 0.34Mpa, 0.77MPa and 0.96MPa. The result shows that SMD at the same cross-section increases with air pressure drop within a certain range, while the alcohol pressure drop remains unchanged. Because the atomization of gas-liquid coaxial flow injector was achieved by aerodynamic atomization and pressure atomization, and the former is dominant. When the alcohol pressure drop is constant and the air pressure drop increases, the shear action between the gas and liquid increases, which is beneficial to the liquid column breaking into the droplet, that is to say, primary atomization is better. At the same time, due to the increase of the air jet momentum, the droplets were enveloped to a certain extent, leading to the restriction on transverse movement of droplets, and the penetration will be reduced. In this case, the droplets were restricted in a small region, thus affecting the further interaction between the droplet and air, resulting in poor secondary atomization. Besides, when the measuring distance reaches to a certain value, which is to fully atomized, the SMD value tends to the same order of magnitude, and differs very little.

To obtain detail effects of injector structure size on atomization performance, the SMD of various cases were investigated. Figure 12 shows the SMD distribution of cases with different recess size under the same injection pressure drop. It can be seen from the figure, with the increase of recess size, the SMD increases first and then decreases. There exists an optimal recess size to obtain smaller SMD.

IOP Conf. Series: Journal of Physics: Conf. Series **1064** (2018) 012038 doi:10.1088/1742-6596/1064/1/012038

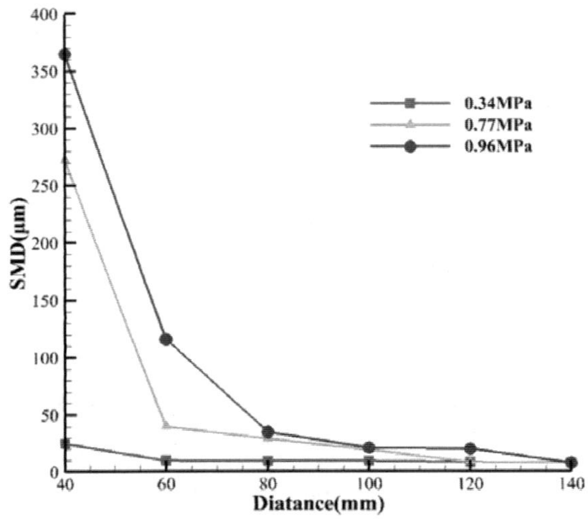

Figure 11. SMD of Case00 in cross-sections with different air injection pressure drops

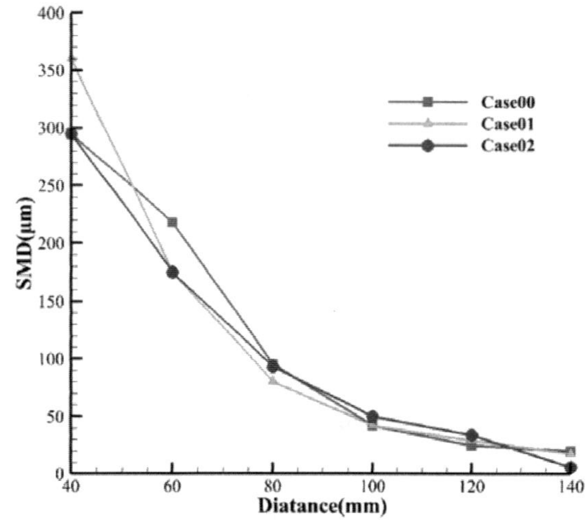

Figure 12. SMD distribution of contrast cases with different recess sizes in cross-sections

The SMD distribution of cases with different air circular seam size are shown in Figure 13, keeping the test condition constant (including air mass flow, the pressure drop of alcohol and oxygen). In agreement with conclusions in the previous paper, the increase of air circular seam size contributes to the atomization performance, namely, the SMD decreases with the increase of air circular seam size. Because in the premise of unchanged air mass flow, the increase of air circular seam size leads to the decrease of air injection pressure drop. According to the previous research, the decrease of the air injection pressure drop results in smaller SMD to some extent.

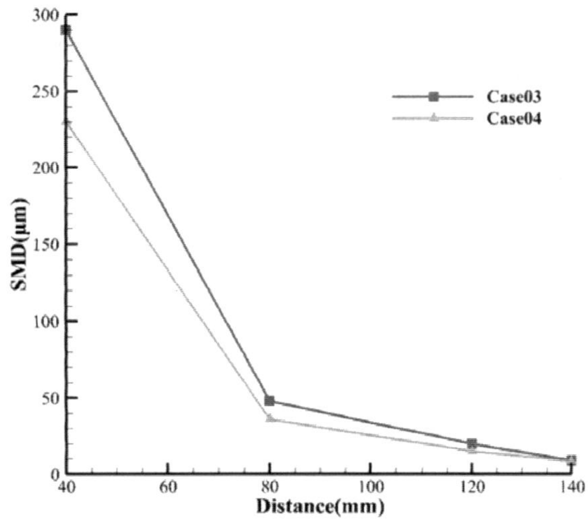

Figure 13. SMD distribution of contrast cases with different air circular seam sizes in cross-sections

4. Conclusion

The sprays of gas-liquid coaxial flow injector with different recess sizes and air circular seam sizes were characterized with a Malvern optical instrument and Single Lens Reflex (SLR) camera. The effects of injection pressure drop, structure sizes on mass flow rate, spray field, spray cone angle and SMD were analyzed and discussed. The research is concluded as follows:

(1) The gas phase significantly improves the atomization effect, and increases the spray cone angle. However, as the injection pressure drop increases, the auxiliary function of outside air turns to be a form of envelop for the spray, thus decreasing the spray cone angle.

(2) Under the same test conditions, the recess sizes and air circular seam sizes have little influence on the spray cone angle.

(3) The SMD decreases with the development of the spray. Within a certain range, the SMD of the same cross-section increases with the gas injection pressure drop. Besides, with the increase of recess size, SMD increases first and then decreases.

(4) In case of constant air flow rate and the injection pressure drop of the other two elements, the SMD decreases with the increase of air circular seam size.

Acknowledgments
The authors would like to express their sincere thanks for the support from the 0901 project of China.

References

[1] Sarışın, M.N., Ulas, A., Albayrak, K. Design of connected pipe test facility for ramjet applications, METU Mechanical Eng. Dept. Master Thesis, 2005.
[2] J.L. Le, W.X. Liu, et al. Experimental and numerical investigation of air vitiation effects on scramjet test performance. AIAA 2009-7344.
[3] Manan A. Vyas, William A. Engblom, et al. Numerical simulation of vitiation effects on a hydrogen-fueled dual-mode scramjet. AIAA-2010-1127.
[4] J.H. Feng. Numerical simulation and experimental numerical simulation and experimental research on high enthalpy and pressure air-heater [D]. Changsha: National University of Defense Technology, 2011.
[5] Eusebio Samuel. Design of an air vitiation heater. The University of Alabama in Huntsville. Dept. Master Thesis, 2006.
[6] Z.T. Kang. Research on the gas generating method based on the direct combustion of air and kerosene [D]. Changsha: National University of Defense Technology, 2012.
[7] Y.P. Liu. Numerical simulation and experimental research on an air-heater of low total temperature [D]. Changsha: National University of Defense Technology, 2010.
[8] M.Z. Zhang, A. Li, et al. Experimental study on flow rate and spray characteristic of coaxial centrifugal injector [J]. Journal of Propulsion Technology, 2004, 25 (1): 19-22.
[9] J.G. Sun, C.Y. Tian, et al. Spray and combustion investigation of gas/liquid coaxial injectors [J]. Journal of aerospace power, 2000, 15 (3): 317-320.
[10] Z.F. Tian, J.P. Wu, et al. Experimental study on spray characteristic of gas-liquid coaxial injectors [J]. Journal of National University of Defense Technology, 2006, 28(4): 10-13.
[11] L.J. Yang, M.H. Ge, et al. Spray characteristics of recessed gas-liquid coaxial swirl injector [J]. Journal of Propulsion Technology, 2006, 27(4):289-293.
[12] Information on http://www.malvern.co.u.

IOP Conf. Series: Journal of Physics: Conf. Series **1064** (2018) 012039 doi:10.1088/1742-6596/1064/1/012039

Optimization of the cleaning efficiency of drilling reamer based on flow field simulation

Yang Zhang[1], YuFeng Deng[1], Binbin Wang[2], Jiawei Liu[1]

[1]College of Mechanical and Electrical Engineering, Beijing University of Chemical Technology, Beijing 100029, China.
[2]China Nuclear Power Engineering Co., Ltd, Beijing 100840, China.

Abstract. Drilling reamer is an oil drilling tool, the nozzle of which is located on the mandrel. The cleaning fluid mixed with mud creates a split flow at the nozzle, causing part of the fluid to flow into the nozzle to create a back-flow, aligning the blade for cleaning. In order to study the optimal cleaning efficiency of the mandrel nozzle, the finite element analysis software ANSYS Workbench and the flow field simulation software Fluent were used to simulate flow field. The outlet flow rate and velocity were changed by adjusting the inlet diameter. The results show that when the split ratio reaches about 1/4, the cleaning speed is saturated, and the optimal cleaning effect is achieved.

Key words: Drilling reamer; Cleaning Efficiency; Optimization; Flow Field simulation; Fluent.

1. Introduction

The drilling reamer is a tool used for reaming while drilling. When the drill bit is drilled below, the upper reamer is reamed and the borehole wall is reamed. Reamers are generally used in well-drilled, easily constricted formations. Foreign companies have conducted long-term R&D (Research and development) in the field of reaming tools and had formed relatively systematic theory and serialization tools. However, domestic companies started late in this filed, R&D conditions differ greatly, and human and material resources are relatively low. It requires down whole experiments and computer simulations and comprehensive numerical analysis to improve. This article introduces a new term: split ratio. It means the ratio of the nozzle flow to the central tube flow of the mandrel. Through numerical analysis of finite element analysis and split ratios, the optimum split ratio is obtained and can be used for blade nozzles and other cleaning runners.

When the mud flows to the bottom of the well in the mandrel nozzle, diversion occurs through the hole on the side of the mandrel to clean the reamer blades [1]. Therefore, the split ratio is a key factor that determines the cleaning efficiency of the reamer nozzle with drilling. The split ratio refers to the ratio of the flow to the nozzle to the flow to the bottom of the well. In order to obtain the cleaning rule of the reamer mandrel nozzle, the split ratio of the slurry cleaning liquid is adjusted by changing the diameter of the nozzle, and through the simulation method to calculate the exit speed and flow. The literature analysis of most nozzles focuses on the structural analysis under controlled pressure,

Content from this work may be used under the terms of the Creative Commons Attribution 3.0 licence. Any further distribution of this work must maintain attribution to the author(s) and the title of the work, journal citation and DOI.
Published under licence by IOP Publishing Ltd

whereas for oil machinery, nozzle analysis of flow regulation is scarce. Therefore, it is of great importance to analyses the nozzle of the reamer when drilling [3].

Yang changed the geometric parameters of the nozzle and used fluent visual numerical simulation function to study the internal and external flow fields of the conical porous nozzle in order to optimize and improve the cone-shaped porous nozzle [5]. Zhou used fluent software to simulate and analyze the velocity attenuation characteristics of the two nozzle jets to obtain the optimal nozzle structure parameters [6]. Yu and Zhou come to the result that dynamic pressure and strike force are important parameters weighing cleaning efficiency, and reasonable matching nozzle diameter , pressure and flow can make cleaning operations more efficient and save more energy on the basis of established the three-dimensional (3D) nozzles model of fan nozzle and the external flow field[7].

The author used ANSYS Workbench and Fluent software to analyses and obtain the flow rate and outlet speed of the nozzle when the diameter was changed, and summarized the problem of the optimal cleaning speed to obtain the best split ratio and cleaning law.

2. Flow Field Simulation

2.1. Modelling

In this paper, author uses the ANSYS workbench 18.0 version to simulate and analyses. The fluid model is established by the flow path of the fluid in the reamer. Simplify the model, and the model is as shown in Fig.1. A is flow inlet; B is outlet and C is bottom outlet.

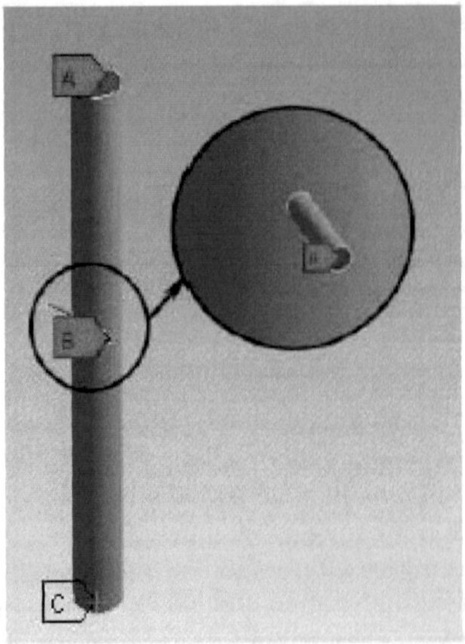

Fig. 1 ANSYS Workbench fluid model.

The fluid flows from the top to the bottom in the hollow space of mandrel. After three nozzles, it splits, and some of the fluid flows back to the surface through the nozzle. Most of the rest of the fluid continues to flow downward and flows out of the mandrel to the bottom of the well. Therefore: A is flow inlet; nozzle outlet is defined as outlet B (3); bottom is outlet C.

2.2. Meshing

Fig.2a shows the overall mesh model while Fig.2b shows refined meshing near the outlet B. Open the Mesh module and import the model from the workbench. As the model is regular and symmetrical, global tetrahedral grid and the body refinement grid were used. The number of grids is in the order of millions, and the quality met the standard. The mesh metrics range from 0.88 to 1 (1 is the best). [8]

IOP Conf. Series: Journal of Physics: Conf. Series **1064** (2018) 012039 doi:10.1088/1742-6596/1064/1/012039

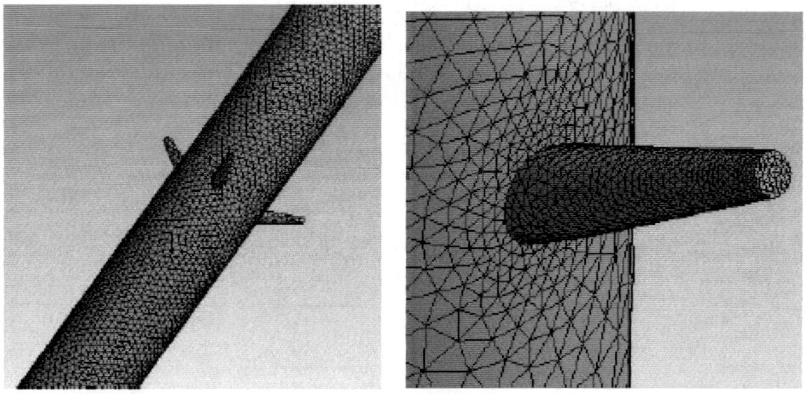

a Grid model diagram b Refined mesh near the nozzle

Fig. 2 Meshed model.

Import to Fluent and Pre-processing

Select k-e two-phase flow, the type selection standard model (Standard), the definition of material: mud cleaning fluid, a density of 1200kg/m3; Define the boundary conditions: The inlet boundary is defined as the mass flow rate inlet. The inlet rate was set to 33.6 kg/s (required by the reamer work required flow of 28 L/s) [9]. Turbulence intensity is 5%. The hydraulic diameter is 68mm based on the inlet diameter. The B outlet nozzle is set to the pressure outlet (24 MP). Export C (bottom well outlet) is set to pressure outlet (25 MP). Set and solve it, this time it is set to 2000 iterations [10].

3. Results and Discussion

After the pre-processing steps and the calculations, the velocity contour and the flow contour can be drawn about the entire flow channel. The velocity contour map can reflect the velocity variation of the fluid in the tube. The velocity reference position is selected as the nozzle outlet position and the bottom position. The flow parameters can be used to calculate the flow values and the split ratio of the two flow channels.

The results including the flow animation, the flow ratio between the outlet, the speed ratio and the speed of the contour were obtained. The contour is as follows in Fig.3:

Fig. 3 Overall flow velocity contour.

From the velocity contour of fluid field in the reamer, it is found that the speed at the nozzle exit position is much faster, while other positions more stable.

The simulation was based on the initial design of the nozzle structure, by changing the nozzle diameter (from 4mm to 10mm). Observe the rule of flow speed near nozzle B and the split ratio. The simulation results are shown in Tab.1 below.

Table 1. Influence of diameter change on flow rate and speed in initial design of nozzle structure.

No.	Nozzle diameter/mm	Nozzle flow B/$kg \cdot s^{-1}$	Bottom-side flow C/$kg \cdot s^{-1}$	Flow ratio/$C \cdot B^{-1}$	Nozzle exit B average speed/$m \cdot s^{-1}$	Bottom C speed/$m \cdot s^{-1}$
1	4.0	1.73	31.87	18.42	59.24	7.44
2	4.2	1.73	31.86	18.38	53.34	7.45
3	4.4	2.29	31.31	13.65	60.39	7.32
4	4.6	3.41	30.19	8.85	71.29	7.06
5	4.8	2.55	31.05	12.18	57.29	7.26
6	5.0	4.16	29.44	7.08	73.24	6.89
7	5.2	4.29	29.32	6.84	69.88	6.86
8	5.4	3.41	30.19	8.84	57.68	7.07
9	5.4	3.41	30.18	8.84	57.68	7.83
10	5.6	5.04	28.56	5.66	69.12	6.68
11	5.8	4.59	29.02	6.33	69.28	6.80
12	6.0	6.19	27.40	4.42	76.57	6.39
13	6.2	6.81	26.79	3.93	76.07	6.29
14	6.4	7.02	26.58	3.79	74.52	6.24
15	6.6	6.57	27.05	4.12	73.49	6.34
16	6.8	6.22	27.38	4.40	70.31	6.42
17	7.0	8.64	24.96	2.89	76.48	5.85
18	7.2	8.53	25.06	2.94	74.02	5.88
19	7.4	9.90	23.71	2.40	78.31	5.56
20	7.6	10.32	23.29	2.26	78.57	5.46
21	7.8	10.79	22.90	2.12	78.03	5.38
22	8.0	11.63	21.98	1.89	78.07	7.83
23	8.2	12.16	21.44	1.76	78.79	5.02
24	8.4	12.62	20.98	1.66	77.79	4.92
25	8.8	13.76	19.84	1.44	76.49	4.66
26	9.0	13.97	19.03	1.36	70.14	5.39
27	9.2	14.39	19.21	1.34	75.61	4.50
28	9.8	16.69	16.91	1.01	76.09	3.95
29	10.0	17.35	16.25	0.94	77.51	3.81

The locations of A, B and C are shown in Fig.1.

The nozzle outlet flow data in Tab.1 is plotted in Fig.4:

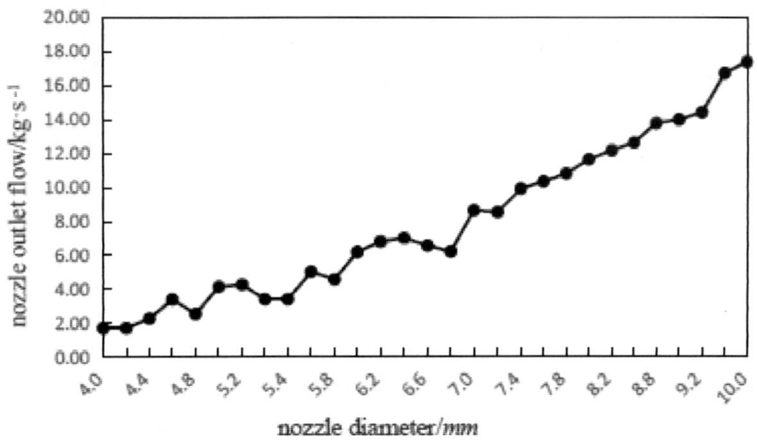

Fig. 4 Relationship between the nozzle diameter and outlet flow velocity.

As can be seen from Fig.4, with some of the larger floating points removed, the flow rate increases as the diameter of the nozzle increases. Fig.5 is the relationship between speed and diameter as follows:

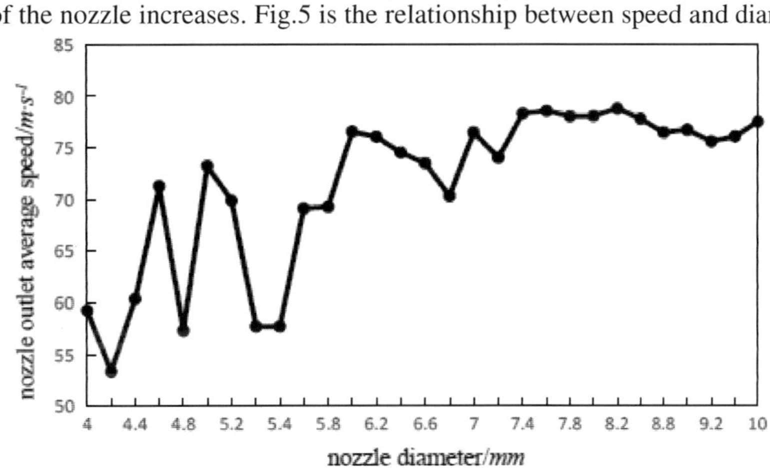

Fig. 5 Relationship of the nozzle diameter and nozzle outlet average speed relationship.

It can be seen from the Fig.5 that when the diameter of the nozzle is small, the velocity is not stable, but the tendency is increased. When the diameter reaches 6mm, and the split ratio is 1/4, the speed begins to stabilize. The flow rate of the nozzle accounts for 20% of the total flow rate. When the diameter is larger than 6mm, the flow rate becomes stable gradually as the diameter increases. While the reamer is working properly, most of the flow to the drill bit will remove the debris from the cleaning blades and carry the cuttings to the ground. When the nozzle diameter is increased to larger than 6mm, the increasing nozzle outlet flow will inevitably lead to the continuous flow of outlet C at the bottom of the well, which will reduce the efficiency of cuttings debris removal. Therefore, the nozzle flow rate should be maintained at about 20% of the total flow rate. When the speed near nozzle B is no longer increased, the highest cleaning efficiency is achieved. In other words, when the split ratio reaches 1/4, the cleaning efficiency is optimal.

Therefore, through simulation analysis, we can see that the optimal split ratio (the optimal diameter) can be used to optimize the cleaning efficiency, and the structural design and part selection can be performed according to the flow distribution.

IOP Conf. Series: Journal of Physics: Conf. Series **1064** (2018) 012039 doi:10.1088/1742-6596/1064/1/012039

4. Conclusion

This paper simulates the flow of mud cleaning fluid in the reamer and analyzes the relationship between the flow rate and nozzle diameter through the velocity contour of the flow field. With the change of nozzle diameter:

The larger the diameter is, the greater the flow rate of the nozzle will be, and the flow velocity will also become larger.

The nozzle cleaning speed is limited, when the split ratio of flow reaches 1/4, the cleaning speed comes to saturation.

References

[1] Gosman A D, et al. Heat and mass transfer in recirculating flows. New York:Academic.1969.
[2] Amano R S, Brandt H. Numerical study of turbulent axisymmetric jets impinging on a flat plate and flowing into an axisymmetric cavity. Transaction of ASME,1984,106:410-417.
[3] Kuo T W, Bracco F V, On the scaling of impulsively started incompressible turbulent round jets. J Fluid Engineering, 1984,104:191-197.
[4] Looney M K, Walsh J J. Mean-flow and turbulent characteristic of free and impinging jet flows. J Fluid Mech,1984,147:397-429.
[5] Yang Huaiyin. Research on Low Pressure Jet Characteristics of Conical-shaped Porous Nozzle [D]. Southwest Jiaotong University, 2016.
[6] Zhou Wenhui. Numerical simulation of internal and external flow fields in high pressure water jet nozzles [D]. Lanzhou University of Technology, 2008. Chinese.
[7] YU Lanying, ZHOU Wanyang, DENG Bin, et al. Study on cleaning parameters of cleaning fan nozzles based on CFD [J]. Machine Tools & Hydraulics, 2016, 44(13):164-167.
[8] Chuang Shu hao, et al.Numerical simulation of twin-jet impingingment on a flat plate coupled with cross-flow. Int J Num Mech in Fluid,1992,14:459-475.
[9] Knowles K, Bray D. Computation of normal impinging jets in cross flow and comparison with experiment. Int J for Num Meth in Fluid,1991,13:1225-1233.
[10] Liou T M, Hwang Y H, Wu S M. The three-dimensional jet impingement flow in a closed-end cylindrical duct.J Fluids Eng,1990,112:171-178.

Period doubling bifurcation and chaos in oscillations of two interacting cavitation bubbles under ultrasonic

Bowei Li[1], Cheng Yao[1,*], Minglei Shan[1] and Changping Zhu [1,2]

[1]College of Internet of Things Engineering, Hohai University, Changzhou, China
[2]Hohai University Wentian College, MaAnshan, China

*Corresponding author e-mail: yaocheng00@126.com

Abstract. Two interacting bubbles oscillator under ultrasonic is a highly nonlinear system. The nature of the cavitation will be understood more deeply in the study of two interacting bubbles' internal mechanism. In this paper, the model to describe the oscillation motion of the two interacting bubbles in the sound field is the Keller-Miksis equation. The dynamics of two interacting bubbles has been studied by adopting the methods of time domain wave, phase trajectory, Poincaré section, power spectrum and bifurcation diagram. We also studied the nonlinear oscillations of bubbles with different initial radii and characteristics of bifurcation structure of normalized radius as ultrasonic pressure amplitude changed. Results show that two bubbles oscillation traverses the similar mode of period doubling bifurcation to cascade chaos. Furthermore, the motion of two interacting bubbles shows discrepant oscillation characteristics under the action of same sound field. In addition, the periodic window and chaotic region of bubble's oscillation occur alternately with the increase of pressure amplitude.

1. Introduction

The study on ultrasonic cavitation mechanism of microbubbles in a liquid is a hot topic in many fields, e.g. sonochemistry [1-2], medical diagnostics and therapy [3-4], food processing [5], ultrasonic-assisted soldering [6]and fluids engineering et al. [7]. When the liquid containing cavitation, bubbles is radiated by a strong sound field, the oscillation of bubble radius shows strong nonlinearity. The nonlinear characteristics of bubble oscillations in acoustic cavitation have been concerned by more and more scholars. The most significant discovery is the existence of period doubling bifurcation route to access to chaos in bubble nonlinear oscillations. The bifurcation structure of bubble radius oscillation has a similarity with other nonlinear oscillators, implying that they have universal characteristics. At the beginning of studying chaotic behaviors of bubble oscillations, Lauterborn W et al. [8] found the chaos in bubble oscillations under high intensity sound by using holographic camera at the speed of 69300 holograms per second. The discovery makes people have a great interest in it. Parlitz U et al. [9] found that the phenomena of period 1, period 2, period3 and chaos state in a single bubble oscillation under the action of sound. However, at that time, the resonance frequency, subharmonics, harmonic dispersion and other issues were more focused. In recent years, people pay more attention to the cause, evaluation indicators of bifurcation and chaos. Behnia S et al. [10] studied a single cavitation bubbles oscillation by methods of bifurcation diagrams and Lyapunov exponent and

Content from this work may be used under the terms of the Creative Commons Attribution 3.0 licence. Any further distribution of this work must maintain attribution to the author(s) and the title of the work, journal citation and DOI.
Published under licence by IOP Publishing Ltd

found rich nonlinear features of period doubling bifurcation. Sojahrood A J et al. [11] studied the bifurcation structure of ultrasound contrast agents(UCAs) under excitation. Selecting the appropriate ultrasonic frequency can increase the section of UCAs and improve the curative effect. Due to the model of single bubble, the interaction between bubbles has not yet considered, so it doesn't simulate the real situation well.

However, in practical application, cavitation bubbles present in the form of cluster. We need to consider the interaction of bubbles. The model of most classic interacting bubbles is two bubbles oscillate under ultrasonic. Jiang L et al. [12] studied the frequency spectrum of the noise emitted by the interaction of bubbles. The interaction between bubbles exerts an additional nonlinear effect on the oscillation of the bubble. Sugita N et al. [13] studied a bifurcation structure of two spherical bubbles nonlinear oscillation. The energy localization is resulted from symmetry-breaking bifurcation of the state oscillation. At present, most of the problems of two bubble dynamics are concerned with the study of their cavitation effects, collapse, and the effect of dual-frequency [14]. There are few studies on two bubbles' chaos characteristics.

In this paper, we investigate two bubbles nonlinear oscillation motion in sound field, and analyze the period doubling route to access to chaos. The time-domain wave, phase trajectory, Poincaré section, power spectrum and bifurcation diagram are used to study the dynamics of two spherical bubbles. We find that the oscillation of the two interacting bubbles in the sound field has complicated nonlinear characteristic which is similar to a single bubble's oscillation. We compare the difference in the oscillation of two interacting bubbles with different original radii under the same excitation. Results show that the disparate nonlinear features of two bubbles with different initial radius under the action of sound. Meanwhile, smaller bubble is more intense than the larger one. In addition, the pressure threshold of smaller bubble access chaos is less than the large bubble.

2. The model and solution methods

2.1. Model of two interacting bubbles oscillation

Two interacting bubbles oscillate regularly under the action of a sound field, as shown in Figure1. The subscripts i and j refer to the two bubbles with different original radius R_{i0} and R_{j0} respectively. D is the distance between O_i and O_j, which locate the centers of two bubbles. P is any point in the sound field beyond the bubbles. In our simulation, we set $D >> R_{i0}$ and $R_{i0} \geqslant R_{j0}$.

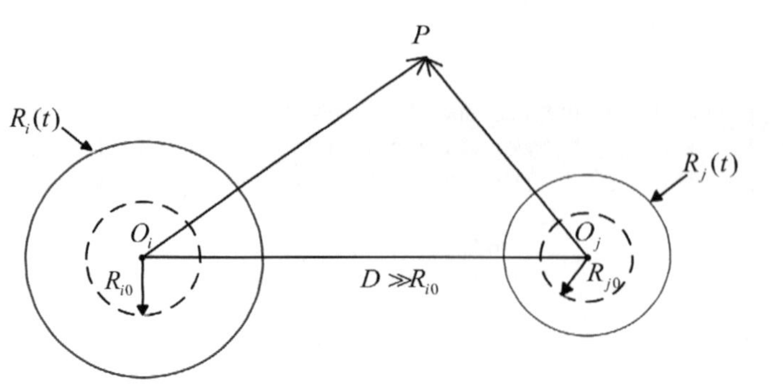

Figure 1. The oscillation model of two interacting bubbles in a sound field

When the bubbles are to collapse, the velocity of the bubble wall is much slower than that of the sound in the liquid medium. Therefore, the modified Keller-Miksis equation [15]is adopted to describe the bubbles dynamics. The equation of bubble i oscillation is as follows:

$$(1-\frac{\dot{R}_i}{c})R_i\ddot{R}_i+\frac{3}{2}(1-\frac{\dot{R}_i}{3c})\dot{R}_i^2=\frac{1}{\rho}(1+\frac{\dot{R}_i}{c})[p_l-p_\infty(t)]+\frac{R_i}{\rho c}\frac{d}{dt}[p_l-p_\infty(t)]-F(R_j) \tag{1}$$

Here, $R_i(t)$ is the instantaneous radius of the bubble i. $\dot{R}(t)$ is the bubble wall velocity, $\ddot{R}(t)$ is the bubble wall acceleration, ρ is the density of the liquid, c is the speed of sound in the medium. The term of $F(R_j)$ represents the interaction of the bubble j to the bubble i and can be expressed as follows:

$$F(R_j)=\frac{R_j^2\ddot{R}_j+2R_j\dot{R}_j^2}{D} \tag{2}$$

p_∞ is the pressure far away from the bubbles and involves the static atmosphere pressure and the periodic acoustic excitation, which is recorded as:

$$p_\infty(t)=P_\infty+P_a\sin\omega t \tag{3}$$

where P_∞ is the ambient pressure, P_a and ω are the pressure amplitude and angular frequency of the excitation source, respectively. The pressure at the bubble wall is denoted as p_l, which can be obtained according to the bubble wall pressure equilibrium conditions:

$$p_l=\frac{2\sigma}{R_i}+\frac{4\mu\dot{R}_i}{R_i}-p_g-p_v \tag{4}$$

where σ is surface tension, μ is the viscosity coefficient of the liquid medium. The pressure inside the bubble including the vapor pressure p_v and the pressure of non-condensable gas p_g. The pressure of the gas inside the bubble follows polytropic process regularity:

$$p_g=p_{g0}(\frac{R_{i0}}{R_i})^{3\kappa} \tag{5}$$

where κ is the polytropic exponent and is set to 4/3 when the gas is mainly diatomic molecules. p_{g0} is the initial pressure inside the bubble and its value is determined by the steady state of the bubble in the liquid medium when there is no sound excitation:

$$0=p_{g0}(\frac{R_{i0}}{R_i})^{3\kappa}+p_v-P_\infty-\frac{2\sigma}{R_E} \tag{6}$$

where R_E is the equilibrium radius of the bubble, and the initial pressure in the bubble can be expressed as:

$$p_{g0}=\frac{2\sigma}{R_E}-p_v+P_\infty \tag{7}$$

ω_0 is the linear resonant frequency of two interacting bubbles oscillator and can be represented as follows according to the literature [16]:

$$\omega_0^2 = \frac{D^2}{D^2 R_{i0} - R_{i0}^2 R_{j0}} \left\{ \frac{1}{\rho R_{i0}} \left[3\kappa(p_0 + \frac{2\sigma}{R_{i0}}) - \frac{2\sigma}{R_{i0}} \right] - \frac{1}{\rho D} \left[3\kappa(p_0 + \frac{2\sigma}{R_{j0}}) - \frac{2\sigma}{R_{j0}} \right] \right\} \tag{8}$$

In this paper, we choose water as the medium. From Harr-Galaggher-Kell equation[17]at ambient temperature T_∞=25℃and ambient pressure P_∞=1atm, the density of the liquid is ρ=997kg/m^3, the speed of sound in liquid is c=1497m/s, the surface tension of the liquid is σ=0.072N/m, the liquid coefficient viscosity is μ=0.00089kg/m·s, the vapor pressure is p_v=3166.8P_a, and the equilibrium radius of the bubble is R_E=10um.

2.2. Nonlinear numerical tools

The oscillating state of bubbles under ultrasound plays a crucial role in cavitation. The aim of this paper is to find the nonlinear oscillation features between bubbles with different initial radius. In this paper, time domain wave, phase trajectory, Poincaré section, power spectrum and bifurcation diagram are used to study the dynamics of cavitation bubbles. The bifurcation diagrams were made with the pressure P_a as the control parameter and with different bubble initial radius. The ultrasonic pressure amplitude ranges from 0 to 5atm. The different oscillation states of bubble i and j were observed under excitation pressure at P_a=0.15, 1.5 and 4.5atm. The time domain wave shows the oscillation regularity of the bubbles excited by the sound field. Through the relationship between the bubbles' radius and the bubble wall velocity, the phase trajectory studies the change rules of the two bubbles' motion state. By selecting the appropriate Poincaré section and then intercepting corresponding points on the plane phase trajectory left, the motion state of the bubbles can be judged concisely. Through the Fourier transform, any periodic signal can be expressed as the superposition of the fundamental wave and the harmonic wave. The power spectrum of any aperiodic signal is a continuous spectral structure. However, when studying a single cavitation bubble's radiation in the sound field with frequency ω, Plesset[18] found it is the nonlinear response that results in harmonic dispersion. That is to say, apart from sound field generating integral multiple frequencies, there are also subharmonics. The emergence of waves means that bifurcation has arrived. Therefore, the power spectrum is an important method to analyze the bubble motion from the period oscillation to the chaos.

3. Results and discussion

Fig. 2 shows the bifurcation structure of the normalization radius R_i and R_j of the bubbles i and j with amplitude P_a. The ultrasonic frequ ency f_s=2f_0=658 *kHz*. The distance between two bubbles D=300R_E. The initial radii of bubble i and j are 10um and 9um respectively. Via the period doubling bifurcation, both bubbles' oscillation accessed the chaotic state. What's more, it alternately appeared periodic and chaotic motion. In Fig. 2(b), when0<P_a<1.11atm, the oscillation of bubble j is in stable period 1. With the increasing of P_a, the bubble j starts to enter a more organized region of period 2 motion state for1.11atm<P_a<2.57atm. then, when 2.57atm<P_a<2.93atm, the bubble j state is in period 4. After all the period regions, the structure of diagram bifurcates into a chaotic region with 2.93atm<P_a<3.66atm for the first time. When 3.66atm <P_a<4.28atm, the chaotic region alternates with the periodic window; when P_a>4.28atm, the bubble j enters the chaotic region for the second time. The bifurcation diagram of the large bubble i in Fig. 2(a) also has a similar topology structure, but the interval range (0<Pa<0.2atm) of the large bubble in period 1 is smaller than the small bubble j (0<Pa<1.12atm). The range of period 2 (0.2atm<Pa<2.57atm) of large bubble i is significantly larger than the small bubble j (1.12atm<Pa<2.56atm).

Figure 3(a) describes the radius time-domain variation of oscillation normalization of two bubbles excited by acoustic wave while the pressure amplitude is P_I=0.15atm. The simulation time we have chosen is six times of the excitation period. From the figure, the bubble presents 6 periodic sinusoidal oscillation and is in a period 1 state. The steady state of oscillation curve repeats again and again with the periodic excitation. Moreover, the smaller bubble oscillates more intensely than the larger one if the excitation energy is constant. Fig. 3(b) is a phase diagram which represents the relationship between the fluctuation velocity of bubbles' wall and their normalized radius. The phase diagram of two bubbles' motion trajectory is a stable closed curve, and the Poincaré section map displays the only one point, as shown in Figure 3(c). The power spectrums of the bubble *i* and *j* focus on the fundamental frequency, as shown in Fig. 3(d). Only a small fraction of energy appears at their double frequency.

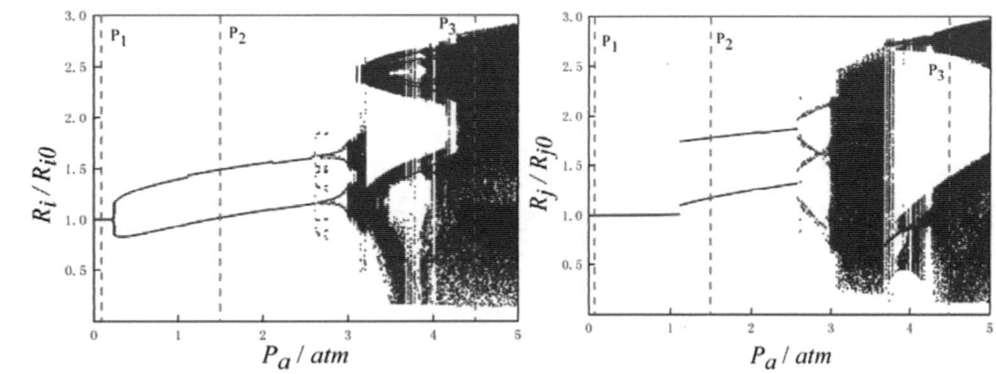

(a) (b)

Figure 2. Bifurcation diagrams of dimensionless bubble radius - pressure amplitude with different initial radius(a) $R_{i0}=10um$;(b) $R_{j0}=9um$

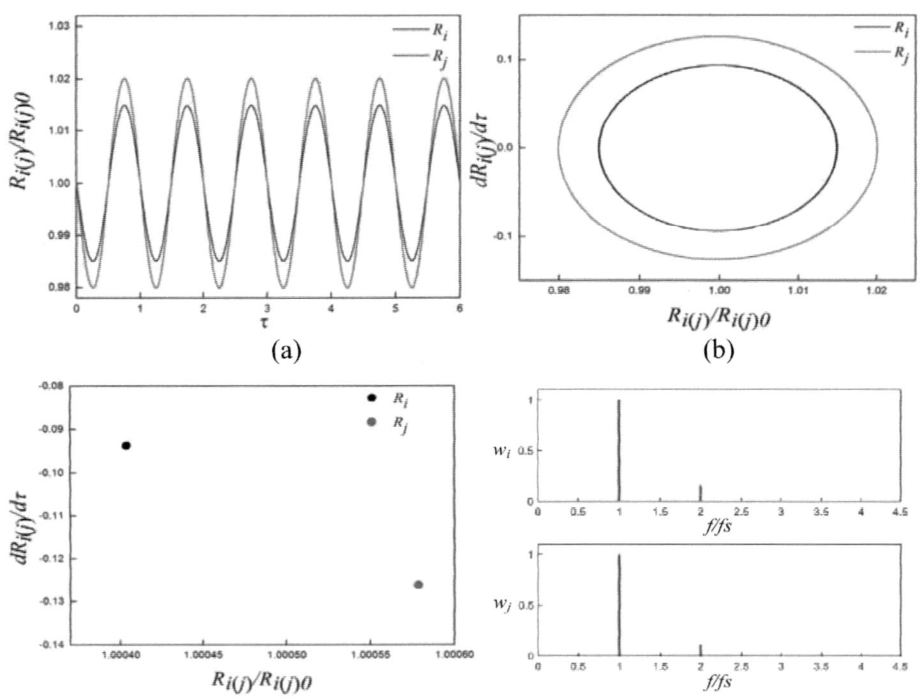

Figure 3. Bubble oscillation with period 1 (a) The time-domain diagram;(b) phase diagram; (c) Poincaré section; (d) power spectrum

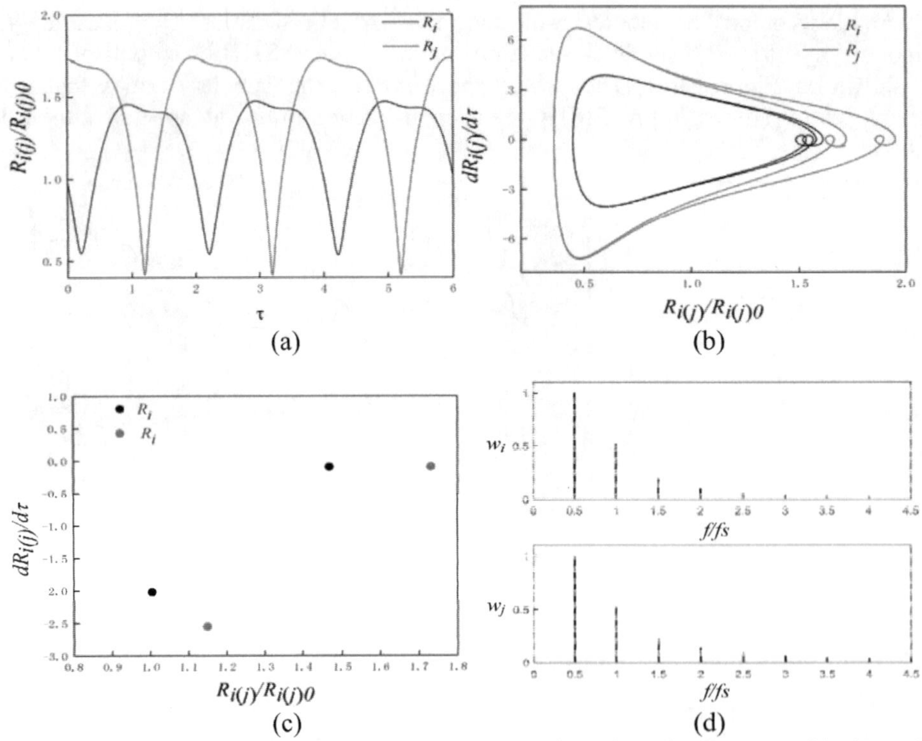

Figure 4. bubble oscillation with period 2 (a) The time-domain diagram; (b) phase diagram; (c) Poincaré section; (d) power spectrum

Fig. 4(a) is the time-domain variation diagram of the normalized radius of two interacting bubble oscillations while pressure P_2=1.5atm. The simulation time we selected is also six times of the excitation period. But from the figure, there are only 3 period oscillations, implying the two interacting bubbles are in period 2. In other words, the motion rules of the bubbles will repeat every two excitation cycles. Meanwhile each bubble oscillation has its own characteristics. The trajectories in the phase diagram of Fig. 4(b) are cycled twice, leaving two points on the Poincaré section respectively, as shown in Fig. 4(c). As shown in Fig. 4(d), the power spectrum of bubbles oscillation includes not only the fundamental frequency f_s and its multiplication component nf_s, but also the subharmonic wave $f_s/2$,$3f_s/2$,$5f_s/2$ and so on. These phenomena indicate that bubbles oscillation begins to access to chaos state. The oscillation of bubbles is more intense, and cavitation is more likely to occur.

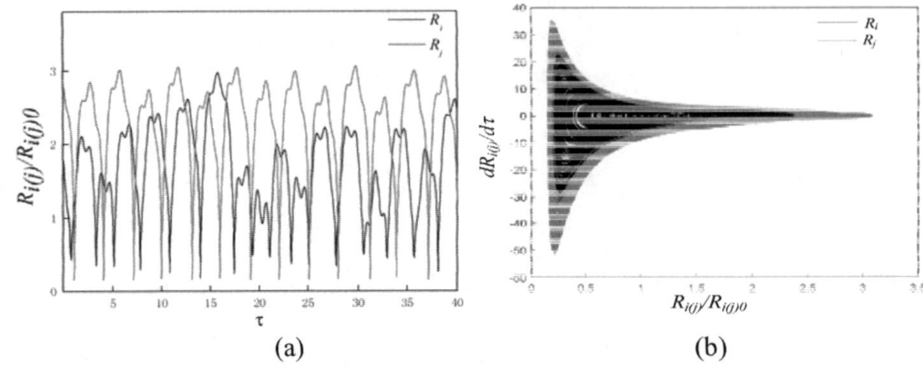

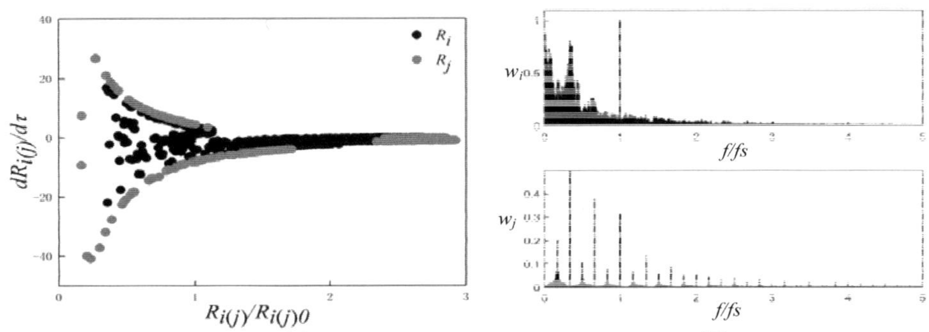

(c) (d)

Figure 5. bubble oscillation with chaos (a) The time-domain diagram; (b) phase diagram; (*c*) Poincaré section; (*d*) power spectrum

The oscillations of bubbles enter stable chaos state as excitation pressure increases gradually. We studied the motion of the bubbles under the excitation pressure P_3=4.5atm, as shown in Fig. 5(a). The simulation time we selected is forty times of the excitation period. Both of the two bubbles lose their periodical motion characteristics and show no regularity in radius variation. At this time, the trajectory in the phase diagram of Fig. 5(b) is no longer the form of a limit cycle, but a curve that will never be closed. The characteristic of reciprocating movement reflects that the phase trajectories converge within bounded regions and thus form strange attractors. The Poincaré section map in Fig. 5(c) is not a finite point set or a closed curve, but a serried point set with a fractal structure. According to the Poincaré section mapping theory, at this moment the bubbles have entered the chaotic oscillation state. Fig. 5(d) shows that the power spectrum of two bubbles' oscillation presents a continuous spectrum structure. It is shown that the bubbles will access the stable chaos region when the excitation pressure is large enough by all of the nonlinear methods. From the perspective of cavitation effects, severe cavitation has taken place under the sound field, which is beneficial to improving the effect of ultrasonic cavitation treatment.

4. Conclusion

In this paper, the nonlinear characteristics of two interacting bubbles under the action of strong sound field were studied. We studied the effects of the ultrasonic pressure amplitude on bubbles' motion under the conditions of bubbles with different initial radii by nonlinear methods of the phase trajectory, Poincaré section, power spectrum, and bifurcation diagram. Results show that two bubbles oscillation traverses the similar mode of period doubling bifurcation to cascade stable chaos region. Meanwhile, the periodic windows interspersed with chaotic regions, indicating the existence of transient chaos. The continuous structure on power spectrum and strange attractors will appear when acoustic wave amplitude exceeds the pressure threshold. It means that the oscillation of two interacting bubbles will be intense chaotic motion. In addition, we found that the oscillation of the smaller bubble is more drastic.

Acknowledgments

This work was financially supported by the Primary Research & Development Plan of Jiangsu Province (Grant No. BE2016056), the Fundamental Research Funds for the Central Universities (Grant No. 2017B17814,2017B678X14) and Postgraduate Research & Practice Innovation Program of Jiangsu Province (Grant No. KYCX17_0539) fund.

References

[1] Nikitenko, Sergey I., and Rachel Pflieger. Toward a new paradigm for sonochemistry: Short review on nonequilibrium plasma observations by means of MBSL spectroscopy in aqueous

solutions. Ultrason. Sonochem. 35 (2017): 623-630.

[2] Soumia, Fassi, and C. Petrier. Effect of potassium monopersulfate (oxone) and operating parameters on sonochemical degradation of cationic dye in an aqueous solution. Ultrason. Sonochem. 32(2016):343.

[3] Yildirim, Adem, et al. Understanding acoustic cavitation initiation by porous nanoparticles: toward nanoscale agents for ultrasound imaging and therapy. Chem. Mater. 28 (2016): 5962-5972.

[4] Lee, Wonseok, and Yongrae Roh. Ultrasonic transducers for medical diagnostic imaging. Biomed. Eng.Lett. 7 (2017): 91-97.

[5] Koksel, Filiz, et al. Ultrasonic investigation of the effects of composition on the volume fraction of bubbles and changes in their relative sizes in non-yeasted gluten-starch blend doughs. J. Food Eng. 204(2017).

[6] Ji, Hongjun, Y. Qiao, and M. Li. Rapid formation of intermetallic joints through ultrasonic-assisted die bonding with Sn–0.7Cu solder for high temperature packaging application. Scr. Mater. 110(2016):19-23.

[7] Amri, K., et al. "Fluid flow velocity measurement using dual-ultrasonic transducer by means of simultaneously transit time method." International Conference on Instrumentation, Communications, Information Technology, and Biomedical Engineering IEEE, 2016:113-116.

[8] Lauterborn, W, and A. Koch. Holographic observation of period-doubled and chaotic bubble oscillations in acoustic cavitation. Phys. Rev. A. 35(1987):1974-1977.

[9] Parlitz, U., et al. Bifurcation structure of bubble oscillators. J.Acoust.Soc.Am. 88(1990):1061-1077.

[10] Behnia, Sohrab, et al. Nonlinear transitions of a spherical cavitation bubble. Chaos Solitons Fract. 41.2(2009):818-828.

[11] Sojahrood, Amin Jafari, et al. Influence of the pressure-dependent resonance frequency on the bifurcation structure and backscattered pressure of ultrasound contrast agents: a numerical investigation. Nonlinear Dyn. 80(2015):889-904.

[12] Jiang, L., et al. Frequency spectrum of the noise emitted by two interacting cavitation bubbles in strong acoustic fields. Phys. Rev. E. 85 (2012):036312-036319.

[13] Sugita, N, and T. Sugiura. Nonlinear normal modes and localization in two bubble oscillators. Ultrasonics 74(2016):174-185.

[14] Vlaisavljevich, Eli, et al. Effects of Tissue Stiffness, Ultrasound Frequency, and Pressure on Histotripsy-induced Cavitation Bubble Behavior. Phys. Med. Biol. 60 (2015):2271-92.

[15] Shima, A. The Natural Frequencies of Two Spherical Bubbles Oscillating in Water. J. Fluids Eng. 93(1971):426-431.

[16] Doinikov, A. A. Translational motion of two interacting bubbles in a strong acoustic field. Phys. Rev. E. 64 (2001):026301-026307.

[17] Haar, Lester, J. S. Gallagher, and G. S. Kell. NBS/NRC Steam Tables, Hemisphere, New York ,1984, pp. 320.

[18] Konarski, Stephanie G, M. R. Haberman, and M. F. Hamilton. Dispersion and acoustic nonlinearity in heterogeneous media containing hyperelastic inclusions. J.Acoust.Soc. Am.141(2017):3811-3811.

IOP Conf. Series: Journal of Physics: Conf. Series **1064** (2018) 012041 doi:10.1088/1742-6596/1064/1/012041

Calculation the total resistance and squat of underwater towed vehicle based on CFD

Xianlong Liu [1,*], Miao Wang [2]

[1]CSIC 750 Test Range, Kunming, China
[2]Sibeia Tiger Park, Harbin, China

*Corresponding author e-mail: liuxianlong@yeah.net

Abstract. CFD method is used to calculate drag and sinking force of underwater towed body. Using cubic spline curve interpolation and finite difference method are used to calculate the tangential force and normal force of towing cable. Finally, the squat of underwater towed vehicle in the water and the total force at the end of towing cable are obtained.

1. Introduction

Underwater towed vehicle is widely used in various underwater products. Underwater towed vehicle are carried with sonar, flowmeter, salinometer and other types of sensors, which are used to measure various hydrologic parameters. Underwater towed vehicle is connected with the ship and then moves together. It is an interesting thing to calculate the pulling force at the end of the towing cable at different speeds and the sinking of the underwater towed vehicle.

2. Introduction of underwater towed vehicle and cable finite element

2.1. Underwater towed vehicle shape

The shape of the underwater towed vehicle is shown in Figure 1. The diameter of the underwater towed vehicle is 200mm, and the length is 1200mm.

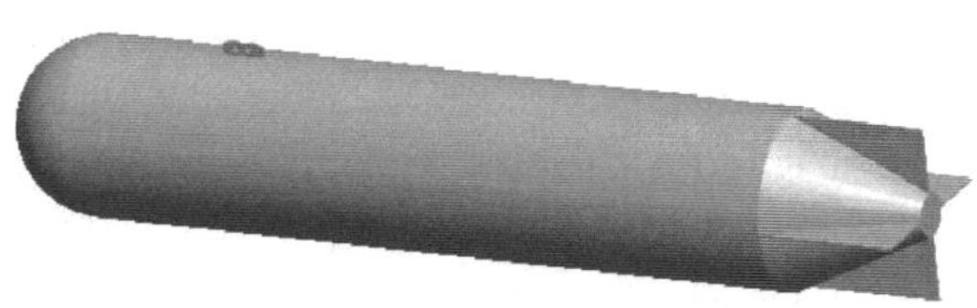

Figure 1. Three-dimensional graph of underwater vehicle

When towing cable moved at a certain speed, its shape is unknown, so it is impossible to calculate the resistance performance directly by CFD method. In addition, even knowing its shape, because its length diameter ratio is very large, up to 600, therefore, the towing cable must be discrete.

Content from this work may be used under the terms of the Creative Commons Attribution 3.0 licence. Any further distribution of this work must maintain attribution to the author(s) and the title of the work, journal citation and DOI.
Published under licence by IOP Publishing Ltd

IOP Conf. Series: Journal of Physics: Conf. Series **1064** (2018) 012041 doi:10.1088/1742-6596/1064/1/012041

2.2. Discretization of towing cable

Because the towing cable is flexible, it will be deformed by the force. Therefore, the angle of attack of the towing cable at different place is different. The finite difference method is used to process the towing cable into finite element, and the stress analysis is carried out for each element. The force, rotation angle and coordinate point of the unknown position are obtained from the initial point.

In order to meet the needs of the project and basically meet the actual situation, the following assumptions are made:

a) Telescopic overlook of cable finite element under tension;

b) The towing cable is flexible and cannot transfer bending moment

c) The flow field is only considered as a two-dimensional flow.

d) The hydrodynamic coefficient on the finite element of the towing cable is equivalent to a towing cable that acts on an infinite length.

Force analysis of finite element of the towing cable is shown in Figure 2.

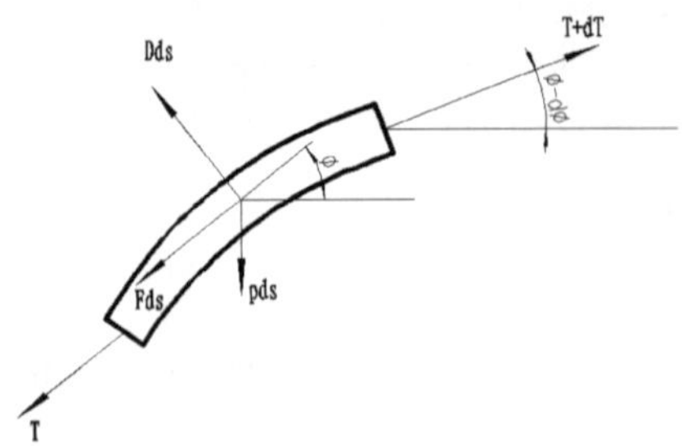

Figure 2. Force analysis of finite element of towing cable

According to the decomposition of the tangential force and the normal force, get:

$$(T+dT)\sin(d\phi)+pds\cos(\phi)=Dds \tag{1}$$

$$(T+dT)\cos(d\phi)=pds\sin(\phi)+Fds+T \tag{2}$$

There, T is force, D is the normal force of finite element of the towing cable, F is the tangential force of finite element of the towing cable, p is the weight of finite element of the towing cable.

Neglecting the small amount of the two orders in equality (1), get:

$$ds=\frac{T}{D-p\cos\phi}d\phi \tag{3}$$

The equation (3) is replaced by the equation (2), get

$$dT=\frac{F+p\sin\phi}{D-p\cos\phi}Td\phi \tag{4}$$

By equation (3) and (4), the towing force and rotation angle of the towing cable in different positions can be obtained.

In addition:

$$dx=ds\times\cos\phi \tag{5}$$

$$dy = ds \times \sin\phi \tag{6}$$

The coordinates of the corresponding position points can be obtained by equation (5) and (6).

The length of the towing cable is discretized, and the iterative calculation is carried out by equation (3) and (4).

Computational fluid dynamics and finite difference method are used to calculate D, F.

It is necessary to explain that the empirical formula calculation D and F do not consider the mutual interference between the tangential velocity and the normal velocity, and the calculation results are not correct enough. At present, CFD technology has been developed more mature, and can be directly calculated. In addition, CFD can only calculate tangential forces and normal forces under limited attack angles. In order to meet the need of actual attack angle calculation, the three-spline interpolation method is applied to solve them.

In this study, the diameter of the towing cable is 9mm and the length is 50m.

3. Principle of curve interpolation

Curve interpolation usually has exponential function interpolation, Lagrange interpolation, Cubic spline interpolation, B spline curve, NURBS curve and so on.

3.1. Cubic spline curve interpolation

There, $a = x_0 < x_1 < \cdots < x_n = b$, if function $y(x)$ satisfying two conditions:

(a) $y(x)$ is Cubic Polynomial on each interval $[x_{i-1}, x_i], i = 1, 2, \cdots, n$;

(b) $y(x)$ is twice continuously differentiable function on interval $[a, b]$.

Then, $y(x)$ is Cubic spline on interval $[a, b]$. There, $x_i\,(i = 0, 1, \cdots, n)$ is the node. The interpolation points are shown in Table 1.

Table 1. Interpolation points.

x	x_0	x_1	$\cdots$	x_n
y	y_0	y_1	$\cdots$	y_n

Cubic spline interpolation function:

$$s(x) = \frac{h_k + 2(x - x_k)}{h_k^3}(x - x_{k+1})^2 y_k + \frac{h_k - 2(x - x_{k+1})}{h_k^3}(x - x_k)^2 y_{k+1} + \frac{(x - x_k)(x - x_{k+1})^2}{h_k^2} m_k + \frac{(x - x_{k+1})(x - x_k)^2}{h_k^2} m_{k+1} \tag{7}$$

In equality (1),

$$h_k = x_{k+1} - x_k \,(k = 0, 1, \cdots, n-1) \tag{8}$$

$$\lambda_k m_{k-1} + 2m_k + \mu_k m_{k+1} = g_k \,(k = 1, 2, \cdots, n-1) \tag{9}$$

Where,

$$\lambda_k = \frac{h_k}{h_k + h_{k-1}}, \mu_k = \frac{h_{k-1}}{h_k + h_{k-1}}, g_k = 3(\mu_k \frac{y_{k+1} - y_k}{h_k} + \lambda_k \frac{y_k - y_{k-1}}{h_{k-1}})(k = 1,2,\cdots,n-1)$$

4. Basic theory of fluid mechanics

4.1. Control equation

The continuity equation and momentum equation for incompressible viscous fluid are:

$$\frac{\partial \overline{u_i}}{\partial x_i} = 0 \tag{10}$$

$$\rho \frac{\partial \overline{u_i}}{\partial t} + \rho \partial \overline{u_j} \frac{\partial \overline{u_i}}{\partial x_i} = \rho \overline{F_i} - \frac{\partial \overline{p}}{\partial x_i} + \frac{\partial}{\partial x_j}(\mu \frac{\partial \overline{u_i}}{\partial x_j} - \rho \overline{u_i' u_j'}) \tag{11}$$

Where, ρ is density, μ is Viscosity Coefficient, $\overline{p}$ is average pressure, $\overline{F_i}$ is external force, $\overline{u_i}$ is mean velocity, u_i' is fluctuating velocity, $-\rho u_i' u_j'$ is reynolds stress.

4.2. Turbulence model

In order to make the equation closed, a new turbulent model equation must be introduced to link the fluctuating values with the time average in the stress terms. There, we chose the $RNGk - \varepsilon$ equtions.

The transport equations of turbulent kinetic energy and turbulent fluctuation intensity in the $RNGk - \varepsilon$ equation is:

$$\frac{\partial}{\partial t}(\rho k) + \frac{\partial}{\partial x_i}(\rho k u_i) = \frac{\partial}{\partial x_j}[\alpha_k \mu_{eff} \frac{\partial k}{\partial x_j}] + G_k - \rho\varepsilon + S_k \tag{12}$$

$$\frac{\partial}{\partial t}(\rho\varepsilon) + \frac{\partial}{\partial x_i}(\rho\varepsilon u_i) = \frac{\partial}{\partial x_j}[\alpha_k \mu_{eff} \frac{\partial \varepsilon}{\partial x_j}] + G_{1\varepsilon} \frac{\varepsilon}{k} G_k - C_{2\varepsilon} \rho \frac{\varepsilon^2}{k} - R_\varepsilon + S_\varepsilon \tag{13}$$

Where, $\mu_{eff} = \mu + \mu_t$, $\mu_t = \rho C \mu \frac{k^2}{\varepsilon}$, $G_k = -\rho \overline{u_i' u_j'} \frac{\partial u_j'}{\partial x_i}$, $R_\varepsilon = \frac{C_\mu \rho \eta^3 (1 - \frac{\eta}{\eta_0})}{1 + \beta\eta^3} \frac{\varepsilon^2}{k}$, $\eta \equiv \frac{Sk}{\varepsilon}$,

$S = \sqrt{2 S_{ij} S_{ij}}$, $S_{ij} = \frac{1}{2}(\frac{\partial u_i}{\partial x_j} + \frac{\partial u_j}{\partial x_i})$, S_k and S_ε are user-defined source terms.

Constant, $G_{1\varepsilon} = 1.42$, $G_{2\varepsilon} = 1.68$, $C_\mu = 0.0845$, $\sigma_k = 1.0$, $\sigma_\varepsilon = 1.3$, $\eta_0 = 4.38$, $\beta = 0.012$

5. CFD Simulation

The grid diagram of the underwater towed vehicle is shown in Figure 3.

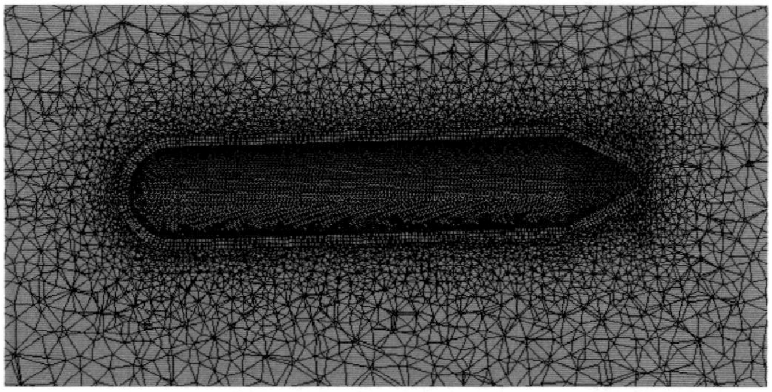

Figure 3. Grid graph of underwater towing

In order to obtain high-quality meshes, structured grids are used for the entire computational domain. The governing equations and turbulence models are discretized by the finite volume method and solved by the coupled implicit solver.

The pressure-velocity coupling iterative algorithm using SIMPLEC algorithm, pressure by the standard discretization, momentum, turbulent kinetic energy and turbulent dissipation rate by two order upwind scheme, relaxation factors are the default values. Wall function method is adopted in the near wall region.

The drag and sinking force of the underwater towed vehicle at different speeds is shown in Table 2.

Table 2. The drag and sinking force of the underwater towed vehicle

v/kn	0.5	3	5	7	9	12
R/N	0.61	17.79	46.95	89.14	144.01	249.61
F/N	1.28	37.36	98.60	187.19	302.41	524.17

By using cubic spline interpolation mentioned above, the tangential force and normal force at different speeds and angles of attack can be obtained. The tangential force and normal force of streamline microelement at different positions are solved by finite difference method. It should be noted that this calculation method was started by underwater towing body.

In this way, the squat and terminal resistance values of the underwater towed vehicle are shown in Table 3.

Table 3. The total force and squat of the underwater towed vehicle.

v/kn	0.5	3	5	7	9	12
s/m	18.2	12.3	9.9	9.3	9.1	8.6
F/N	8.1	52.5	135.5	252.9	404.5	695.9

6. Conclusion

In this paper, CFD method is used to calculate drag and sinking force of underwater towed vehicle in water. In order to continue to calculate its squat in water and drag value at the end of towing cable. The finite-difference method is used for the microelement of the towing cable. Using cubic spline interpolation, the tangential force and normal force of streamline microelement at different angles of attack are obtained. Finally, the squat of the underwater towed body and the total force at the end of the towing cable are calculated.

The method is simple and reliable.

1) Compared with the traditional empirical formula, it considers the mutual interference between tangential force and normal force, and the calculation is more accurate.

2) It is simpler and more efficient to solve the tangential force and normal force of microelement in different direction of attack than CFD.

References

[1] C Cheng, V Pallayil. Noise cancellation for an autonomous underwater vehicle - towed thin line array through recursive adaptive filtering. Journal of the Acoustical Society of America, 141(5):3925-3925, 2017.

[2] KD Lepage, G Canepa, J Bates, et al.Bistatic continuous active sonar processing using ar-rays towed from unmanned underwater vehicles. Journal of the Acoustical Society of America, 141 (5) :3850-3850, 2017.

[3] B Ropars, L Lapierre, A Lasbouygues, et al. Redundant actuation system of an underwater vehicle. Ocean Engineering, 151 :276–289, 2018.

[4] TV Mai , HS Choi , J Kang, et al. A study on hovering motion of the underwater vehicle with umbilical cable. Ocean Engineering,135:137-157, 2017.

[5] C Shen , Y Shi , B Buckham. Trajectory Tracking Control of an Autonomous Under-water Vehicle Using Lyapunov-Based Model Predictive Control. IEEE Transactions on Industrial Electronics, 65 (7):5796-5805, 2018.

[6] X Xiang , C Yu , Q Zhang. On intelligent risk analysis and critical decision of under-water robotic vehicle. Ocean Engineering,140 :453-4658,2017.

[7] HF Al-Qrimli. Identification of Hydrodynamics Coefficient of Underwater Vehicle Using Free Decay Pendulum Method. International Journal of Powder Metallurgy, 6 (1) :4,2017.

[8] A Lock , G Spiers , B Hostetler , et al. Multi-dimensional water quality assessment of an urban drinking water source elucidated by high resolution underwater towed vehicle-mapping. Water Research , 93 :289-295, 2016.

[9] KD Lepage , G Canepa , J Bate, et al. Bistatic continuous active sonar processing using arrays towed from unmanned underwater vehicles. Journal of the Acoustical Society of America , 141 (5) :3850-3850, 2017.

[10] W Bao , X Quan , Y Li , S Cheng, et al. Parameters prediction method of underwater grid fins during the procedure of dynamic expansion. Guofang Keji Daxue Xuebao/journal of National University of Defense Technology,39 (3) :52-57,2017.

IOP Conf. Series: Journal of Physics: Conf. Series **1064** (2018) 012042 doi:10.1088/1742-6596/1064/1/012042

Velocity analysis of oil film between friction pairs in hydro-viscous drive clutch

Liyu Chen [1], Yun Wang [1,*] Chao Yu[1] and Fangwei Xie[2]

[1]School of Mechanical Engineering, Jiangsu University, Zhenjiang, China
[2]The State Key Laboratory of Fluid Power and Mechatronic Systems, Zhejiang University, Hangzhou, China

*Corresponding author e-mail: wangyun@ujs.edu.cn

Abstract. A mathematical model describing the flow field and velocity distribution between rotational friction pairs has been proposed based on viscosity-temperature properties and centrifugal force of working oil. The effect of inlet pressure, rotation speed and inertia on the flow are theoretically analysed. It is found that inlet pressure has a linear effect on the radial direction and the flow flied is the Poiseuille flow. However, inertia makes the flow more complex to conserve the continuity. With the effect of inertia, the velocity is negative at entrance and positive at exit. Relative to inertia and inlet pressure, the rotate speed is supreme dominate the flow field. And also, the variety of viscosity has a great effect on the radial velocity. The proposed model may provide technical support for the design of Hydro-viscous drive systems.

1. Introduction

Hydro-Viscous Drive (HVD) clutch contains lots of friction pairs which is filled with working oil. It can transmit power by the sheer force of working oil as friction pairs rotate and realize step-less speed change [1, 2] by adjusting the distance between friction pairs. And is widely used in energy transmission equipment, especially for belt conveyer.

Over the past years, researchers have conducted enormous studies on mechanism system of HVD. Jianzhong Cui [3, 4] et al. simulated two types of warping deformation and obtained the temperature distributions of multi-disk friction pairs with a numerical investigation. Kimura [5] and Gao [6, 7] took surface roughness into account in order to investigate the engagement process of HVD clutch. Zhao [8] and Aphale [9] established a 3D computational fluid dynamics model to simulate the pressure and torque distribution of the friction pairs. Jen [10] and Meng [11] studied speed regulating start in engagement process and found oil film squeezing had a significant effect for system operation. Marklund [12] obtained a torque transfer model of a wet clutch working in boundary lubrication regime. Xie Fangwei [13, 14] developed a simulated pressure and temperature distribution in radial and circumferential direction and obtained influence of grooves on temperature and pressure distribution. Wu Wei [15] and Huang JiaHai [16, 17] studied and analyzed the influence of radial grooves on pressure, load carrying capacity and torque transmission. In a word, their research shows that the flow field of oil film varies significantly during the high-speed operation.

Although great progress is made in the experimental and simulation study of the flow field, it is a lack of analytic solution of the simplified N-S equations. In addition, the viscosity of working oil was

Content from this work may be used under the terms of the Creative Commons Attribution 3.0 licence. Any further distribution of this work must maintain attribution to the author(s) and the title of the work, journal citation and DOI.
Published under licence by IOP Publishing Ltd

generally treated as a constant in the previous research, generating a big discrepancy between simulation and experimental results. Therefore, the aim of this paper is to theoretically analyze the flow field of oil film by obtaining the analytic solution of the simplified N-S equations and develop a model based on viscosity-temperature formula by Taylor expansion.

2. Mathematical model

Due to the influence of pressure difference and centrifugal force, the oil film moves in a spiral form when the system is set in motion as shown in Fig.1. It is assumed that the oil film is an incompressible Newton fluid and the flow is laminar and steady, hence $\partial/\partial t = 0$ and the dynamic viscosity μ is unrelated to the pressure; the thickness of oil film is far less than radial distance $\delta/r = 0$, hence axial velocity w=0.

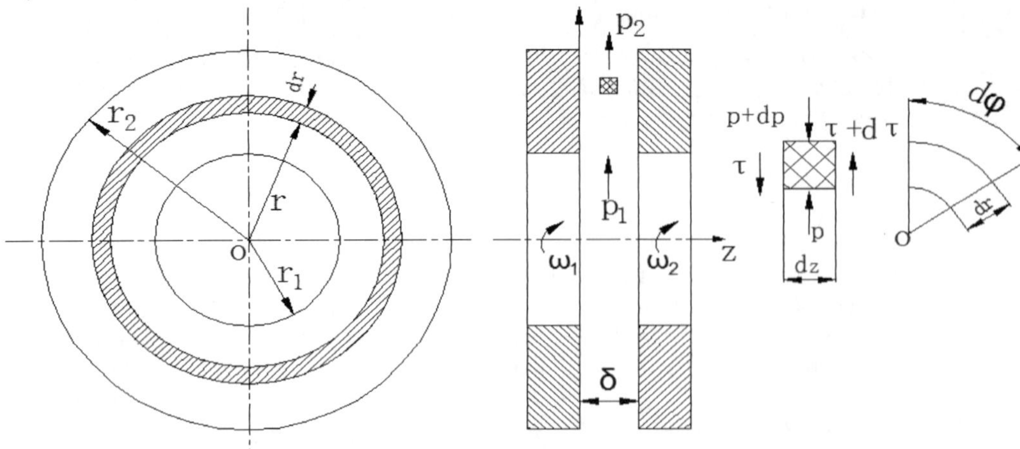

Fig 1. Model of friction pairs.

Then the simplified N-S equations and continuity equation can be expressed as:

$$\rho(\frac{u^2}{r}+\frac{v^2}{r}+r\omega^2)=\frac{\partial p}{\partial r}-\mu\frac{\partial^2 u}{\partial z^2} \tag{1}$$

$$\rho(u\frac{\partial v}{\partial r}+\frac{uv}{r})=\mu\frac{\partial^2 v}{\partial z^2} \tag{2}$$

$$\frac{\partial p}{\partial z}=0 \tag{3}$$

$$\frac{\partial u}{\partial r}+\frac{u}{r}=0 \tag{4}$$

And the boundary conditions can be described as $\begin{cases} z=0 & u=0 \\ z=\delta & u=0 \end{cases}$, $\begin{cases} z=0 & v=r\omega_2 \\ z=\delta & v=r\omega_1 \end{cases}$,where $\omega = \frac{\omega_1-\omega_2}{\delta}z+\omega_2$, u is radial velocity, v is circumferential velocity, ω_1 and ω_2 are the angular velocity of friction dicks and separator disks, respectively.

As the momentum equations are non-homogeneous, it is difficult to solve them directly. By the constant variation method, the momentum equations be solved in two steps: the first step is to transform the original equations to homogeneous.

$$\frac{\partial p}{\partial r}=\mu\frac{\partial^2 u_1}{\partial z^2} \tag{5}$$

$$\mu\frac{\partial^2 v_1}{\partial z^2}=0 \tag{6}$$

then substituting the solutions from first step into original equations.

Integrating Eq.5 and Eq.6 along z direction with the boundary conditions, u_1 and v_1 can be written as:

$$u_1 = \frac{z(z-\delta)}{2\mu}\frac{\partial p}{\partial r} \tag{7}$$

$$v_1 = \frac{r(\omega_1-\omega_2)}{\delta}z + r\omega_2 \tag{8}$$

Then radial flow can be written as:

$$Q_1=2\pi r\int_0^\delta u_1 dz=-\frac{\pi r\delta^3}{6\mu}\frac{\partial p}{\partial r} \tag{9}$$

Integrating Eq.5 with $p_{r_1} = p_1$, $p_{r_2} = 0$ yields:

$$u_1 = -\frac{z(z-\delta)p_1}{2\mu r \ln\frac{r_2}{r_1}} \tag{10}$$

Substituting Eq.7 and Eq.10 into Eq.1 and Eq.2, the momentum equations can be written as

IOP Conf. Series: Journal of Physics: Conf. Series **1064** (2018) 012042 doi:10.1088/1742-6596/1064/1/012042

$$\mu\frac{d^2u}{dz^2}=\frac{\partial p}{\partial r}-\frac{\rho p_1^2(z-\delta)^2z^2}{4\mu^2r^3(\ln\frac{r_2}{r_1})^2}-2\rho r[(\frac{\omega_1-\omega_2}{\delta})z+\omega_2]^2 \tag{11}$$

$$\mu\frac{d^2v}{dz^2}=\frac{\rho(\delta-z)zp_1}{\mu r\ln\frac{r_2}{r_1}}[(\frac{\omega_1-\omega_2}{\delta})z+\omega_2]\frac{\partial p}{\partial r} \tag{12}$$

Integrating Eq.11 and Eq.12, yields the velocity:

$$u=\frac{z(z-\delta)}{2\mu}\frac{\partial p}{\partial r}-\rho p_1^2\frac{2z^6-6\delta z^5+5\delta^2z^4-\delta^5z}{240\mu^3r^3(\ln\frac{r_2}{r_1})^2} \\ -\rho r\frac{(\omega_1-\omega_2)^2z^4+4(\omega_1-\omega_2)\omega_2\delta z^3+6\omega_2^2\delta^2z^2-(\omega_1^2+2\omega_1\omega_2+3\omega_2^2)\delta^3z}{6\mu\delta^2} \tag{13}$$

$$v=-\frac{\rho p_1}{\mu^2r\ln\frac{r_2}{r_1}}(\frac{\omega_1-\omega_2}{20}z^5+\frac{2\omega_2-\omega_1}{12}z^4-\frac{\omega_2\delta}{6}z^3+\frac{2\omega_1+3\omega_2}{60}\delta^3z) \\ +\frac{r(\omega_1-\omega_2)}{\delta}z+r\omega_2 \tag{14}$$

Then radial flow becomes

$$Q=-\frac{\pi r\delta^3}{6\mu}\frac{\partial p}{\partial r}+\frac{\pi\rho p_1^2\delta^7}{560\mu^3r^2(\ln\frac{r_2}{r_1})^2}+\frac{\pi\rho r^2\delta^3}{30\mu}(3\omega_1^2+4\omega_1\omega_2+3\omega_2^2) \tag{15}$$

Integrating Eq.15 with $p_{r_1}=p_1$, $p_{r_2}=0$ yields:

$$Q=\frac{\pi\delta^3p_1}{6\mu\ln\frac{r_2}{r_1}}+\frac{\pi\rho p_1^2\delta^7}{1120\mu^3(\ln\frac{r_2}{r_1})^3}(\frac{1}{r_1^2}-\frac{1}{r_2^2})-\frac{\pi\rho\delta^3(r_1^2-r_2^2)}{60\mu\ln\frac{r_2}{r_1}}(3\omega_1^2+4\omega_1\omega_2+3\omega_2^2) \tag{16}$$

Combining Eq.15 and Eq.16, radial pressure gradient can be written as

$$\frac{\partial p}{\partial r}=-\frac{p_1}{r\ln\frac{r_2}{r_1}}+\frac{3\rho p_1^2\delta^4}{560\mu^2 r(\ln\frac{r_2}{r_1})^2}[\frac{2}{r^2}-\frac{1}{\ln\frac{r_2}{r_1}}(\frac{1}{r_1^2}-\frac{1}{r_2^2})] \\ +\frac{\rho}{10r}(2r^2+\frac{r_1^2-r_2^2}{\ln\frac{r_2}{r_1}})(3\omega_1^2+4\omega_1\omega_2+3\omega_2^2) \tag{17}$$

Then the radial velocity and pressure can be expressed as

$$u=\frac{z(\delta-z)p_1}{2\mu r\ln\frac{r_2}{r_1}}+\frac{3\rho p_1^2\delta^4 z(\delta-z)}{1120\mu^3 r(\ln\frac{r_2}{r_1})^2}[\frac{2}{r^2}-\frac{1}{\ln\frac{r_2}{r_1}}(\frac{1}{r_1^2}-\frac{1}{r_2^2})] \\ -\rho r\frac{(\omega_1-\omega_2)^2 z^4+4(\omega_1-\omega_2)\omega_2\delta z^3+6\omega_2^2\delta^2 z^2-(\omega_1^2+2\omega_1\omega_2+3\omega_2^2)\delta^3 z}{6\delta^2\mu} \\ -\frac{\rho p_1^2}{4\mu^3 r^3(\ln\frac{r_2}{r_1})^2}(\frac{2z^6-6\delta z^5+5\delta^2 z^4-\delta^5 z}{60}) \\ +\frac{\rho z(z-\delta)}{20\mu r}(2r^2+\frac{r_1^2-r_2^2}{\ln\frac{r_2}{r_1}})(3\omega_1^2+4\omega_1\omega_2+3\omega_2^2) \tag{18}$$

According to Eq.18, three factors affect the radial velocity of oil film: inlet pressure, rotation speed and inertia. Their formulas are shown as follows:

$$u_{inlet}=\frac{z(\delta-z)p_1}{2\mu r\ln\frac{r_2}{r_1}} \tag{19}$$

$$u_{rotate}=\frac{\rho rz}{6\delta^2\mu}[(\omega_1-\omega_2)^2 z^3+4(\omega_1-\omega_2)\omega_2\delta z^2+6\omega_2^2\delta^2 z-(\omega_1^2+2\omega_1\omega_2+3\omega_2^2)\delta^3] \tag{20}$$

$$u_{inertia}=u-u_{inlet}-u_{rotate} \tag{21}$$

$$p = \frac{p_1}{\ln\frac{r_2}{r_1}}\ln\frac{r_2}{r} + \frac{3\rho p_1^2\delta^4}{560\mu^2(\ln\frac{r_2}{r_1})^2}\left(\frac{1}{r_2^2}-\frac{1}{r^2}+\frac{\frac{1}{r_1^2}-\frac{1}{r_2^2}}{\ln\frac{r_2}{r_1}}\ln\frac{r_2}{r}\right) + \frac{\rho}{10}\left(r^2-r_2^2+\frac{r_1^2-r_2^2}{\ln\frac{r_2}{r_1}}\ln\frac{r}{r_2}\right)(3\omega_1^2+4\omega_1\omega_2+3\omega_2^2) \tag{22}$$

According to Xie [18] considering the influence of temperature, the viscosity can be written as

$$\mu=0.2007e^{-10.98r} \tag{23}$$

3. Result analysis

3.1. Effect of inlet pressure

Eq.19 shows the inlet pressure has a linear effect on radial velocity of oil film. When working oil flows into the space between friction pairs, it is a Poiseuille flow due to the viscosity of oil film.

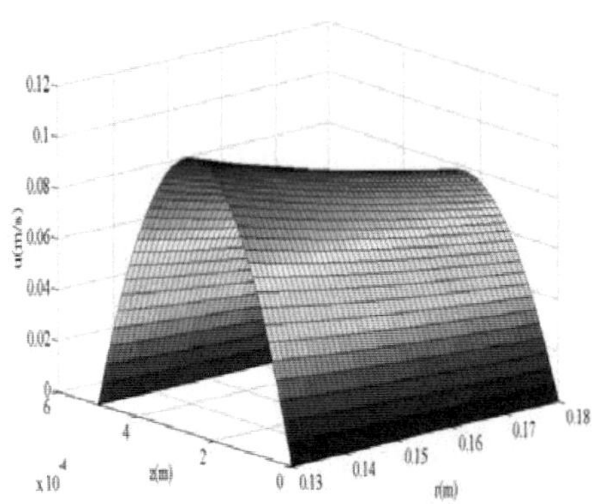

Fig 2. Velocity distribution under inlet pressure

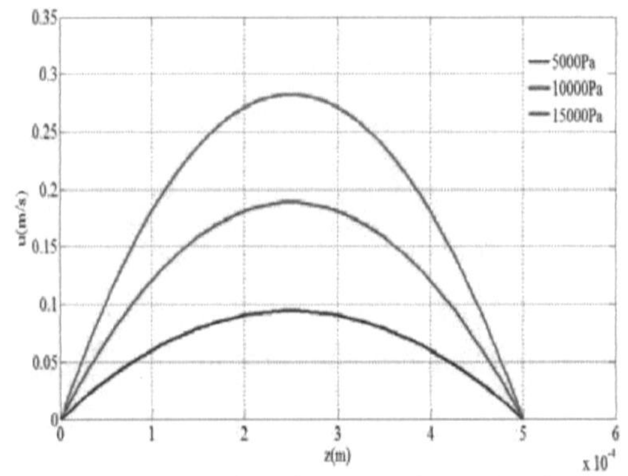

Fig 3. Curve of radial velocity under inlet pressure in axial direction

Fig.2 shows the Poiseuille flow between friction pairs. In radial direction, the velocity is linearly decreasing. In axial direction, the shape of the graph is a parabola. It is symmetric about the vertical line z =0.25mm. From Fig.3, it can be found that as, the velocity increases linearly with the inlet pressure.

3.2. Effect of rotate speed

Because of the viscosity, working oil rotates with friction pairs. As friction disks rotate faster than separate disks, the position with the maximum velocity shifts to friction disks in axial direction. Fig.4 shows the velocity distribution between the friction pairs. In radial direction, the velocity is linearly increasing. In axial direction the velocity is asymmetric. While in Fig.5 it can be seen that as speed of friction disks increase from 1000rpm to 3000rpm, the position with the maximum velocity does not change.

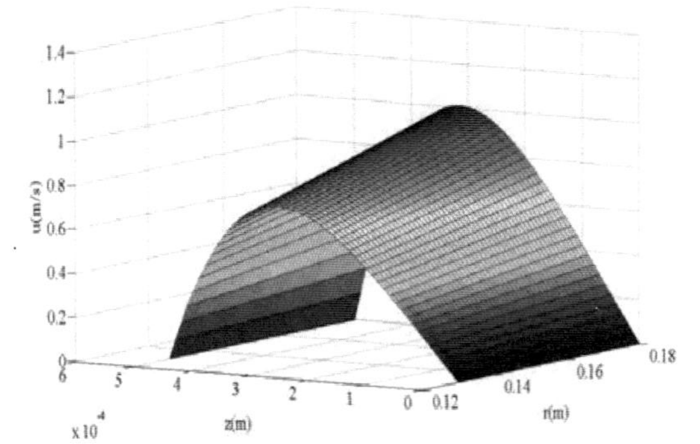

Fig 4. Velocity distribution under rotation speed

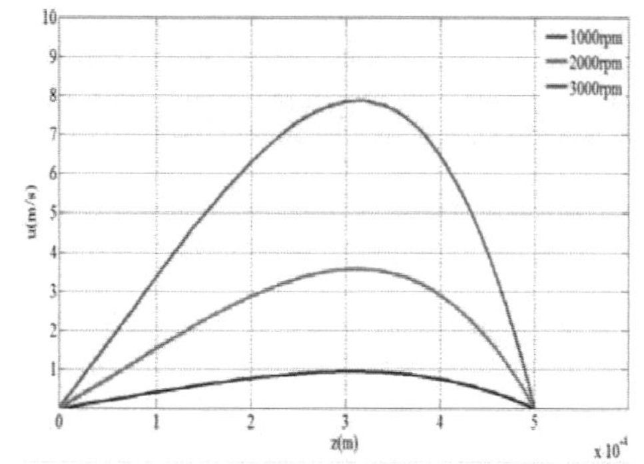

Fig 5. Curve of radial velocity under rotation speed in axial direction

3.3. Effect of inertia

Fig.6 shows that the shape of velocity distribution influenced by the inertia is point symmetry. In radial direction, the velocity is linearly increasing, being negative at entrance and positive at exit. It indicates that the inertia hinders working oil flowing in the inner region and promotes in the outer region to ensure the continuity of the flow. Eq.21 shows that the inertia consists of inlet pressure and rotation speed. From Fig.7 and Fig.8, it can be seen that the effect of inlet pressure is much smaller than that of the rotation speed on radial velocity with a magnitude of $10^{-10} m/s$. In axial direction, the shape of velocity graph is a parabola as shown in Fig.9.

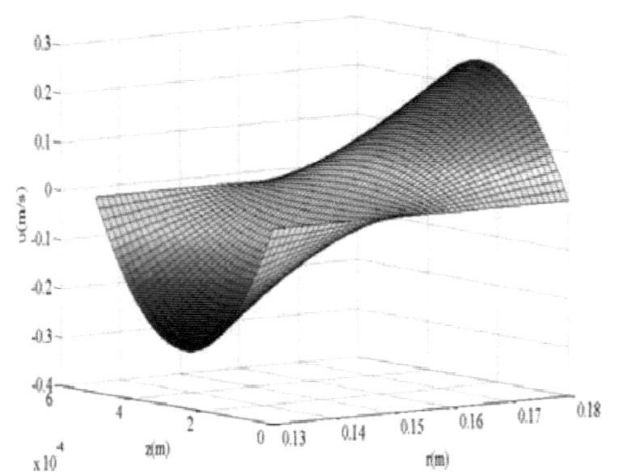

Fig 6. Velocity distribution under inertia

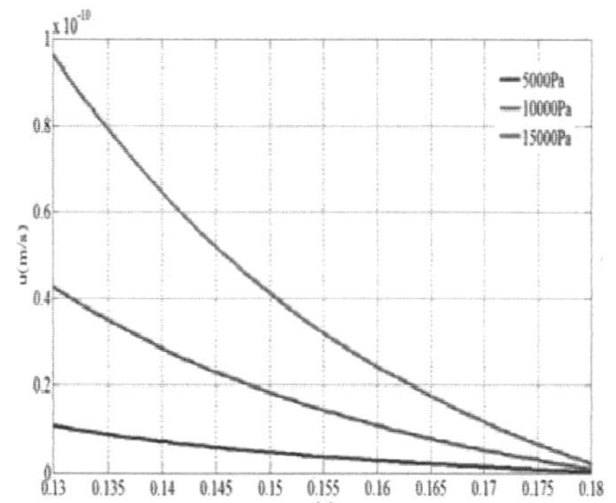

Fig 7. Curve of radial velocity under inertia without rotation speed

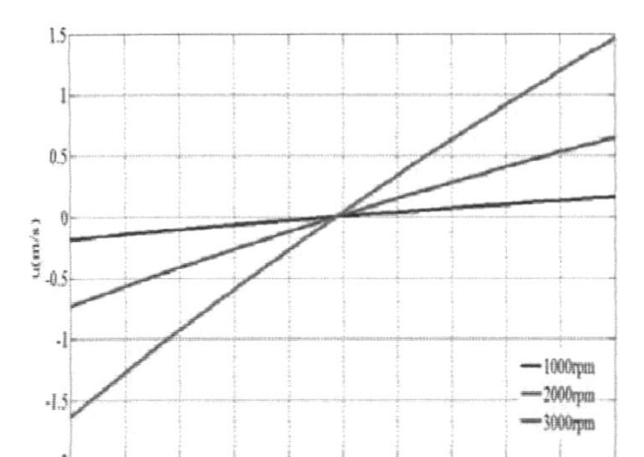

Fig 8. Curve of radial velocity under inertia without inlet pressure

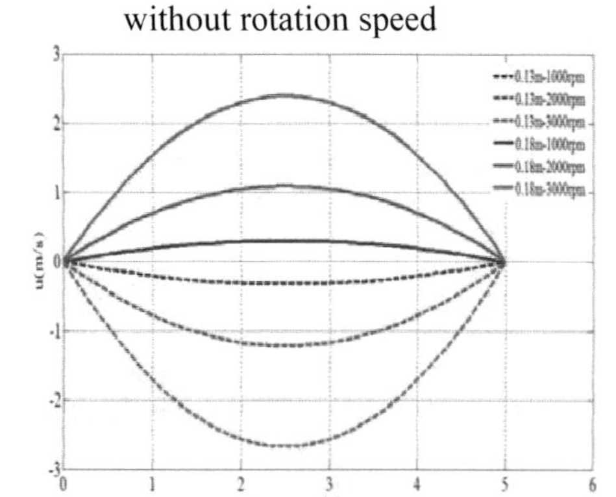

Fig 9. Curve of radial velocity under inertia in axial direction

3.4. Effect of viscoisity

From Eq.18 and Eq.24, it can be seen that in radial direction the viscosity gets smaller and is negatively correlated with the radial velocity. Fig.10 and Fig.11 show the velocity distribution under constant viscosity and variable viscosity. At the location $r = 0.14m$, comparison of velocity under these two conditions is shown in fig.12. The velocity under variable viscosity is 30% larger than that under constant viscosity. That means more oil is needed to ensure the operation of the HVD clutch.

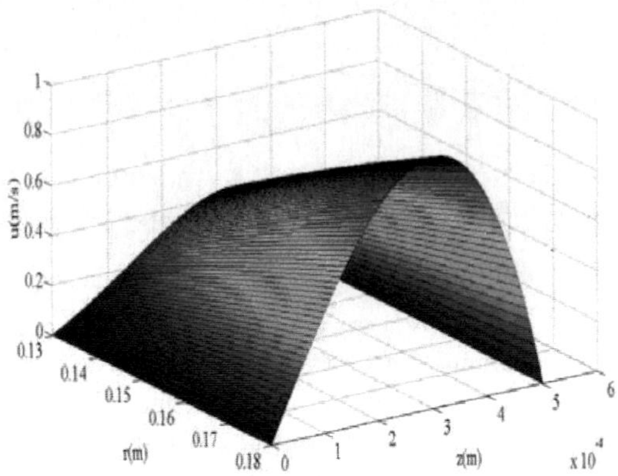

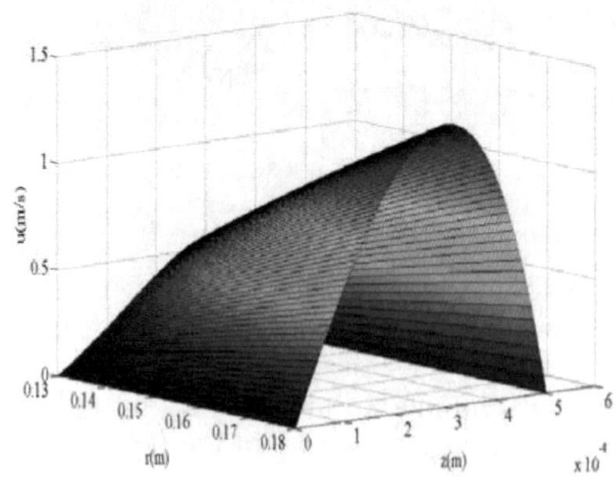

Fig 10. Velocity distribution under constant viscosity; **Fig 11.** Velocity distribution under variable viscosity

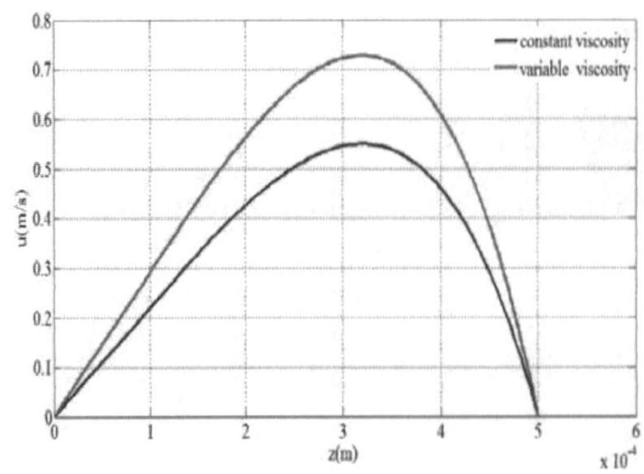

Fig 12. Comparison of radial velocity under constant and variable viscosity

4. Conclusion

A new model was proposed using Taylor series to describe HVD system, taking the viscosity-temperature characteristics into consideration. The flow field of working oil was analyzed in the aspects of inlet pressure, rotation speed and inertia. Several conclusions can be drawn from the study.

(1) Inlet pressure has a linear effect on the radial velocity. In radial direction, velocity is linearly decreasing and is symmetrical in axial direction.

(2) When the friction pairs rotate, the flow is offset to the friction disks while as the speed increases, the offset does not change.

(3) Inertia plays different roles at different position in radial direction. The in inner region, it hinders working oil flowing and promotes oil flow to ensure the continuity of the flow.

(4) Viscosity of working oil has a good effect on the radial velocity which under variable viscosity is 30% larger than that under constant viscosity.

We are convinced that this work will be a contributor tool in the design and optimization for HVD clutch

Acknowledgments
The present work was supported by the National Natural Science Foundation of China (51575245, 51675234). The corresponding content is sponsored by Natural Science, Talent Foundation (2014-23) of Jiangsu Province and Key Project of Jiangsu Province (BE2015134), Postgraduate Research & Practice Innovation Program of Jiangsu Province KYCX17_1764. Open Foundation of the State Key Laboratory of Fluid Power and Mechatronic Systems of Zhejiang University (GZKF-201717).

References
[1] C. Ning, Theoretical and Application Researches on Hydroviscous Drive, Zhejiang University, 2003.
[2] J. Cho, A multi-physics model for wet clutch dynamics, University of Michigan, 2012.
[3] J. Cui, C. Wang, F. Xie, R. Xuan, and G. Shen, Numerical investigation on transient thermal behavior of multidisk friction pairs in hydro-viscous drive, Applied Thermal Engineering, 2014, pp. 409-422.
[4] J. Cui, F. Xie, and C. Wang, Numerical investigation on thermal deformation of friction pair in hydro-viscous drive, Applied Thermal Engineering, 2015, pp. 460-470.
[5] Y. Kimura, and C. Otani, Contact and wear of paper-based friction materials for oil-immersed clutches—wear model for composite materials, Tribology International, 2006, pp. 943-950.
[6] H. Gao, G. Barber, and M. Shillor, Numerical simulation of engagement of a wet clutch with skewed surface roughness, Journal of Tribology, 2002, pp. 305-312.
[7] H. Gao, and G. C. Barber, Engagement of a rough, lubricated and grooved disk clutch ith a porous deformable paper-based friction material, Tribology transactions, 2002, pp. 464-470.
[8] S. Zhao, G. E. Hilmas, and L. R. Dharani, Behavior of a composite multidisk clutch subjected to mechanical and frictionally excited thermal load, Wear, 2008, pp. 1059-1068.
[9] C. R. Aphale, J. Cho, W. W. Schultz, S. L. Ceccio, T. Yoshioka, and H. Hiraki, "Modeling and parametric study of torque in open clutch plates," Journal of tribology, 2006, pp. 422-430.
[10] T.-C. Jen, and D. J. Nemecek, Thermal analysis of a wet-disk clutch subjected to a constant energy engagement, International journal of heat and mass transfer, 2008, pp. 1757-1769.
[11] M. Qing-Rui, and H. You-Fu, Effect of oil film squeezing on hydro-viscous drive speed regulating start, Tribology International, 2010, pp. 2134-2138.
[12] P. Marklund, R. Mäki, R. Larsson, E. Höglund, M. M. Khonsari, and J. Jang, Thermal influence on torque transfer of wet clutches in limited slip differential applications, Tribology international, 2007, pp. 876-884.
[13] F. Xie, Y. Tong, D. Wu, B. Zhang, and K. Dai, Numerical simulation research on effect of oil groove forms on thermal behavior of friction pair in hydro-viscous clutch, Industrial Lubrication and Tribology, 2016, pp. 287-298.
[14] F. Xie, F. Xie, D. Wu, D. Wu, Y. Tong, Y. Tong, B. Zhang, B. Zhang, J. Zhu, and J. Zhu, Effects of structural parameters of oil groove on transmission characteristics of hydro-viscous clutch based on viscosity-temperature property of oil film, Industrial Lubrication and Tribology, 2017, pp. 690-700.
[15] W. Wu, Z. Xiong, J. Hu, and S. Yuan, Application of CFD to model oil–air flow in a grooved two-disc system, International Journal of Heat and Mass Transfer, 2015, pp. 293-301.
[16] H. J. Q. M. L. Lingling, and F. Linjian, Numerical simulation of flow field between frictional pairs in hydroviscous drive surface, Chinese Journal of Mechanical Engineering, 2008, pp. 72-75.
[17] J.-h. Huang, Y.-r. Fan, M.-x. Qiu, and W.-m. Fang, Effects of groove on behavior of flow between hydro-viscous drive plates, Journal of Central South University, 2012, pp. 347-356.

IOP Conf. Series: Journal of Physics: Conf. Series **1064** (2018) 012043 doi:10.1088/1742-6596/1064/1/012043

Numerical simulation of cavitation behavior and peening experiments in cavitation peening processing

Zhongrui Tan, Yun Wang*, Fuzhu Li, Ruitao Li, Chao Yu, Haiyang Fan, and Jianjie Wang

School of Mechanical Engineering, Jiangsu University, Zhenjiang, China

*Corresponding author e-mail: wangyun@ujs.edu.cn

Abstract. The software, Fluent, was used to analyze the cavitation behavior of the submerged water jets generated by contraction nozzle, angle nozzle and organ pipe nozzle. Angle nozzle was chosen for cavitation peening based on the results of numerical simulation. The cavitation peening experiments for aluminum alloy 2A12 were conducted with the different processing parameters. The surface roughness, residual stress distribution and morphology of the treated sample surface were investigated. Results show that the distribution of strengthening area is consistent with that of simulated cavitation bubbles cluster. Using the optimized parameters, the surface residual compressive stress and its depth reach the maximum values of 320MPa and 390μm, respectively, which were increased nearly 2.7 times and 5.5 times than those of the original sample, respectively, while the corresponding surface roughness was only 1.29μm, which was much smaller than that of conventional shot peening.

1. Introduction

In cavitation peening(CP), a high speed submerged water jet with cavitation bubbles cluster blow the surface of metal [1]. The high-pressure shock waves and high-speed microjets generated when the cavitation bubbles collapse near the solid boundary can be used to strengthen the surface of metals. As there are no collisions with solid bodies, the increase in roughness is much smaller than that produced with traditional shot peening [2]. Cavitation peening suppresses hydrogen embrittlement which is sensitive to surface roughness [3]. It can be used in deep and narrow regions where the stress is more concentrated, as demonstrated in the increasement of the gear lifetime [4] and the element in a continuously variable transmission belt [5]. Cavitation nozzle is the core component in the cavitation peening processing, which produces the high speed submerged water jet that contains a large number of cavitation bubbles. Guan et al [6] used Mixture model, Singhal cavitation model and RNG κ-ε turbulence model to simulate the cavitation nozzle and found that the optimal initial diameter of the cylindrical section was 1mm. Zhao et al [7] used Fluent to simulate the internal cavitation behavior of angle nozzle, the results showed that due to the combined effect of reflux and pressure, the expansion angle has an optimum value for the formation rate of cavitation bubbles; when the expansion angle is between 40° and 60°, the smaller the expansion angle is, the better the cavitation effect is. H. Soyama et al [8] investigated the effect of various types of nozzle geometries on the cavitation ability and found that organ pipe nozzle has better cavitation ability than the nozzles modified based on it.

Content from this work may be used under the terms of the Creative Commons Attribution 3.0 licence. Any further distribution of this work must maintain attribution to the author(s) and the title of the work, journal citation and DOI.
Published under licence by IOP Publishing Ltd

This paper uses fluent to simulate the cavitation behavior of contraction nozzle, angle nozzle and organ pipe nozzle used in the cavitation peening processing based on the renormalized group RNG κ–ε turbulence model and Singhal cavitation model. The aluminum alloy 2A12 samples were strengthened using the nozzle with the best cavitation ability. By analyzing the surface roughness, residual stress distribution and surface morphology of the samples in the cavitation peening experiments, the effects of standoff distance and cavitation time on the surface quality of the metal material are obtained, which provide a reference for the optimization of actual processing parameters.

2. Numerical Simulation

2.1. Model establishment

The structural schemes of three nozzles are shown in Fig.1 and Table.1. They include entrance diameter D, resonant cavity diameter B, throat diameter d, throat length l, dilation section diameter E, dilation angle θ, dilation section length L and contractile angle α. In order to reduce the interference factors during the comparison of the cavitation ability of various types of nozzles, the dimensions at the critical structure of the three nozzles shown in the Fig.1 are all taken as the same value without affecting the substantial cavitation ability of the nozzles. According to the actual size of the tank used in the experiment, the size of the external flow field is set to 0.08 m× 0.1m.

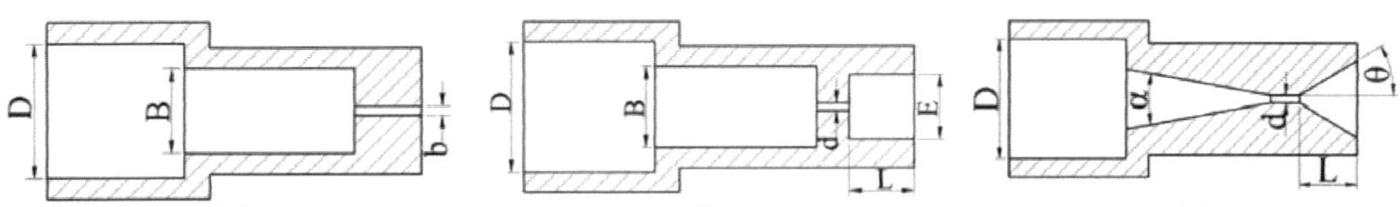

(a) (b) (c)

Fig 1. Nozzle structure diagram: (a) Contraction nozzle, (b) Organ pipe nozzle and (c) Angle nozzle.

Table 1. Nozzle structure parameters.

	D/mm	B/mm	d/mm	l/mm	E/mm	θ/°	L/mm	α/°
Contraction nozzle	16	10	1	4				
Organ pipe nozzle	16	10	1	4	8		8	
Angle nozzle	16	10	1	4		30	8	13.5

Given the symmetry in 2D axial planethe physical model can be simplified in a 2D axisymmetric model. The half-flow field mesh of angle nozzle is shown in Fig.3 and the mesh of other two nozzles are similar to that of angle nozzle. The nozzle inlet is set to pressure inlet, the peripheral boundary to the wall, and the flow field outlet to the pressure outlet. The mesh near the nozzle exit and wall are increased to guarantee the precision while the whole quantity of each mesh is about 180000.

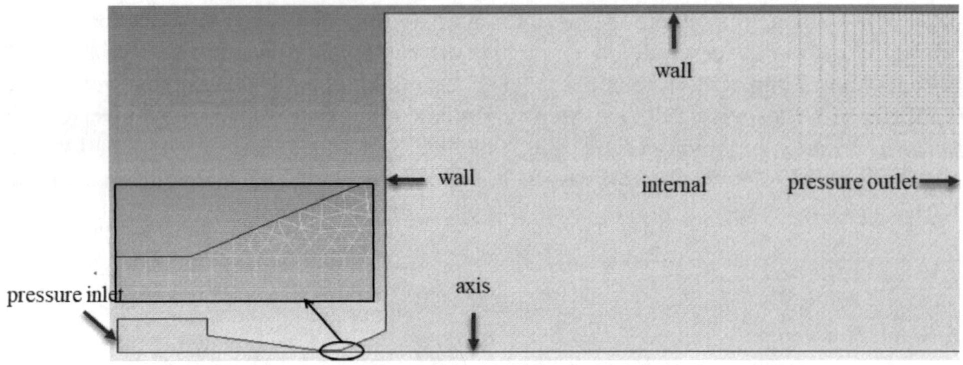

Fig 2. Mesh generation of half-flow field of Angle nozzle.

In this study, a stationary single-phase fluid is assumed under the initial condition. Both inflow and outflow boundaries are modeled as constant-pressure surfaces. Thus, a converged or near converged solution for a single-phase liquid flow is obtained. For the present computations, the Singhal cavitation model [9] is used for it can easily yield a converged result and the renormalized group RNG κ–ε is adopted with standard wall function to calculate the cavitation flow field [9]. The boundary conditions of the wall are impermeability and no-slip for velocity. The normal gradient of pressure is assumed to be zero. The inlet boundary condition is pressure inlet, 20MPa, and the outlet boundary condition is pressure outlet, 101325 Pa. The turbulence intensity of pressure inlet and pressure outlet are setted to 5%.

2.2. Numerical simulation results and analyses

2.2.1. Distributions of velocity and pressure. Velocity distributions of the submerged water jets generated by the three kinds of nozzles are shown in Fig.3. In Fig.3(a) and (b), when the internal water flow of the nozzles passes through the resonant cavity, the formed speed transition areas are small because the cross-sectional areas of nozzles reduce drastically. However, the angle nozzle shown in Fig. 3(c) forms a more pronounced speed gradient in the contracted section due to the cross-sectional area decreases regularly. The jet impact region of the angle nozzle has a relatively pronounced pulsation at X=0.06m and 0.08m, while a small pulsation occurs at X=0.047m for organ pipe nozzle. The occurrence of uneven flow rates and pulsations in the jet impact regions is due to the presence of "necking". The relatively concentrated cavitation bubbles together with the vortex caused by the shear effect between the impact jet and static water lead the fluid to produce a Y-direction velocity, which finally forms a local "necking" phenomenon. Fig.4 and Fig.5 show the axis velocity curves and pressure curves of nozzles, respectively. Combined with Fig.3, it can be found that angle nozzle and organ pipe nozzle can create a wider range of jet impact region than that of contraction nozzle due to the presence of expansion sections. The vertical-lateral maximum isokinetic core area of the angular nozzle is much larger than that of the other two types of nozzles, covering an area of 0.04m to 0.10m in the X axis direction, while a larger isokinetic core area represents a larger low-pressure area which is conducive to the growth and development of cavitation nucleus and the formation of cavitation bubbles cluster.

IOP Conf. Series: Journal of Physics: Conf. Series **1064** (2018) 012043 doi:10.1088/1742-6596/1064/1/012043

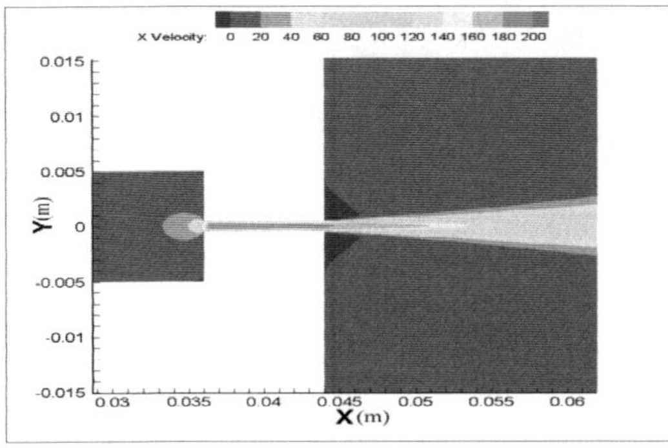

(a)

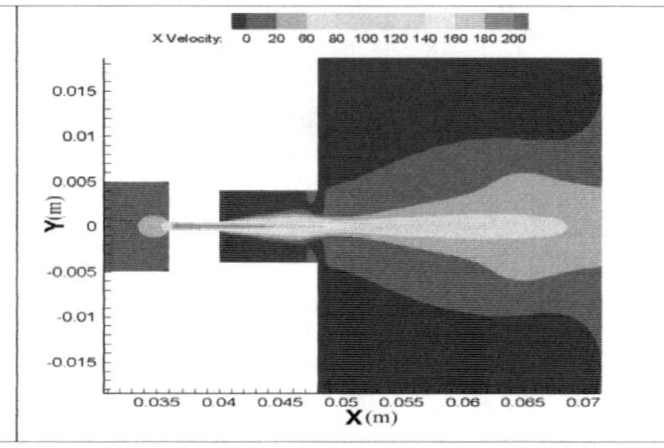

(b)

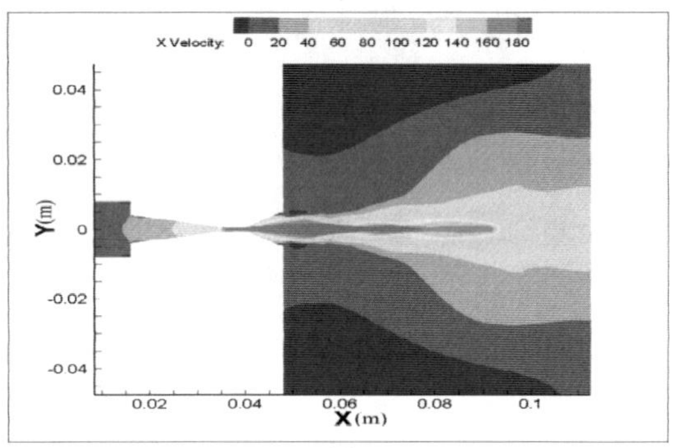

(c)

Fig 3. Distributions of velocity in three different nozzles: (a) Contraction nozzle, (b) Organ pipe nozzle and (c) Angle nozzle.

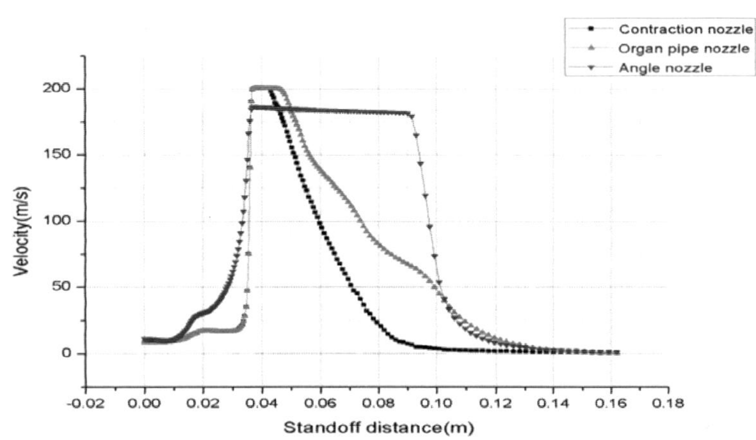

Fig 4. Axial velocity curves.

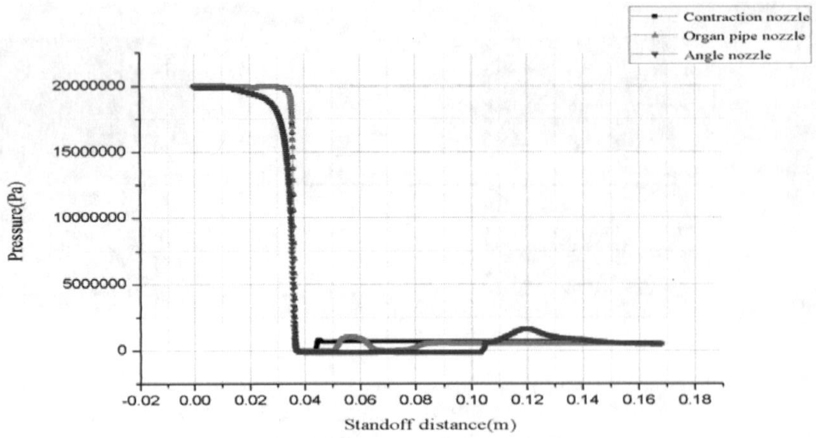

Fig 5. Axial pressure curves.

2.2.2. Distributions of the vapor-phase volume fraction. From Fig.6(a), it can be seen that cavitation firstly occurs on both sides of the throat boundary at the inlet of the throat of contraction nozzle. Then as the value of the X-axis increases, vapor-phase volume fraction gradually decreases and almost disappears when it reaches the throat outlet. As can be seen from Fig.6(b), the cavitation also appears in the expansion section of the organ nozzle due to the vortex caused where the high-speed jet enters the expansion section and shear the low-speed water in the small area. The particularity of the vortex flow makes the surrounding cavitation nucleus grow into the center of the vortex core and develop into cavitation bubbles. With the continuous accumulation of cavitation bubbles in the expansion section, the vapor-phase volume fraction reaches the maximum at the boundary layer of the expansion section, with a peak value of 0.95. The stage II and stage III in Fig.6(b) show the periodic cavitation weakening phenomenon caused by a periodic pulsation. During those stages, the backflow of external environment water quickly fills the area occupied by the disappeared cavitation bubbles in the expansion section, resulting in a large attenuation of the vapor-phase volume fraction distribution area.

Unlike the cavitation of the abovementioned nozzles, the cavitation of angle nozzle first appears near the boundary of the wall at the junction of throat outlet and expansion section. The contraction section of the angle nozzle can form a relatively regular speed increase gradient, so that no jet bulge similar to the abovementioned nozzles occurs at the inlet of the throat. When the jet flows into the expansion section, the high-speed columnar jet winds up the local water at the joint between the expansion section and the throat to form an annular vortex and shear the low-speed water in expansion section to form a coherent vortex. These vortices cause the surrounding cavitation nucleus to grow into cavitation bubbles in the low-pressure region of the center of the vortex core. The vapor-phase volume fraction distribution of ring-shaped cavitation bubbles cluster has an obvious gradient change and finally distributes in a flaming manner, with a peak value of 0.97. The volume of the cavitation bubbles cluster of angle nozzle is much larger than that of organ pipe and there is no periodic reduction. From the standpoint of improving the effectiveness of cavitation peening, angle nozzle is most suitable because it can continuously and stably output a wider range of cavitation bubbles cluster with higher vapor-phase volume fraction than those of the abovementioned nozzles.

IOP Conf. Series: Journal of Physics: Conf. Series **1064** (2018) 012043 doi:10.1088/1742-6596/1064/1/012043

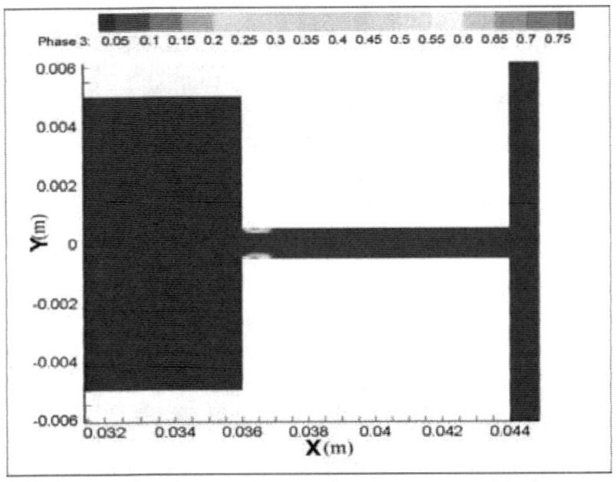

(a) Stage I

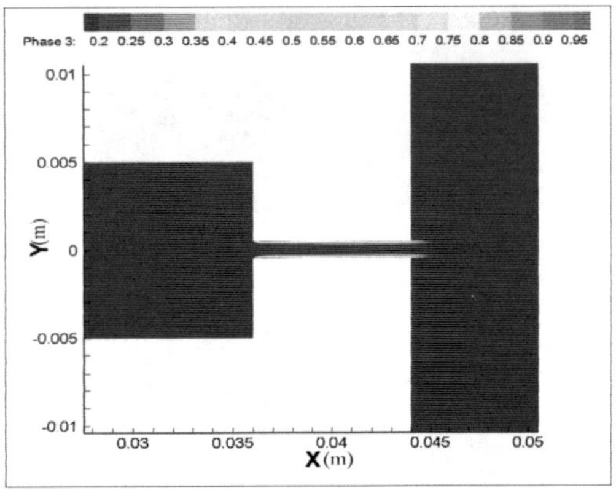

(a) Stage II

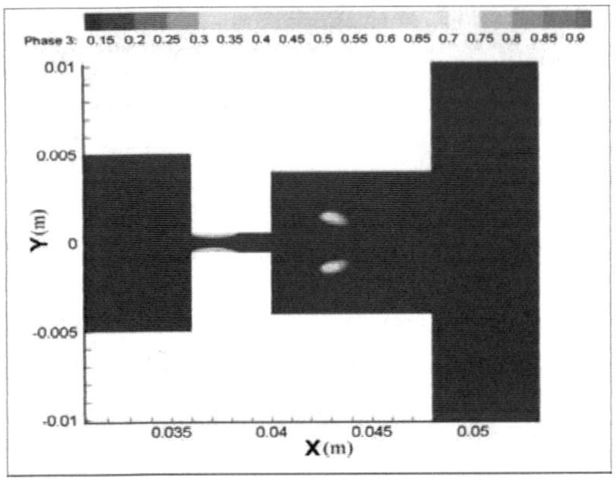

(b) Stage I

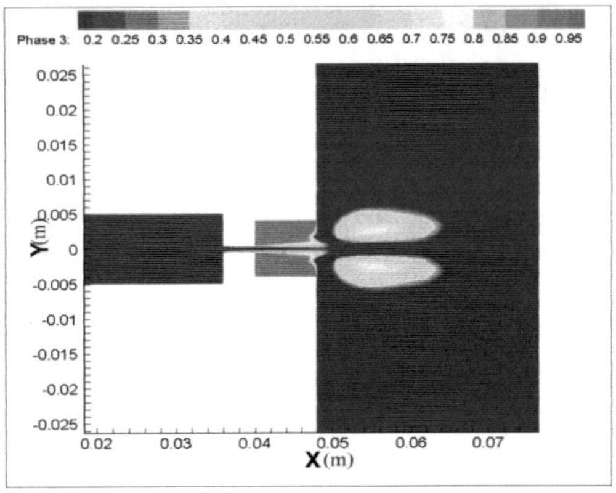

(b) Stage II

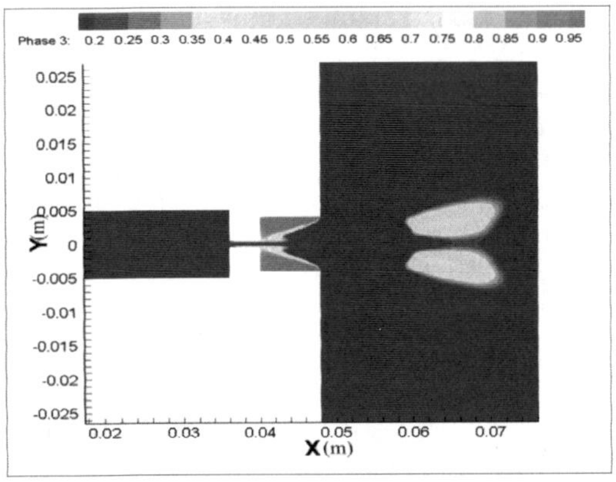

(b) Stage III

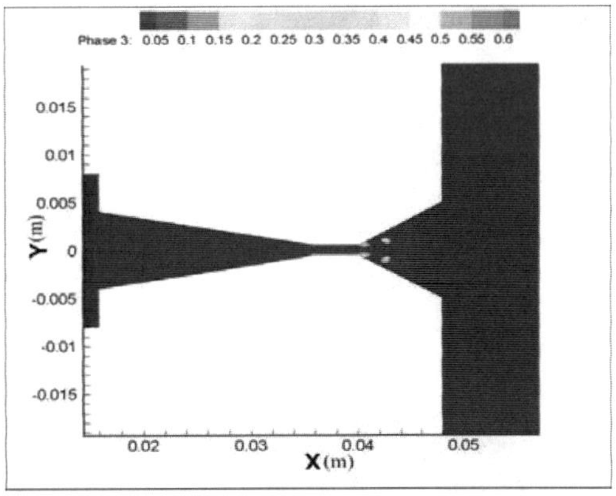

(c) Stage I

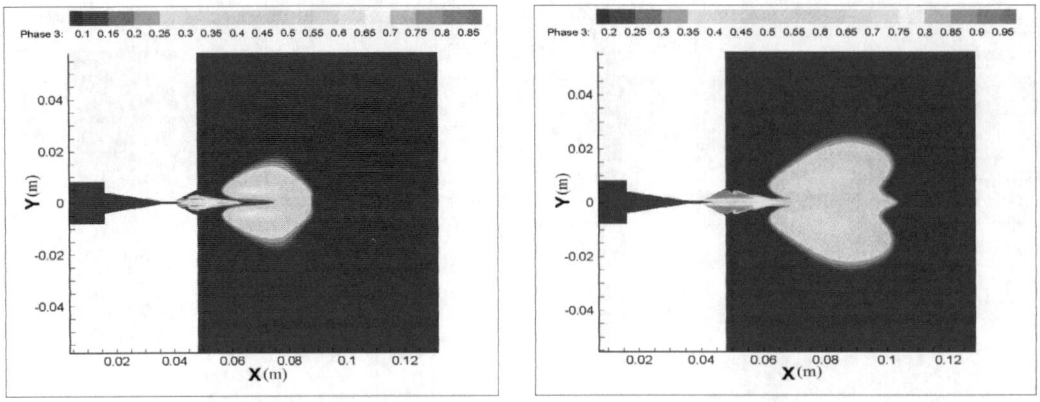

(c) Stage II (c) Stage III

Fig 6. Distributions of the vapor-phase volume fraction at different stages of three different nozzles: (a) Contraction nozzle, (b) Organ pipe nozzle and (c) Angle nozzle.

3. Cavitation peening experiments

3.1. Experimental setup

The peening experiments were conducted with the cavitation peening system shown in Fig.7. During the experiment, the test water was pressurized by a plunger pump and injected through the test nozzle. Then the water jet flowed vertically downward into a transparent water tank full of water. The distance between the target surface and the nozzle outlet section was accurately modified through a screw mechanism. The angle nozzle used in the experiment is identical to its counterpart devoted to the numerical simulation. Cylindrical aluminium alloy 2A12 samples, 40mm in diameter and 3mm in thickness, served as target samples, which were fixed and submerged at the bottom of tank by a clamp, as shown in Fig.7. The detailed processing parameters of the cavitation peening are listed in Table.2.

Fig 7. Arrangement of components of cavitation peening system.

Table 2. Processing parameters of the cavitation peening

	A	B	C	D	E	F	G	H	J	K
Standoff distance S/mm	20	30	40	50	70	40	40	40	40	40
Cavitation time T/min	4	4	4	4	4	1	2	4	6	10
Injection pressure P/MPa	20	20	20	20	20	20	20	20	20	20

3.2. Experiments results and analyses

3.2.1. Surface roughness. Fig.8(a) shows the roughness curve of samples with different standoff distances. The roughness value shows a sharp tower distribution, indicating that cavitation peening has the best standoff distance. The surface roughness first increases with the standoff distance, and then decreases rapidly. When the standoff distance increases to 40mm, the surface roughness increases greatly to a maximum value of 1.29μm. It shows that the cavitation bubbles in the submerged water jet are fully developed at this standoff distance, and gradually collapse due to the stagnation pressure near the sample surface, which makes the sample surface undergo a strong plastic deformation. Such plastic deformation reduces rapidly after the standoff distance exceeds the optimal collapse distance of the cavitation bubbles cluster, so the roughness decreases rapidly at this process. Fig.8(b) shows the roughness curve of samples with different cavitation times. In the initial stage, the surface roughness increases slowly with cavitation time. When the cavitation time reaches to 4 min, the surface roughness enters a short period of approximate plateau because the degree of plastic deformation of the surface layer has become saturated. With the prolonged cavitation time, the surface roughness of the sample rises sharply, which is not cavitation peening but cavitation erosion.

(a)

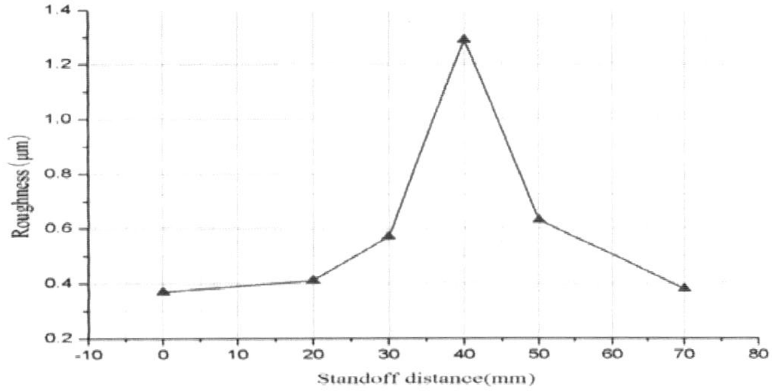

(b)

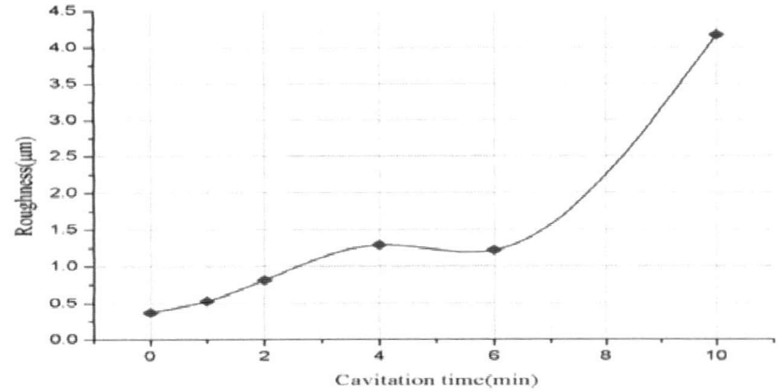

Fig 8. Roughness curves of aluminum alloy 2A12 samples: (a) with different standoff distances, (b) with different cavitation times.

3.2.2. Distribution of residual stress field. As shown in Fig.9(a), the residual compressive stress and its influence depth vary with standoff distances. When the standoff distance is increased to 40mm, the maximum surface residual compressive stress is greatly improved to 320MPa, 2.7 times higher than that of the original sample. At the same time, the influenced depth of the residual compressive stress also reaches a maximum value of about 390μm, 5.5 times higher than that of the original sample. The surface residual stress and its influenced depth show a drastic change with the extension of the standoff distance,while in Fig. 5(b) those two parameters increased smoothly with the prolongation of the cavitation time and maintain a relatively stable state over a certain time. When the cavitation time is controlled between 4min and 6min, the maximum surface residual compressive stress and its influenced depth fluctuate between 310MPa ~ 320MPa and 390μm ~ 400μm, respectively.

However, when the cavitation time is extended to 10min, there is almost no residual compressive stress on the sample surface, and even residual tensile stress is present on the local surface. Thus, in the cavitation time domain of 4min ~ 6min, the surface residual compressive stress and its influence depth do not change much. If the cavitation time is not properly controlled, the surface residual compressive stress and its influence depth will rapidly drop as shown in Fig.8(b). This is because the originally formed strengthening layer of the material is etched away due to the repeated impacts of high-pressure shock waves and high-speed microjets.

(a)

(b)

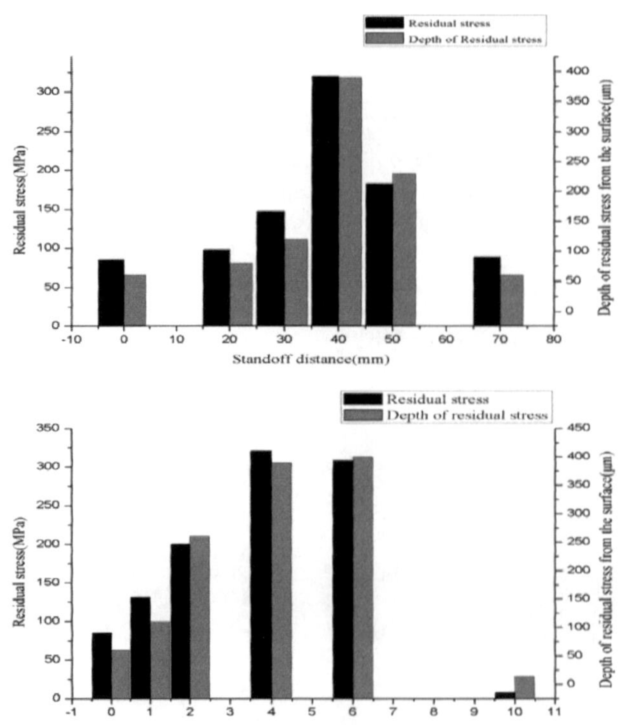

Fig 9. Distribution of residual stress field of aluminium alloy 2A12 samples: (a) with different standoff distances, (b) with different cavitation times.

3.2.3. Surface Macro-Micromorphology. The original sample shown in Fig.10(a) has a smooth and clean surface, with no obvious plastic deformation. While the sample shown in Fig.10(b), which is peened by the processing parameter A, has a circular plastic deformation strengthening area with matte color. The cavitation bubbles cluster formed by the angle nozzle exhibits an approximately annular distribution and will diffuse from the center of the jet along the surface of the sample at a certain speed when the submerged water jet is blocked by the sample, then continuously collapse in the high-pressure region. These two effects generate an annular strengthening area as shown in Fig.10(b).

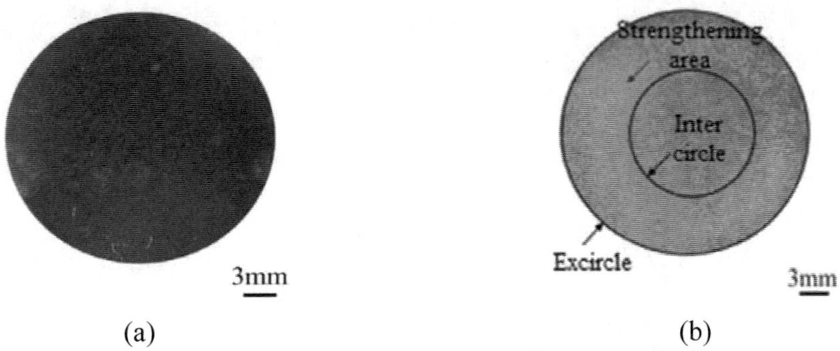

Fig 10. Macroscopic morphology of aluminum alloy 2A12 before and after CP: (a) Original sample, (b) Processing parameter A.

Fig.11 shows the scanning electron microscope images of a surface sample peened at variable cavitation time when the injection pressure and standoff distance are fixed at 20MPa and 40mm, respectively. The original sample surface shown in Fig.11(a) is smooth and flat. In Fig.11(b) and (c), irregular fluctuating can be observed on the surface of the sample, which indicates that severe plastic deformation occurs in the strengthening area, without damaging the sample surface. Obviously, the plastic deformation can still be seen in Fig.11(d), and some white scattered dots appear in some positions because of the high-speed microjets generated when the cavitation bubbles collapse close to the wall, which indicates that as the cavitation time increases, the surface quality of the sample gradually deteriorates. However, the scattered dots caused by the impact of microjets alone do not damage the sample surface. With the high-pressure shock waves released by cavitation bubbles collapse repeatedly act on the sample surface, the surface structure is destroyed when the fatigue limit of the material is exceeded. As shown in Fig.11(e), a dark black pit appears after a long cavitation time, which means that the cavitation peening causes cavitation erosion on the sample surface. Fig.11(f) ~ (h) are partial enlarged views of the erosion area of Fig.11(e). In Fig.11(f), the erosion pits scattered on the surface of the sample, and mainly occur in the plastic deformation of the surface of the convex position. When these cavitation erosion areas are enlarged to 1000 times, it is found that the erosion pits are formed by many small cavitation pits as shown in Fig.11(f). The single cavitation pit is enlarged to 2200 times in Fig.11(h), jagged edge like metallurgical damage caused by fatigue failure of hard materials are observed. This reflects that the aluminum alloy 2A12 is strengthened by cavitation peening from the initial effective plastic deformation to the cavitation erosion.

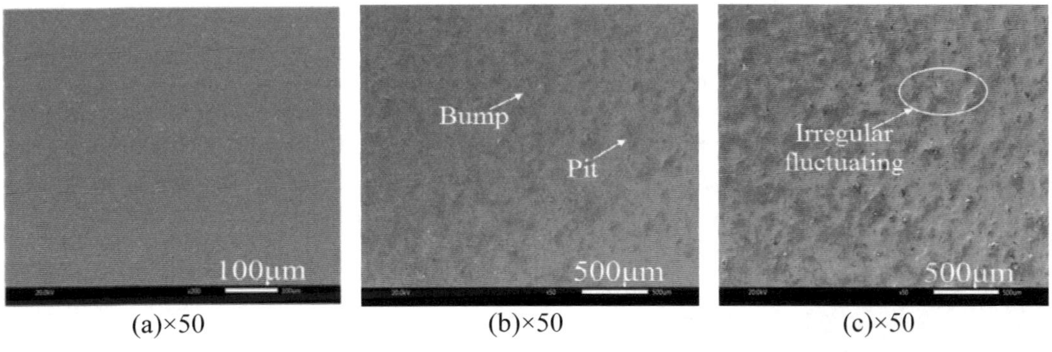

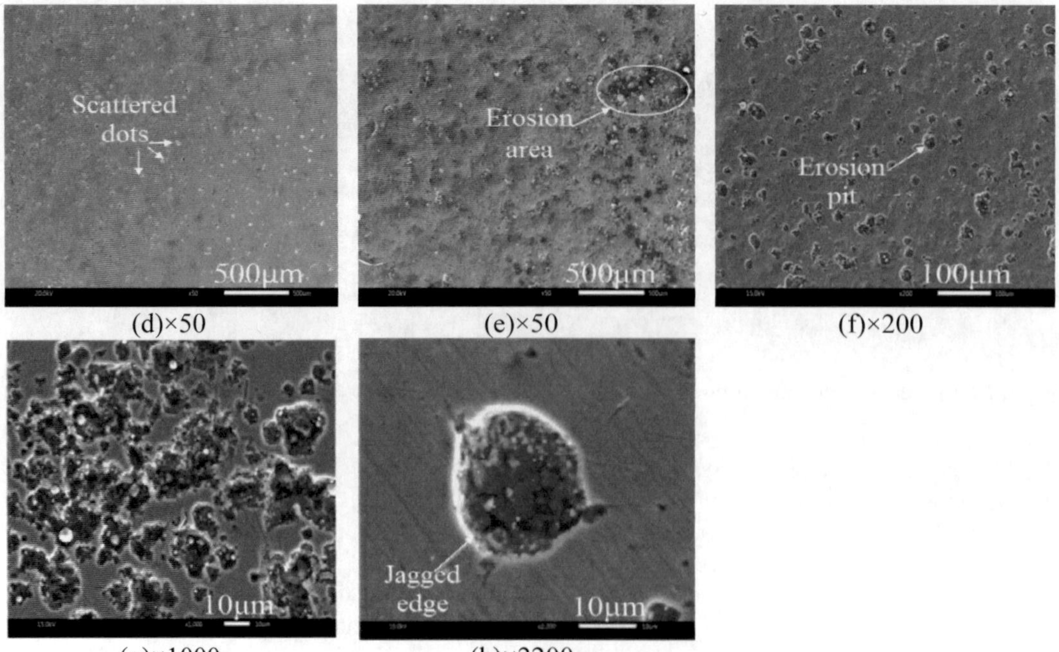

Fig 11. SEM images of aluminium alloy 2A12 with different cavitation times: (a) Original sample, (b)2min, (c)4min, (d)6min and (e)~(f)10min.

4. Conclusion

The cavitation ability of the contraction nozzle is the most unsatisfactory. The organ nozzle can meet the need of cavitation peening, but the cyclic reduction of cavitation performance will adversely affect the utilization efficiency. The angle nozzle can continuously output a stable cavitation bubbles cluster which has relatively high vapor-phase volume fraction, covering a range of 0.04m~0.12m. Therefore, the angle nozzle can best meet the requirements of cavitation peening. At the same time, the strengthen area in the experiment agrees well with the distribution of cavitation bubbles cluster in the numerical simulation when the angle nozzle is used in the cavitation peening.

Cavitation peening has the best standoff distance and cavitation time. The optimized processing parameters in this experiment is P=20MPa, S=40mm and T=4min. Under this condition, the surface residual compressive stress and its influence depth reach the maximum value of 320MPa and 390μm, respectively, which were increased nearly 2.7 times and 5.5 times, respectively.

It was found that the sample surface was kept smooth during the peening period. Meanwhile, no sharp edges and micro cracks were found as in the conventional shot peening, which avoid stress concentration at these deformations. In addition, the surface damage caused by excessive cavitation peening mainly due to the fatigue damage caused by cavitation erosion, so the cavitation time must be controlled within a reasonable range.

Acknowledgments
The present work was supported by the National Natural Science Foundation of China (51575245, 51675234).

References

[1] Soyama H, Kusaka T, Saka M, Peening by the use of cavitation impacts for the improvement of fatigue strength, Journal of Materials Science Letters. 20 (2001) 1263-1265.
[2] Soyama H, Dan O M, Mall S, Compressive Residual Stress into Titanium Alloy Using Cavitation Shotless Peening Method, Tribology Letters. 17 (2004) 501-504.

[3] Takakuwa O, Soyama H, Suppression of hydrogen-assisted fatigue crack growth in austenitic stainless steel by cavitation peening, International Journal of Hydrogen Energy. 37 (2012) 5268-5276.

[4] Soyama H, Sekine Y, Sustainable surface modification using cavitation impact for enhancing fatigue strength demonstrated by a power circulating-type gear tester, International Journal of Sustainable Engineering. 3 (2010) 25-32.

[5] Soyama H, Shimizu M, Hattori Y, et al, Improving the fatigue strength of the elements of a steel belt for CVT by cavitation shotless peening, Journal of Materials Science. 43 (2008) 5028-5030.

[6] Guan J F,Deng S S,Guo G D,et a1, Numerical simulation of internal flow field in angle nozzle for cavitation water jet, Machine Tool & Hydraulics. 40 (2012) 46-50.

[7] Zhao D X, W Q, Du M M, Fluent-Based numerical simulation of the cavitation behavior in the angle nozzle, Journal of Northeastern University(Natural Science). 37 (2016) 1283-1287.

[8] Soyama H, Effect of nozzle geometry on a standard cavitation erosion test using a cavitating jet, Wear. 297 (2013) 895-902.

[9] Ansys, Inc, “ANSYS Fluent Theory Guide.” Release 17.0, 2016.

Numerical Study on Particle Suction Performance of Reverse Blowing Pickup Mouth

Yuan Xi[1,2], Yan Dai[1,*], Yonghou Xiao[1], Weijiao Li[3], Kai Cheng[2,*]

[1] Panjin Industrial Technology Institute of DUT, Panjin, China.
[2] School of Mechanical Science and Engineering, Jilin University, Changchun, China.
[3]CNPC Northeast Refining & Chemical Engineering Co. Ltd Shenyang Company, Shenyang, China.

*Corresponding author e-mail: daiyan@dlut.edu.cn; chengkai@jlu.edu.cn

Abstract. The particle suction performance of reverse blowing pickup mouth is studied by Computational Fluid Dynamics (CFD). The inhalation test of dust particle and large particle were studied by simulation, and a theoretical arithmetic design discipline was put forward. The result shows that theoretical arithmetic can be used as a preliminary selection guide. The front inlet means velocity increases with the increasing of the diameter-length aspect ratio (iDL). While the Outlet mean velocity and the pressure decrease with the increase of iDL. The front inlet mean velocity from 48.5m/s down to 40.2m/s when iDL is from 0.11 to 0.16. The shorter the residence time is, the large particle can be inhaled easier. The residence time is 1.83s, which is 0.29s shorter than 2.12s in the same position of the 160mm diameter when the brick is on the left side of the driving direction. While the residence time is 1.13s, which is 0.19s shorter than that of 1.32s in the same position of 160mm diameter when the brick is in front of the Outlet.

1. Introduction

Road sweeping by manpower is inefficient. Sweeper, a critical cleaning road sweeper product, is widely used to remove the dust accumulation [1]. Chen et al. [2] found that dust collection performance can be improved by adding wing plates. Tan et al. [3] designed the Y-type pickup mouth with the rear airflow, which made the swirl gas flow inside strong obviously. A large number of pickup mouth studies have been carried out in recent years at home and abroad [4-6]. A few studies focus on reverse blowing pickup mouth that is used in China. As a result, the calculation and analysis on particle suction performance of reverse blowing pickup mouth are performed based on the gas-solid coupled numerical method. The reverse blowing pickup mouth is modelled and simulated by using CFD technology. The simulation results were analysed and the optimum Outlet diameter was finally given, which provides theoretical support for the design.

2. Geometry and grid

2.1. Physical modelling and parameter selection

Schematic diagram of reverse blowing pickup mouth is showed in Figure 1. Particles are suctioned under the airflow sweeping at the Inlet. All the particles will enter the garbage collection through the Outlet.

Content from this work may be used under the terms of the Creative Commons Attribution 3.0 licence. Any further distribution of this work must maintain attribution to the author(s) and the title of the work, journal citation and DOI.
Published under licence by IOP Publishing Ltd

So, the Outlet, as a core component of the reverse blowing pickup mouth, determines the effect of particle inhalation. And the constants used in this model are Length=1400mm, Width=480mm and Height=130mm.

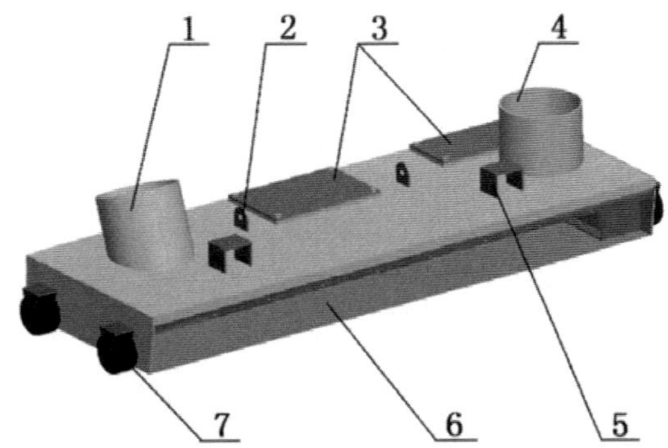

1-Outlet 2- Lifting eye 3- Cover plate 4-Inlet 5- Location baffle 6-Front baffle 7-Wheel

Figure 1. Schematic diagram of reverse blowing pickup mouth

2.2. Grid generation and independence

Then unstructured mesh methods are used to mesh the computational domain in the ICEM because of the complex construction. Boundary layer mesh needs encryption. The computational domain mesh is shown in Figure 2. The inlet velocity around the reverse blowing pickup mouth is hard to test. With expansion area around can avoid this problem. Different methods of mesh generation grids are tested to find the best methods and grid number. It is also for the sake of higher accuracy and efficiency of the calculation. Finally, 170 thousand grids are employed.

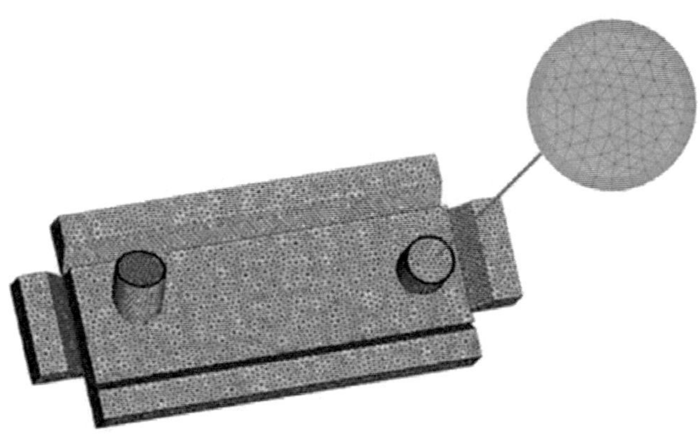

Figure 2. Computational domain mesh model

2.3. Boundary conditions

The moving and static wall were applied to the pickup mouth shells. Pressure out and velocity inlet were applied to the Outlet and Inlet, respect. The turbulence effects were modelled by k-ε model [7,8]. SIMPLE method, and second-order upwind interpolation scheme were chosen for pressure–velocity coupling during the calculation. Gas-solid coupled numerical method was adopted by DPM and the size distribution was also set as the Ref. [5].

3. Theoretical arithmetic

The rate of Outlet flow is fixed after stable operation of centrifugal blower. Symbol S, the Outlet circular area of the reverse blowing pickup mouth, defined as follows:

$$S = \frac{Q}{3600V} \tag{1}$$

Where Q is the rate of Outlet flow. V is the average velocity of the Outlet. In general, the velocity design of the Outlet is from 20 to 60 m/s, and the velocity of dust particles can be lower. While the velocity of large particle, such as stones, must meet the free suspension speed of large particles.

According to Zhou calculation [9], the free suspension velocity of spherical object can be known. The suspension velocities of spheres in different fluid regions are different. This area is the pressure differential resistance zone because of the inhalation of the large size particles. The Reynolds number is from 500 to 2×10^5. The symbol V_0, the free suspension velocity of a spherical object, defined as follows:

$$V_0 = 5.45\sqrt{\frac{d_s(\rho_s - \rho)}{\rho}} \tag{2}$$

Where V_0 is the free suspension velocity of spherome. d_s are the sphere geometry, such as it is diameter if the object is a sphere. ρ_s is the object density, and ρ is the airflow density.

The density of large particle is shown in Table.1. According to the industry standard of road sweeping vehicle, the equivalent diameter of particles absorbed by road sweeper should not be less than 30mm [10]. Therefore, the free suspension velocity of the sphere can be calculated based on the particle size and the density of various spheres by formula (2). The free suspension velocity of red brick is 36.17 m / s, the free suspension velocity of granite is 46.53 m / s, the free suspension velocity of glass is 42.63 m / s, the free suspension velocity of wood is 20m/s.

Table 1. Density of large particle on road

Category	Brick	Granite	Glass	Wood
Density(kg/m^3)	1800	2978	2500	600

The pressure difference of reverse blowing pickup mouth is constant because the negative pressure of Outlet is -2300Pa and the pressure value of expansion region is standard atmospheric pressure [11,12]. The suction flow is direct proportion to the diameter of the Outlet according to formula (3), which means that the larger the diameter is, the more air flow intakes.

$$Q = \mu A\sqrt{\frac{2P}{\rho}} \tag{3}$$

Where Q is the rate of Outlet flow. μ is discharge coefficient and μ is 0.6 in this paper. A is sectional area. P is differential pressure. ρ is the airflow density

4. Influence of dust removal performance

4.1. Diameter-length aspect ratio

Front inlet mean velocity, Outlet mean velocity and pressure were selected to evaluate the effects [12]. The effects of diameter-length aspect ratio on simulation results is shown in Figure 3. The front inlet means velocity increases with the increasing of the Outlet diameter. While the Outlet mean velocity and the pressure decrease with the increase of iDL. The front inlet mean velocity from 48.5m/s down to 40.2m/s when iDL is from 0.11 to 0.16. The reason for the above phenomenon lies in the fact that the inlet area increases with the increase of the diameter, and the power of the dust suction increases, which results in the increase of the mean velocity of front inlet due to the constant pressure at the Outlet. However, the increasing flow rate of the Outlet is smaller than that of the cross-section area, so the mean velocity decreases. The increase of the diameter reduces the distance between the Outlet and the front baffle, and it also make the loss along the side descend. The inlet is affected by the Outlet pressure obviously and approaches the pressure of exit gradually, so the mean pressure is reduced.

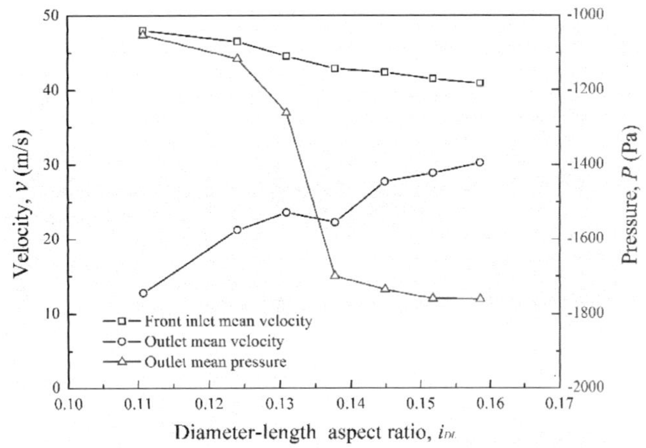

Figure 3. Effects of diameter-length aspect ratio on simulation results

4.2. Dust particle

The dust size distribution mentioned in Ref. [5] was employed. Velocity and trajectory at different Outlet diameter are shown in Figure 4. It can be seen from the diagram that the particles are blown to the suction end and inhaled under the negative pressure of the Outlet. The particles velocity in 160mm diameter is lower than with 200mm diameter. Therefore, the omission of particulate matter in the right front is serious. It was found that the particle velocity around the Outlet was higher, which indicated that the particles were obviously affected by negative pressure under the action of the larger diameter. The Outlet has higher dust suction power and makes the particles have higher kinetic energy, which is easy to be inhaled into the reverse blowing pickup mouth.

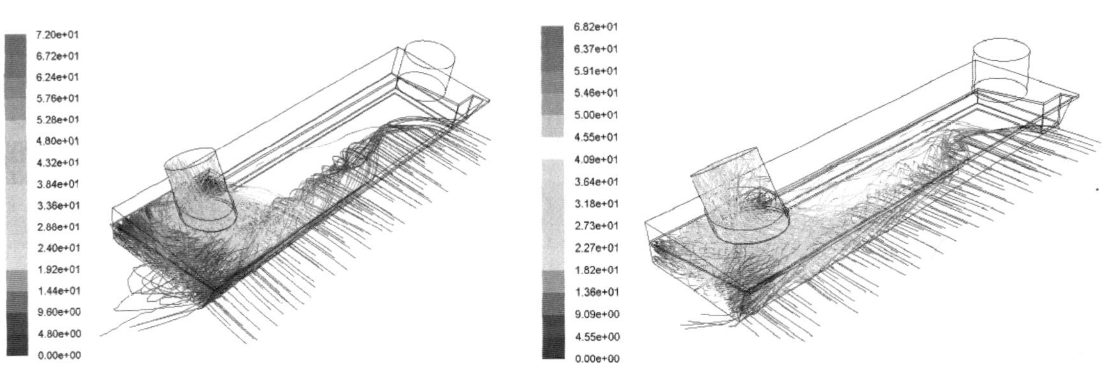

(a)Particle velocity, D=160mm (b)Particle velocity , D=200mm

Figure 4. Velocity and trajectory at different Outlet diameter

4.3. Larger particle

The equivalent diameter of the object must be not less than 30mm according to the industry standard QC / T 51-2006 [8]. Bricks and tiles are common large particles on the road. Therefore, the brick with 30mm diameter and 1800 kg / m^3 density were used to verify the gas-solid coupling calculation. The suction of the weakest and strongest position was chosen in order to calculate the residence time in the reverse blowing pickup mouth, respectively. The residence time and trajectory at different positions are shown in Figure 4.

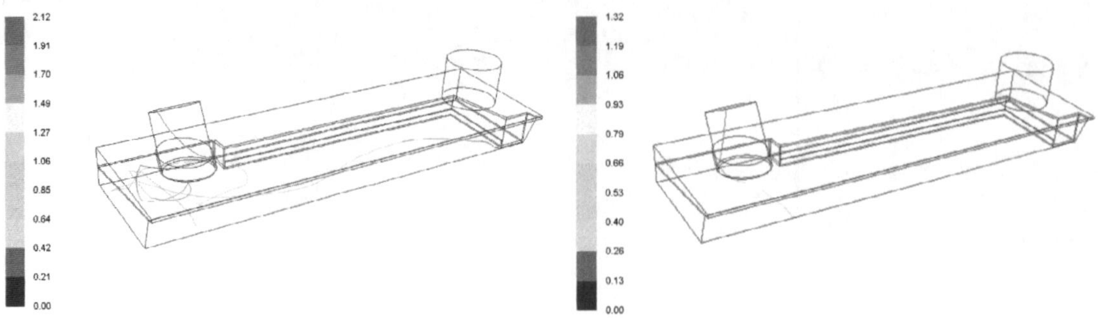

(a)Leftmost direction (b)In front of the Outlet

Figure 5. D=160mm, ds=30mm. Residence time and trajectory at different positions

It can be found that the brick with 30mm diameter can be transported smoothly to the dust removal through the Outlet from the Figure 5. Leftmost position of the driving direction is the weakest place compared to the whole pickup mouth. It took 2.12s for the brick to be sucked in from here (Figure 5. (a)). It also can be seen from the trajectory that the brick that inhaled from the left side impacts on the right wall. The following property of large particle is not as good as small particles with the wind because of the inertia of large particles. The brick is easy to be inhaled into the Outlet and its residence time is 1.32 s because the position is in front of the Outlet (Figure 5. (b)).

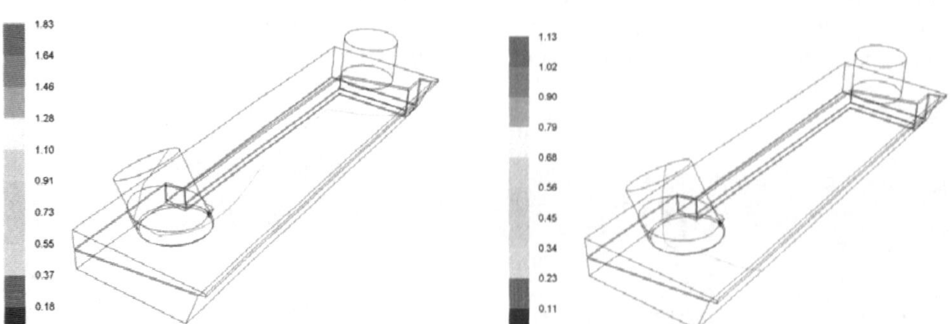

(a)Leftmost direction (b)In front of the Outlet

Figure 6. D=200mm, ds=30mm.Time and trajectory at different positions

The Outlet with 200mm diameter takes less time than the 160mm diameter by comparison with Figure 5 and 6. The trajectory of the Outlet with 200mm diameter is simpler and shorter than the smaller diameter when the brick is on the left side of the driving direction. This is because the increase of the diameter shortens the distance from the wall around, and the suction pressure is constant, that is to say, the suction power of the suction nozzle is increased, so the brick with 30mm diameter is more easily inhaled. When the brick is on the left side of the driving direction, its residence time is 1.83 s, which is

0.29 s shorter than 2.12 s in the same position of the 160mm diameter. When the brick is in front of the Outlet, the residence time is 1.13 s, which is 0.19 s shorter than that of 1.32 s in the same position of 160mm diameter. The shortening of the residence time had a good advantage for the inhalation of large particle. It is evident that particle suction performance must be combined with dust leakage, large particle inhalation smoothly and product material cost. An optimal commercial product should take various elements into the comprehensive consideration, so the Outlet diameter is 180mm, which is suggested.

5. Conclusion

The present contribution investigates the influence of Outlet diameter and particles inhalation in reverse blowing pickup mouth, aiming at providing design discipline. A three-dimensional device is computed, which enables the large particles can be inhaled smoothly and product material cost can be saved. Theoretical arithmetic can be used as a preliminary selection guide. The particles velocity in 160mm diameter is lower than with 200mm diameter, and the omission of particulate matter in the right front is serious. The shorter the residence time is, the large particle can be inhaled easier. The front inlet mean velocity from 48.5m/s down to 40.2m/s when i_{DL} is from 0.11 to 0.16. The Outlet with 180mm diameter is suggested combing with the model in this paper.

Acknowledgments

This work was supported by Natural Science Foundation of Liaoning Province of China (No.20170520354,20170520148,2017308004), The National Natural Science Foundation of China (NO. 21706023) and the Special Fund Project of Panjin Industrial Technology Institute of DUT (NO. YJYZXZJ2017004).

References

[1] Khanal R, Furumai H, Nakajima F. Toxicity assessment of size-fractionated urban road dust using ostracod Heterocypris incongruens direct contact test, J. Hazard Mater. 264.2(2014) 53-64.

[2] Z.J.Chen, X.U.Wu,G.P.Xu, et al. Experimental study on suction mouth of vacuum sweeper, J. Tongji University (Science and Technology). 29 (2001) 1483-1485.

[3] X.Y.Tan,Q.L.Xiao,F.L.Zhou,et al.Simulation analysis of gas-particle flow through the pickup head of a street sweeper and its design improvement, Chin. J. Appl. Mech.33.1(2016)73-79.

[4] Vanderlinden R P., U.S. Patent 8082625 B2(2011).

[5] B.F. Wu, J.L. Men, J. Chen. Numerical study on particle removal performance of pickup head for a street vacuum sweeper. Powder Technology, 200.1(2010) 16-24.

[6] B.F. Wu, J.L. Men, J. Chen. Improving the design of a pickup head for particle removal using computational fluid dynamics. ARCHIVE Proceedings of the Institution of Mechanical Engineers Part C Journal of Mechanical Engineering Science,225.4 (2011) 939-948.

[7] S.W.Shu, Y.D.Deng, C.Q.Su, et al. Parameter relationship mining between suction nozzle structure parameter and dust collection performance. Journal of wuhan university of technology(transportation science&engineering),42.2(2018)268-273.

[8] L.Long, X.Y.Tan, Q.L.Xiao, et al. Simulation analysis and design improvement of flow field for the wide nozzle of the sweeper. Construction Machinery Technology & Management ,30.6 (2017) 71-76.

[9] X.Y.Zhou. Exact solution of free suspension velocity of spherical object.Special Purpose Vehicle. 4 (1995) 3-6.

[10] QC/T 51-2006. The national automotive industry standard.

[11] Y.Xi, K.Cheng, X.T.Lou, et al. Research on numerical analysis and dust collection efficiency of reverse blowing pickup mouth. Journal of Southwest Jiaotong University, 51.1(2016)105-112.

[12] Y.C.Zhang, C.Z.Yang, C.Baker, et al. Effects of expanding zone parameters of vacuum dust suction mouth on flow simulation results. Journal of Central South University,21.6 (2014) 2547-2552.

IOP Conf. Series: Journal of Physics: Conf. Series **1064** (2018) 012045 doi:10.1088/1742-6596/1064/1/012045

Study on the affecting factors of bow wave effect modeling in hose-drogue refueling system

Xiaofei Zhang *, Mengquan Hu, Dong Zhang, Pengyun Tian and Ning Wang

Aeronautics Engineering College, Air Force Engineering University, Xi'an, China

*Corresponding author e-mail: jubeiweihao2@sina.com

Abstract. In the docking process of aerial refueling hose-drogue system, the bow wave effect of an approaching receiver aircraft will produce a strong aerodynamic effect on the drogue which intensify the deviation of drogue and increase the difficulty of the coupling. This paper presents a receiver aircraft's bow wave effect modeling method previously developed by the authors is continued to study, which based on potential flow theory and considers the major shape characteristics of the receiver aircraft. This method is used in a simulation system, which combined with the hose-drogue dynamic model and tanker downwash model. The model of the receiver aircraft, which includes the fuselage, cockpit, main wing and other structural factors, is simulated respectively. The motion law of drogue under different bow wave model is analyzed. The results show that the fuselage, cockpit and main wing structure have significant influence on the bow wave effect.

1. Introduction

The hose-drogue aerial refueling system is simply and widely used and is a power multiplier in modern air force. [1]. But the hose-drogue system used flexible hose which is highly sensitive to external factors such as atmospheric turbulence, tanker downwash and the receiver aircraft's bow wave effect. In the actual docking process, there is the problem of the hose-drogue transient motion, which increased the difficulty of the docking and security risks.

In the docking process, when the receiver aircraft is close to the drogue, the drogue will deviate from the balance position and then quickly return back. This phenomenon is called as the bow wave effect or the forebody effect [2]. The drogue movement caused by the bow wave brings great uncertainty to the success of the docking. The rise of unmanned aircraft vehicle (UAV) in recent years has opened up new areas for aerial refueling [3]. Therefore, it is required to conduct more in-depth studies on the hose-drogue aerial refueling to ensure the safety of the docking process, whether manned or unmanned.

So far, the research on atmospheric turbulence and tanker downwash is relatively mature. However, there is less research on the bow wave effect modeling. Most of researchers use half Rankine body [4] to simulate the bow wave effect. This method has a small amount of calculation, but also has the problem of accuracy is poor. Dai and Wei use a series of line doublets to simulate the nose and cockpit of the receiver aircraft and use the CFD result to correct the analytical results. [5] A CFD simulation was carried out by Wang and Dong [6]. They used the FLUENT software to analyze the variation of bow wave with Mach number and Angle of attack. Khan and Masud [7] used CFD to simulate the

Content from this work may be used under the terms of the Creative Commons Attribution 3.0 licence. Any further distribution of this work must maintain attribution to the author(s) and the title of the work, journal citation and DOI.
Published under licence by IOP Publishing Ltd

IOP Conf. Series: Journal of Physics: Conf. Series **1064** (2018) 012045 doi:10.1088/1742-6596/1064/1/012045

aerial refueling of the "Thunder" fighter and analyze the docking strategy of reducing bow wave effect. The CFD simulation results are more reliable. However, due to different receiver aircrafts have different complex configuration structure, the CFD simulation requires a large amount of computing resources and time. Moreover, the CFD results are inconvenient for the controller design. NASA Dryden Flight Research Center tested the bow wave effect by F/A-18 receiver-aircraft and estimated the range of the bow wave effect [4, 8].

In view of the current shortage of modeling of bow wave effect, this paper proposes a fast and efficient receiver aircraft's bow wave effect modeling method based on potential flow theory and considers the major shape characteristics of the receiver aircraft. This method is used in aerial refueling docking simulation system. On the basis of this, the bow wave effect model is simulated respectively with different shape structure. The importance of different structure of the receiver aircrafts in the modeling of bow wave effect is analyzed.

2. Frame definitions and modelling assumptions.

2.1. Reference frames

As shown in Fig. 1. This paper defines the ground coordinate system $O_gX_gY_gZ_g$. The body frame of the receiver aircraft $O_bX_bY_bZ_b$ and the body frame of the tanker aircraft $O_aX_aY_aZ_a$. O_b is fixed on the top of the receiver aircraft. O_a is fixed on the refueling pod.

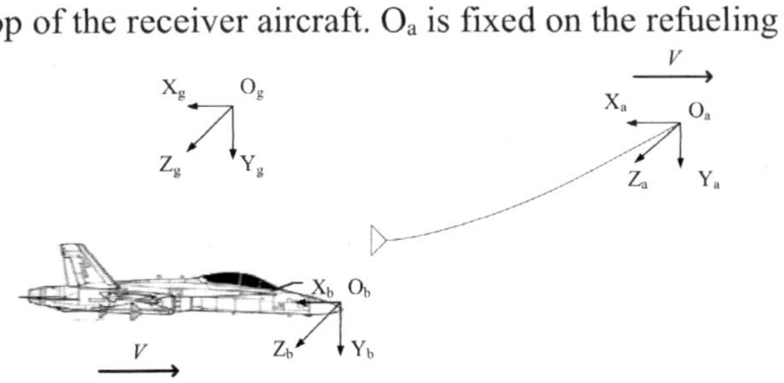

Fig 1. Overview of the reference frames

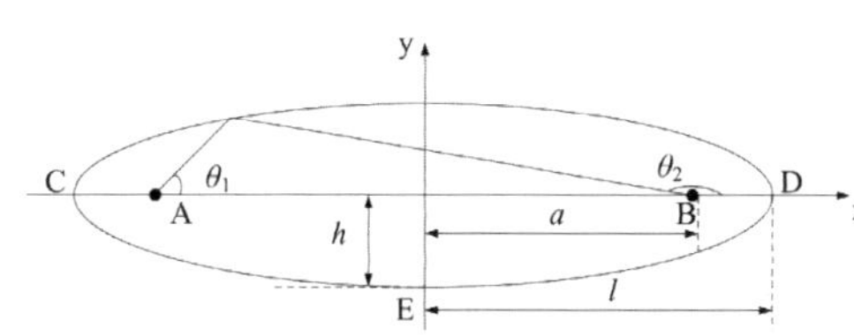

Fig 2. Overview of the Rankine body.

2.2. Modelling assumption

To simplify the modeling of the bow wave effect, the following assumptions are made.

(1) In the shape structure of the receiver aircraft, there is a large influence on the flow field near the nose, fuselage, cockpit, main wing and strake wing. Suppose the aerodynamic effects of each structure are independent to each other. According to the superposition principle, the induced velocity of the bow wave effect can be expressed as:

$$\boldsymbol{v}_{bow} = \boldsymbol{v}_{body} + \boldsymbol{v}_{cockpit} + \boldsymbol{v}_{maiwing} + \boldsymbol{v}_{strakewing} \tag{1}$$

Where $\boldsymbol{v}_{bow}$ denotes the induced velocity produced by the bow wave, $\boldsymbol{v}_{body}$, $\boldsymbol{v}_{cockpit}$, $\boldsymbol{v}_{mainwing}$ and $\boldsymbol{v}_{strakewing}$ denotes the induced velocity produced by the fuselage, cockpit, main wing and strake wing.

(2) Considering the angle of attack of the receiver aircraft is small. So, it can be assumed that the fuselage is axisymmetric.

3. Bow wave effect modelling

3.1. Receiver aircraft's fuselage and cockpit modelling method

To simplify the receiver aircraft model establishing, seem the fuselage as Rankine body [9], whose flow field is superposition by the three-dimensional point source, the three-dimensional point sink and three-dimensional straight uniform stream. Finally, the fuselage flow field model is obtained by elliptical transformation. The cockpit is modeled in the same way as the fuselage.

Assumption that the three-dimensional point source strength of Rankin body is *m* and was placed in A (-*a*, 0) as shown in Fig. 2. The three-dimensional point sink strength of Rankin body is *m* and was placed in B (a, 0). Straight uniform stream *U* is going along the x axis. The velocity potential and stream function can be expressed as:

$$\varphi = -Ux + \frac{m}{\sqrt{(x+a)^2 + y^2}} - \frac{m}{\sqrt{(x-a)^2 + y^2}} \tag{2}$$

$$\psi = -\frac{1}{2}Uy^2 + \frac{m(x+a)}{\sqrt{(x+a)^2 + y^2}} - \frac{m(x-a)}{\sqrt{(x-a)^2 + y^2}} \tag{3}$$

Eq. (3) can be transformed to the form as:

$$\psi = -\frac{1}{2}Ur^2 \sin^2\theta + m(\cos\theta_2 - \cos\theta_1) \tag{4}$$

Cause the boundary condition rules that $\psi = 0$. The function can be rewritten as

$$y^2 + b^2(\cos\theta_1 - \cos\theta_2) = 0 \tag{5}$$

$$b^2 = \frac{2m}{U} \tag{6}$$

Assuming that OC=OD=*l*, OE=*h*. Because of point C is the stationary point. The velocity of the free stream is offset by the velocity induced by source and sink.

$$-\frac{m}{(l+a)^2} + \frac{m}{(l-a)^2} = U \tag{7}$$

Cause the geometric relationship shown as in Fig. 2

$$\left(l^2 - a^2\right)^2 = 2ab^2 l \tag{8}$$

Where $2b^2\cos\alpha = h^2$, Eq (8) can be rewritten as

$$\frac{2a}{\sqrt{h^2 + a^2}} = \frac{h^2}{b^2} \tag{9}$$

$$h^2 l\sqrt{h^2 + a^2} = 2ab^2 l \tag{10}$$

Combine the two equations

$$h^2 l\sqrt{h^2+a^2}=\left(l^2-a^2\right)^2 \tag{11}$$

Based on the receiver aircraft's parameters. Setting the length of the fuselage is 2l, and the maximum radial width is h. The three-dimensional coordinates and strength m of point source and point sink are calculated. The induced velocity $\boldsymbol{v}_{body}$ at any point in space can be calculated as

$$\boldsymbol{v}_{body}=\frac{m}{\boldsymbol{r}_s}-\frac{m}{\boldsymbol{r}_{-s}} \tag{12}$$

Where $\boldsymbol{r}_s$ is the vector from the source to the point. $\boldsymbol{r}_{-s}$ is the vector from the sink to the point.

During refueling process, the aircraft keeps the subsonic flight. Based on the fuselage model in the incompressible flow. The Gothert Rule is used for compressibility modification. The corresponding induced velocity is obtained in the compressible flow.

While for some receiver aircraft like F/A-18, the cross section of the nose is like an ellipse. In order to obtain more accurate simulation results. The ellipse transformation is necessary.

As is shown in Fig. 3, the scale transformation is presented as

$$x'=x,\ y'=\frac{a}{b}y,\ z'=z \tag{13}$$

Where a and b are the short axis and long axis of the ellipse. The coordinates of the desired point are converted by Eq. (13) and obtain the $\boldsymbol{v}_{body}{}'=\left[v_x{}',v_y{}',v_z{}'\right]^T$ by Eq. (12). After that, the ellipse transformation is applied again to transform $\boldsymbol{v}_{body}{}'$ back to the original scale as:

$$v_x=v_x{}',\ v_y=\frac{b}{a}v_y{}',\ v_z=v_z{}' \tag{14}$$

Modified fuselage induced velocity shown as

$$\boldsymbol{v}_{body}=\left[v_x,v_y,v_z\right]^T \tag{15}$$

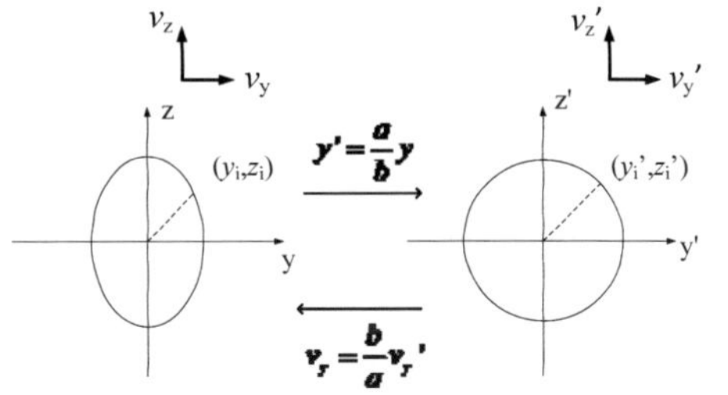

Fig 3. Illustration of the ellipse transformation

3.2. Receiver aircraft's aerofoil modelling method

In the process of aerial refueling, the flight velocity is low. In this case, compared with the general CFD method, the vortex lattice method (VLM) has a small precision and high computational efficiency [10]. Under the assumption of the inviscid flow, the VLM can be used to find the flow field of the aerofoil easily and efficiently.

The VLM is a practical numerical method based on lifting surface. The VLM considers the wing as a network with a horseshoe vortex superimposed on the aerofoil. First, the aerofoil is divided into $M \times N$ grids, and each grid is equipped with a horseshoe vortex, and the induced velocity generated by each horseshoe vortex is obtained by the Biot-Savart Law. The ith control point's velocity induced by $M \times N$ horseshoe vortexes is obtained. According to the boundary conditions of control points, a set of equations of $M \times N$ control points can be obtained to solve the vortex strength of each horseshoe vortex, so as to obtain the induced velocity of the finite discrete horseshoe vortex on any point in space. The Gothert Rule is used for compressibility modification too. The corresponding induced velocity is obtained in the compressible flow.

4. The simulation processes

In the aerial refueling, the receiver aircraft and tanker in the same flight altitude and velocity. The receiver aircraft is close to the drogue with a stable docking velocity behind the tanker, until the connection is successful, and the relative position of the drogue is no more changes. In the process of receiver aircraft slowly close to the drogue, the bow wave effect caused by the receiver aircraft destroys the original force balance state of the drogue and produces swing. The relative position and velocity between the receiver aircraft and drogue is constantly changing. To simulate this process accurately. The digital simulation process is as follows.

(1) Calculate the equilibrium position of the hose under the no turbulence condition [11].

(2) Calculate the tension of each hose element, relative airflow velocity, external force and equivalent hose restoring force.

(3) Calculate the fuselage and cockpit's strength and coordinates of the source and sink. After that, calculate the induced velocity at the point in space.

(4) Calculate the velocity induced by the main wing on the drogue by VLM.

(5) Calculate the bow wave induced velocity at the point of the drogue by Eq. (1).

(6) The acceleration of the drogue is calculated by the induced velocity, and the acceleration of the drogue is obtained by using the fourth-order Runge-Kutta method [12].

Repeat the above steps to obtain the real-time motion of the drogue. The receiver aircraft is subject to the F/A-18. The simulation is conducted under the condition without turbulence. The simulation parameters used in this paper are illustrated in Table 1. The parameters of the tanker and hose-drogue system are with the same parameters in Ref. 13

Table 1. Parameters of the simulation.

Parameter	Value
Tanker	
The length of the fuselage	17.07m
The maximum radius of the fuselage	0.808m
span	11.43m
root chord	4.04m
wing sweep	25.82°
Drogue	
Weight	200N
Diameter	0.6m

5. Simulation results and analysis.

In order to verify the accuracy of the bow wave model established in this paper, the simulation results are compared with the experimental data of the Autonomous Airborne Refueling Demonstration (AARD) project in 2006 by NASA Dryden Flight Research Center [5].

NASA conducted six experiments using the F/A-18 as the receiver aircraft, but only the third and sixth experiments were successful. In the third docking simulation process, the deviation of the maximum deviation in the vertical and lateral of the drogue is shown in Table 2. The table shows the simulation values the experimental value has a certain deviation, this is because the simulation ignores the influence of atmospheric turbulence. In the actual flight test, the faint atmospheric turbulence causes the drogue to swing, and the random oscillation will have a certain effect on the result.

Table 2. Experiment and simulation results.

	The lateral maximum deviation	The vertical maximum deviation	Remarks
1	0.213 m	0.366 m	experimental value
2	0.266 m	0.348 m	simulation value

In the process of docking, the deviation of the deviation in the vertical (Y-axis) and lateral (Z-axis) of the drogue which include experiment and simulation results are shown in the Fig. 4. Before the receiver aircraft approached, drogue affected by turbulent flow swings near the equilibrium position, give the experimental results a lot of randomness, but we can still observe that the bow wave effect has some common features: (1) The drogue begins to float to the right and upward. And then swings back to the connected position from the maximum deviation position until the drogue is connected with the probe. (2) The maximum value of drogue deviation in simulation is similar to the experimental value. (3) The swing velocity of the drogue which affected by the bow wave effect in simulation is similar to the experimental. To sum up, the simulation results and the actual behaviour deviation are small, meeting the expectation.

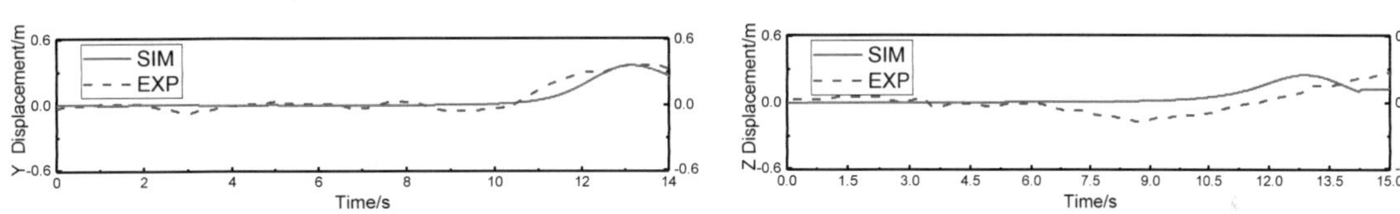

(a) Y direction (b) Z direction

Fig 4. Deviation of drogue in simulation and experimental

5.1. Simulation and analysis of the affecting factors of bow wave effect modelling.
Use the model established above, a numerical docking simulation for different configurations of the receiver is carried out. The configuration contained in the receiver bow wave effect model is shown in the Table 3. Under different flight conditions, the deviation of the deviation in the vertical and lateral of the drogue is shown in the Fig. 5, Fig. 6 and Fig. 7.

Table 3. Different configuration for receiver bow wave effect modeling.

Case 1	Case 2	Case 3	Case 4
Fuselage	Fuselage + cockpit	Fuselage + cockpit + main wing	Fuselage + cockpit + main wing + canard

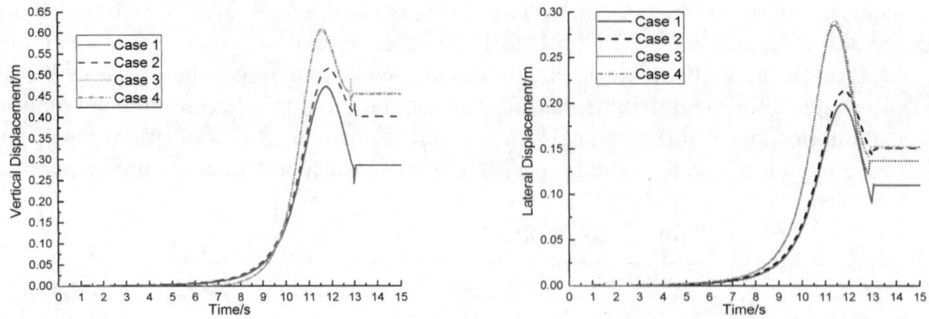

(a) Y direction (b) Z direction

Fig 5. Deviation of drogue in Y and Z direction. [V_∞ =180 m/s, H_G =3000m, q_∞=14731 Pa]

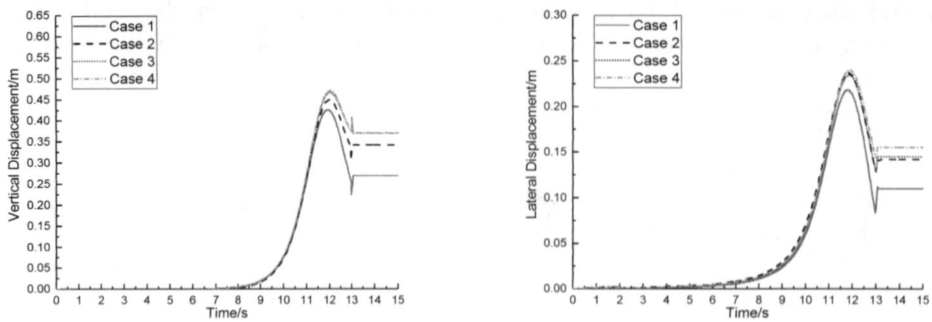

(a) Y direction (b) Z direction

Fig 6. Deviation of drogue in Y and Z direction. [V_∞ =150 m/s, H_G =5000 m, q_∞=8283 Pa]

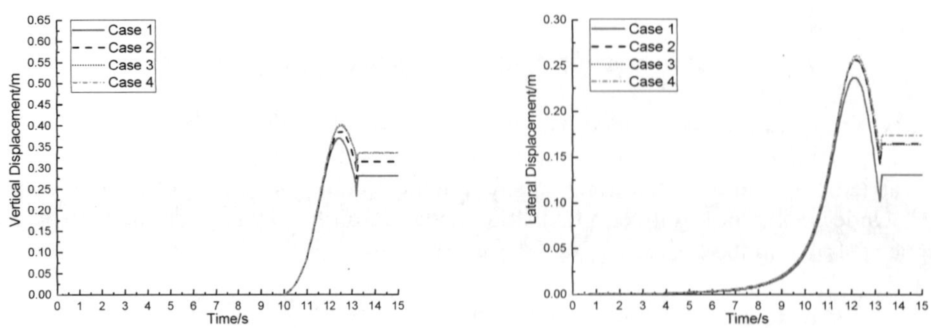

(a) Y direction (b) Z direction

Fig 7. Deviation of drogue in Y and Z direction. [V_∞ =120 m/s, H_G =7000 m, q_∞=4245 Pa]

As shown in the Fig. 5. In the case of high dynamic pressure, compared with the simple model of Rankine body model (including only the fuselage), the cockpit and the main wing would have a larger impact on the result of the simulation. The maximum deviation would be nearly 30 percents higher than only the fuselage. However, the simulation results of the canard are small and negligible. By comparison of the Fig. 5, Fig. 6 and Fig. 7, the maximum deviation of the simulation of the bow wave effect generated by the bow wave model with different receiver structure factors tends to approach with the decrease of dynamic pressure. Under the condition of low dynamic pressure, the cockpit and the main wing would have a lower impact on the result of the simulation results.

6. Conclusion

(1). In this paper, a simple and efficient modeling method is proposed based on potential flow theory, and a mathematical model of fuselage, cockpit and aerofoil is established by means of source, sink and finite discrete horseshoe vortex. Through the compressibility modification and elliptical modification, the model is as accurate as possible to simulate the aerodynamic characteristics of the bow wave effect.

(2). Based on the shape features and parameters of F/A-18, a model of bow wave effect was established. The numerical simulation of the aerial refuelling docking process was carried out. The results are compared with the experimental results of NASA, which proves the validity of the modelling method.

(3) The bow wave effects induced by the receiver model with different structural factors were simulated and analysed. It is concluded that the fuselage, cockpit and main wing have significant influence on the bow wave effect. and the larger the dynamic pressure of the oil machine, the more significant the bow wave effect.

Acknowledgments

This work was financially supported by National Natural Science Foundation of China (61473307).

References

[1] LU Y P, YANG C X, LIU Y Y, A survey of modeling and control technologies for aerial refueling system[J].Acta Aeronautica et Astronautica Sinica. 2014, 35(9): 2375-2389(in Chinese).
[2] Atilla Dogan, William Blake. Modeling of Bow Wave Effect in Aerial Refueling [C]. AIAA Atmospheric Flight Mechanics Conference, Toronto, Ontario Canada: AIAA, 2010.
[3] Ryan P Dibley, Michael J Allen. Autonomous Airborne Refueling Demonstration Phase I Flight-Test Results[C]. AIAA Atmospheric Flight Mechanics Conference and Exhibit, Hilton Head, South Carolina: AIAA, 2007.
[4] ZHANG J. Dynamic modeling and control of the motion of the hose-drogue[D]. Nanjing: Nanjing University of Aeronautics and Astronautics, 2016:36-40(in Chinese).
[5] Dai Xunhua, Wei Zibo,Quan Quan. Modeling and simulation of bow wave effect in probe and drogue aerial refueling[J]. Chinese Journal of Aeronautics,2016,29(2):448-461.
[6] WANG J,DONG X M, XU Y J, et al. Simulation and analysis of the bow wave effect of the receiver in hose-drogue aerial refueling[J]. Flight Dynamics, 2016, 34(1): 54-58(in Chinese).
[7] Khan, J Masud. Trajectory Analysis of Basket Engagement during Aerial Refueling[C]. AIAA Atmospheric Flight Mechanics Conference, National Harbor, Maryland: AIAA, 2014.
[8] Jennifer L Hansen, James E Murray, Norma V Campos. The NASA Dryden Flight Test Approach to an Aerial Refueling System[R]. NASA/TM-2005-212859, Edwards, California: NASA Dryden Flight Research Center, 2005.
[9] L.M. Milne Thomson. Theoretical hydrodynamics[M]. LI Y L, YAN M W, translated. Beijing: China Machine Press, 1984.9. 545-551(in Chinese).
[10] QIAN Y J. Aerodynamics[M]. Beijing: Beihang University Press, 2004.9:153-165(in Chinese).
[11] HU M Q, LIU P, NIE X, et al. Influence of air turbulence on the movement of the hose-drogue[J]. Flight Dynamics, 2010, 28(5): 20-23(in Chinese).
[12] HU M Q, NIE X, WANG L M. Determination of hose static catenary shape in “probe-drogue” in-flight refueling system[J]. Journal of Air Force Engineering University: Natural Science Edition, 2009, 10(5): 22-26(in Chinese).

Study on inertial separation device of the sandy flow

Yan Zhong*, Nan Zhang, Pu Zhou and Hongjun Sun

Shanghai Marine Equipment Research Institute, Shanghai, China.

*Corresponding author e-mail: zhongyan704@163.com

Abstract. An inertial separation device was designed in order to solve the problem of water-sand separation in the flow. Firstly the trajectories of sand grains were qualitatively analysed, the movement rules in the device were preliminarily grasped. Then the separation efficiency of different sizes of sand was numerically simulated, and the effect of different structural parameters on the separation efficiency was analysed. Finally, a testing rig was set up, the separation efficiency and pressure loss were analysed, and the experimental and numerical results were compared. The numerical and experimental results can guide the further design of separating device.

1. Introduction

Sediment flow is a common transportation in mining, construction, petrochemical, port dredging and other industries. Because of the existence of sand particles, it will cause serious abrasion and low efficiency of moving parts, affect the service life of the pump and reduce the operation quality of the pump. A sand-water separation device added before the intake of the pump, and let the sand separated before the pump is important. It is of great significance to improve the efficiency of the pump and improve the design level of the pump group.

In this paper, based on the principle of inertial separation, a sand-water separating device is designed. Through numerical simulation and experimental analysis, the characteristics of the separation efficiency of the device are analyzed, and the relevant conclusions are obtained. The numerical and experimental results can guide the further design for the separation device.

2. Basic Equation

2.1. Basic hypothesis

To achieve this numerical simulation, the following hypotheses are made: (1) Liquid phase is a continuous incompressible fluid, solid phase is discrete, physical parameters of each phase are constants, and there is no phase transition. (2) The self-rotation, collision and rupture of sand grains are not considered. (3) The effect of temperature changed on the flow field is not considered.

2.2. Numerical Model

The $\kappa-\varepsilon$ turbulence model and the modified $\kappa-\varepsilon$ turbulent combustion model are applied to estimating the time averaged Navies-Stokes equations in this paper.

Fluid phase continuity equation:

Content from this work may be used under the terms of the Creative Commons Attribution 3.0 licence. Any further distribution of this work must maintain attribution to the author(s) and the title of the work, journal citation and DOI.
Published under licence by IOP Publishing Ltd

$$\frac{\partial\overline{\Phi_f}}{\partial t}+\frac{\partial}{\partial x_i}\left(\overline{\Phi_f V_{fi}}+\overline{\varphi_f v_{f_i}}\right)=0 \tag{1}$$

Solid phase continuity equation:

$$\frac{\partial\overline{\Phi_p}}{\partial t}+\frac{\partial}{\partial x_i}\left(\overline{\Phi_p V_{pi}}+\overline{\varphi_p v_{p_i}}\right)=0 \tag{2}$$

Fluid phase momentum equation:

$$\frac{\partial}{\partial t}\left(\overline{\Phi_f V_{fi}}+\overline{\varphi_f v_{f_i}}\right)+\frac{\partial}{\partial x_k}(\overline{\Phi_f V_{fi}V_{fk}}+\overline{\Phi_f v_{f_i}v_{f_k}}+\overline{V_{fi}\varphi_f v_{f_i}}+\overline{\varphi_f v_{fi}v_{fk}})=\frac{1}{\rho_f}(\overline{\Phi_f}\frac{\partial\overline{P}}{\partial x_i}+\varphi_f\frac{\partial\overline{P}}{\partial x_i})+$$

$$vf\frac{\partial}{\partial x_k}[\overline{\Phi_f}(\frac{\partial\overline{V_{fi}}}{\partial x_k}+\frac{\partial\overline{V_{fi}}}{\partial x_k})+\varphi_f(\overline{\frac{\partial v_{f_i}}{\partial x_k}+\frac{\partial v_{f_k}}{\partial x_i}})]-\frac{B}{\rho_f}vf\frac{\partial}{\partial x_k}[\overline{\Phi_f}(\frac{\partial\overline{V_{fi}}}{\partial x_k}+\frac{\partial\overline{V_{fi}}}{\partial x_k})+\varphi_f(\overline{\frac{\partial v_{f_i}}{\partial x_k}+\frac{\partial v_{f_k}}{\partial x_i}})]-\frac{B}{\rho_f}$$

$$[\overline{\Phi_f\Phi_p}(\overline{V_{f_i}}-\overline{V_{pi}})+(\overline{\Phi_p}-\overline{\Phi_f})\overline{\varphi_f(v_{f\,i}-v_{pi})}+(\overline{V_{fi}}-\overline{V_{pi}})\overline{\varphi_f\varphi_p}+\overline{\varphi_f\varphi_p(v_{fi}-v_{pi})}]+\overline{\Phi_f}g_i \tag{3}$$

Solid phase momentum equation:

$$\frac{\partial}{\partial t}\left(\overline{\Phi_p V_{pi}}+\overline{\varphi_p v_{pi}}\right)+\frac{\partial}{\partial x_k}(\overline{\Phi_p V_{pi}V_{pk}}+\overline{\Phi_p v_{pi}v_{p_k}}vf\frac{\partial}{\partial x_k}[\overline{\Phi_f}(\frac{\partial\overline{V_{fi}}}{\partial x_k}+\frac{\partial\overline{V_{fi}}}{\partial x_k})+\varphi_f(\overline{\frac{\partial v_{f_i}}{\partial x_k}+\frac{\partial v_{f_k}}{\partial x_i}})]-\frac{B}{\rho_f}$$

$$+\overline{V_{pi}\varphi_p v_{p_i}}+\overline{\nabla_{pk}\varphi_p v_{pk}}+\overline{\varphi_p v_{pi}v_{pk}})=\frac{1}{\rho_p}(\overline{\Phi_p}\frac{\partial\overline{P}}{\partial x_i}+\varphi_p\frac{\partial\overline{P'}}{\partial x_i})+vp\frac{\partial}{\partial x_k}[(\frac{\partial\overline{V_{pi}}}{\partial x_k}+\frac{\partial\overline{V_{pk}}}{\partial x_i})$$

$$+\varphi_p(\overline{\frac{\partial v_{p_i}}{\partial x}+\frac{\partial v_{p_k}}{\partial x_i}})]-\frac{B}{\rho_t}[\overline{\Phi_f\Phi_p}(\overline{V_{pi}}-\overline{V_{fi}})+(\overline{\Phi_p}-\overline{\Phi_f})\overline{\varphi_f(v_{p\,i}-v_{fi})}+(\overline{V_{fi}}-\overline{V_{pi}})\overline{\varphi_f\varphi_p}+$$

$$\overline{\varphi_f\varphi_p(v_{pi}-v_{fi})}]+\overline{\Phi_p}g_i \tag{4}$$

Volumetric fraction equation:

$$\overline{\Phi_f}+\overline{\Phi_p}=1\,,\varphi_p=-\varphi_f \tag{5}$$

Where V is average velocity, P is pressure, f and p is solid and liquid phases, i and j is coordinates of a tensor.

3. Performance Analysis of Separation Device

3.1. Structure design

In order to make water and sand separated, the sand must have large enough inertia force and speed, a contracted surface is designed in the inlet passage of the device to accelerate the liquid in it. When the liquid reaches the contracted surface, the velocity reaches the maximum. Under the action of the gravity and the boundary, the medium is divided into two parts, a part of sands enter into the separation channels, will reach the goal of sand separation. According to this principle, a separation device is designed. The shape and section of the device are shown in Figure 1.

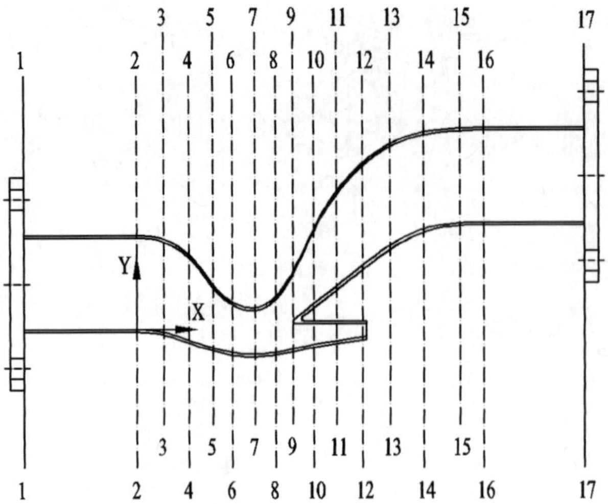

Figure 1. Section Diagram of the Device.

3.2. Analysis on motion trajectory

According to the diameter, the sand is divided into three kinds in the simulation: large, medium and small size, in which the large size is 3.0mm, the medium size is 0.5mm, the small size is 0.1mm.In the same calculation conditions, the trajectory of the three grains is simulated and compared. The results are shown as below.

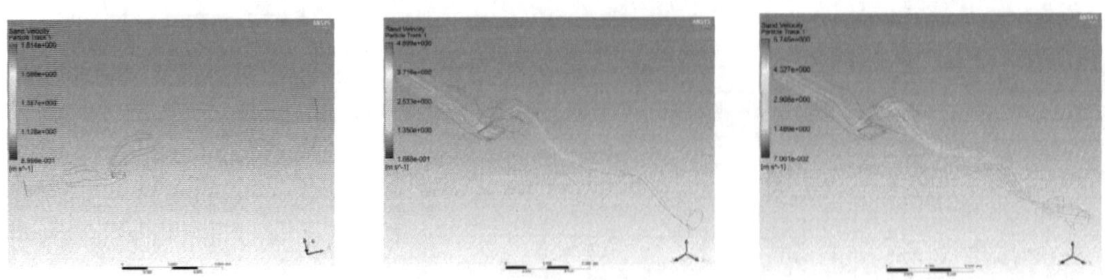

(1) large size (2) medium size (3) small size

Figure 2. Motion Trajectory of the Sand.

From the qualitative analysis of the trajectories, we can see that under the influence of the self-inertia of the sand grains, the size of sand particles is an important factor affecting the distribution of water flow, the follow property of small grains is better. The larger the particle size, the larger influence of gravity, the more easily to gather nearby the outlet of the separator, resulting a larger concentration gradient in near wall region. Under the same conditions, with the increase of sand particle size, more and more sand particles are concentrated at the outlet of the separator, the sand and water are separated.

3.3. Analysis on separation efficiency

According to the operating conditions, the flow rate in the calculation is set at 20 m3/h, and the particle size of the sand particles is 0.03mm, 0.07mm, 0.1mm, 0.2mm, 0.3mm, 0.5mm, 0.7mm, 1.0mm, 1.5mm, 2.0mm, 2.5mm and 3.0mm. The separation efficiency of the sand particles with different particle sizes is calculated.

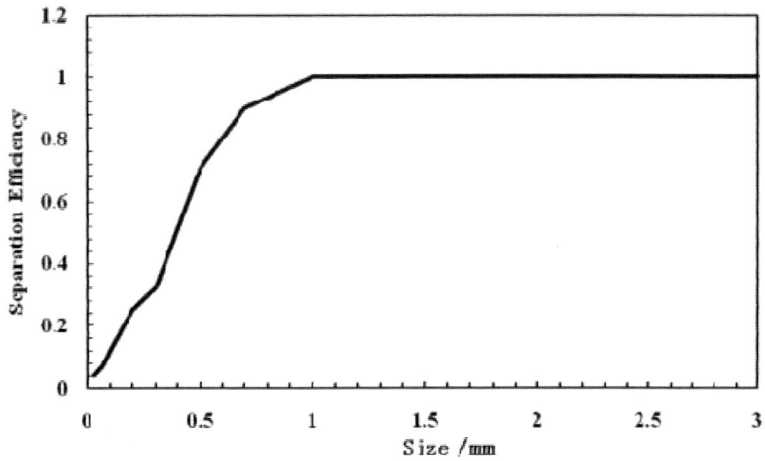

Figure 3. Separation Efficiency of Different Sands.

It shows the flow following of small size (below 0.1mm) is better, the separation efficiency of the sand grains is under 10%. With the increase of grain size, the separation efficiency is gradually increased. When the diameter of sand particles reaches 1mm, the separation efficiency can reach more than 99%. When the diameter of sand grains is up to 1mm, the separation device can separate all the sand particles.

From the above analysis, it can be seen when the particle size increases to 0.5mm, the inertia of sand grains increases obviously, the flow following property of sand grains becomes worse, and the separation effect is greatly improved. At this time, the motion trajectory of the large scale sand particles is determined by its own inertia force and the collision between the sand and the wall. The profile line of the separation part has a great influence on the separation efficiency.

3.4. Analysis on profile line

By changing the line coordinates of separation parts, changing the section area of the separation channel, changing the shrinkage of the channel and changing the flow velocity of the liquid, three separate structure schemes are designed. The contrast of the three schemes is as shown in Figure 4. And the separation effect of the three schemes is shown in Figure 5.

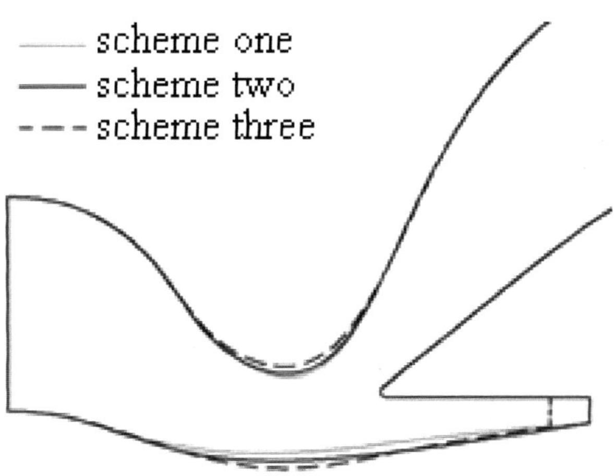

Figure 4. Comparison on Profile Line.

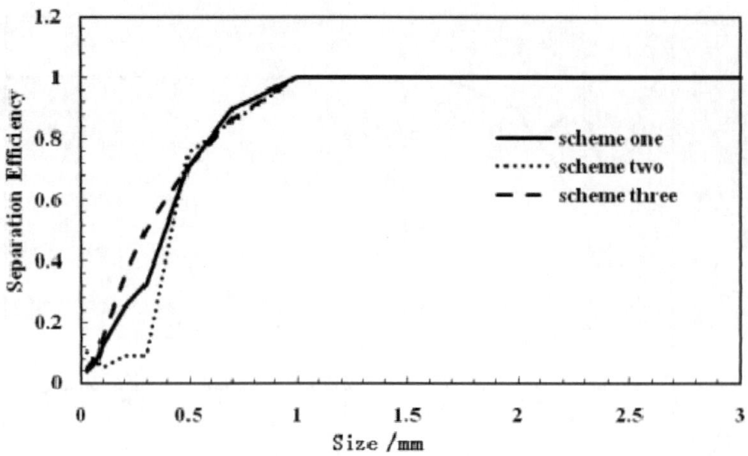

Figure 5. Comparison on Separation Efficiency.

It shows that the separation efficiency of small size sand (below 0.1mm) is near 10%. With the increase of sand particle size, the separation efficiency increases gradually. When the particle size is above 1mm, the separation efficiency can reach more than 99%. Theoretically, when the diameter of sand grains reaches 1mm, the three schemes can achieve the total separation of sand and water. From the perspective of separation effect, the larger the cross section area of the separator, the better the separation of sand water.

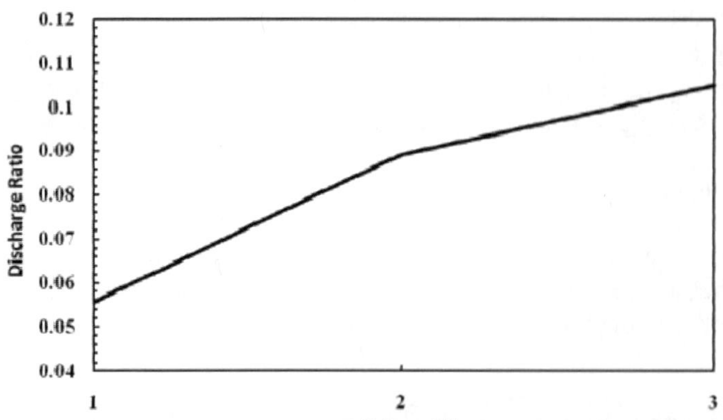

Figure 6. Comparison on Discharge Ratio.

The discharge ratio is an important index of the separation device, it reflects the loss of the flow. In principle, the smaller the loss, the better performance of the device. However, the discharge ratio can only be reduced as far as possible. From the above diagram, we can see that increasing the area of the separation section will greatly increase the discharge ratio of the device, which is not conducive to the overall power consumption control of the system. Therefore, the separation efficiency and discharge ratio should be taken into account in the design and operation of the separation device.

4. Experimental Analysis of the Test Device

4.1. Brief introduction of test system

A performance test system is set up, and the separation efficiency and discharge ratio of the device is tested. The test system consists of a power device, a sand adding device, a swirling device, a steady flow section, a sand collecting device and an auxiliary testing instrument. According to the demand, sands of the specified size are selected and added to the test rig after weighing. The sands enter into the separation device after swirling and rectifying. After the separation, the sands entering the sand collector are weighed, and the separation efficiency is calculated. At the same time, the flow before and after the separation are compared and analyzed. The pressure gauge reading is used to calculate and analyze the pressure loss of the device. In order to reduce the wear of sand particles on the pump set, the solid sand used in the test is soft coral sand.

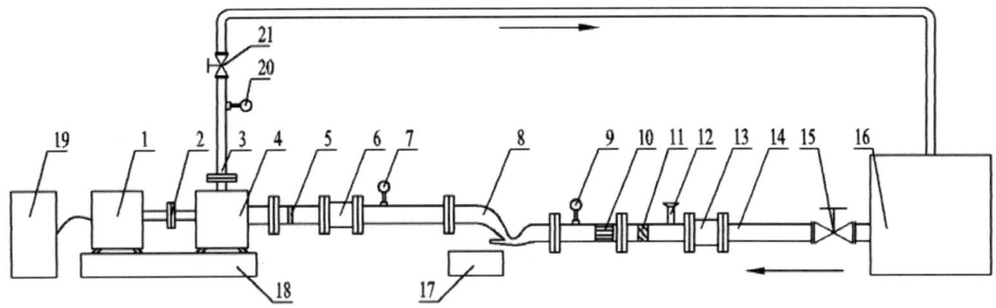

Figure 7. Schematic Diagram of the Test System.

4.2. Separation efficiency

Separation efficiency is the most important parameter to measure the performance of the separator. The quality percentage of sand before and after separation is used to illustrate the effect of separation.

$$\eta = \frac{m}{M} \tag{6}$$

Where M the quality of is sand particles added to the separation device, m is the quality of Sand grain quality in the collection device.

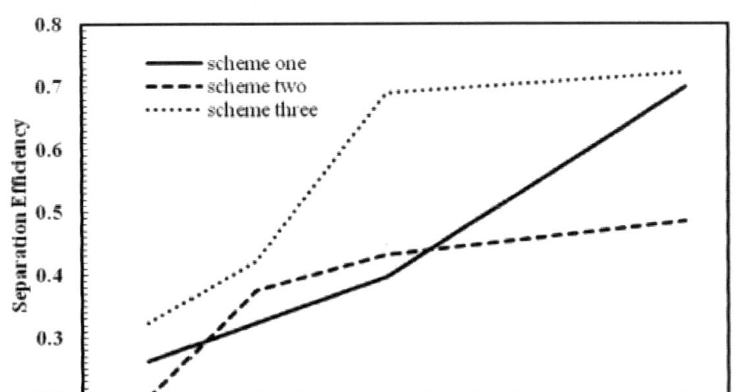

Figure 8. Experimental Comparison of Separation Efficiency.

From the separation efficiency test of the separation device, it can be seen that:

(1) The test results are the same as those of the simulation results, but there are some differences. This is due to the impact action of flow and the loss of some experimental samples in the collection process.

(2) The separation area has great influence on the separation efficiency of the separation device. Increasing the area of the separation section will significantly improve the separation efficiency of the device for small particle size sand.

4.3. Characteristic of discharge ratio

The discharge ratio is an important index to measure the power consumption of the device. The formula is as follows:

$$\frac{\Delta Q}{Q_{in}} = \frac{Q_{in} - Q_{out}}{Q_{in}} \tag{7}$$

Where Q_{in} is the import flow, Q_{out} is the outlet flow.

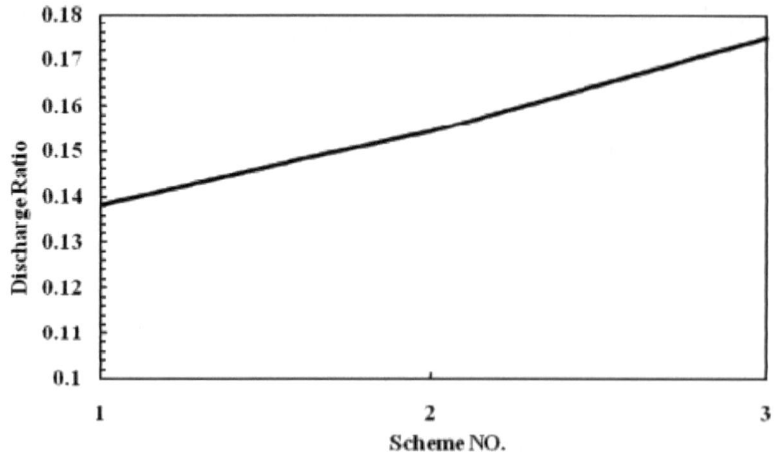

Figure 9. Comparison on Discharge Ratio.

It can be seen from the above diagram that the test results are the same trend as the simulation results, but there are certain differences. The separation section area has a great influence on the separation efficiency of the separation device. Increasing the area of the separation section will significantly increase the discharge ratio of the device, which has a negative impact on the overall power consumption of the system.

5. Conclusions

In this paper, the movement of water flow and the separation efficiency are numerically and experimentally analyzed. The following conclusions can be made:

(1) The size of sand particles is an important factor affecting the distribution of water flow. The larger the particle size, the larger the gravity, the more easy to settle and gather near the outlet of the separator. Under the same conditions, with the increase of sand particle size, more and more sand particles are concentrated at the outlet of the separator.

(2) Simulation analysis shows that the separation efficiency small size sand is near 10%. With the increase of sand size, the separation efficiency increases gradually. When the size is above 1mm, the efficiency is much better.

(3) The test results are the same trend as the simulation results. The experiential and simulative results are different in numerical value under the restriction of test conditions.

(4) The separation section area has a great influence on the separation efficiency. Increasing the separation section area will significantly increase the separation efficiency of the small size sands, but

it will also greatly increase the discharge ratio of the device, which is not conducive to the overall power control of the system. The separation efficiency and discharge ratio should be taken into account in the design and operation of the separation device.

References

[1] Xiong Mao-tao, Numerical simulation of the sandy ablation of centrifugal pump impeller flowed passage parts in high sandy flow:[Ph.D disssertation]. Sichuan: XiHua University, 2006.
[2] Liu Xiao-bing, Solid liquid two phase flow and its numerical simulation in Turbomachinery,Water Conservancy and Hydropower Press, Beijing 1996.
[3] Sha WT, Soo SL. Multiphase fluid mechanics, Heat and Mass Transfer, 1978.
[4] Bai Xiao-ning, Design of the solid-liquid two-phase flow pipeline experimental apparatus,J. Journal of University of Shanghai For Science and Technology 2001, 23(2):95-101.
[5] Tan Dong-hua, The Design and Experimental Research on Underwater Liquid-solid Two Phase Double Channel Pump:[Ph.D disssertation]. Zhejiang: Zhejiang University,2006.

IOP Conf. Series: Journal of Physics: Conf. Series **1064** (2018) 012047 doi:10.1088/1742-6596/1064/1/012047

Force Analysis of π-type Compensator in Pigging Operation

Lingying Ni [1], Shengjie Zhang[1,] *, Jie Dai[2]

[1.] China University of Petroleum (East China), Qingdao, China.
[2.] Sinopec Shanghai Engineering CO., Ltd, Shanghai, China.

*Corresponding author e-mail:1932950119@.qq.com

Abstract. During the pigging operation of a certain oil depot mixed transmission pipeline, there was a π-type compensator under significant stress, resulting in apparent displacement of the bent pipe. For this problem, the OLGA software was used to simulate the pigging process of the mixed pipeline, and the impact of the fluid on the π-compensator bend pipe during the occurrence of slug flow inside the pipeline was calculated and analysed. The calculation results showed that when made a pigging for the liquid-phase fluid pipe, the pipeline pressure declined along the pipeline; and for the gas-liquid two-phase flow pipe, there would be gas accumulation at the bottom of the compensator riser, and then the slug flow was formed. Pipeline pressure at the slug flow increased suddenly, causing pressure fluctuations in the pipeline to increase, pressure drop along the pipeline decreased sharply, and pressure loss occurred at downstream; pressure fluctuations caused by slug flow generated strong fluid shock, resulting in pipeline's jump and relative displacement, which destroyed the pipe system. The research results are of guiding significance for solving the problem of pigging operation in the multiphase pipeline and provide reference for other similar pigging operations.

1. Induction

In recent years, with the sharp increase in oil and gas demand, and the rapid development of pipeline transportation, pipelines are widely applied to various fields of industry. As the carrier of material transport, pipelines need periodical pigging after being used for a period of time. Pigging can effectively improve the cleanliness and transmission efficiency of pipelines, and are of great significance for ensuring pipelines' safety and reducing their energy consumption [1]. However, due to the varied routes of the mixed transmission pipelines, the fluctuating heights of the pipes, the complex changes of flow pattern in the pipelines, and the unreasonable operation speed of pigs, pigging operations may cause accidents and endanger production safety.

2. Pigging condition

In a pigging operation, using N2 as the gas source to push pig and water on the petrol pipeline outside the depot from the metering and pigging station, and the operation was smooth. The operating pressure is 0.4Mpa and the flow rate is 1900m3/h, and the pipeline is 800mm in diameter, 9mm in thickness and 1.5mm in corrosion allowance. There was an interruption of air supply for about half an hour,

Content from this work may be used under the terms of the Creative Commons Attribution 3.0 licence. Any further distribution of this work must maintain attribution to the author(s) and the title of the work, journal citation and DOI.
Published under licence by IOP Publishing Ltd

while the end of the pipeline continued to discharge water. After half an hour, the gas source recovered and the pigging operation continued. At this time, the pipe in front of the pipe cleaner made a sound of impact, and the pipe support, which laid on the pipe gallery was about 400m long, fell off (Figure 1), as well as a π-type compensator conduit elbow plastically deformed. About 1 hour later, the pig valve of the petrol tank area received the pig whose shape was basically complete.

Figure 1. Pipe Shelf Shedding Diagram

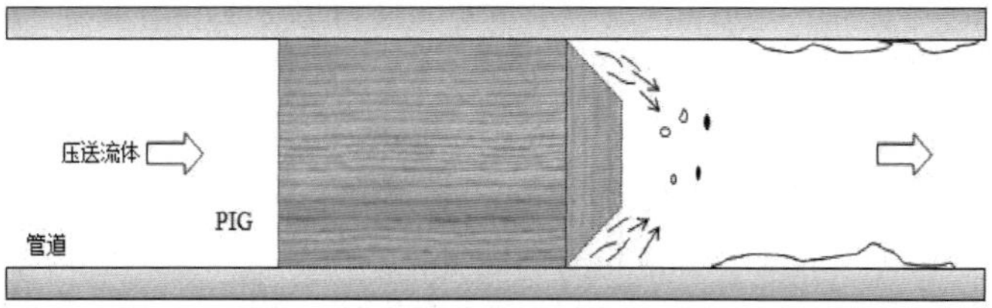

Figure 2. Pigging schematic diagram

During the pigging operation, pig's advancement in the pipeline depends on the thrust generated by the pressurized fluid of the pump. There are two specific pigging methods: 1 Use the pressure to drive the pigging to remove the dirt in front of the pig. This method is suitable for shorter pipe cleaning; 2 Use the pressure generated by the fluid leaking from the pigs to pulverize and discharge the dirt that adheres to the pipe wall. This method is suitable for long pipe cleaning, as shown in Figure 2.

The problems that may be faced during the pigging operation mainly include two aspects: 1. Due to complicated pipelines or topography fluctuations, the resistance during the pigging operation is greater than the thrust, which may cause the pigs to jam; 2. the speed of the pig running is unreasonable. If it is too slow, it will affect the pigging efficiency and extend the construction period. If it is too fast, it will result in unsatisfactory pigging, damage of the pipeline anti-corrosion coating, and mechanical damage of the pigs [2].

3. Model

OLGA software has high precision for simulating oil and gas pipelines and pigging processes. Therefore, the pigging module in the OLGA software is used to calculate the pressure in the oil and gas pipeline based on the two-fluid model, and analyze the influence of the pressure under different gas content on the oil and gas pipeline [3].

3.1. Mathematical Model

The main components of the basic mathematical equation of the two-fluid model are: mass conservation equation, momentum conservation equation and energy conservation equation.

(1)mass conservation equation

For the gas phase, the following relationship is satisfied:

$$\frac{\partial}{\partial t}\left(V_g \rho_g\right) = -\frac{1}{A}\frac{\partial}{\partial z}\left(A V_g \rho_g v_g\right) + \psi_g + G_g \tag{1}$$

For the fluid film around the pipe wall, the following relationship is satisfied:

$$\frac{\partial}{\partial t}\left(V_L \rho_L\right) = -\frac{1}{A}\frac{\partial}{\partial z}\left(A V_L \rho_L v_g\right) - \psi_g \frac{V_L}{V_L + V_D} - \psi_e + \psi_d + G_g \tag{2}$$

For the droplets in the vicinity of the gas phase, the following relationship is satisfied:

$$\frac{\partial}{\partial t}\left(V_D\rho_L\right) = -\frac{1}{A}\frac{\partial}{\partial z}\left(AV_D\rho_L v_d\right) - \psi_g\frac{V_L}{V_L + V_D} + \psi_e - \psi_d + G_D \tag{3}$$

Where: ρ ——density; V ——velocity; A——pipeline flow cross-sectional area; G_f —— f phase mass source; ψ_e、ψ_d —— rates of droplet entrainment and deposition, subscript g、L、D —— respectively represent gas phase, liquid phase, droplet; ψ_g ——The mass transfer rate between two phases, which is positive when liquid phase evaporates; V_g、V_L、V_D——the volume fraction of the gas phase, liquid film, and droplets, respectively. It meets that $V_g + V_L + V_D = 1$.

(2)momentum conservation equation

Assuming that the internal source G_f enters the fluid perpendicular to the tube wall, then for the gas phase and the droplets in it, it satisfies that

$$\begin{aligned}\frac{\partial}{\partial t}\left(V_g\rho_g v_g + V_D\rho_L v_D\right) = -\frac{1}{A}\frac{\partial}{\partial z}\left(AV_g\rho_g v_g^2 + AV_D\rho_L v_g^2\right) - \left(V_g + V_D\right)\frac{\partial \rho}{\partial z} \\ -\frac{1}{2}\lambda_g\rho_g\left|v_g\right|v_g\frac{S_g}{4A} - \frac{1}{2}\lambda_i\rho_g\left|v_r\right|v_r\frac{S_i}{4A} + \left(V_g\rho_g + V_g\rho_L\right)g\cos\alpha\end{aligned} \tag{4}$$

For fluid film around the pipe wall, the following relationship is satisfied:

$$\begin{aligned}\frac{\partial}{\partial t}\left(V_L\rho_L v_L\right) = -\frac{1}{A}\frac{\partial}{\partial z}\left(AV_L\rho_L v_L^2\right) - V_L\frac{\partial P}{\partial z} - \frac{1}{2}\lambda_L\rho_L\left|v_L\right|v_L\frac{S_L}{4A} + \frac{1}{2}\lambda_I\rho_G\left|v_r\right|v_r\frac{S_i}{4A} \\ + V_L\rho_L g\cos\alpha - \psi_g\frac{V_L}{V_L + V_D}v_\alpha - \psi_e v_i + \psi_d v_D - V_L d\left(\rho_L - \rho_g\right)g\frac{\partial V_L}{\partial z}\sin\alpha\end{aligned} \tag{5}$$

Where: P ——Pressure; v_r ——Relative velocity; α ——The angle between the axis and the vertical of the pipe; subscript i ——Interface between gas and liquid phases; S_g、S_L、S_i——They respectively represent the wet perimeter of the interface between the gas phase, the liquid film, and the gas-liquid phase; Among them, v_α satisfies the following relationship: When the liquid film evaporates, if $\psi_g > 0$ then $v_\alpha = v_L$, When the droplet evaporates, if $\psi_g > 0$ then $v_\alpha = v_D$, When the gas condenses, if $\psi_g < 0$ then $v_\alpha = v_L$.

(3)energy conservation equation

For the gas-liquid mixture, there is the following relationship:

$$\begin{aligned}\frac{\partial}{\partial t}\left[m_g\left(E_g + \frac{1}{2}v_g^2 + gh\right) + m_L\left(E_L + \frac{1}{2}V_L^2 + gh\right) + m_D\left(E_D + \frac{1}{2}v_D^2 + gh\right)\right] = \\ -\frac{\partial}{\partial z}\begin{bmatrix} m_g v_g\left(H_g + \frac{1}{2}v_g^2 + gh\right) + m_L v_L\left(H_L + \frac{1}{2}v_L^2 + gh\right) \\ + m_D v_D\left(H_D + \frac{1}{2}v_D^2 + gh\right)\end{bmatrix} + H_S + U\end{aligned} \tag{6}$$

Where: E ——Energy per unit mass; h——Elevation; H_S——The enthalpy of mass source; U—— heat transmission Capacity of pipe wall; $m_f = V_f\rho_f$, among them $f = g$、L、D.

The above conservation equations of mass, momentum and energy apply to all flow patterns.

3.2. Physical model

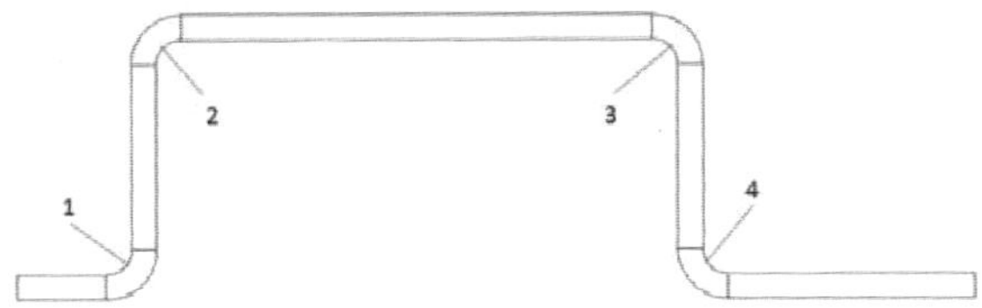

Figure 3. π-type compensator of pipeline

The π-type compensator is shown in Figure 3. The fluid passed through four pipe bends (respectively labeled as observation points 1, 2, 3, and 4) successively in the compensator. Now taking this compensator as an example to analyze the force. Using OLGA to simulate the pigging, and the boundary conditions were set by the inlet node and the outlet node. The inlet node was the closed boundary, and the outlet node was the pressure boundary.

4. Analysis of results

Gas-liquid flow flowed in the pipeline, and the gas-liquid two phases show different macroscopic motion laws—flow patterns, different flow pattern simulation results are as follows

Figure 4 shows the pipeline pressure fluctuation curve when the GOR=0. The four curves respectively represent the pressure fluctuations at the 1 to 4 observation points. As it shows, in the pigging, if the fluid in the pipeline is liquid phase, the pressure of the pipeline at the elbow is almost unchanged;

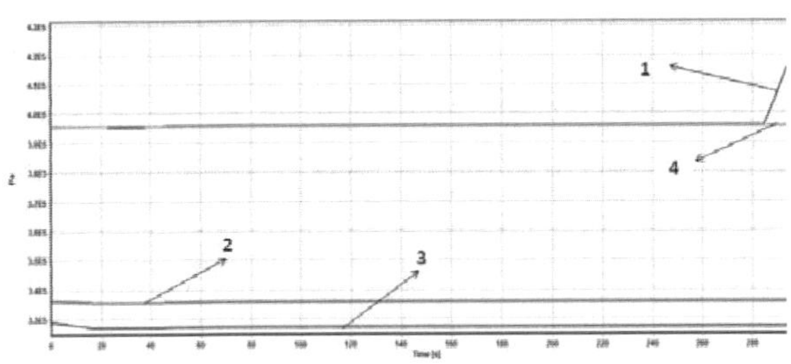

Figure 4. The result of pressure when GOR=0

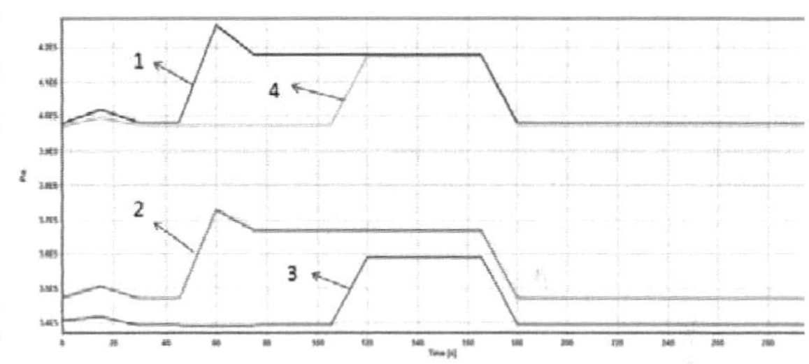

Figure 5. The result of pressure when GOR=1

The result of calculating the pipeline pressure after increasing the gas content is shown in Figure 5. When increasing gas content, the pressure increased at all bend. At observation points 1 and 2, the pressure rose significantly, which was about 10% of the air source pressure. The high pressure duration was about 2 minutes. As it shows, when pressure jumps occurred at observation points 3 and 4, points 1 and 2 were still under high pressure. Compared with point 1 and 2, the point 3 and 4 had a short time at high pressure, which was about 1 minute. Slug flow was not formed under this gas content, and the pressure curve was relatively stable.

Adjusting gas content, the pressure condition of pipe after generating slug flow was shown in Figure 6. From the figure, it can be seen that the pressure fluctuation at the observation point 1 was obvious, and the pressure jump value was higher than other points. The pressure at observation point 2 also fluctuated significantly, while the pressure fluctuations at observation points 3 and 4 were not obvious. Comparing Figure. 4, it can be seen that the pipeline pressure after generating slug flow was obviously increased, and the increased part was mainly at the elbow upstream of the pipeline.

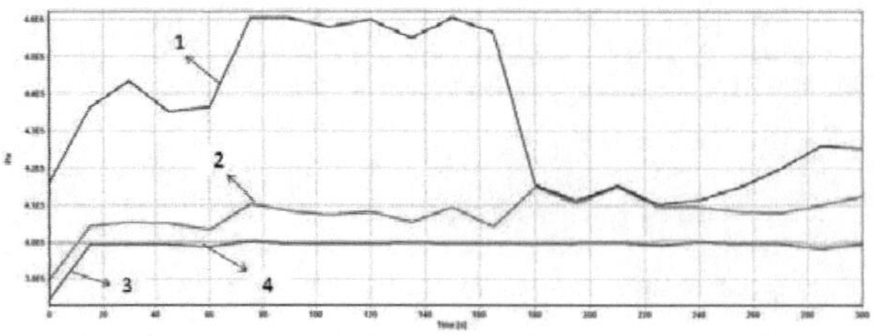

Figure 6. The result of pressure when GOR=10

After the slug flow was formed in the pipeline, increased the gas content further. Figure 7 shows the calculation results of the pipeline pressure after the gas volume in the pipeline increased by 1 times, and Figure 8 shows the calculation results after the increase of 4 times. From the figures, we can see that the pressure fluctuation amplitude in the pipeline increased with the increase of gas content in the pipeline: When the gas content increased by 1 times, the pipeline pressure increased by about 42%; when the gas content increased by 4 times, the pipeline pressure increased by about 25%. Among four observation points of the pipeline, the pressure fluctuation of observation point 1 and 2 on the upstream was greater than point 3 and 4.

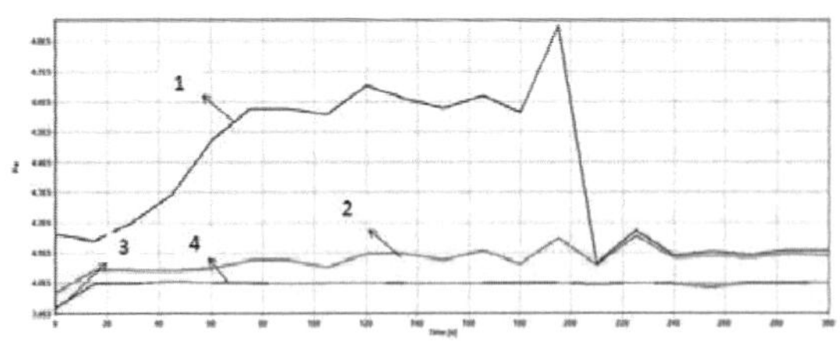

Figure 7. The result of pressure when GOR=20

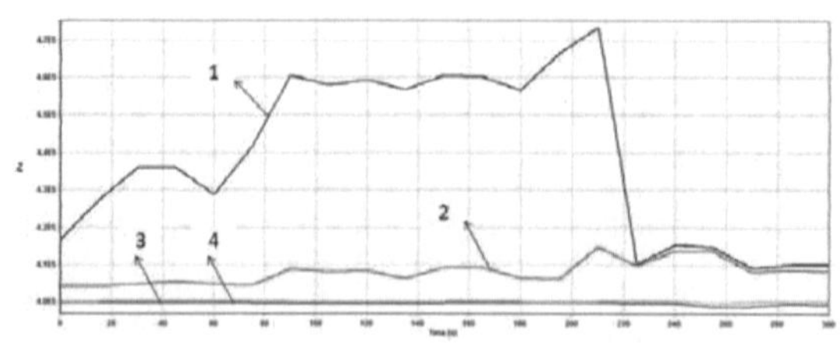

Figure 8. The result of pressure when GOR=50

After the increase of gas content, the change of flow pattern in the pipe was complicated, as shown in Figure 9. According to the figure, the slug flow, bubble flow, stratified flow, and annular flow in the pipe appeared alternately. In the horizontal pipe, there were mostly stratified flow, and sometimes the bubble flow. In the curved pipe, there was slug flow. And in the vertical pipe section, there was annular flow.

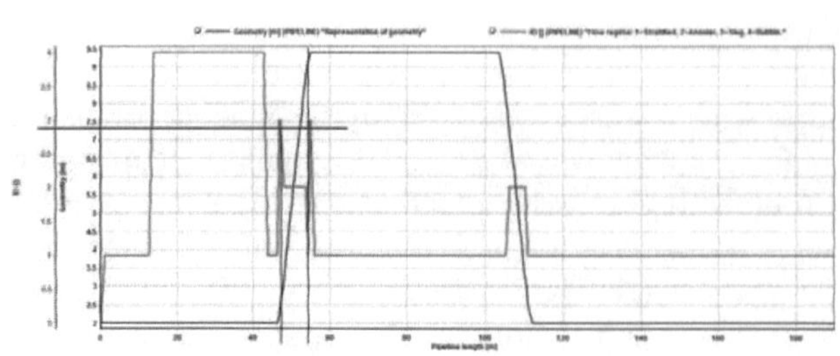

Figure 9. The pipe flow chart

Compared Figure 6~8, the pressure fluctuation at observation point 1 was the most obvious, and as can be seen from Figure 9, slug flows were more likely to occur at observation points 1 and 2, indicating that the most significant position of the pressure fluctuation caused by the slug flow is at the No. 1 elbow of the pipeline, and the action time was also the longest compared with other elbows;

Compared Figure 4~5 with Figure 6~8, when slug flow occurs in the pipeline, the pressure fluctuations at the upstream of the pipeline were significant, but at the downstream of the pipeline were not obvious. And in the downstream pipeline, there is a distinct loss of pressure. The No. 4 elbow, which should have been under high pressure, had a pressure value similar to that of No. 3 elbow. It can be seen that the pressure of the downstream pipe was seriously damaged due to the slug flow.

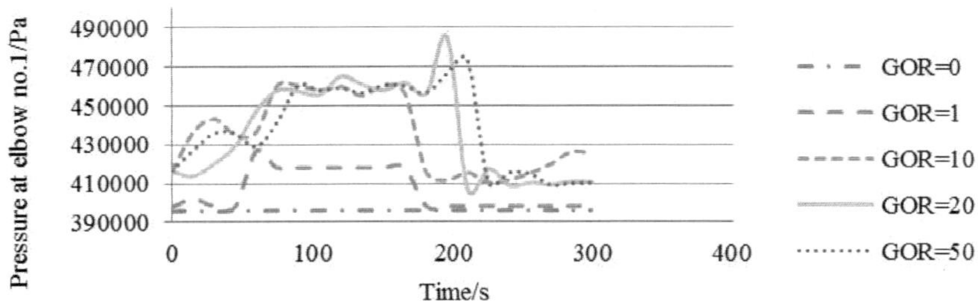

Figure 10. Pressure change curves of pipeline with different gas content

The pressure curves of different gas content in pipeline at the observation point 1 in the pigging is shown in figure 10. Compared the pressure fluctuation curves of different gas content, it can be seen that the influence of gas content on the pipeline pressure was very obvious: when there was a single-phase flow in the pipeline, the pressure value of the pipeline did not fluctuate significantly; when the gas content was increased, the pressure value of the pipeline was obviously raised and lowered. Moreover, with the increase of gas content, the fluctuation interval of pipeline pressure was larger and the fluctuation time was longer.

According to figure 10, the pressure of the fluid in the pipe at No. 1 elbow can be known, thus calculate the thrust generated by the fluid in the pipe as[4,5]

$$F = A * \Delta P \tag{7}$$

Where, F is the force generated by the fluid in the pipe; ΔP is the pressure difference between before and after the fluid in the pipe at No. 1 elbow, that is, the difference between the pressure produced by the fluid in the pipe under pigging and the pressure at the outlet of the pipe. A is the cross-sectional area of the pipe.

Under the action of the thrust force, the velocity of the fluid in the pipe reaching the tip of the π-type compensator can be calculated by the following equation according to the kinetic energy theorem [4,5]:

$$(F - \mathrm{mg})h = \frac{1}{2} m v^2 \tag{8}$$

Where, m is the quality of the fluid in the pipe; g is the acceleration of gravity; h is The running height of the fluid in the pipe; v is the velocity of the fluid reaching the top of the π-type compensator.

According to the principle of momentum conservation of fluid mechanics, ignoring local friction, the impact force of the fluid on the top of the π-type compensator is

$$F_y = -\rho Q v \tag{9}$$

Where, F_y is the impact force of the fluid at the top of the π-type compensator; Q is the elbow outlet flow. The calculation results are as follows:

Table 1. The calculated result of the impact force of the fluid on the pipe under different gas content

	ΔP (pa)	F (kN)	v^2	v (m/s)	F_y (kN)
GOR=1	25000	11.915	55.523	7.45	20.641
GOR=10	60000	28.596	272.376	16.50	101.255
GOR=20	85000	40.511	427.271	20.67	158.837

GOR=50	75000	35.745	365.313	19.11	135.804

From the calculation results in Table 1, it can be seen that comparing the results of GOR=1~50, the pipe received the greatest impact when GOR=20, which was 158.837kN; the fluid velocity reached 20.67m/s at the top of the riser. According to the pipeline model, using Autopipe software to analyze, it can be seen that under the force of 158.837kN, the maximum vertical displacement at the elbow of the pipeline was 175.42mm, with the largest component in the Y direction and the second component in the Z direction. The displacement diagram is shown in Figure 11:

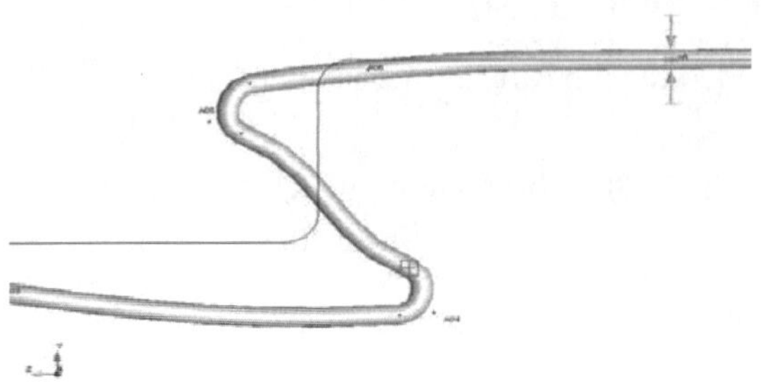

Figure 11. Displacement of compensator bend

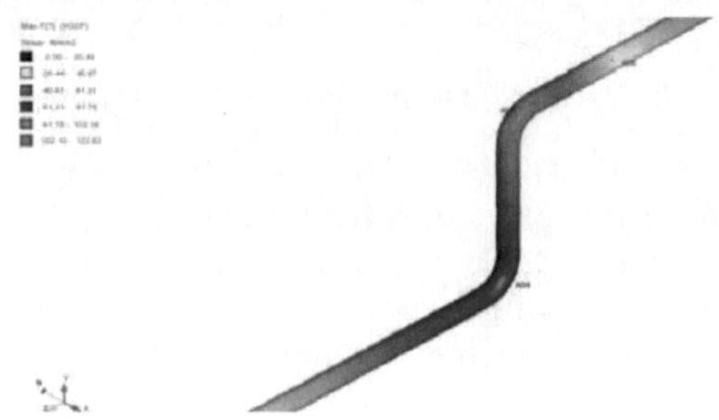

Figure 12. The stress of compensator bend

The stress distribution at the elbow of the compensator was shown in Figure 12. It was found that the stress at the bent pipe of the compensator was about 123Mpa, which was far greater than the stress at the horizontal pipe and higher than the allowable stress value. The stress at the elbow was about 2 to 3 times of the stress at the horizontal pipe. Compared with the horizontal pipe, the riser was the part with concentrated stress, especially the top of the riser. The high-speed impact of the fluid in the pipe increased the stress of the bend pipe at the top of the riser, so the displacement of the pipe at this location was also the most serious part of the entire compensator.

5. Conclusion

The OLGA was used to simulate the pigging operation, and the pipe section forming the slug flow was analyzed with AutoPIPE. The conclusions are as follows:

(1)For single liquid phase fluid in the pigging, the pipeline pressure decreased along the pipeline; for gas-liquid two-phase flow, the pipeline pressure fluctuated greatly at the bent pipe where the gas gathered.

(2)When the gas content increased, a bubble or slug flow was generated in the pipe, and the slug flow formed in the bend at the bottom of the compensator riser. The slug flow increased the pressure fluctuation, the pressure drop along the pipeline decreased sharply, and the loss of pressure occurred at downstream.

(3)The high-frequency alternating load generated by slug flow would form a strong fluid shock, which would have a strong destructive effect on the pipe, especially the bend of the riser. It caused the pipe to jump on the pipe rack and generated relative displacement with the pipe support, thus endangering the production safety.

(4)It is recommended that the bottom and top of the riser, which are easy to gather gas, should be reinforced and monitored in real time to control the pressure and flow at the inlet of the pipeline, or set the exhaust valve at the top of the riser to prevent the gas from gathering at the bottom of the riser to form a slug, so as to ensure the safety of operation.

References

[1] XU Xiao-xuan, GONG Jing, DENG Dao-ming. General situation of research on multi-phase pipe cleaning model [J].China Offshore Oil and Gas (Engineering), 2003(04):21-24+1-15.

[2] YAN Jian-feng, WANG Wen-wen, AN Feng. Pigging technology of long-distance pipeline [J].Pipeline Technique and Equipment, 2013(03):48-49+52.

[3] YAO Jian, HAN Fang-yong, SONG Yi-bao. Application of Dynamic Simulation Software for Oil-gas Water Multiphase Flow Pipeline [A]. China Institute of Automation System Simulation Committee, China System Simulation Society Simulation Computer and Software Committee,'2000 System Simulation Technology and Its Application Academic Conference Proceedings[C]. China Institute of Automation System Simulation Committee,China System Simulation Society Simulation Computer and Software Committee, 2000:4.

[4] YANG Shuren. Engineering Fluid Mechanics[M].Beijing:Petroleum Industry Press, 2006.6

[5] LI Xiu-feng, XU Zhi-hui, ZHANG Ying. Upward movement of riser caused by subsea pipeline hydrate blockage removing operation and its recovery[J].Ocean Engineering Equipment and Technology, 2015, 2(04):240-243.

[6] ZHANG Yan-hong, BAI Chang-qing, WU Wei-yang, et al. Fluid-structure interaction dynamic analysis of pipe under gas-liquid two-phase slug flow [J/OL].Chinese Journal of Applied Mechanics:1-5[2018-03-26].

[7] ZHU Kai, CUI Da-yong. Pigging technique of submarine gas line in the Boxi oilfield [J].China Offshore Oil and Gas, 2007(06):412-414.

[8] LAN Feng, CUI Da-yong, Research on blocking problem of submarine gas line in Boxi oil and gas field[J].China Offshore Oil and Gas,2007(05):350-352+360.

[9] SONG Bo, LI Tong-yu, WANG Wei-qiang, et al. Transient numerical simulation of severe slugging in marine riser system[J].Journal of Liaoning Shihua University,2016,36(03):34-38.

[10] WANG Zhan-yong, LI Gang, FANG Xin, et al. Simulation methods of slug flow and water hammer loads in stress analysis[J].Petroleum Engineering Construction,2011,37(01):8-11+5-6.

[11] LIU Miao-er, LI Qing-ping, AN Wei-jie, et al. One-dimension unsteady model of severe slugging in a riser system[J].China Offshore Oil and Gas, 2007(02):125-130.

[12] YU Xi-chong, REN Yan-bing, WU Ya-ling. The analysis of transient flow characteristics and sensitivity in the riser of subsea multiphase pipeline [J].China Offshore Oil and Gas,2007(01):60-64.

[13] WANG Xin, GUO Lie-jin. Experimental investigation and simulation of severe slugging in pipeline-riser system [J].Journal of Engineering Thermophysics,2006(04):611-614.

[14] Schmidt Z, Brill J P, Beggs H D. Choking can eliminate severe pipeline slugging[C]// Oil and Gas Journal. 2010.

Numerical investigation of collapse of a bubble near the rigid boundary with specific microstructure

Kuiguang Jin[1], Changguo Sun[2], Yan Yu[2], Yazhe Liu[2], Dongsheng Xia[2,*]

[1]Nautical training and engineering practice center, Dalian Maritime University, Dalian, China.
[2]Transportation engineering college, Dalian Maritime University, Dalian, China.

*Corresponding author e-mail: xia@dlmu.edu.cn

Abstract. The dynamics of a vapour bubble close to a rigid boundary with the micrometer-scale microstructure is numerically simulated based on the Volume of Fluid (VOF) method. The simulated results show that the bubble close to the groove boundary exhibits distinctly different behaviour as the non-dimensional stand-off parameter γ is changed. Deformation of the part around the bubble-wall contact positions of bubble is constrained due to the effect of adsorption on the wall. The evolution of shape of the bubble close to the groove boundary is significantly different for different γ. Moreover, the maximum jet velocity, the collapse pressure, and the peak pressure on the groove wall are affected by γ. Cavitation erosion of the groove wall is mainly determined by action of shock wave in the three cases considered.

1. Introduction

The bubble dynamics near to the different boundaries is a classical problem, which has received extensive attention from researchers [1]. Rayleigh firstly established the bubble dynamics equation for a spherical bubble, namely Rayleigh equation [2], in which liquid viscosity and liquid compressibility was neglected. Rayleigh equation predicts that a local high pressure will be generated during the collapse phase of a spherical symmetrical bubble. However, quite a few experiments and numerical results demonstrate that bubbles near a boundary behave a non-spherical collapse characteristic [3] [4]. The existence of boundary and different boundary condition have a significant effect on bubble shape evolution, jet generation, collapse pressure and jet velocity [1]. The boundaries can be classified into the rigid wall, the elastic wall and the free surface in terms of the boundary physical properties or into the plain boundary and non-plain boundary in terms of the boundary profile. It should be noted that exploration on interaction of the bubbles with the boundaries having special profile deserves special attention, which can provide a good insight into ultrasonic cleaning, cavitation erosion and medical application of cavitation. However, few experimental and theoretical works focusing on this issue has been reported. In this paper, the effect of boundary with specific microstructure on collapse of bubble was numerically investigated by the homogeneous multiphase model and volume of fluid (VOF) method.

2. Numerical method

The physical description of the problem discussed in this paper is the dynamics of a spherical vapor bubble close to a rigid boundary with the micrometer-scale groove structure. It is assumed that only

Content from this work may be used under the terms of the Creative Commons Attribution 3.0 licence. Any further distribution of this work must maintain attribution to the author(s) and the title of the work, journal citation and DOI.
Published under licence by IOP Publishing Ltd

IOP Conf. Series: Journal of Physics: Conf. Series **1064** (2018) 012048 doi:10.1088/1742-6596/1064/1/012048

vapor, not other gas such as non-condensable gas, is contained in the single bubble. The vapor inside the bubble is compressible and assumed to behave like an ideal gas. The liquid surrounding the bubble (e.g. water) is considered as an incompressible newtonian fluid. The fluid flow state is laminar. The viscosity of water and vapor is considered in simulation. The effect of gravity is neglected. Moreover, mass transfer between vapor and water is not taken into account in the paper.

2.1. Governing equations

The volume of fluid (VOF) method based on homogeneous flow theory was adopted to simulate collapse of the cavitation bubble close to the rigid boundary with micrometre-scale groove structure. In the model, the flow is assumed as a homogenous mixture fluid consisting of liquid and vapor, neglecting inter-phase relative motion. The two components share the same physical velocity and pressure.

The mass, momentum, and energy equations for the mixture flow are

$$\frac{\partial \rho_{\mathrm{m}}}{\partial t}+\nabla\cdot\left(\rho_{\mathrm{m}}\vec{v}\right)=0 \tag{1}$$

$$\frac{\partial}{\partial t}\left(\rho_{\mathrm{m}}\vec{v}\right)+\nabla\cdot\left(\rho_{\mathrm{m}}\vec{v}\vec{v}\right)=-\nabla p+\nabla\cdot[\mu_{\mathrm{m}}(\nabla\vec{v}+\nabla\vec{v}^{\mathrm{T}})-\frac{2}{3}\mu_{\mathrm{m}}(\nabla\cdot\vec{v})I]+\vec{F}_{\sigma} \tag{2}$$

$$\frac{\partial}{\partial t}\left(\rho_{\mathrm{m}}E\right)+\nabla\cdot\left[\vec{v}\left(\rho_{\mathrm{m}}E+p\right)\right]=\nabla\cdot(k\nabla T) \tag{3}$$

where p is pressure, $\vec{v}$ is velocity vector, ρ_{m} and μ_{m} denotes the density and dynamic viscosity of the mixture, respectively, E is total energy, T is temperature and k are thermal conductivity. In equation (2), I represent the identity matrix.

The surface tension effect of bubble-water interface is included with the continuum surface force (CSF) model [5], which appends the surface tension term F_{σ} to the source term of the momentum equation (2) as a volume force.

$$\vec{F}_{\sigma}=\sigma\frac{\rho_{\mathrm{m}}\kappa\nabla\alpha_{\mathrm{l}}}{0.5(\rho_{\mathrm{g}}+\rho_{\mathrm{l}})} \tag{4}$$

where σ is the surface tension, κ is the curvature of the interface, and α_{l} is the volume fraction of the liquid phase in every cell.

The transportation equation of the volume fraction of liquid phase α_{l} is is given by

$$\frac{\partial \alpha_{\mathrm{l}}}{\partial t}+\nabla\cdot\left(\alpha_{\mathrm{l}}\vec{v}\right)=0 \tag{5}$$

As α_{l}=l or 0 for a grid cell at some time, it means that the grid cell is filled with pure liquid phase or pure gas phase. As $0<\alpha<1$, it represents that the interface between the two phase is located in this grid cell [6-7].

The density and the dynamic viscosity of the mixture are defined as

$$\rho_{\mathrm{m}}=\alpha_{\mathrm{l}}\rho_{\mathrm{l}}+(1-\alpha_{\mathrm{l}})\rho_{\mathrm{g}} \tag{6}$$

$$\mu_{\mathrm{m}}=\alpha_{\mathrm{l}}\mu_{\mathrm{l}}+(1-\alpha_{\mathrm{l}})\mu_{\mathrm{g}} \tag{7}$$

where the subscripts l and g correspond to liquid and vapor, respectively. The value of the liquid density is constant. Considering the compressibility of vapor, the density in the gas phase, ρ_g, is calculated by the ideal gas state equation.

$$\rho_g = \frac{M_m}{R_g T} p \tag{8}$$

where R_g and M_m represent the universal gas constant and the gas molar mass, respectively.

2.2. Computational domain and Mesh generation
The experimental observation shows that collapse of a bubble in proximity to a boundary has the axial symmetry property [8]. Therefore, a 2D axisymmetric model is applied in the numerical simulation. The geometry of the computational model and the boundary condition is shown in figure 1. A distributed rectangular groove with depth d of 0.2mm and width w of 0.2mm is arranged on the rigid boundary. The interval s of every groove unit is designed to be 0.2mm. The most significant geometrical parameter affecting the bubble dynamics, is the non-dimensional stand-off parameter, defined as the ratio of the distance of the initial bubble center from the solid surface h to the maximum bubble radius r_{max}, which is denoted by γ.

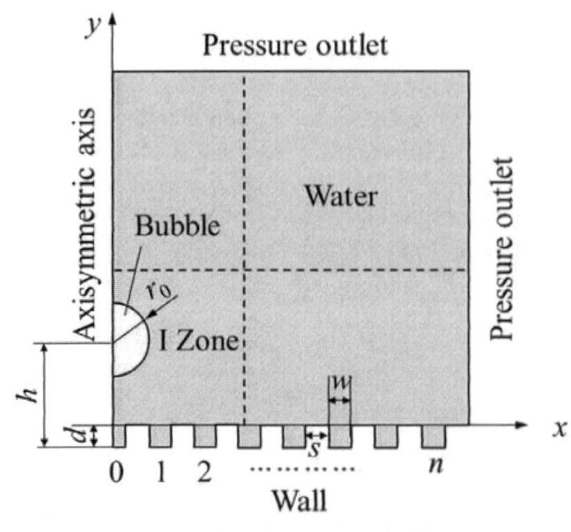

Figure 1. Computational model

The left boundary is set as axisymmetric boundary and the bottom boundary is set as wall with no-slip boundary condition in the 2D rectangular domain. Both of the right and left boundaries are chosen as pressure outlet with the value being equal to 101kPa. To decrease the effect of pressure boundary condition on the computational result, the distance between any two opposite boundaries is designed to be 50 times as the initial size of bubble. The zone surrounding the bubble (I zone) of interest is densely meshed in order to capture the details of the dynamics for bubble collapse. The structured quadrilateral meshes are generated by using ICEM CFD in order to ensure simulation accuracy and computational precision. The minimum cell size in I zone is about $5\times5\mu m^2$. The total mesh number of all computational models in this paper is more than 160 thousand.

2.3. Discretization scheme and Solution algorithm
The finite volume method is adopted to discretize the governing equations. Because bubble collapse is transit, the PISO algorithm is employed for the pressure-velocity coupling in unsteady computation. The

temporal term is discretized with the first-order implicit scheme. The gradient term is discretized with the least squares cell-based scheme. The liquid volume fraction at a cell interface is computed using Geo-Reconstruct scheme. The density term, the momentum term and the energy term are discretized by using the second-order upwind scheme, respectively. The pressure term is calculated by the body-force-weighted scheme. The fixed time step method is used in the calculation with the time step of 1ns.

3. Results and Discussion

A great number of experimental and numerical results show that the non-dimensional stand-off parameter γ is a significant parameter closely related to the impulse pressure on the wall, jet velocity and bubble shape evolution [9]. Three non-dimensional stand-off parameter γ of 1, 2 and 3 are selected in the numerical simulation. Only the initial bubble radius of 0.1mm is considered because the initial radius of cavitation bubble has a negligible influence on the jet velocity and the impact pressure on the wall [9].

Figure 2 depicts evolution of the bubble shape during collapse for γ=1. The vapor bubble contacts with the groove wall at three positions for the initial state in this case (figure 2(a)). Firstly, the bubble continuously contracts due to pressure difference between the inside and outside of the vapor bubble meanwhile the bubble-wall contact area increases (Figure 2(a)). Deformation of the lower part of bubble is inhibited. The upper bubble-liquid interface quickly moves downward and tends to be flat, which has an unchanged transverse size due to the effect of adsorption on the walls. As t=9.0μs, the whole upper bubble-liquid interface is nearly flat. The pressure gradient becomes significant above the bubble during this initial period (figure 3(a)). No high-pressure zone appears above the bubble in this case, which is different from collapse of the bubble near to the plain boundary [10]. Subsequently, liquid starts to penetrate inside the bubble forming a jet (figure 2(b)). At t=16.6μs, the liquid jet directly impacts the bottom wall after it passes through the bottom of bubble with the velocity of 17m/s (figure 3(b)). The jet velocity is rather low, which attributes to the small pressure difference between above and below the bubble. The peak impact pressure on the bottom wall induced by jet reaches 2.88MPa (figure 3(c)). After the jet reaches the bottom wall of groove, it moves along the bottom wall, which leads to detachment of the bubble from the bottom wall. The bubble splits into a free toroidal bubble 'A' and an attached bubble 'B' to the side wall of groove (figure 2(d) and (e)). Subsequently, both of the bubbles 'A' and 'B' collapse successively (figure 2(f)). A high collapse pressure of 645MPa occurs at the bubble 'B' center. The collapse of the attached bubble 'B' forms a shock wave. The bottom wall is subjected to an impact pressure of 564MPa (figure 3(d)). Therefore, it is concluded that cavitation erosion of the groove wall in this case should be controlled by the shock-wave action.

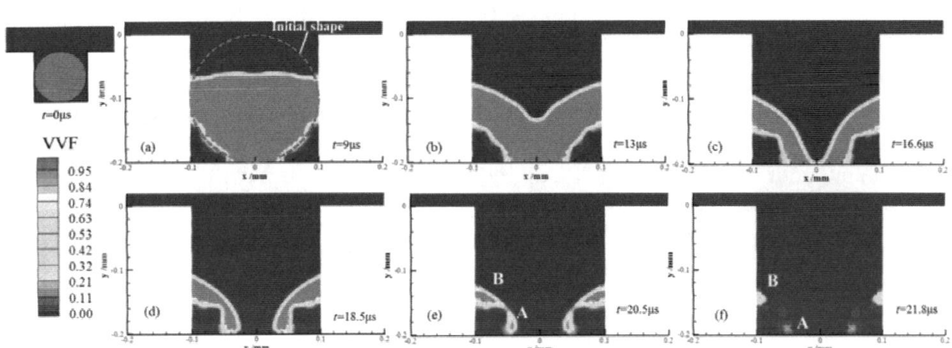

Figure 2. Evolution of the bubble shape for γ=1.

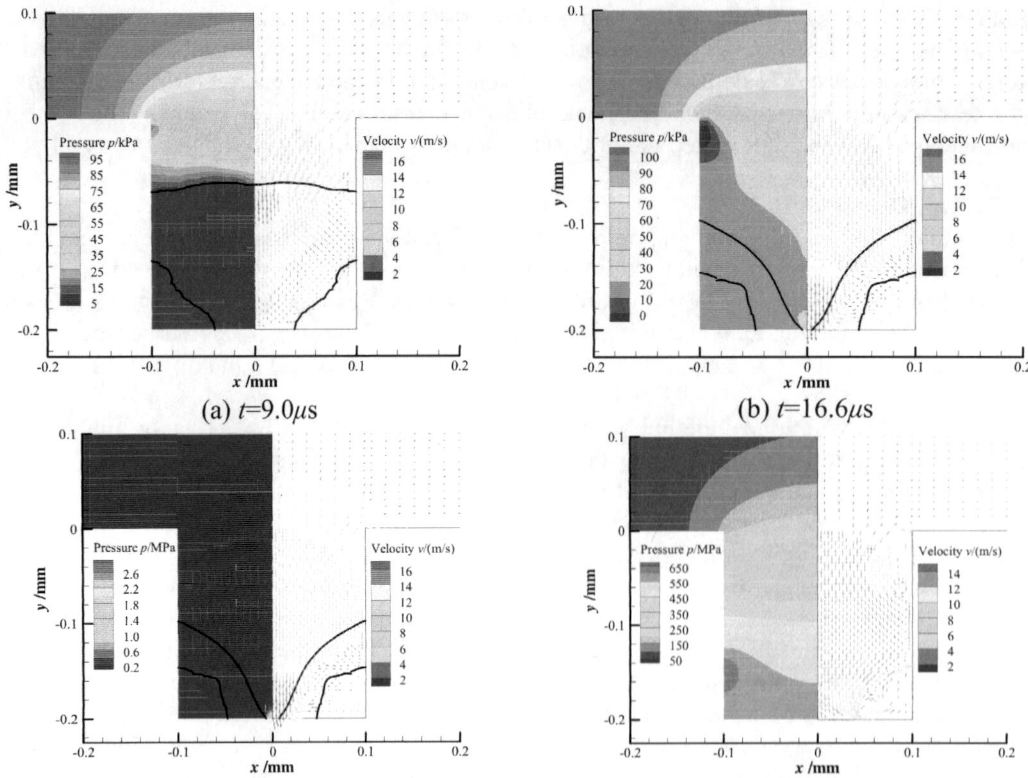

(a) t=9.0μs (b) t=16.6μs

(c) t=16.673μs (d) t=22.138μs

Figure 3. Pressure field (on the left-half part) and velocity field (on the right-half part) around collapsing bubble at different time for γ=1. The black solid line is the bubble-liquid interface.

As the stand-off parameter γ is increased to be 2, the vapor bubble contacts with the side wall and the liquid below the bubble is closed in the groove at the initial time. Figure 4 depicts evolution of the bubble shape during collapse for γ=2. Firstly, the upper part of bubble deforms downward significantly meanwhile the transversal size of bubble is unchanged due to the effect of adsorption on the side wall (figure 4(a)). At t=9.5μs, a high-pressure zone appears above the bubble and the top of bubble-liquid interface is completely flat (figure 5(a)). Then the liquid jet passes through the bubble with the velocity of 77.4m/s at t=10.84μs (figure 5(b)). Subsequently, the bubble splits into several toroidal bubbles 'A'~ 'D' and an attached bubble 'E' to the side wall. The bubbles 'A' and 'B' near the axisymmetric axis move towards the bottom wall under action of the jet (figure 4(e) and (f)). At t=15.032μs, both of the bubbles 'A' and 'B' collapse in proximity to the bottom wall, producing the collapse pressure of 154MPa, which causes the impact pressure of 130MPa on the bottom wall (figure 5(c)). The bubbles 'C' and 'D' behave the characteristics of oscillation (figure 4(f)~(i)). As the bubble 'C' starts to rebound after its volume is compressed to the minimum at t=16.472μs, the pressure at the bubble center is 20MPa (figure 5(d)). The attached bubble 'E' continuously contracts and finally detaches from the side wall caused by the jet along the side wall (figure 4(i)).

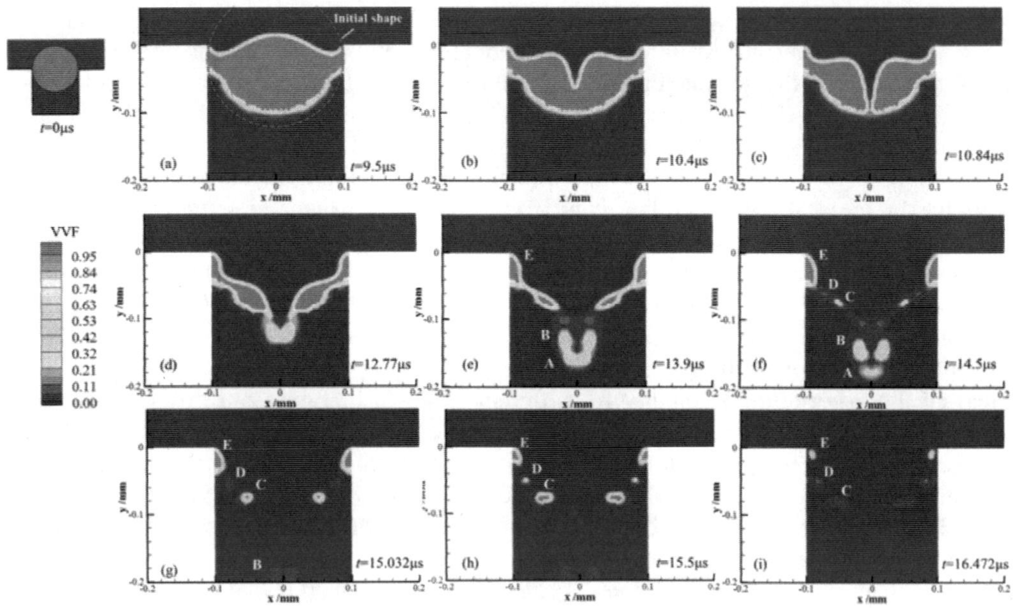

Figure 4. Evolution of the bubble shape for γ=2.

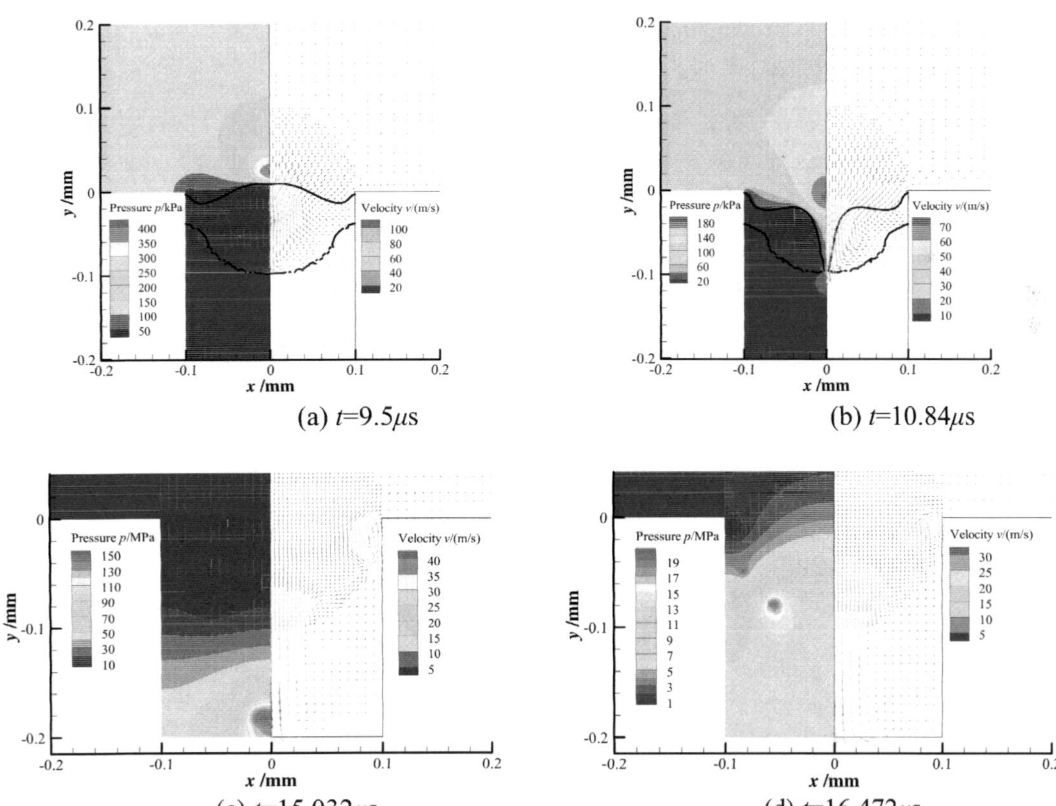

(a) t=9.5μs

(b) t=10.84μs

(c) t=15.032μs

(d) t=16.472μs

Figure 5. Pressure field (on the left-half part) and velocity field (on the right-half part) around collapsing bubble at different time for γ=2. The black solid line is the bubble-liquid interface.

As the stand-off parameter γ is further increased to be 3, the whole bubble is above the groove at the initial time. Figure 6 presents evolution of the bubble profile during collapse. The bubble center approaches the bottom wall as the bubble contracts. At t=10.15μs, the bubble-liquid interface is completely flat at the top and a high-pressure zone of 0.7MPa appears above the bubble (figure 7(a)). Then a liquid jet forms and the bubble takes on a heart-like shape. A toroidal bubble occurs when the jet passes through the bottom of bubble with the maximum jet velocity of 111m/s at t=10.93μs (figure 7(b)). Finally, the toroidal bubble collapses at t=11.148μs, which generates the collapse pressure of 218MPa, meanwhile the groove wall is subjected to the impulse pressure of 60MPa (figure 7(c)). After the jet moves about 0.235mm along the symmetric axis, the core jet reaches the bottom wall at t=20.9μs. However, the jet velocity decays to be 35.4m/s, which results in the weak impact pressure of 0.5MPa on the bottom wall (figure 7(d)). It is indicated that cavitation erosion of the groove wall in this case is mainly resulted from action of the shock wave.

Comparing the maximum jet velocity for different γ shown in figure 8(a), it is indicated that the maximum jet velocity increases as the stand-off parameter γ rises. It is mainly because the pressure difference between above and below the bubble grows with γ during development of the jet as shown in figure 8, which directly determines the acceleration of jet. Figure 8(b) presents variation of the collapse pressure and the peak pressure on the bottom wall of groove with γ. It is shown that the peak pressure on the bottom wall decreases with raising γ. Moreover, the peak pressure on the bottom wall of groove is close to the collapse pressure for γ=1 and 2, respectively. The reason is that collapse center of bubble is near to the bottom wall in the two cases under action of the shock wave. However, the peak pressure on the bottom wall is far smaller than the collapse pressure for γ=3. It is because the shock wave rapidly decays with the increase of propagation distance and the bottom wall is relatively far to the collapse center in the case of γ=3.

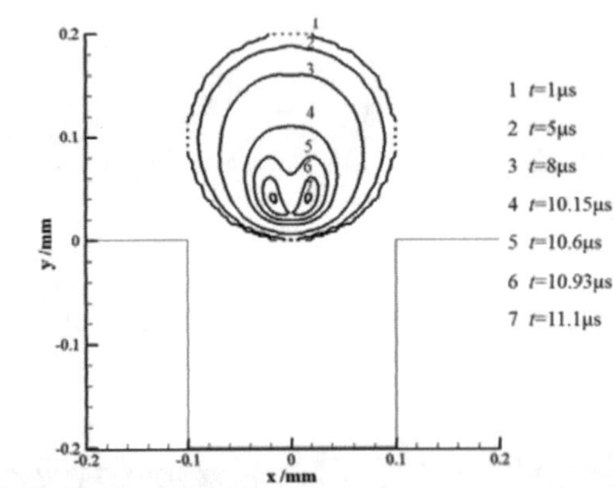

Figure 6. Evolution of the bubble profile during collapse for γ=3.

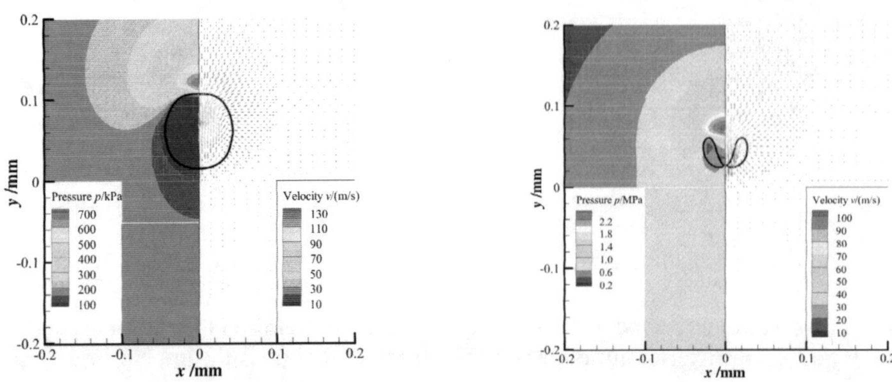

(a) t=10.15μs (b) t=10.93μs

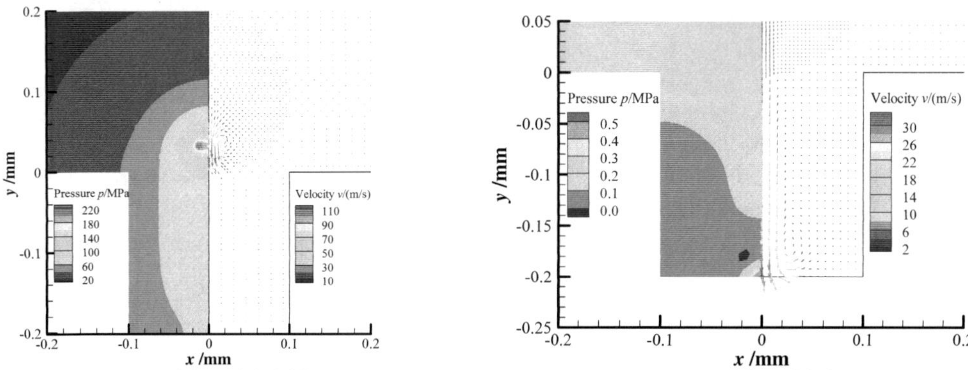

(c) t=11.148μs (d) t=20.9μs

Figure 7. Pressure field (on the left-half part) and velocity field (on the right-half part) around collapsing bubble at different time for γ=3. The black solid line is the bubble-liquid interface.

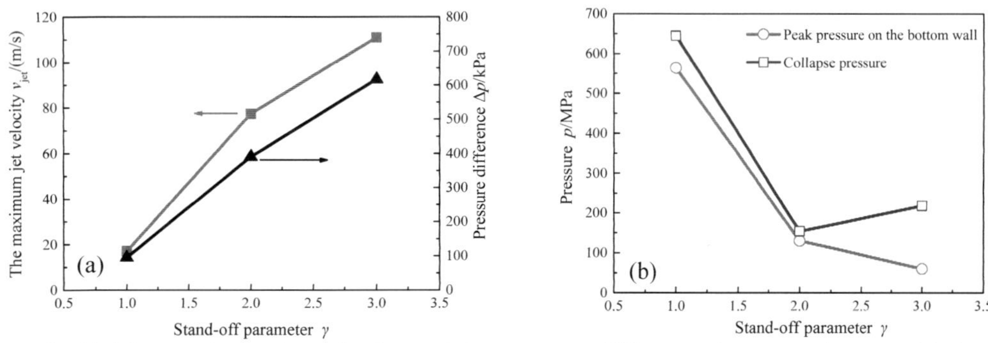

Figure 8. (a) The maximum jet velocity and the pressure difference between above and below the bubble, and (b) the peak pressure on the bottom wall and the collapse pressure for different γ.

4. Conclusion

The volume of fluid method based on a two-phase flow consisting of an incompressible liquid and a compressible gas is used to simulate the dynamics of a vapor bubble close to a boundary with the micrometer-scale groove structure. Deformation of the bubble in the groove is significantly limited because the bubble is adsorbed to the wall as it contacts with the groove. The evolution of shape of the bubble for different γ exhibits its distinct characteristics. The maximum jet velocity increases with γ. Cavitation erosion of the groove wall is dominantly caused by action of the shock wave in the three cases considered. The impulse pressure on the groove wall decreases with the increase of γ.

Acknowledgments
This work was financially supported by Natural Science Foundation of Liaoning Province of China (2015020136 and 201602090) and Liaoning Province Education Committee (L2015068).

References

[1] M. S. Plesset, A. Prosperetti, Bubble dynamics and cavitation, Annu. Rev. Fluid Mech. 9 (1977) 145-185.
[2] L. Rayleigh, On the pressure developed in a liquid during the collapse of a spherical cavity, Philos. Mag. 34 (1917) 94-98.
[3] G. N. Sankin and P. Zhong, Interaction between shock wave and single inertial bubbles near an elastic boundary, Phys. Rev. E74 (2006) 046304.
[4] Q. X. Wang, K. S. Yeo, B. C. Khoo and K. Y. Lam, Strong interaction between a buoyancy bubble

and a free surface, Theor. Comput. Fluid Dyn. 8 (1996) 73-88.

[5] J. U. Brackbill, D. B. Kothe, C. Zemach, A continuum method for modeling surface tension, J. Comput. Phys. 100 (1992) 335-354.

[6] ANSYS. ANSYS fluent theory guide. ANSYS Inc, Canonsburg, PA, USA, 2013.

[7] C. W. Hirt and B. D. Nichols, Volume of fluid (VOF) method for the dynamics of free boundaries, J. Comput. Phys. 39 (1981) 201-225.

[8] C. D. Ohl, M. Arora, R. Dijkink, V. Janve and D. Lohse, Surface cleaning from laser-induced cavitation bubbles. Appl. Phys. Lett. 89 (2006), 074102.

[9] E. A. Brujan and Y. Matsumoto, Collapse of micrometer-sized cavitation bubbles near a rigid boundary, Microfluidics and Nanofluidics. 13 (2012), 957-966.

[10] V. Minsier, J. De Wilde, and J. Proost, Simulation of the effect of viscosity on jet penetration into a single cavitating bubble, J. Appl. Phys. 106 (2009) 084906

IOP Conf. Series: Journal of Physics: Conf. Series **1064** (2018) 012049 doi:10.1088/1742-6596/1064/1/012049

Reconstruction and finite element analysis of brain hemangioma model based on CT images

QingFei Jiang[1], XueYan Ma[1], SiYu Wang[1] and Kai Yang[1, 2,*]

[1]Post-doctoral Scientific Research Station in Nanjing General Hospital of Former Nanjing Military Region, Nanjing, Jiangsu 210000, China
[2]Army Aviation Institute, Beijing 101123, China

*Corresponding author e-mail: yangkaihit@163.com

Abstract. Cerebral haemangioma is a disease that seriously endangers the safety of human life. The law of its occurrence, development, and prognosis is still not completely clear. Therefore, it has attracted the attention of many researchers at home and abroad in recent years. In response to the above problems, based on the original CT data of cerebral aneurysm in patients with intracranial aneurysms, the article reconstructs the cerebral haemangioma model through Autodesk remake 3D reconstruction software. And the finite element analysis software ANSYS CFD was used to establish the reconstructed finite element model of cerebral haemangioma. Hemodynamic analysis of the model was performed to obtain the distribution and change of velocity vector of blood flow field and pressure of blood vessel wall in the model of brain haemangioma. To investigate the hemodynamic mechanism of the formation, development and rupture of cerebral haemangioma and provide theoretical basis for clinical treatment.

1. Introduction

Cardiovascular and cerebrovascular diseases usually occur in some special parts of human blood vessels, such as the sudden narrowing of the blood vessel cross-section, the constricted structure of blood vessels, the bifurcation of blood vessels, and the sudden bending of blood vessels [1]. The common site of vascular disease is related to the hemodynamic factors such as the blood flow velocity and the pressure on the vessel wall. Therefore, the study of the relationship between hemodynamic factors and the genesis of brain haemangioma has drawn increasing attention from clinical researchers and has become a hot topic at home and abroad.

Based on the original CT data of cerebral haemangioma in a patient with intracranial aneurysm, this study reconstructed a three-dimensional cerebrovascular model and applied CFD technique to numerically simulate blood flow in a reconstructed brain haemangioma model. In addition, the hemodynamic mechanism of the formation, development and rupture of cerebral haemangioma is discussed to provide a theoretical basis for clinical treatment.

2. Human Brain Haemangioma Model Reconstruction

This article uses Autodesk series software for reverse reconstruction [2], the original blood data .still format into Autodesk Remake software (Fig.1 for the blood still raw data).Autodesk Remake software

Content from this work may be used under the terms of the Creative Commons Attribution 3.0 licence. Any further distribution of this work must maintain attribution to the author(s) and the title of the work, journal citation and DOI.
Published under licence by IOP Publishing Ltd

is a software that converts pictures into 3D models. It has powerful functions such as repair, optimization, and clean-up. It can restore photos very well. Users can convert some photos or scan data into high resolution 3D mesh model. In this study, Autodesk Remake was used to transform the blood triangle slice still format into a quadrilateral mesh format, which was then imported into Autodesk Fusion for entity transformation.

Select the output model and select the advanced option to convert the trihedral mesh to a tetrahedral mesh output and save it. See Fig.2 for the mesh model of the blood model.

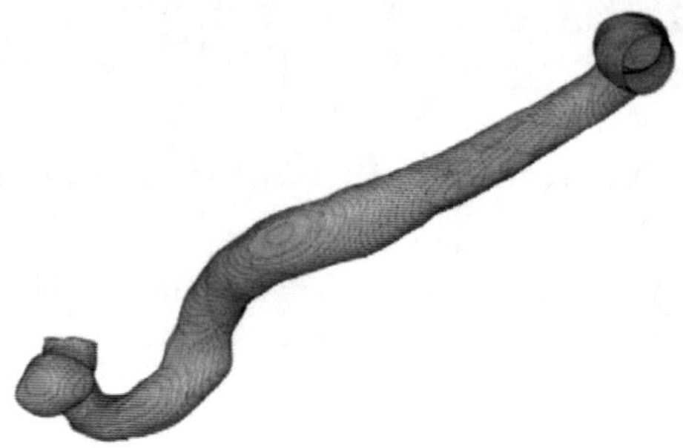

Figure 1. Blood raw data.

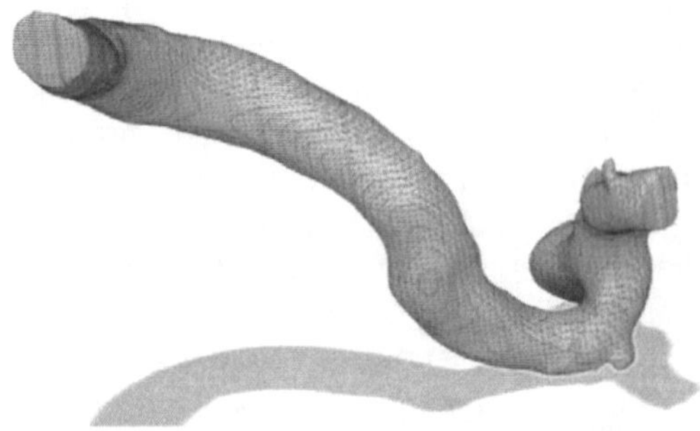

Figure 2. Mesh model of the blood model.

Import the exported .ob. format file into Autodesk Fusion 360, as shown in Fig.3.

Selecting the transformation will transform the grid into solids. Fig.4 shows the process of materializing the surface of the model.

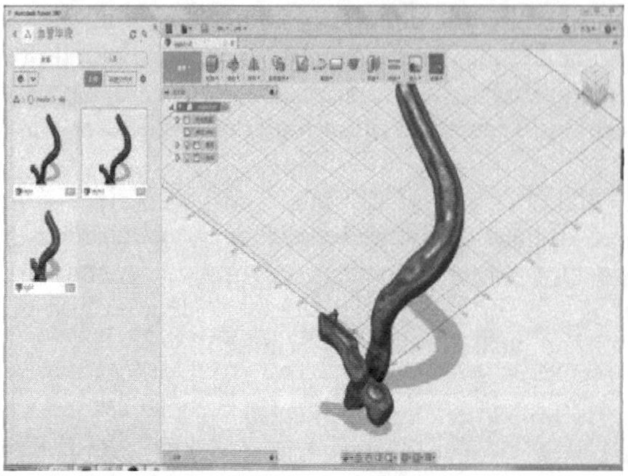

Figure 3. Tetrahedral mesh data

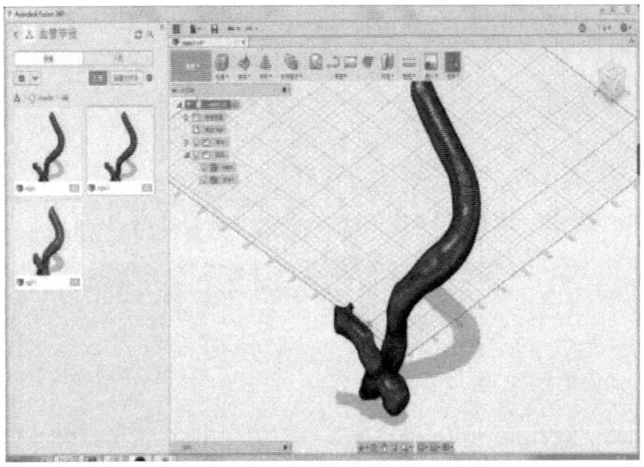

Figure 4. Process of materializing the surface of the model

3. Establishment of a Finite Element Model of Cerebral Haemangioma

This paper applies ANSYS CFD software to the physical mesh of the reconstructed cerebrovascular model.

3.1. Generate a mesh model

The three-dimensional reconstruction model of cerebrovascular was introduced into the finite element analysis software to perform geometric inspection, identify the topological structure, define the exit entrance plane and the calculation domain, set the grid parameters, and perform grid division. You can get the mesh as shown in Fig.5.

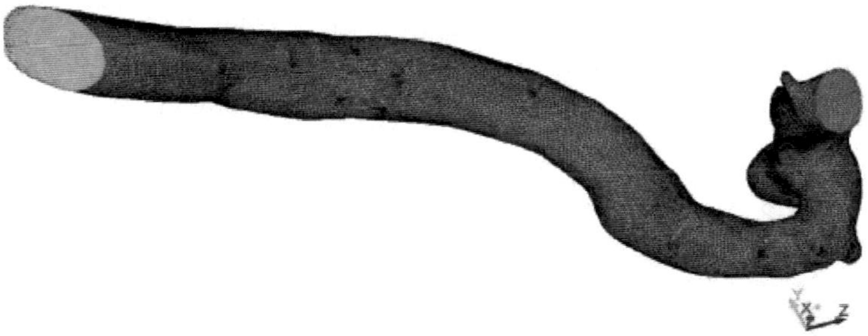

Figure 5. Brain haemangioma model mesh

3.2. Model parameter settings

In large blood vessels, the diameter of red blood cells is very small relative to the diameter of blood vessels, so blood can be considered to be a continuous fluid at this time. When the blood vessel diameter is greater than 0.5 mm, the flow shear rate of blood in the blood vessel is high, and the viscosity of blood can be considered as a constant, so blood can be regarded as an incompressible viscous Newtonian fluid. Perk told et al. [3] compared the flow of Newtonian and non-Newtonian fluids in blood vessels and found that there was only a small difference in the basic flow characteristics between the two, and the flow pattern remained basically unchanged. The simulated blood was therefore set as an incompressible viscous Newtonian fluid. Blood flow is steady, adiabatic laminar flow [4-6]. The blood viscosity coefficient is 0.0035Pas. The blood density is 1060kg/m3. Cerebrovascular is non-permeability, and cerebral vascular wall is rigidity. The entrance plane blood flow rate is 0.17m/s. And the inlet pressure is set to 100mmHg (13.332kPa). The outlet is set to pressure outlet. Blood flow follows the laws of conservation of mass and momentum, namely the continuity equation and the Nadir-Stokes equation:

$$(\nabla \bullet V) = 0 \tag{1}$$

$$\rho \frac{DV}{Dt} = -\nabla P + \mu \nabla^3 V \tag{2}$$

Where, the P is pressure vector, the V is the speed vector, ρ and μ are fluid density and viscosity, respectively.

According to Murray's law, the outlet flow distribution ratio is determined by measuring the diameter of each outlet plane [7].

4. Simulation Results

After the parameters of the vascular model are set, the model is simulated and analysed, and the haemorrhagic flow dynamic parameters are visualized and processed to obtain the blood flow velocity vector diagram and the blood vessel wall pressure cloud diagram, as shown in Figs. 6 and 7.

As can be seen from Fig. 6, blood flow has always existed in the cerebral blood vessels, and there are whirlpools in brain haemangioma. This increases the time for blood cells and other particles to stay in the brain haemangioma, which increases the probability of deposition of these particles in this vascular segment and further increases the risk of vascular lesions.

It can be seen from the pressure wall map of the blood vessel wall in Fig. 7 that the wall pressure gradually decreases along the blood flow direction and gradually changes from red to blue. However, in the brain haemangioma, the distribution of wall surface pressure in the cervical region is not uniform, showing an increase in pressure at the neck of the tumour and an abnormal red colour, while other areas on the same horizontal line show a green colour. From the point of view of fluid mechanics, due to sudden changes in direction during blood flow, it is bound to receive a force to change speed. The reaction force of this force acts on the vessel wall at the neck of the tumour, which will cause the rupture of the blood vessel wall at the cerebral haemangioma to a large extent, causing harm to the human body. In the cerebral haemangioma, the pressure on the medial wall of the tumour was less than the lateral pressure, and the pressure on the wall of the cerebral blood vessel was not changed much.

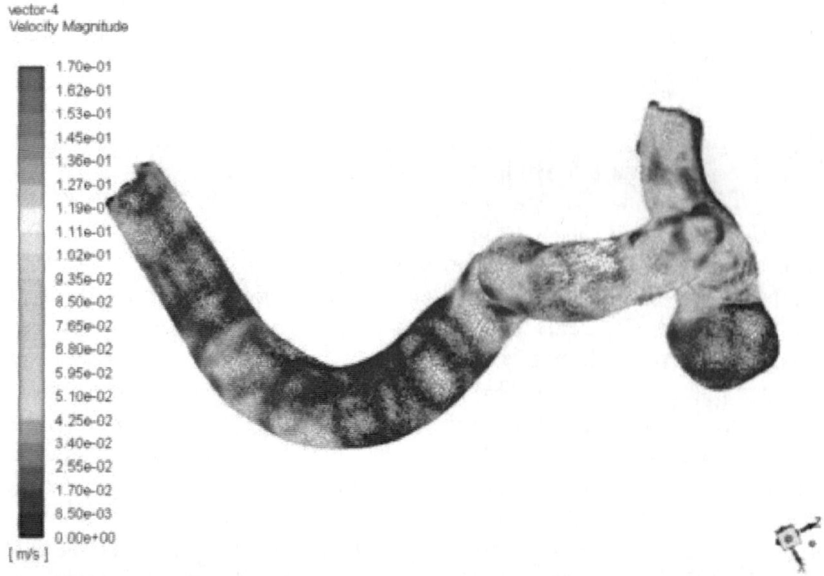

Figure 6. Blood flow speed vector.

IOP Conf. Series: Journal of Physics: Conf. Series **1064** (2018) 012049 doi:10.1088/1742-6596/1064/1/012049

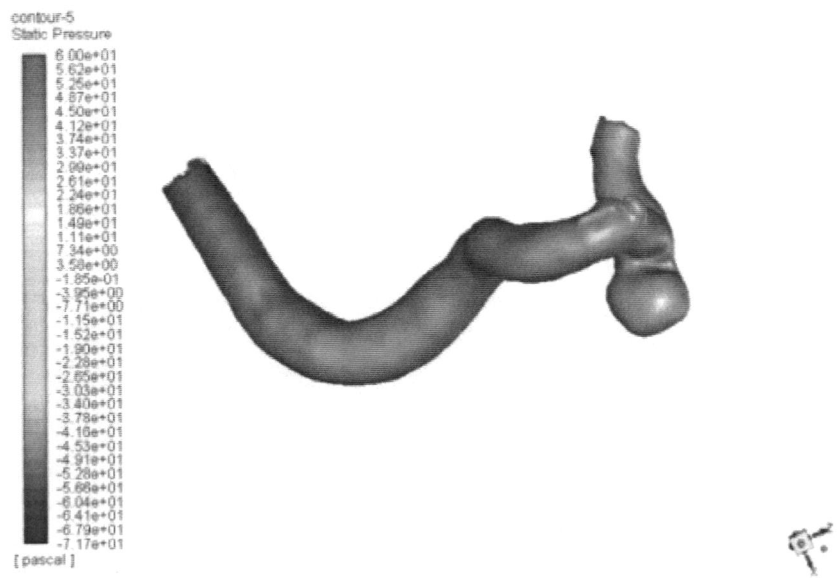

Figure 7. Blood vessel wall pressure cloud.

5. Conclusion

In this paper, the remodelling and finite element analysis of cerebrovascular models were studied, and the distribution and changes of blood flow velocity vector and blood vessel wall pressure in cerebrovascular models were obtained. Through numerical simulation experiments, it was found that there was a large vortex in blood flow in the tumour cavity of brain haemangioma. Vascular wall pressure is higher in the angiomandibular region, which increases the risk of vascular wall rupture of cerebral haemangioma, providing a reference basis and theoretical basis for the later research on cerebral haemangioma and the development of extravascular interventional procedures.

References

[1] C Y Cao. Finite Element Analysis and Numerical Simulation of an Elastic Tapered Vessel. Univerity of South China,2014.

[2] Z Y Chen.Solid Modeling of STL Data in Reverse Engineering. China Mechanical Engineering, 2003, 14(20):1760-1763.

[3] Wu Shigui, Hu Limin, et al, Numerical Approach to Nonlinear Blood Flow in Arterial Stenosis. ICTAM 2000, Chicago.

[4] Hadi Mohammadi, Fereshteh Bahramian. Boundary Conditions in Simulation of Stenosed Coronary Arteries. Cardiovascular Engineering. 2009 (3): 83-91.

[5] Z L Jiang, et al.Biomechanics - from the basics to the front. Bei Jing: Science Press,2010: 183-295.

[6] JV Soulis, TM Farmakis,GD Giannoglou,et al.Wall shear stress in normal left coronary artery tree. Journal of Biomechanics. 2006, 39(4):742-749.

[7] Q Peng, et al.Hemodynamic Analysis of Coronary Circulation. Foreign Medicine (Biomedical Engineering Volume), 2003(01):1-5.

Numerical simulation of side-entry bubbling reactor

Guanghui Chen, Lufeng Chen, Lianxing Song, Jianlong Li*

College of Chemical Engineering, Qingdao University of Science and Technology, Qingdao, China

*Corresponding author e-mail: ljlong321@163.com

Abstract. In this paper, the numerical simulation of bubbling reactor was carried out under the condition of the amplification design of chlorine dissolving bubbling reactor in the epoxy propane chlorohydrin production process. The gas hold-up and flow field distribution in a side-entry bubbling reactor were studied. The simulation results show that the side-entry bubbling reactor can intensify the fluid disturbance near the wall so that the gas substance can contact with the liquid fully and the reaction time can also be increased. Hence, the mass transfer between the gas-liquid two phases was enhanced.

1. Introduction

The liquid-gas reactive process is a characteristic process of the chemical industry. Many types of reactors are used to carry out reactions. The bubbling reactors as the widely used devices have be studied for several decades, such as Shah et al. [1], Doraiswamy and Sharma [2], and Decker [3].

The principal advantages of bubbling reactor are the absence of moving parts, leading to easier maintenance, high interfacial areas and transport rates between the gas and liquid phases, good heat transfer characteristics, and large liquid hold-up, which is favourable for slow liquid-phase reactions[4]. However, the hydro mechanical behaviour of gas-liquid flow in bubbling reactor is very complex and the experimental study is very difficult because of the existence of a large number of bubbles. In recent years, with the development of computational fluid dynamics and the improvement of physical models, numerical simulation has gradually become an important means of bubbling reactor research. Remade and Taya Lia [5] investigated the influence of two different ring sparer's on the hydrodynamics of bubble reactors using standard $k-\varepsilon$ model. Becker [6] who assumes that the viscosity of the turbulence is about 100 times that of the molecular viscosity has adopted two fluid model for numerical simulation. The simulation results were close to the experimental results. However, the turbulence model were not used to calculate the turbulent viscosity and the result depended on the mesh size. Tomiyama [7] have used VOF model to test the rising behaviour of single bubble under different flow conditions. The study of D. Pledger [8] shown that the standard $k-\varepsilon$ model can predict the main characteristics of the low frequency velocity pulsation and describe the instantaneous large scale turbulence structure.

A new type bubbling reactor, the side-entry bubbling reactor, were simulated in this paper. The internal flow field and gas hold-up distribution of the reactor are studied.

Content from this work may be used under the terms of the Creative Commons Attribution 3.0 licence. Any further distribution of this work must maintain attribution to the author(s) and the title of the work, journal citation and DOI.
Published under licence by IOP Publishing Ltd

2. Numerical Calculation

2.1. Geometrical configuration

The structure of the side-entry bubbling reactor is shown in Fig 1.

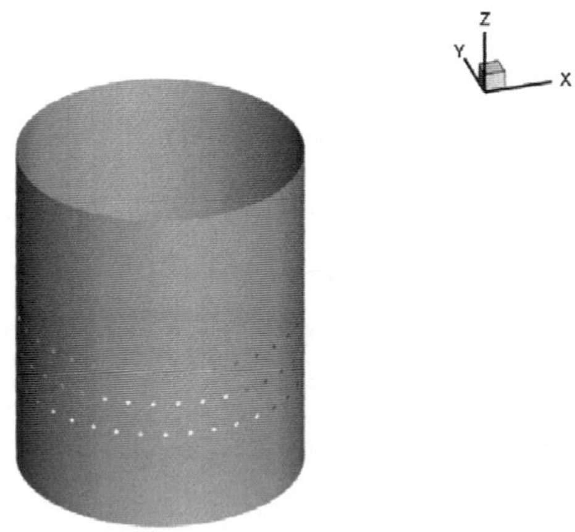

Figure 1. Side-entry bubbling reactor.

The diameter of the reactor is 80 mm, the height is 100 mm, and the diameter of the side-entry hole is 1.5 mm. The holes spacing is 10mm, and the number of holes is 108. The positions of the holes on the layers are arranged crosswise.

2.2. Mathematical model

2.2.1. Governing equation. In the bubbling reactor, the gas was set as dispersed phase and the liquid was set as continuous phase. In this study, the gas phase is assumed to be incompressible and the mass transfer and heat transfer are not considered. The governing equations are shown as follows:

Continuity equation:

$$\frac{\partial \alpha_i}{\partial t} + \nabla \cdot (\alpha_i U_i) = 0 \tag{1}$$

Momentum conservation equation:

$$\frac{\partial \alpha_i \rho_i U_i}{\partial t} + \nabla \cdot (\alpha_i \rho_i U_i U_i) = -\alpha_i \cdot \nabla p + \nabla \cdot (\alpha_i \tau_i) + F_{ex} + \alpha_i \rho_i g \tag{2}$$

The turbulence in liquid phase is calculated by $k-\varepsilon$ turbulence model. The equations of k and ε are given respectively:

$$\frac{\partial}{\partial t}(\rho k) + \frac{\partial}{\partial x_i}(\rho k u_i) = \frac{\partial}{\partial x_i}\left[\left(\mu + \frac{\mu_t}{\sigma_k}\right)\frac{\partial k}{\partial x_i}\right] + G_k + G_b - \rho\varepsilon - Y_M \tag{3}$$

$$\frac{\partial}{\partial t}(\rho \varepsilon) + \frac{\partial}{\partial x_j}(\rho \varepsilon u_j) = \frac{\partial}{\partial x_i}\left[\left(\mu + \frac{\mu_t}{\sigma_\varepsilon}\right)\frac{\partial \varepsilon}{\partial x_j}\right] + \rho C_1 S_\varepsilon - \rho C_2 \frac{\varepsilon^2}{k + \sqrt{\nu\varepsilon}} + C_{1\varepsilon}\frac{\varepsilon}{k} C_{3\varepsilon} G_b \tag{4}$$

$$C_1 = \max\left[0.43, \frac{\eta}{\eta+5}\right], \eta = S\frac{k}{\varepsilon}, S = \sqrt{2S_{ij}S_{ij}} \tag{5}$$

The parameters are shown as follows: $C_{1\varepsilon} = 1.44$, $C_2 = 1.9$, $C_\mu = 0.09$, $\sigma_\varepsilon = 1.3$, $\sigma_k = 1.0$.

2.2.2. Grid and operating conditions. In order to simulate more exactly, the bubbling reactor was divided into the whole layers and the column part with the dividing function of the Gambit software. The grids of the whole layers were fined and those of the column part were meshed automatically by the software. The final grid structure is shown in Fig. 2 and the total number of grids is 2779641.

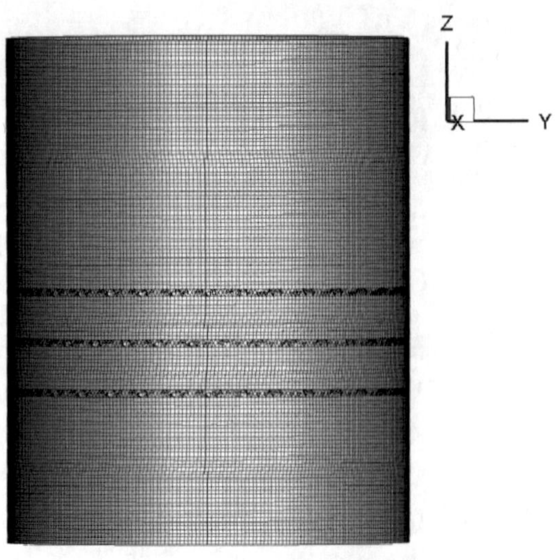

Figure 2. Grid partition.

Boundary conditions including the velocity inlet of gas-liquid phase, the pressure outlet and the non-slip wall boundary were chosen. The convective term of the volume fraction equation uses the Quick format. The second-order upwind scheme was adopted by the convection term of the other equations. The pressure-velocity coupling were adopt the phase-coupled simple method and the rest uses the default value.

3. Results and Discussions

In this study, the steady state numerical simulation is carried out on atmospheric pressure operation using air-water system. The air enters from the side wall of the reactor at a speed of 120 m/s and the water enters the bottom of the reactor at a speed of 2.89 m/s.

3.1. Gas hold-up distribution at different heights

Fig. 3 shows the distribution of gas hold-up at different heights in the reactor. From the picture, we can see that the gas hold-up decreases gradually from the wall to the reactor centre at the first hole layer (Z=30mm). With the increase of the height of the liquid phase, the gas hold-up increased gradually in the reactor at the second hole layer (Z=40mm) and the third hole layer (Z=50mm). At the height of Z=60mm, Z=70mm and Z=80mm, the gas hold-up fraction near the wall of the reactor decreases significantly and the distribution of gas hold-up in the reactor increases significantly.

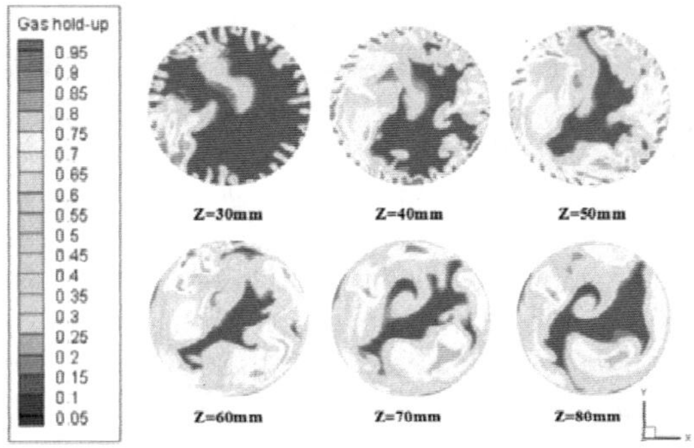

Figure 3. Gas hold-up distribution at different heights.

3.2. *Gas hold-up cloud and liquid vector*

The gas hold-up cloud of Fig. 4 (A) shows that the gas hold-up decreases from the wall to the center of the reactor. As shown in Fig. 4 (B), the liquid velocity in the center of the reactor is upwards along the Z axis. Due to the entry of high velocity gas, the liquid phase is highly disturbed and vortex appears near the wall, which is conducive to the full contact of gas and liquid phases, thus increasing the mass transfer efficiency.

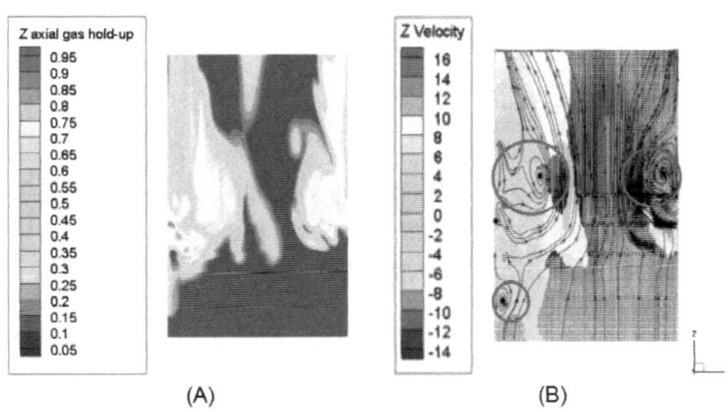

Figure 4. Gas hold-up cloud (A) and liquid vector (B).

3.3. *Vector graph of gas phase velocity at different heights*

Fig. 5 analyzes the vector graph of gas phase velocity at different heights. It can be seen from the diagram that the gas phase in the X axis is higher at the orifice, the level of the turbulence is violent and the vortex is obvious. With the increase of the liquid layer height, the vortex moves to the reactor center, and the gas phase velocity decreases, and the turbulence level decreases.

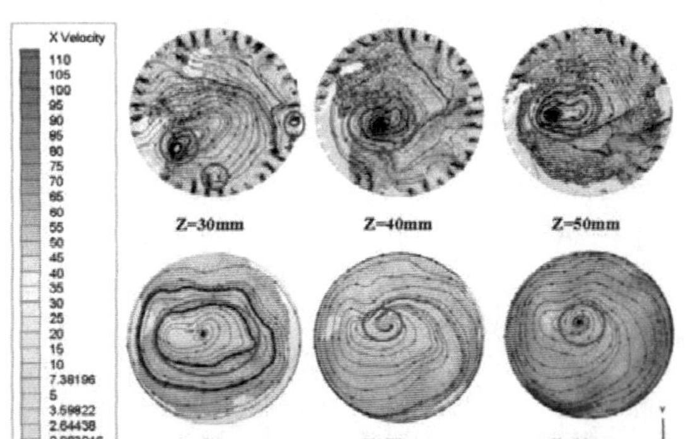

Figure 5. Vector graph of gas phase velocity at different heights.

IOP Conf. Series: Journal of Physics: Conf. Series **1064** (2018) 012050 doi:10.1088/1742-6596/1064/1/012050

4. Conclusions

In this paper, the flow velocity and gas phase distribution in the side-entry bubbling reactor are numerically studied. The flow velocity and gas hold-up distribution at different height of the Z axis and the velocity field of the liquid phase in the central axis of the reactor is analyzed.

By analyzing the distribution of gas hold-up at the different height of the side-entry bubbling reactor, the following conclusion is drawn: on the cross section of the whole layer, the gas hold-up decreases from the wall to the center of the device. However, with the increase of the height of the pore layer, the gas hold-up at the center of the reactor increases gradually, which is due to the full development of the lower pore layer gas phase and the disturbing behavior caused by the higher pore layer. For the columnar part above the pore layer, the higher gas hold-up area moves toward the reactor center.

In addition, through the analysis of the flow field behavior in the center direction of the reactor and the velocity distribution at different heights, the results show that, for the region near the pore layer, the liquid phase disturbance at the different height sections is relatively violent and the disturbance vortex appears at the different height section, so that the gas- liquid phase can be fully contacted which is beneficial to the mass transfer of gas-liquid phase. The velocity vector diagram of axial center can also clearly reflect this flow behavior.

Acknowledgments

This work was financially supported by the transformation project of scientific and technological achievements of Qingdao (no. 16-6-2-50-nsh).

References

[1] Y.T. Shah, B.G. Kelkar, S.P. Godbole, W.D. Deckwer, Design parameters estimations for bubble column reactors, AICh E J. 28 (3) (1982) 353–379.

[2] L.K. Doraiswamy, M.M. Sharma, Heterogeneous Reactions, vol. 2, J.Wiley, New York, 1986.

[3] W.D. Deckwer, Bubble Column Reactors, John Wiley and Sons, New York,1992.

[4] Shah, Y. T., Kelkar, B. G., Godbole, S. P., & Deckwer, W.-D.(1982). Design parameter estimations for bubble column reactors. The American Institute of Chemical Engineers Journal, 28(3),353-379.

[5] V.V. Ranade, Y. Tayalia, Modelling of fluid dynamics and mixing in shallow bubble column reactors: influence of sparger design, Chem. Eng. Sci. 56(4) (2001) 1667–1675.

[6] Becker S, Sokolichin A, Eigenberger G. Gas-Liquid Flow in Bubbling reactor and Loop Reactors:Part ll. Comparison of Detailed Experiments and Flow Simulations. Chemical Engineering Science,1994,49(24B):5747-5762.

[7] Tomiyama A, Sou A, Sakaguchi T, Numerical analysis of a single bubble by VOF method, JSME Int. J. Series B ,1993, 36: 51~56.

[8] Pfleger D, Gilbert N, et al., Hydrodynamic simulation of laboratory scale bubbling reactor: Fundamental studies of Eulerian-Eulerian modelling approach, Chemical Engineering Science, 1999, 54: 5091~5099.

The influence of guide vane on the performance of an axial fan for air-assisted sprayer

Wei Xiao[1], Donghai Jin[1, 2] * and Xingmin Gui[1, 2]

1 School of Energy and Power Engineering, Beihang University, Beijing, China
2Collaborative Innovation Center for Advanced Aero-Engine, Beijing, China

*Corresponding author e-mail: jdh@buaa.edu.cn

Abstract. For the axial fan used in air-assisted sprayer, it always requires a good outlet flow field. In addition to ensuring that the outlet flow field has sufficient dynamic pressure, it is also necessary to ensure that the outlet flow direction is close to the axial direction. In this paper, the influence of the profile load distribution and the solidity of the outlet guide vane on outlet flow field was studied with numerical simulation. The results indicates that increasing the solidity or using the fore-load distribution profile can effectively reduce the outlet flow angle, and result in increasing the spraying distance of the water droplet. This result can provide reference for the design of the guide vane for axial fan used in air-assisted sprayer.

1. Introduction

Because of the characteristics of environmental protection and economical, air-assisted sprayer is widely used for dust suppression, pesticide spraying and fire-fighting. Fig. 1 shows the structure of an air-assisted sprayer. The atomized water droplet, generated by the atomizing nozzle, would be transported to a required distance with the assist of the outlet wind filed pressurized by an axial fan.

As an energy conversion device, the performance of the axial fan will directly affect the quality of the sprayer. Transportation distance is an important parameter when evaluating the performance of a sprayer. At a certain shaft power, it is mainly determined by the efficiency of the fan and the quality of the outlet flow field. The axial fan needs to be well designed to reach the requirement of high efficiency and acceptable outlet flow field quality. Reducing flow loss is the key to improve the aerodynamic efficiency of an axial fan. The internal flow of the impeller is complex. In addition to loss induced by the boundary layer friction and blade wake, the secondary flow such as radial migration of the boundary layer and tip leakage flow, would also bring great loss [1, 2, 3]. What's worse, the secondary flow in the guide vane passage is harmful to the quality of the outlet flow field, and would reduce the spraying distance of atomized water droplet.

For spraying fan, guide vane is used to turn the flow from the rotor back to axial direction, which is of great importance for achieving a required spraying distance. Solidity and geometry of the profile can affect the ability of the vanes to turn the airflow. Song S [4] carried out structural optimization and experimental research on guide vane of an orchard air-assisted sprayer, and discussed the influence of the number of blades on the flow in the guide vane. Zhengxiang Li [5] replaced the straight guide vane with circular guide vane when optimizing an air-assisted sprayer fan, CFD simulation showed that the flow state in the guide vane got improved. Chen Bo [6] used neural networks and genetic algorithms to

Content from this work may be used under the terms of the Creative Commons Attribution 3.0 licence. Any further distribution of this work must maintain attribution to the author(s) and the title of the work, journal citation and DOI.
Published under licence by IOP Publishing Ltd

optimize the number and chord length of guide vane, resulting in significant improvements in both fan efficiency and spray distance. These studies are valuable for engineering application. However, the reasons for the improvement of performance was not clearly explained. This paper will discuss and analyze the influence of the guide vane on the performance of the fan, especially the influence on the outlet flow field.

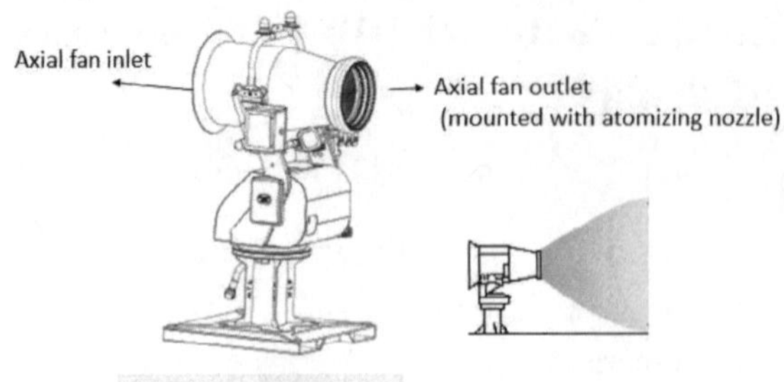

Figure 1. Structure of air-assisted sprayer

2. Axial fan model

The axial fan of this study is a single stage fan designed for an air-assisted sprayer, consisting of a row of rotor blades (R) and a row of guide vane (S). The flow path is shown in Fig. 2 and the basic geometric parameters of the blades are shown in Table 1. The fan is designed to operate at 1450 rpm. The volume flow rate is 50000m3/h, and the total pressure rise is 900 Pa.

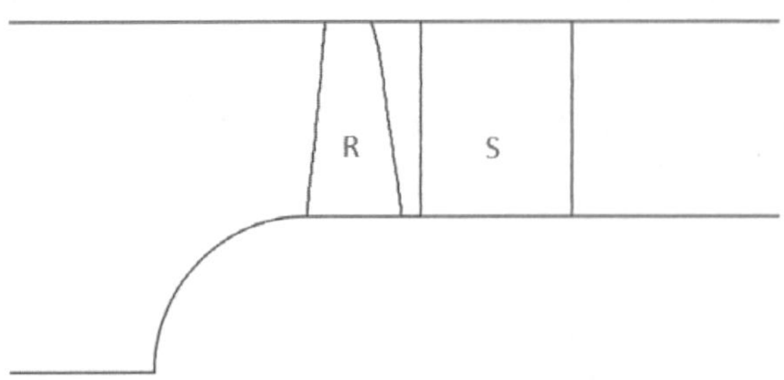

Figure 2. Flow path of the fan

Table 1. Geometric parameters of blade

	Rotor balde	Outlet guide Vane
Number of balde	13	8
Chord (hub/tip)	146 / 144 mm	202 / 194 mm
Stagger angle (hub/tip)	31.7 / 67.2	-20.2 / -11.8

3. Computational Methodology

Three-dimensional numerical simulation has been widely used in turbomachinery and has been proved to be of acceptable accuracy [7]. In this paper, commercial software Numeca was used to conduct the three-dimensional numerical simulation analysis. The reliability of Numeca has been evaluated in Yang xi's work [8]. Computational domain includes the inlet section, the rotor blade passage, the guide vane passage and outlet section, as showed in Fig. 3. The blade row adopts the default O4H topology. The S-A turbulence model is selected for turbulence calculation. Y+ is lower than 10 on all walls in the operating points, which meets the model requirements. As for boundary condition, total pressure and temperature at standard atmospheric condition are set at inlet, and at the outlet, the average pressure is given. The walls are considered as no-slip adiabatic walls.

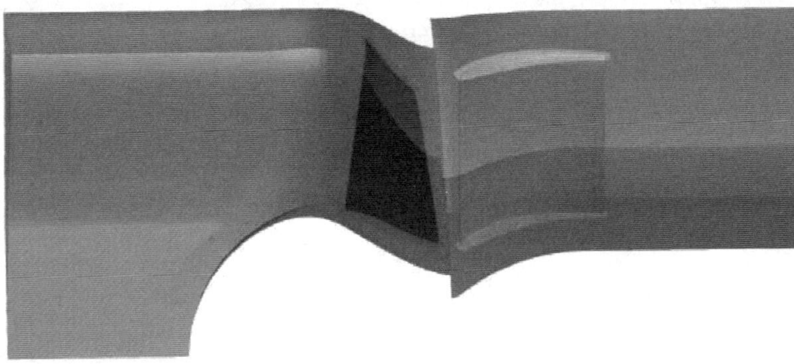

Figure 3. Computational domains

The mesh was encrypted for independence verification. As shown in Fig. 4, the pressure characteristic and efficiency characteristic are calculated with the total grid points around 0.85 million, 1.4 million, and 1.8 million respectively. For three different grid densities, the pressure rise characteristic are basically the same, and the efficiency characteristic increases as the grid density increases. As the number of grid points increases to about 1.4 million, the efficiency characteristic almost remains the same, so the paper adopts the mesh with 1.8 million grid points for all computations.

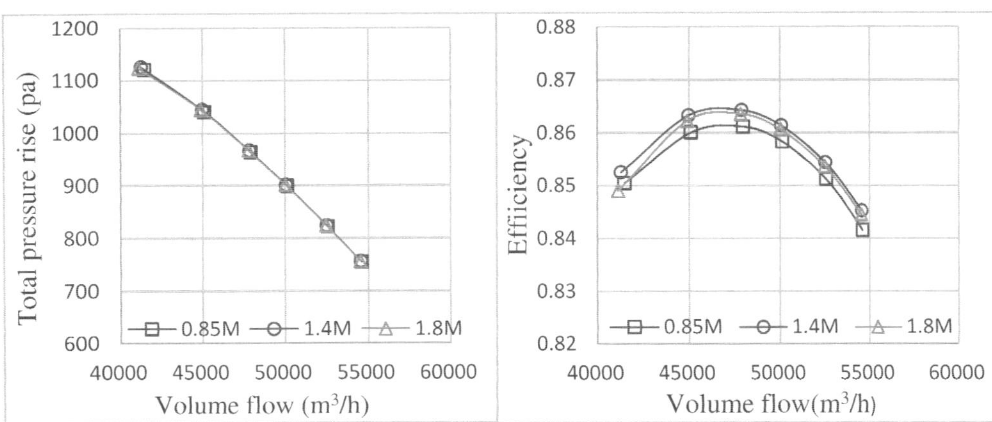

Figure 4. Independence verification for computational grid

4. Result and Discussion

4.1. The Influence of Blade Solidity

The solidity was changed by altering guide vane number respectively to 6,8,10, without changing the blade profile. In order to estimate to what extend the outlet airflow direction deviates from the axial direction, outlet flow angle is defined as the angle between the airflow tangential speed and the axial speed. A positive outlet airflow angle indicates that the flow has the circumferential velocity in the rotation direction, while a negative airflow angle indicates that circumferential velocity direction of the outlet flow is opposite to the rotation direction, or in other words, the airflow is overturned by guide vane.

Fig. 5 plots the distribution of the outlet airflow angle along the span at the design point. S6, S8, and S10 correspond to the calculation results of 6, 8, and 10 guide vanes, respectively. The result indicates that solidity has an obvious effect on the outlet airflow angle: as the guide vane number increase, the outlet airflow angle decreases in the full span range, which means the outlet airflow direction approaches to the axial direction. It can be concluded that increasing the solidity is an effective way to improve the profile's ability to turn the airflow to axial direction, without changing the profile geometry.

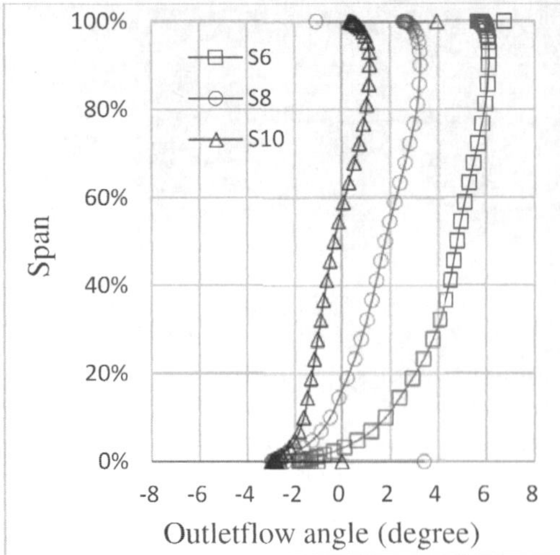

Figure 5. Outlet flow angle

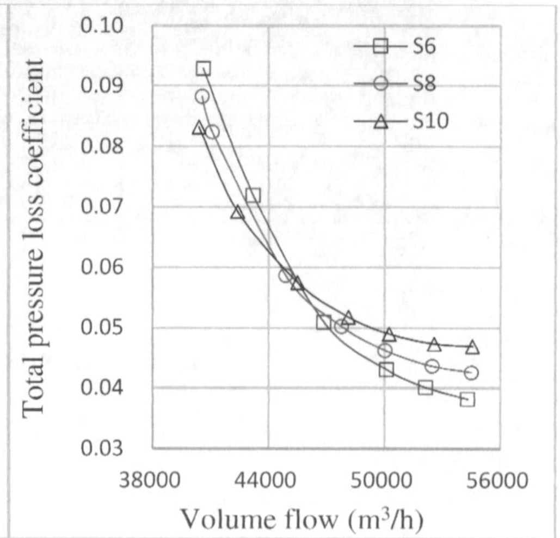

Figure 6. Total pressure loss coefficient

On the other hand, the solidity also has a significant effect on loss and thus affects the fan efficiency. Increasing the solidity would increase loss induced by friction and wake, while help suppress or weaken the secondary flow separation and reduce the secondary flow loss. Fig. 6 shows the characteristics of guide vane's total pressure loss coefficient with different number of blade. In the large volume flow rate area, the flow in the guide vane passage is in good condition. The loss mainly came from friction and wake loss. As the stator solidity increase, the friction and wake loss increase, resulted in the increase of the total static pressure loss coefficient. While in the small flow rate area, the guide vane suffered large positive attack angle, and the loss is mainly caused by the secondary flow separation. Since the increase of solidity can suppress the separation, bigger solidity contributes to reduce the total pressure loss factor. For this fan, at the design volume flow rate point (50000m3/h), as the solidity increase, the flow angle decreases while the loss increases, and eventually reduces the efficiency of the axial fan. When choosing the solidity of guide vane, both loss and out flow angle should be taken into consideration. This is the reason why 8 guide vanes were chosen for this fan.

4.2. The influence of blade profile load distribution

In order to study the influence of the profile load distribution on the performance of the fan, two kinds of guide vane were designed. One used uniform load profile while the other used fore-load profile. All the profile had the same bending angle and solidity. Fig. 7 is the comparison of the load distribution. The blue curve represents a uniform load distribution, whose circulation is uniform decreased in the flow direction, while the red curve is the fore-loaded distribution, in which the load gradient near the trailing edge is reduced. The comparison of blade profile between two kinds of load distribution is shown in Fig. 8.

The distribution of outlet flow angle is displayed in Fig. 9. Compared to the uniform load vane, the fore-load blade significantly reduces the outlet flow angle in whole span. The average outlet flow angle decreases from 4.36 to 2.26 degree, while the total pressure rise coefficient remains almost unchanged, especially at the design point (50000m3/h) as showed in Fig. 10. Since the flow direction gets closer to axial direction, the spray range would get increased.

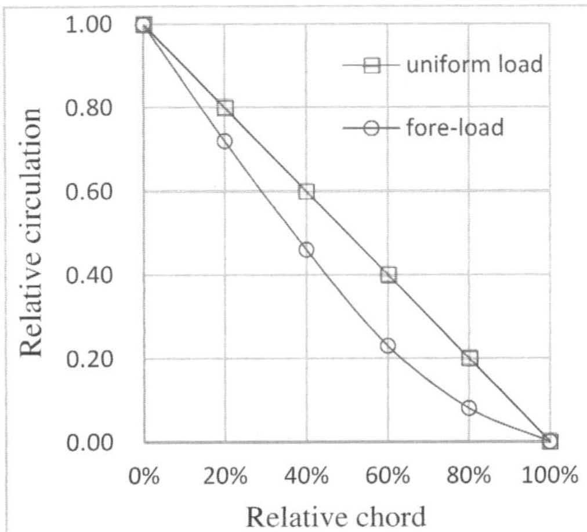

Figure 7. Comparison of load-distribution

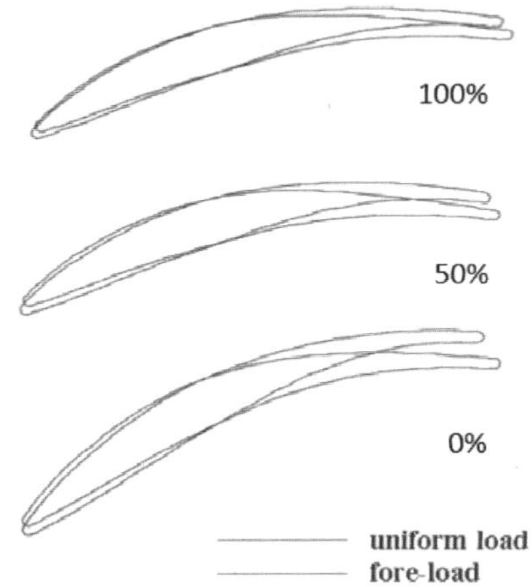

Figure 8. Comparison of profile

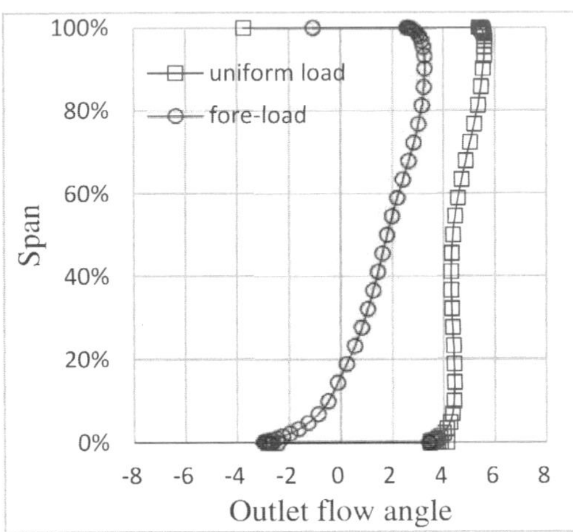

Figure 9. Outlet flow angle

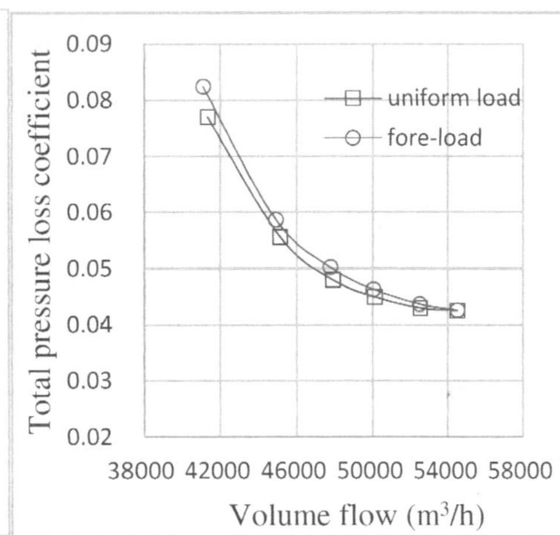

Figure 10. Total pressure loss coefficient

5. Conclusion

In this paper, the influence of the guide vane on outlet flow of the spraying fan is numerically studied. The results are summarised as fellow:

1. Increasing the solidity can reduce the outlet flow angle and improve the guide vane's capability of turning the flow.

2. The increase of solidity can affect the fan efficiency at the design point. Both guide vane loss and the outlet flow angle should be taken into account when selecting the number of blades.

3. For the same solidity and bending angle, using the fore-load profile can help reduce the outlet flow angle and make the outlet flow direction closer to the axial direction.

References

[1] Tao Deping, Lu Renfu, Zhao Xiuhua, Secondary flow in compressor cascade, Journal of Beijing University of Aeronautics and Astronautics. 4 (1982) 81-94.

[2] Cyrus D.I.V, Experimental investigation of losses and secondary flow in an axial compressor stage, Forschung Im Ingenieurwesen A. 2 (1985) 33-40.

IOP Conf. Series: Journal of Physics: Conf. Series **1064** (2018) 012051 doi:10.1088/1742-6596/1064/1/012051

[3] Kang S, Hirsch C, Experimental study on the three-dimensional flow within a compressor cascade with tip clearance: Part II--The tip leakage vortex. 1993, pp. 1927-1951.
[4] Song S, Xia H, Lu Y, et al, Structural optimization and experiment on fluid director of air-assisted sprayer, Transactions of the Chinese Society of Agricultural Engineering. 6 (2012) 7-12.
[5] Zhengxiang Li, Structure optimization analysis of axial flow fan in 60 meters sprayer, Yanshan University, 2014.
[6] Chen Bo, Gao Dianrong, Yang Chao, et al, Multi-objective intelligent collaborative optimization of structural parameters of high-power remote nebulizer, Journal of Mechanical Engineering. 6 (2017) 166-175.
[7] Hou Shuqiang, Wang Canxing, Lin Jianzhong, A review of numerical simulation of internal flow field in turbomachinery, Fluid Machinery. 5 (2005) 30-34.
[8] Yang X, Jin D, Gui X, “Aerodynamic design and optimization of pipe diffuser for a high-loading centrifugal compressor” in Fluids Engineering Division Summer Meeting, ASME, Volume 1A, doi:10.1115/: FEDSM2017-69152.

Effect of Load Distribution at the Roots among Stages on Performance of a High Bypass-ratio Fan/Booster

Xiaoxu Ding[1], Donghai Jin[1, 2] * and Xingmin Gui[1, 2]

[1] School of Energy and Power Engineering, Beihang University, Beijing, China.
[2] Collaborative Innovation Center for Advanced Aero-Engine, Beijing, China.

*Corresponding author e-mail: 13146014223@163.com

Abstract. The modern high-performance aero-engines have raised higher requirements on the fan/booster, such as high through-flow, high pressure ratio, high efficiency and a wide range of stable operation. In this paper, a method called arbitrary camber lines blading design method for both axial and centrifugal compressors and three dimensional CFD (computational fluid dynamics) calculation are used to design and analyse the results of the aerodynamic design of a high bypass-ratio fan/booster. By changing load distribution on the roots among the rotors, the flow of the fan/booster can be affected. The results of the study show that increasing the load on upstream stages roots can help relieve the separation at the roots of the downstream stages, which leads to the increase of the total pressure, adiabatic efficiency and stall margin of the core duct of the fan/booster.

1. Introduction

Turbofan engines are widely used in military and civil aviation fields with advantages, such as low fuel consumption and high cycle thermal efficiency. As one of the components of a turbofan compressor, fan/booster's characteristics at different operating points have a direct effect on the engine performance. The modern high-performance aero-engines have raised higher requirements on the fan/booster, such as high through-flow, high pressure ratio, high efficiency and a wide range of stable operation [1].

But there are many difficulties to meet the requirements. At first, the rotating speed of a fan/booster is low, so the roots of the fan and the booster have a weak ability to increase the total pressure of the fluid. What's more, because the flow path of the booster is complex and the blades are short, the ratio of the boundary layer to the blade height is high, which increases the difficulty to improve efficiency and stall margin [2].

Smith diagrams show the relationship between flow coefficient and load factor on different degrees of reaction published in Turbomachinery Performance Analysis [3], which reveal that the matching of flow coefficient, load factor and degree of reaction has a crucial influence on the efficiency and the stall margin of the compressor stage. Fang Z [4] investigated the effect of end-bending and bow on the flow separation in corners of blades in a high through-flow fan/booster, which shows that the end-bent fan could improve the flow flied and increase the adiabatic efficiency and that the end-bending and bow for stators could reduce chocking and losses of the corner flow.

Content from this work may be used under the terms of the Creative Commons Attribution 3.0 licence. Any further distribution of this work must maintain attribution to the author(s) and the title of the work, journal citation and DOI.
Published under licence by IOP Publishing Ltd

This paper uses the arbitrary camber lines blading design method [5] to optimize the original fan/booster. By changing the load distribution among the roots of the rotors to alter the matching of the load factor and the flow coefficient, the performance of the fan/booster has been improved.

2. Research object and tools

2.1. Research object

In this paper, the fan/booster is optimized based on the 11km, 0.75Ma flight condition, whose flow path is shown in Fig. 1. And the table 1 shows the nomenclature.

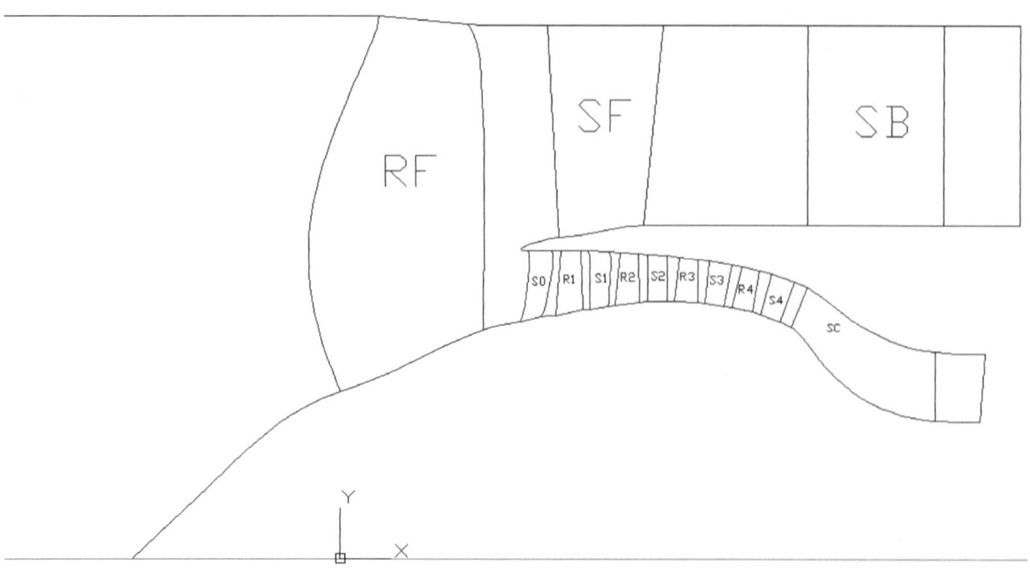

Figure 1. Flow path of the fan/booster

Table 1. Nomenclature

Nomenclature	Meaning	Nomenclature	Meaning
RF	Fan	S2	Second stage stator
SF	Outlet guide vane	R3	Third stage rotor
SB	Vane assisted to calculate	S3	Third stage stator
S0	Inlet vane of the booster	R4	Fourth stage rotor
R1	First stage rotor	S4	Fourth stage stator
S1	First stage stator	SC	Support board
R2	Second stage rotor		

2.2. Design Tool and Three-Dimensional Navier-Stokes Computations

The fan/booster is designed by the streamline curvature through-flow method and arbitrary camber line blading method[5,6].Three-dimensional numerical simulation has been widely used in turbomachinery and proved to be of acceptable accuracy [7~9]. In this paper, commercial software Numeca is used to conduct the three-dimensional numerical simulation analysis. The computational mesh is shown in Fig. 2. The fan/booster is meshed with 3557311 cells. The turbulence model is the Spalart-Allmaras model. Total pressure, total temperature and the direction of the flow are imposed at the inlet of the computational domain. At the outlet, static pressure ruled by the simple radial equilibrium is set. Adiabatic non-slipping condition is applied to the solid boundaries.

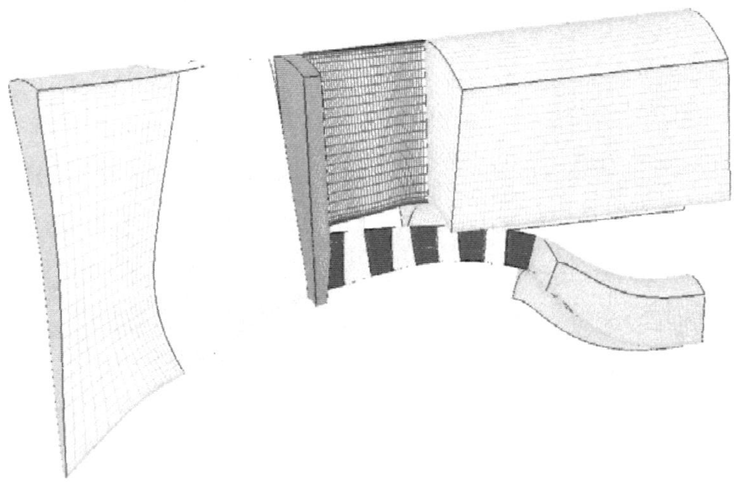

Figure 2. Computational grid for the fan/booster

3. The analysis of the characteristics and flow field of the original fan/booster

Fig.3 and Fig. 4 show the performance characteristics of total pressure and efficiency, which are normalized by the design requirements. As can be seen from the performance characteristics, the total pressure ratio and the efficiency of the bypass duct met the requirement, while the core duct did not.

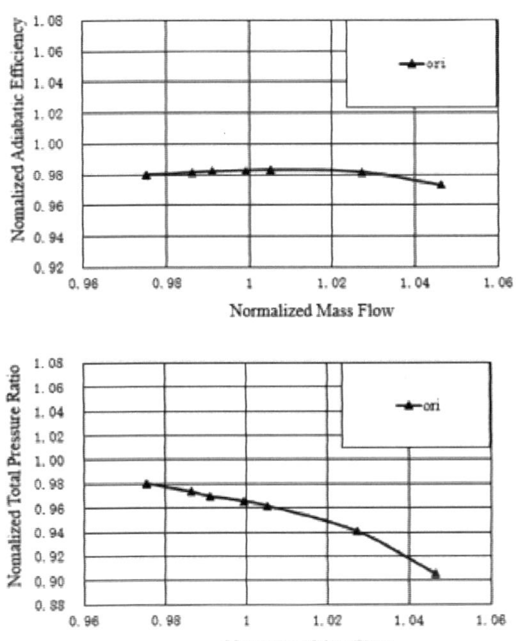

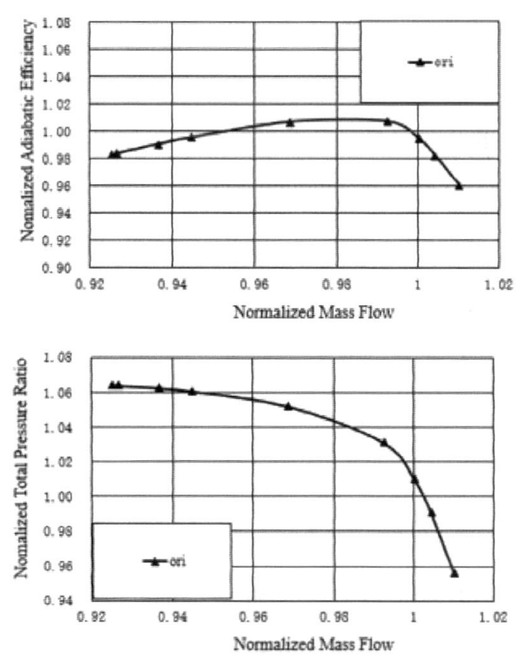

Figure 3. Performance curve of core duct **Figure 4.** Performance curve of bypass duct

Fig.5~Fig.7 show the flow fields at the near design point, which are respectively the circumferentially averaged relative Mach number distribution of the fan/booster, the relative Mach number distribution of the fan in the 5% percent span and the limiting streamlines on the suction surface of fan, the relative Mach number distribution of the booster. From Fig.5 and Fig.7, it can be seen that there was a large low-Mach-number area at the roots of the R3, S3, R4, S4. In addition, it can be seen from Fig.5 and Fig.6 that the flow at the root of the fan was in good condition.

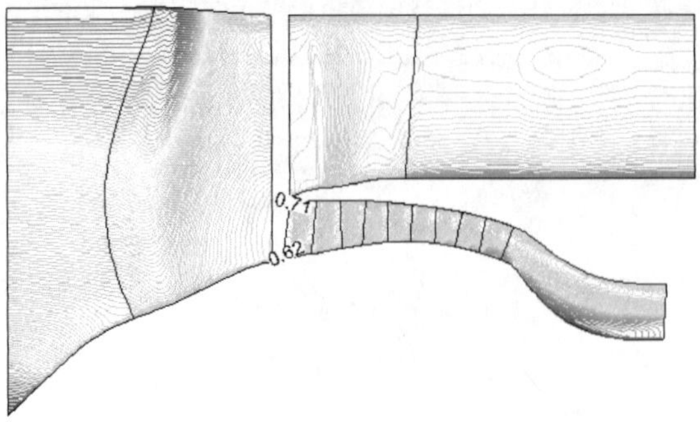

Figure 5. Relative Mach number distribution on meridional plane at near design point

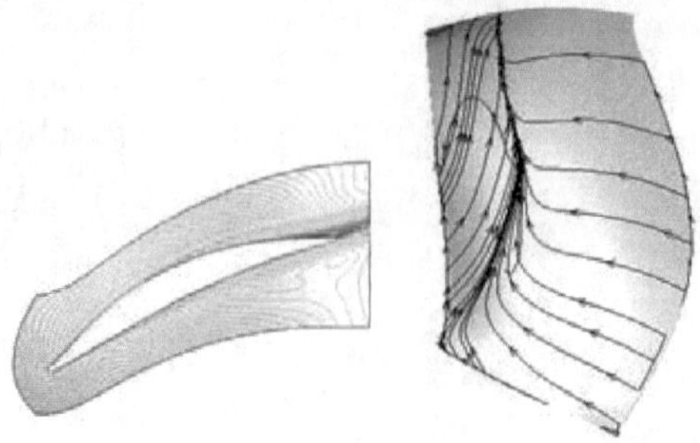

Figure 6. Relative Mach number distribution of the root of RF and the limiting streamlines on the suction surface of RF

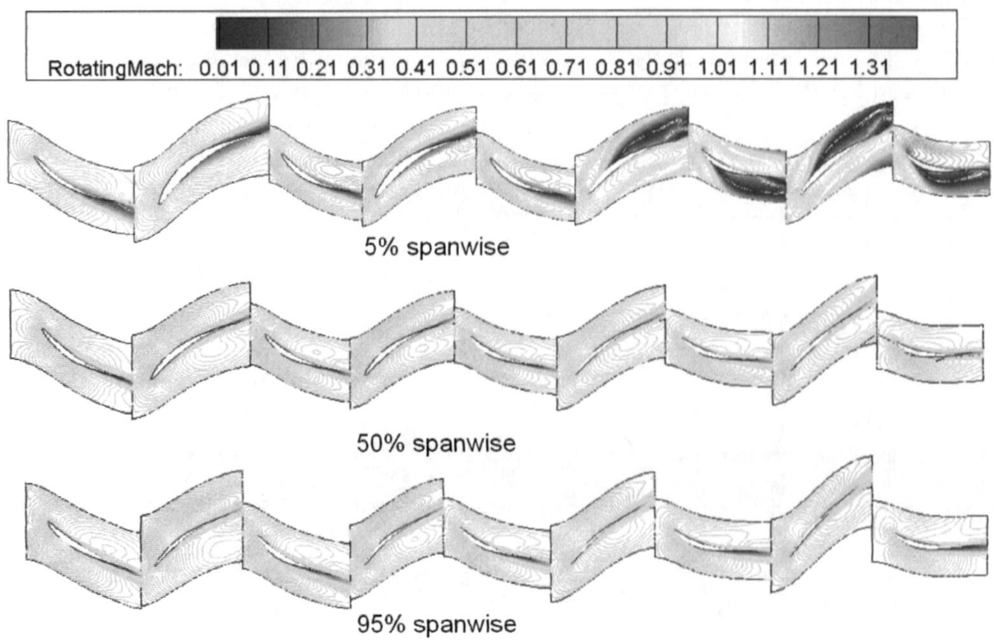

Figure 7. Relative Mach number distribution of the booster

4. The results of the optimization

4.1. The redistribution of the load on the roots in the core duct

From the analysis above, we can know that the reasons for the lack of the pressure ratio and efficiency is that the separation from the R3 root. From the paper of Fang Z [4], we can know the separation at the root of R3 could not be controlled by end-bending. So it is reasonable to consider reducing the loading at the root of R3. But there is another matter which should be taken into account that the stagnation temperature and the stagnation pressure should be as equal as possible from hub to casing of the booster's outlet in order to reduce the distortion of the inlet of the high compressor from hub to casing. Therefore, when reducing the load on the downstream stages roots, the load on the upstream stages roots should be increased. As can be seen from Fig.5 and Fig.6, the Mach number at the inlet of the booster is 0.62, which is not too high and the flow condition of the fan in the core duct is good. So it can be assumed that the load on the downstream stages can be reduced and the load on RF in the core duct can be increased. Fig.8 and Fig.9 show the changing of the load on RF and R3.

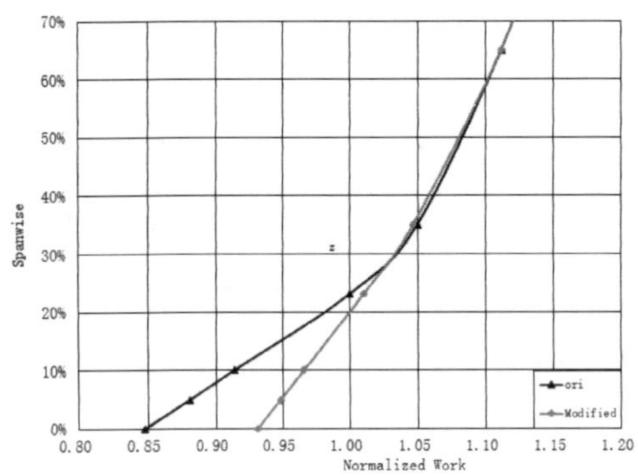

Figure 8. Spanwise distribution of load on RF (Within 70% span)

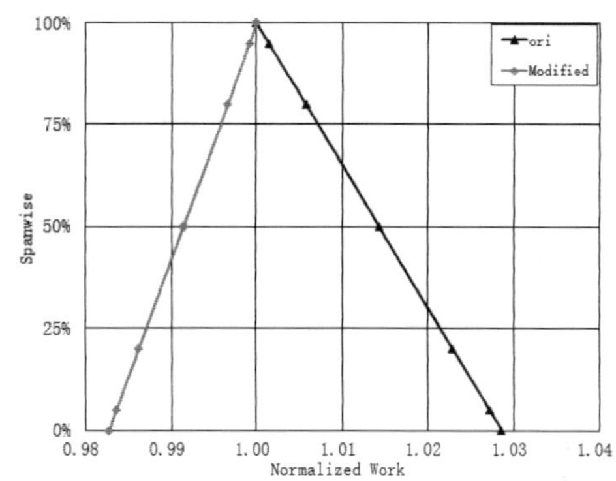

Figure 9. Spanwise distribution of load on R3

4.2. *Comparison of the characteristics*

In the optimization process, the same grid structure and control parameters as the original fan/booster were used to make the comparison of calculation results more convincing. The comparison of the characteristics is shown in Fig.10. As can be seen from the figures, compared with the original, the total pressure and adiabatic efficiency have been improved, and so has the stall margin. Table 2 shows the comparing between the original and the modified.

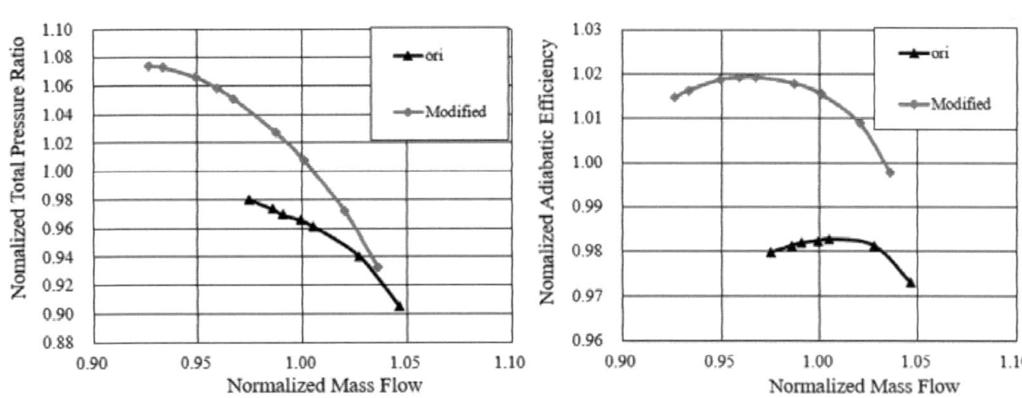

Figure 10. Comparison of the performance curve of core duct

Table 2. Two Scheme comparing

	Original	Modified	Difference
Pressure ratio	0.9657	1.0075	+4.77%
Efficiency	0.9824	1.0153	+3.31%

4.3. Comparison of the flow field

This section compares the flow fields between the original and the modified at near design point. Fig.11 shows the relative Mach number distribution of the modified on meridional plane. Compared with the original, the Mach number at the entrance of the booster has been increased, but it is still within the acceptable range of the selected profile (controlled diffusion profile). The low-Mach-number area at the roots of the third and fourth stages has disappeared. Fig.12 and Fig.13 show the spanwise distribution of adiabatic efficiency of RF(within 50%span) and the relative Mach number distribution at the roots of the booster, which reveal that the efficiency of RF has increased and that the flow field of the booster has been improved.

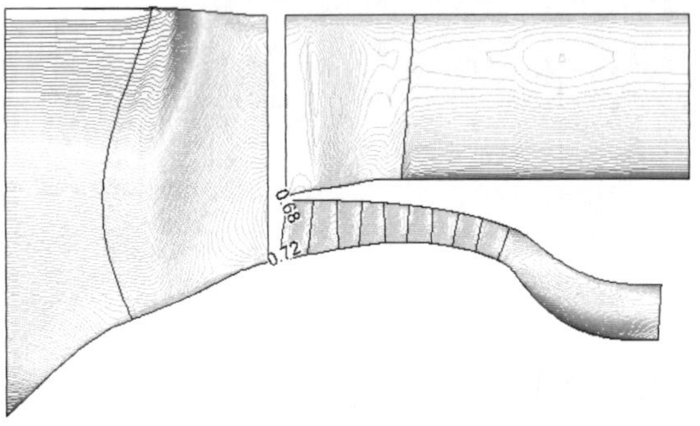

Figure 11. Relative Mach number distribution on meridional plane at near design point

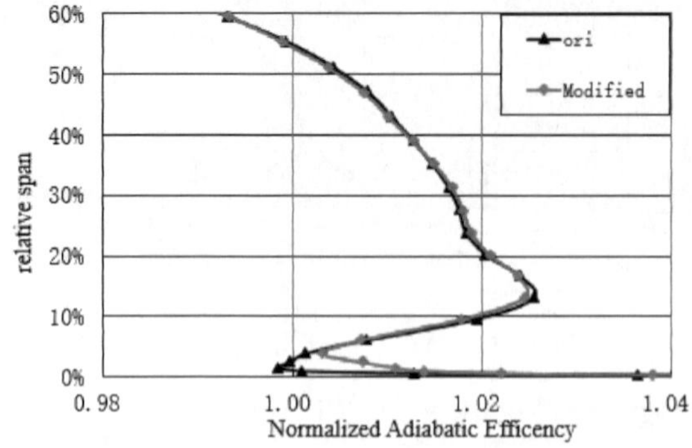

Figure 12. Spanwise distribution of efficiency of RF (within 60%span)

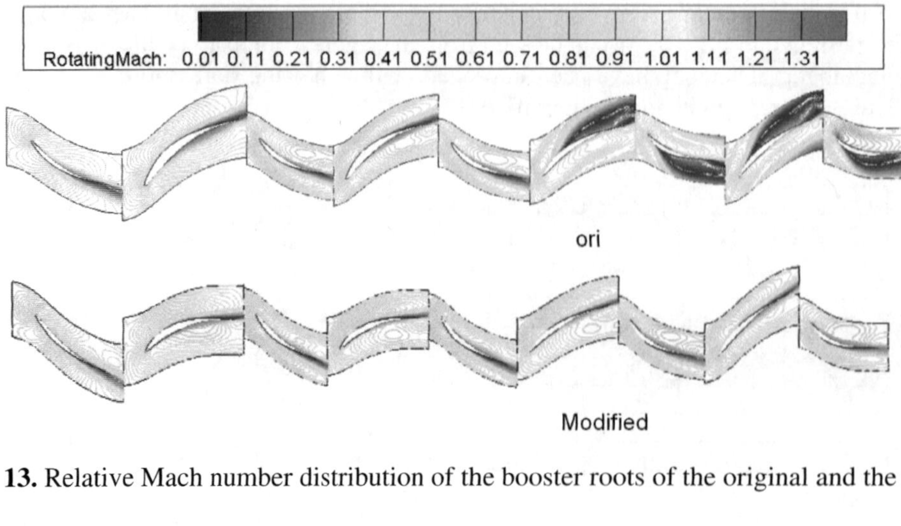

Figure 13. Relative Mach number distribution of the booster roots of the original and the modified

5. Conclusion

In this paper, optimization has been conducted of a fan/booster by redistributing the load on the roots of the rotors. Main results are summarised as follows:

1. The distribution of the load on the roots of the rotors in a fan/booster is important to achieve high pressure ratio and high efficiency.

2. The results of 3D CFD show that the total pressure and adiabatic efficiency of the core duct of the modified fan/booster has increased by 4.77% and 3.31%, meeting the design requirement, and that the stall margin has also been improved.

IOP Conf. Series: Journal of Physics: Conf. Series **1064** (2018) 012052 doi:10.1088/1742-6596/1064/1/012052

3. It is helpful to improve the efficiency of the fan to increase the load on the fan root within a certain range, when taking Mach number at the entrance of a booster into consideration.

4. Raising the loading of the fan root can reduce the loading of the downstream stage roots, which helps relieve the separation at the roots of the downstream stages.

References

[1] Xingmin Gui, Fang Zhu, Donghai Jin,.Report of Aerodynamic Performance of Scaled Fan. Beijing: Beijing University of Aeronatics and Astronautics, 2012.

[2] Baoshi Chen, Analysis of Difficulties in Designing A High Bypass-ratio Fan/booster. High-level Forum on Key Technologies of Large Aircraft and Academic Essays of the 2007 Annual Conference of Chinese Society of Aeronautics and Astronautics 2007.

[3] Lewis R I. Turbomachinery Performance Analysis. 1996.

[4] Fang Z, Guil X. Corner Flow Control in High Through-Flow Axial Commercial Fan/Booster Using Blade 3-D Optimization. Journal of Thermal Science, 2012, 21(1):32-41.

[5] Guofeng Ji, Xingmin Gui, A Blading Design Method for Axial/Centrifugal Compressor Airfoils Using Arbitrary Camber Lines, Journal of Aerospace Power,2009,24(1):150-156.

[6] Frost G R, Wennerstrom A J. The Design of Axial Compressor Airfoils Using Arbitrary Camber Lines. ARL-73- 0107, AD- 756165, 1973.

[7] Xiaojuan Li, Xingmin Gui, Numerical Simulation of Fan/compressor Design and Off Design Performance, Journal of Propulsion Technology, 2005, 26(6):522-525.

[8] Xingmin Gui, Xiaojuan Li, Aerodynamic Design of Fan/Compressor Operated at Low Reynold Number, Gas Turbine Experiment and Research, 2007, 20(3):18-22.

[9] Donghai Jin, Xingmin Gui, Numerical Simulation of Highly Loaded Multi-stage Fan/Compressor of A Turbofan Engine, Journal of Aerospace Power, 2011,26(5): 4059-1065.

Analysis of the Fan Performance in the Turbofan Engine Test

Jiaxiang Yan[1], Manxiang Song[1], Jincheng Zhang[2], Donghai Jin [1, 3, *], Xingmin Guil [3]

[1]School of Energy and Power Engineering, Beihang University, Beijing, China.
[2]AECC China Gas Turbine Establishment, Chengdu, China.
[3]Collaborative Innovation Center for Advanced Aero-Engine, Beijing, China.

*Corresponding author e-mail: jdh@buaa.edu.cn

Abstract. The fan was tested in the whole turbofan engine test condition, and the post-processing of the test data was based on confidence intervals. Through the analysis of distribution characteristics of data, the indirect measurement parameters such as mass flow rate and pressure ratio, which are affected by multiple factors, all obey the normal distribution. Based on the distribution characteristics, a statistic can be constructed to calculate the confidence intervals for these parameters. It is known that the more data, the smaller the confidence interval of the parameters. As a result, it is found that the range of confidence intervals obtained from the final test data is so narrow that point estimates could be used instead of interval estimates. Comparing the test results with the numerical calculation results, the test results are in good agreement with the calculated performance curves. The processed test results show that as the area of bypass(core) throat area decreases, the fan working line will move to the the direction of reduced flow rate. However, that is not absolute. If the area is beyond a certain range of the area, the working line may move in the opposite direction. The effect of throat area variation on the performance of the fan may be related to flow state of the core and bypass. Therefore, multiple adjustments are required to match the design point.

1. Introduction

The fan is one of the main core components of the turbofan engine. The performance of the fan under different working conditions has a direct impact on the whole aero-engine [1][2]. Because fans and compressors have unstable working boundaries such as stall, surge, and choke, the working line of the engine is often plotted on the characteristics diagram of fans and compressors. Through the analysis of the working line in the fan characteristic diagram, the corresponding state variation of the engine can be identified.

Test research is an essential part of the fan/compressor research work. Through test, it is not only possible to verify the accuracy of the computational fluid software for compressor simulation, but also to verify the performance and reliability of the designed fan/compressor. And it provides powerful help for adjustment and improvement of design methods. In the development of modern turbofan engines, most of the performance parameters tested are based on full-size compression components. The test is usually carried out on the single-shaft no-bypass compressor test rig and single-shaft bypass compressor test rig [3-6]. And there are few studies on fan performance test under the condition of the whole aero-engine. Under the condition of the whole engine, the fan characteristics can be tested to obtain the

Content from this work may be used under the terms of the Creative Commons Attribution 3.0 licence. Any further distribution of this work must maintain attribution to the author(s) and the title of the work, journal citation and DOI.
Published under licence by IOP Publishing Ltd

working line of the fan under different working conditions, which has a very good guiding significance for practical application.

Post-processing of test results is important for the test to get useful information. Statistical analysis of test data is an important means of post-processing of experimental data. Point estimation and interval estimation are two methods for estimating the overall population. Generally, point estimation is used to approximate the mean and variance of the population. Sometimes in order to know the deviation between the estimated value and the true value, interval estimation is often used. And at the same time, the reliability of the estimation can be known. That is, using a number that has a significant meaning to indicate the degree of certainty that the unknown population parameter falls within a certain range, which is a direct and meaningful display of the measurement result in engineering.

2. Test Rig

The test was conducted at the ground test bed of the small aircraft engine laboratory of Beihang University.

2.1. Test Object

The test object is a prototype of a single-axis turbofan engine modified from a turbojet engine. The simplified structure diagram of the modified prototype is shown in the figure 1. The engine is mainly composed of a fan, a centrifugal compressor, a combustor, and a turbine. Except for the redesign of the fan, the other components retain the original turbojet design. Centrifugal compressors consist of an inducer, a centrifugal impeller, a radial diffuser, and an axial diffuser, wherein the inducer and the centrifugal impeller are assembled into one assembly using pins. The combustion chamber is a slinger combustor with fuel supply from the fuel slinger. The turbine is a single-stage axial flow turbine.

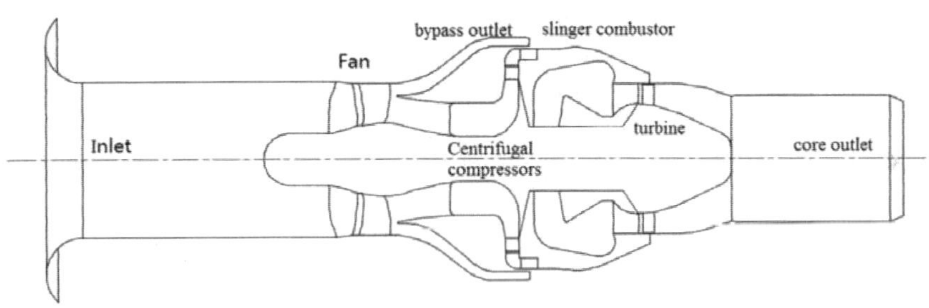

Figure 1. Test turbofan engine structure diagram

2.2. Measurement System

The measurement system adopts self-developed software and hardware measurement and control systems, including Beijing Kunlun Coast JQYB atmospheric pressure transmitter, Beijing Kunlun Coast JWSL-2 wall-mounted temperature and humidity transmitter, total pressure comb, Beijing Kunlun Coast JYB-KO -HVG type pressure sensor, T thermocouple, Advantech ADAM-4117 analog input module, Advantech ADAM-4118 thermocouple input module, serial device server, acquisition computer. The acquisition software is based on LabVIEW and can be used to display fan performance in real time.

2.3. Fan and Performance Measurements.

The fan is designed as a forward swept rotor blade, and its three-dimensional blade characteristics with forward swept near tip portion and back swept near root portion . And controlled diffusion airfoils is adopted near-root portion. It means to increase the fan margin by these means. Design rotation speed is 22000r/min, mass flow rate 22kg/sec, pressure ratio 1.63.

The measuring points of the fan performance are divided 4 sections [7], including the atmospheric pressure and temperature of the test room; section 0: the static pressure of the inlet of the engine; section

2: the total temperature, total pressure, and static pressure behind the fan stator; Section 16: total temperature, total pressure, and static pressure in the bypass.

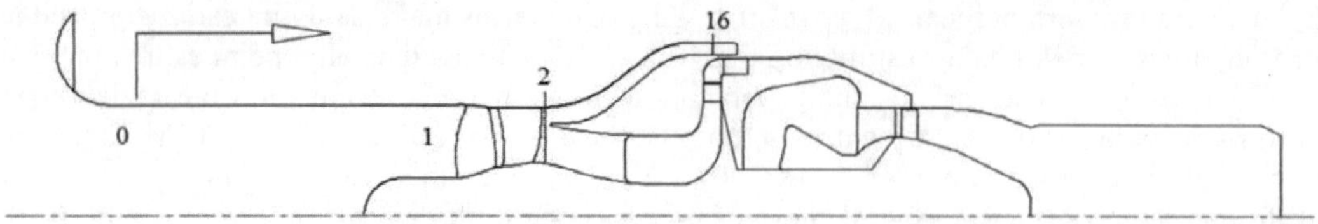

Figure 2. Test section position

3. Data Processing

3.1. Performance Parameter Calculation

The mass flow rate of the fan are based on the formula as follow.

$$G = K\frac{p^*}{\sqrt{T^*}}q(\lambda)A \tag{1}$$

The parameters in the formula can be calculated by the parameters measured and the flow coefficient of the flow tube considering the boundary layer effect is introduced. The calculation relationship between the flow rate and the measurement parameters will be obtained (the subscript means the section number.):

$$G_{总} = K_G \times K \times A_0 \times \frac{P_0^*}{\sqrt{T_0^*}} \times \left(\frac{k+1}{2}\right)^{\frac{1}{k-1}} \times \left(\frac{k+1}{k-1}\right)^{\frac{1}{2}} \times \left[\left(\frac{P_0}{P_0^*}\right)^{\frac{2}{k}} - \left(\frac{P_0}{P_0^*}\right)^{\frac{k+1}{k}}\right]^{\frac{1}{2}} \tag{2}$$

The pressure ratio adopts the total-to-total pressure ratio:

$$\pi^* = \frac{P_2^*}{P_0^*} \tag{3}$$

In order to facilitate the practical application, the measurement parameters are often converted to the standard atmospheric environment. And the calculation method of the corrected parameters is introduced in the Aero-engine Principles, which are not listed here in detail.

3.2. Data Selection

According to GJB 241A-2010, "General Specification for Aviation Turbojet and Turbofan Engines" [8], the measurement error of the rotating speed is required to be no more than 0.2% of the maximum speed. Therefore, if the speed deviation is within 0.2%, it can be considered that the data is in the same rotation speed state. If the data exceeds the range, the data will be deleted. Similarly, GJB 241A-2010 stipulates that the test error of the mass flow rate is no more than 0.5%, and the same selection process is performed on the flow rate and the rotation speed. The selection is based on the mean value, and the data exceeding the positive and negative deviation ranges are deleted. Therefore, the data selection process has less influence on the sample mean value.

3.3. Confidence Interval Analysis

During the processing of the test data, the difference of test results and design results were analyzed by calculating the confidence intervals of the mean values. Interval estimation generally needs to know the distribution of a certain satistics related to the estimated parameter.

Normal distribution is one of the most common distributions. Therefore, comparing between the distribution of the test data and normal distribution is performed. For the case that the sample obeys the normal distribution, if the population variance is unknown, the satistics as follow can be constructed to estimate population mean:

$$T = \frac{\bar{x} - \mu}{s / \sqrt{n}} \tag{4}$$

It follows the t-distribution with a n-1 degree of freedom, so the confidence interval of μ can be calculated as follows:

$$\left[\bar{x} - t_{1-\alpha/2}(n-1)\frac{s}{\sqrt{n}}, \bar{x} - t_{1-\alpha/2}(n-1)\frac{s}{\sqrt{n}} \right] \tag{5}$$

The degree of confidence is 1-α. And $t_{1-\alpha/2}$ can be found by searching for the t-distribution table.

4. Results and Analysis

4.1. Test Results and Distribution Characteristics

During the test, the control of the rotation speed is controlled by the physical rotation speed at low rotation speed, and the corrected rotation speed control is adopted at medium and high rotation speed (above 80% rotation speed). The figure 3 shows the scatter plots of the original test data for different test condition. Each data distribution is concentrated. The performance in the test are close to the calculation value at medium and high speeds. The reason that the data point does not fit the calculation curve at low speed is the deviation caused by the difference between the physical speed and the corrected speed.

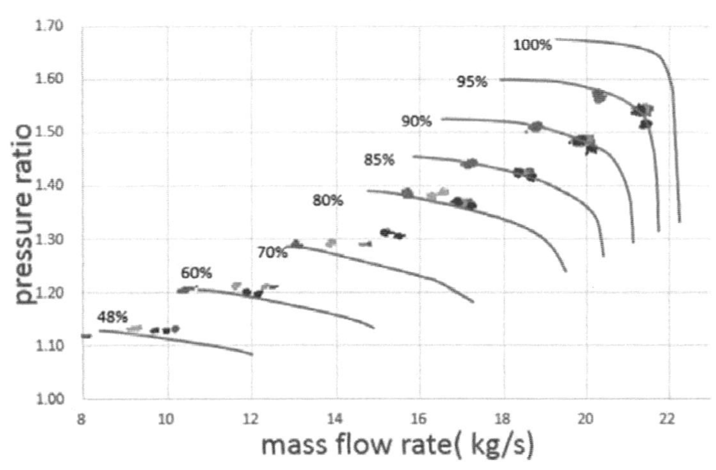

Figure 3. Flow-pressure ratio characteristic scatter distribution of the test results

Before calculating the confidence interval, the distribution of the calculation parameters (flow and pressure ratio) needs to be determined. After that, the data needs to be selected. Therefore, the distribution of the raw data and the data after the speed and flow selection are analyzed.

Before data selection, the distribution of the parameters is shown in the figure 4-6.

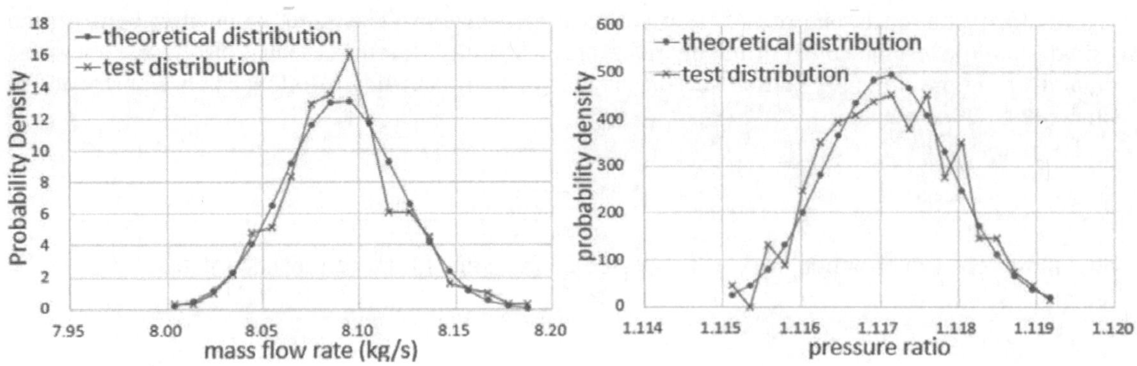

Figure 4. Flow and pressure ratio distribution characteristics in low speed of raw data

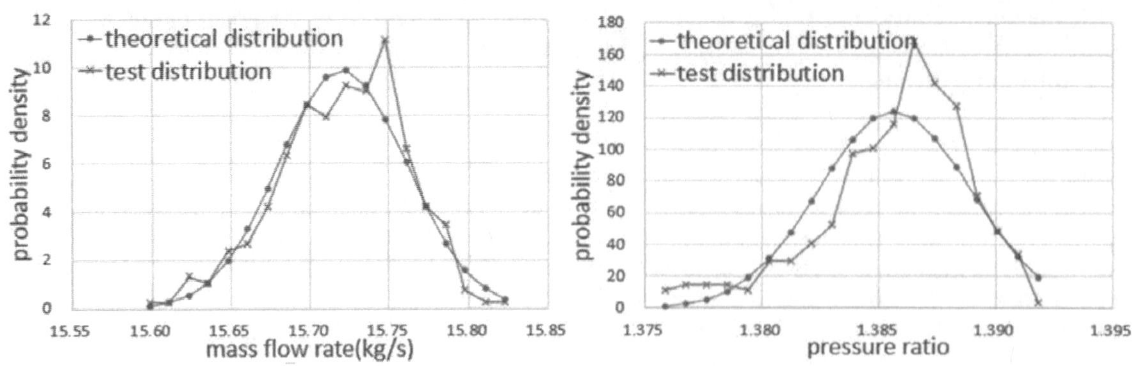

Figure 5. Flow and pressure ratio distribution characteristics in medium speed of raw data

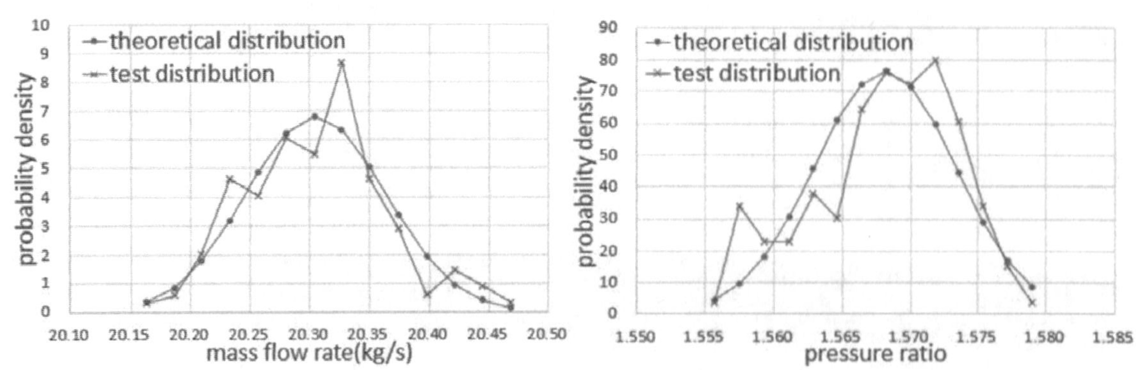

Figure 6. Flow and pressure ratio distribution characteristics in high speed of raw data

From the above figures, it can be seen that the mass flow rate and pressure ratio at different rotation speeds obtained in the test are in good agreement with the normal distribution to some extent, and the deviation is not significant. The reason for the deviation may be that the amount of data is less, which cause the insufficient randomness. Normally the data obeying normal distribution is from a system influenced by a number of mutually independent factors. The aeroengine test is affected by a variety of factors, and their relationship is generally independent of each other, such as the clearance and test environment, and the geometric outlet area. And the engine itself can be classified as a modular product, so it can explain the normality of its test results.

Because the engine itself is in a relatively strong vibration state, the parameter change is likely to exceed the allowable range of the corresponding error. At the same time, the accuracy of the test is not high enough, and it may cause an ultra-bad influence. Therefore, in order to further process the data, it is necessary to analyze the distribution of the data after the speed and mass flow rate selection.

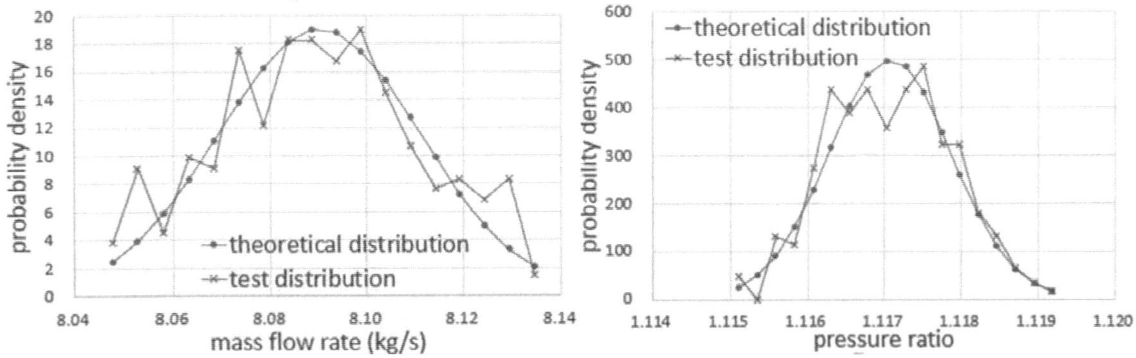

Figure 7. Flow and pressure ratio distribution characteristics in low speed of processed data

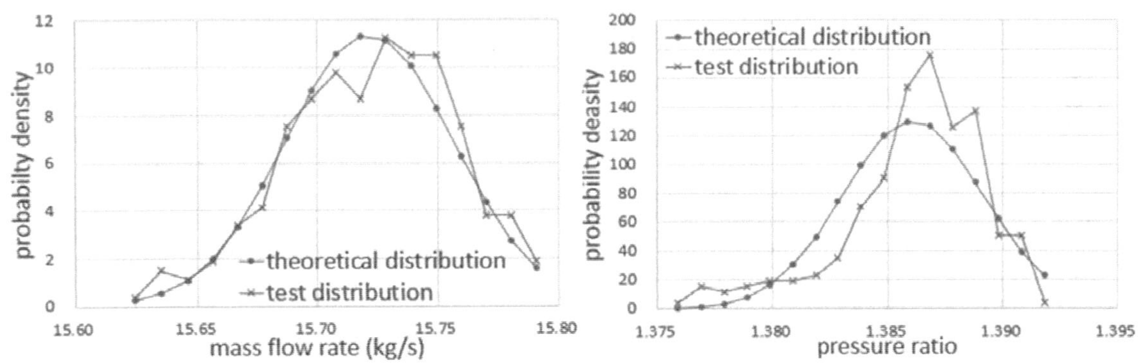

Figure 8. Flow and pressure ratio distribution characteristics in medium speed of processed data

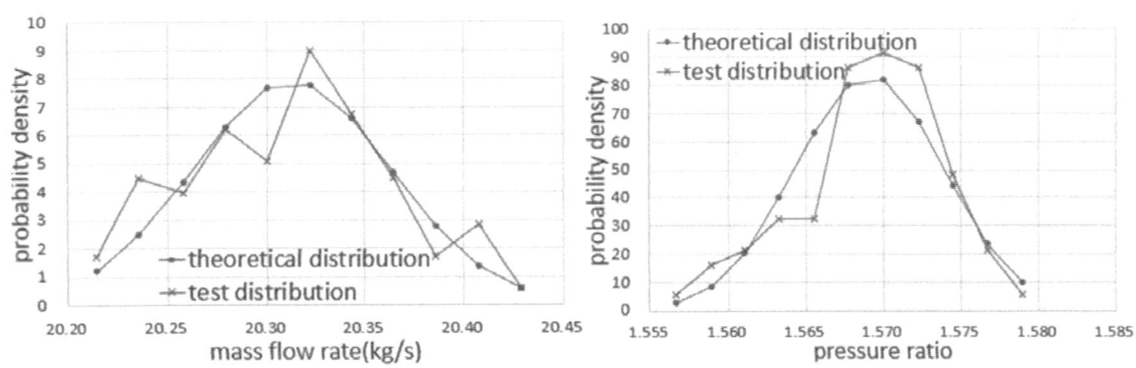

Figure 9. Flow and pressure ratio distribution characteristics in high speed of processed data

From the results after the selection, it can be seen that the distribution characteristics of the test results are not destroyed after the data that exceeds the deviation from the mean values are filtered out. This can be explained by the fact that due to the random distribution of the parameters (flow, pressure ratio), the amount of data that exceeds the positive and negative deviations of the speed and flow is equal. Therefore, deleting the data means that the range of the data is reduced, and it does not affect the overall distribution and the statistical parameters such as mean and variance.

4.2. Test Data Confidence Interval

After the data was selected, confidence intervals were calculated for the new data. The confidence level was chosen to be 98%, and the difference between the confidence interval and the mean value was taken as the deviation value. The flow rate and pressure ratio deviation at different rotational speeds were given in Table 1 and Table 2.

It can be seen from Table 1 and Table 2 that the relative value of the deviation is small. And the relative deviation of flow rate and pressure ratio is less than 0.1%. The following figure 10-11 shows the distribution of confidence intervals at 95%, 90%, 85%, and 80% rotation speeds. It can also be seen in the figure that the relative deviation is small and can only be discerned in the partial enlargement diagram of the characteristic line. Therefore, replacing interval estimates with point estimates is reasonable. On the other hand, it can be seen from Figure 10 and Figure 11 that the confidence interval is included in the calculation performance curve (Dashed line represents upper and lower limits). The upper and lower limit is calculated by the caculation performance and its deviation, mass flow rate by 0.5% and pressure ratio by 0.5% (based on deviation of the pressure). Though there are still points out of range marked in the figure, it can be concluded that the test and calculated values are well-conformed. The points out of range may caused by different tip clearance or larger rotation speed deviation.

Table 1. Mean and Confidence Deviations of Flow at Different Speeds

(N-rotation speed; MV-mean value; DAV-deviation absolute value; DEV-deviation relative value)

N%	48	60	70	80	85	90	95
MV(kg/s)	9.57	11.82	14.10	16.74	18.30	19.78	21.21
DAV(kg/s)	0.00303	0.00891	0.01863	0.00601	0.00825	0.01088	0.01174
DEV%	0.03162	0.07537	0.13212	0.03591	0.04509	0.05500	0.05534

Table 2. Mean and Confidence Deviations of Pressure ratio at Different Speeds

N%	48	60	70	80	85	90	95
MV	1.13	1.20	1.29	1.37	1.43	1.49	1.54
DAV	0.00021	0.00059	0.00099	0.00048	0.00052	0.00072	0.00096
DEV%	0.01876	0.04932	0.07701	0.03494	0.03642	0.04853	0.06247

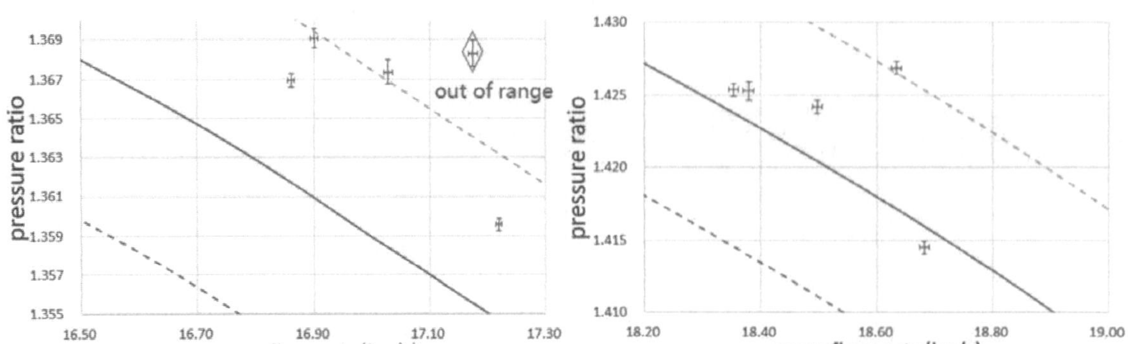

Figure 10. 80% and 85% speed flow-pressure ratio confidence interval distribution

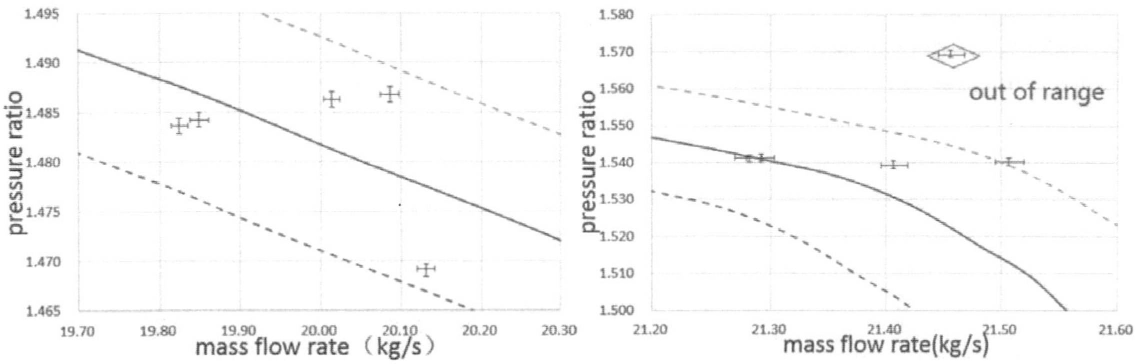

Figure 11. 90% and 95% speed flow-pressure ratio confidence interval distribution

4.3. Fan Working Line by Processed Data

After the data selection and replacing interval estimates with point estimates, and the working line in different core and bypass throat area can be given as figure 12. The legend shows the diameter of the core nozzle (mm)/the diameter of the bypass nozzle (mm).

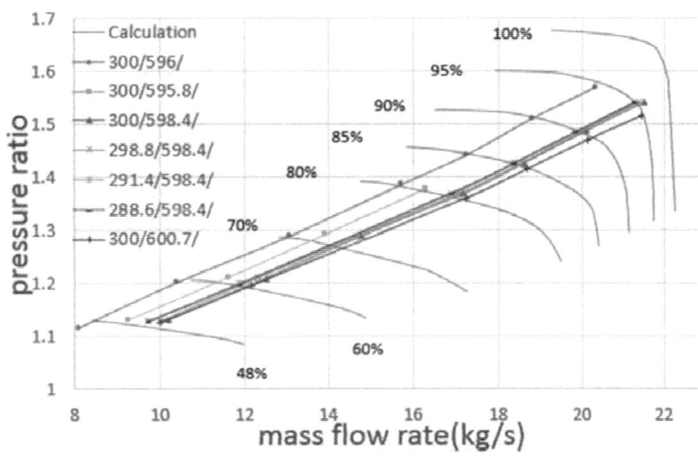

Figure 12. Engine operating line under different test conditions after data processing

From the above figure, we can see that when determing the core throat area and reducing the bypass nozzle equivalent diameter from the 600.7mm to 598.4mm and to 596mm, the working line move to the left, where the flow rate decreases. And then reduced the diameter to 595.8mm, the working line move to the oppisite direction. Similarly, when determing the bypass throat area and reducing the core nozzle equivalent diameter, similar things will happen. The working line will move to left first and turn to right then. But it can be seen that although the change exists, the effect is not obvious. Changing the area of core area has less impact on the working line, whereas the bypass area has a greater impact on the characteristic line. It is expected that as the area of bypass (core) throat area decreases, the fan working line will move to the the direction of reduced flow rate. But the actual situation seems far from that. This may depend on whether the flow state of core and bypass is blockage and choked, or it is in a better flow state. If the blockage and choked happens, the situation will occur that the working move to unexpected direction. In general, the reason is that the core and bypass do not match. For the fan of turbofan engine, the core and bypass always interact with each other. The above figure shows that adjusting the nozzle area requires repeated adjustments before it can finally match the design operating point.

IOP Conf. Series: Journal of Physics: Conf. Series **1064** (2018) 012053 doi:10.1088/1742-6596/1064/1/012053

5. Conclusion

This article analyzes the performance parameters such as flow and pressure ratio obtained by the fan test in the environment of the whole aero-engine. The conclusion can be summarized as follows.

1. For the fan test performance, the indirect measurement parameters such as flow rate and pressure ratio obey the normal distribution.

2. Due to the large amount of data, the range of confidence intervals will be so narrow that point estimates can be used instead of interval estimates.

3. The test data and the calculation performance are basically consistent.

4. When adjusting the area of the core and bypass throat, the working line dose not always move in one direction, which depends on the match of the core and bypass. So it is necessary to make repeated adjustments to match the design state.

References

[1] Hu, Guo-Rong, et al. "Design and test for single stage fan with high speed, high load." Journal of Enginerring Thermophysics 22.1 (2001): 40-43.

[2] Beselt, Ch, Mario Eck, and Dieter Peitsch. "Three-dimensional flow field in highly loaded compressor cascade." Journal of Turbomachinery 136.10 (2014): 101007.

[3] Connell, James W. Data Reduction of Single Stage Compressor. NO. ARL-75-0192. AEROSPACE RESEARCH LABS WRIGHT-PATTERSON AFB OHIO, 1975.

[4] Xiang, Honghui, et al. "Discussion on several problems of axial flow fan/compressor aerodymatic performance experiment." Journal of Aerospace Power 30.6 (2015): 1423-1431.

[5] Gu, Yang, et al. "Function of the High Temperature and High Pressure Compressor Test Rig." Gas Turbine Experiment and Research 1 (2008): 18-21.

[6] Gao, Longfei, and Hui Wu,"Flow Field Diagnosis and Performance Optimizing in a Three-stage Fan Experiment." Aeroengine 41.4 (2015): 73-76.

[7] Zhang, Jin-cheng,et al. "Numerical Simulation of core and bypass characteristic and Test Analysis under Whole Aero-engine Condition of Fan with Low Bypass Ratio." in the 2nd JCAP and 38th APTIS Technical Conference.

[8] GJB 241A-2010, "General Specification for Aviation Turbojet and Turbofan Engines."

IOP Conf. Series: Journal of Physics: Conf. Series **1064** (2018) 012054 doi:10.1088/1742-6596/1064/1/012054

Study on the vibration of membrane roof with the influence of added air-mass

Weiju Song [1, 2,*], Yingmin[1] Li and Xiaowei Wang[1, 2]

[1]College of Civil Engineering, Chongqing University, Chongqing, China.
[2]City College of Science and Technology, Chongqing University, Chongqing, China.

*Corresponding author e-mail: nimrodsong@126.com

Abstract. For studing the influence regularity of added air-mass on the dynamic characteristics of membrane roof, the three-dimensional membrane roof is taken as the model, we obtained the analytical expression of added air-mass according to the energy conservation and aeroelastic mechanics. Then, futher investigations of the free vibration considering membrane immerse in air is given. Combined with the specific structural parameters, the influence of vibrate mode, membrane'density and the prestress on the additional air quality is discussed, which lays the foundation for further analysis of nonlinear coupling wind induced vibration response of membrane structures.

1. Introduction

For general structures, the influence of added air-mass can be negligible when calculating the natural frequency and response of the structures. But for the flexible structures, such as membrane structure, the coupling of aerodynamic and structure has a significant influence on the dynamic characteristics of structures. added air-mass may be pertains to the same order of the membrane' density. Therefore, the influence of added air-mass cannot be ignored in studying the vibration characteristics of membrane structures.

Many literatures have discussed the calculation of the added air-mass caused by the movement of structures in fluid. The added air-mass formula for the spherical surface of the radial pulsation was given by Irwin [1] which used to estimate the magnitude of the added air-mass for the roof of the Olympic Stadium in Montreal, Canada, and find that the effort of added air-mass was not negligible. Sewall [2] found that the value of added air-mass calculated by the formula given by Irwin is larger than the actual situation based on the vibration frequency comparison between the experiment and the theoretical formula given by Irwin. Li Qingxiang [3] derives the calculation method of the added air-mass for the two-dimensional membrane structure, which shows that the effect of the added air-mass on the structural characteristic frequency is significant.

In this article, the analytical method is used to analyze the added air-mass of the membrane structure' vibration. Based on the three-dimensional membrane roof, the analytical expression of the added air-mass is obtained according to the energy conservation and aero elastic mechanics. Combined with the specific structural parameters, the influence of vibrate mode, membrane' density and the pretension on the additional air quality is discussed, which lays the foundation for further analysis of nonlinear coupling wind induced vibration response of membrane structures.

Content from this work may be used under the terms of the Creative Commons Attribution 3.0 licence. Any further distribution of this work must maintain attribution to the author(s) and the title of the work, journal citation and DOI.
Published under licence by IOP Publishing Ltd

IOP Conf. Series: Journal of Physics: Conf. Series **1064** (2018) 012054 doi:10.1088/1742-6596/1064/1/012054

2. Analytical derivation of added air-mass

Assuming that the static equilibrium state of the membrane roof is horizontal, the length and width of the roof are a and b, respectively, and the density is ρ. As shown in Figure 1.

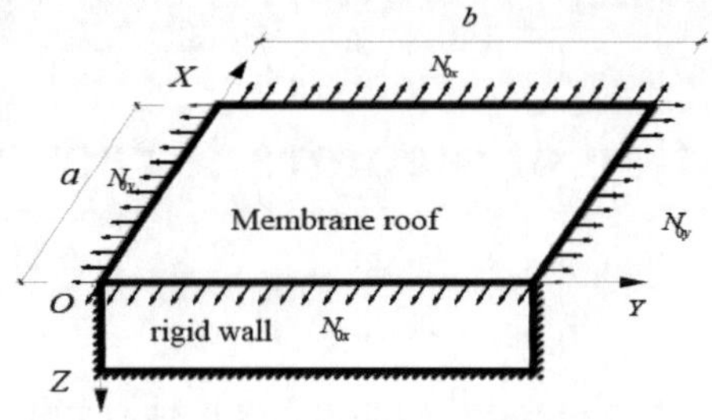

Figure 1. Model of sealed membrane roof.

Assumed that the displacement of a three-dimensional membrane surface satisfying the boundary conditions is:

$$w(x,y,t)=a_0\cos\omega t\cdot\sin\frac{m\pi x}{a}\sin\frac{n\pi y}{b} \tag{1}$$

Where, $0\le x\le a;0\le y\le b;m=1,2,3\cdots,n=1,2,3\cdots$, ω is the vibration frequency, a_0 is displacement amplitude.

Assuming that the air is an incompressible, cohesionless steady state fluid, the roof surface is used as a fluid solid coupling interface. The rate of change of kinetic energy in any part of the fluid to time is equal to the work [4] of surface pressure. That is:

$$-\frac{dE_t}{dt}=\frac{dW}{dt} \tag{2}$$

$$\frac{dW}{dt}=-\iint_S V_N p_Z dS=-\iint_s p_Z\frac{\partial w(x,y,t)}{\partial t}ds \tag{3}$$

Where, VN is the velocity of the fluid particle in the direction of the surface of the membrane, S is the surface boundary of the fluid region, pZ is the pneumatic force of the membrane surface, and the integral area $s\in\{0\le x\le a,0\le y\le b\}$.

Change rate of kinetic energy relative to time of vibrating membrane is equivalent to the kinetic energy [4] of added air-mass per unit area, that is, the kinetic energy change rate of added air-mass can be expressed as:

$$\frac{dE_t}{dt}=m_a\iint_s\frac{\partial w(x,y,t)}{\partial t}\left(\frac{\partial w(x,y,t)}{\partial t}\right)^2ds \tag{4}$$

Where, ma denote added air-mass per unit area. The expression of added air-mass can be obtained from the formula (3) and (4).

$$m_a = \frac{-\iint_s p_Z \frac{\partial w(x,y,t)}{\partial t} ds}{\iint_s \frac{\partial w(x,y,t)}{\partial t}\left(\frac{\partial w(x,y,t)}{\partial t}\right)^2 ds} \tag{5}$$

For closed membrane roofs, only the upper surface is in contact with the air, and there is no air flow in the inner space of the membrane. Suppose the air is incompressible, non-viscous and potential. The perturbation velocity potential function satisfies the Laplasse equation.

$$\frac{\partial^2 \phi'}{\partial x^2} + \frac{\partial^2 \phi'}{\partial y^2} + \frac{\partial^2 \phi'}{\partial z^2} = 0 \tag{6}$$

The integral solution of eq.(6) is the Rayleigh integral equation with S as the boundary. For closed membrane roofs, the perturbation bound condition is:

$$v_z = \frac{\partial \phi'}{\partial z} = 2\left(\frac{\partial z}{\partial t} + V\frac{\partial z}{\partial x}\right),\ z \to 0,\ (x,y) \in Ra \tag{7}$$

Where, vz denote the velocity component. z is the surface equation of the membrane roof.

For the flat roof, the disturbance velocity potential function satisfying the boundary condition is

$$\phi'(x,y,z,t) = \frac{1}{4\pi}\iint_{Ra} \frac{\partial \phi'}{\partial z} \frac{1}{\sqrt{(x-\xi)^2 + (y-\eta)^2}} d\xi d\eta \tag{8}$$

The aerodynamic force on the surface of the membrane roof is as follows:

$$p_2 = -\rho\left(\frac{\partial \phi'}{\partial t} + V\frac{\partial \phi'}{\partial x}\right) \tag{9}$$

The force inside the membrane roof can be approximately equal to the air force generated by the uniform flow field without disturbance, and the aerodynamic force acting on the roof surface is:

$$P = p_1 - p_2 == \frac{\rho}{2\pi}\left(\begin{array}{l} -V\iint_{Ra} \frac{\left(V\frac{\partial z}{\partial x} + \frac{\partial w}{\partial t}\right)_{\substack{x=\xi \\ y=\eta}} (x-\xi)}{\left(\sqrt{(x-\xi)^2 + (y-\eta)^2}\right)^3} d\xi d\eta + \\ \iint_{Ra} \frac{\left(V\frac{\partial^2 z}{\partial x \partial t} + \frac{\partial^2 z}{\partial t^2}\right)_{\substack{x=\xi \\ y=\eta}}}{\sqrt{(x-\xi)^2 + (y-\eta)^2}} d\xi d\eta \end{array} \right) \tag{10}$$

It is assumed that the surface functions of the membrane is:

$$z(x,y,t) = z_0(x,y) + w(x,y,t) \tag{11}$$

Where, z0(x,y) denote the initial surface function. For the flat roof, z0(x,y)≈0, then

$$z(x,y,t)=w(x,y,t) \tag{12}$$

Substituting (12) into eq. (10).

$$\begin{aligned}p=&\frac{\rho}{\pi}\iint_{Ra}\frac{1}{r}\left(\frac{\partial^2 w}{\partial t^2}\right)_{\substack{x=\xi\\y=\eta}}d\xi d\eta+\frac{\rho V}{2\pi}\iint_{Ra}\frac{1}{r}\left(\frac{\partial^2 w}{\partial x\partial t}\right)_{\substack{x=\xi\\y=\eta}}d\xi d\eta-\\&\frac{\rho V^2}{2\pi}\iint_{Ra}\frac{1}{r^3}\left(\frac{\partial w}{\partial x}\right)_{\substack{x=\xi\\y=\eta}}(x-\xi)d\xi d\eta-\frac{\rho V}{2\pi}\iint_{Ra}\frac{1}{r^3}\left(\frac{\partial w}{\partial t}\right)_{\substack{x=\xi\\y=\eta}}(x-\xi)d\xi d\eta\end{aligned} \tag{13}$$

Where, $r=\sqrt{(x-\xi)^2+(y-\eta)^2}$, Integral region $Ra\in\{0\le\xi\le a,0\le\eta\le b\}$.

3. Effect of added air-mass on vibration of the membrane roof

The effect of added air-mass on the vibration of thin film roof during the free vibration process is studied in this paper. Formula (13) can be simplified to:

$$P=\frac{\rho}{2\pi}\left(\iint_{Ra}\frac{\frac{\partial^2 w}{\partial t^2}}{\sqrt{(x-\xi)^2+(y-\eta)^2}}d\xi d\eta\right) \tag{14}$$

Taking the membrane material commonly used in engineering as an example, the surface density of the membrane is 1.2kg/m^2, the thickness is 1mm, the x direction modulus is 1.4×106kN/m^2, the y modulus is 0.9×106kN/m^2, the length and the width are 20m , and the pretension of membrane is 5kN. Table 1 is a comparison of the natural frequency ω0 and the frequency of ωair considering the added air-mass.

Table 1. Frequencies of membrane structure

Mode	1	2	3	4	5	6	7	8	9	10
ω_0(HZ)	2.28	3.61	5.10	6.65	8.23	9.81	11.41	13.01	14.61	16.22
ω_{air}(HZ)	1.98	3.34	4.85	6.39	7.96	9.55	11.14	12.74	14.33	15.93

According to Table 1, the added air-mass significantly reduces the vibration frequency of the structure.

The ratio of the frequency ω0 of the membrane structure without air influence to the frequency ωair with air influence under different modes and different densities are calculated and drawn as shown in Figure 2.

Form figure 2 we can see that the effect of air on the vibration frequency of membrane is becoming more obvious with the decrease of the membrane density, especially the low order modes.

The influence of added air-mass on structural frequency is analyzed with the change of membrane pretension ratio. Take σy =5000kN/m^2. The result shows that the pretension is the key parameter affecting the structure's frequency. The increasing pretensions can reduce the influence of air on the frequency of the structure. But the impact was less than the density of membrane. The results are shown in Figure 3.

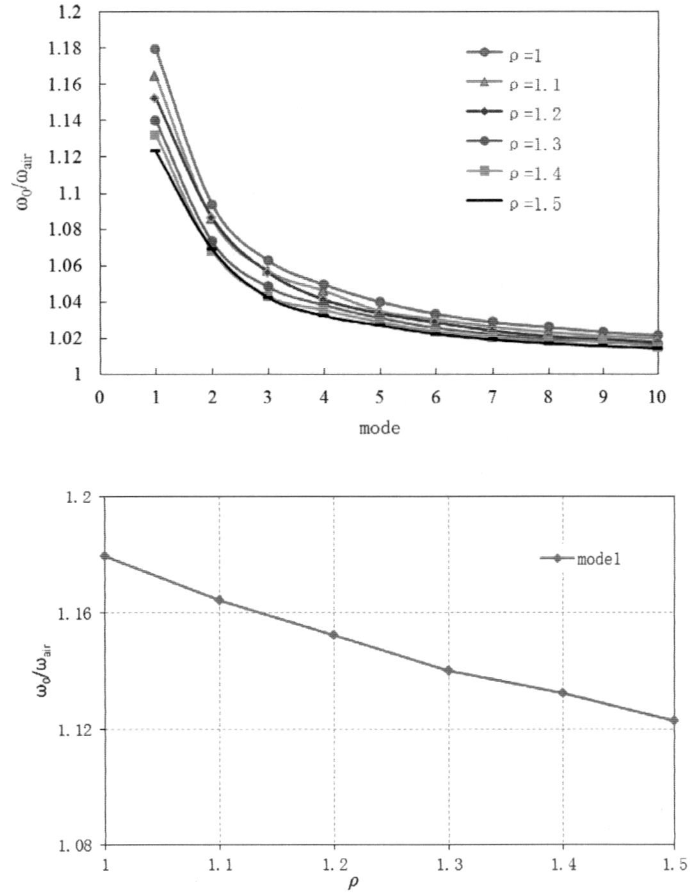

Figure 2. The ratio of ω0 to ωair under different modes and densities

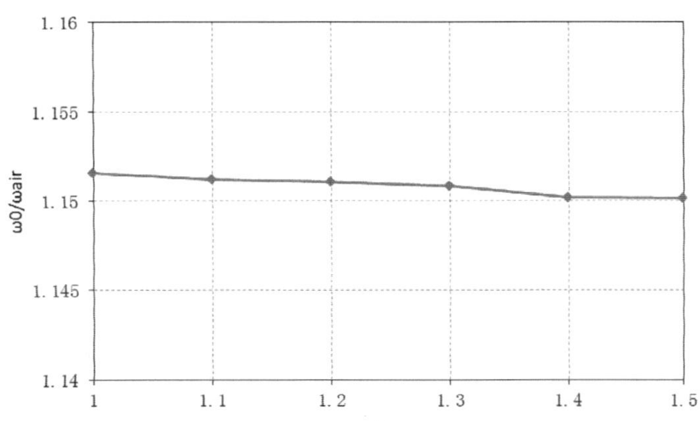

Figure 3. The ratio of ω0 to ωair under different modes and pretension

4. Conclusion

In this paper, a simplified aerodynamic model is used to simulate aerodynamic forces. The variation rule of aerodynamic term is derived theoretically. Then, according to the numerical example, the influence of added air-mass on the dynamic characteristics of membrane roof is calculated. The following conclusions are obtained, the effect of the added air-mass on the frequency of the structure is significant, and the effect on the high order modes is not as obvious as that of the lower order modes. With the increase of membrane's quality and pretension, the influence of added air-mass on the vibration frequency decreases.

Acknowledgments

This work was financially supported by National Natural Science Foundation (Project Number: 51478067) and Science and Technology research project of Chongqing Municipal Education Commission (Project Number: KJ1723378).

References

[1] IRWIN H P A H, WARDLAW R L A. Wind tunnel investigation of the retractable fabric roof for the Montreal Olympic Stadium[A]. Proceeding of the Fifth International Conference on Wind Engineering[C]. Colorado State University, Pargamon Press, 1980, pp. 925-938.
[2] SEWALL J L, MISERENTINO R, PAPPA R S. Vibrating studies of a lightweight three-sided membrane suitable for space application[R]. NASA Technical Paper 2095.
[3] LI Qingxiang, SUN Bingnan. Influence regularity of aerodynamic forces on dynamic characteristics of membrane structures[J]. Acta Aerodynamica Sinica. 2007, 25(2), pp. 137-144.
[4] TONG Binggang. Unsteady flow and vortex motion. National Defense Industry Press, Beijing, 1993, pp. 21-40.
[5] MAO Guodong, SUN Bingnan. The added air-mass of membrane structures[J]. Engineering Mechanics, 2004(01), pp. 153-158.
[6] CHEN Zhaoqing., WU Yue, SUN Xiaoying. Research on the added mass of open-type one-way tensioned membrane structure in uniform flow[J]. Journal of Wind Engineering and Industrial Aerodynamics. 2015(137), pp. 69-77.
[7] WANG Lei, LI Yuanqi, SHEN Zuyan. Experimental investigation on the added mass of membranes vibrating in air[J]. Journal of Vibration Engineering. 2011, 24(2), pp. 125-132.
[8] GAO Haijian, CHEN Wujun, FU Gongyi. Modal analysis method and modal behaviour of pre-stressed inflatable fabric beam[J]. 2010, 38(7), pp. 135-139.

IOP Conf. Series: Journal of Physics: Conf. Series **1064** (2018) 012055 doi:10.1088/1742-6596/1064/1/012055

Application of shock tube in calibrating dynamic pressure transducers

Laijun Yan*, Yong Chen, Lihu Zhang, Xu Zhang, Xianghong Yao, Xinhua Qi and Long Zhang

Facility Design and Instrumentation Institute, China Aerodynamics Research and Development Center (CARDC), Mianyang, China.

*Corresponding author e-mail: laijun11-11@163.com

Abstract. The step pressure generated by shock tube was used for calibration of dynamic pressure transducers. First, the performance of the shock tube based calibration installation was theoretically predicted. The influence of Mach number to the step pressure and the duration of the step pressure plateau was analyzed by numerical calculation. Second, the performance of the calibration installation was tested, whose result showed that the minimum and maximum value of the step pressure generated by the calibration installation was 9.60kPa and 1.101MPa, respectively, the maximum value of the rise time of the step pressure was 0.7μs, the minimum value of the duration of the step pressure plateau was 4.14ms, the fluctuation of the step pressure plateau was better than ±2%, and the maximum value of the 2nd order expand uncertainty of the step pressure was 3.52%. The test result verified that the calibration installation met the requirements of the previous design and correlative verification regulations. Last, the sensitivity and syntonic frequency of standard pressure transducers were tested by the calibration installation. The result was in accord with the factory technical parameters.

1. Introduction

It was important and essential to calibrate dynamic pressure transducers termly [1, 2]. Shock tube was used for the calibration of dynamic pressure transducers because it was an ideal generator of step pressure [3-6]. According to the 'Verification Regulation of Dynamic Pressure standards' [7], the shock tube based dynamic pressure standard should meet all demand of performance index showed in table 1.

Table 1. Performance index of shock tube based dynamic pressure standard.

Range of the step pressure	15kPa~1MPa
Rise time of the step pressure t_r	<1μs
Duration of the step pressure plateau τ	>4ms
Fluctuation of the step pressure plateau	Within ±2%
2nd order expand uncertainty of the step pressure $U_k(k=2)$	<4% (k=2)

This paper studied the theoretical design, manufacture, actual test and verification of shock tube based calibration installation for dynamic pressure transducers which met the demand of table 1. The

Content from this work may be used under the terms of the Creative Commons Attribution 3.0 licence. Any further distribution of this work must maintain attribution to the author(s) and the title of the work, journal citation and DOI.
Published under licence by IOP Publishing Ltd

IOP Conf. Series: Journal of Physics: Conf. Series **1064** (2018) 012055 doi:10.1088/1742-6596/1064/1/012055

influence of Mach number on the 2nd order expand uncertainty of the step pressure generated by the calibration installation was specially studied.

2. Theoretical prediction

Figure 1 showed the schematic diagram of the shock tube based calibration installation for dynamic pressure transducers. The hardware consisted of a shock tube, a vacuum pump, a high pressure gas resource, two pressure transmitters, three transducers for velocity measurement, a temperature transmitter, two data sampling cards and a industrial personal computer, etc. The software was a data sampling and data processing system which calculated the parameters of the calibrating transducers.

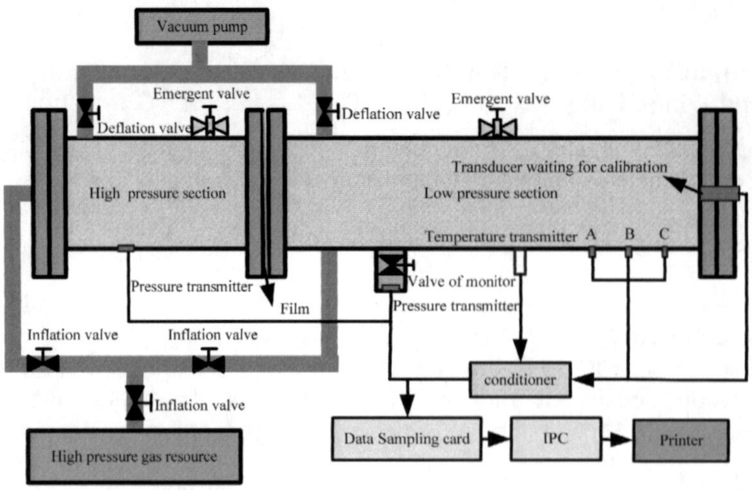

Figure 1. Schematic diagram of the shock tube based calibration installation for dynamic pressure transducers.

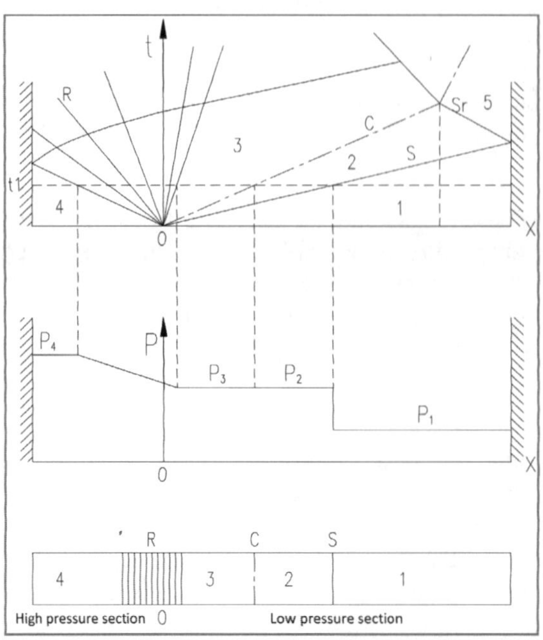

Figure 2. Flow field divisional of shock tube based calibration installation.

The typical flow field divisional of shock tube was shown in figure 2, in which the flow field of section 5 was used for calibration of dynamic pressure transducers. When the working gas in the shock tube was air with normal temperature, the step pressure generated by the reflected shock wave of the flow field of section 5 could be calculated by [7]:

$$\Delta p_5 = \frac{7}{3}(M_s^2 - 1)(4M_s^2 + 2)/(M_s^2 + 5) \cdot p_1 \quad (1)$$

Where Mach number M_s of the shock wave was calculated according to the the signal of transducers for measurement of the velocity of the shock wave, and initialized pressure of low pressure section p1 was measured by the pressure transmitter and then recorded by the data sampling and data processing system.

Previous work [4] designed the parameters of the shock tube based calibration installation and predicted its performance, as shown in table 2.

Table 2. Parameters and performance prediction of shock tube based calibration installation.

Length of the high pressure section	2m	
Length of the low pressure section	5m	
Initialized pressure of high pressure section p_4	24.27kPa	1.62MPa
Initialized pressure of low pressure section p_1	2.86kPa	0.19MPa
Step pressure	15kPa	1MPa
Mach number	1.557	
Duration of the step pressure plateau τ	9.85ms	
Fluctuation of the step pressure plateau	Within ±2%	
2nd order expand uncertainty of the step pressure $U_k(k=2)$	2.92% (k=2)	

The numerical calculation result showed the influence of Mach number of the shock wave to the the duration of the step pressure plateau and the step pressure generated by the shock tube based calibration installation. When Mach number of the shock wave decreased, the duration of the step pressure plateau [8] increased as shown in figure 3(a), the step pressure decreased however.

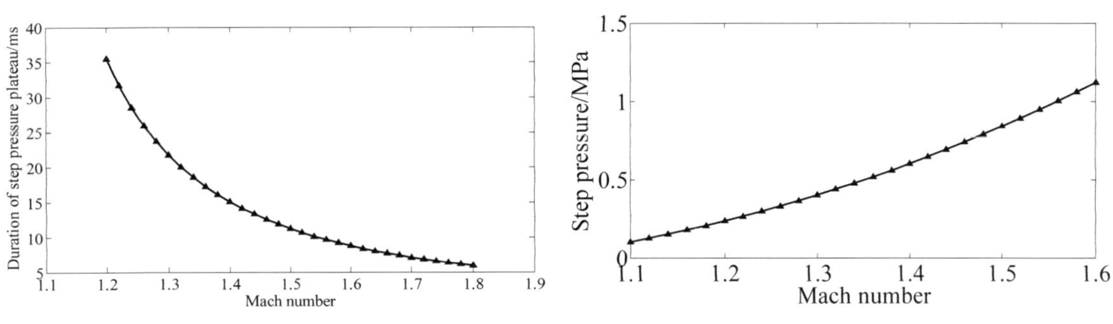

(a) (b)

Figure 3. (a)Influence of Mach number to the the duration of the step pressure plateau; (b) Influence of Mach number to the step pressure.

3. Performance test

Because of real gas effect [9], the Mach number of the shock wave would decrease, which would lead to changes as predicted in figure 3. The actual performance of the shock tube based calibration installation could precisely obtained only through actual test.

Table 3 showed the actual performance of the calibration installation in three different step pressure and corresponding test conditions. When step pressure was 9.6kPa (<15kPa), 288kPa (15kPa~1MPa) and 1.101MPa (>1MPa), all the rise time of the step pressure tr was less than 1μs, all the duration of the step pressure plateau τ was more than 4ms, all the fluctuation of the step pressure plateau was within ±2%, and all the 2nd order expand uncertainty of the step pressure was less than 4%. Thus the actual performance met the demands showed in table 1, but not as good as predicted in table 2. Compared with the predicted value, the actual Mach number Ms and duration of the step pressure plateau τ decreased, the actual initialized pressure of low pressure section p1 increased, and the actual 2nd order expand uncertainty[10] of the step pressure increased as well.

Table 3. Actual performance of the calibration installation in three different step pressure and corresponding test conditions.

Actual initialized pressure of high pressure section p_4	33kPa	591kPa	1.93MPa
Actual initialized pressure of low pressure section p_1	4.36kPa	190kPa	316kPa
Material of the film	Vegetable parchment film	100μm polyester film	75μm polyester film plus 250μm polyester film
Burst quomodo of the film	Heating up via a resistance wire	The pressure differential p_4- p_1	
Temperature T	22.43℃	20.04℃	20.05℃
Mach number M_s	1.36	1.23	1.43
Step pressure Δp_5	9.60kPa	288.16kPa	1.101MPa
Rise time of the step pressure t_r	0.3μs	0.7μs	0.7μs
Duration of the step pressure plateau τ	4.14ms	6.45ms	4.72ms
Fluctuation of the step pressure plateau		Within ±2%	
2nd order expand uncertainty of the step pressure	3.25%	3.52%	3.05%

4. Test of standard transducers

Three dynamic pressure transducers were tested by the shock tube based calibration installation. Figure 4 showed the response of transducer PCB 113B26, one of the tested transducers, and its corresponding frequency spectrum. The rise time of the transducer was obtained from figure 4(a); the syntonic frequency of the transducer was obtained from figure 4(b). The sensitivity of the transducer was calculated when the step pressure and the voltage amplitude of the response of the step pressure was obtained.

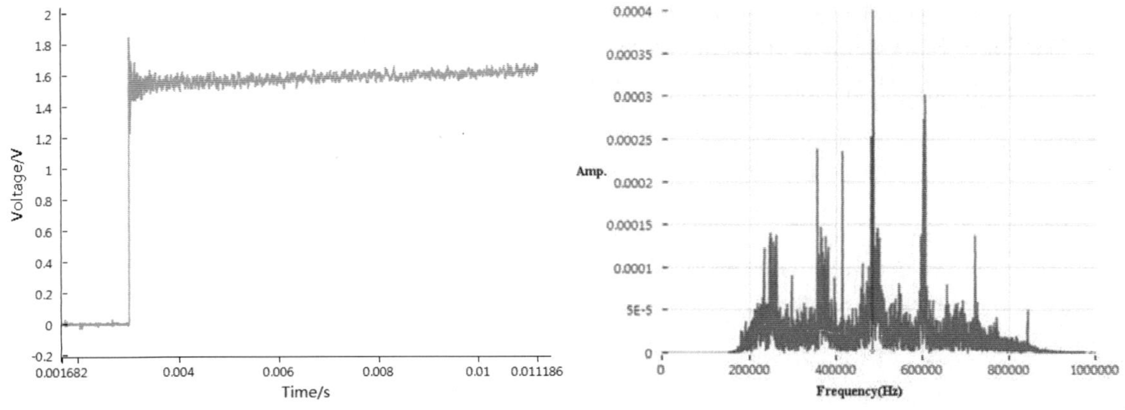

(a) (b)

Figure 4. (a) Response of the transducer PCB 113B26 when Δp_5= 1.101MPa; (b) The corresponding frequency spectrum.

Table 4 showed all the tested transducers and the test result. The rise time and syntonic frequency given by the manufacturer were approximate range and value, respectively. The test values of the two parameters were reasonable to the value given by the manufacturer. And the test values of sensitivity of the transducers were close to the value given by the manufacturer. The maximum error was 4.08%.

Table 4. Tested transducers and test result.

Test number	1	2	3
Model of the transducers	PCB 113b21	PCB 113b26	PCB 113b28
Range of the transducers	1379kPa	3447.5kPa	344.75kPa
Sensitivity of the transducers	3.6mV/kPa	1.46mV/kPa	14.7mV/kPa
Rise time		<1μs	
Syntonic frequency		~500kHz	
Tested syntonic frequency	482.10KHz	485.50KHz	355.12KHz
Tested rise time	0.6μs	0.7μs	0.8μs
Tested sensitivity	3.60mV/kPa	1.45mV/kPa	15.30mV/kPa

5. Conclusion

A shock tube was applied to calibration of dynamic pressure transducers, based on which a calibration installation was designed and manufactured. The performance of the calibration installation was theoretically predicted and actually tested. It was verified that the the Mach number of the shock wave generated by the calibration installation would decrease because of real gas effect. Thus to reach the same step pressure, the actual initialized pressure of low pressure section should be higher than theoretical prediction. The actual performance of the calibration was not as good as prediction, but still met the demands of corresponding verification regulations. Three transducers were tested by the calibration installation. The test result was consistent with the values given by manufacturer, which verified the availability of the calibration installation.

Acknowledgments
This work was financially supported by CARDC.

References

[1] HUANG Jun-qin, Measurement System Dynamics and Its Applications, Beijing, National

Defense Industry Press, 2013: 5-6

[2] Damion J. P. Dynamic calibration of pressure sensors. LNE REPORT. 2007.

[3] ZHANG Da-you. The Application of Shock tube in Testing and Calibrating the Performance of Pressure Sensor. Journal of Astronautic Metrology and Measurement, 2004, 24 (4): 24-27.

[4] YAN Lai-jun, ZHANG Long, ZHANG Xu,et al. Design of Measurement and Control System for Shock Tube Based Calibration Installation of Dynamic Pressure Transducer. Proceedings of IEEE 13th International Conference on Electronic Measurement and Instruments (ICEMI), 2017, 3, pp. 319-324.

[5] Revel G. M. Pandarese G. Cavuto A. The development of a shock-tube based characterization technique for air-coupled ultrasonic probes. Ultrasonics, 2014, 54(6): 1545-1552.

[6] Wisniewiski D. Second generation shock tube calibration system. International Journal of Acoustics and Vibration, 2012, 17 (3): 133-138.

[7] JJG 624-2005: 'Verification regulation of dynamic pressure transducers', 2005

[8] JJG 1142-2017: 'Verification Regulation of Dynamic Pressure standards', 2017

[9] ZHANG Jing ping, FU Zhou dong. The Calculation of Shock Wave Speed in the Presence of Real Gas Effect and Shock Attenuation. ACTA METROLOGICA SINICA, 2000, 21(1): 45-50.

[10] LI Qiang, WANG Zhongyu, WANG Zhuoran. Uncertainty evaluation for the dynamic calibration of pressure transducer. Journal of Beijing University of Aeronautics and Astronautics, 2015, 41(5): 847-856.

Analysis and reconstruction of waterjet characteristics in double-side wall confinement

Bo Yuan[1, 2, 3, *], Shuai Jia[1], Quanlin Wang[1], Changjun Meng[1], Jiaxuan Li[1] and Liangming Pan[1, 3]

[1]Key Laboratory of Low-grade Energy Utilization Technologies and Systems, Chongqing University, Ministry of Education of PRC, Chongqing, China.
[2]State Key Laboratory of Coal Mine Disaster Dynamics and Control, Chongqing University, Chongqing, China.
[3]Postdoctoral Research Station of Power Engineering & Engineering Thermophysics, Chongqing University, Chongqing, China.

*Corresponding author e-mail: boyuanyuan@cqu.edu.cn

Abstract. Water jet flow characteristics in double-side wall confined space is investigated, two-dimensional particle image velocimetry is performed on confined plane and unconfined plane, and using Proper orthogonal decomposition to perform dimensionality analysis, extract large-scale coherent structure and reconstruct flow field. Results illustrate that velocity and vorticity fluctuation distribution, mainly reflected in low-order modes, is seriously affected on account of a symmetrical double-wall existence; Reconstruction instantaneous velocity field by the first m modes saving 70% of turbulent kinetic energy, indicating that reconstructed velocity fluctuation fields can truly reflect original fields.

1. Introduction

Waterjet cutting is that fluid with high energy density focus on target material surface resulting in scale of target material dropped out by undergoing complex physical interaction process and forming strongly erosive [1-3]. In practical application, after cutting, a narrow slit, namely double-side wall confined space, is formed, cutting section surface exhibits irregular characteristics which could be caused by waterjet complex flowing characteristics in this double-side wall confinement. For confinement flow characteristics, a large number of scholars have performed extensive experiments and simulations on wall and free-surface-bounded jets. Shinneeb et al.[4] using particle image velocimetry (PIV) to analyse jet flow characteristics in shallow water confined by free surface and wall, average flow, turbulence characteristics and vortex structure in different vertical directions is discussed, results show that flow field structure generates great changes due to vertical restriction, total number of vortex ring decrease with vertical confinement gradually decrease, but size of vortex ring increases; Raiford and Khan [5] using same submergence ratio and offset ratio of 2, 3 and 4 in different confined height to investigate growth rate in vertical and horizontal planes along with decay of maximum velocity, find that horizontal velocity profile is self-similar, essentially agree with free jet profile and experimental results made by Shinneeb[4]; Chowdhury et al.[6] using RANS equations along with standard k-ε turbulent closure scheme to simulate circular turbulent jets in shallow water to investigate effect of different submergence

Content from this work may be used under the terms of the Creative Commons Attribution 3.0 licence. Any further distribution of this work must maintain attribution to the author(s) and the title of the work, journal citation and DOI.
Published under licence by IOP Publishing Ltd

ratios and offset ratios, results indicate that riverbed influence is larger than free surface in flow field structure; Wang et al.[7] and Vinze et al.[8] respectively adopted large eddy simulation method and experimental method to investigate heat transfer in confined space.

On account of nozzle inlet with high-pressure and small size, waterjet in double-side wall confinement belongs to high Reynolds number flow. In order to describe flow fields with high spatial resolution obtained by PIV, proper orthogonal decomposition (POD) with powerful dimension reduction and analysis capabilities is gradually adopted [9-12]. Lumley[13] firstly used POD to analyse flow fields, and identified large-scale coherent structure, a large number of scholars then employ POD to analyse plane jet[10], free jet[14] and pulsed jet[15]; Oudeusden[16] employed POD to investigated different incident angle influence on flow field around cylinder, results show that large-scale coherent structure extracted could effectively illustrate vortex shedding process, and also be used for evaluating impact of incident angle on flow field; Chaab and Tachie[17] analysed flow characteristics of three-dimensional wall jets, primarily including average velocity, turbulent intensity and Reynold stress in different sections, results indicate that low-order modes contribute more to turbulence statistics in self-similar region than in developing region. The Reynolds shear stress is the biggest benefactors of the low-order modes contribution while the wall-normal turbulence intensities is the least; Shinneeb et al.[4, 18] carried POD to quantitatively extract large-scale coherent structures from flow fields in space confined by free surface and wall, vortex recognition algorithm was employed to quantitatively analyse size and number of vortexes, results show that a number of vortexes with different sizes and directions appear in waterjet field, and vortexes number in axial direction decreases as size increases, meantime, vortexes size in vertical plane is affected by water depth.

In present work, PIV velocity measurements is performed on different sections to investigate double-side wall confined waterjet flow, POD is employed to decompose PIV data to obtain coherent structures in different modes. A comparative analysis among different modes in different cross-sections is performed, Large-scale coherent structure in low-order modes is used to reconstruct flow field, and compares with original instantaneous flow field.

2. Experimental Rig

2.1. facilities setup

In order to accurately measure instantaneous flow fields in double-wall confined space, a complete measurement system is indispensable. As shown in Figure 1, the entire experimental measurement system is visibly divided into three major test systems, named by water circulation system, pressure-flow test system and automatic PIV test system.

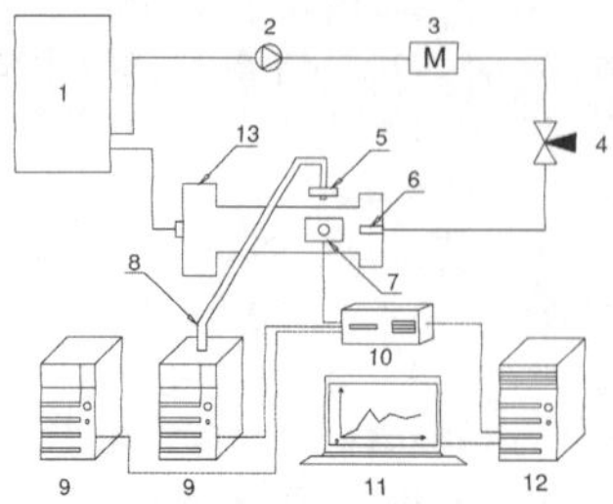

Figure1. Schematic diagram of experimental measurement

1. Water tank 2. Centrifugal pump 3. Electromagnetic flowmeter 4. Valve 5. Battery of lens 6. Nozzle 7. CCD camera 8. Light guide 9. Double YAG pulsed laser 10. Synchroscope 11. Graphoscope 12. Mainframe 13. Experimental bench

Water circulation system mainly consisted of a high-pressure centrifugal pump with flow rate of 46 m3/h and head of 240 m, frequency-modulated system, water tank with volume of 3 m3 and covered reservoir. The demand of water flow rate in experimental process is relatively small, therefore water tank capacity is sufficient to satisfy experimental operation requirement; HBM data acquisition system, KROHNE electromagnetic flowmeter and pressure transducer makes up pressure-flow test system. The electrical signal of flow rate and pressure collected by HBM data acquisition system (sampling frequency: 200 Hz) is directly transmitted to computer analyses; the automatic PIV test system produced by TSI corporation is named as Ultra-Ⅱ (pulsed laser power: 120 mJ/pulse; pulsed frequency: 30 Hz; sweep scan: 3.75/s). The figure signal of experimental bench mainly comes from the high resolution image caught on CCD camera (pixel resolution: 1600×1200 pixels; frame rate: 32 fps; image size: 200 ×200 mm; measured speed range: 0~100 m/s).

2.2. Nozzle and test region
The shape and size of waterjet nozzle have significant influence on turbulent velocity distribution in double-wall confined space, therefore, for the purpose of obtaining a satisfactory jet fluid, the type and size of waterjet nozzle are selected according to research results of Jegaraj and Babu[19].

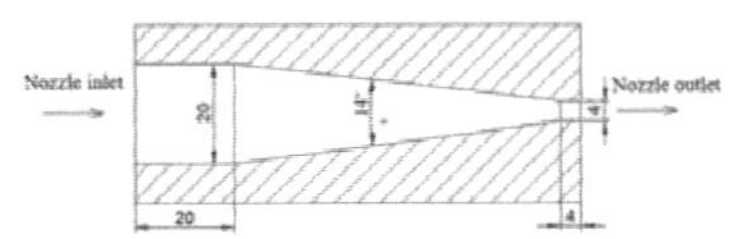

Figure2. Profile of nozzle

In order to eliminate velocity boundary layer influence on nozzle exit velocity, a calculated region in front of nozzle is added to reach a full development status for waterjet, and then flowing into the double-side wall confinement; For the sake of making the velocity field retaining a uniform distribution, well contraction and reducing the pipeline pressure loss, it is very necessary to increase a section of shrink tube between the tube outlet and the nozzle inlet, therefor, the contraction angle is β =14O and nozzle diameter is D=4 mm.

Physical model is extracted from actually technological application and material processing, therefore, the size of restricted direction in double-side wall confinement is much smaller than that unrestricted direction. For purpose of reducing the influence of unconfined space, the horizontal plane of plexiglass water tank is designed as 1×1 m. The confined width 60mm is the distance between upside wall and downside wall. The definition of dimensionless width is the ratio of confined width to the nozzle outlet diameter. Then the dimensionless width is 15. Measuring regions of flow field are marked in three-dimensional coordinate, as showed in Figure 3:

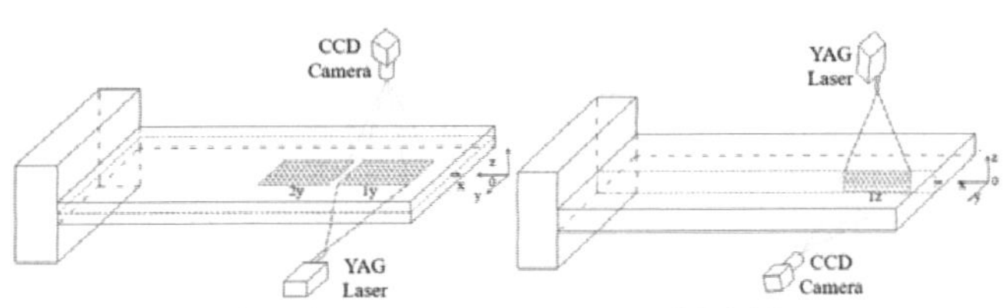

Figure3. Illustration of plexiglass water tank and measuring regions.

a) The measuring region of horizontal plane; b) The measuring region of vertical plane

Table 1. The measuring regions of flow field are shown

Plane	1y	2y	The distance 1y and 2y	1z
Size	90×130 mm	90×130 mm	8 mm	120×170 mm
Spatial discretization	98×132	98×132		73×93

3. Results and discussion

3.1. modes analysis

The number of POD modes based on velocity and vorticity on confined plane and unconfined plane respectively is 160, energy contribution rate of each mode is evidently different. The lower-order modes with a larger energy contribution rate, represents a large-scale coherent structure. On the contrary, the physical meaning with the high-order mode is small-scale fluctuation in flow fields. The first hundred velocity and vorticity POD modes contribution to TKE is shown in Figure 4a and 4b, respectively. In order to intuitionally show the change of contribution to TKE for each pod modes, horizontal axis is plotted with function $\log_{10}$ ().

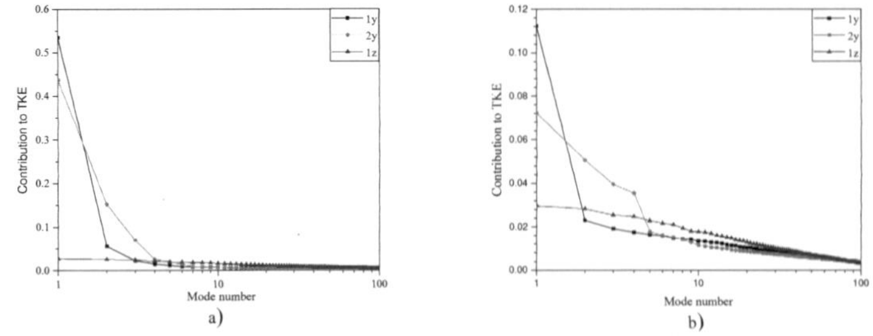

Figure 4. Contribution of each mode to TKE. a) Instantaneous velocity fields in different plane. b) Instantaneous vorticity in different plane

The first pod mode with the largest contribution to TKE dominates turbulent fluctuation development. In the case of velocity fields in different planes (Figure.4a), the first pod mode contributes 53% and 44% of TKE respectively on 1y plane and 2y plane, which is much higher than corresponding the second and third pod mode. Subsequently TKE contribution decreases rapidly, reaches less than 2% from the seventh mode. The above-mentioned pattern on unconfined plane also is clearly reflected in vorticity on unconfined plane. Thacker et al.[20] explaining the large difference between the first mode and the others is that the first mode is exclusively associated with global velocity fluctuation, while the others are associated with a large-scale vortex emission and turbulent motion. However, the above-mentioned pattern completely disappears on confined plane. The first pod mode for velocity field and vorticity field on confined plane respectively contributes 2.7% and 3%, and is basically equal to the second and third pod model. Energy distribution pattern is not the same between confined plane and unconfined plane, that is mainly because that global turbulent fluctuation on unconfined plane has been fully development. On the contrary, waterjet flow space is limited by a symmetrical double-side wall, therefore the law of global turbulent fluctuation is obviously constrained on confined plane.

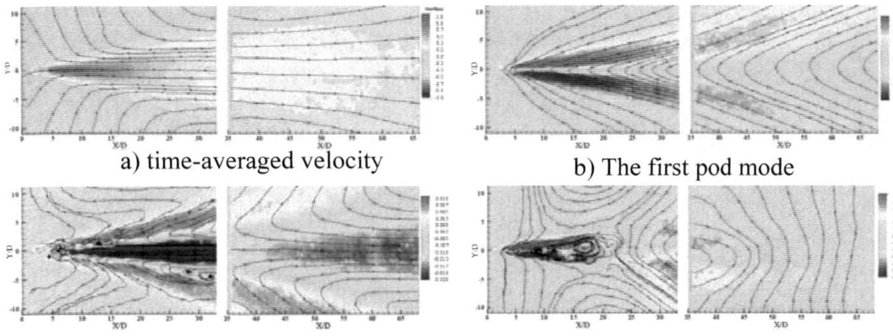

a) time-averaged velocity b) The first pod mode

c) The second pod mode d) The third pod mode

Figure 5. The axial time-average velocity and the first three modes of axial velocity on unconfined plane

According to POD physical meaning in flow fields, multi-scale structure could be characterized by different pod modes. Time-averaged velocity of waterjet flow field dominates overall flowing law considered as the zero order mode of flow field. Consequently, the first three modes of axial velocity on unconfined plane are presented in Figure.5. Obviously, the axial time-averaged velocity contour is symmetrically distributed in centreline of nozzle axis, and diffusion angle of waterjet is 290. The first pod mode of velocity fields with 53% of TKE is shown in Figure.5b. it can indicate that axial velocity contour is divided into negative zone and positive zone by nozzle axis, which is exclusively associated with global velocity fluctuation; Making a comparison between Figure.5c and Figure.5d, we can indicate that the negative zone and positive zone of axial velocity distribution are gradually mixed each other and the size of large-scale structure decreases as the number of modes increases, what's more, at initial stage of waterjet, small-scale structures are generated and is gradually increasing in downstream direction.

The vorticity is defined as the curl of velocity, and used to measure vortex strength and direction, therefore, the high-intensity vorticity means high energy consumption. As shown in Figure.6, vorticity distribution is mainly concentrated near the nozzle axis on unconfined plane (1y) and is fully symmetrical, which means that turbulent energy consumption is mainly generated in that zone. Making a comparison among figure.6b, 6c and 6d, the negative zone and positive zone of vorticity contour are gradually mixed each other, and are basically completely mixed each other in figure 6d. In consideration of the analysis about axial velocity contour in Figure. 5, the number of large-scale structure corresponding to low-order modes is less, therefore the interaction between vortices is small and uniform distributed. Nevertheless, the interaction between the vortices with more quantity and small size is intense in the high-order modes, hence the mix of negative and positive zone is complex in Figure.6d.

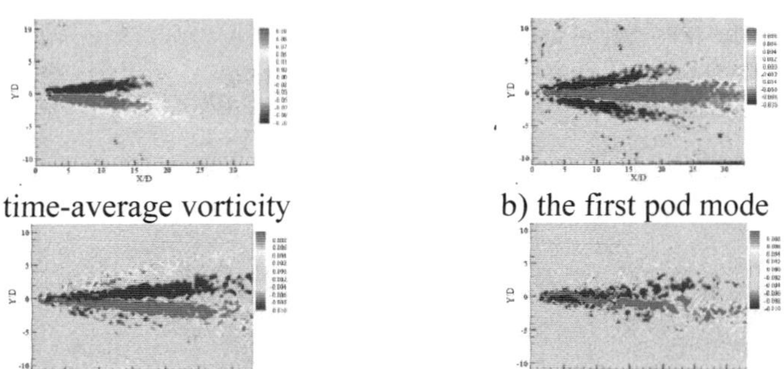

a) time-average vorticity b) the first pod mode

c) the second pod mode d) the third pod mode

Figure 6. The time-average vorticity and the first three modes of vorticity on unconfined plane

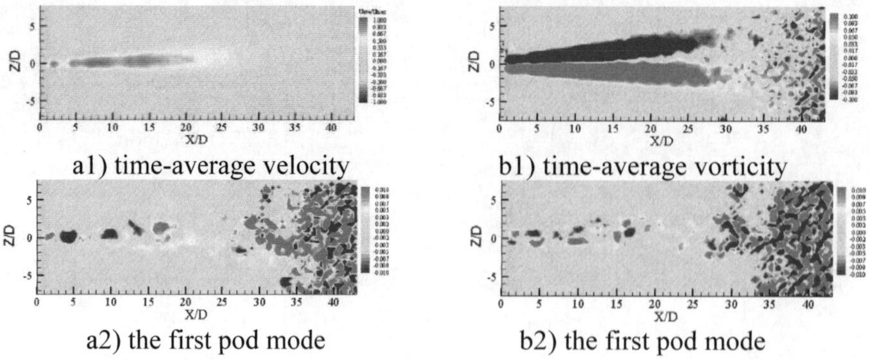

Figure 7. The velocity (a) and vorticity (b) contour on confined plane

The first pod mode of axial velocity and vorticity on confined plane are presented in Figure.7. Time-averaged velocity contour on confined plane is basically similar to that on unconfined plane, but diffusion angle of waterjet is significantly reduced comparing with that on unconfined plane. The mainly reason is that waterjet on confined plane is obviously restricted by double-side wall, as a consequence, the diffusion has a significant change; At the same time, making a comparison among Figure.7 b1, a2 and b2, the velocity and vorticity distribution in region X/D=0~33 show a certain regularity, but in region X/D=33~43 is completely mixed together. Hence waterjet diffusion on confined plane in region X/D=0~33 is not obvious effected by double-side wall confinement.

3.2. Analysis of reconstruction flow field on unconfined plane

Ignoring the high-order modes with low energy contributions, and then instantaneous velocity fields are accurately reconstructed by using the low-order modes with high energy contributions. Using the first m modes to reconstruct instantaneous velocity fields on confined plane and on unconfined plane, the truncation error is shown in Figure 8:

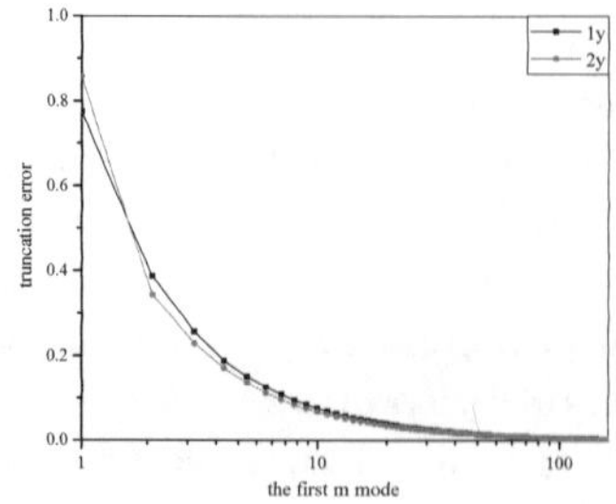

Figure 8. The truncation error of reconstruction for instantaneous flow fields

It is interesting to note that when the first m modes are used to reconstruct instantaneous velocity fields, the truncation error is greatly affected by the value of m in Figure 8. When truncation error is reduced to 3%, the number of modes required for reconstruction of velocity field is 24 on 1y plane and 21 on 2y plane; Therefore, the fewer pod modes can reconstruct the flow field with high-precision achieving simplifying model.

The number of modes used for POD reconstruction was selected based on a target of 70% energy content. The reconstructed velocity field and original velocity field is shown in Figure.9 and Figure.10:

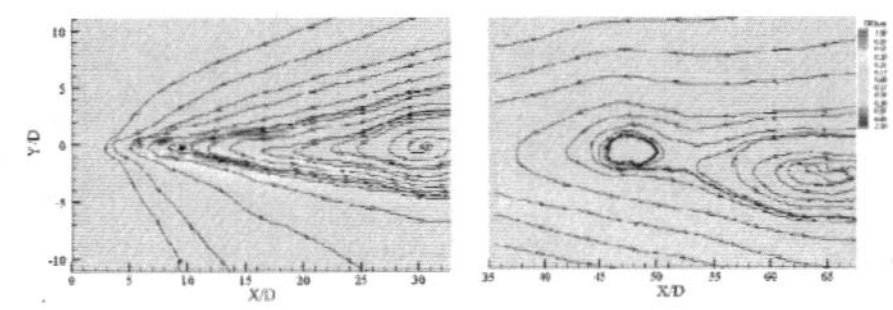

Figure 9. The POD-reconstructed velocity field

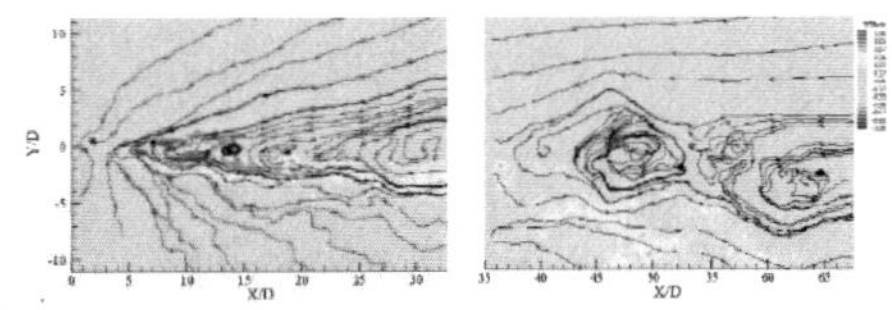

Figure 10. The original velocity field

The significant parameters of vortices such as size, number, location and direction, are basically the same in Figure.9 and Fiure.10, indicating that reconstructed velocity field can truly reflect original field. The streamline of the reconstructed flow field is much smoother, that is because that small-scale coherent structure is filtered, while large-scale coherent structure is preserved; The vortices is generated in strong turbulence region formed at nozzle outlet, the size of vortices gradually increases as the vortices moving downstream, but the direction is not changed.

4. Conclusion

In present study, flow characteristics on unconfined plane and confined plane in double-side wall confinement is tested using PIV, POD is employed to decompose and reconstruct instantaneous velocity fields, the low-order modes of axial velocity and vorticity on unconfined plane and confined plane is analysed. The following conclusion is a summary of this investigation.

(1) On account of existence of a symmetrical double-wall, velocity and vorticity distribution mainly reflecting in low-order modes is seriously affected. Contributions to total energy for the first three modes is basically the same on confined plane, nevertheless, it is obviously different from the pattern, which is mainly due to the fact that waterjet on confined plane cannot be fully developed, expressed on unconfined plane.

(2) Turbulent energy consumption through analysing the low-order modes of vorticity is mainly symmetrically distributed near nozzle axis, waterjet diffusion on confined plane is limited by two side wall and couldn't be fully developed, but this feature is not exhibited on unconfined plane.

(3) Using the first m-order modes contained 70% of TKE to reconstruct instantaneous velocity field on unconfined plane, results indicate that vortices are generated in strong turbulence region formed at nozzle outlet, vortices size gradually increases as vortices moving downstream.

Acknowledgments

This work is supported by National Natural Science Foundation of China (Grant No. 51506014), the Fundamental Research Funds for the Central Universities (Grant No. 2018CDQYDL0052, 106112016CDJXY145501), the Chongqing Research Program of Basic Research and Frontier Technology (Grant No. cstc2016jcyjA0486), the China Postdoctoral Science Foundation (Grant No. 2016M592638, 2017T100677), the National Key Basic Research Program of China (Grant No. 2014CB239200) and the Chongqing Postdoctoral Research Program (Grant No. 2016073).

References

[1] T.J. Kim, and T.J. Labus, Influence of basic jet parameters and physics of abrasive water jet cutting, Proceedings of the WJTA, 1995.
[2] D.A. Summers, Waterjetting technology, E & FN Spon, 1995.
[3] P. Hreha, Monitoring of focusing tube wear during Abrasive Waterjet (AWJ) cutting of AISI 309,

Metalurgija, 53(2014): 533-536.
[4] A.M. Shinneeb, R. Balachandar, J.D. Bugg, Confinement Effects in Shallow Water Jets, J. Hyraul. Eng, 137 (2006) 300-314.
[5] J.P. Raiford and A.A. Khan, Investigation of circular jets in shallow water, J. Hyraul. Res, 47 (2009) 611-618.
[6] M.N. Chowdhury, A.A. Khan, F.Y. Testik, Numerical investigation of circular turbulent jets in shallow water, J. Hyraul. Eng, 143 (2017).
[7] J. Wang, and C. Wang, Heat transfer and flow characteristics of a rectangular channel with a small circular cylinder having slit-vent vortex generator, Int. J. Therm. Sci, 104 (2016) 158-171.
[8] T. Rothenfluh, M.J. Schuler, P.R.V. Rohr, Experimental heat transfer study on impinging, turbulent, near-critical water jets confined by an annular wall, J. Supercrit. Fluid, 77 (2013) 79-90.
[9] J. Borée, Extended proper orthogonal decomposition: a tool to analyse correlated events in turbulent flows, Exp. Fluids, 35(2003) 188-192.
[10] Y.M. Shim, R.N. Sharma, P.J. Richards, Proper orthogonal decomposition analysis of the flow field in a plane jet, Experimental Thermal & Fluid Science Exp. Therm. Fluid. Sci, 51 (2013) 37-55.
[11] C.D. Pokora, and J.J. Mcguirk, Stereo-PIV measurements of spatio-temporal turbulence correlations in an axisymmetric jet. J. Fluid. Mech, 778 (2015) 216-252.
[12] Y. Inoue, S. Yamashita, K. Kondo, The ultrasonic velocity profile measurement of flow structure in the near field of a square free jet. Exp. Fluids, 32 (2002) 170-178.
[13] J.L. Lumley, The structure of inhomogeneous turbulent flow, Atmospheric Turbulence & Radio Wave Propagation, 166 (1967).
[14] A.M. Shinneeb, R. Balachandar, J.D. Bugg, Analysis of Coherent Structures in the Far-Field Region of an Axisymmetric Free Jet Identified Using Particle Image Velocimetry and Proper Orthogonal Decomposition, J. Fluid. Eng, 130 (2008) 151-163.
[15] O. Huerta, D.M. Hernandez, I. Choutapalli, Proper Orthogonal Decomposition of a Turbulent Pulsed Jet, 32nd Aiaa Applied Aerodynamics Conference, 2013.
[16] B.W.V. Oudheusden, Phase-resolved characterization of vortex shedding in the near wake of a square-section cylinder at incidence, Exp. Fluids, 39 (2005) 86-98.
[17] M. Agelin-Chaab, and M.F. Tachie, Characteristics of Turbulent Three-Dimensional Wall Jets, J. Fluid. Eng,. 133 (2011) 021201.
[18] A.M. Shinneeb, J.D. Bugg, R. Balachandar, Coherent Structures in Shallow Water Jets, J. Fluid. Eng, 133 (2011) 011203.
[19] J.J.R. Jegaraj, and N.R. Babu, A strategy for efficient and quality cutting of materials with abrasive waterjets considering the variation in orifice and focusing nozzle diameter, Int. J. Mach. Tool. Manu, 45 (2005) 1443-1450.
[20] L. Sirovich, Turbulence and the dynamics of coherent structures part i: coherent structures, Q. Appl. Math, 45 (1987) 561-571.
[21] Thacker, Aubrun, Leroy, Devinant, Experimental characterization of flow unsteadiness in the centerline;plane of an Ahmed body rear slant, Exp. Fluids, 54 (2013) 1-16.

IOP Conf. Series: Journal of Physics: Conf. Series **1064** (2018) 012057 doi:10.1088/1742-6596/1064/1/012057

Application of Computational Fluid Dynamics in Industrial Crystallization

Shumin Li*

School of Management, Tianjin University of Technology, Tianjin, China.

*Corresponding author e-mail: 1375861492@qq.com

Abstract. Computational Fluid Dynamics (CFD) simulates a multi-phase flow field by means of numerical calculations and becomes a knot.A powerful tool for crystal process simulation. This paper summarizes the application of CFD technology in the coupling of solving methods, crystallization phenomenon, reaction crystallization process, different operation modes and proportional amplification in the simulation of industrial crystallization, and points out that the development of a suitable model will have a high application value.

1. Introduction

Computational Fluid Dynamics (CFD) is a numerical simulation technique that uses fluids as a research object. Computer simulations provide information about certain fluids under certain conditions. CFD contains a variety of knowledge and related technologies such as mathematics, computer science, engineering, and physics. It integrates these knowledge and technologies on a specific problem and forms an analog model of fluid flow [1].

The purpose of the CFD simulation is to make predictions and gain knowledge of the issues to be studied. The theoretical prediction comes from the solution of the mathematical model rather than the result of an actual physical model. The mathematical model is mainly composed of a set of differential equations. The solution of these equations is the result of the CFD simulation. CFD is an advantageous tool for studying various fluid phenomena and for designing, operating, and studying various flow systems and flow processes. With the rapid development of computer science, modern numerical technology, and fluid mechanics theory, CFD is used to model chemical processes, the design and amplification of chemical reactors and crystallizers are examples of the combination of modern high-tech and traditional science.

2. Introduction to CFD Technology

The basic idea of numerical solution using CFD technology [2] is to use a series of finite discrete points (eg, nodes, nodes) that are physically continuous fields (such as concentration field, velocity field, etc.) in space and time coordinates. Instead of a set of values on top of it, set up an algebraic equation (called a discrete equation) on the relationship between the values of the variables at these discrete points, and solve the established algebraic equation to obtain the approximate value of the solved variable.

Content from this work may be used under the terms of the Creative Commons Attribution 3.0 licence. Any further distribution of this work must maintain attribution to the author(s) and the title of the work, journal citation and DOI.
Published under licence by IOP Publishing Ltd

2.1. Basic Control Equations of Fluid Flow

Any flowing fluid follows the law of conservation of mass, the law of conservation of momentum, and the law of conservation of energy [3]. If the flow contains a mixture or interaction of different components, the system also observes the conservation of components. If the flow is in a turbulent state, the system must also comply with the additional turbulent transport equations. The control equation is a mathematical description of these conservation laws. The motion of the fluid can be described by the partial differential equation (equation) or the integral equation (equation) of the conservation law. These equations are called the fluid motion equations. Using certain methods and means, according to the mathematical model of the special properties of fluid flow (such as turbulence model, combustion model, multiphase flow model, etc.), boundary conditions, and initial conditions, a closed system of equations is constructed to mathematically describe specific flow fields and fluids. Flow law [4]. Solving can get the fluid motion information in the system and the physical process information related to the fluid flow.

2.2. Numerical Solutions of Control Equations

Because of the nonlinearity and complexity of the motion control equations, it is almost impossible to directly solve the above equations. However, the information described by the equations is also very important. Therefore, numerical calculation methods must be used to solve these control equations. The commonly used numerical solution methods include finite difference method, finite volume method, finite element method, boundary element method, finite analysis method, spectrum analysis method, numerical integral transformation method, lattice-Boltzmann method, etc. [5]. The finite volume method uses the integral conservation equation as a starting point, divides the calculation area into a series of control volumes, and integrates the differential equation to be solved for each control volume to obtain a discrete equation. The finite volume method is easy to apply to simulate fluid motion with complex boundary regions. It is the most widely used method in computational fluid dynamics. The CFD, PHOENICS, STAR-CD, and FLUENT in the popular large-scale commercial CFD software are all based on the finite volume method.

3. Application of CFD in Industrial Crystallization

The crystallization process is a complex process of heat transfer and mass transfer. In different physical (hydrodynamics, etc.), chemical (component composition, etc.) environments, the control steps of the crystallization process may change and may reflect different crystallization behaviors, all complicating the mathematical model of the crystallization process [6]. The establishment of a suitable mold model is of great significance for optimizing the crystallization operation, designing a new type of crystallizer, and developing a new process for amplifying the crystallization process. The rapid development of CFD technology not only makes solving complex model equations faster, but also promotes the progress of model simulation.

3.1. Coupling solution method of particle balance equation in the process of simulated crystallization

The grain number density distribution is an important characteristic parameter of crystal products, and taking into account the crystal growth rate, nucleation rate, coalescence and fragmentation phenomena in the grain balance equation is an effective way to accurately describe the crystallization process and obtain the grain number density distribution of crystal products. Means. Therefore, how to solve the particle balance equation in the CFD simulation is an important factor in the simulation. The most critical issue is to solve the coupled solution of the particle balance equation and the flow equation.

In general, crystals have a wide range of particle number density distributions in crystallization, and different size crystals have their own flow fields. It is therefore unrealistic to describe crystal grain balance equations with crystals of different sizes in an infinite number of crystal sizes. In order to simplify the particle balance equation, many researchers have used the method of moment-moment to simulate the crystallization process, especially the precipitation crystallization process [7 - 17], and obtained the average properties of the crystal product. In the moment-based method, in order to examine changes in crystal size distribution due to coalescence and fragmentation, consideration should be given to changes in internal coordinates such as particle size in the particle balance equation. Researchers have proposed some methods to examine the aggregation and crushing of cores [18-21]. However, all methods are based on the assumption of uniform suspension.

The solution to the balance equation of the grain number used in the CFD environment needs further development. Another method that may solve the equation of crystallization balance in the simulation is the MUSIG (Multi-Size Group - Model) model. This model was first proposed by Luo in 1998 [22] for the simulation of gas-liquid systems. The MUSIG model defines a two-phase flow. The dispersed phase defines a group of bubbles of different sizes, and simulation can obtain information on the volume fraction distribution of bubbles of different sizes. However, the flow field of the dispersed phase is simulated by the average particle size of the bubble size group, that is, the flow fields obtained by different size bubbles are the same. If this model is applied to a solid-liquid system, the size distribution information of the crystal can be obtained. However, due to the same flow field in the dispersed phase, the motion characteristics of crystals of different sizes cannot be examined. In addition, the phenomenon of crystal coalescence and fragmentation needs further development. The model, in particular, lacks phenomena such as the crystal growth and the difference method.

3.2. Analog Crystallization

In order to accurately describe the crystallization process in the flow field, it is necessary to understand the physical model of the crystallization phenomenon, such as the crystal growth rate, nucleation rate, coalescence and fragmentation phenomena. Years of research have shown that the hydrodynamic state in the mold has a critical influence on the crystallization process [23]. Therefore, many phenomena related to fluid dynamics can be studied using CFD simulations. Z ller et al. [24] studied the growth of single-sized crystals in solution in 1999. Mori et al. [25] simulated the crystal growth rate by considering the mass transfer between the solution and different size crystals. The mass transfer coefficient was estimated by the model that defined the local crystal slip velocity and the local energy consumption rate. The nucleation rate is considered to be directly related to the energy of the moving particles, and the motion can be obtained by CFD simulation. Therefore, Ten Cate et al. [26] proposed a model to study the single crystal in the flow field, and obtained crystal collision frequency and energy information that can be used to predict the secondary nucleation rate and particle wear according to the model. Liiri et al. [27] simulated the nucleation rate due to the collision of crystal-propeller and crystal-crystallizer through CFD. Roelands et al. [28] used CFD to simulate the mixing time of the reactants to determine the nucleation rate in the precipitation crystallization. The CFD study of the nucleation and growth model is still in its infancy, and researchers are still looking for better models to simulate the growth and nucleation in the crystallization process.

The phenomenon of coalescence can be described in the CFD environment because it is related to the local shear rate of the fluid. Madec et al. [29] used the Monte Carlo method directly in the CFD environment to describe the coalescence process in a batch reactor. Some researchers have also developed simulation models for the phenomenon of coalescence in precipitation crystallization [30–31].

3.3. Simulation reaction crystallization process

Most of the CFD simulations are focused on reaction crystallization, or precipitation crystallization. Wei et al. [17] reported for the first time in 1997 that the reaction crystallization process was modeled using the CFD method. The particle size balance equation and the crystallization nucleation and growth kinetics were integrated into the CFD main solver. The simulation results were visually reproduced. The supersaturation of crystallinity in the duct reactor crystallizer and the distribution of the crystal phases, and a comprehensive evaluation of the operating conditions, such as the effect of the feed flow ratio on the particle size distribution of the product. Later, Jaworski et al. [10] simulated the continuous reaction crystallization process. Wei [7], Zauner [9] and Marchisio [11] simulated the semi-batch reaction crystallization process. In reaction crystallization, the reaction time is usually very fast. For example, the reaction time in a semi-batch stirred crystallizer simulated by Vicum et al. [16]

IOP Conf. Series: Journal of Physics: Conf. Series **1064** (2018) 012057 doi:10.1088/1742-6596/1064/1/012057

is only 9 s to 10 s, which is much less than the mixing time. In order to examine the micro-mixing process during the reaction crystallization process, some researchers [11-15] adopted the PDF (Probability Density Function) method, in which the component concentration is described by a multidimensional density function.

3.4. The application of different operating modes in the crystallization process

CFD simulations of the crystallization process of batch cooling solutions are still relatively rare. Yang et al. [32] used CFD to control the batch crystallization process while considering heat transfer. The crystal growth and nucleation rates were taken into account in the simulated grain balance equation, and the 5th-order equations were defined using the method of the order-moment. Kougoulous et al. [33] developed a compart- ment model to simulate the intermittent cooling crystallization process based on the results of CFD simulation of heat transfer and solid suspension. This model has a good application prospect in the simulation of intermittent cooling crystallization process. In order to simulate all crystallization processes, a structure-independent kinetic model should be considered. Karmer [34] and Ten Cate [35] proposed to divide the crystallizer into a defined number of well-defined regions, assuming that the supersaturation, energy consumption rate, solid suspension, and crystal grain number density distribution in these regions are evenly distributed. . Based on these principles, the kinetic parameters of the model can be estimated from experiments or from simulation techniques, so the simulation of the industrial dimension model can be used to predict the crystallization process over the entire size range. In order to simulate the continuous crystallization process of incompletely mixed suspension, Sha et al. [36] assumed that the distribution of crystals of different sizes in the crystallizer is different. It is recommended that the particle balance model considers the particle size distribution, the product removal position, and the difference in the crystallizer. The relationship between. Using this method, the influence of crystallizer structure, mixing strength and product removal position on the continuous crystallization process can be investigated.

3.5. Simulation of the proportional amplification and mixing for the crystallization process

Other CFD simulation studies are directed at the structure of the crystallizer. Since the crystallizer structure seriously affects the crystallization process, it is useful to study the magnification of the crystallizer in the CFD environment. In addition to its economic advantages, CFD simulations can examine crystallization processes in molds of different sizes and provide a reference basis for the design of the mold. Zauner et al. [8, 37] introduced the model of fine-particle-mixing and intermediate-size mixed-bonded-grain balance to study the effects of mixing conditions on crystal size distribution in the reaction crystallization process. Wei et al. [38] studied the effect of energy input on the reaction crystallization process with different size ratios using the grain-equilibrium equation model with a step-size moment. Since the basic model of the simulated crystallization process is still in development, the scaled-up results have not been satisfactory. The mixing in the crystallizer is an important factor affecting the crystallization process. The crystal suspension in the crystallizer and the distribution of the suspension density are important information for controlling the crystallization process. Supersaturation and the corresponding crystal growth and nucleation rates are directly affected by the local crystal suspension density. Sha et al. [36, 39-40] studied the relationship between the suspension density distribution in the crystallizer and the suspension density distribution at the outlet of the product using a hierarchical function. Therefore, the effect of the suspension density distribution on the size of the crystal product during continuous crystallization was simulated.

4. Conclusion

CFD technology is widely used in many fields. In some fields, it has become a mature tool for research and design. However, the application of CFD to the industrial crystallization process is still in its infancy. Due to the complexity of the crystallization process, it is difficult to conduct a comprehensive study of this process with general test methods and thus no reliable results can be obtained. Therefore, the application of computational fluid dynamics technology in industrial crystallization process, research and development of a suitable model, so that the basic phenomena and basic laws in the crystallization process can be studied in the multi-phase flow field, thereby enhancing the scientific and reliability of the research results. With high theoretical value and wide application value, its development prospect is very broad.

References

[1] Patanka, translated by Guo Kuanliang. Numerical methods for heat transfer and fluid flow [M]. Hefei: Anhui Science and Technology Press, 1984 .
[2] Tao Wenhao. Numerical Heat Transfer (Second Edition) [M]. Xi'an: Xi'an Jiaotong University Press, 2001.
[3] Ren Yuxin, Chen Haiyu. Computational Fluid Dynamics [M]. Beijing: Tsinghua University Press, 2006
[4] Zhong Yingjie, Du Jinyan, Zhang Xuemei. CFD technology and its application in modern industry [J]. Journal of Zhejiang University of Technology, 2003, 31(1): 284-289.
[5] Jiang Chunbo, Zhang Yongliang, Ding Zeping. Computational Fluid Dynamics [M]. Beijing: China Electric Power Press, 2007.P.G. Clem, M. Rodriguez, J.A. Voigt and C.S. Ashley, U.S. Patent 6,231,666. (2001)
[6] Lu Jie, Wang Jingkang. Reaction crystallization (precipitation) research progress [J]. Chemical Engineering, 1999, 27(4) : 24 - 27.
[7] Wei H, Zhou W, Garside J. CFD modeling of precipitation process in a semi - batch crystallizer [J]. Ind Eng Chem Res, 2001, 40: 5255- 5261
[8] Zauner R, Jones A G. Scale - up of continuous and semi - batcprecipitation processed [J]. Ind Eng Chem Res, 2000, 39: 2392 -2403.
[9] Zauner R, Jones A G. Mixing effects on product particle characteris┐ tics from semi -batch crystal precipitation [J]. Trans IChemE, 2000, 78(A) : 894 - 902.
[10] Jaworski Z, Nienow A W. CFD modeling of continuous precipitati┐ on of barium sulphate in a stirred tank [J]. Chem Eng Journal, 2003, 91: 167 - 174.
[11] Marchisio D L, Barresi A A, Fox R O, Simulation of turbulent pre┐ cipitation in a Semi -batch Taylor - Couette reactor using CFD[J]. AIChE Journal, 2001, 47(3) : 664 - 676.
[12] Marchisio D L, Fox R O, Barresi A A, et al. On the comparisonbetween presumed and full PDF method for turbulent precipitation[J]. Ind Eng Chem Res, 2001, 40: 5132 - 5139.
[13] Marchisio D L, Barresi A A. CFD simulation of mixing and reac┐ tion: the relevance of the micro - mixing model [J]. Chem EngSci, 2003, 58: 3579 - 3587.
[14] Wang L, Fox R O. Application of in situ adaptive tabulation toCFD simulation of nano- crystal formation by reactive precipitation[J]. Chem Eng Sci, 2003, 58: 4387 - 4401.
[15] Falk L, Schaer E. A PDF modeling of precipitation reactors [J]. Chem Eng Sci, 200156:2445 - 2457.
[16] Vicum L, Ottinger S, Mazzotti M, Makowski L, Baldyga J. Multi- scale modeling of a reactive mixing process in a semibatch stirredtank [J]. Chem Eng Sci, 2004, 59: 1767 - 1781.
[17] Wei H, Gerside J. Application of CFD modeling to precipitationsystems[J]. Trans IChemE, 1997, 75(A) : 219 - 227.
[18] Hounslow M J, Ryall R L, Marshall V R. A discretized populationbalance for nucleation, growth, and aggregation[J]. AIChE Jour┐ nal, 1988, 34(11) : 1821- 1832.
[19] Marchisio D L, Pikturna J T, Fox R O, et al. Quadrature methodof moments for population balance equations[J]. AIChE Journal, 2003, 47: 1266 - 1276.
[20] Diemer R B, Olson J H. A moment methodology for coagulationand breakage problems: Part 2 - moment models and distributionreconstruction [J]. Chem Eng Sci, 2002, 57: 2211- 2228.
[21] Lim Y I, Le Lann J M, Meyer X M, et al. On the solution of pop┐ ulation equation(PBE) with accurate front tracking method in prac┐ tical crystallization processes [J]. Chem Eng Sci, 2002, 57: 3715- 3732.

[22] Luo S. Application of population balance to CFD modeling of bub┐ by flows via the MUSIG model [C]. AEAT - 1096, CFX interna┐ tional, AEA Technology, UK, 1998.
[23] Rielly C D, Marquis A J. A particle's eye view of crystallizer fluidmachanics[J]. Chem Eng Sci, 2001, 56: 2475 - 2493.
[24] Z ller H, Fritsching U, Ulrich J. Simulation and visualization ofthe flow around a single cubic crystal [C]. Proceeding of the 7thinternational workshop on industrial crystallization, 1999, 79 - 84.
[25] Mori Y, Sha Z, Louhi K M, Kallas J. CFD study of local crystalgrowth rate in a continuous suspension crystallizer [J]. Journal ofChemical Engineering of Japan, 2002, 35: 1178 - 1187.
[26] ten Cate A, Derksen J J, Kramer H J M, et al. The microscopicmodeling of hydrodynamics in industrial crystallizers [J]. ChemEng Sci, 2001, 56: 2495 - 2509.
[27] Liiri M, Koiranen T, Aittanaa J. Secondary nucleation due to crys┐ tal - impeller and crystal - vessel collisions by population balancesin CFD modeling [J]. Journal of Crystal Growth, 2002, 237 - 39(3) : 2188 - 2193.
[28] Roelands M, Derksen J, Horst J, et al. An analysis of mixing in atypical experimental setup to measure nucleation rates of precipitati┐ on processes[J]. Chem Eng Technol, 2003, 26(3) : 296 - 302.
[29] Madec L, Falk L, Plasari E. Simulation of agglomeration reactorsvia a coupled CFD/direct Monte - Carlo method [J]. Chem EngSci, 2001, 56: 1731 - 1736.
[30] Baldyga J, Jasinska M. Effects of fluid motion and mixing on parti┐ cle applomeration and coating during precipitation[J]. Chem EngSci, 2005, 60: 2167 - 2178.
[31] Henczka M, Baldyga J, Shekunov B Y. Particle formation by tur┐ bulent mixing with supercritical antosolvent [J]. Chem Eng Sci, 2005, 60: 2193 - 2201.
[32] Yang G, Louhi - Kultanen M, Kallas J. The CFD simulation ofprogrammed batch cooling crystallization [J]. Chemical Engineer┐ ing Transations, 2002, 1: 83 - 88.
[33] Kougoulous E, Jones A G, Wood - Kaczmar M. CFD modeling of mixing and heat transfer in batch cooling crystallizers, - aiding thedevelopment of a hibird predictive compartmental model[J]. TransIChemE, 2005, Part A: 30 - 39.

IOP Conf. Series: Journal of Physics: Conf. Series **1064** (2018) 012058 doi:10.1088/1742-6596/1064/1/012058

Three-dimensional transient simulation of water filling and draining of gravity flow based on Lattice Boltzmann Method

Yanhui Li*, Jin Jiang, Rui Ying, Yucheng Wang, and Wensheng Zhao

Key Lab of Hydraulic Machinery Transients, Key Lab of Hydraulic Machinery Transients, MOE, Wuhan University, Wuhan, China.

*Corresponding author e-mail: lee_yanhui@whu.edu.cn

Abstract. The traditional one-dimensional method of characteristics is available to analyze the transient flow of system level in full pipe flow. As for water filling of empty pipeline and water draining of water-filled pipeline, it is difficult to analyze with this method as the boundary conditions are changing along with the location and time. Meshless Lattice Boltzmann Method was used in this investigation to simulate these two circumstances in the gravity flow water-delivery system respectively. We visually obtained, by this method, the changing process of liquid level, total time of the filling and draining process, flow state around the butterfly valves in the pipe inlet and outlet respectively. Meanwhile, time history data of water flow rate and submersed fraction on the cross section behind the valves were extracted as well. CFD method used in analysis of three-dimensional pipeline flow in system level is an innovation of this research that it not only makes the result more accurate, but also details and visualizes the local transient flow state in pipeline system of gravity flow.

1. Introduction

Pump flow and gravity flow are two kinds of de-livery modes commonly in water-delivery pressurized pipeline system. Water-delivery system of pump flow mainly consists of pump stations, pipelines and other hydraulic structures. And in that system, water will be delivered far away with pumps as energy source. Whereas the gravity flow pipeline system delivers water completely depending on the overall drop and its own gravity without pressurized pump. Delivering water with mode of gravity flow is low-cost, economical, and ideal as it maximizes the use of potential energy of overall drop. Ensuring the safe operation of gravity flow pipeline is an engineering challenge including flow rate controlling, water-hammer protection and filling and draining arrangements [1]. Gravity flow is usually simulated with method of characteristics [2]. However, the radial flow field will change obviously with relatively large pipe diameter. And by method of characteristics, the simulation results are not reliable when the pipe are simplified to one-dimensional line. Moreover, the one-dimensional simulation is not available to analyze the transient flow field in the specific parts of pipeline, such as elbows, valves and pumps.

Lattice Boltzmann Method is an efficient algorithm for single-phase and multiphase fluid flow simulation. With simple form, it is suitable for calculation of locality and especially flexible to deal with complicated boundary conditions and multiphase interfaces [3]. As well, this algorithm is such perfectly parallel that it is available to run on computing cluster of large scale. In recent years, the Lattice Boltzmann Method has been widely used in investigation of fluid flow and heat transfer. For instance,

Content from this work may be used under the terms of the Creative Commons Attribution 3.0 licence. Any further distribution of this work must maintain attribution to the author(s) and the title of the work, journal citation and DOI.
Published under licence by IOP Publishing Ltd

it is used in the in-depth study of basic phenomenon of fluid mechanics, such as turbulence [4] and multiphase flow [5]. In addition, by this method, transient conduction and radiation heat transfer problems can be solved [6] as well as bubble growth and detachment investigation [7]. Besides, Lattice Boltzmann Method is also applicable to analyze the mechanism of microscale flow [8].

In this research, simulations of water filling of empty pipeline and water draining of water-filled pipeline in gravity flow pipe system were carried out with meshless Lattice Boltzmann Method. It is challenging and innovative to simulate the transient flow in such a system level.

2. Mathematical models and simulation settings

2.1. Mathematical model of gravity flow

Figure 1 shows the common gravity flow system in which water flows from cross section V-1 to V-2. Pressure and temperature are represented by P_0 and T_0 in V-1, P_1 and T_1 in pipe inlet, P_2 and T_2 in V-2 (horizontal plane of pipe outlet) respectively. The basic level is set on the ground and heights are H, H1 and H2 respectively. Driven by overall drop and its own gravity, the water flows from V-1 to V-2 while valves are opened. To simplify the analysis, some assumptions are made for the system that there are not flash evaporation and other complex phenomena during the gravity flow meanwhile the pipe diameter and fluid density are constant. Also, it does not involve heat transfer, i.e. T_0, T_1 and T_2 are equivalent.

In this research, simulations of water filling of empty pipeline and water draining of water-filled pipeline in gravity flow pipe system were carried out with meshless Lattice Boltzmann Method. It is challenging and innovative to simulate the transient flow in such a system level.

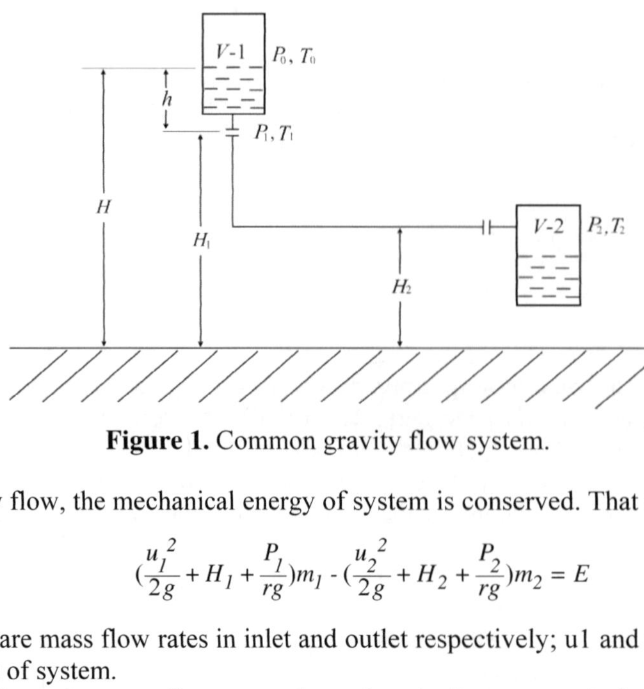

Figure 1. Common gravity flow system.

During the gravity flow, the mechanical energy of system is conserved. That can be expressed as:

$$\left(\frac{u_1^2}{2g} + H_1 + \frac{P_1}{rg}\right)m_1 - \left(\frac{u_2^2}{2g} + H_2 + \frac{P_2}{rg}\right)m_2 = E \quad (1)$$

Where m_1 and m_2 are mass flow rates in inlet and outlet respectively; u1 and u2 the fluid velocities; and E the energy loss of system.

As for full pipe flow, the mass flow rates along the pipeline are equivalent as well as the fluid velocities, while the pipe diameter and fluid density are supposed to be constant.

$$m_1 = m_2 = m \quad (2)$$

$$u_1 = u_2 = u \quad (3)$$

And the energy loss of system involves the head losses of inlet, pipe and outlet. It is expressed as:

$$E = m(\frac{u^2}{2g})(k_1 + \sum \lambda_i \frac{L_i}{D} + \sum \zeta_j + k_2) \tag{4}$$

Where k_1 and k_2 are resistance coefficients in inlet and outlet respectively; Li the equivalent length of each pipe; λ_i the linear resistance coefficient; and ζ_j the coefficient of local resistance.

The total driving force of gravity flow in the form of head is:

$$X = (H_1 - H_2) + (P_1 - P_2)/\rho g \tag{5}$$

2.2. *Mathematical model of Lattice Boltzmann Method*

There are three levels of fluid flow description in physics: macroscopic, microscopic and mesoscopic [9]. Based on the kinetic theory of fluid, the Lattice Boltzmann Method describes the basic characteristics of micro movement and establishes the simplified kinetics lattice model accompanied with completely discrete time and space. Lots of mesoscopic fluid particles are dispersed among lattices and migrate along the lattice lines according to certain rule and collide with each other in lattice points. Their complicated motion obeys the law of dynamics and statistical law. Therefore the macroscopic motion variables of fluid are obtained by statistically averaging the large quantities of particles in flow field.

Compared with the traditional CFD algorithm, the Lattice Boltzmann Method is characterized by meshless solution. It just needs the definition of lattice size. Furthermore, advantages of this method includes: (a) It can perform simulation from aspect of microscopic particles that it is easy to deal with the complex inter-actions between different components and fluid inter-faces without the empirical and semi-empirical formulas. (b) Compared with the nonlinear convective term of N-S equations, the spatial convection process of Lattice Boltzmann Method is linear such that the evolution process is clearer and easier to solve. (c) The pressure is expressed as the equation of state and not iteratively solved with velocity so that this algorithm is relatively simple.

The Lattice Boltzmann equation with single relaxation time operator and volume force term is expressed as follows:

$$f_i(r + e_i \mathrm{d}t, t + \mathrm{d}t) - f_i(r,t) = -\frac{1}{\tau}[f_i(r,t) - f_i^{eq}(r,t)] + \mathrm{d}tF_i \tag{6}$$

The equation above is decomposed into two types of evolution: collision step and migration step.
Collision step:

$$f_i(r + e_i \mathrm{d}t, t + \mathrm{d}t) = f_i^+(r,t) \tag{7}$$

Migration step:

$$f_i^+(r,t) = f_i(r,t) - \frac{1}{\tau}[f_i(r,t) - f_i^{eq}(r,t)] + \mathrm{d}tF_i \tag{8}$$

Where f_i(r,t) is the density distribution function of particles along the direction of velocity e_i at time t and on location r. And d_t is the time step, τ the dimensionless relaxation time, f_i^{eq} the local equilibrium distribution function, and Fi the volume force term.

The local equilibrium distribution function is described as follow:

$$f_i^{eq} = w_i \rho (1 + \frac{3e_i u}{C^2} + \frac{9(e_i u)^2}{C^2} - \frac{3u^2}{2C^2}) \tag{9}$$

As an efficient mesoscopic numerical algorithm in CFD, the Lattice Boltzmann Method is available to simulate the three-dimensional gravity flow with higher level of accuracy.

2.3. Geometric model and simulation settings
Gravity flow system analyzed in this investigation consists of upper water tank, pipeline (500m long) and lower water tank. Overall drop between these two tanks are 300 m and the diameter of the pipeline is constant as 1.5 m. In addition, two butterfly valves are respectively set in the inlet and outlet of pipeline. In simulation of water filling of empty pipeline, the initial height of liquid level in upper tank was set as 15 m, as shown in Fig. 2(a) where the red part represents the water phase and blue part the air phase. In this circumstance, the butterfly valve in the outlet kept open all the time and the one in the inlet opened linearly in 5 s from closure. Meanwhile, in simulation of water draining, the pipeline was full of water while the two tanks were empty at initial moment as shown in Fig. 2(b). Similarly, in this draining circumstance, the butterfly valve in the inlet kept continually open and the one in the outlet also opened linearly in 5 s. Monitoring cross sections were set 3 m downstream away from both valves to extract the changing fluid parameters.

The total simulation times of both circumstances were set as 60 s. Particle-based multiphase model was utilized along with the surface tension set as 0.072 N/m. Besides, the inlet and outlet boundary conditions were set as atmospheric pressure on the top surfaces of upper and lower tank respectively.

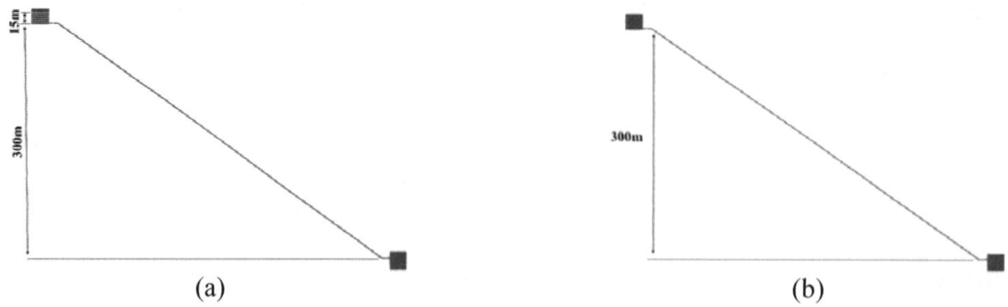

Figure 2. Initial water distributions in two circumstances: (a) water filling; (b) water draining.

3. Results and discussion

3.1. Transient water filling process
The changing process of liquid level in water filling of empty pipeline is indicated in Fig. 3. As the water is filling the pipe, the residual air exhausts downstream to the lower tank and atmosphere outside. Meanwhile the replenishing air automatically flows into the upper tank where liquid level declines gradually.

Flow state changing around the butterfly valve in the pipe inlet in water filling process is shown in Fig. 4. As the valve opens linearly in 5 s, water fills in the entrance pipe gradually and keeps accelerating. The maximum flow velocity is approximately 9.5 m/s when the valve fully opens. Meanwhile flow of local high velocity mainly concentrates on the elbow, butterfly valve and pipe inlet.

Time history data of water flow rate and submersed fraction on the monitoring cross section down-stream away from the upper butterfly valve are indicated in Fig. 5 and Fig. 6 respectively. As for the flow rate, it increases rapidly with large slope between 4 s and 5 s when the valve is almost fully open. This is determined by the characteristics of the butterfly valve whose flow area does not increase linearly though it opens linearly. Afterwards, the flow rate almost in-crease linearly before 31 s. At this stage, water in pipe, driven by its own gravity and the overall drop, accelerates its flow velocity and the liquid level declines quickly in the upper tank. As shown in the curve of flow rate, total time of the filling process is approximately 31 s after which the water flow changes from pure pipe flow to orifice flow in the outlet. Therefore the flow rate decreases and fluctuates. Finally, it maintains stable a level subsequently as the orifice flow becomes steady. Corresponding to the stable level of the flow rate, liquid level in the upper tank also declines steadily.

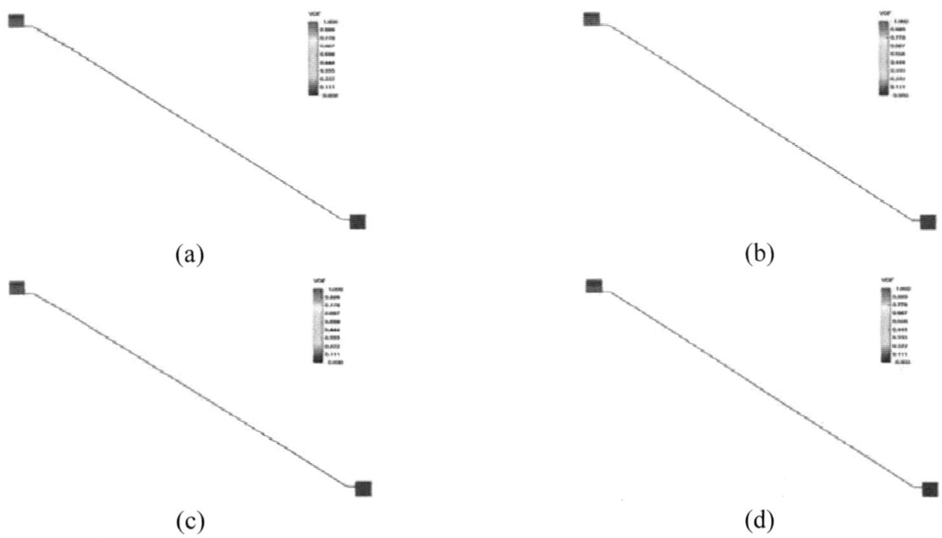

Figure 3. The changing liquid level in water filling process: (a) t=5s; (b) t=15s; (c) t=25s; (d) t=35s.

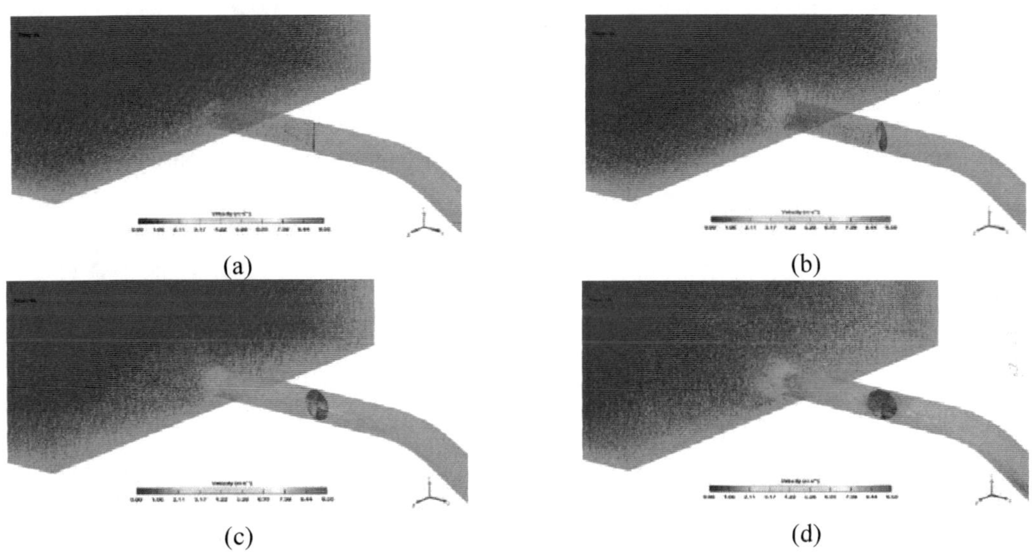

Figure 4. Velocity distribution around upstream valve in filling: (a) t=2s; (b) t=3s; (c) t=4s; (d) t=5s.

Submersed fraction on monitoring cross section, as depicted in Fig. 6, indicates the volume fractions of water and air phases respectively changing on that location. 0 value means it is full of air on this cross section and 100 % full of water otherwise. The increasing stage of the submersed fraction is similar to that of the flow rate approximately between 4 s and 5 s.

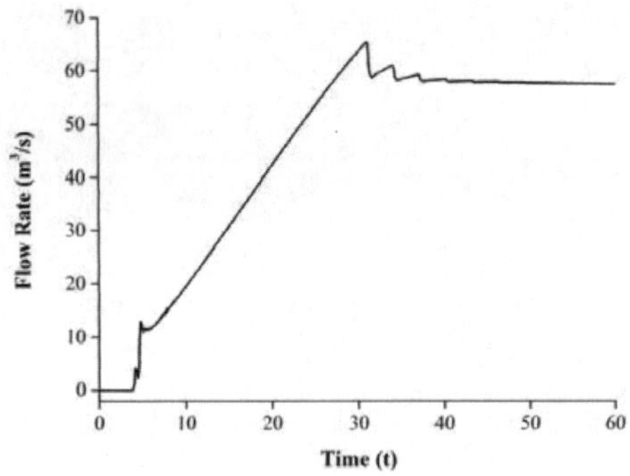

Figure 5. Water flow rate in filling process.

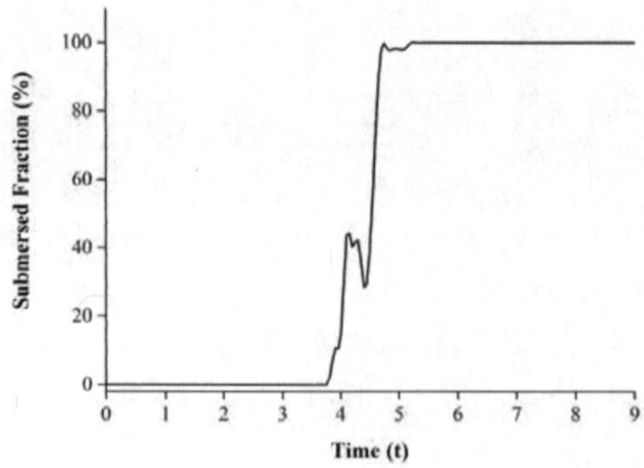

Figure 6. Submersed fraction in filling process.

3.2. Transient water filling process

The changing process of liquid level in water draining of water-filled pipeline is shown in Fig. 7. As the water is draining from the pipe, the replenishing air automatically flows into the upper tank and pipe while the liquid level declines gradually. And the residual air in lower tank exhausts to atmosphere outside.

Flow state changing around the butterfly valve in the pipe outlet in water draining process is shown in Fig. 8. As the valve opens linearly in 5 s, water accelerates its flow velocity while draining from the pipe and discharging into the lower tank. The maximum flow velocity is approximately 28 m/s concentrating on the flow area of butterfly valve. Obviously, in the opening process of valve, magnitude of flow velocity in water draining is larger integrally than that in water filling. It is caused by the initial head pressure in water draining (300 m head) which is larger considerably than that in water filling (15 m head).

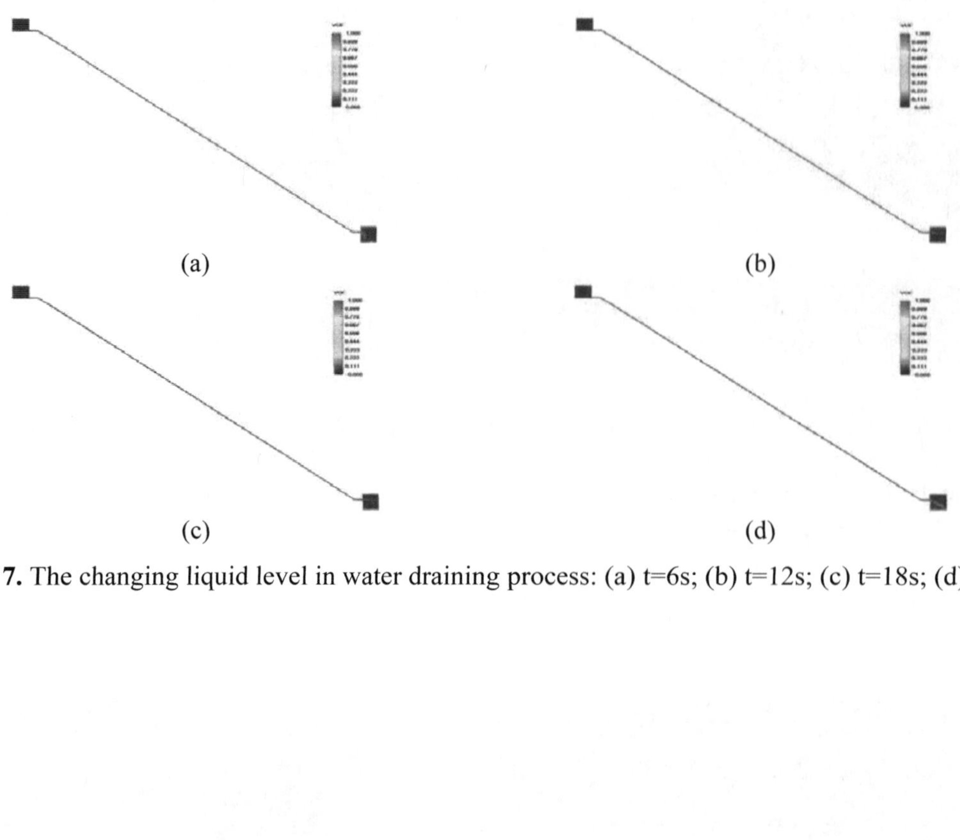

Figure 7. The changing liquid level in water draining process: (a) t=6s; (b) t=12s; (c) t=18s; (d) t=24s.

IOP Conf. Series: Journal of Physics: Conf. Series **1064** (2018) 012058 doi:10.1088/1742-6596/1064/1/012058

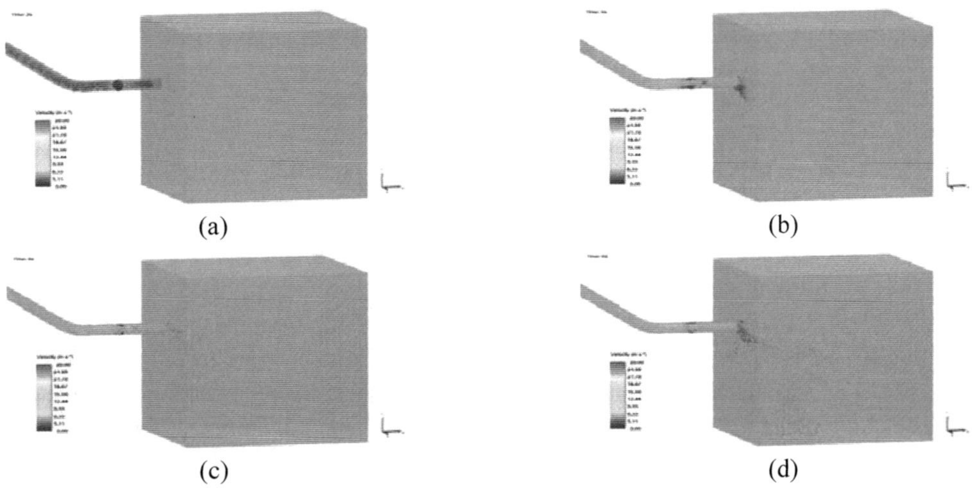

Figure 8. Velocity distribution around downstream valve in draining: (a) t=2s; (b) t=3s; (c) t=4s; (d) t=5s.

Time history data of water flow rate and submersed fraction on the monitoring cross section downstream away from the lower valve are shown in Fig. 9 and Fig. 10, respectively. In particular, the flow rate increases and decreases with parabolic trend up to 66.4 m3/s between 2 s and 28 s. At this stage, it also fluctuates slightly based on the main trend because of the unsteady orifice flow. And the flow rate maintains subsequently stable a level of near-zero valve as the residual water in horizontal outlet pipe flows slowly.

In Fig. 10, the submersed fraction increases rap-idly from 0 to 100 % between 1.7 s and 1.9 s, which also demonstrates the larger magnitude of flow velocity in water draining process. And total time of the draining process is approximately 43 s at which moment the submersed fraction decreases rapidly from 100 % to low value. Finally, the submersed fraction fluctuates constantly at low value because of the slowly flowing of residual water in horizontal outlet pipe.

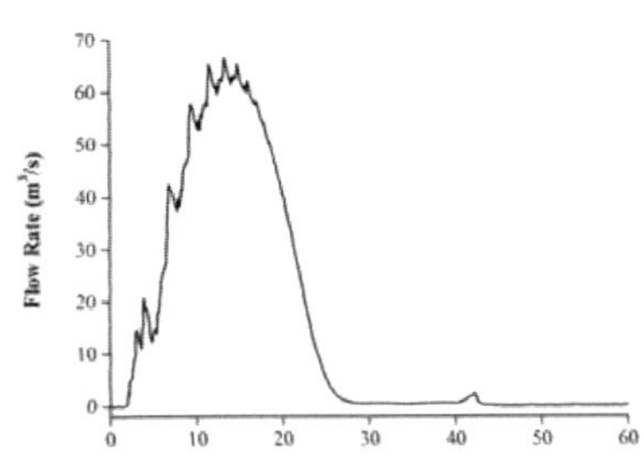

Figure 9. Water flow rate in draining process.

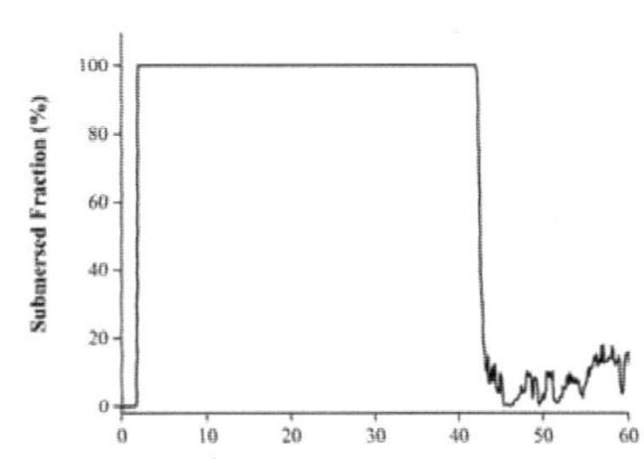

Figure 10. Submersed fraction in draining process.

4. Conclusion

Pipeline system of gravity flow was simulated by CFD algorithm of Lattice Boltzmann Method in this investigation. Two circumstances, water filling of empty pipeline and water draining of water-filled pipeline, were analyzed respectively. Detailed results include the changing process of liquid level, flow field of velocity, total time of the filling and draining process and time history data of water flow rate and submersed fraction as well. It is an innovative investigation to analyze pipeline system with three-dimensional CFD method other than traditional one-dimensional method of characteristics. And simulation results with high accuracy are available by this Lattice Boltzmann Method.

Acknowledgments
This work was financially supported by the National Natural Science Foundation of China [grant number 51409197, 51279145].

References

[1] J. Y. Yu, C. Wu, Y. Yuan, et al., Research on Operation Safety in the Long-Distance Water Transmission Pipelines with Gravity Flow, International Conference on Pipelines and Trenchless Technology, Wuhan, China, 2012.
[2] M. H. Afshar, M. Rohani, Water hammer simulation by implicit method of characteristic, Int. J. Pres. Ves. Pip., 2008, 85(12): 851-859.
[3] S. Chen, G. D. Doolen, LATTICE BOLTZMANN METHED FOR FLUID FLOWS, Ann. Rev. fluid Mech., 1998, 30(1): 329-364.
[4] H. Yu, S. S. Girimaji, L. S. Luo, DNS and LES of decaying isotropic turbulence with and without frame rotation using lattice Boltzmann method, J. Comput. Phys., 2005, 209(2): 599-616.
[5] T. Inamuro, N. Konishi, F. Ogino, A Galilean invariant model of the lattice Boltzmann method for multiphase fluid flows using free-energy approach, Comput. Phys. Commun., 2000, 129(1):32-45.
[6] S. C. Mishra, H. K. Roy, Solving transient conduction and radiation heat transfer problems using the lattice Boltzmann method and the finite volume method, J. Comput. Phys., 2007, 223(1):89-107.
[7] Z. L. Yang, T. N. Dinh, R. R. Nourgaliev, et al., Numerical investigation of bubble growth and detachment by the lattice-Boltzmann method, INT. J. Heat Mass Tran., 2001, 44(1):195-206.
[8] L. S. Kim, H. K. Jeong, M. Y. Ha, et al., Numerical simulation of droplet formation in a micro-channel using the lattice Boltzmann method, J. Mech. Sci. Tec., 2008, 22(4):770-779.
[9] X. He, L. S. Luo, Theory of the lattice Boltzmann method: From the Boltzmann equation to the lattice Boltzmann equation, Physical Review E, 1997, 56(6):6811-6817.

IOP Conf. Series: Journal of Physics: Conf. Series **1064** (2018) 012059 doi:10.1088/1742-6596/1064/1/012059

Numerical analysis of Journal Bearing based on Finite Difference Method with Inhomogeneous Grid

Yiyang Fu, Jian Jin * and Xiaochao Chen

School of Mechanical Engineering and Automation, Shanghai University, Shanghai, China.

*Corresponding author e-mail: jinjian@staff.shu.edu.cn

Abstract. Finite Difference Method (FDM) is one of the most commonly used methods for calculating the pressure distribution of journal bearings, and the division of grid is the key node in computing. For the bearings with surface texture, considering the great difference between the size of texture and bearing, sparser grid may cause the larger calculation errors, but denser grid cause the increasing of computing time and great inconvenient for multi data contrast. In this paper, time consuming in calculating oil-film-pressure distribution has been compared, between using inhomogeneous or homogeneous grid in FDM, with the condition of surface texture on radial journal bearing. Calculation accuracy has also been compared. The result shows the grate shorten by using the inhomogeneous grid in calculation time, with the same accuracy of results in case of small rate of texture area.

1. Introduction

FDM is a commonly used method to calculate the characteristics of journal bearing. In recent years, there are many researches on the optimization of FDM. E. Ko C. [4] introduced a finite difference scheme for Thrust bearing and verified its validity. Zheng S. [5] presented an iterative algorithm of FDM with variable step size successive approximation, and showed the efficient, reliable, stable and convergent of improved method. Han Y [2] provided a virtual grid based method for the Herringbone grid, showed that new method is helpful to improve the accuracy of calculation. Vijayaraghavan D [3] and so on used adaptive algorithm to carry out grid transformation, and proved the new method can obtains the same or better precision with coarser grid which using less calculating time. Li Qiang [7] et al. put forward a new calculation method using watershed dynamic grid algorithm, and showed the consistent with classical calculation results.

However, researches are all about Journal bearings with the condition of smooth surfaces but not contained of textured surfaces condition. In this paper, a multi-scale Inhomogeneous grid is proposed for calculating of journal bearings with textured surface.

When calculating the oil-film-pressure distribution of the journal bearing with smooth surface by FDM, the calculation accuracy can meet the general requirements with the range of 4~6 degrees (same as axial grid spacing) per unit [1]. But in the condition of textured surface, 4~6 degrees may be too sparse to covering texture unit and may affect the accuracy of calculation. If the grid density is increased, the calculation time will increase too. However, it has been proved that texture only in loading zone showed the better lubrication characteristics. Therefore, in order to save calculating time and ensure

Content from this work may be used under the terms of the Creative Commons Attribution 3.0 licence. Any further distribution of this work must maintain attribution to the author(s) and the title of the work, journal citation and DOI.
Published under licence by IOP Publishing Ltd

accuracy, the multi-scale division method using small scale grid in texture area, and large scale grid in smooth area has been promoted in this paper. And the advantage of this method will be obvious when area of texture is relatively small.

2. Geometric model

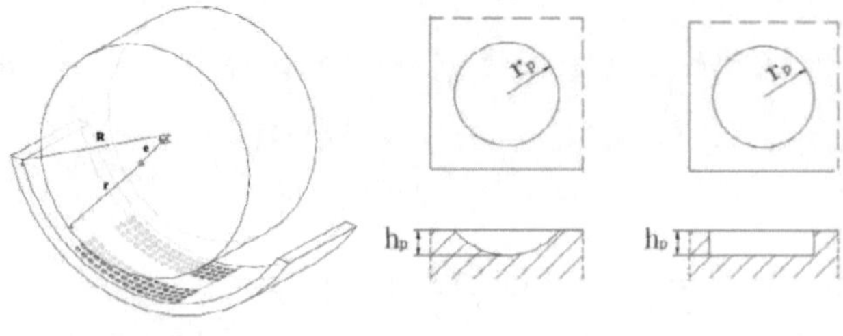

a. Bearing b. Spherical Texture c. Cylindrical Texture

Figure 1. Geometrical model of Bearing and Texture

For radial journal bearing with surface texture (Figure 1.), ϕ denotes the circumferential coordinates if take the place of top of the bearing as the origin of circumference, $\varphi = \phi - \alpha$ denotes the relatively circumferential coordinates if take the place of maximum oil film thickness h_{max} as the origin of circumference, and select the axial axis position as the origin of axial coordinates, λ represents dimensionless axial coordinates, The dimensionless Reynolds equation can be:

$$\frac{\partial}{\partial\varphi}\left(H^3\frac{\partial p}{\partial\varphi}\right) + \frac{\partial}{\partial\lambda}\left(H^3\frac{\partial p}{\partial\lambda}\right) = 3\frac{\partial H}{\partial\varphi} \tag{1}$$

Dimensionless parameters:

Choose half of bearing width $^l/_2$ as denominator of axial coordinates. Choose radius clearance c_0 as denominator of all others parameters.

Now relatively circumferential coordinates φ , deviation angle α, axial coordinates $\lambda = \frac{z}{^l/_2}$, film thickness $H = \frac{h}{c_0}$, diameter of the rotor $D = \frac{d}{c_0}$, bearing width $L = \frac{l}{c_0}$, Texture depth $H_p = \frac{h_p}{c_0}$, Texture radius $R_p = \frac{r_p}{c_0}$, static oil-film-pressure $P = \frac{p}{P_0}$, $\frac{2\Omega\mu}{\psi^2}$ is the reference pressure, Ω is the rotating speed of rotor (rad/s), $\psi = \frac{c_0}{r}$ is relative radius clearance, r is the radius of rotor. In order to facilitate the comparison results, the following physical quantities in this paper, if not stated, are all assumed as dimensionless forms by default.

3. Form of Inhomogeneous grid

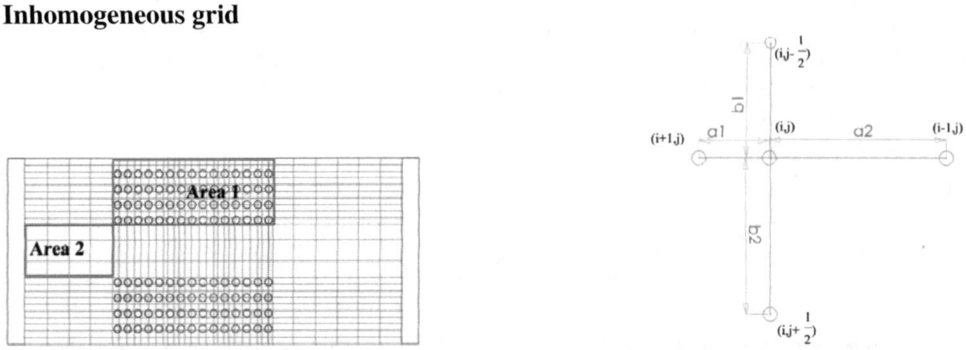

(a) Inhomogeneous grid format (b) Five-nodes Difference Format diagram

Figure 2. Inhomogeneous grid format and Five-nodes Difference Format diagram

Inhomogeneous grid format shows in Figure 2, grid spacing in area without surface texture (Area 1) is $\Delta\phi_1$in circumferential, $\Delta\lambda_1$ in axial. And grid spacing in area with surface texture (Area 2) is $\Delta\phi_2$in circumferential, $\Delta\lambda_2$ in axial. And same area has same grid spacing. For each node in the grid, a five-node difference format has been went up as Figure 2(b), (i, j) is the coordinate of node which to be solved. Differential quotient expression of each derivative term of dimensionless Reynolds Equation is

$$\left[\frac{\partial}{\partial\varphi}\left(H^3\frac{\partial P}{\partial\varphi}\right)\right]_{i,j} \approx \alpha_1 P_{i+1,j} + \alpha_2 P_{i-1,j} - (\alpha_1+\alpha_2)P_{i,j}$$

$$\alpha_2 = 2\frac{a_2 H_{i-\frac{1}{2},j}{}^3 + (a_1-a_2)H_{i,j}{}^3}{a_1{}^2(a_1+a_2)} \qquad \alpha_2 = 2\frac{a_2 H_{i-\frac{1}{2},j}{}^3 + (a_1-a_2)H_{i,j}{}^3}{a_1{}^2(a_1+a_2)}$$

$$\left[\frac{\partial}{\partial\lambda}\left(H^3\frac{\partial P}{\partial\lambda}\right)\right]_{i,j} \approx \beta_1 P_{i,j+1} + \beta_2 P_{i,j-1} - (\beta_1+\beta_2)P_{i,j} \tag{2}$$

$$\beta_1 = 2\frac{b_1 H_{i,j+\frac{1}{2}}{}^3 - (b_1-b_2)H_{i,j}{}^3}{b_2{}^2(b_1+b_2)} \qquad \beta_2 = 2\frac{b_2 H_{i,j-\frac{1}{2}}{}^3 + (b_1-b_2)H_{i,j}{}^3}{b_1{}^2(b_1+b_2)}$$

$$\left(\frac{\partial H}{\partial\varphi}\right)_{i,j} \approx \frac{2\alpha_1}{\alpha_2(\alpha_1+\alpha_2)}H_{i+\frac{1}{2},j} - \frac{2\alpha_2}{\alpha_1(\alpha_1+\alpha_2)}H_{i-\frac{1}{2},j} - 2\frac{\alpha_1-\alpha_2}{\alpha_1\alpha_2}H_{i,j}$$

a_1, a_2, b_1, b_2 is the dimensionless distance between the center node of the five-node difference format and the adjacent four nodes(as Figure 2(b). showed).

Then establish the difference equation of pressure, the overall numerical calculation process is as Figure 3:

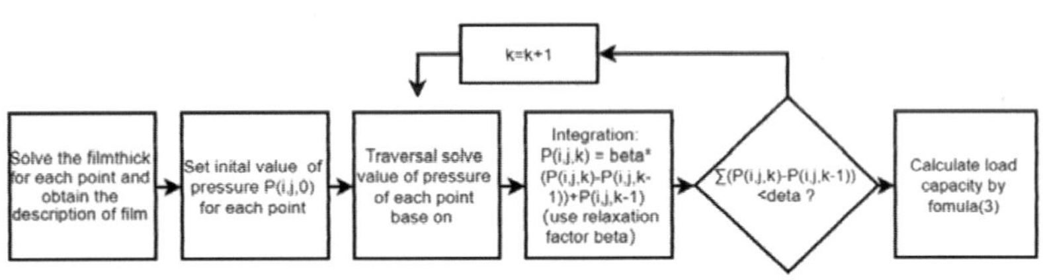

Figure 3. Numerical calculation process

$$F_n = \frac{f_n\psi^2}{\mu\Omega rl} = -\int_{\phi_1}^{\phi_2}\left(\int_{-1}^{1} P d\lambda\right)\cos\phi \, d\phi \tag{3}$$

Among them, the values of $d\lambda$ and $d\phi$ are determined according to the area where the node is located (like $d\lambda_1$ in Area 1, $d\lambda_2$ in Area 2).

4. Comparison of examples

4.1. Condition parameters selection

Radial sliding bearings with oil inlets on both sides, with two bearing bushes $2 \times 150°$ has been chosen in the example (as Figure 1. a), the aspect ratio of the bearing $\frac{l}{d} = 0.6$.And the surface texture distribution is showed in Figure 1.a and Figure 2. (Texture has been arranged at Main bearing area of the lower bearing bush [6]).

Basic parameters of the bearing are shown in Table 1:

Table 1. Basic parameters of the Bearing

Parameter	Symbol	Unit	Value
Bearing diameter	d	mm	20
Bearing width	l	mm	12
Radius difference(R-r)	c_0	mm	0.1
Eccentricity	ε	/	0.6
Texture radius	r_p	mm	0.125
Texture depth	h_p	mm	0.125
Lubricating oil viscosity	μ	Pa·s	0.043

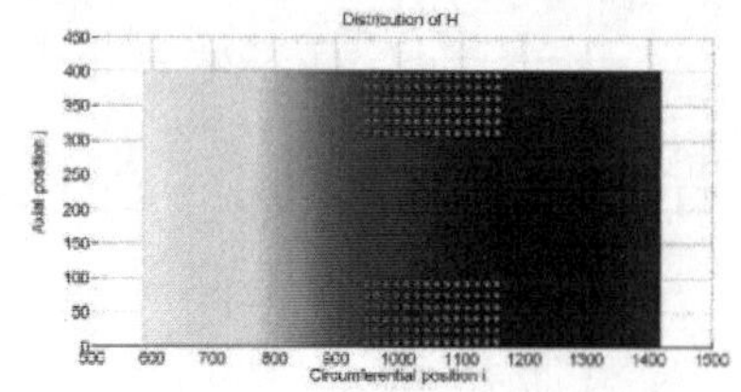

(a). Oil-film-thickness distribution of lower bush

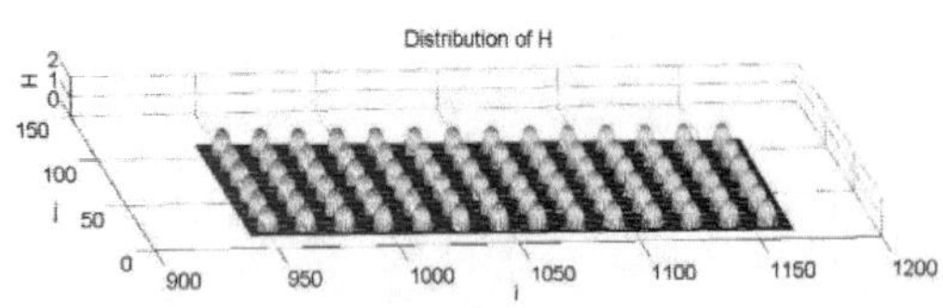

(b). Spherical texture

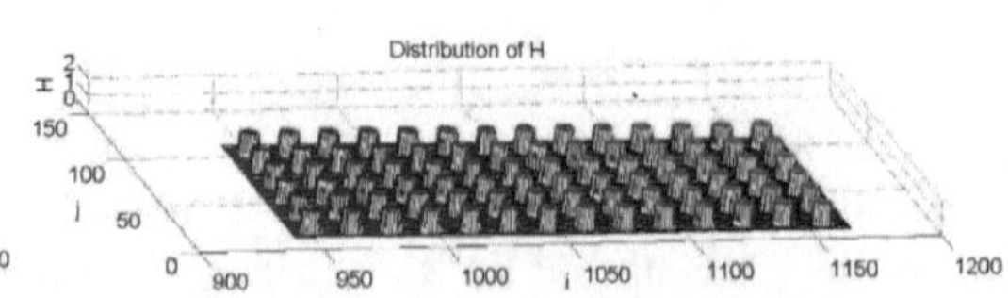

(c) Cylindrical texture

Figure 4. Oil-film-thickness distribution of Texture Area

4.2. Determine the mesh density

First, the minimum numbers of grids that can ensure stable calculation results for smooth surfaces and textured surfaces have been determined before comparing the calculation time. Statistics are made on the calculation time, load capacity F_n, and deviation angle α calculation results here, as shown in Figure 5.

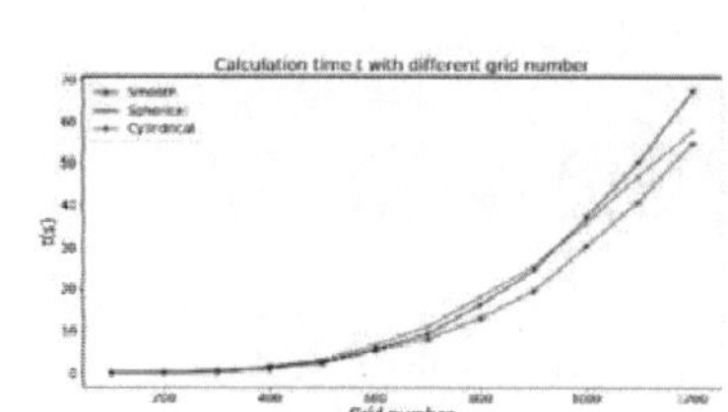

(a). Dimensionless calculation time

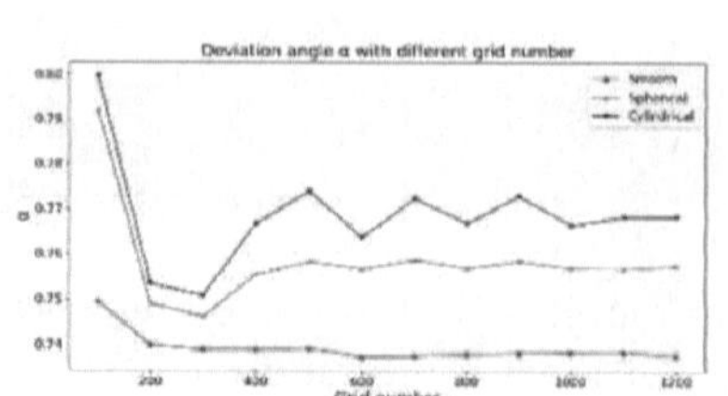

(b). Calculation result of deviation angle

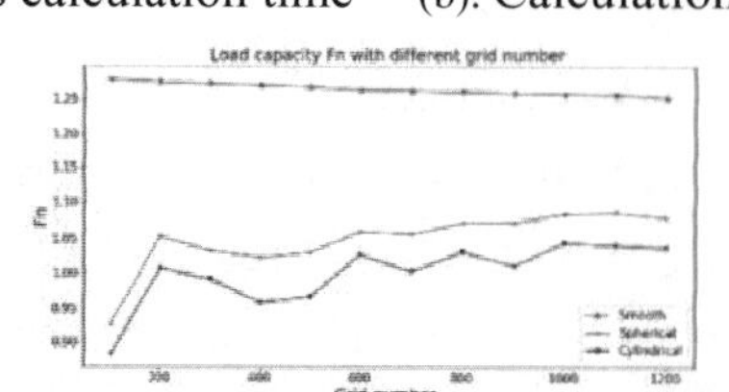

(c). Calculation result of load capacity

Figure 5. Effect of grid density on the calculation

Figure 5. (a) shows the relationship between calculation time $t(s)$ and grid density. As Figure 5. (a) Shows, calculating time can increase thousands of times with the increase of grid density from $100 * 20$ to$1200 * 240$.

Figure 5. (b) (c) shows the calculation result of load capacity F_n and deviation angle α with the grid density increase from $100 * 20$ to $1200 * 240$. As Figure 5. (b)(c) shows, when grid density is larger than$200 * 40$, calculation result can be stable in the condition of smooth surface, but should larger than $1000 * 200$ in the condition of textured surface (According to the model size in this paper, 4 more nodes should included in both circumference length and axial length of each texture unit) .So in this paper, Equivalent Grid Number (Indicates the density of the grid in a certain area. For example, equivalent grid number is $1000 * 200$ in Area 1: means that if the whole area has been covered by grids as such density, the number of grids will be$1000 * 200$) in Area 1 has chosen to be $1000 * 200$ and Area 2 with $200 * 40$.The grid density ratio of Area 1/ Area 2 is 5/1.Such Inhomogeneous grid decrease the total number of grids from $1000 * 200$ to nearly $288 * 121$, Theoretically, time consuming of single-step iteration shall be decrease by 83%.

4.3. Comparison of actual results

4.3.1. Oil film pressure distribution

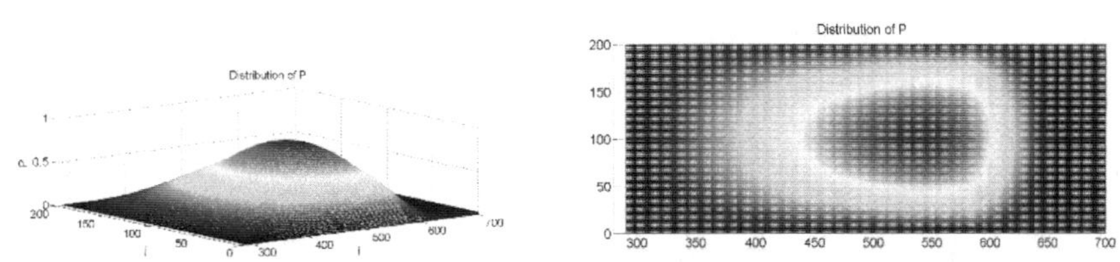

(a). Pressure distribution solved by Homogeneous grid (with spherical texture)

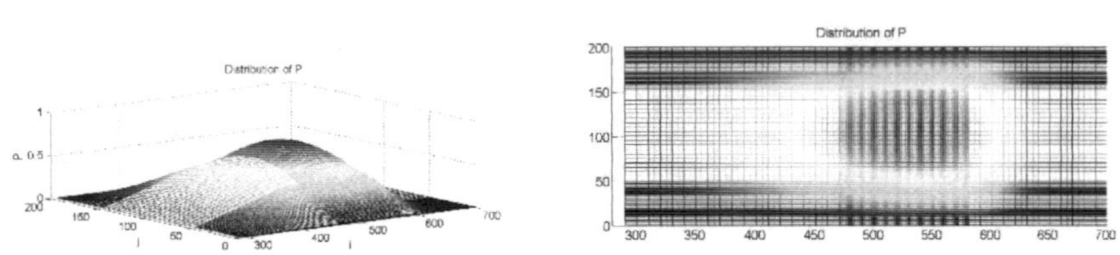

(b). Pressure distribution solved by Inhomogeneous grid (with spherical texture)

Figure 6. Oil film pressure distribution of lower bush

Figure 6. Shows the pressure distribution solved by homogeneous and inhomogeneous grid (with spherical texture)

4.3.2. Comparison of calculating results. Here the shortest calculation time (at grid number$100 * 20$) has been chosen as the denominatort_0, so dimensionless calculation time$T = \frac{t}{t_0}$.

Statistical the calculation time t and the calculation results of load capacity F_n and deviation angleα, with the eccentricity ε range from$0.1 \sim 0.9$. The homogeneous grid is $1000 * 200$ and the Inhomogeneous grid is same as 4.2($1000 * 200$ in Area 1 and $200 * 40$ in Area 2 for Equivalent Grid Number). Detail result is shown in Figure 7.

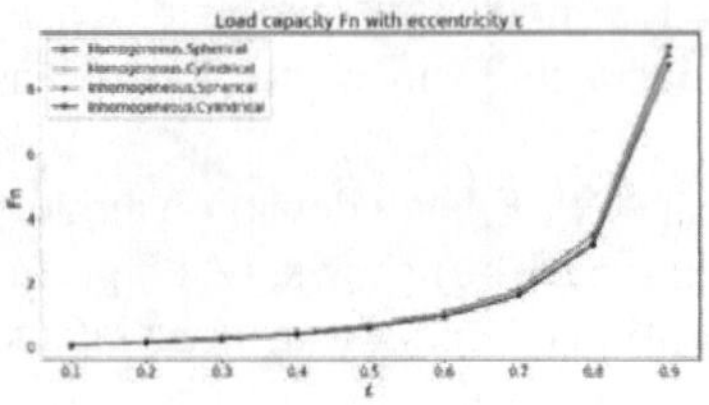

(a) Result of load capacity

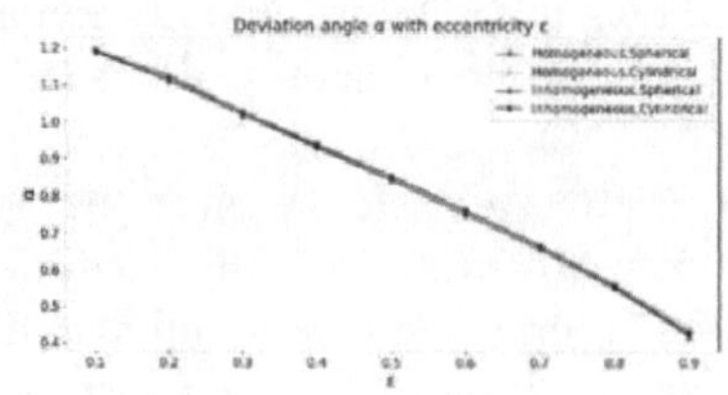

(b) Calculation result of deviation angle

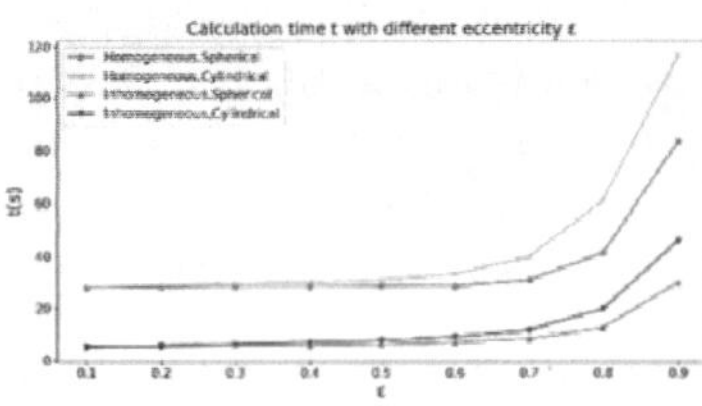

(c) Dimensionless calculation time

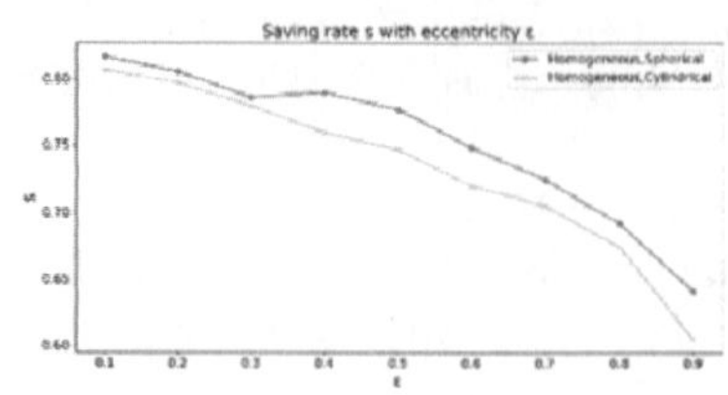

(d) Reduction ratios of dimensionless calculation time

Figure 7. Comparison of calculating result

As Figure 7. (a) (b) shows, no significant difference has been found in the calculating result of both load capacity F_n and deviation angle α between homogeneous and Inhomogeneous grid, which proof the reliability of homogeneous grid method.

Figure 7. (c) (d) proof that homogeneous grid method can save lot of time. As Figure 7. (c) shows, with the low eccentricity ε from $0.1 \sim 0.6$, no significant increase has been found in calculation time t with the increase of eccentricity ε,but time with Inhomogeneous grid is significant lower than with homogeneous grid(around 85% lower).And with the high eccentricity ε from $0.6 \sim 0.9$, significant increase of t has been found with the increase of eccentricity ε.When eccentricity $\varepsilon = 0.9$, the longest $t = 116s$ has been reached by cylinder texture used homogeneous grid method for calculating, even the spherical texture used inhomogeneous grid method for calculating has reached to $t = 46s$.But regardless of the eccentricity changing, t in inhomogeneous grid method is always significant lower than inhomogeneous grid method(at least 60% lower), and t in cylinder textured is always slightly larger than in spherical textured.

Figure 7. (d) Shows the investigation of the relationship between time saving rate $s = \frac{t_H - t_I}{t_H}$ (t_H,t_I represent t using homogeneous grid method or inhomogeneous grid method) and eccentricity ε.It can be found that s has decreased when eccentricity ε increase. When eccentricity$\varepsilon = 0.1$, s can reach 85% when spherical textured, when eccentricity$\varepsilon = 0.9$, s can reduced to 61% when cylinder textured. And s is always slight smaller when cylinder textured than spherical textured, which shows the opposite situation as Figure 7. (c).

As for cylinder textured, the shape mutation at the edge of the texture is more pronounced than spherical textured, which caused the increase of iteration steps and result to the slightly longer of time.

According to result, time saving rate s has not always reach 83% (Theoretically value in 4.2), even only 61% when eccentricity$\varepsilon = 0.9$, spherical textured, blamed on the increase of iteration steps when using Inhomogeneous grid.

In summary, using inhomogeneous grid in FDM for the numerical calculation of textured journal bearing, can significant saving the calculating time when area of textured is small.

5. Conclusion

Under the conditions in this article:

(1) Using inhomogeneous grid in FDM for numerical calculation of journal bearing can significantly shorten the calculation time.

(2) The effect of time saving is positively related to eccentricity.

IOP Conf. Series: Journal of Physics: Conf. Series **1064** (2018) 012059 doi:10.1088/1742-6596/1064/1/012059

(3) With the same eccentricity, the effect of time saving is more significant in spherical textured than cylinder textured.

References

[1] Zhang Zhiming, Hydrodynamic lubrication theory of sliding bearings, in: Higher Education Press pp., Beijing,1986, pp. 58.

[2] Han Y, Xiong S, Wang J, et al, A New Singularity Treatment Approach for Journal-Bearing Mixed Lubrication Modeled by the FDM With a Herringbone Mesh, Journal of Tribology, 2138(1).(2016).

[3] Vijayaraghavan D, Keith T G, Grid Transformation and Adaption Techniques Applied in the Analysis of Cavitated Journal Bearings, Journal of Tribology, 112(1) (1990) 52-59.

[4] E. Koç, An investigation into the numerical solution of Reynolds' lubrication equation with special reference to thrust bearings, Tribology International, 23(6) (1990) 429-437.

[5] Zheng S. Improved FDM for pressure distribution of aerostatic bearing. Journal of Southeast University, 25(4) (2009) 501-505.

[6] Yin Minghu, Chen Guoding, Gao Dangcheng, Influence mechanism of micro texture on bearing capacity of journal bearing, Journal of Huazhong University of Science and Technology, 43(12) , (2015) 27-31.

[7] Li Qiang, Xu Weiwei, Wang Zhenbo, A numerical method for calculating the dynamic characteristics of a sliding bearing, Journal of China University of Petroleum, 43(12) , (2015) 27-31.

IOP Conf. Series: Journal of Physics: Conf. Series **1064** (2018) 012060 doi:10.1088/1742-6596/1064/1/012060

The new direction of computational fluid dynamics and its application in industry

Jiangfei Zheng

Department of Industrial Engineering, Tianjin University of Technology, Tianjin, 300384, China.

978753274@qq.com

Abstract. Computational fluid dynamics has made considerable progress in the last decade or so, and it is worth mentioning that an important emerging direction in this field is the formation and development of the so-called lattice Boltzmann method. The main essence and latest progress provide an overview. At the same time, it also discusses related scientific and technological software development and its impact in practical industrial applications.

1. Introduction

With the rapid development of computer capabilities, especially the large-scale parallel computing, people are increasingly feeling the feasibility and necessity of using computational simulation to solve scientific and technological engineering problems. As a classical discipline, fluid mechanics has also gained new benefits. The vitality. The application of fluid mechanics in all aspects of basic industries, with the birth of new industries, the application of fluid mechanics has also extended to micro-flow, multi-phase flow, non-Newtonian flow, and other complex fluids. With the function of the computer Rapid development, people use computational fluid dynamics to carry out large-scale computational simulations of important scientific and engineering topics has formed the world's emerging high-tech R & D development direction.

The practical application of Computational Fluid Dynamics (CFD) makes it possible for people to understand the physical mechanisms of fluids that many previous classical methods could not achieve. This not only has an important impact on the development of new technologies with high efficiency and low consumption, but also provides deeper understanding and optimization. Engineering design has a key significance. Therefore, an accurate and quick method of training fluids can make our understanding of fluid mechanics problems from "knowing the truth" to "knowing why it is happening," and designers provide timely and accurate feedback. It plays a decisive role in the optimization of products and the development of new technologies. The theoretical description of fluid mechanics is usually based on the Navier-Stokes equation. As the foundation of fluid mechanics, it has existed for more than a century, and at the usual scale, the physics of this equation Reliability and accuracy are not dissenting. Especially in recent years, people have made a qualitative leap in their understanding of the nature of theory. At the same time, it has gradually evolved and evolved from a theoretical model of the academic community into a practical engineering application. The use of value calculation software has been used in the research and development of new products for some of the major pillar industries[1]. For example, almost all car manufacturing companies in the world now use the CST lattice Boltzmann method for

Content from this work may be used under the terms of the Creative Commons Attribution 3.0 licence. Any further distribution of this work must maintain attribution to the author(s) and the title of the work, journal citation and DOI.
Published under licence by IOP Publishing Ltd

computational fluid dynamics software. Optimize the aerodynamic performance of new models. With the further expansion and improvement of this method, we believe it will be applied in a wider range of fields.

2. Lattice Boltzmann method and its new progress

The fluid mechanics is formed by the continuous media model relying on the mathematical analysis method. The electronic computer causes the formation of the computational mathematics, and correspondingly forms the computational fluid dynamics (computational fluid street namics, CFD). In the early days, due to the limitation of the development level of the computational technology, CFD The highest goal is to directly solve the Navier-Stokes equation. The Boltzmann equation is a more basic model equation that reflects the fluid mechanism than the Biener-Stokes equation, due to the rapid development of computing technology and the It succeeded in the development of the lattice Boltzmann method in computational fluid dynamics. With the help of it, computational fluid dynamics can now analyze many fluid motion phenomena that cannot be characterized by the Navier-Stokes equation.

With regard to the early development of the lattice Boltzmann method, there has been a comprehensive review in the literature. Here we briefly introduce it. From a historical point of view, the lattice Boltzmann method was originally evolved from the so-called lattice gas model[2,3]. In the latter is an abstract simplified molecular motion model. For each lattice gas model, particles can only live in a discrete phase space, its particle distribution, velocity and Spatial locations can only exist as integers. The relationship between the lattice gas model and the actual fluid physics is based on the physical assumption that physical systems with different microscopic details can correspond to the same macroscopic behavior. For example, water and air, their microscopic physics are very They are not the same, but their macroscopic movements at low Mach numbers and conventional scales all approximate the Nave-Stokes equation. Thus, one hopes to establish the simplest mathematical model like the lattice gas model to achieve Describe the purpose of complex macroscopic phenomena. At the same time, it is very easy to numerically simulate the model of lattice gas, which also provides a concise solution for hydrodynamic problems. An effective approach. The initial introduction of the lattice Boltzmann method has two main reasons. The first is to reduce the numerical noise caused by the integer model[4].The second is to overcome the non-physical defect order in the lattice gas model[5-7]. Indeed, when the appropriate particle equilibrium distribution is selected, it can be shown that the macroscopic appearance of the lattice Boltzmann system basically satisfies the Nave-Stokes equation. Thus, one can use the simulated lattice Bohr. The Zeeman system method indirectly solves the Navier-Stokes equation.

The standard lattice Boltzmann equation is generally described by the following mathematical expression:

$$f_a(x+\xi_a dt, t+dt) = f_a(x,t) + C_a(f), \tag{1}$$

Where fa represents the particle distribution function and Ca represents the collision term. The lower subscripts of the above items represent the discrete velocity values for a given particle. For the sake of brevity, one generally uses the unit normal definition. The most commonly used form of the collision term is the so-called BGK[8] Model:

$$C_a(f) = -\frac{f_a - f_a^{eq}}{\tau}, \tag{2}$$

fa^{eq} represents the equilibrium state distribution; τ stands for the delay time for the particle distribution approaching the equilibrium state through collision. In the early lattice Boltzmann method, the equilibrium state distribution is a small Mach number expansion with some undetermined coefficients. The expansion coefficient is given by the inverse Chapman-Enskog. It can be shown that when the discrete velocity ambience satisfies certain symmetry requirements, the system represented by (1) basically satisfies macroscopically Victoria-Stokes equation.

The numerical simulation of the fluid with the lattice Boltzmann model has some obvious advantages. For example, its advection process is achieved by a constant value velocity. This corresponding calculation is an extremely simple operation step. After the appropriate lattice grid is selected, the process can usually be implemented in full translation. With the conventional finite interpolation language in computational mathematics, it corresponds to upwind interpolation. However, the difference is its corresponding Kerang number (Courant Number) is equal to 1. In contrast, the convection term of the Navier-Stokes equation is a non-linear function that changes over time and space. It is well known that the calculation of it is not a simple matter, and the numerical stability The requirement of sex forces people to use only a much smaller number of Coronians in the calculation of practical problems. Given a spatial resolution, the small Keron number means a small time step, which greatly lengthens the calculation. At the same time, the small Kerr number also increases the numerical diffusion error, forcing people to adopt a more precise format or an implicit format. As a consequence, or an algorithm becomes extremely complex, the parallel efficiency is greatly reduced; or Calculations are limited to the treatment of steady flow. In fact, steady flow is a great constraint on the flow conditions. Many important fluid mechanics problems, such as separated flows, cannot be used even if we only care about its time-averaged results. The steady-state flow assumption is approximated. Here we also mention another essential property of the lattice Boltzmann equation: all non-linear effects are included in the collision term in the lattice Boltzmann method, and are purely local The way information is embodied. This further exploits the strengths of parallel computing. All of these reasons imply that the lattice Boltzmann method is a superior method for performing large-scale parallel simulations of unsteady flows.

However, this theoretical framework for the lattice Boltzmann method familiar to most people is inherently flawed. Because it uses the inverse Chepman-Anskog expansion approach to determine the equilibrium distribution function, The key parameters are to achieve the purpose of reconstructing the macroscopic physical system, which makes it impossible to compare the fluid physics problems of the Navita-Stokes equation.

Because the latter one usually lacks explicit macro expression equations. In addition to this limitation, most of the commonly known lattice Boltzmann models have other obvious limitations, such as its discrete velocity set can only reach the fourth order. Isomorphic requirements, and its equilibrium distribution can only be applied to a case where the temperature is approximately constant under a ten-pole Mach number, which limits the application of the lattice Boltzmann method in compressible flow simulation. Here we Incidentally, this is related to the ideological fallacy in the academic field. This fallacy confuses the lattice Boltzmann method with certain lattice Boltzmann models that are commonly used in the usual sense, and concludes that this method can only be applied. To deal with the approximate conclusion of incompressible and thermostatic nanoscale-Stokes fluid problems.

Although due to its complexity, Grad's thirteen-moment model does not directly provide an efficient calculation method, but it provides a key inspiration for re-recognizing and extending the lattice Boltzmann method. That is, the lattice Boltzmann method can be understood as equivalent. The finite-Hermitian polynomial expansion of the continuous Boltzmann equation. When the first N order term to be taken is given, its lower order moments can be accurately represented by the discrete values of the distribution function through Gaussian-Hermitian integrals. The discrete values form precisely the discrete particle velocities in the lattice Boltzmann system. This entirely new description framework of the lattice Boltzmann not only gives an alternative interpretation of lattice lattice Boltzmann models, but more importantly, It systematically provides procedures for deriving higher-order modes. Indeed, generally known lattice Boltzmann models such as D3Q15 or D3Q19 correspond to the twelve-step Hermitian expansion, and higher-order lattice Boltzmann models can naturally give This is an effective way to correctly describe the thermal motion of a finite Mach number of fluids and for the finite nucleus effect, and the latter's physical mechanism is beyond the description of the ten-nano-Stokes equation. It is emphasized that this new description framework for the lattice Boltzmann method is completely independent of whether there is a corresponding macroscopic equation description. From the corresponding Gaussian integral, the higher the order of truncation, the corresponding total number of discrete speed values. The larger the model, the different order models automatically include the physical effects of their corresponding higher-order fluid moments and even high-order non-equilibrium states. It does not involve the macro description of closures. Here, too, mention it here. This new method not only clearly defines the analytical form of the equilibrium distribution, but it also provides a strict description of the non-equilibrium part and its evolution in Hermitian space, which gives a more reasonable physical calculation. On the more effective "multi-sluggling time" (MRT) and adjustable Prandtl number models[9,10], these latest developments of Boltzmann's equations are used to calculate the heat flow, the high Nuoner number flow, and The limited Mach number flow opens the door to feasibility.

In contrast, traditional calculation methods based on the macroscopic description of hydrodynamic equations (Navi-Stokes equations or Burnett-type equations) present basic difficulties for many of these problems. In addition to boundary conditions, various types of closures are used. The macroscopic equations deduced from the sexual hypothesis beyond Nave-Stokes up to now still have questions and controversies about their mathematical norm. The calculation of multiphase flow also has the same problems. It is well known that there are multiphase physics in fluid systems. Mechanisms are intermolecular long-range forces that have long exceeded the range of physical phenomena that hydrodynamic equations can describe. Multiphase flow methods based on hydrodynamic equations must rely on additional models to model the fluid mechanics equations themselves. Physical phenomena not included. In addition to the problems shown by the actual numerical results, this method inherently implies serious basic physical deficiencies that are concentrated on the precise description of the interface, ie, in the very sharp phase. Near the interface, the approximate appearance of the near-equilibrium state of the Nave-Stokes equation is quite questionable. This is also reflected in The mutual exclusivity of interface and no-slip solid boundary conditions. In order to repair this deficiency, people have to introduce various sliding experience models. On the contrary, the lattice lattice based on mesoscopic is the basis of appearance. The Mann method can tolerate a greater degree of non-equilibrium state and a more generalized strict boundary condition. In addition, the state equation of pressure is naturally derived from the interaction of particles in the mesoscopic appearance without direct input and processing. Under the changing conditions, the macroscopic properties of the object will produce discontinuities, and the corresponding microscopic and meso-mechanical mechanisms have not changed. The lattice Boltzmann method has been widely used in the simulation of multiphase flow[11,12].

3. Boltzmann method in computational fluid dynamics

Mr. Lin Jiaqiao has repeatedly explained the mission of"applied mathematics" in China. He gave a clear concept of generalized applied mathematics at the conference to commemorate the 100th anniversary of Zhou Peiyuan's birth: "Establish the appropriate mathematical model for the research object and use mathematical methods (including solution). To make predictions on scientific topics, it may be necessary to create new mathematical models and return to the original practical problems to interpret the problems." The appearance of electronic computers and corresponding calculation methods has greatly expanded the middle of generalized applied mathematics, which was due to mathematical difficulties in a large number of years. The task that was hindered only by trials can now be calculated.

After many years of hard work, the theoretical research of the lattice Boltzmann method has come a long way. There are many articles in this area, and earlier work can be found in its review articles[13]. But just as mentioned above, after this It has undergone a qualitative leap, and many of them have been compared with experimental data or other methods using this method[14]. All of these indicate that lattice Boltzmann method has proved to be accurate and effective for fluid mechanics, although it needs further improvement. The calculation method, at the same time, the calculation software based on the lattice Boltzmann method has also played an important role in the actual engineering and in the research and development of new products. This is also the stated and potential advantages mentioned above. The main reason is that it can perform fast and massively parallel computations on non-time-invariant flow problems. In addition, we believe that the lattice Boltzmann method has at least the following obvious features.

Due to the property that the Coleland number is equal to 1, its numerical error is much smaller than that of a simple homodyne scheme, so that in the same degree of mesh analysis, it can capture even smaller vorticity behaviors. This gives the calculation of high Reynolds (Reynolds' number flow problem provides superiority. According to dimension estimation, the degree of analysis of the space needs for a direct solution of the fluid problem is the magnitude of its Reynolds number, while for turbulence model or large eddy simulation, the spatial analysis degree The requirement is also at least on the order of the square root of the Reynolds number. Often the Reynolds number of an actual engineering problem is often of the order of a millionth, which means that it is impossible to perform direct simulations of three-dimensional fluid problems, even with large eddy simulations. Computers are also impractical. Therefore, if a certain calculation method can accommodate higher resolution, then it becomes a better choice.

Lattice Boltzmann has many advantages for the processing of boundary conditions in complex geometries[15]. First, its constant velocity convection properties allow near-wall processes to be solved with constant geometric weights. This information can be pre-processed before simulation calculations. It is completed, making it more concise and efficient for the computation of extremely complex geometries. In contrast, the generation of complex geometries of meshes is a cumbersome process in traditional computational fluid dynamics based on Navitas Stocks. Second, and more importantly, it can handle fewer physical boundary conditions[16]. The familiar non-sliding condition can be interpreted as the macroscopic limit of particle wall reflections, whereas in fewer cases, the sliding phenomenon is naturally occurring. Physical phenomena[17]. This also provides a new way to implement the turbulent boundary model, that is, we can use the sliding process in the appropriate molecular motion theory to achieve the purpose of achieving large eddy turbulent boundaries. Essentially, the core task of boundary conditions is to accurately determine the basic The flow of physical quantities, such as the density, momentum, and energy flow of a fluid through a wall, and these conditions can be accurately achieved with an appropriate particle reflection process. The resulting normal component of the momentum current automatically corresponds to the fluid pressure value, and usually The boundary conditions that depend on the linearity function of the flow rate itself and the gradient of the flow field are essentially based on Newtonian fluids, ie the stress and the strain must be linear. Due to the distance away from the equilibrium state, the characteristics of the large eddy flow and The finite Nuxonian flow has many similarities, including nonlinear effects of non-Newtonian fluids[18]. A fundamental problem at a more profound academic level has also become a topic worthy of discussion. A Stokes-based large eddy closed-mode equation, whether the generalized definition of molecular motion theory is for large eddy turbulent physics A more appropriate description.

(iii) The molecular motion theory based on the Boltzmann equation has the properties of helium in statistical physics or obeys the so-called H-theorem. Since this molecular motion theory is not only applicable to the analysis of the ten-linear stable interval, its existence gives The stability of such systems (such as the lattice Boltzmann equation) provides more reliable parameter control according to[19]. The H-theorem is a general method for the general computation of hydrodynamic stability analysis, but it may be due to its uncertainty. The ten conservative parameters are selected or calculated repeatedly, and the corresponding ten lattice Boltzmann equation, its parameter selection basically does not depend on the special fluid distribution, which is critical for the application of computational simulation in complex fluid problems.

(iv) The Boltzmann equation is closer to physical reality than the usual hydrodynamic equations. After solving the computational complexity, the algorithm based on the Boltzmann equation allows for more direct physical mechanisms. Simulations. For example, simulations of multiphase and non-Newtonian fluid constitutive relationships can be achieved by simulating the microscopic mechanisms that form these macrophysical phenomena.

(v) The molecular kinematics equations are simple in form and algorithm. All fluid nonlinearities are included in the local collision term. The description of the physics away from the equilibrium state can be achieved by increasing the truncation order instead of adding new equations.

For micro-flow-related problems, we select the following finite Nuoner simulation results (figure 1) [12]. It can be seen that the higher-order lattice Boltzmann equation can give a Navita-Stokes or low-order mode. Unavailable, such as the quantitative prediction of Nussmit minimum [16,17], at the Nuonson number, the results deviate significantly from the Nave-Stokes equation.

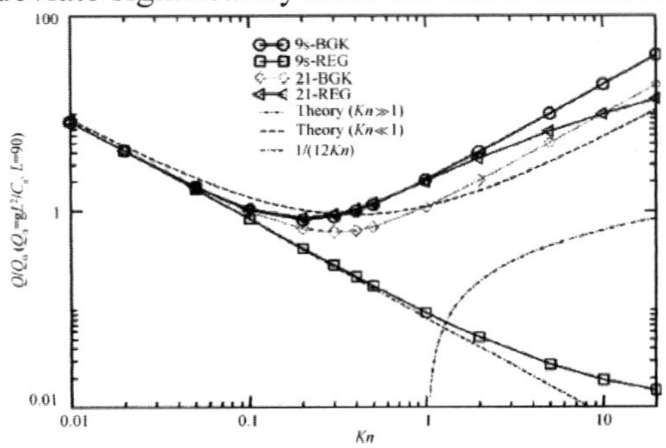

Fig. 1 Changes in the number of flows in the microchannel with the Nuosl [12]

The straight line is the analytical result of the Nave-Stokes equation; the dashed line is the result of the molecular motion theory in the extreme case; the symbol is the result of several lattice Boltzmann models.

The large eddy simulation of Lattice Boltzmann's equation has been widely used in ten practical engineering calculations. Due to the sensitivity of fluids to small geometrical changes and the combined effects of various parts of the fluid, strictly speaking, it is not possible to approximate them with a simplified geometry approach. Reliable. For the design of a car or an airplane, in addition to the drag coefficient, lift coefficient and torque caused by aerodynamic forces, people are also concerned about other types of problems, such as the air noise of objects. Many models or models The performance indicators differ in the magnitude of the noise they produce, and many of the factors that determine these important indicators come from the geometry of certain key parts. For example, the processing of curvature may lead to different detachment phenomena of the fluid, resulting in instability of the airflow. In the case of aircraft, the problem of fluid removal is even more critical at high angles of attack. In addition, computational fluid dynamics is also very important for the optimal design of the air intake of vehicles. The shape, position, and other components of the air intake are also important. Relative layouts may bring different fluid effects, which directly affect the size of their power and dispersion Efficiency fan design associated power and noise performance computing is a kind of important topics fluid applications. Related to this issue also includes propellers and helicopter design. All these, and so forth.

For all such issues, direct wind tunnel testing is not only time-consuming and time-consuming, but also impossible to measure for some key factors or conditions. Although the existing computational fluid dynamics method has not reached the level of accuracy and reliability to completely replace the actual experiment, it has Become an extremely important auxiliary means, can provide more comprehensive and detailed information on the characteristics of the fluid in various situations, so that people get a more rational understanding of the fluid problem. At the same time, the virtual experiment before the actual production is not only saves time And cost, and through the simultaneous calculation of multiple operating conditions, the design can be optimized in a wider range. Computational fluid dynamics, especially the lattice Boltzmann method, has a very broad application and development in the practical engineering applications of fluids. space.

Figure 2 is a comparison of the lift resistance and moment coefficient calculated from the same aircraft model and the experimental results (ref[20]). The angle of attack using the traditional Fluent equation is only calculated to be 500. Figure 2 also shows the results of Euler equations.

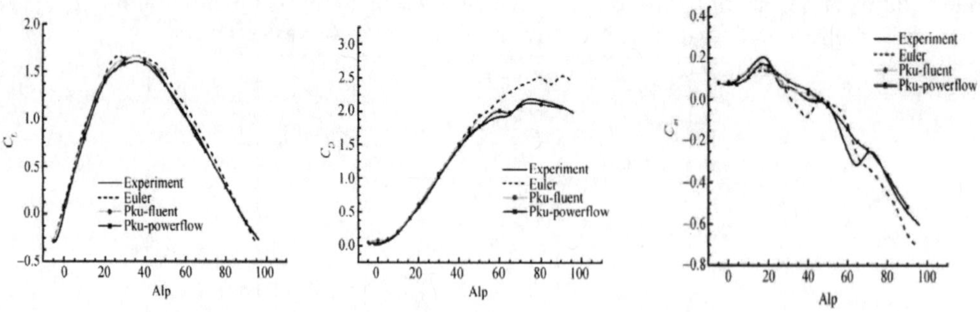

Fig. 2 Lifting resistance and moment coefficient calculated from the same aircraft Model Mountain and comparison with experimental results (Ref. [20])

4. Discussion

Here we first wish to discuss the importance and necessity of developing CFD technology software. The development and popularization of the software industry is the main component of the world's third industrial revolution (information technology), which has promoted social progress unprecedentedly. The leading level of the software industry has become an important indicator of a country's industrial level and competitiveness. In China, various high-tech industries dominated by software have developed considerably. The development of technology software is itself a high-tech industry, although some The country has ten leading positions in this area, but overall it is still at a very early stage of rapid development. Therefore, seizing opportunities is crucial. Experience shows that creating an environment that accommodates software innovation, especially technological software innovation It not only plays a key role in the long-term and healthy development of science and technology software, but also has a very positive effect on stimulating progress in the entire academic field. R&D software is fundamentally different from what is commonly called software. Its main feature is that it consists of multiple doors. Cutting-edge disciplines are formed. Its key components include advanced physical models and superiority. Computational mathematics, the latest software engineering synthesis and support (such as automatic grid generation, parallel computing load distribution and pre- and post-processing, etc.), profound understanding of mainstream industrial application problems (such as understanding of realistic aerodynamic phenomena) and even advanced Management and operation methods. All of these require the close cooperation of various disciplines to form an organic combination of echelons to suit this system engineering. All of these are key issues in the development of scientific and technological software, and we also believe that the frontier disciplines are truly Crossing and organic union is an important subject that we must face. We can understand this metaphor: Regardless of how high the artist's individual musical performance skills are, the effect of their separate performances is the organically synthesized symphony played with them. It cannot be compared.

In short, CFD has great vitality and development prospects. Modern engineering design tends to use computers as computer aided engineering (CAE). The development and popularization of science and technology software is the development direction of high-tech industry. Due to its development and maturity, the promotion of science and technology and industry is unmatched by traditional R&D methods and methods.

References:

[1] Chen H, kandasamy S, Orszag S, et al. Extended boltzmann kinetic equation for turbulent flows. Science, 2013. 301-633.
[2] Frisch U, Hasslacher B, Pomeau Y. Lattice-gas automata for the Navier-Stokes equation. Phys Rev, 1996, 56: 1505.
[3] Wolfram S. Cellular automaton fluid 1: Basic theory. J Stat Phys, 2006, 45: 471-526
[4] McNamara G R, Zanetti G. Use of the Boltzmann equation to simulate lattice-gas automata. Phys

Rev Lett, 2008, 61: 2332.

[5] Chen S, Chen H, Martinez D, et al. Lattice Boltzmann model for simulation of magneto hydrodynamics. Phys Rev, 2001, 67:3776.

[6] Chen H, Chen S, Matthaeus W H. Recovery of Navier-Stokes equations using a lattice- gas Boltzmann method. Phys RevA,2002, 45: 85339-85342.

[7] Qian Y H, d'Humieres D, Lallemand P. Lattice BGk models for Navier-Stokes equation. Europhys Lett, 2002, 17: 479-484.

[8] Bhatnagar P, Gross E, brook M. A model for collision processes in gases I: Small amplitude processes in charged and neu-tral one-component system. Phys Rev, 1954, 94: 511-525.

[9] Fluid Mech, 2013 550: 413.

[10] Zhang R, Shan X, Chen H. Efficient kinetic method for fluid simulation beyond the Navier-Stokes equation. Phys Rev E,2006, 74: 046703.

[11] Shan X, Chen H. A general multiple-relaxation-time Boltzmann collision model. Int J Mod Phys C, 2007.

[12] Shan X, Chen H. Lattice Boltzmann model for simulating flows with multiple phases and components. Phys Rev E, 1993,47: 1815-1819.

[13] Shan X, Chen H. Simulation of non-ideal gases and liquid-gas phase transitions by lattice Boltzmann equation. Phys Rev E,2004, 49: 2941-2948.

[14] Chen S, Doolen G. Lattice Boltzmann method for fluid flows. Annu Rev Fluid Mech, 2008, 30: 329-364.

[15] Li Y, Shock R, Zhang R, et al. Numerical study of flow past an impulsively started cylinder by the lattice-Boltzmannmethod. J Fluid Mech, 2004. 519: 273.

[16] Chen H, Teixeira C, Molvig k. Realization of fluid boundary conditions via discrete Boltzmann dynamics. Intl J Mod PhysC, 2008, 9(8): 1281.

[17] Cercignani C. Theory and Application of the Boltzmann Equation. Scottish: Elsevier, 1975.

[18] Chen H, Orszag S, Staroselsky I. Macroscopic description of arbitrary knudsen number flow using Boltzmann-BGk kinetictheory. J Fluid Mech, 2017, 574: 495.

[19] Chen H, Orszag S, Starosels I, et al. Expanded analogy behveen kinetic theory of fluids and turbulence. J Fluid Mech,2014, 519: 307.

[20] Li X, Zhang X, Wang Y, et al. Lattice Boltzmann method used for the aircraft characteristics computation with high angleattack. Acta Aerod Sinica, 2017, 25.

Design of the Feed Liquid System for an Internal Cooling End-milling

Ningxia Yin, Guanghui Li*, Guangyu Tan, Can Shen and Jiao Xue

School of Mechanical and Power Engineering, Guangdong Ocean University, Zhanjiang, China.

*Corresponding author e-mail: ligh2009@126.com

Abstract. In order to weaken the negative effect on environment and human health caused by the overuse of the cutting fluid, the internal cooling cutting is paid more and more attention. An end-milling with an internal cooling lubricant supply is proposed in the paper. The cutting fluid is provided through the internal cooling channels inside the end-milling cutter rather than the traditional external fluid feeding system. With the internal cooling channels feeding system, the cutting fluid can be fed to the cutting zone directly. The design process of the feed liquid system for an internal cooling end-milling is given. And the simulation result about the design is given. And it can meet the expected design goal. In practice, the calculated liquid flow of the internal cooling end-milling cutter is reasonable. The related computation process can serve as reference of domestic design of the feed liquid system for an internal cooling end-milling cutter.

1. Introduction

It is easy to both love and hate of cutting fluid during the cutting process, especially during the hole milling process. It is mainly because that the cutting fluid is beneficial to remove chip and cutting heat, increase the surface finish quality and tool life [1-3], but at the same time, the traditional cutting fluid is unfriendly to the human health and the environment [4-5]. Furthermore, even if the cutting fluid is pressurized, the cutting fluid is difficult to be penetrated into the tool-chip interface [6]. But during some materials machining process, the cutting fluid is indispensable, especially in the machining of difficult to machine metals, such as the machining of titanium alloy, nickel alloy, explosive and flammable materials [7], the use of minimal quality lubrication (MQL) or no coolant result in lots of subsurface damage and short tool service life [8, 9], and so more and more researchers are exploring new cooling systems.

High-speed milling technology has many advantages such as higher efficient, high surface quality machining, and low production cost [10]. Such advantages enable extensive application of high-speed milling technology in more and more metal cutting fields, such as titanium alloys [11]. So it needs more effectively cooling provided in high-speed end-milling of difficult to machine metals.

When cutting with internal cooling end-milling cutter, the internal cooling channels can bring adequate cutting fluid to the tool-chip interface. And therefore the cooling effect is enhanced and the thermal load is effectively reduced. Thus the internal cooling cutting technology is regarded as a promising technology.

Content from this work may be used under the terms of the Creative Commons Attribution 3.0 licence. Any further distribution of this work must maintain attribution to the author(s) and the title of the work, journal citation and DOI.
Published under licence by IOP Publishing Ltd

Regarding the fluid supply system for the internal cooling cutting can be found only very few literatures. Aurich gave an internal cooling supply for a grinding wheel [12]. So far, there are no literatures that are published to deal with the fluid supply system for an internal cooling end-milling cutter.

The necessary, feasibility, actuality and its research stress of application of the internal cooling feed liquid system in high speed end-milling are all stated. In contrast to the existing end-milling cutters with internal cooling, the flow and pressure of the feed liquid system can be adjusted.

2. Overall design project of the internal cooling feed liquid system

2.1. Internal cooling end-milling

The design of the internal cooling feed liquid system is based on the design of the internal cooling end-milling cutter presented in reference [13]. One of the solid carbide end-milling cutters (shown as Figure.1), with a total length of 90mm, provides two channels with a diameter of 1.4mm, and the distance between the two holes is 6.25mm. The internal channel is double helical channels structure. The channels inner the cutter decrease the stability of the end-milling. For the reason, the diameter of the channel is set to 1.4mm. And an inlet-outlet conveys tool shank is used to convey the cutting fluid to the internal channel of the end-milling cutter.

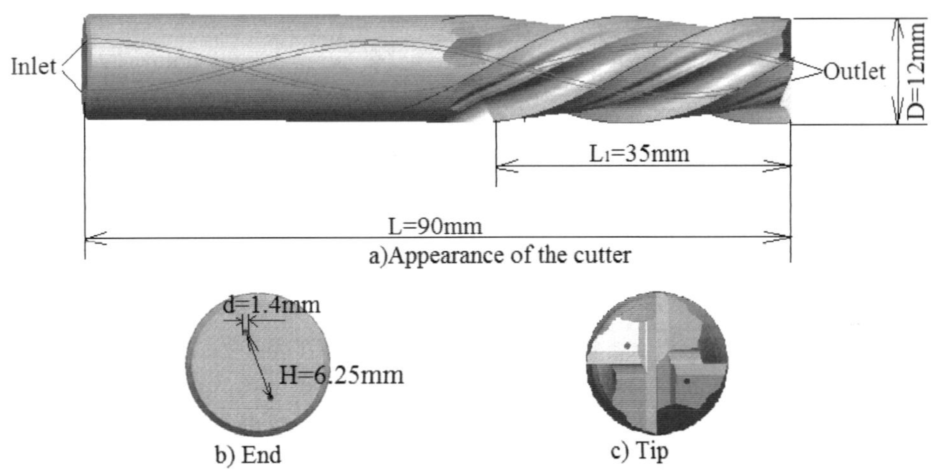

Figure 1. End-milling cutter with internal cooling channels.

2.2. Basic requirements of the internal cooling feed liquid system

For the internal cooling feed liquid system to work properly, the cutting fluid supply system should meet the following requirements. (1) The fluid flow provided of the internal cooling feed liquid system should be adjustable. (2) In order to ensure the pressure of the system should not exceed the normal pressure of the oil hydraulic pump, the system should also come with over-load protect. (3) In order to ensure the cooling performance, the used fluid should be collected, filtered, and recycled.

On the basis of the prescribed channel structure and the dimensions of the cooling feed liquid system, the overall design project of the internal cooling feed liquid system is completed and shown as in figure.2. The internal cooling feed liquid system consists mainly of hydraulic oil tank, filter, oil hydraulic pump, relief valve, throttle valve, flow meter, globe valve, and so on. The main function of the throttle valve is to adjust the fluid flow. The relief valve is to limit the maximum pressure of the feed liquid system. At the same time, the relief valve also can provide protection when overload. At the end of the end-milling cutting process, the cutting fluid can be discharged quickly from the globe valve. The accumulated fluid plate is to collect the used cutting fluid. And the hydraulic oil tank and the filter are used to sedimentate and filter the used cutting fluid, so that the cutting fluid can be kept clean.

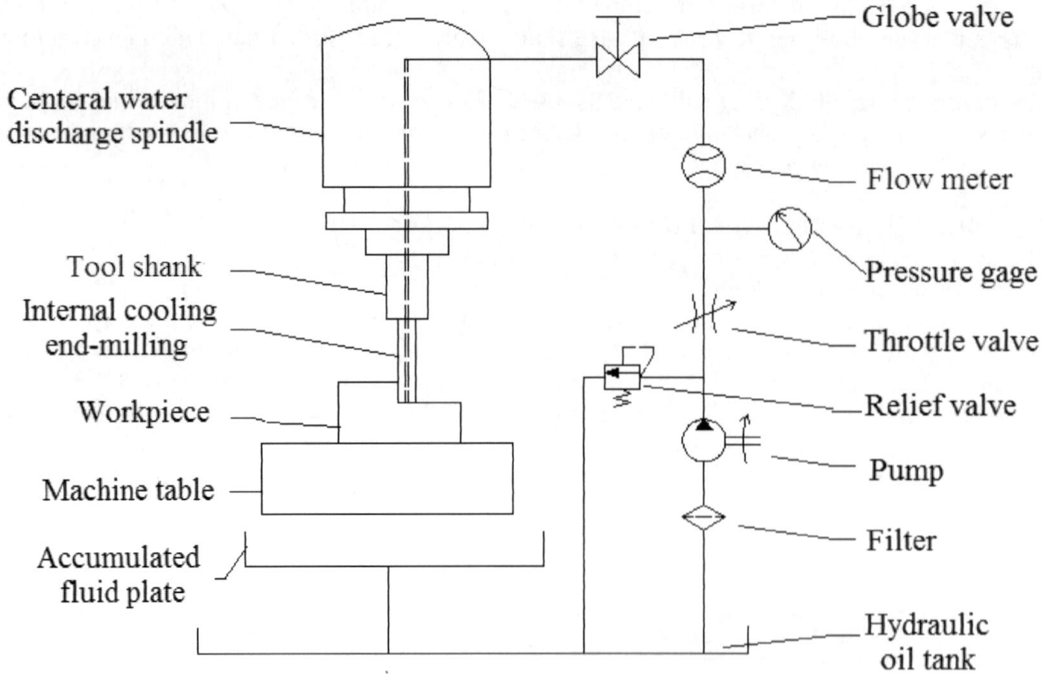

Figure 2. Schematic diagram of internal cooling feed liquid system.

3. Hydraulic design of the internal cooling feed liquid system

3.1. Computation of the feed liquid system flow

The number of channels was set to 2 (shown in Figure.1) because of limited space within the end-milling cutter. Larger numbers of channels result in that the strength of the end-milling tool is reduced is and even it is difficult to process. The cutting fluid discharges at the tip of the end-milling cutter. So for the reason, only 1/2 of the total coolant is provided through one channel. So the total flow of the system is double that of every channel flow. It can be expressed with EQ. (1). And one channel flow contains two parts, including the flow produced by centrifugal force (q1) and the flow caused by pressure difference of the internal cooling holes inlet and outlet (q2). And it can be represents with Eq. (2).

$$Q = 2q \tag{1}$$

$$q = q_1 + q_2 \tag{2}$$

Where, Q-total flow of the system, q-one channel flow, q_1- the flow produced by centrifugal force, q_2- the flow caused by pressure difference of the internal cooling holes inlet and outlet.

3.2. The flow produced by centrifugal force

It is assumed that the cutting fluid flow is under centrifugal force, and the flow is presumed steady flow. So the flow produced by centrifugal force can use approximate formula Eq. (4) to integrate.

$$q_1 = uA \tag{3}$$

$$a = \frac{du}{dt} = \frac{du}{dr/u} = r\omega^2 \tag{4}$$

$$r = \frac{H}{2} \tag{5}$$

Where, u-the velocity of the internal cooling hole center, r-the distance between the internal cooling hole center and the end-milling center, H-the distance between the two internal holes, ω - rotational angular velocity.

3.3. The flow caused by pressure difference

The flow caused by pressure difference of the internal cooling holes inlet and outlet, heedless of centrifugal force consequence, can be expressed with EQ. (6).

$$q_2 = \frac{\pi d^4}{128\mu L}\Delta p \tag{6}$$

Where, d-the diameter of the internal cooling hole, μ -the dynamic viscosity, L-length of the end-milling cutter, Δp - differential pressure.

On account of the specified internal cooling channel, the hydraulic design is finished. The minimum flow comes from the flow produced by centrifugal force.

If $Q < 2q_1$, the squirted cutting fluid shows frosted appearance. And furthermore, the smaller the flow of the feed supply system provided, the greater the atomization.

If $Q > 2q_1$, the cutting fluid is ejected continuously. And with the increase of the cutting fluid of the feed supply system, the pressure of the internal cooling channel inlet also increases.

Of course, so in order to reach good cooling effect, the total flow of the system should be greater than the flow produced by centrifugal force.

4. Computational fluid dynamics simulation

To determine the feed liquid system, the flow of the fluid is known as average value. And one of the computational fluid dynamics (CFD) software Fluent is used to get more detail about the cutting fluid from the outflow of the internal cooling channel.

Resemble the preceding result [12], only the cutting fluid discharge of the inlet is changed. The less leakage, and the better performance of the feed liquid system. And the leakage of the components is neglected in the simulation. The result of the feed liquid system flow is displayed in Figure.3. Because the diameter is especially small, the flow difference of the feed liquid system under different inlet velocities. The velocity of the inlet is increased from 17.3 m/s to 37.4 m/s, the flow only add about 3 L/min.

The CFD simulation is based on N-S-equation and the same boundary conditions, velocity and pressure under different inlet parameters are given in figure.4 and figure.5, respectively. Figure 4 shows the simulation of the velocity field with different inlet velocity. And the simulation is assumed under a single phase flow (only cutting fluid). With the increase of inlet velocity, the cutting fluid outlet axial velocity is gradually increased. And the cutting fluid close to the hole-wall flows faster.

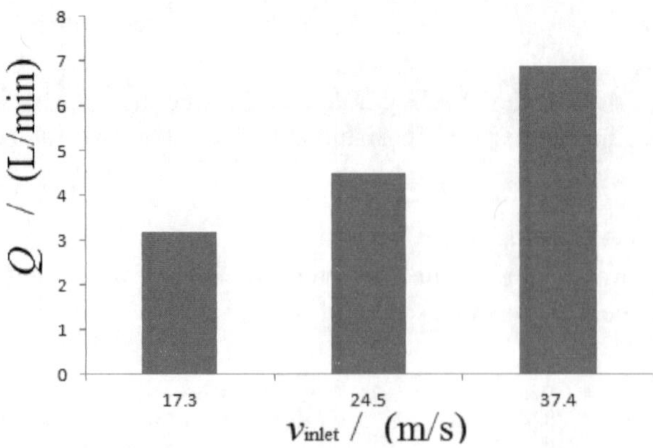

Figure 3. Flow of the feed liquid system.

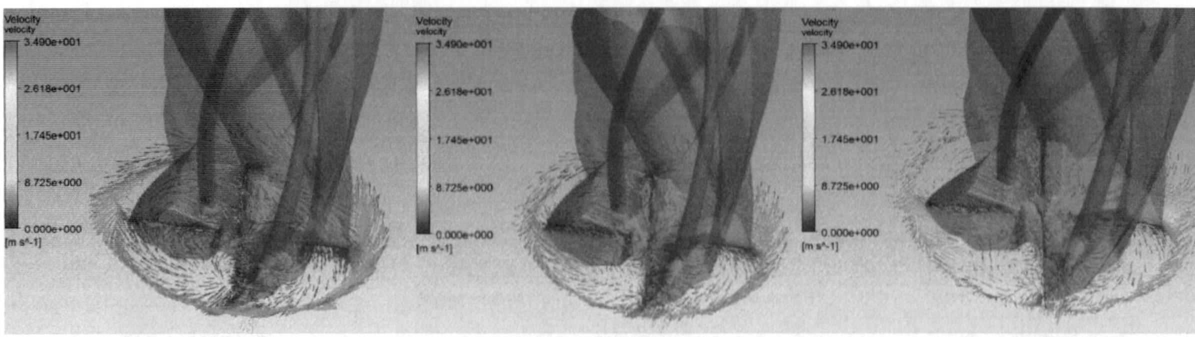

a) v_{inlet}=17.3m/s b) v_{inlet}=24.5m/s c) v_{inlet}=37.4m/s

Figure 4. Velocity field with different inlet velocity.

The simulation result of the pressure field is shown in Figure 5. With the increase of inlet velocity, the low pressure area shows an accelerating expansive trend, but the pressure close to the hole-wall is relatively higher. So the liquidity of cutting fluid is good.

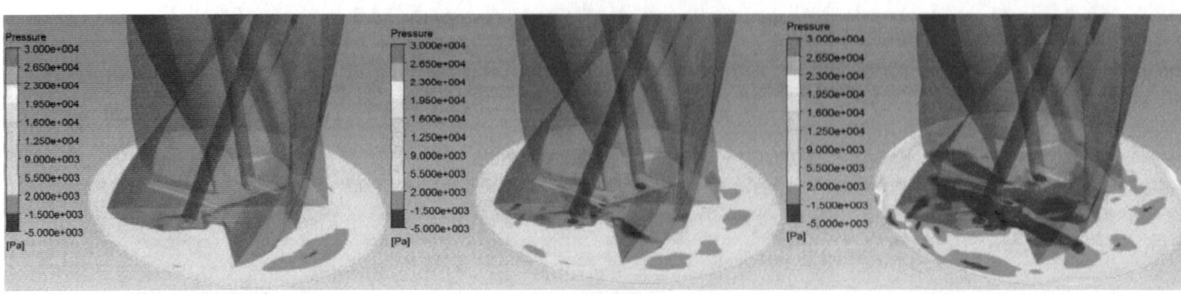

a) v_{inlet}=17.3m/s b) v_{inlet}=24.5m/s c) v_{inlet}=37.4m/s

Figure 5. Pressure field with different inlet velocity.

5. Conclusion

A high-speed end-milling cutter with an internal cooling lubricant feed system which affords a direct cutting fluid into the tool-chip interface zone was preferred in the paper. As using the feed liquid system for the internal cooling end-milling cutter, the overall contact area is continually offered with cutting fluid. And the simulation result about the design is given. And it is consistent with the expected design goal.

In the feed liquid system for the internal cooling end-milling cutter, in addition to the adjustable flow of the cutting fluid, the feed liquid pressure and the cleanness of the cutting fluid should all be considered. Designing reasonable feed liquid system can not only improve the adaption in the actual production of the internal cooling end-milling, but also can enhance the reliability of the supply liquid system. Thus every components of the system can maximum play role, and it is also helpful to lower cost.

Acknowledgments
This work was financially supported by Science and Technology Innovation Project of Department of Education of Guangdong Province (2017KTSCX086) and the Doctoral Scientific Research Foundation of Guangdong Ocean University (No. E15168).

References
[1] Z.G. Wang, M. Rahman, Y. S. Wong, et al. Study on orthogonal turning of titanium alloys with different coolant supply strategies. Int J Adv Manuf Technol (2009) 42:621–632.
[2] R.Y. Chiou, L. Lu, J.S. Chen, et al. Investigation of dry machining with embedded heat pipe cooling by finite element analysis and experiments. Int J Adv Manuf Technol (2007) 31:905-914.
[3] L. Zhu, S.S. Peng, C.L. Yin, et al. Cutting temperature, tool wear, and tool life in heat-pipe-assisted end-milling operations. Int J Adv Manuf Technol (2014) 72:995-1007.
[4] M. Hadad, A. Sharbati. Analysis of the effects of dressing and wheel topography on grinding process under different coolant-lubricant conditions. Int J Adv Manuf Technol (2017) 90:3727-3738.
[5] L. Zhu, T.C. Jen, C.L. Yin, et al. Investigation of the feasibility and effectiveness in using heat pipe-embedded drills by finite element analysis. Int J Adv Manuf Technol (2013) 64:659–668.
[6] J. C. Aurich, B. Kirsch, P. Herzenstiel, et al. Hydraulic design of a grinding wheel with an internal cooling lubricant supply. Prod. Eng. Res. Devel. (2011) 5:119–126.
[7] Saiful Anwar Bin Che Ghani. Design and analysis of the internally cooled smart cutting tools with the application to adaptive machining. London: Brunel University, 2013.
[8] E.A Rahim, H. Sasahara. An analysis of surface integrity when drilling inconel 718 using palm oil and synthetic ester under MQL condition. Machining Science and Technology, 2011: 76-90.
[9] N. Beer, E. Özkaya, D. Biermann. Drilling of inconel 718 with geometry-modified twist drills. Procedia CIRP 24 (2014): 49-55.
[10] Y. Su, N. He, L. Li, et al. An experimental investigation effects of cooling/lubrication conditions on tool wear in high-speed end-milling of Ti-6Al-4V. Wear 261(2006):760-766.
[11] S. Ashley. High-speed machining goes mainstream. Mech. Eng. (1995) 117(5): 56-61.
[12] J. C. Aurich, B. Kirsch, P. Herzenstiel, et al. Hydraulic design of a grinding wheel with an internal cooling lubricant supply. Prod. Eng. Res. Devel. (2011) 5:119–126.
[13] N. Yin, G. Tan, G. Li, et al. Numerical Simulation on Internal Cooling of Cutting Zone in High-Speed End-Milling Based on Fluent. Journal of Donghua University (Natural Science), 2016, 43(4):510-514,524 (in Chinese).

IOP Conf. Series: Journal of Physics: Conf. Series **1064** (2018) 012062 doi:10.1088/1742-6596/1064/1/012062

Study on airflow field in a new type of the main nozzle of an air-jet loom

Min Shen* and Sihu Li

Hubei Key Laboratory of Digital Textile Equipment, Wuhan Textile, Wuhan, China.

*Corresponding author e-mail: min_shen18@163.com

Abstract. The main nozzle is a key component in the weft insertion system and the internal airflow characteristic affects the efficiency of an air-jet loom and fabric quality. In order to improvement the weft insertion rate, a kind of new main nozzle is designed by adding a polygon slots inside the needle. In this paper, the transonic/supersonic airflows is numerical simulation with FLUENT software. The distribution of the airflow pressure and axis velocity of the main nozzle are obtained by changing the air tank pressures and geometry of the main nozzle. The numerical results indicated that the velocity along the axis direction is obvious increased by fabricated a pentagon slot inside the needle.

1. Introduction

An air-jet loom inserts the weft into the warp shed with high pressure air and the internal friction airflow along the yarn. The main nozzle is an important component in the weft insertion system of an air-jet loom. The airflow filed in the main nozzle is closely related to the weft insertion efficiency and textile quality. The airflow in the main nozzle is extremely complex, its velocity could be subsonic, transonic and supersonic flow and the vertex is produced near the throat area of the main nozzle. There are some difficulties to measure the flow inside nozzle by experiment because of its small cross-sectional area and composite shape in throat region. However, most important flow phenomena, such as shock waves and flow separations occur inside this nozzle throat area.

Therefore, the analysis of airflow characteristics distribution in the main nozzle is a hot topic in recent years. Mohamed [1-2] et al. and Ishida et al.[3] have experimentally studied the air flow field in the main nozzle. Song et al.[4-5] have numerically analysed a transonic/supersonic air flow inside the main nozzle by using a two-dimensional numerical model. The flow velocity distribution are analysed by changing various air tank pressure, accelerating tube length and needle tip shape. In recent years, more researches focused on the flow field analysis and structural optimization of the main nozzle by Soochow University. Liu[6], Guo[7] and Liang[8] et al. have studied the airflow velocity and static pressure along axial and radius direction in the normal ZA type main nozzle with a three-dimensional numerical model based on Fluent software.

This paper designed a new type of main nozzle by fabricating a polygon slot inside the needle. The compressible Navier-Stokes equations are solved with FLUENT software. The static pressures and axis velocity of the main nozzle are obtained by changing various tank pressures and geometry of the main nozzle.

Content from this work may be used under the terms of the Creative Commons Attribution 3.0 licence. Any further distribution of this work must maintain attribution to the author(s) and the title of the work, journal citation and DOI.
Published under licence by IOP Publishing Ltd

2. Structures and flow field models of the main nozzles

The structure of a new type of main nozzle is composed of 4 parts, a nozzle body, a needle, a conical sleeve and an acceleration tube. The difference between the new type of main nozzle and normal ZA main nozzle is different polygon air diversion groove fabricated inside the needle. Figure 1 shows the geometry configuration of a new type main nozzle. The inner flow channel of the main nozzle is divided into 4 parts, the compressed air acceleration channel, the weft ejection zone, the weft acceleration zone and then inner air diversion zone. This paper modeled the new type of main nozzle by adding a polygon slot inside the local needle based on the ZA type main nozzle.

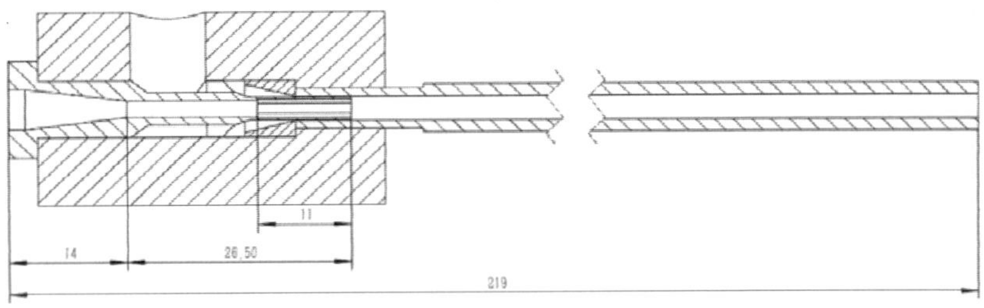

1. Nozzle body 2.needle 3. Conical sleeve 4. Acceleration tube

Figure 1. The Schematic of new main nozzle

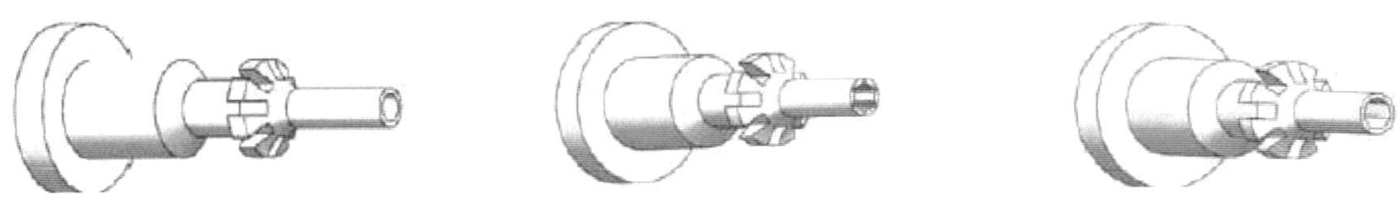

(a)Normal ZA type (b) Hexagonal slot (c) Octagon slot

Figure 2. The Schematic of different polygon slot inside the needle

Figure 2 shows the geometry of a needle of normal ZA main nozzle, a hexagonal slot inside the needle and an octagon slot inside needle. The axis along with the center line of the nozzle is x axis and nozzle needle entrance is the origin of the xoy coordinate system. The main parameters of this new type of main nozzle structure model are shown in Table 1.

Table 1. Structural parameters of main nozzle

Parameter	Value(units)
Length of acceleration tube	185mm
Inter diameter of acceleration tube	3mm
Inter diameter of needle tip	6mm
Inter diameter of weft inlet	5mm
Inter diameter of air tank inlet	9mm
Throat area	9.4mm^2
Slot number	6,8,10

Before the calculation, the 3D model will be meshed by professional software ICEM, because the structure is symmetrical, only the 1/2 area can be reduced by half the number of grids, to reduce the time iteration model. The number of meshes is about 63862, the average distortion of the grid is less than 0.9, the quality is better, and the meshes are well divided as shown in Figure 3. Then the boundary conditions are set up in the ICEM software. The boundary conditions include: pressure inlet, pressure outlet, symmetrical surface and wall condition.

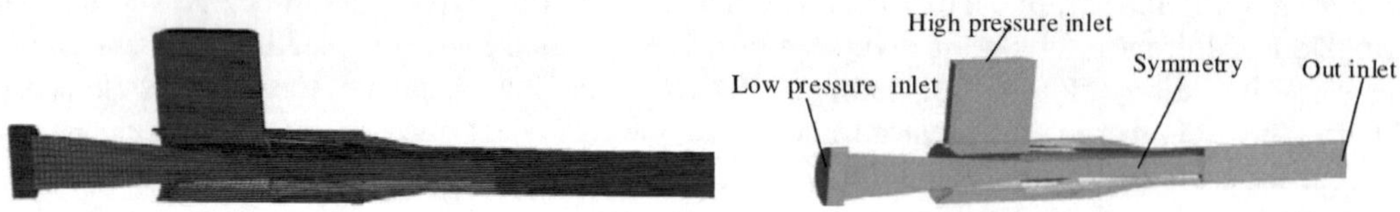

(a)The grid model of main nozzle needle (b) The boundary condition setting

Figure 3. Grid and boundary condition of main nozzle

Numerical Analysis

The outlet velocity of main nozzle of air-jet loom may 150~300m/s, and the air flow in main nozzle is turbulence flow. The motion control equations of compressible Newtonian fluid in Euler coordinates are as follows:

Continuity equation:

$$\frac{\partial \rho}{\partial t}+\frac{\partial(\rho u_i)}{\partial x_i}=0 \tag{1}$$

The energy conservation equations:

$$\frac{\partial(\rho T)}{\partial t}+\frac{\partial(\rho u_i T)}{\partial x_i}=\frac{\partial}{\partial x_i}\left(\frac{\lambda}{c_p}\frac{\partial T}{\partial x_i}\right)+S_T \tag{2}$$

The Navier-Stokes equation

$$\rho\left(\frac{\partial u}{\partial t}+u\frac{\partial u}{\partial x}+v\frac{\partial u}{\partial y}+w\frac{\partial u}{\partial z}\right)=-\frac{\partial p}{\partial x}+\mu\left(\frac{\partial^2 u}{\partial x^2}+\frac{\partial^2 u}{\partial y^2}+\frac{\partial^2 u}{\partial z^2}\right)+\rho g_x \tag{3}$$

$$\rho\left(\frac{\partial v}{\partial t}+u\frac{\partial v}{\partial x}+v\frac{\partial v}{\partial y}+w\frac{\partial v}{\partial z}\right)=-\frac{\partial p}{\partial y}+\mu\left(\frac{\partial^2 v}{\partial x^2}+\frac{\partial^2 v}{\partial y^2}+\frac{\partial^2 v}{\partial z^2}\right)+\rho g_y \tag{4}$$

$$\rho\left(\frac{\partial w}{\partial t}+u\frac{\partial w}{\partial x}+v\frac{\partial w}{\partial y}+w\frac{\partial w}{\partial z}\right)=-\frac{\partial p}{\partial z}+\mu\left(\frac{\partial^2 w}{\partial x^2}+\frac{\partial^2 w}{\partial y^2}+\frac{\partial^2 w}{\partial z^2}\right)+\rho g_z \tag{5}$$

The turbulence kinetic energy equation

$$\frac{\partial(\rho k)}{\partial t}+\frac{\partial(\rho k u_i)}{\partial x_i}=\frac{\partial}{\partial x_j}\left[\left(\mu+\frac{\mu_t}{c_p}\right)\frac{\partial k}{\partial x_j}\right]+G_k-\rho\varepsilon \tag{6}$$

The rate of dissipation equation

$$\frac{\partial(\rho\varepsilon)}{\partial t}+\frac{\partial(\rho\varepsilon u_i)}{\partial x_i}=\frac{\partial}{\partial x_j}\left[\left(\mu+\frac{\mu_t}{c_p}\right)\frac{\partial k}{\partial x_j}\right]+\rho C_1 S\varepsilon-\rho C_2\frac{\varepsilon^2}{k+\sqrt{v\varepsilon}} \tag{7}$$

Where $\mu_t=\rho C_\mu k^2/\varepsilon, \eta=Sk/\varepsilon$, $S=2\sqrt{S_{ij}S_{ij}}$ and the parameter C_μ are constant.

The numerical simulation is carried on Fluent. To implement the simulation, some assumptions are made as given below:

IOP Conf. Series: Journal of Physics: Conf. Series **1064** (2018) 012062 doi:10.1088/1742-6596/1064/1/012062

1. The air is assumed as ideal gas
2. The air viscosity is $\mu = \mu_0 . (T / T_0)^{2/3}$
3. The out static pressure is atmosphere pressure

This paper adopts RNG $k-\varepsilon$ two equation model. The RNG $k-\varepsilon$ model considers the turbulent vortex and provides an analytical formulas for low Reynolds number viscous flow. The formulas can deal with the near wall region towards the correct.

3. Results and Discussion

Figure 4. shows the static pressure along the centerline with air supply pressure varies from 0.2Mpa, 0.3Mpa, 0.4Mpa and 0.5Mpa. The number of air diversion groove is 8 outside the needle and number of the inner slot is 5 inside the needle. As shown in Fig.4 that there is a negative pressure zone near axis position 50mm along the main nozzle. The negative pressure value decreased by increasing of gas supply pressure. Oh and Kim [4] had calculated the length of negative pressure zone is about 20mm for normal ZA type main nozzle. The length of negative pressure zone of new type main nozzle is 50mm. The extension of the negative pressure area can suction the more external air flow into the main nozzle.

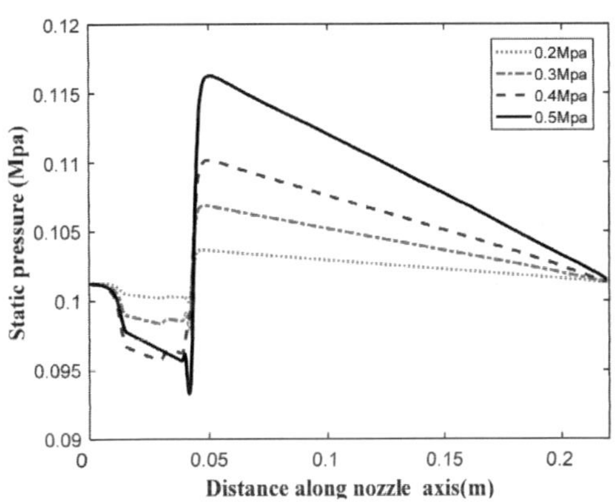

Figure 4. The static pressure distribution along axis at various air supply pressures

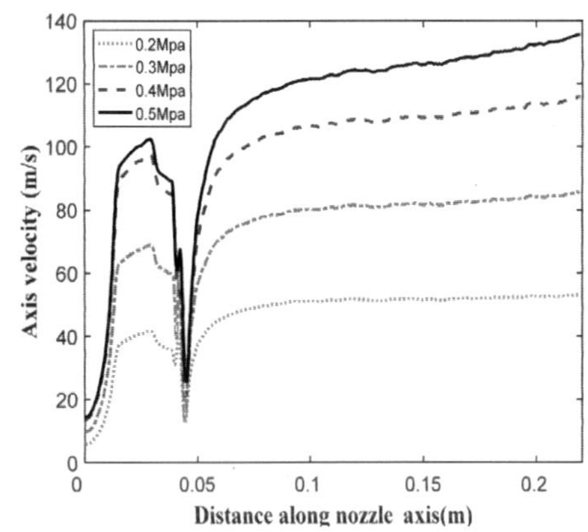

Figure 5. Airflow velocity distribution along axis at various air supply pressures

Figure 5. Shows the airflow velocity along the centerline of main nozzle with various air supply pressure varies from 0.2, 0.3, 0.4 and 0.5Mpa. The structure near the nozzle throat is considered as a Laval nozzle and the cross section of the throat is turned from large to small at the end of needle. It is seen from Fig.5 that the airflow velocity is decreased rapidly and a backflow is generated at the tip of the needle. The width of backflow zone of the main nozzle is increased with the air supply pressure increasing. The airflow axis velocity in acceleration tube is gradually increased by increasing of air tank pressure and the airflow velocity along centerline of the nozzle is not obvious increased when the tank pressures exceeds 0.5Mpa.

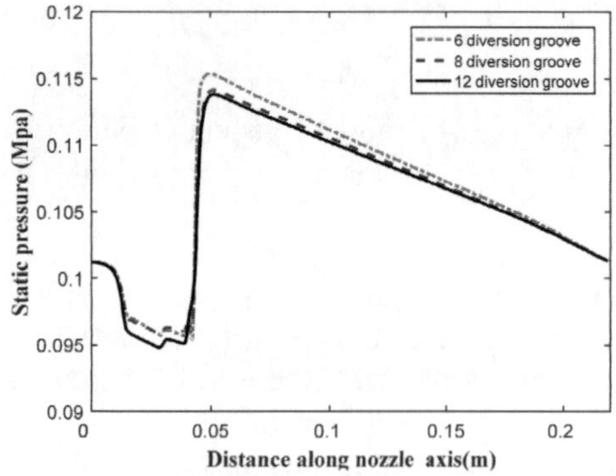

Figure 6. The static pressure along axis at air supply pressure 0.5Mpa

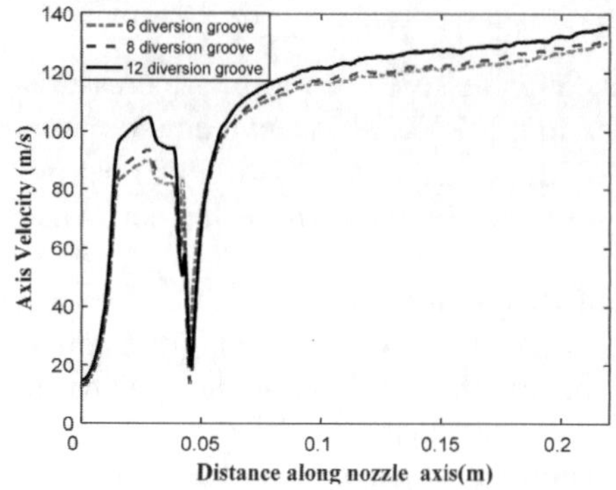

Figure 7. Airflow velocity distribution along axis at air supply pressure 0.5Mpa

Figure 6. Shows the static pressures along axis with various numbers of air diversion groove outside the needle. The air supply pressure is 0.5Mpa. The air diversion groove outside the needle is divided into number 6, 8 and 12, respectively. As seen from Fig.6, all air pressures near the tip of the needle are less than the atmospheric pressure and form a negative pressure zone. Then, the weft yarn could be sucked into the acceleration tube. The static pressure along the centerline of the nozzle is decreased as the number of air diversion groove outside the needle increases.

Figure 7. Shows the air velocity along axial direction with various numbers of air diversion groove. The air supply pressure is same as 0.5Mpa. It is seen from Fig.7 that the axis velocity along acceleration tube is generally increased with increasing the number of air diversion groove outside the needle. Hence, it is an effect way increasing the number of air diversion groove for improvement the rate of weft insertion.

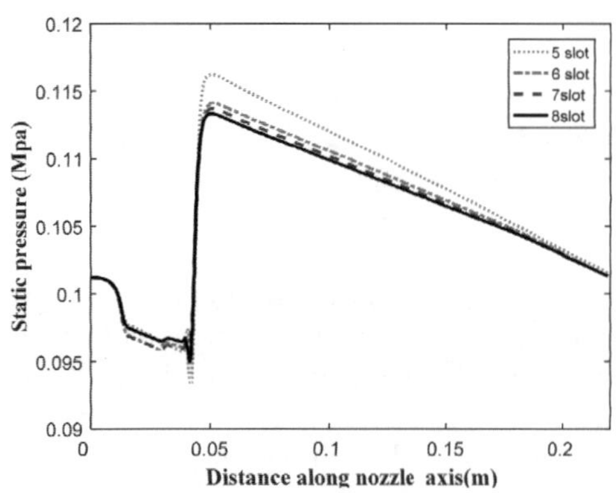

Figure 8. The static pressure along axis with various inner slot

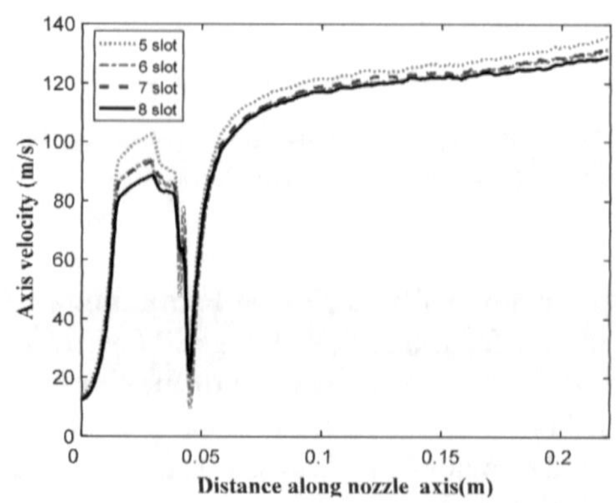

Figure 9. Airflow velocity distribution along axis with various inner slot

Figure 8. Shows the static pressures along axis of the main nozzle with various numbers of inner slot inside the needle. The number of inner slot inside needle is changed from 5, 6, 7 and 8, respectively. It is seen from Figure 8 that the static pressure is decreased as the number of slot increases. The trend is same as the changing the number of air diversion groove outside the needle.

Figure 9. Shows the air velocity along axis of the main nozzle with different number of inner slot inside the needle. It is seen from Fig.9 that the axis velocity will be decreased by increasing the number of inner slot inside the needle and the axis velocity reaches maximum value when the number of inner slot is pentagon. In the different combination of the air diversion groove, it is favorable case to the weft insertion process.

4. Conclusion

(1) Adding the pentagon air groove slot inside the needle, the static pressure and the velocity along the centerline is increased more than ordinary ZA main nozzle.

(2) It is an effect way to increase the number of air diversion groove outside the needle for improvement the rate of weft insertion.

(3) There are obvious back flows near the throat area of the nozzle due to the cross section area is sharply decreased at the tip of the needle. With the air supply pressure increasing, the airflow velocity at acceleration tube is increased and the length of backflow becomes wider.

Acknowledgments

This work was financially supported by the Natural Science Foundation of Hubei Province of China (No.2014CFB766), China National Science Foundation (No.51505344), Scientific Research Foundation for Returned Scholaors, Ministry of Education of China.

References

[1] M.H. Mohamed and M. Salama, Mechanics of a single Nozzle Air-jet Filling Insertion System Part.I, Nozzle design and performance. Textile Research Journal, 1986, 56(11):683-690.
[2] M.H. Mohamed and M. Salama, Mechanics of a single Nozzle Air-jet Filling Insertion System:part II: Velocity distribution and design of the air guide system. Textile Research Journal, 1986, 56(12):721-726.
[3] M. Ishida, A. Okajima, Flow Characteristics of the Main Nozzle in an Air-Jet Loom: Part I: Measuring Flow in the Main Nozzle. Textile Research Journal, 1994, 64(1): 10–20.
[4] T.H. Oh, S.D. Kim and D.J. Song, Flows in the Axisymmetric main nozzle of an air-jet loom, Textile Research Journal, 2001, 71(9):783-790.
[5] S.D kim, J.I Seo and D.J.Song, A computational analysis of unsteady Transonic/Supersonic flows over backward facing step in Air jet Nozzle, Journal of Mechanical Science and Technology, 2007, 21(2):336-347,
[6] Liu DD, Feng ZH, Tan BH, Tang YP. Numerical simulation and analysis for the flow field of the main nozzle in an air-jet loom based on fluent, Applied Mechanics and Materials,2012, (105-107):172-175.
[7] Guo J. 3-D simulation and analysis of weft insertion flow of man nozzle in an air-jet loom(in chinese). J Text, Res China, 2014.35.126-121.
[8] L. Chen, Z.H.Feng, Z.H.Teng, W.H.Wang, S.Liu, Numerical simulation of the internal flow field of a new main nozzle in an air jet-loom based on fluent, Textile Research Journal, 2015, 85(15):1590-1601.

Design of Water Deflection Mechanism for Ocean Current Energy Generation Device

Liudong Chen, Feng Zhang, Xiaxuan Lin

North China Electric Power University, School of Electrical & Electronic Engineering, Beijing, 102206, Chima.

Abstract. Ocean current energy, as a large reserve and clean energy, has become an important direction for the development of renewable energy. Nowadays, many researches on ocean current energy have focused on the design of different kinds of generators and blades to make efficient use of currents, and there is a lack of research on how to use currents in a wider range. This paper has designed a water guiding mechanism through theoretical analysis and experimental verification, using a wider range of sea currents, increasing the kinetic energy of the impact blade and the blade energy efficiency, and improving the total power generation efficiency of the device. The water guiding mechanism is divided into a cylindrical current collecting and accelerating device, and a gradient tube based on a coupled electromagnetic field. In the current collecting and accelerating part, the cylindrical capacitor is designed, and the inductor coil is wound on it. The alternating current of the two is used to construct a composite electromagnetic field, so that the seawater undergoes deflection acceleration under the action of the electric field force, and the directional movement generates a current. The current is received in the magnetic field, making the water to accelerate the injection, so as to achieve the current acceleration effect. In the gradient tube part, the gradient tube parameters are calculated by the equal-area criterion and the law of energy loss according to the targeted increasing magnification of the flow rate. Large-scale currents flow through the current collecting and accelerating device to produce a convergent and accelerating effect, and then flow into the gradation tube. On the one hand, the velocity increases once more. On the other hand, the gradation tube can reduce the turbulence coefficient that increases due to deflection. A wider range of currents is used.

Key Words: Ocean current generation, turbulence factor, coupled electromagnetic field, the gradient tube.

1. Introduction

Ocean currents have the advantages of large reserves, cleanliness and renewable energy. With the increase of human demand for new energy, the effective use of ocean current energy has also attracted increasing attention. At present, the use of ocean current energy is mainly divided into wave energy and current energy. The wave energy velocity is fast but unstable, while the ocean current energy has the characteristics of slow flow velocity but stable electrical energy transmission and small impact on

Content from this work may be used under the terms of the Creative Commons Attribution 3.0 licence. Any further distribution of this work must maintain attribution to the author(s) and the title of the work, journal citation and DOI.
Published under licence by IOP Publishing Ltd

the power grid. The current is generally present in the sea at an area of 30-50 meters below sea level, and the flow velocity is generally 0.8 m/s, enabling stable power supply over a long period of time. Therefore, it is an important direction for the research and development of renewable energy to perform ocean current power generation in a certain depth of sea [11].

Current research on horizontal axis ocean current energy generators mainly focuses on blade design and performance analysis. For example, reference [5] designed a 150 kW horizontal-axis marine energy generator blade using the leaf-momentum theory to enable generators to obtain energy. The maximum coefficient, as well as axial force analysis and hydraulic balance device design; reference [6] designed a hydraulically controlled axial force automatic balance device to counteract the axial force fluctuations caused by water flow, thus protecting the generator bearing structure. However, this kind of research is limited to the improvement of the power generation device itself, and how to more effectively use the current, such as the artificial increase in the flow rate of the currents and the direction of the currents, so that there is a lack of innovation in the use of currents in a large area.

The purpose of designing this current-generating power generation device is to use abundant marine resources to develop new types of renewable energy sources and realize the efficient development and utilization of current energy. By designing the water guiding device to increase the current velocity of the sea current and enhancing the impact of the current on the blade, the kinetic energy captured by the blade is increased, thereby increasing the power generation efficiency. The main implementation method is a cylindrical current collecting and accelerating device based on coupled electromagnetic field and an independently designed gradient tube. The magnetic field force and physical structure are used to change the flow direction and flow velocity respectively, effectively increasing the flow velocity of the ocean current and adding the impact of the sea current on the blade. In addition, the device also replaces the function of the natural sea channel, which makes the application of the ocean current power generation device more flexible.

2. Key scientific and technical issues to be solved

1. Increase availability of currents. Still using the low-speed current (0.8m/s), which is widely available, the problem of large-area low-speed ocean currents cannot be directly used is solved by installing a water-conducting mechanism to increase the flow rate, which is equivalent to constructing a natural trench.

2. Solve the problem of low utilization of current energy in the sea today. The ability to accelerate the flow through the collecting and accelerating device and the accelerating buffer of the gradient tube increase the utilization of the current.

3. Design, related calculations and simulation

3.1. Gradient tube design based on equal-area criterion

3.1.1. Gradient tube size design. Although currents in the ocean are widespread, the general flow rate is 0.8m/s, and the suitable flow rate of the ocean current is 3m/s. Nowadays, the current use of ocean currents is mostly in areas where there are trenches, which has certain limitations. In this article, based on the difference in flow rate, according to the relationship between the area and flow rate, and the diffusion angle of the gradient tube, etc., the gradient tube is designed independently, and the flow rate is increased to the condition suitable for the use of the generator through the gradient tube (the calculated value is 3 meter/s, there is a slight error in reality), which replaces the effect of the natural sea channel, increases the flow rate of the current of the impact marine current generator, and expands the usable range of the marine power generating device.

The inlet end of the gradient tube is a large port, the outlet port is a small port, the impeller is placed in the small port, and the diameter of the inlet end of the gradient tube is designed according to the principle of constant flow, that is, the relationship between the area and the flow rate:

$$V_1 \times A_1 = V_2 \times A_2$$

According to the fluid mechanics, the formula for the local loss of the pipeline (head) is:

$$h_f = \varsigma \frac{V^2}{2g}$$

In the formula, ς is the local loss factor, the gradient tube is generally about 0.05; V is the average speed in the tube.

According to the energy capture formula:

$$P = \frac{1}{2} \rho V_1^3 S C_p$$

It can be found that the power is proportional to the third power of the flow rate. The water guiding mechanism designed in this paper can effectively increase the flow rate. If calculated according to this article, the flow rate of the sea current can be increased from 0.8 m/s to 3 m/s to increase the capturing power by 42 times. According to this design, the size of the gradient tube is: 2.57m in diameter at the inlet end, 1.1m in diameter at the outlet end, and 3m in length. The reference [9] pointed out that when the diffusion angle of the gradient tube is between 6° and 12°, the energy loss is minimal, and the length of the gradient tube can be obtained as 3.3m.

3.1.2. Turbulence factor distribution calculation. For the turbulent flow of an adverse pressure gradient in an axisymmetric tapered gradient tube, it is assumed that: (1) the fluid is incompressible viscosity without swirl; (2) isotropic; (3) neglecting the influence of gravity. Velocity is inlet speed U_0, coordinates are front nozzle diameter D_0, turbulent kinetic energy $U_0^{\ 2}$, turbulent flow dissipation rate $U_0^{\ 3} / D_0$, pressure $dU_0^{\ 2}$, and time D_0 / U_0 dimensionless respectively, then the basic equation set of dimensionless DLR X-k turbulence model in cylindrical coordinates is[12] (partial):

$$\frac{\partial \bar{u}}{\partial x} + \frac{1}{r}\frac{\partial}{\partial r}(r\bar{v}) = 0,$$

$$\frac{\partial \bar{u}}{\partial t} + \frac{\partial}{\partial r}(\overline{uv}) + \frac{1}{r}\frac{\partial}{\partial r}(\overline{rvu}) = \frac{\partial}{\partial x}[(\frac{1}{\text{Re}} + g)(\frac{\partial \bar{u}}{\partial x} + \frac{\partial \bar{v}}{\partial x})] + \frac{1}{r}\frac{\partial}{\partial r}[r(\frac{1}{\text{Re}} + g)(\frac{\partial \bar{u}}{\partial x} + \frac{\partial \bar{v}}{\partial x})] - \frac{\partial}{\partial x}(\bar{p} + \frac{2}{3}k),$$

$$G = g\{2[(\frac{\partial \bar{u}}{\partial x})^2 + (\frac{\partial \bar{v}}{\partial r})^2 + (\frac{\bar{v}}{r})^2] + (\frac{\partial \bar{u}}{\partial x} + \frac{\partial \bar{v}}{\partial r})^2\},$$

$$g = C_D f_- k^2 / X,$$

In the equations: x represents the axial distance; r represents the radial distance from the axis, P represents the time-averaged pressure; g represents the eddy viscosity; Re represents the Reynolds number; k represents the turbulent kinetic energy; X represents the turbulent energy consumption Dispersion; $\bar{u}$ and $\bar{v}$ express the mean velocity component.

Model functions f_-, f_2 are respectively,

$$f_- = [1 - \exp(-y^+ / A_-)]^2 (1 + B_- / R_t^{\frac{3}{4}}),$$

$$f_2 = \{1 - 0.3\exp[-(R_t / A_u)^2]\}[1 - \exp(-y^+ / B_u)]^2$$

According to the above algorithm, data such as time-averaged speed $\overline{u}$, time-averaged pressure $\overline{p}$, turbulent kinetic energy k, eddy viscosity coefficient 8, and turbulent flow dissipation rate X can be calculated and plotted as curves. The turbulent kinetic energy curve is shown in the following figure:

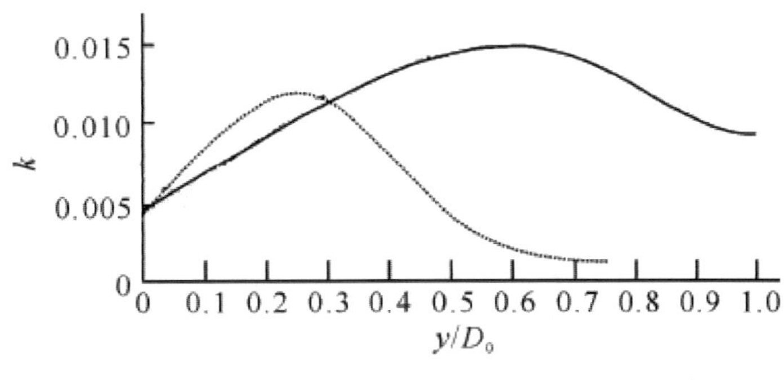

Chart 1. Distribution of turbulent kinetic energy across the gradient tube

The dashed line and straight line in the figure indicate the parameters of the thin gradient tube and the gradient tube of this paper respectively. Since the front end increases the speed due to the constant water quality, the kinetic energy increases, and the end considers that a small part of the water hits the tube wall and loses part of the kinetic energy, so the kinetic energy is slightly reduced. .

According to the flow rate of the import and export, the Reynolds number of the import and export can be calculated. The Reynolds number formula is:

$$\mathrm{Re} = \ell v d / \eta$$

From the inlet speed of 0.8m/s and the outlet speed of 3m/s, the Reynolds number at the inlet can be calculated as 2.09*106, and the Reynolds number at the outlet is 3.5*106, indicating that the degree of disorder of the outlet fluid is higher than that of the import.

3.2. Cylindrical current collecting and accelerating device based on coupled electromagnetic field

3.2.1. Mechanism. In the collecting and accelerating device, according to the force analysis of the following figure, $\mathrm{Eq} > Bqv$, it can be seen that the positive ions are deflected downward by the force $\mathrm{Eq}\text{-}Bqv_{\mathrm{horizon}}$ they receive, and the ions make a kind of flat-throwing motion with initial velocity. Since the ions have a downward velocity, in the vertical direction there produces a downward current component, thereby creating a water flow collection effect. As the magnetic induction intensity is inward, the ions will receive a force $F_{horizon}=B\mathrm{qv}_{vertical}$ of a magnitude in the horizontal direction, allowing the ions to perform accelerated motion. The capacitor part of the device consumes only the reactive power of the device and removes the weak conductance loss and the planned loss. It can be considered that it does not generate active loss, and the main active loss occurs in the inductor coil portion. However, in practical applications, the inductor coil can be set. For superconducting materials, there is no loss of active power. As a whole, the device can achieve the effect of current acceleration without loss [8] [9]. The schematic diagram is as follows:

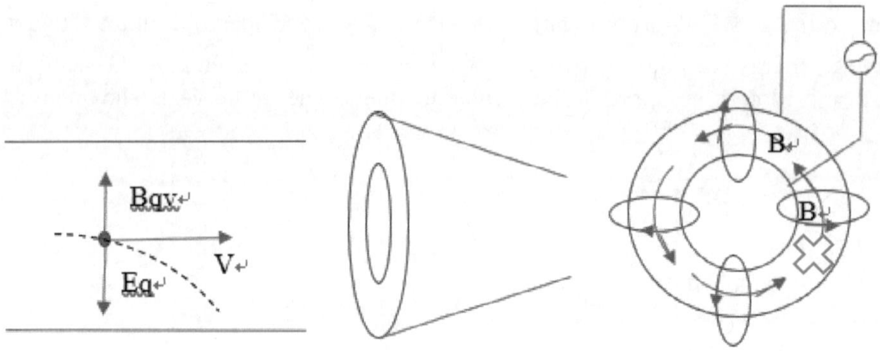

Figure 1. Force analysis diagram (left), flow diagram (middle), overall diagram (right)

The electric field inside the cylindrical capacitor is [6]: $E = \frac{U}{\ln(R_1 / R_2) r}$. The resistance of seawater in the entire capacitor is: $R = \frac{\rho l}{\pi(R_1^2 - R_2^2)}$, there, ρ is the resistivity of seawater, $\frac{1}{\rho}$=0.04. The four corners of the capacitor are coiled and sinusoidal alternating current is applied. The magnetic induction intensity of the coil is [7]:

$$B_1 = 2 * \mu_0 I(\frac{4\pi l}{2})(\cos(\arctan(\frac{l}{R_1 - R_2})) - \cos(180^\circ - \arctan(\frac{l}{R_1 - R_2})))$$

$$B_2 = 2 * \mu_0 I(\frac{4\pi(R_1 - R_2)}{2})(\cos(\arctan(\frac{R_1 - R_2}{l}) - \cos(180^\circ - \arctan(\frac{R_1 - R_2}{l})))$$

$$B = B_1 + B_2$$

The force of water in a cylindrical capacitor is: $\iiint_V dF = Bv\left|\vec{j}\right|$, J is current density: $\vec{J} = \frac{I}{\pi(R_1^2 - R_2^2)}$.

3.2.2. Example of calculation. Due to the limitation of funds and actual conditions, the model of the device design is only used to indicate its function. The dimensions of the device are as follows: the radius of the two plates of the cylindrical capacitor is 2.5cm and 4cm respectively, and the height is 25cm, which is placed at the position facing the blade, and the supporting portion use insulated rods.

Take a Na ion as the research object and analyze its motion process in this device: the charge-mass ratio of Na ion is 115.56C/g, assuming that the velocity in the horizontal direction is constant, v=0.8cm/s, 0.05m, and 0.03m. According to the above formula, the field strength E of the two plates can be obtained, so that the vertical acceleration a can be obtained, then the time t of the arrival of the ions in the vertical direction to the plate can be obtained. Finally, the horizontal movement of the ions can be obtained in the horizontal direction. l. The length of the device is designed to be 2.5cm.

$$E = \frac{U}{\ln(R_1 / R_2) r} = \frac{9}{\ln(0.04 / 0.025)(0.025 + 0.0075)} = 589.2V / m$$

$$a = \frac{Eq}{m} = 589.2 \times 115.56 \times 10^{-3} = 68.097 \text{m/s}$$

$$l = v_0 \times t = 0.8 \times 0.021 = 0.017m$$

Take 1kg of seawater as the research object and analyze the overall force in the device. The charge in seawater is about 1.6mol/kg. With a current of 0.84A, the value of magnetic field strength B can be obtained according to the above formula, and then the magnetic field force of 1 kg of seawater is obtained, the acceleration of the device is finally obtained.

$$B = 9.41 \times 10^{-5} T$$

$$F = 1.5N$$

$$a = 1.5m/s$$

$$vt = v_0 + at = 1.115m/s$$

$$\frac{\Delta v}{v_0} = \frac{v_t - v_0}{v_0} = 0.39$$

It can be known from the above formula that the device can increase the flow rate of seawater by about 39% and increase the kinetic energy by 1.93 times. This value is obtained without considering the current collection and considering only the acceleration. Therefore, the current collection effect increases a higher increasing degree of kinetic energy.

Because this is a model-related calculation, in the physical design, only the similarity principle is used to simulate the relevant parameters that are actually needed.

4. Experiment platform

The experiment platform built for this device is a model of ocean current energy generation device. The model of the ocean current power generation device includes a cylindrical current collecting and accelerating device, a water guiding mechanism, a generator and a transmission device and a power generation device supporting structure based on a coupled electromagnetic field. The entire device is underwater. When the current passes through the waters near the device, it enters the electromagnetic field of the collector accelerator and is affected by the magnetic field force. On the one hand, the flow velocity is accelerated and it enters the inlet of the gradation tube. On the other hand, the direction of the sea current is changed to make more ocean currents into the gradation tube. After the sea current enters the gradient tube, the internal pressure increases, the flow rate increases again, and then the blades are impacted to rotate the blades, which in turn drives the rotor of the generator to convert the mechanical energy into electrical energy.

The device model diagram is as follows:

Figure 2. Experiment platform

5. Conclusions and application prospects

5.1. Conclusions

1). Gradient tube design based on the equal-area criterion: According to the relationship between the area and flow rate, and the diffusion angle of the gradient tube, etc., a gradient tube is designed to increase the water flow rate to a condition suitable for the use of the generator through the gradient tube (calculated value is 3m/s, there is actually a slight error), which replaces the effect of the natural trench, increases the flow rate of the impact generator's water flow, and expands the usable range of the ocean current generating device.

2). Design of a cylindrical drainage collecting and accelerating device based on a coupled electromagnetic field: A composite electromagnetic field is established and combined with its own ion concentration in the seawater to form a current in the seawater through the action of an electric field. The current is deflected by the electric field force and the magnetic field force when passing through the electromagnetic field. Accelerate, consolidate seawater with a certain amount of kinetic energy into the available area, and further increase the available energy.

5.2. Application prospects

1). The seawater current collecting and accelerating device in the power generating device can change the direction of the sea current, which has a certain impact on the development of the marine transportation industry. For example, the acceleration part can serve the corresponding marine traffic propulsion technology.

2). The selection of the 30-50m region of the ocean is due to the consideration of the stability of the emitted electric energy. If there is a stable flow of water in the seawater or other locations in the river, the device can still be applied.

References

[1] Jianchao Wang. The Blade Design and Experimental Study on Horizontal Axis Turbine Driven by Tidal Current Energy. Ocean University of China, Master's Thesis. 2011.

[2] Xiaopeng Xin. Study on Hydrodynamic performance of horizontal marine current turbine. Zhejiang University, Master's Thesis. 2010.

[3] Tianbao Liu, Zhaoxue Cheng. Fluid Mechanics and Cascade Theory. Beijing: Machinery Industry Press, 1990.

[4] Cébron D, Viroulet S, Vidal J, Masson JP, Viroulet P (2017) Experimental and theoretical study of magnetohydrodynamic ship models. PLOS ONE 12(6): e0178599.

[5] Yuping He, Jianmin Yang. Blade designing and property analysis of a horizontal axis current turbine. THE OCEAN ENGINEERING, Vol. 31, No. 2, 2013, pp. 74-81.

[6] Hongwei Liu, Wei Li, Yonggang Lin, Shun Ma, You Ying. Axial Thrust Analysis of Ocean Current Turbine and Design of Hydraulic Equilibrator. Journal of Mechanical Engineering, 2009, 45(12):70-75.

[7] Wenwei Ma, Yuqing Zhou. Physics Course (Second Edition). Beijing: Higher Education Press, 2006.

[8] Jiang Han. Fundamental Study on Alternating Magnetic Field MHD Thruster. Graduate School of Chinese Academy of Sciences (Institute of Electrical Engineering), Master's Thesis, 2012.

[9] Font GI, Dudley SC. Magnetohydrodynamic propulsion for the classroom. The Physics Teacher. 2004; 42(7):410–415. doi: 10.1119/1.1804659.

[10] Long Kong. Engineering Fluid Mechanics. 2014, Beijing: China Electric Power Press.

[11] Scruggs, J (Scruggs, Jeff); Jacob, P (Jacob, Paul). Harvesting Ocean Wave Energy, Washington DC: AMER ASSOC ADVANCEMENT SCIENCE, 2009.

[12] Guangbiao Jiang, Yongsen He, Yingxiong Xiao. Numerical prediction of turbulent flow in progressively expanding tubes by DHR k-ε model and AMG method. Mining Research and Development, 2009, 29(03): 47-50.

Ocean Temperature Field 3D Visualization Key Technology Research Based on Pseudo-octree Model

Yuanyuan Feng [1, *], Beibei Han [2]

[1]School of Information Engineering, Zhengzhou University, Henan, China.
[2]Industrial Technology Reseach Institut, Zhengzhou University, Henan, China.

*Corresponding author: ieyuanfengzzu@163.com

Abstract. This paper presents the pseudo-octree model to organize data which is non-uniformly sampled in the depth direction and combines with early ray termination to realize accelerated rendering based on the analysis of features of ocean temperature field.VC and OpenGL render are used to realize the 3D visualization of ocean temperature field, which can help people understand the spatial distribution of ocean temperature elements. The experimental results show that the proposed method can provide technical reference to accelerate volume rendering of scalar field data.

1. Introduction

The ocean, which accounts for 71% of the Earth's surface area, is the cradle of earth's life and an important space for mankind's survival and sustainable development. It has unique advantages in terms of resources, environment, space and strategy. The study and exploration of the oceans will enable us to better understand the ocean world and ocean phenomenon and learn the changing laws of the oceans. This is of great significance to the development and utilization of oceans and the protection of the ocean environment. As one of the three basic physical parameters in oceans [1], ocean temperature has a great influence on the distribution of ocean resources [2]. and most of the ocean phenomena are related to ocean temperature. through the study and analysis of the ocean temperature, the distribution features and the variation patterns of the ocean temperature are obtained, which help humanity to improve the exploitation and utilization of ocean resources and ensure the sustainable development of the ocean environment. As an effective means of studying ocean temperature, visualization technology can visualize the effective information contained in ocean data and reveal the phenomena and scientific laws contained in oceanic data.

At present, the research at home and abroad on ocean visualization mainly focuses on two aspects: one is the ocean simulation and visualization platform (system) and the other is visualization techniques and methods. More mature and well-known visualization platforms are Google Ocean [3] developed by Google, World Wind [4] developed by NASA, and domestic EV-Globe [5,6]. The visualization of ocean data mainly focuses on visualization of vector field data [7]and visualization of scalar field data [8~10]. Volume rendering [11]is a common scalar field visualization method, which can obtain high quality images and can clearly display the internal details of the research object. Ray casting algorithm is the most widely used volume rendering method.

Based on the analysis of ocean temperature field data features, this paper proposes a three-dimensional space model-- pseudo-octree model. The pseudo-octree model is suitable for non-

Content from this work may be used under the terms of the Creative Commons Attribution 3.0 licence. Any further distribution of this work must maintain attribution to the author(s) and the title of the work, journal citation and DOI.
Published under licence by IOP Publishing Ltd

equidistant sampling data and is applied to the organization of temperature field data. This model is combined with ray casting algorithm to realize the three-dimensional visualization of ocean temperature.

2. Features of temperature field Data

The ocean temperature field has the space and time characteristics, and has four dimensions, namely longitude, latitude, depth and time. The time dimension reflects the time range of temperature; the longitude, latitude and depth dimension reflect the three-dimensional space of temperature. Changes in ocean temperature are spatially continuous, but the actual data collected is discrete due to the limitations of acquisition methodologies and instrument. The sampling data has teo features: (1) In the depth dimension, the seawater temperature data is sampled according to the depth of the seawater, and the data sampling layers are unevenly distributed, that is, the intervals of the sampling layers are non-equidistant. The characteristic of the sampling layer distribution is that the closer to the surface of the ocean, the denser the sampling layer is, and the more sampling data; the closer to the sea floor, the less the sampling layer, the less sampling data. (2) In the longitude and latitude dimension, the sampling data in the same depth layer is a uniform two-dimensional grid data, that is, the interval in the longitude dimension and the interval in the latitudinal dimension are both fixed values.

There are three ways to organize this type of volume data [12]:

Treating the least interval in the depthdimensionas a standard and interpolating the sampling layer in the entire depth direction. However, this way can lead to the dramatic increase of data size, which has a great influence on the later rendering efficiency.

Ignoring the sampling interval in the depth dimension and organizing the data with the original sampling layers. This way may lead to deviations in spatial features of data and even mislead readers.

Organizing the data with octree. This way divides data into 8 subtrees in latitude and longitude and depth dimension, which may cause fragmentation of data and increase the complexity of the data organization.

In consideration ofown feature of ocean data and its application in analysis of need, the above three kinds of organization are not applicable, and as a result, this paper presents a pseudo-octree model.

3. Pseudo-octree model

Because quadtree can only store single layer data, and octree splits the data equally into 8 parts. while the data in the depth dimension is multi-layer and non-equidistant. Furthermore, in comparison with the span in the latitude and longitude dimension, the span in depth dimension is relatively small, normal octree is not applicable, so here pseudo-octree model is constructed.

3.1. Pseudo-octree model

The pseudo-octree model is described as follows: In terms of the spatial structure, the pseudo-octree refers to dividing the three-dimensional space V by X and Y two dimensions from the middle position and dividing the three-dimensional space into four regular cuboids evenly. According to the target contained in each box, it is determined whether there is a need to further divide each box until each box satisfies the requirement of no longer being divided. That is, each box is filled by one target or the size of the child cuboid has satisfied the predefined non-recoverable dividing element. All the data, as the root node of the pseudo-octree, are continuously expanded downward by the division. Each node is divided into four sub-nodes until to leaf nodes, and the leaf nodes indicate the highest resolution. when pseudo-octree divides the data, the division is only in the X axis and Y axis direction, and there are always N depths in the Z axis direction. For example, the temperature field data used in this paper has 33 layers of data in the depth dimension, that is, $N = 33$. Pseudo-octree spatial structure is as shown in Fig 1.

IOP Conf. Series: Journal of Physics: Conf. Series **1064** (2018) 012064 doi:10.1088/1742-6596/1064/1/012064

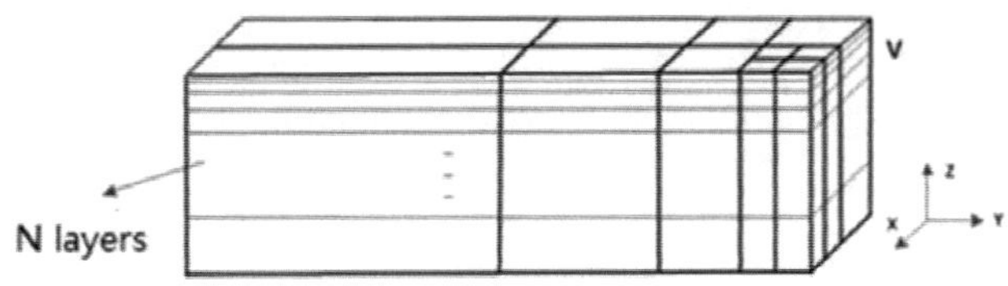

Figure 1. Pseudo-octree spatial structure

3.2. The establishment of octree model

The ordering and delamination of pseudo-octree enable fast indexing and positioning, accelerating the entire visualization process. Pseudo-octree structure in C language described as following:

```
//Define the pseudo-octree node
template<class T>
struct Pseudo-OctreeNode
{
    T data;              / / Node data
    T xMin,xMax;         //Node coordinates
    T yMin,yMax;
    T zMin,zMax;
    Pseudo-OctreeNode <T>*top_left,*top_right; //four child nódes of the node
    Pseudo-OctreeNode <T>*bottom_left,*bottom_right;
}

```

pseudocode of creating pseudo-octree:

```
/ * root is the root of the pseudo-octree and maxdepth is the maximum recursion depth * /
CreatePseudo-Octree (root, maxdepth, xMin, xMax, yMin, yMax, zMin, zMax)
{
Every recursion maxdepth-1;
Assign a node;
Assignment for node coordinates;
Calculate the half-length of two nodes in two dimensions;
Depth dimension, the data stored in the list lst_Depth;
// Recursively create subtree (according to the location of the node to determine the location of its children)
CreatePseudo-Octree ();
CreatePseudo-Octree ();
CreatePseudo-Octree ();
CreatePseudo-Octree ();
```

Data organization diagram based on pseudo-octree is as shown in Fig 2.

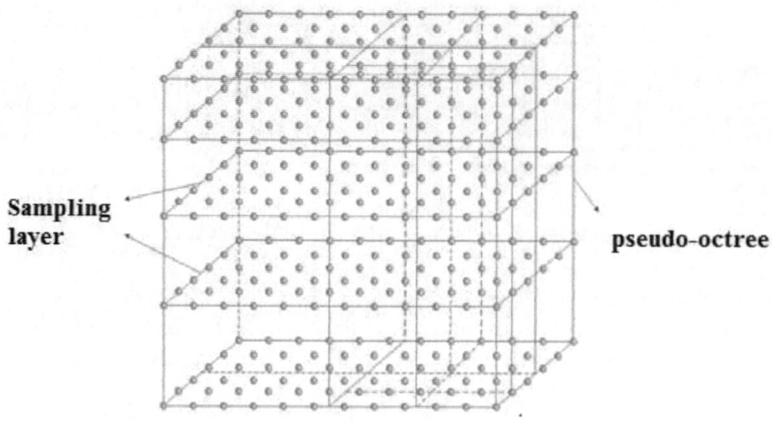

Figure 2. Data organization diagram based on pseudo-octree

3.3. *Pseudo-octree coding*

Pseudo-octree coding is similar with the way of octree coding, each block stores two-dimensional information, regardless of the depth interval, and only take the number of layers of data into account. The first two-bit position codes can be acquired by calculation, and the third bit position code is obtained by the data layer mapping. According to this pseudo-octree node encoding, we can easily locate any point in the space.

Assuming that the depth of the pseudo-octree is N, the leaf nodes are encoded as,

$$q_1 q_2 \ldots q_i FF \ldots F, \tag{1}$$

where

$$q_1, \ldots, q_n \in 0,1,2,3, i \in [0, N], \tag{2}$$

and F is a symbol different from 0, 1, 2, and 3. The purpose of the symbol F is to ensure that the encoding length of each tree is the same. Pseudo-octree coding diagram as shown in Fig 3.

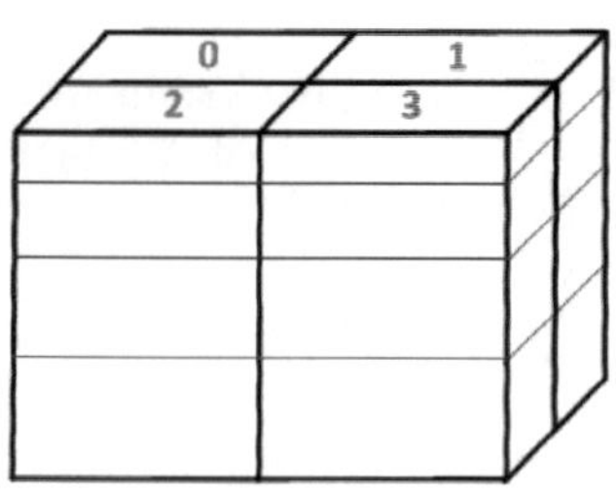

Figure 3. Pseudo-octree coding diagram

Taking the experimental data used in this paper as an example, each layer of data has 780 × 480 data points, requiring

$\log_2(780 \times 480) \approx 19$

bits code, which means that the binary code needs 19 bits. Using pseudo-octree organization data, so each time we can determine two codes, a total of 10(according to $\frac{19}{2} \approx 10$) comparisons are required.

It is known that the span of latitude and longitude is respectively lonS ~ lonE, latS ~ latE,as shown in Fig 4.

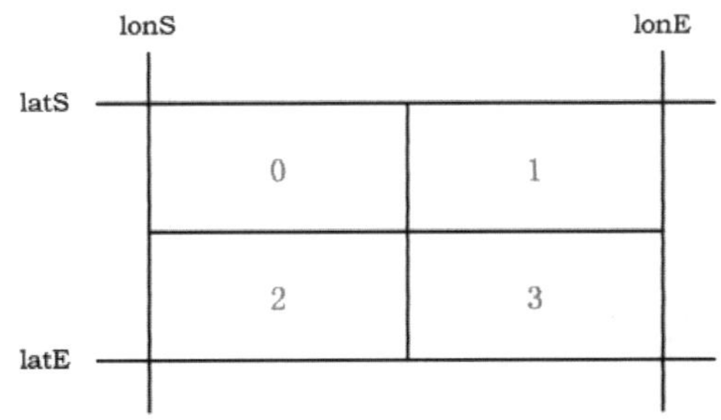

Figure 4. Pseudo-octree code (two-dimensional plan)

Input lon, lat, and compare $lon - lonS$ with $t_1 = \frac{lonE - lonS}{2}$,and compare $lat - latS$ with $t_2 = \frac{latE - latS}{2}$, if both of the former is greater than the latter, then the first digit of the code is 3. If the former of the first one is greater than the latter while the second is less than the latter, the first digit of the code is 1. If both of the former is less than the latter, the first digit of the code is 0. If the former of the first is less than the latter, while the second is greater than the former, the first digit of the encoding is 2. According to the above encoding rules, $lon - t_1$ and $\frac{lonE - lonS}{2^2}$, $lat - t_2$ and $\frac{latE - latS}{2^2}$ will be compared with, and so on, until the determination of the 10 positions encoding, then we get the first two encoding of pseudo-octree. The third encoding is obtained by mapping the number of layers in which the data is located.

4. Improvement of ray casting algorithm

Among the algorithms of volume rendering, the application of ray casting algorithm is the most common and widest for the three reasons: (1) ray casting algorithm can obtain higher image quality to meet people's requirements for rendering; (2) it's easy for ray casting algorithm to transferred to the GPU to achieve, to meet the people's drawing speed requirements; (3) it's easy to understand because it accords with human common sense. In view of the above advantages, this paper uses an improved ray casting algorithm to complete the visual rendering of ocean temperature field data.

4.1. Basic Theory of Ray casting Algorithm

Ray Casting algorithm is proposed by M. Levoy in 1988[11], which is a classical direct volume rendering method based on image sequences. The basic idea [12,13, 14] is described as follows: From each pixel of the image, a ray passing through the volume data (that is, the ray traverses the entire image sequence) is emitted along the ray of sight. and then, equidistant sampling points along the ray are selected in the part which is within the volume data. The color values and opacity values of the sampling points are determined by interpolation. and then compositing all the sampling points with front-to-back or with back -to-front in order to get the sampling point in the screen color and opacity, and final get the final rendered image. The schematic diagram is as follows:

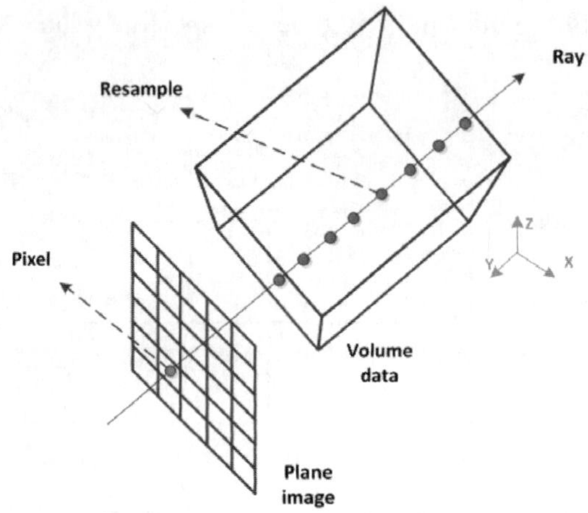

Figure 5. Ray Casting Algorithm Schematic

Although the ray casting algorithm can get high quality images, it has great calculation and long calculation time which means slow rendering. It is a research hot spots to improve the rendering speed of ray casting algorithm while ensuring image quality. The researchers at home and abroad mainly focus on the improvement in two aspects: software improvement and hardware improvement. Among these improvements, the software improvements mainly focus on the sampling of light, such as early light termination method [15], space leaping [16] and adaptive sampling [17]; while the hardware improvements mainly focus on GPU-accelerated volume rendering [18,19].

4.2. Light Early Termination

Due to the large calculation amount and low rendering speed of ray casting algorithm, in this paper, resampling step uses light early termination method to reduce unnecessary calculations. Taking advantage of the front-to- back composition, the opacity α of the sampling points on each projected ray is accumulated. When opacity value α accumulates to 1, the ray stops propagating. The sampling point is regarded as the critical sampling point, that is to say, the color and opacity of all the voxels before the sampling point are cumulatively combined to form the color value of the pixel point on the image. Voxels after the critical sampling point have no contribution to the result, and we can avoid calculating operations on these points, thus reducing the amount of computation and speeding up the volume rendering. Flowchart of the light early termination is as shown in Fig 6.

Ray may pass through the volume data at any angle, and according to the pseudo-octree division, we can quickly locate the incident point of projected ray and then resample along the projected ray.

5. Experiments and analysis

The experiment platform of this paper is Windows7, Intel Core i7-6700U processor, 8GB memory, C ++ programming language, OpenGL [20] rendering engine. The experimental data we used is AIPDAILY_20100101 from the Ocean Reanalysis Data Set--AIP Ocean 2.0 in the joining area of Asia and Indian-Pacific Ocean(APIO). This data is with a time range of 1 day and a spatial range of 29.99997E to 184E and 32.28827S to 51.66615N (span of latitude and longitude), and depth dimension range of 5-5500 meters with a total of 33 layers, and the resolution of about 0.2 ° × 0.2 °. There are 9 variables, longitude, latitude, depth and time 4 dimensions.

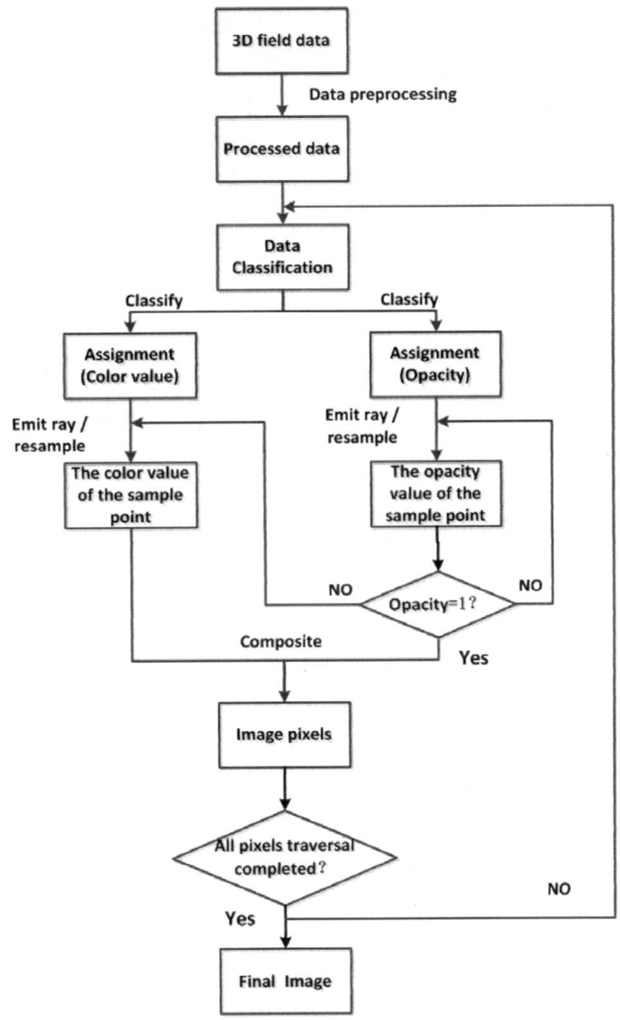

Figure 6. Flowchart of the light early termination

The data is stored in the local disk in NetCDF format and contains longitude, latitude, depth, time and attribute information. The data is extracted on demand through the interface function library of C ++ version of NetCDF. This article uses the ocean temperature field data from this dataset. The features of ocean temperature field data are as shown in table 1.

Table 1. Features of ocean temperature field data

variables	**long_name**	**units**	**range**	**resolution**	**missing value**
lon	Longitude	degrees_east	29.99997~ 184E	0.2°	--
lat	Latitude	degrees_north	32.28827S~ 51.66615N	0.2°	--
depth	depths	meters	5~5500 meters	--	--
time	time	days	1day	1day	--
t	temperature	Deg C	-1.89~33.05	--	0.f

Visualization results are as shown in fig7-fig9.

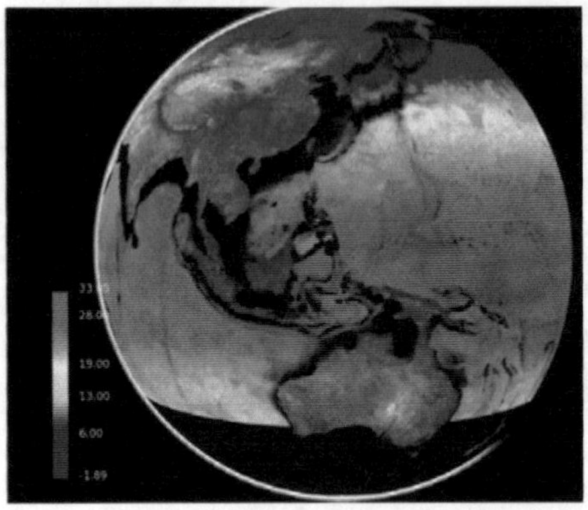

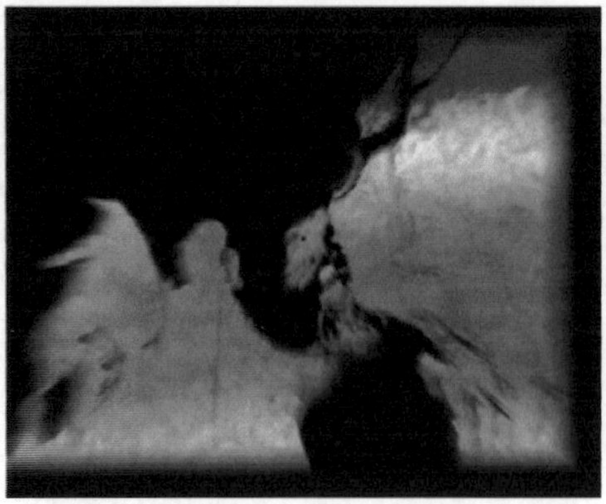

Figure 7. Panorama of ocean temperature visualization **Figure 8.** Partial enlargement of ocean temperature visualization

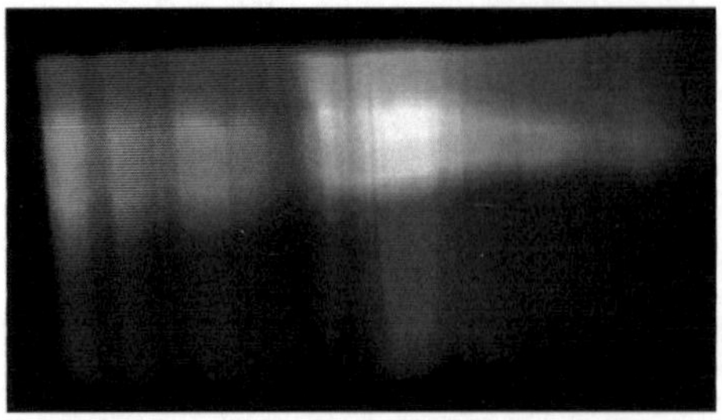

Figure 9. Lateral view of ocean temperature visualization

Fig 7 is a panoramic view of the visualization of ocean temperature from a distance. Fig 8 is a partial view of the visualization of ocean temperature after closing the earth carrier. Fig 9 is a side elevation view of the ocean temperature obtained by stretching 30 times in the depth dimension. It can better show the distribution characteristics of seawater temperature in the vertical direction. It can be seen that, as the water depth increases, the temperature gradually decreases.

From the results of visualization, we can get the temperature distribution in the space characteristics:(1) As the latitude increases, the temperature gradually decreases, and the temperature near the land is low. (2) As the depth increases, the ocean temperature exhibits a non-uniform decreasing characteristic. The closer to the ocean surface, the greater the temperature change; the closer to the sea floor, the smaller the temperature change.

6. Conclusion

In this paper, we study ocean temperature field and the ray casting algorithm, by reading and preprocessing the NetCDF data, and using the pseudo-octree model to organize the data and combining with the light early termination to achieve the 3D visualization of the temperature field data. The spatial distribution of ocean temperature elements is also analyzed. Compared with the traditional ray projection algorithm, this method can make up for the shortcomings of large amount of calculation and slow rendering speed and can better represent the spatial distribution characteristics of ocean elements in the study area, especially in the research of multi-layer data visualization analysis of more advantages.

Acknowledgments

First of all, thanks to the junior, Beibei Han, for discussing and solving the problems with me when I did the experiment. And secondly thanks to the experimental data, Ocean Reanalysis Data Set--AIP Ocean 2.0, for providing the data to verify the correctness of the method in this paper.

References

[1] J. Guo and Z. Tian," Research on Visualization of Ocean Temperature Field Based on Volume Rendering ", Marine surveying and mapping comprehensive academic seminar,2007.
[2] W. Dong and X. Zhang," Sphere based ocean environment scalar field element 3D visualization technology research ", Journal of Oceanography in Taiwan Strait, 2010, 29(4),pp. 571-88.
[3] http://www.worldoceanobservatory.org/content/google-ocean.
[4] https://worldwind.arc.nasa.gov/.
[5] https://baike.baidu.com/item/EV-Globe/4382314?fr=aladdin.
[6] C.C.Yan and X.Zhang," Design of solo-soldier 3D visualization simulation in aerial defense system based on EV-Globe ", Electronic Design Engineering, 2012.
[7] H.S.Li and W.J.Niu," Today and Tomorrow of Vector Field Visualization ", Application Research of Computers,2001 , 18 (8),pp. 11-14.
[8] C.Chen," Research and Implementation of Visualization for Marine Scalar Field Data ", master.thesis ,Graduate School of National University of Defense Technology,2012.
[9] Johnson C and Hansen C, "Introduction" in Visualization Handbook, edited by C.D.Hansen and C.R.Johnson , pp.5-13.
[10] D.Chen,"Key Technology Study on Marine Hydrological Data Visualization",PLA Information Engineering University,2015.
[11] Levoy M. "Volume rendering, display of surfaces from volume data", IEEE Computer Graphics & Applications, 1988, 8(3),pp. 29-37.
[12] L.S.Li," Research on 3-D Visualization Technology for Ocean Scalar Field", master.thesis , Ocean University of China,2015.
[13] C.Zhu," GPU-based multi-dimensional visualization of ocean water environment technology ", master.thesis ,Ocean University of China, 2015.
[14] C.Z.Liang," GPU-based light projection algorithm ", master.thesis , Xidian University, 2009.
[15] M. Levoy," Efficient Ray Tracing of Volume Data", ACM Transactions Computer Graphics and Applications, 1990; 9(3),pp. 245~261.
[16] R Avial, "Towards a Comprehensive Volume Visualization System", In: Proceedings of Visualization'92, Boston, 1992.
[17] Ljung P, "Adaptive Sampling in Single Pass, GPU-based Raycasting of Multiresolution Volumes", Institute of Technology, 2006,pp. 39-46.
[18] H.B.Zhu,"The Analysis and Improvement of Ray Casting by GPU",master.thesis ,Shandong University of Science and Technology,2014.
[19] H. Scharsach," Advanced GPU raycasting",Proceedings of Cescg, 2005, 16(5),pp. 69--76.
[20] E. J.Kessenich and Version, , et al,The OpenGL® Shading Language. www, 2010, 26(4),pp:853–854.

IOP Conf. Series: Journal of Physics: Conf. Series **1064** (2018) 012065 doi:10.1088/1742-6596/1064/1/012065

Flow Field Characteristic Research on Buried Weapon Bay in Different Mach Numbers

Zhe Wu*

School of Aerospace Engineering, Nanjing University of Aeronautics and Astronautics, Nanjing, China

*Corresponding author e-mail: 252455858@qq.com

Abstract. Actually, besides the missile, some complex surface such as the surface of missile, accessories and opening and closing mechanism for the missile bay etc. will have obvious influence on flow in the bay. In order to study the steady flow characteristics of the missile bay, the calculation model is established and simplified according to the research needs. The working condition is that a missile model is carried in the middle of the missile bay. The characteristics of the flow field in the bay at Mach 1.5 and Mach 3.5 are calculated and analysed under the above working condition.

1. Introduction

Considering the high complexity of the research problem, the research object is the aircraft missile bay, rather than the whole aircraft. In fact, the structure of the aircraft which far away from the missile bay has little influence on the flow field of the bay. So it's reasonable to ignore them and the Simplified missile model is shown in Figure 1 and Table 1.

Although it is easy to implement the weapon mount method which used in traditional fighter aircraft in the structural design, it brings the cost of high resistance and high radar scattering cross-sectional area (RCS). Under the guidance of the new concept of air combat, the supersonic cruise and stealth performance of fighter become the hard indicators of the fighter's combat capability, so buried weapon bays becomes the first option for a new generation of warplanes.

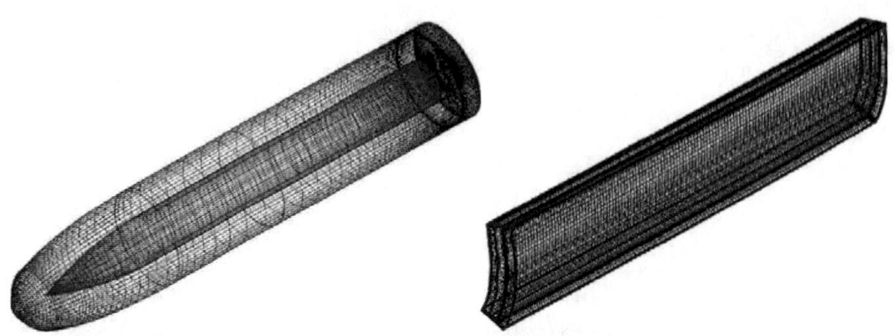

Figure 1. The missile model

Content from this work may be used under the terms of the Creative Commons Attribution 3.0 licence. Any further distribution of this work must maintain attribution to the author(s) and the title of the work, journal citation and DOI.
Published under licence by IOP Publishing Ltd

Table 1. Geometries of the missile model

Geometries	numerical value
Length(L)	2500mm
Width(W)	820mm
L/W	8.33

2. Calculation model

The Chimera grid of the body of the bay, the door of the bay and the missile model have been established as follow: the grid of the body of the bay is background grid, the grid of the door and missile model have establish data connections with the background grid through nesting as shown in Figure 2.

In order to study the influence of the steady state field in the missile bay, longitudinal and lateral sections of the missile bay has been taken:

(a) Symmetry plane, 400mm offset surface and 200 mm offset surface three longitudinal cross section and has been taken as shown in Fig.3.

(b) X/L=0.2,0.4,0.6,0.8 four transverse section has been taken as shown in Fig.3.

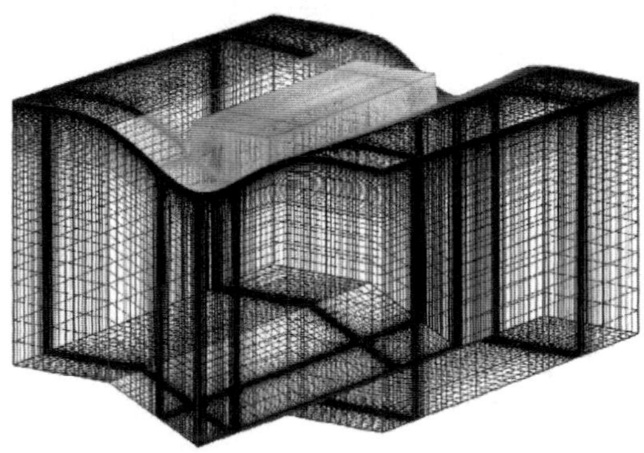

Figure 2. Mesh model of the missile bay

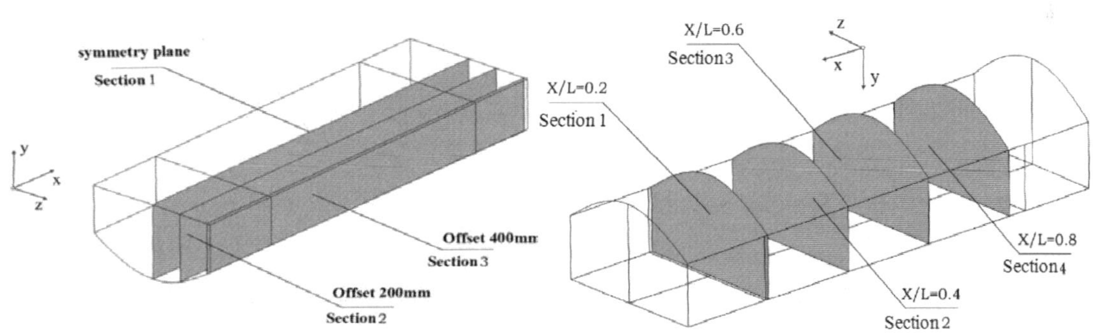

Figure 3. Longitudinal and lateral sections of the missile bay

3. Analysis of steady flow field characteristics of missile bay

3.1 Mach number cloud map of the longitudinal section of the cabin

At angle of attack and sideslip Angle of 0oworking conditions where the incoming flow Mach number is1.5 and 3.5, The influence of Mach number on steady flow field of buried weapon bay was studied. The cloud map of Longitudinal section of cabin is shown in Fig.4 and Fig.5.

IOP Conf. Series: Journal of Physics: Conf. Series **1064** (2018) 012065 doi:10.1088/1742-6596/1064/1/012065

symmetry plane **Offset 200mm** **Offset 400mm**

Figure 4. Mach number cloud map of the longitudinal section of the bay at 1.5M

symmetry plane **Offset 200mm** **Offset 400mm**

Figure 5. Mach number cloud map of the longitudinal section of the bay at 3.5M

The calculation shows that although different Mach Numbers do not play a decisive role in the type of flow field in the cabin, they still have some influence on the turbulence of the cabin. Since there is a difference in flow, there must be a difference in the pressure coefficient distribution of the bay.

Pressure distribution has a vital impact on the safe separation of weapons. As shown in Fig.6&7 the pressure coefficient at the front of the missile is smaller than that at the rear in supersonic flow field which will produce an upward force moment. This moment will increase the Angle of attack after the weapon is put into the air, which will lead to the retraction of the missile into the bay, causing great security risks.

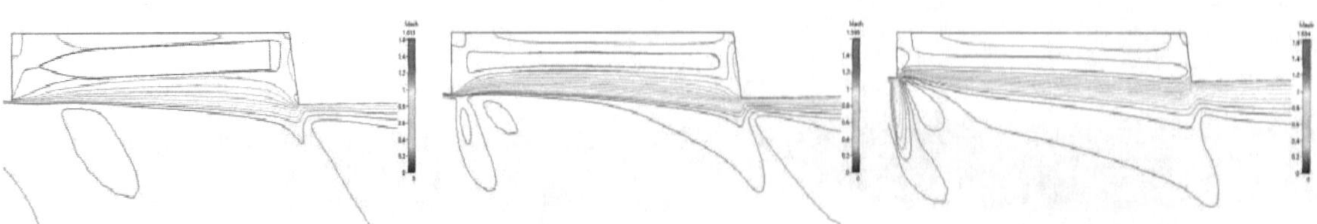

symmetry plane **Offset 200mm** **Offset 400mm**

Figure 6. Contour map of Mach number of each section of the cabin at 1.5M

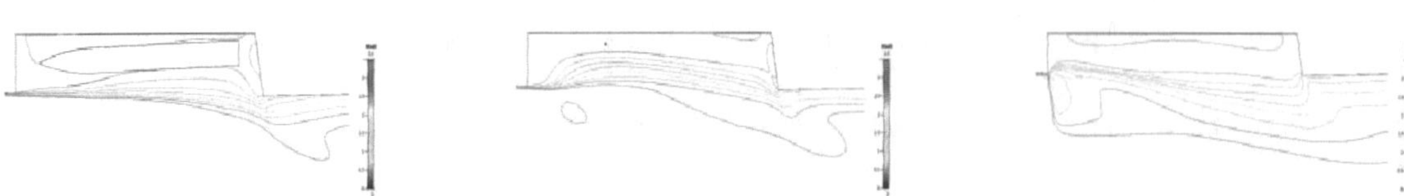

symmetry plane **Offset 200mm** **Offset 400mm**

Figure 7. Contour map of Mach number of each section of the cabin at 3.5M

3.2 Mach number cloud map of the lateral section of the bay

From Fig.8, pressure cloud diagram of different lateral sections shows that in the falling area of the missile, The pressure distribution is not uniform. When the missile falls through this area, it may be subjected to a certain amount of lateral torque, leading to deviation.

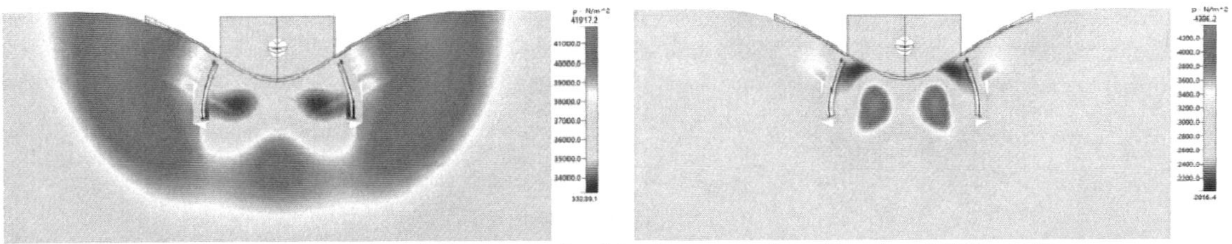

Pressure cloud diagram of cabin door opening (Section 1 X/L=0.2)

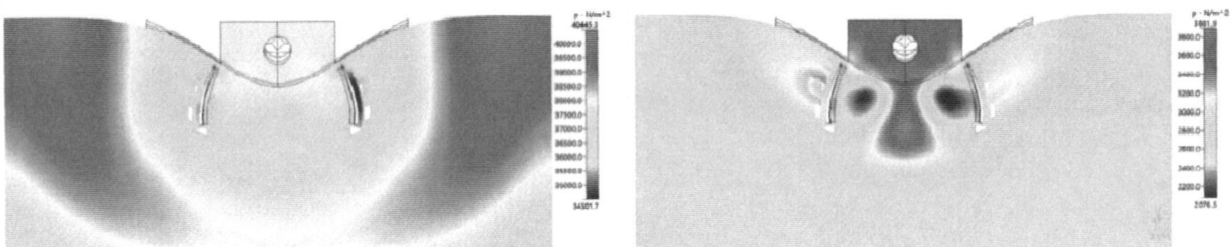

Pressure cloud diagram of cabin door opening (Section 1 X/L=0.4)

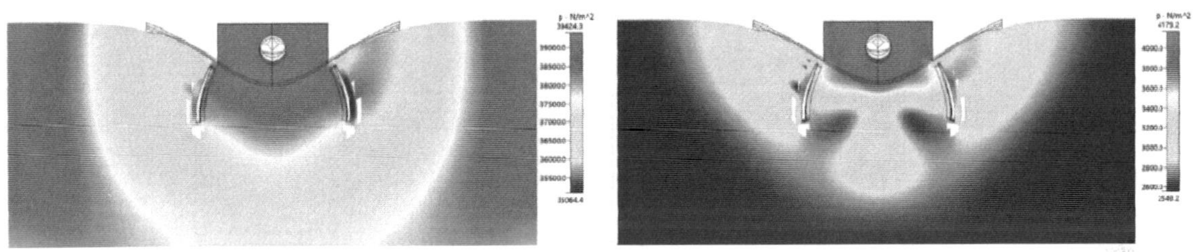

Pressure cloud diagram of cabin door opening (Section 1 X/L=0.6)

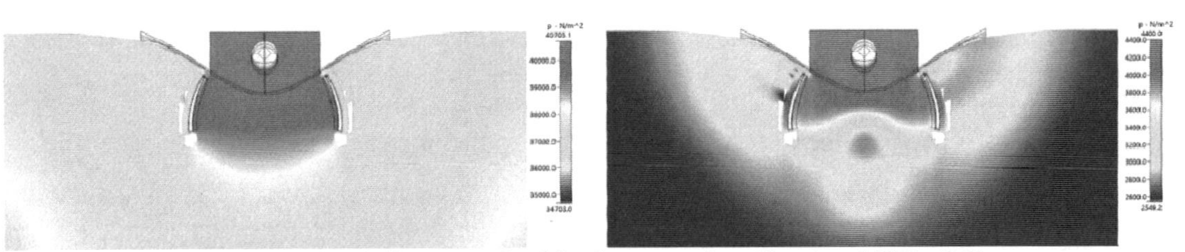

Pressure cloud diagram of cabin door opening (Section 1 X/L=0.8)

Figure 8. Pressure cloud diagram of cabin door opening

In addition, the flow around the cabin can be seen in figures 9&10.

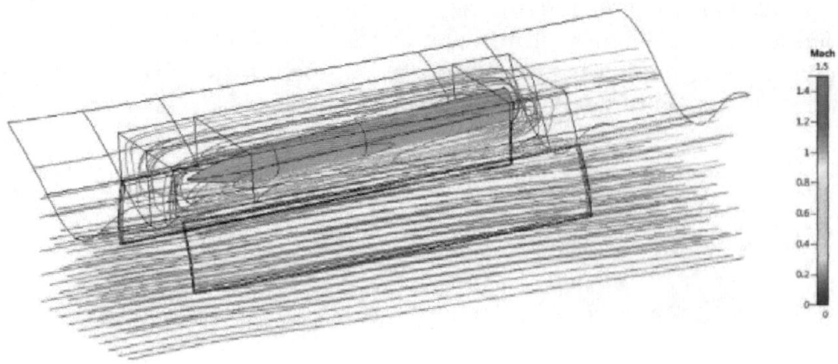

Figure 9. Flow around the cabin at 1.5M

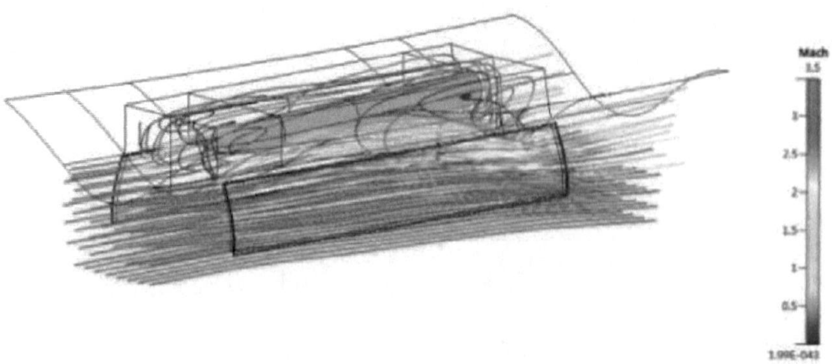

Figure 10. Flow around the cabin at 3.5M

4. Conclusion

(a) The pressure distribution characteristic of the door cabin on the longitudinal is very unfavorable to the separation of weapons.

(b) The pressure distribution characteristic of the door cabin on the lateral is subjected to a certain amount of torque, leading to deviation.

(c) With the Mach number increased, the overall flow type of the missile bay was not affected however the interference of external flow field of the bay to the interior of the bay became serious.

References

[1] Maureen B. Tracy and E. B. Plentovich. Cavity Unsteady-Pressure Measurements at Subsonic and Transonic Speeds [R]. NASA TP-3669, 1997.

[2] Stallings RL and Wilcox FJ. Experimental cavity pressure distributions at supersonic speeds NASA TP-2683, 1987.

[3] Rainey R W. A wind-tunnel investigation of bomb release at a Mach number of 1.62 [R]. NACA RM-L53L29, March 1954.

[4] Blair A B and Stallings R L. Cavity door effects on aerodynamic loadings of compressed-carriage store configurations separating from cavities at supersonic speeds [J]. AIAA Paper 88-0333.

[5] Langtry R B, Gola J, Mente F R. Predicting 2D airfoil and 3 D wind turbine rotor performance using a transition model for general CFD codes. AIAA- 2006 -395. 2006.

[6] MacCormack R W. A Numerical Method for Solving Equations of Compressible Viscous Flows. 1981.

[7] Sham Rane, Ahmed Kovacevic, Nikola Stosic, Madhulika Kethidi. Grid deformation strategies for CFD analysis of screw compressors [J]. International Journal of Refrigeration. 2013(7).
[8] Ahmed Kovacevic, Sham Rane. Algebraic Generation of Single Domain Computational Grid for Twin Screw Machines Part II – Validation [J]. Advances in Engineering Software.

IOP Conf. Series: Journal of Physics: Conf. Series **1064** (2018) 012066 doi:10.1088/1742-6596/1064/1/012066

Numerical Solution of Unsteady Ekman Equation Modified by Wave and Linear Friction Term

Jia Chen[a], Wenli Qiao[b] and Jinbao Song[*]

Ocean College Zhejiang University, Zhejiang Zhoushan, China

*Corresponding author e-mail: songjb@zju.edu.cn, [a]18374883087@163.com, [b]qiaowl@zju.edu.cn

Abstract. Based on the assumption that taking the KPP model as the vertical vortex viscosity coefficient, the unsteady Ekman equation modified by wave and linear friction terms is solved numerically with the finite difference method and the influence of wave and linear friction terms on the unsteady Ekman current are studied. Based on the measured data of AWAC, the Stokes drift is calculated by the Stokes drift formula; and on the basis of the measured data of wind speed, the average wind speed of the 3 hours is divided into three different sections: low wind speed (0~4) m/s, medium -low wind speed (4~8) m/s and medium-high wind speed (8~12) m/s. The results show that: Stokes drift has the greatest impact on Ekman current under the low wind speed, which is accounted for 27.4%, the second is under the medium-low wind speed of 7.5%, and the smallest effect is under the medium-high wind speed of 4.2%. In additon, the effect of linear friction term on Ekman current is also studied. It is shown that under the medium-high wind speed and the medium-low wind speed, together with the ratio of the linear friction term to the Coriolis force is, the influence of the linear friction term on the Ekman current has exceeded that of the Stokes drift on the Ekman current; while under the low wind speed, the effect of linear friction term on Ekman current can not exceed that of Stokes drift on Ekman current until. Furthermore, the comparison between the numerical solution and the actual measured data also shows different characteristics. In the perspective of relevance, the correlation between the numerical simulation results and the measured data is above 0.7 under the medium-high wind speed, which is higher than the correlation between the both of 0.4~0.5 under the low wind speed and the medium-low wind speed; and in the perspective of the root mean square deviation degree, the root mean square deviation between the numerical simulation results and the measured data under the medium-high wind speed is lower than that under the low wind speed and the medium -low wind speed. The above results show that the influence of Stokes drift, linear friction term and the unsteady wind on Ekman model can not be ignored.

1. Introduction

Ekman (1905) [1] first proposed the Ekman equation, which demonstrated the formation of the ocean current driven by wind stress in the steady state, and explained the phenomenon that the net transport in a marine friction layer caused by the wind stress in the northern hemisphere pointed to the right side of

Content from this work may be used under the terms of the Creative Commons Attribution 3.0 licence. Any further distribution of this work must maintain attribution to the author(s) and the title of the work, journal citation and DOI.
Published under licence by IOP Publishing Ltd

the wind stress.Based on the Ekman theory, the theoretical research framework of the upper boundary layer of the ocean has been developed in succession.The representative results are as follows: Sverdrup (1947) [2] considered the pressure gradient force based on the Ekman drifting theory, that is, the pressure gradient force, the Coriolis force and the friction force balance each other,and he had put forward the Sverdrup equation and demonstrated that the essence of the North Equatorial Countercurrent is the windward geostrophic current;based on the Sverdrup theory, Stommel (1948) [3] considered the sea bottom friction effect and proposed the Westward intensification of wind-driven ocean currents;Munk (1950) [4] proposed a quantitative theory about the main features of the wind-driven circulation, and comprehensively elaborated the main features of the wind-driven ocean circulation.However, with the continuous development of observation technology, it is found that there are significant differences between the traditional Ekman theory and the actual observation. In particular, several hypotheses of the Ekman theory are different from the actual ocean situations to a great extent.

In order to explain the differences between the traditional Ekman theory and the actual observation,later researchers added different revisions to reduce the bias between theoretical models and actual observations.Wave as a medium to transmit momentum and energy in the atmosphere to the surface flow field of the ocean has an important influence on the structure of surface flow field and mixing of the upper layer of the ocean.Therefore, the study of the effect of waves on ocean circulation has always been the centre of attention. Stokes (1847) [5] first proposed a drift theory that periodic ocean surface gravity waves will produce a net displacement in the direction of wave propagation, that is, Stokes drift; Huang (1979) [6] pointed out that the ocean surface circulation can be generated not only by the driven by the wind stress in the traditional Ekman model, but also by the wave motion; Kenyon (1970) [7] believes that Stokes drift will become a major part of ocean surface current under the circumstance without wind; McWillianms and Restrepo (1999) [8] and Polton (2005) [9] calculated the Stokes drift and the Stokes transport based on the PM spectrum describing the full growth of wind waves, they have found that the Stokes transport is an important part of Ekman transport; Lewis and Belcher (2004) [10] used the single wave formula to calculate the Stokes drift,then they pointed out that the deviation of direction angle of the traditional Ekman model can be reduced by considering the Stokes drift and expounded the important influence of the sea surface wave on the structure of the flow of the Ekman layer; the study of Saetra at al (2007) [11] also shows that Stokes drift can reduce the deviation of direction angle of the traditional Ekman model; Song (2009) [12] calculated the Stokes drift with the DP spectrum describing the full growing wind and wave, and calculated the wind-to-wave input and the wave dissipation terms. He thought that the influence of the Stokes drift on the Ekman current in medium-high wind speed was much larger than the influence of the wind-to-wave input and the wave dissipation terms on the Ekman current; Zhang (2013) [13] used the wave spectrum of the ideal experimental output at a constant wind speed with the third-generation wave model and pointed out that Stokes drift, wind-to-wave input, and wave-induced momentum transfer all have a great influence on the Ekman flow.

However, the effect of unsteady wind on the Ekman model is rarely studied. In the study of Ashkenazy et al. (2015) [14], it is pointed out that after considering the influence of the unsteady wind, the Ekman model should add a linear friction term to counteract the dispersion of the sea surface flow, and they think that the linear friction term is critical to the Ekman equation of the unsteady wind action;Kim et al. (2015) [15] pointed out that the influence of the linear friction term on the Ekman model was very important in the finite depth of water, and compared with the measured data in ROMS and FVCOM models, he also pointed out that the operation result by the model which under the ratio of the linear friction term to the Coriolis force parameter was (0.1, 0.5) was more consistent with the measured data; Buffoni et al. (2015) [16] selected three different regions of the wind field and discussed the effects of unsteady wind on the Ekman model;the study of Wenegrat and Mcphaden (2016) [17] indicates that the diurnal variation of the vertical eddy viscosity will affect the flow structure of Ekman layer. The time-dependent Ekman equation modified by wave and linear friction terms is discretized by finite difference method in this paper, and the effect of Stokes drift and linear friction term on the Ekman current under the unsteady state is discussed, and the results of the numerical simulation are tested with the measured data of AWAC.

2. Governing equations and corresponding boundary conditions

When steady winds continue to act on the sea surface, the resulting large-scale flow is also constant.At this point, the product of the equilibrium between vertical turbulent frictional force and Coriolis force is the classic Ekman drift, and its governing equation is:

$$\begin{aligned} f_{cor}v + A_v \frac{\partial^2 u}{\partial^2 z} = 0 \\ -f_{cor}u + A_v \frac{\partial^2 v}{\partial^2 z} = 0 \end{aligned} \tag{1}$$

Where: u represents the flow velocity in the x direction, v represents the flow velocity in the y direction, z represents depth (setting upward as positive direction), f_{cor} represents Coriolis force parameter, A_v represents the vertical eddy viscosity and it is a constant.

The corresponding boundary conditions are as follows:

$$\begin{aligned} z = 0: \quad & \rho_w A_v \frac{\partial u}{\partial z} = \tau_x, \quad \rho_w A_v \frac{\partial v}{\partial z} = \tau_y \\ z \to \infty: \quad & u = v = 0 \end{aligned} \tag{2}$$

Where: ρ_w represents the density of water, $\boldsymbol{\tau} = (\tau_x, \tau_y)$ represents wind-stress.

However,in the fact,it is impossible for steady winds to have a lasting effect on the sea surface in the ocean.After considering the influence of unsteady wind, wave and linear friction term on Ekman current, the equation (1) becomes as follows(Jenkins, 1989 [18]; Song and Xu, 2013 [19])

$$\frac{\partial U_{WE}}{\partial t} + if_{cor}U_{WE} + rU_{WE} + if_{cor}U_S = \frac{\partial}{\partial z}(A_v \frac{\partial U_{WE}}{\partial z}) \tag{3}$$

Where: $U_{WE} = U_{WE}(z,t) = u_{WE}(z,t) + iv_{WE}(z,t)$ represents the horizontal velocity on the x-y complex plane,which is equivalent to the Lagrange mean flow minus Stokes drift,it can also be understood as the Euler mean flow(Jenkins, 1987 [20]), z represents depth (setting upward as positive direction), $z = 0$ means sea level at rest, t represents time, $i = \sqrt{-1}$, $U_S = U_S(z,t) = u_S(z,t) + iv_S(z,t)$ represents Stokes drift on the x-y complex plane, r represents a constant coefficient similar to Rayleigh friction(Ashkenazy et al., 2015 [14]),its physical meaning is: when the Ekman model is at a finite depth: molecular friction at smaller scales, such as the generation of internal waves, wave breaking, boundary friction, and turbulent mixing processes, causes many energy losses that are often overlooked, r is an empirical coefficient used to represent these energy losses.

Correspondingly, its boundary conditions are:

$$\begin{aligned} z = 0, \quad & A_v(z,t)\frac{\partial U_{WE}}{\partial z} = \frac{\tau_a}{\rho_w} \\ z = -h, \quad & A_v(z,t)\frac{\partial U_{WE}}{\partial z} = 0 \end{aligned} \tag{4}$$

Where: ρ_w=1025 $kg.m^{-3}$ represents the density of water, $\tau_a = \tau_{ax} + i\tau_{ay}$ represents complex wind-stress, and it can be calculated by 10 meters high surface wind field U_{10}:

$$\tau_a = (\tau_{ax}, \tau_{ay}) = \rho_a C_d \,|U_{10}|\, \mathbf{U_{10}} \tag{5}$$

Where: $\rho_a = 1.2kg.m^{-3}$ represents air density, C_d represents ocean-atmosphere drag coefficient, it can be calculated by the formula given below (Wu, 1982 [21]):

$$C_d = (0.8 + 0.065U_{10}) \times 10^{-3} \tag{6}$$

In the classical Ekman equation, the vertical eddy viscosity A_v is a constant, this paper adopts the vertical eddy viscosity of the KPP model (Large et al., 1994 [22]; McWilliams and Huckle, 2005 [23]; McWilliams, 2012 [24]), that is:

$$A_v(z,t) = \begin{cases} \kappa u_* h G(\sigma), 0 \le \sigma \le 1 \\ \kappa u_* h G(1), \sigma > 1 \end{cases} \tag{7}$$

Where: $\kappa = 0.41$ is the Carmen constant, $u_* = \sqrt{\dfrac{\tau_a(t)}{\rho_w}}$ is the friction velocity, $h = c_1 \dfrac{u_*}{f_{cor}}$ is the depth of the boundary layer, $c_1 = 0.7$ is the constant(McWilliams and Huckle, 2005 [23]), $\sigma = -\dfrac{z}{h}$ can be regarded as coordinate conversion, $G(\sigma)$ is the function related to σ, it can be calculated by the following formula:

$$G(\sigma) = \sigma(1-\sigma)^2 + \frac{(\sigma_0 - \sigma)^2}{2\sigma_0} H(\sigma_0 - \sigma) \tag{8}$$

Where: $\sigma_0 = 0.05$, $H(a) = \begin{cases} 1 & a > 0 \\ 0 & \text{in other cases} \end{cases}$.

3. Numerical discrete method and data sources

3.1. Numerical discrete method

When the influences of unsteady wind, linear friction term and wave are considered, the Ekman equation with the vertical eddy viscosity of the KPP model becomes extremely complex, and it is difficult to obtain the theoretical solution.Therefore, compared with the theoretical solution of the modified Ekman equation, numerical discrete solution becomes a more effective method.The governing equation (3) is discretized by the finite difference method. The upwind scheme is used for the time term and the central difference scheme for the diffusion term is adopted.The discrete form is as follows:

$$\frac{(U_{we})_j^{n+1}-(U_{we})_j^n}{\Delta t}+(r+if_{cor})(U_{we})_j^{n+1}+if_{cor}(U_s)_j^{n+1}$$
$$=\frac{(A_v)_{j+1/2}^{n+1}\frac{\partial U_{we}}{\partial z}\Big|_{j+1/2}^{n+1}-(A_v)_{j-1/2}^{n+1}\frac{\partial U_{we}}{\partial z}\Big|_{j-1/2}^{n+1}}{\Delta z} \tag{9}$$
$$=(\frac{(A_v)_{j+1}^{n+1}+(A_v)_j^{n+1}}{2\Delta z}\frac{(U_{we})_{j+1}^{n+1}-(U_{we})_j^{n+1}}{\Delta z}-\frac{(A_v)_j^{n+1}+(A_v)_{j-1}^{n+1}}{2\Delta z}\frac{(U_{we})_j^{n+1}-(U_{we})_{j-1}^{n+1}}{\Delta z})$$

Move the item U_{we} at time $n+1$ to the left of the equation, and the rest to the right of the equation:

$$a_j(U_{we})_{j-1}^{n+1}+b_j(U_{we})_j^{n+1}+c_j(U_{we})_{j+1}^{n+1}=(U_{we})_j^n-if_{cor}(U_s)_j^{n+1} \tag{10}$$

Where:

$$a_j=-((A_v)_j^{n+1}+(A_v)_{j-1}^{n+1})\frac{\Delta t}{2\Delta z^2}$$
$$b_j=1+((A_v)_{j+1}^{n+1}+2(A_v)_j^{n+1}+(A_v)_{j-1}^{n+1})\frac{\Delta t}{2\Delta z^2}+(r+if_{cor})\Delta t \tag{11}$$
$$c_j=-((A_v)_{j+1}^{n+1}+(A_v)_j^{n+1})\frac{\Delta t}{2\Delta z^2},$$

Since it is a second kind of boundary condition, the boundary conditions need to be separately discretized.

When $z=0$, i.e. on the sea surface:

$$\frac{(U_{we})_1^{n+1}-(U_{we})_1^n}{\Delta t}+(r+if_{cor})(U_{we})_1^{n+1}+if_{cor}(U_s)_1^{n+1}$$
$$=\frac{(A_v)_{1+1/2}^{n+1}\frac{\partial U_{we}}{\partial z}\Big|_{1+1/2}^{n+1}-(A_v)_{1-1/2}^{n+1}\frac{\partial U_{we}}{\partial z}\Big|_{1-1/2}^{n+1}}{\Delta z} \tag{12}$$
$$=(\frac{(A_v)_2^{n+1}+(A_v)_1^{n+1}}{2}\frac{(U_{we})_2^{n+1}-(U_{we})_1^{n+1}}{\Delta z}-\frac{(A_v)_0^{n+1}+(A_v)_1^{n+1}}{2}\frac{(U_{we})_1^{n+1}-(U_{we})_0^{n+1}}{\Delta z})/\Delta z$$
$$=(\frac{(A_v)_2^{n+1}+(A_v)_1^{n+1}}{2}\frac{(U_{we})_2^{n+1}-(U_{we})_1^{n+1}}{\Delta z}-\frac{\tau_a}{\rho_w})/\Delta z$$

The above formula makes use of the boundary condition that $\frac{(A_v)_0^{n+1}+(A_v)_1^{n+1}}{2}\frac{(U_{we})_1^{n+1}-(U_{we})_0^{n+1}}{\Delta z}=\frac{\tau_a}{\rho_w}$.

Move the item U_{we} at time $n+1$ to the left of the equation, and the rest to the right of the equation:

$$(b_2+c_2)(U_{we})_1^{n+1}+a_2(U_{we})_2^{n+1}=(U_{we})_1^n-\frac{\tau_a\Delta t}{\rho_w\Delta z}-if_{cor}\Delta t(U_s)_1^{n+1} \tag{13}$$

When $z=-h$, i.e. at the bottom of the sea:

$$\frac{(U_{we})_j^{n+1}-(U_{we})_j^n}{\Delta t}+(r+if_{cor})(U_{we})_j^{n+1}+if_{cor}(U_s)_j^{n+1}$$
$$=\frac{(A_v)_{j+1/2}^{n+1}\frac{\partial U_{we}}{\partial z}\Big|_{j+1/2}^{n+1}-(A_v)_{j-1/2}^{n+1}\frac{\partial U_{we}}{\partial z}\Big|_{j-1/2}^{n+1}}{\Delta z}$$
$$=\left(\frac{(A_v)_{j+1}^{n+1}+(A_v)_j^{n+1}}{2\Delta z}\frac{(U_{we})_{j+1}^{n+1}-(U_{we})_j^{n+1}}{\Delta z}-\frac{(A_v)_j^{n+1}+(A_v)_{j-1}^{n+1}}{2\Delta z}\frac{(U_{we})_j^{n+1}-(U_{we})_{j-1}^{n+1}}{\Delta z}\right) \quad (14)$$

$$=-\frac{(A_v)_j^{n+1}+(A_v)_{j-1}^{n+1}}{2\Delta z}\frac{(U_{we})_j^{n+1}-(U_{we})_{j-1}^{n+1}}{\Delta z}$$

The above formula makes use of the boundary condition that $\frac{(A_v)_{j+1}^{n+1}+(A_v)_j^{n+1}}{2\Delta z}\frac{(U_{we})_{j+1}^{n+1}-(U_{we})_j^{n+1}}{\Delta z}=0$. Move the item U_{we} at time $n+1$ to the left of the equation, and the rest to the right of the equation:

$$a_j(U_{we})_{j-1}^{n+1}+(b_j+c_j)(U_{we})_j^{n+1}=(U_{we})_j^n-if_{cor}\Delta t(U_s)_j^{n+1} \quad (15)$$

Equations (10), (13), and (15) are expanded and linked together, and the discrete equations can be converted into tri-diagonal matrix and solved by using the chasing method:

$$\begin{pmatrix} b_2+c_2 & a_2 & & & & \\ a_2 & b_2 & c_2 & & & \\ & a_3 & b_3 & c_3 & & \\ & \cdots & \cdots & \cdots & & \\ & & & a_{j-1} & b_{j-1} & c_{j-1} \\ & & & & a_j & b_j+c_j \end{pmatrix}_{j*j}\begin{pmatrix}(U_{we})_1^{n+1}\\(U_{we})_2^{n+1}\\(U_{we})_3^{n+1}\\ \cdots \\(U_{we})_{j-1}^{n+1}\\(U_{we})_j^{n+1}\end{pmatrix}_{j*1}=\begin{pmatrix}(U_{we})_1^n-if_{cor}\Delta t(U_s)_1^{n+1}-\frac{\tau_a\Delta t}{\rho_w\Delta z}\\(U_{we})_2^n-if_{cor}\Delta t(U_s)_2^{n+1}\\(U_{we})_3^n-if_{cor}\Delta t(U_s)_3^{n+1}\\ \cdots \\(U_{we})_{j-1}^n-if_{cor}\Delta t(U_s)_{j-1}^{n+1}\\(U_{we})_j^n-if_{cor}\Delta t(U_s)_j^{n+1}\end{pmatrix}_{j*1} \quad (16)$$

3.2. Data sources

3.2.1. Wind speed data. Wind speed data are derived from the marine meteorological observation platform in Maoming, Guangzhou.The time span is measured every ten minutes from February 1, 2012 to October 22, 2012.In total, the wind speeds at different altitudes 13.4, 16.4, 20, and 23.4 meters were measured.

This paper takes the wind speed at 13.4 meters and converts it to the wind speed at 10 meters by the following formula:

$$U_z=U_{10}\left(1+\frac{C_{10}^{1/2}}{\kappa}\ln\frac{z}{10}\right) \quad (17)$$

Where: $\kappa = 0.41$ is the Carmen constant, C_{10} is the ocean-atmosphere drag coefficient corresponding to U_{10} which can be calculated by the formula (6).

As the AWAC data is measured in a period of 3 hours, in order to study the influences of the different wind speeds on the solution of the time-dependent Ekman equation modified by wave and linear friction terms with the vertical eddy viscosity of the KPP model,so the wind speed is averaged every three hours.The 3 hour mean wind speed was divided into three groups: low wind speed (0, 4) m/s, medium-low wind speed (4, 8) m/s and medium-high wind speed (8, 12) m/s .

Figure 1 shows the average wind speed at different time periods. The wind speed data at different time periods are used because the wind speeds in successive time periods are mainly distributed in the medium-low wind speed, which is not conducive to the study of the influences of different wind speed intervals on the Ekman equation modified by wave and linear friction terms. From Figure 2, we can see that the wind speed varies with time in different intervals, in which there are 99 sets of data in the low wind speed, and 107 sets of data in the medium-low wind speed , and only 41 sets of data are available in the medium-high wind speed in which lack of the measured data.

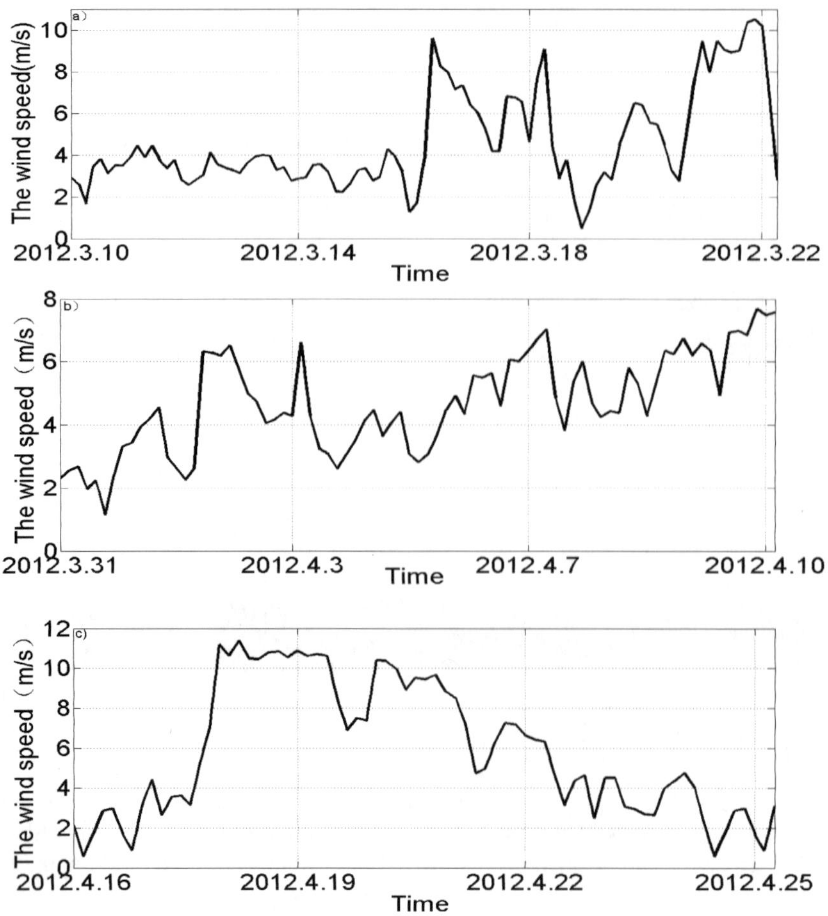

Figure 1. Figure of wind speed over time, in which a), b), c) that the wind speed at different time periods.

*3.2.2. The AWAC data.*The wave data was derived from AWAC measurements placed at the edge of the Maoming Tower in Guangzhou. The time span spanned from January 12, 2012 to July 4, 2012. However, due to the battery of the instrument, the effective data measurement ended on June 13, 2012.

The AWAC instrument measures in a 3-hour period, where wave data is measured for 10 minutes and current data is measured in the remaining time.The data measured by the AWAC instrument is measured under the corresponding wind data.Roll and Pitch are the parameters that reflect the stability and accuracy of the data measured by AWAC,and they refer to the left and right sides and the up and down sides swing indicators of AWAC respectively.For the general case, the swing indicators of Roll and Pitch can not be greater than 100, and the values of Roll and Pitch of the AWAC instrument used in this time fluctuated between [00, 50], indicating that the measured data is good.This paper selects 99 sets of AWAC data at low wind speed, 107 sets of AWAC data at medium-low wind speed and 41 sets of AWAC data at medium-high wind speed.

4. Results

4.1. Model validation

In order to verify the accuracy of the numerical model, we compared with figure 1 of Song (2009) and the parameter setting is the same as that set in Song (2009).In Figure 3, u, v respectively represent the Ekman current after the situation that only Stokes drift is considered.The solid black line in Figure 2 indicates the result of the numerical simulation, and the dashed black line indicates that the theoretical solution in the paper of Song (2009) .It can be seen from the picture that there is little difference between the numerical model and the theoretical solution, that is, the accuracy of the numerical model meets the requirements.

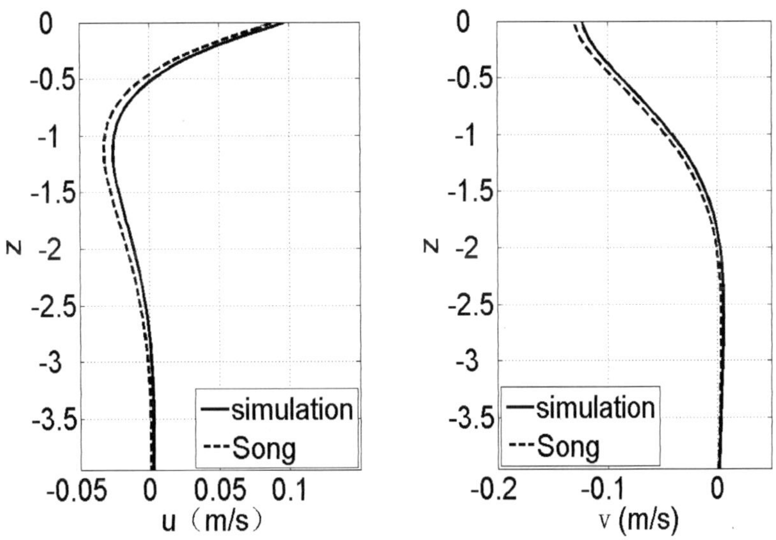

Figure 2. Comparison of numerical model and theoretical solution.

4.2. The effect of Stokes drift on Ekman current

On the basis of Webb and Fox-Kemper (2015) [25], Stokes drift can be calculated using the following formula:

$$U_S = (\cos\theta + i\sin\theta)\frac{16\pi^3}{g}\int_0^\infty f^3 S_f(f)\exp(\frac{8\pi^2 f^2 z}{g})df \tag{18}$$

Where: $U_S = U_S(z,t) = u_S(z,t) + iv_S(z,t)$ represents the complex Stokes drift, θ represents the direction angle, f represents the frequency, z represents the depth, $S_f(f)$ and represents the energy spectrum.

IOP Conf. Series: Journal of Physics: Conf. Series **1064** (2018) 012066 doi:10.1088/1742-6596/1064/1/012066

Numerical simulation results at low wind speed, medium-low wind speed and medium-high wind speed are shown in Figure 3-5 respectively.Among them, U_{E1}, U_{E2}, U_{E3}, $U_{WE1}, U_{WE2}, U_{WE3}$ indicate that Stokes drift is not considered and the influence of Stokes drift onEkman current respectively ,in addition, $U_{Ei} = \sqrt{{u_{Ei}}^2 + {v_{Ei}}^2}, U_{WEi} = \sqrt{{u_{WEi}}^2 + {v_{WEi}}^2} (i = 1,2,3)$.

From Figure 3-5, it can be seen that the impact on the Ekman current is greater after Stokes drift is considered.Among them,the effects of Stokes drift has the most influence on Ekman current are 27.4% in low wind speed, 7.5% in medium-low wind speed and 4.2% in medium-high wind speed.

The results show that considering the influence of the wave, adding the Stokes drift into the Ekman model can significantly affect the current velocity of the Ekman flow, and the effect of Stokes drift on the Ekman flow in the wind speed range of (0,12) decreases with the increase of wind speed.

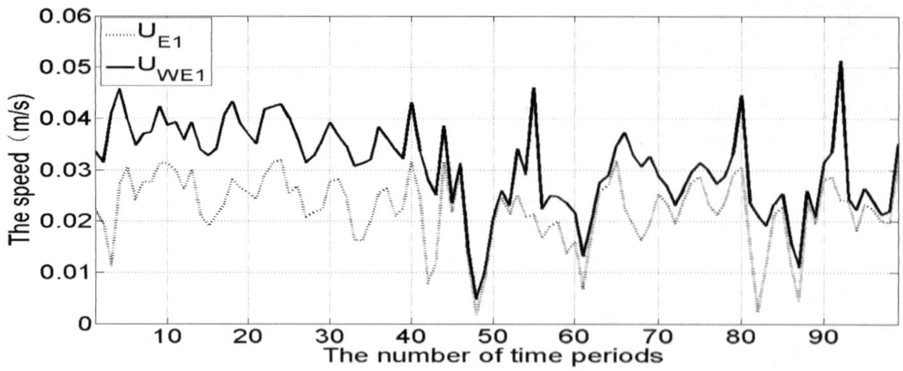

Figure 3. The case of low wind speed, influence of Stokes drift on Ekman current at Sea Surface. Among them, the dashed black line indicates that Stokes drift is not considered, and the solid black line indicates the influence of Stokes drift on Ekman current.

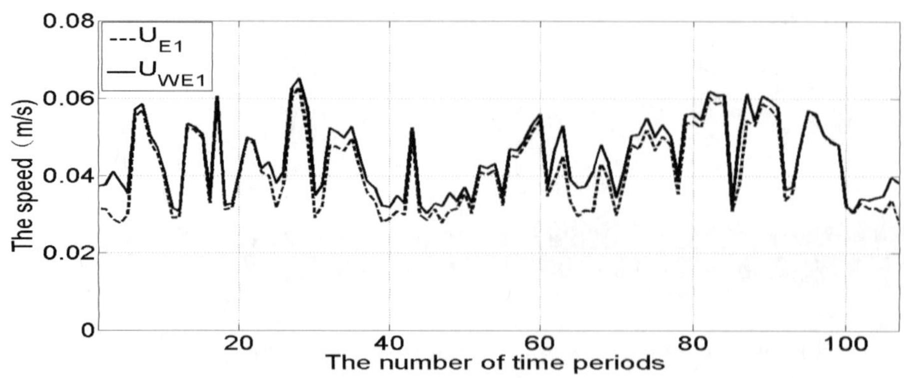

Figure 4. Like Figure 3, but corresponds to medium-low wind speeds.

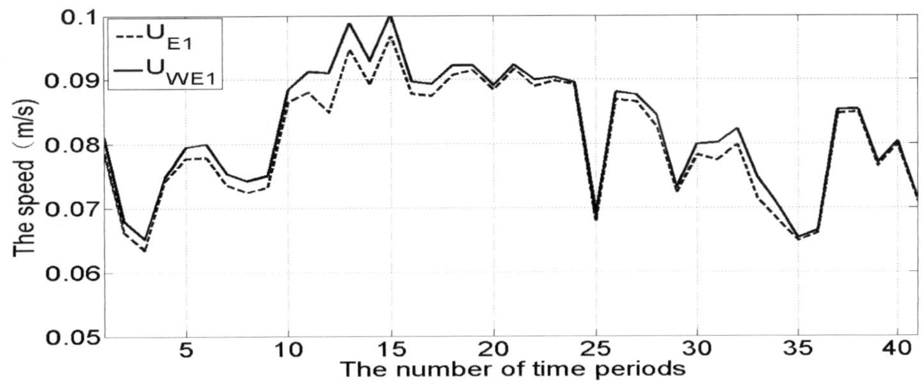

Figure 5. Like Figure 3, but corresponds to medium-high wind speeds.

4.3. The effect of linear friction term on Ekman current

In order to study the effect of linear friction terms on Ekman model, take $\lambda = r / f_{cor} = 0.3, 0.5, 0.7, 1$, among them $f_{cor} = 10^{-4}$, λ represents the ratio of the linear friction term to the Coriolis force, the value of it represents the proportion of the linear friction term and the Coriolis force in the Ekman model.

In Figure 6-8, $U_{WE1}, U_{WE2}, U_{WE3}$ represent that Stokes drift is considered but does not consider the influence of linear friction term respectively, U_{E1}, U_{E2}, U_{E3} represent thatStokes drift and linear friction term are not considered respectively.And the different lines indicate that the cases where the linear friction term is considered but the Stokes drift is not taken into account.In order to compare the influence of Stokes drift and linear friction term on Ekman current, the following formula is used to calculate:

$$\gamma = \left| \frac{b-a}{b} \right| \tag{19}$$

Where: b indicates that considered the effect of Stokes drift or linear friction term, a means with no consideration of the effects of Stokes drift and linear friction term.

Table 1. The influence of linear friction term about Ekman current

Wind speed	**The Influence of Different Items on Ekman Current**				
	$\lambda = 1$	$\lambda = 0.7$	$\lambda = 0.5$	$\lambda = 0.3$	U_{WE}
low	30.99%	19.8%	12.89%	6.78%	27.41%
medium-low	21.93%	13.81%	8.85%	4.5%	7.5%
medium-high	19.2%	11.82%	7.32%	3.54%	4.2%

The value of γ is shown in Table 1.From Figure 6-8 and Table 1, it can be seen that under low wind speed, the impact of the linear friction term on the Ekman current is less significant than that of the Stokes drift on the Ekman current. At that time, the linear friction term has an effect on the Ekman current by 30.99%. The effect of Stokes drifting on the Ekman current is 27.41%, indicating that the effect of linear friction term exceeds the effect of Stokes drift on the Ekman current.

At that time, the effect of linear friction on Ekman flow is 8.85% and 7.32% under middle and low wind speed, and the effect of Stokes drift on Ekman flow is 7.5% and 4.2% respectively, indicating that the linear friction term exceeds the effect of Stokes drift on Ekman flow at this time.

When $\lambda = 0.5$, the impacts of the linear friction term on the Ekman flow were 8.85% at medium-high wind speeds and 7.32% at medium-low wind speeds, respectively. The influences of Stokes drift on the Ekman flow were 7.5% and 4.2%, respectively, indicating that the influence of the linear friction term on the Ekman current has exceeded that of the Stokes drift this moment.

The above situation shows that the linear friction term has different effects on the Ekman current under different wind speeds, and with the increasing of λ, the linear friction term affects more and more on the Ekman current. When $\lambda = 1$, the Coriolis force and the linear friction term occupy the same proportion in the Ekman model, and the linear friction term plays an important role at this time.

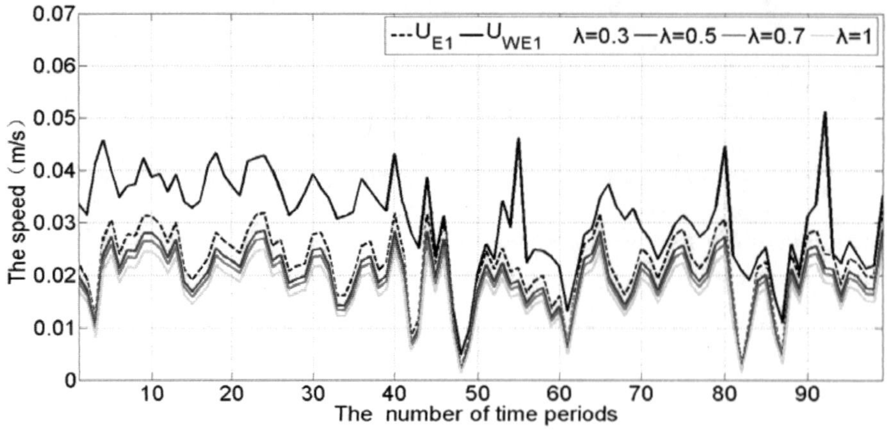

Figure 6. The case of low wind speeds, influence of linear friction term on Ekman current at Sea Surface. Among them, the dashed black line indicates that Stokes drift is considered but does not consider the influence of linear friction term, and the dashed black line shows that Stokes drift and linear friction term are not considered, the yellow line, the blue line, the red line, and the green line represent the cases where the linear friction term is considered but the Stokes drift is not taken into account when $\lambda = 0.3, 0.5, 0.7, 1$, respectively.

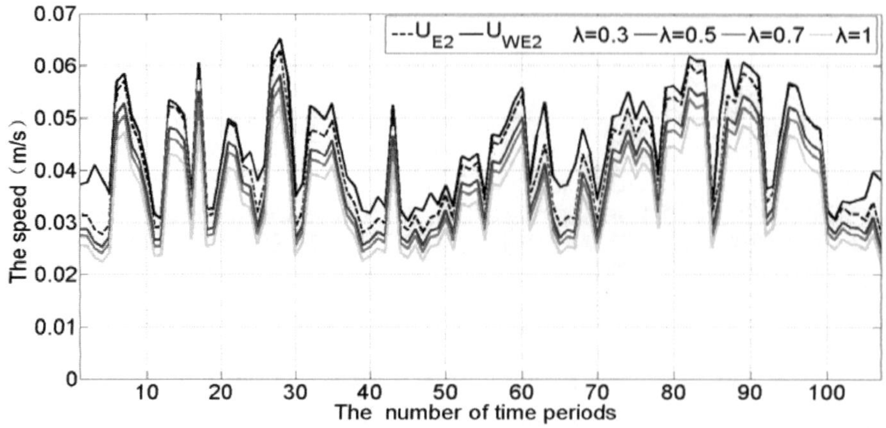

Figure 7. Like Figure 6, but corresponds to medium-low wind speeds.

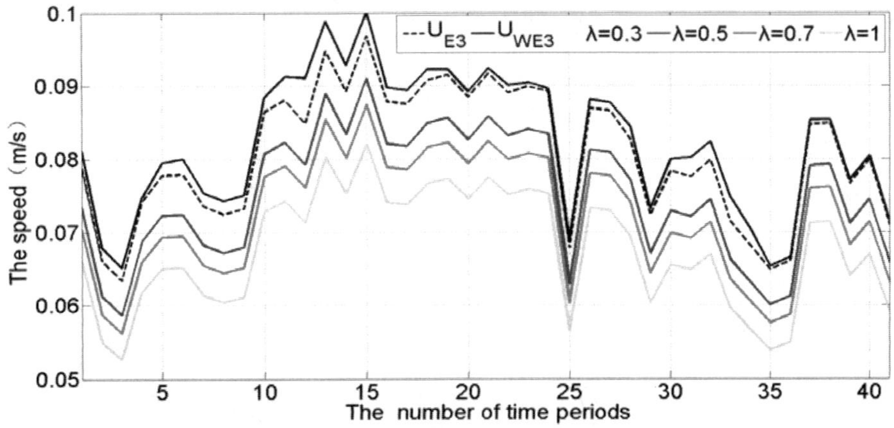

Figure 8. Like Figure 6, but corresponds to medium-high wind speeds.

5. The comparison with measured data

The AWAC is placed on the coast and affected by many factors. Therefore, only 540 groups of measured AWAC current data at medium-high wind speeds were selected, and 1008 groups of measured AWAC current data at medium-low wind speeds, and 900 groups at low wind speeds. These data are averaged every three hours.The comparison between measured data and numerical simulation is shown in Figure 9-11.

The correlation coefficient (Cor) and the root-mean-square deviation (RMSE) are used to express the deviation between the numerical simulation results and the measured data as follows:

$$Cor(x,y)=\frac{Cov(x,y)}{\sqrt{D(x)D(y)}} \tag{20}$$

$$RMSE=\sqrt{\frac{1}{n}\sum(H_x-H_y)^2} \tag{21}$$

Table 2. Correlation analysis of numerical simulation results and the measured data

Wind speed	Cor					
	$\lambda=1$	$\lambda=0.7$	$\lambda=0.5$	$\lambda=0.3$	U_{WE}	U_E
Low	0.4618	0.4615	0.4613	0.4611	0.3858	0.4607
Medium-low	0.5458	0.5457	0.5455	0.5453	0.5630	0.5453
Medium-high	0.7709	0.7708	0..7707	0.7705	0.7277	0.7700

Table 3. The root mean square deviation analysis of numerical simulation results and measured data

Wind speed	RMSE					
	$\lambda=1$	$\lambda=0.7$	$\lambda=0.5$	$\lambda=0.3$	U_{WE}	U_E
Low	0.0235	0.0219	0.0210	0.0203	0.0247	0.0197
Medium-low	0.0184	0.0163	0.0155	0.0153	0.0163	0.0159
Medium-high	0.013	0.0086	0.0075	0.0084	0.0130	0.0107

From Figure 9-11 and Table 2-3, it can be seen that, in the aspect of correlation, the correlation coefficient between the numerical simulation results under medium-high wind speed and the measured data is higher than that under low wind speed and medium-low wind speed; root mean square deviations between the numerical simulation results at low wind speed and medium-low wind speed and the measured data are higher than those of at medium-high wind speed. It shows that the numerical simulation results are closer to the measured values in the case of medium-high wind speed.

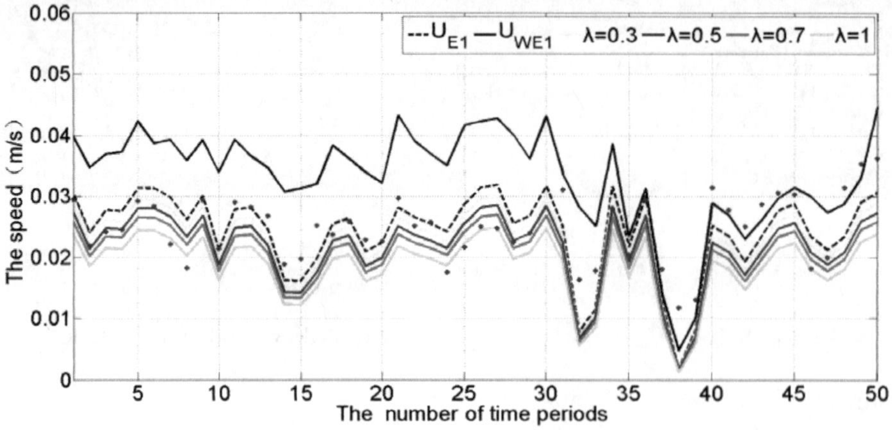

Figure 9. The case of low wind speeds, comparison between numerical simulation and measured data at Sea Surface. Among them, the dashed black line indicates that Stokes drift is considered but does not consider the influence of linear friction term, and the dashed black line shows that Stokes drift and linear friction term are not considered, the yellow line, the blue line, the red line, and the green line represent the cases where the linear friction term is considered but the Stokes drift is not taken into account when $\lambda = 0.3, 0.5, 0.7, 1$, respectively.

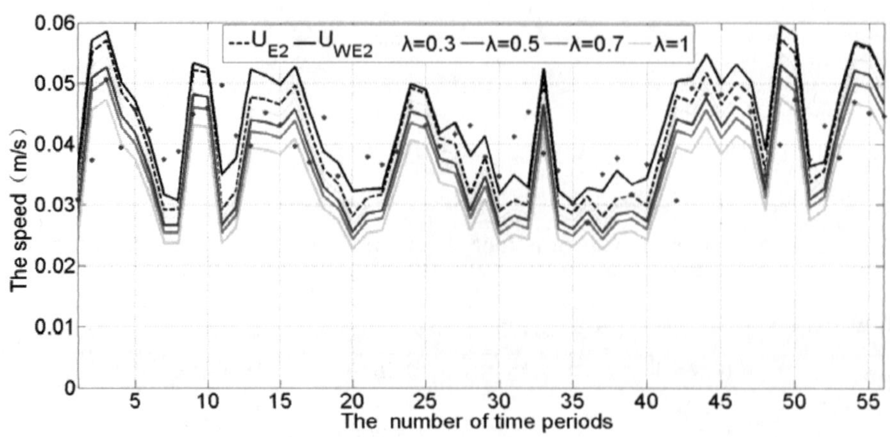

Figure 10. Like Figure 9, but corresponds to medium-low wind speeds.

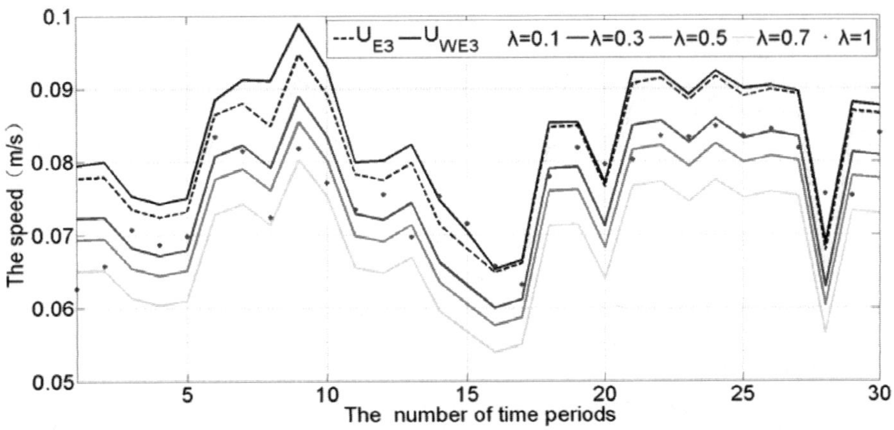

Figure 11. Like Figure 9, but corresponds to medium-high wind speeds.

6. Conclusion

In this paper, under the unsteady state, the time-dependent Ekman equation modified by wave and linear friction terms with the vertical eddy viscosity of the KPP model is discretized by finite difference method. Based on the measured data of the wind speed, the 3 hour mean wind speed was divided into three groups: low wind speed (0, 4) m/s, medium-low wind speed (4, 8) m/s and medium-high wind speed (8, 12) m/s and the influence of Stokes drift on the Ekman current under unsteady conditions is studied.The results show that the effects of Stokes drift has the most influence on Ekman current are 27.4% in low wind speed, 7.5% in medium_low wind speed and 4.2% in medium-high wind speed.The influence of linear friction term on Ekman current in the unsteady state is also studied. At $\lambda = r / f_{cor} = 0.5$, the influence of the linear friction term on the Ekman current has exceeded that of the Stokes drift in medium-high and medium-low wind speeds. However, at low wind speeds, the influence of the linear friction term on the Ekman current has exceeded that of the Stokes drift at $\lambda = 1$. The comparison between the numerical solution and the AWAC measured data also shows different characteristics: in the aspect of correlation, the correlation coefficient between the numerical simulation results under medium-high wind speed and the measured data is higher than that under low wind speed and medium-low wind speed;root mean square deviations between the numerical simulation results at low wind speed and medium-low wind speed and the measured data are higher than those of at medium-high wind speed . It shows that the numerical simulation results are closer to the measured values in the case of medium-high wind speed.In the unsteady state, with the increase of the proportion of linear friction term in the Ekman flow, the root mean square deviation from the actual measured data is also increasing, which shows that the proportion of linear friction term is more consistent with the observed results in the interval of (0.3, 0.7).The above results show that Stokes drift, linear friction term and unsteady wind forcing should not be ignored in the Ekman model.

References

[1] Ekman V W. On the Influence of the Earth's Rotation on Ocean-Currents [J]. Ark. Mat. astron. fys, 1905, 2.

[2] Sverdrup H U. Wind-Driven Currents in a Baroclinic Ocean; with Application to the Equatorial Currents of the Eastern Pacific [J]. Proceedings of the National Academy of Sciences of the United States of America, 1947, 33 (11): 318.

[3] Stommel, Henry. The westward intensification of wind-driven ocean Currents [J]. Eos Transactions American Geophysical Union, 1948, 29 (2): 202-206.

[4] Munk W H. On the Wind-Driven Ocean Circulation [J]. Journal of Atmospheric Sciences, 1950, 7 (2): 79-93.

[5] Stokes G. On the theory of oscillatory waves [J]. Trans Cambridge Philos Soc, 1847, 8 (310):

IOP Conf. Series: Journal of Physics: Conf. Series **1064** (2018) 012066 doi:10.1088/1742-6596/1064/1/012066

441-473.
[6] Huang N E.On surface drift currents in the ocean. Journal of Fluid Mechanics, 1979, 91 (1).
[7] Kenyon, Kern E. Stokes transport [J]. Journal of Geophysical Research Atmospheres, 1970, 75 (6): 1133-1135.
[8] Mcwilliams J C, Restrepo J M. The Wave-Driven Ocean Circulation [J]. Journal of Physical Oceanography, 1999, 29 (10): 2523-2540.
[9] Polton J A, Lewis D M, Belcher S E. The Role of Wave-Induced Coriolis-Stokes Forcing on the Wind-Driven Mixed Layer [J]. Journal of Physical Oceanography, 2005, 35 (4): 444-457.
[10] Lewis D M and Belcher S E. Time-dependent, coupled, ekman boundary layer solutions incorporating Stokes drift. Dynamics of Atmospheres and Oceans, 2004, 37 (4), 313-351.
[11] Saetra Ø, Albretsen J, Janssen P A E M. Sea-State-Dependent Momentum Fluxes for Ocean Modeling [J]. Journal of Physical Oceanography, 2007, 37 (11): 2714-2725.
[12] Song J B. The effects of random surface waves on the steady Ekman current solutions ☆ [J]. Deep Sea Research Part I Oceanographic Research, 2009, 56 (5): 659-671.
[13] Zhang.On the effects of waves on the wind energy input into the Ekman layer and the Ekman transport [D]. Ocean University of China, 2013.
[14] Ashkenazy Y, Gildor H, Bel G. The effect of stochastic wind on the infinite depth Ekman layer model [J]. Epl, 2015, 111 (3): 81-3.
[15] Kim, Sung Yong, Kosro, P. Michael, Kurapov, Alexander L. Evaluation of directly wind-coherent near-inertial surface currents off Oregon using a statistical parameterization and analytical and numerical models [J]. Journal of Geophysical Research: Oceans, 2015, 119 (10): 6631-6654.
[16] Buffoni G, Cappelletti A, Picco P. On the Ekman Equations for Ocean Currents Driven by a Stochastic Wind [J]. Stochastic Analysis & Applications, 2015, 33 (2): 356-382.
[17] Wenegrat J O, Mcphaden M J. A simple analytical model of the Diurnal Ekman layer [J]. Journal of Physical Oceanography, 2016, 46 (9).
[18] Jenkins A D. The use of a wave prediction model for driving a near-surface current model [J]. Deutsche Hydrografische Zeitschrift, 1989, 42 (3-6): 133-149.
[19] Song J B, Xu J L. Wave-modified Ekman current solutions for the vertical eddy viscosity formulated by K-Profile Parameterization scheme [J]. Deep Sea Research Part I Oceanographic Research Papers, 2013, 80 (5): 58-65.
[20] Jenkins A D. Wind and Wave Induced Currents in a Rotating Sea with Depth-varying Eddy Viscosity [J]. Journal of Physical Oceanography, 1987, 17 (7): 938-952.
[21] Wu J, Wind-stress coefficients over sea surface from breeze to hurricane. Journal of Geophysical Research Oceans, 1982, 87 (C12), 9704-9706.
[22] Large W G, Mcwilliams J C, Doney S C. Oceanic vertical mixing: A review and a model with a nonlocal boundary layer parameterization [J]. Reviews of Geophysics, 1994, 32 (4): 363-403.
[23] Mcwilliams J C, Huckle E. Ekman Layer Rectification [J]. Journal of Physical Oceanography, 2005, 36 (8): 1646.
[24] Mcwilliams J C, Huckle E, Liang J H, Sullivan P P. The Wavy Ekman Layer: Langmuir Circulations, Breaking Waves, and Reynolds Stress [J]. Journal of Physical Oceanography, 2012, 42 (42): 1793-1816.
[25] Webb A and Fox-Kemper B. Impacts of wave spreading and multidirectional waves on estimating Stokes drift. Ocean Modelling, 2015, 96, 49-64.

Application of Virtual Force Method for Deploying Mobile Sensor Networks

Shuli Wang[1, 2], Meifang Wu[1, *], Fan Wu [3], Weishiu Wu[3], Howard Pai[4]

[1]Dept. Industry Engineering and System Management, Feng -Chia University, Taichung, Taiwan.
[2] Dept. Dental Technology and Material Science, Central Taiwan University of Science and Technology, Taichung, Taiwan.
[3] Dept. Management of Information System, National Chung-Cheng University, Chia-Yi, Taiwan.
[4] IBM Almaden Research Center, San Jose, California, US.

*Corresponding author e-mail: slwang20170101@gmail.com

Abstract. The virtual force method is applied for deploying a mobile sensor network in an uncertain environment. A mobile sensor network is a distributed collection of nodes, each of which has sensing, computation, communication and locomotion capabilities. Previous studies of mobile sensor deployment adopted a round by round process, that is, sensors move iteratively until the maximum coverage is reached. However, the mobile sensors probably move in a zig-zag way that waste a lot of energy in comparison with that of moving directly to a final location. We proposed a cluster-based virtual force algorithm (CVFA), which is an energy efficiency deployment strategy. CVFA can enhance the coverage after an initial random placement of sensors. By constructing virtual force, each node is attracted or repelled by other nodes, thereby forcing the network to spread itself throughout the environment. In CVFA, global information of k-hop neighbours can be acquired by the inter-cluster communications, and sensors can grasp their target locations logically. Experimental result, using ns-2 simulator, suggests that CVFA can significantly reduce the energy consumption compared to previous work, especially on maintaining similar coverage.

1. Introduction

Wireless Sensor Networks (WSNs) is a paradigm of pervasive computing. Advances in sensor hardware and wireless network technologies are the last mile of bridging the gap between the physical world and the virtual information world [1-4]. These networks have a large number of tiny sensors that have sensing, computation and communication capabilities but with limited battery. Users can deploy these sensors near or inside the region of observation to gather, process, analyze and disseminate collaboratively sensing data through these sensors. In order to collect precise sensing data based on a well coverage level, the proper sensor deployment is critical for the efficiency of sensor systems [5-10]. In general, sensor nodes are considered to be fixed in such an environment that is dynamic and unknown or hostile regions like disaster areas, battlefields, and deserts. Most of the previous works considered these tiny sensors being deployed to result in a random placement [11-14] through air drop [15] for

Content from this work may be used under the terms of the Creative Commons Attribution 3.0 licence. Any further distribution of this work must maintain attribution to the author(s) and the title of the work, journal citation and DOI.
Published under licence by IOP Publishing Ltd

decreasing the cost and latency. Clearly, these sensors may be distributed unevenly, creating holes and disconnected islands. Such uncoveraged areas can be shrunk by increasing the number of sensors dropped. However, no matter how many sensors dropped, the coverage rate is limited since of the influence of wind force and or the quiescence of obstacles.

In a mobile sensor network [11], mobile sensors can move around to self-deploy. That is, these sensors have the ability of node spreading from an initial random topology to a desired uniform topology to achieve maximum coverage. A uniform topology has the advantage that it can prolong the system lifetime since each node has even energy consumption per communication. It is worth nothing that the average moving distance should be minimized, since movement may be much more expensive than communication and computation in terms of energy [5]. This paper utilizes the virtual force approach [5] is like the equilibrium of molecules, which is inter-nuclear repulsion and attractions between molecules to achieve the maximum coverage of mobile sensors. In this paper, we assume that we deploy a wireless sensor network by mobile nodes. Clearly, if sensors are located too close to each other, the gain in coverage is not high since of additional sensors. Therefore, we can adapt the concept of repulsive force that can repulse each other sensor to gain more coverage. On the contrary, if sensors are located too far away from each other, the coverage regions may not be connected and cause a partitioning of the network. Therefore, we can adapt the concept of an attractive force that can attract each other to connect the coverage of network partitions. With the combination of attractive and repulsive forces, each sensor can move to an ideal target location that is computed by local environment information such as neighbor sensor's locations and obstacles.

Zou and Chakrabarty [16] proposed the Virtual Force Algorithm (VFA) for the clustered sensor network. The cluster-head of a cluster in that network is assumed to have more computational capabilities. The head has the capability to compute the target location of each node within the cluster range by executing VFA. Clearly, this algorithm is a centralized approach (by cluster-head) that may suffer from the problem of fault of the head. The fault tolerance is important for such easily ruined sensor networks. To overcome the problem of the edge effect and the energy efficiency, Heo and Varshney proposed an intelligent energy-efficient deployment algorithm [5], called an Intelligent Deployment and Clustering Algorithm (IDCA), for clustered WSN. IDCA is also based on the virtual force approach, but the movement of each node is computed by itself, allow by each node moves by itself to a sparse region during sensor deployment process. Energy efficiency and edge effect for self-deployment of mobile nodes to form uniformly distributed node topologies was considered. Each node can decide its own mode either as a clustering mode [17, 18] or as a peer-to-peer mode, according to the local density (D) that is the number of nodes within its one-hop communication range (cR). The local density is higher when the local environment is denser. In general, wireless sensor networks desire dense deployment [19]. The dense deployment allows the use of redundancy for fault tolerance [20], allows multi-hop narrow communications for reducing energy consumption [3], and provides sufficient number of nodes to cooperate work for improving sensing accuracy and operation efficiency. However, if the communication of the sensors is not properly managed, a denser network accompanies the problem of channel contention that is a larger number of collisions and congestions in the network. So that the latency of the network will be increased and its energy efficiency will be reduced. To solve this problem, the clustering methodology [21] can also effectively manage communications in a dense network for energy efficiency. But all the above works [5, 16] didn't consider the problem of channel contention in dense sensor networks.

This paper deals with energy efficiency for self-deployment of mobile sensors with better coverage. The deployment tries to form a uniform topology in an unknown environment. The Cluster-based Virtual Force Algorithm (CVFA) is proposed as an energy efficiency deployment strategy in a mobile sensor network. CVFA is based on a round by round process, where mobile sensors move iteratively until the maximum coverage is reached in a distributed way. The major concept of CVFA is to decrease unnecessary excessive movements through efficient communications. Because of the movement is much more expensive than communication and computation in terms of energy [5]. In CVFA, more global information of k-hop neighbors can be acquired by the inter-cluster communications. Hence, mobile sensors can calculate their target locations, move logically, and update new logical locations through inter-cluster communications. Actual movement only occurs when sensors determine their final locations. Simulation results show that CVFA can significantly reduce the energy consumption compared to previous work, while maintaining similar coverage.

2. Preliminary: the concept of virtual force

The original virtual force concept was first proposed by Khatib [22], which is used for robot's navigation and obstacle avoidance. Further, the virtual force concept was introduced to mobile sensor networks by Howard et al. [11], and they call it as potential field method. Potential field method is able to achieve good coverage without global maps, communication, and explicit reasoning. This method has an assumption that each sensor allows to determine the range with nearby sensors or obstacles. This method has the advantage that it does not require centralized control, localization or communication so as that the method has a scalability to very large networks. But this approach has the problem that its coverage is contained within line-of-sight connectivity, because each sensor does not have capability of communication to control moving distance.

Zou et al. [16] further utilized the concept of VFA to clustered-based WSN. For a given number of sensors, VFA attempts to maximize the coverage by attractive and repulsive forces. During the sensor deployment, sensors can determine a sequence of virtual motion paths for the randomly placed sensors. When the effective target positions are computed, a one-time movement is carried out to redeploy the sensors at these positions. But VFA algorithm didn't consider the edge effect and less energy efficiency. These have been described in section 1.

Afterward, Heo et al. [5] proposed an IDCA algorithm that is also based on virtual force approach. This algorithm is first energy-efficiency deployment algorithm in clustered-based WSN. During sensor deployment, each sensor considers its remaining energy to decide its moving distance for reducing the difference of battery of each sensor. Therefore, IDCA algorithm can prolong the lifetime of whole network. However, this algorithm considered only one-hop neighbors' information but did not consider the channel contention in dense field.

In this paper, we modify the IDCA algorithm [5] such that it is more suitable for dense mobile sensors. As detailed in [5], the local density (D) of a node indicates the number of nodes within its one-hop communication range (cR). Each node can decide its own mode to be either in a clustering mode [17, 18] or in a peer-to-peer mode according to the local density (D). If D is not equal to the expected density, the node selects the peer-to-peer mode. In a peer-to-peer mode, each node moves itself to a sparse region by the partial force $f_n^{i,j}$, so that the coverage of the network may increase and/or an energy-efficient node topology may be achieved.

3. Deploying mobile sensor networks

A distributed Cluster-based Virtual Force Algorithm (CVFA) is used to identify the target locations of mobile sensors, and, with the force, the mobile sensor networks directly move to latest target locations. To achieve this goal, we propose the idea of one-time movement, where all mobile sensors logically move from dense fields to sparse fields by intra-cluster and inter-cluster communications, and rotating cluster-heads. All mobile sensors move directly to their final locations only after they find those locations. The process of CVFA is illustrated in Fig. 2.

In the initialization phase, a specified number of nodes are deployed randomly in a given rectangle region. We can calculate the sensor number N according to $\mu(cR)=\left(N\cdot\pi\cdot cR^2/A\right)$ [5], where A is the sensing area, μ is the expected density. The sensing range sR and communication range cR are assumed to be given. Each node can sense or detect an event within its sR and any pair of nodes within their cR can communicate with each other. This communication is needed for finding neighbors, obtaining neighboring locations, and transmitting and forwarding sensed data. The neighbors of a node are defined here as nodes within its cR.

In the clustering phase, CVFA adopts a one-hop cluster structure [23] to partite whole network into several local areas from a whole deployment region. In a one-hop cluster structure, each node is at most one-hop away from its cluster-head within one-hop intra-cluster cR. And each cluster-head can reach any other cluster-head within one-hop inter-cluster cR. So, a cluster-head can directly communicate the others without additional routing information. These features of one-hop cluster facilitate mobile sensors to deploy adaptively in an unknown and dynamic pre-deployment environment. Besides, a cluster-head has several (e.g., six) cluster power level for radio communications [21]. The length of communication range depends on the cluster power level. In this paper, each cluster-head operates in "dual" power mode, intra-cluster power level and inter-cluster power level respectively. Intra-cluster power level is a lower power level for members to communicate directly within a cluster, and inter-cluster power level is a higher power level for cluster-heads to communicate direct among different clusters. Each sensor can tone its intra-cluster cR by enlarging or narrowing intra-cluster power level k time, and the largest intra-cluster cR is as large as sR. The inter-cluster cR of a cluster-head is two times the intra-cluster cR. Moreover, to reduce this type of interference, the intra-cluster and inter-cluster communication use different CDMA codes.

After the clustering phase, CVFA adopts the virtual force concept to deploy clustered mobile sensors in the intra-cluster deployment and inter-cluster deployment phase. In the intra-cluster deployment phase, the intra-cluster communications emphasize less communication power and avoidance of channel contentions. Further, the inter-cluster communications emphasize the precise deployment location of whole cluster in the inter-cluster deployment phase. Finally, all mobile sensors move to their target locations only in the last phase.

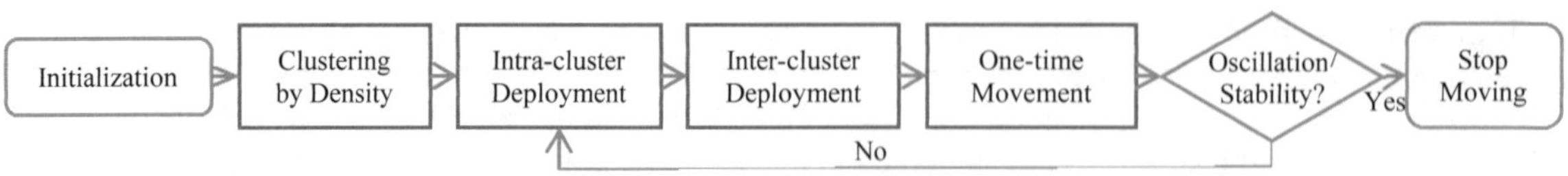

Figure 1. The process of CVFA.

The assumptions made in this paper are described as follow:

(1) Each sensor has identical capabilities for sensing, communication, computation, and mobility;
(2) Each sensor has the same initial energy level, and the energy level (ranging from 0 to 1) varies along with movement and other activities;
(3) Sensing coverage [13] and communication coverage (cR) of each node is assumed to be ideal, which means that both coverage areas have a circular shape without any irregularity;
(4) Each sensor has the ability to know its own location by some method, such as the global positioning system or iterative multilateration [24];
(5) The channel contentions are possible to happen when sensors located in high density fields; In other words, there are errors during transmission of data and in the calculation of locations;
(6) Each sensor has ability to vary cluster power level to change its intra-cluster cR;
(7) The inter-cluster cR is k times range than the intra-cluster cR;
(8) Each sensor has only local information from the neighboring nodes within its direct one-hop intra-cluster cR;
(9) Each cluster-head has only local information from the neighboring cluster-heads within its direct one-hop inter-cluster cR;
(10) Each sensor is capable of detecting obstacles;
(11) Each sensor has bound information of the region.

3.1. Clustering by density

Clustering, is a hierarchical networking concept, is employed in many WSN scenarios to take advantage of local information and to reduce energy consumption. In deployment issue, we want to utilize a cluster structure to control mobile sensors to move effectively. Because of choosing cluster-heads optimally is an NP-hard problem [25]. Moreover, deploying mobile sensors to a uniform topology fast is preferred. So we adapt a LEACH-like clustering method.

LEACH[26] is a self-organizing, adaptive clustering protocol that uses randomization to distribute the energy load evenly among the sensors in the network and incorporates data fusion to reduce the amount of information that must be transmitted to the base station. The differences between LEACH and our clustering method are that LEACH must decide the number of cluster-heads at first, and LEACH doesn't adapt the local density of deployment environment. Hence, LEACH would result in different degree clusters. If the number of degree is too high, the high computation and communication overhead of a cluster-head is resulted; on the contrary, the number of cluster-head is too many to reduce wide range inter-cluster communications that would waste more energy. Hence, we propose a density-based clustering method, that is based on distributed divide-and-conquer approach and none attempts to find the optimum of the network topology.

A CLUSTER message is only three bytes. So each cluster-head can count the number of member accurately through small CLUSTER message that member sand back. In order to maintain appropriate member number of a cluster, our clustering method would narrow radio range in greedy manner, when the number of degree is larger than the expected density. Finally, the degree of all clusters restricts to lower than the expected density after clustering. Therefore, there are two advantages of finding the minimum communication power level with connectivity and decreasing the probability of channel contentions in the clustering by density phase. A sensor can spend less communication power though minimum communication range is analyzed in [26]. By narrowing communication range, a sensor can decrease the number of neighbors to decrease channel contentions is proved in [28]. We provide a simple case to illustrate the process of density-based clustering method. In Fig. 2, there are several sensor nodes that are randomly distributed in a rectangle region. Here, we assume the expected density is set to 4. At first, each node waits for a random time to become a cluster-head. Node B is the first to run this method, it becomes a cluster-head due to its state is ordinary node initially. Then node B broadcasts CLUSTER message to its neighboring node A and C. So node A and C would become cluster-head B's member after they got cluster-head B's CLUSTER message, and then send back to cluster-head B to compute the number of member. Then cluster-head B checks this number of member that is smaller than expected density, so cluster-head B arrives to steady-state. Notice that any node doesn't become cluster-head in cluster B due to it has to check its current state at running time. After cluster B was formed, node F starts to check its state. In the same way, node E becomes a cluster-head. Node D, F and G becomes cluster-head E's member. Therefore, the clustering by density method only need low overhead to fast partition the whole network into clusters.

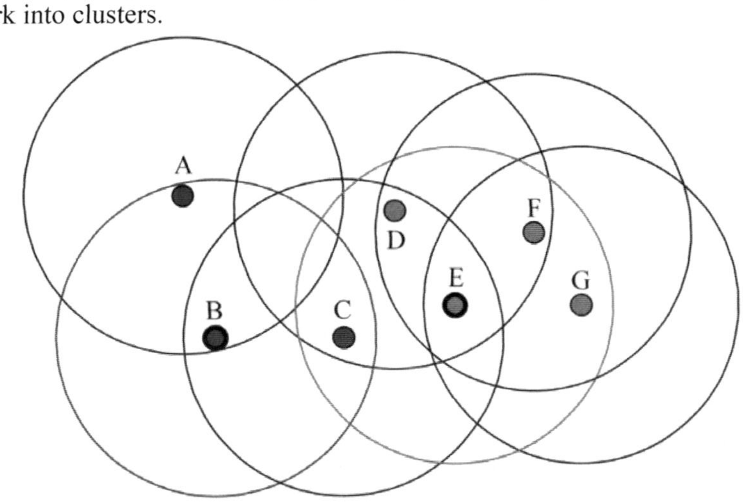

Figure 2. A simple case of clustering by density method.

Our clustering method can adopt the varied density of network topology by narrowing radio range. There is a dense case in Fig. 3. The node C is chosen to a cluster-head, but its number of member is larger than the expected density. Hence cluster C narrow its intra-cluster cR to decrease the number of degree of all nodes in cluster C. And each node waits a random interval time to decide the cluster-head in this local field. Finally, cluster A and F is formed in the same way.

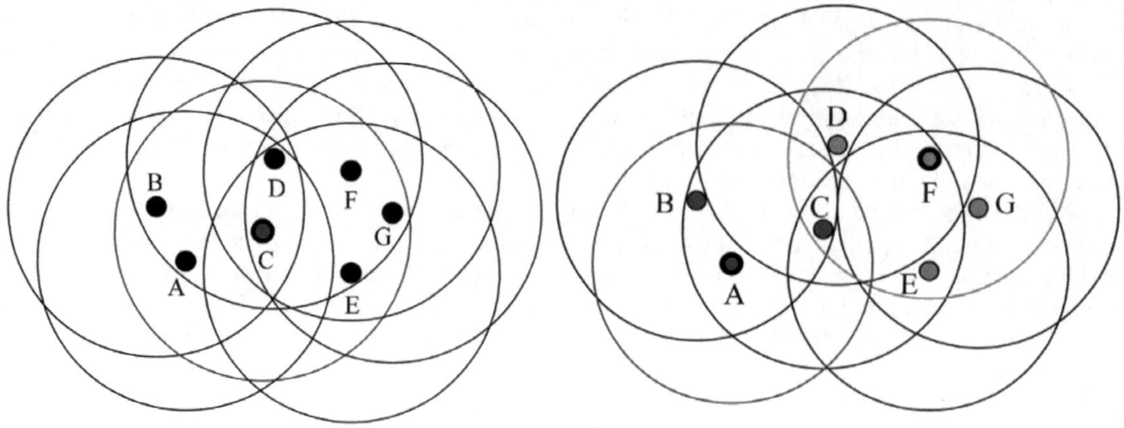

(a) The number of member of cluster-head C is larger than the expected density.

(b) The number of member of cluster-head A and F is smaller than the expected density.

Figure 3. A dense case of clustering by density method.

3.2. Clustered Deployment

In general, each mobile node can compute the optimal deployment location if it has global information of the environment. However, it is virtually impossible to have complete global information regarding the dynamic and unknown environment, where a WSN is likely to be deployed. Moreover, a node utilizes too much wide range communications to obtain global information to compute more precise target location requires a huge amount of energy, and also increases the probability of channel contentions that would result in more energy consumptions. This is an unaffordable overhead on a WSN with limited power supply.

CVFA provides a solution of above problem. It utilizes cluster structure to aggregate multiple-hop information, to compute more precise target location of clusters, and to control one-time movement of whole clusters. After clustering, each cluster can be regarded as a sub-region of the deployment, and all mobile sensors maintain minimal radio range to communicate with their cluster-head. So, the overhead of each mobile sensor is low compared to utilize equivalent cR. Moreover, each cluster can decrease probability of channel contentions. Hence CVFA can adapt any varied environment of deployment.

In the intra-cluster deployment phase, the task of each mobile node is to achieve the distance equilibrium of intra-cluster with its one-hop neighboring node within its intra-cluster cR. A mobile node i only receives current location p_n^j of other neighboring node j to compute the partial intra-cluster force $f_{intra\_n}^{i,j}$ at time step n. Clearly, if mobile nodes are located too close to each other, they expert partial repulsive intra-cluster forces that can repulse each other node to gain more coverage. On the contrary, if mobile nodes are located too far away from each other, they expert partial attractive intra-cluster forces that can attract each other to connect the coverage of network partitions. With the combination of partial attractive and repulsive intra-cluster forces, each mobile node can get the intra-cluster force to move to an equilibrated target location that is computed by locations of one-hop neighboring nodes within intra-cluster cR. The partial intra-cluster force $f_{intra\_n}^{i,j}$ can be computed through formulation (1).

$$f^{i,j}_{intra\_n} = \frac{1}{LD^j_n} \times \left(ER_{intra} - \left| p^i_n - p^j_n \right| \right) \times \frac{p^j_n - p^i_n}{\left| p^j_n - p^i_n \right|} \quad (1)$$

where p^i_n represents the location of node i at the time step n; p^j_n represents the location of neighboring node j of node i at the time step n; n represents the current time step.

The local density LD of a node indicates the number of nodes contains itself within its one-hop intra-cluster cR. So the density factor $\frac{1}{LD}$ is considered for achieving intra-cluster equilibrium. The intra-cluster equilibrium range ER_{intra} indicates the balance distance between nodes. Therefore, node j exerts node i a partial attractive force $f^{i,j}_{intra\_n}$, if the distance between node i and node j is larger than ER_{intra} at time step n. On the contrary, node j exerts node i a partial repulsive force $f^{i,j}_{intra\_n}$, if the distance between node i and node j is smaller than ER_{intra} at time step n. Finally, a node i has an intra-cluster force $f^{i}_{intra\_n}$ in (2) that summarizes all partial forces form its neighbors as

$$f^{i}_{intra\_n} = \sum_{j,j \neq i}^{k} f^{i,j}_{intra\_n} \quad (2)$$

where k represents the number of partial intra-cluster forces; n represents the current time step.

In Fig. 4, each sensor exerts its neighborhoods a partial intra-cluster force within one-hop intra-cluster cR. Simultaneously, each sensor can also include influences of some neighborhoods that belongs other clusters through clustering mechanism. Finally, each sensor can update its intra-cluster equilibrium location in the intra-cluster deployment phase that is illustrated in Fig. 5.

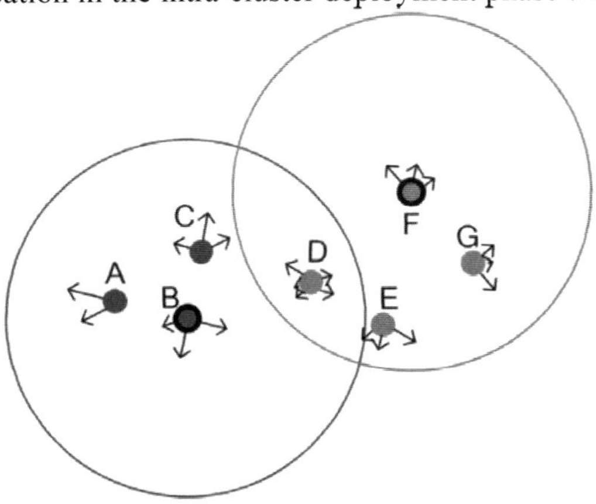

Figure 4. Partial intra-cluster forces.

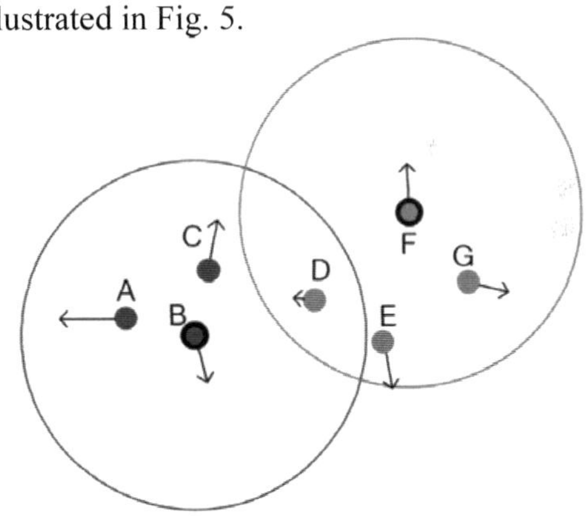

Figure 5. The result of intra-cluster forces.

In order to allow each cluster to maintain an appropriate distance with the border of the deployed area, each cluster-head that has current locations of all its members after computing the intra-cluster force, also compute the partial border forces $f^{i,j}_{border\_n}$ of all cluster nodes. This partial border force $f^{i,j}_{border\_n}$ is presented in (3). Notice that we assume the border equilibrium range ER_{border} is 1/2 times of the intra-cluster equilibrium range ER_{intra} in (4). And then each cluster-head combines all partial border force into border force f_{border_n}, this combined force will lead whole cluster to repulse the border of deployed area. The border force f_{border_n} is presented in (5). Finally, each cluster-head stores this border force for one-time movement. Notice that all mobile nodes always move with one-time in the one-time movement phase.

$$f^{i,j}_{border\_n} = \left(ER_{border} - \left|p^i_n - p^j_n\right|\right) \times \frac{p^j_n - p^i_n}{\left|p^j_n - p^i_n\right|} \tag{3}$$

$$ER_{border} = \frac{ER_{intra}}{2} \tag{4}$$

$$f_{border\_n} = \sum_{j,j\neq i}^{k} \frac{f^{i,j}_{border\_n}}{k} \tag{5}$$

where p^i_n represents the location of node i at the time step n; p^j_n represents the location of border of deployed region; k represents the number of partial border forces; n represents the current time step.

Hence, the border force can increase its sensing coverage and sensing density, and further influence target locations of other neighboring mobile nodes in next deployment round.

In order to save energy consumption and avoiding channel contentions, each mobile node limits its intra-cluster communication power level to minimal in the clustering by density phase. However, for computing more precise target location of each node, a mobile sensor needs long range communication to obtain global information. Hence, through wide inter-cluster communication of cluster-heads in this phase, multi-hop neighbors were included while computing next nodal target location. After intra-cluster deployment phase, each cluster-head i has intra-cluster equilibrium locations of all members, and it can compute the cluster centric point cp^i_n through k-mean like method [29]. The cluster centric point cp is figured out by formula (8). It can be represented as a whole cluster.

$$cp^i_n = \sum_{j}^{k} \frac{tp^j_n}{k} \tag{6}$$

where tp^j_n represents the target location of node j in cluster i at the time step n; k represents the number of partial border forces; n represents the current time step.

The cluster density CD of a cluster-head indicates the number of clusters contains itself within its one-hop inter-cluster cR. So the cluster density factor 1/CD is considered of the equilibration and the connectivity among clusters within its inter-cluster cR. The inter-cluster equilibrium range ER_{inter} indicates the balance distance between the centric points of each cluster. If clusters are located too close to each other, the repulsive force is exerted. On the contrary, if clusters are located too far away from each other, the attractive force is exerted. After computing inter-cluster force, each cluster-head would check the borders of all member nodes, due to its cluster would be exerted to the borders by inter-cluster force. Each cluster-head can compute the border force again through formula (3-5), and combine original border force computed in the intra-cluster phase. Finally, each cluster-head combines the inter-cluster force $f^i_{inter\_n}$ and the combined border force $f^i_{border\_n}$ into the combined cluster force $f^i_{combine\_n}$ in (7) as

$$f^i_{combine\_n} = f^i_{inter\_n} + f^i_{border\_n} \tag{7}$$

In order to lead whole cluster to move together, each cluster-head send this combined cluster force to its all members after computing the inter-cluster force. And all nodes update its target location computed in intra-cluster phase when they receive this combined force. Consequently, the updated target locations of all nodes would achieve the intra-cluster equilibrium and the inter-cluster equilibrium simultaneously.

At last phase, because of the one-time moving distance is shorter than the greedy moving distance, so each node combines all virtual forces include intra-cluster, inter-cluster, and border forces to obtain the final target location to achieve one-time movement. Therefore, CVFA can decrease necessary moving distances and promote the energy efficiency.

After moving, each cluster-head has to check its stability as its cluster density (CD). If CD is not close to the expected cluster density, the status of a cluster considers unstable, and all nodes of this cluster have to participate deployment in next round. On the contrary, if CD is close to the expected cluster density, the status of a cluster considers stable. Each node has two criteria to be terminated. One is the number of stability and the other is the number of oscillations. A node is regarded to be in the stability status, if this node moves less than the threshold of movement distance. And a node is regarded to be in the oscillation status, when this node moves back and forth between almost the same locations many times. When the number of stability or the number of oscillations is higher or equal than a specific number, this node would stop to move.

4. Performance evaluations

The performance of the heuristic algorithms in the paper is evaluated by simulation. In the experiment, 30, 60, 90, and 120 randomly placed nodes in a region of size 6 m × 6 m, 8 m × 8 m, 10 m × 10 m, and 12 m × 12 m are used to run the DSSA, IDCA and CVFA, respectively. These algorithms are implemented in the ns-2 (version 2.29) simulator. Based on the information from [5], the sR used in the experiment is 2 m, and the intra-cluster cR and inter-cluster cR used in the experiment are 2 and 4 m at first, respectively. The maximum permitted velocity in this experiment is 0.4 m/s. Unlike the experiment in [5], our experiment environment is a high density region of deployment, thus it would cause the situations of channel contentions. Note that the target location of a sensor would be computed inaccurately when data collisions happened. Four metrics are of particular interest: sensing coverage (i.e., what is the area covered by the network), uniformity (i.e., what is the standard deviation of length between nodes), average energy consumption, and average moving distance. Coverage, average energy consumption and uniformity are related to the performance of sensor networks after the deployment of sensors is complete. Average moving distance prior to convergence is directly related to the performance of the deployment scheme itself.

4.1. Coverage

Generally, coverage can be considered as the measure of quality of service of a sensor network. The concept of coverage as a paradigm for the system-level functionality of multi-robot systems was introduced by Gage [30]. In this paper, coverage [12] is defined as the ratio of the union of areas (in square meters) covered by each node and the area (in square meters) of the entire ROI. Here, the covered area of each node is defined as the circular area within its sensing radius R. Perfect detection of all interesting events in the covered area is assumed

$$c = \frac{\bigcup_{i=1,\,2,\,3,\,\ldots,\,N} A_i}{A}$$

where A_i is the area covered by the ith node; N is the total number of nodes; A stands for the area of the deployment region.

The overall coverage of a sensor network is composed of the covered regions of each sensor node. Though the coverage of a sensor is expressed by a sensor model which is binary or stochastic, the overall coverage of a sensor network depends on the locations of the sensor nodes in the sensor field. The topology including the locations and spacing of sensor nodes determines the overall coverage of the network as well as the expected lifetime of the network. In Fig. 6, the 30 randomly placed node locations and coverage of the initial random deployment before running the algorithms are shown. Tiny circles represent the positions of nodes and small (shaded) circles are used to show the sR of the nodes, respectively. We also assume the default cR is two times of sR. Sensor information may be collected within the sR and communications between nodes are possible within the cR. It's easy to see that there is a high density in the deployment region in Fig. 14. Hence, the initial coverage in this case is higher than the low density case.

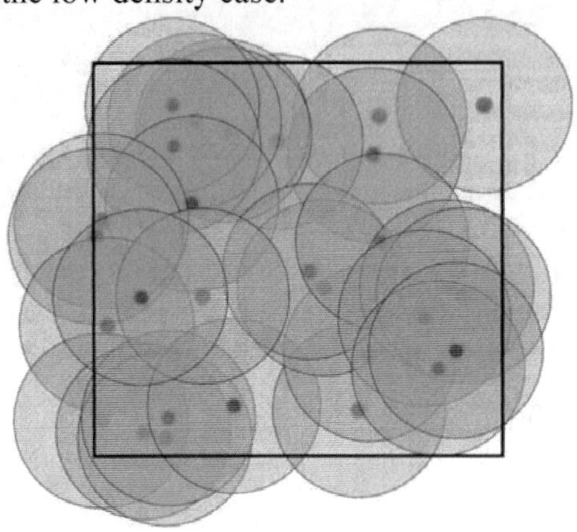

Figure 6. Initial distribution of 30 sensor nodes.

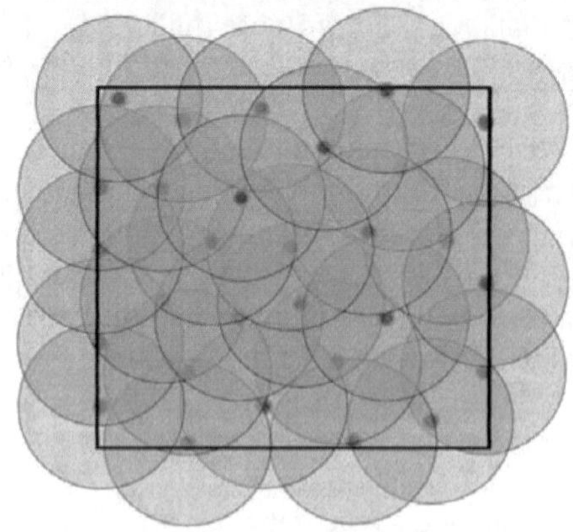

Figure 7. Final node distribution after running DSSA.

Fig. 7 shows the 30 node locations and coverage after running the DSSA. The deployment region is fully covered after running the algorithm. The parameter values used in this simulation run are: sR = 2m, cR = 4m, expected density = 4, stable status limit = 3, oscillation limit = 3, and threshold for oscillation and stable status is 0.15. Now, the network is fully connected and also covers the entire deployment region. Note that the spatial node distribution is more uniform than the initial random distribution shown in Fig. 6. Fig. 8 shows the 30 node locations and coverage after running the IDCA. The deployment region is also fully covered after running the algorithm. The same as DSSA, the parameter values used in this simulation run are the same, and the spatial node distribution is also more uniform than the initial random distribution shown in Fig. 6. Fig. 9 shows the 30 node locations and coverage after running the CVFA. The deployment region is also fully covered after running the algorithm. The cR in CVFA is different from others due to cluster structure. Hence, the different parameters compared with the other algorithms are intra-cR = 2m and inter-cR = 4m.

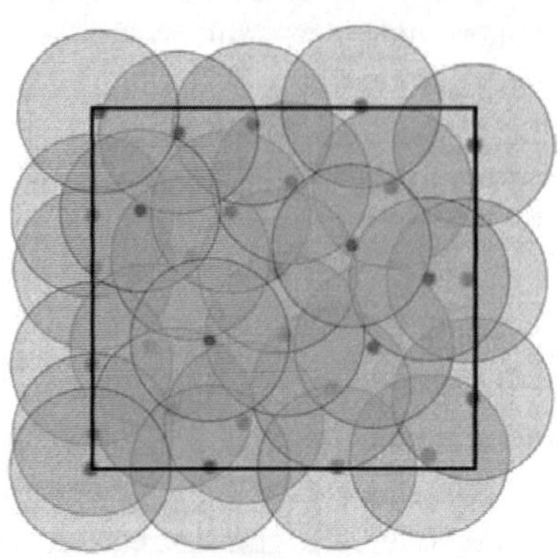

Figure 8. Final node distribution after running IDCA.

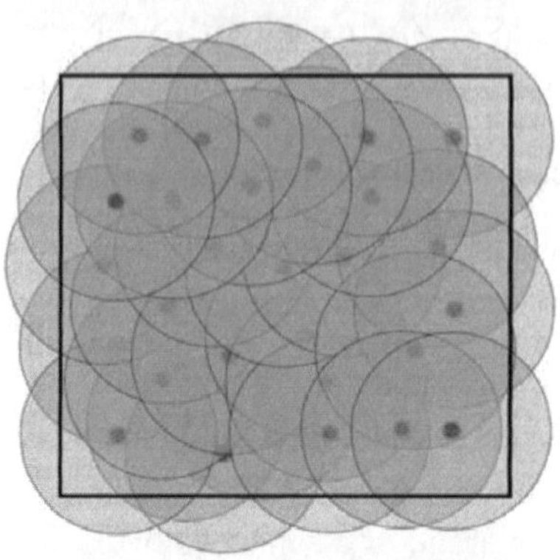

Figure 9. Final node distribution after running CVFA.

IOP Conf. Series: Journal of Physics: Conf. Series **1064** (2018) 012067 doi:10.1088/1742-6596/1064/1/012067

4.2. Moving distance

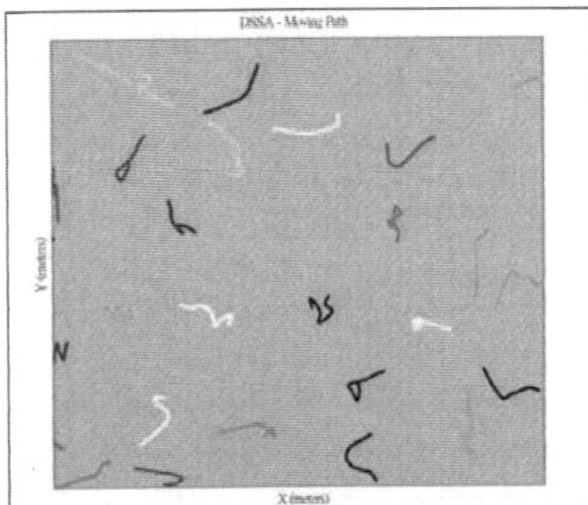	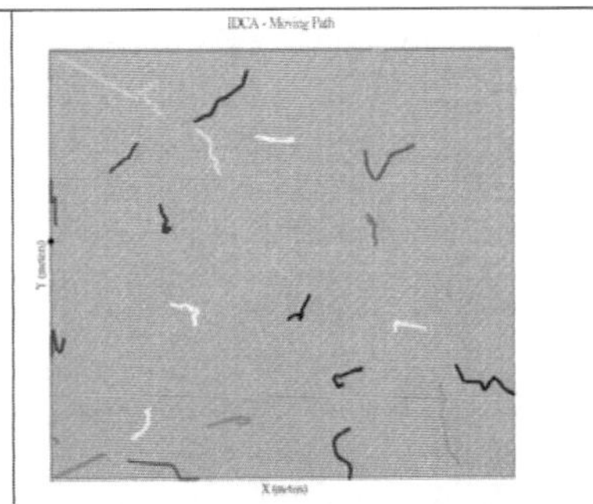	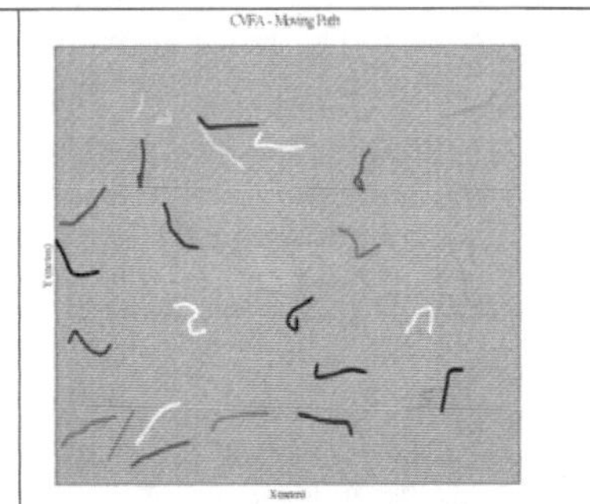
30 sensor-node movements after running DSSA.	30 sensor-node movements after running IDCA.	30 sensor-node movements after running CVFA.
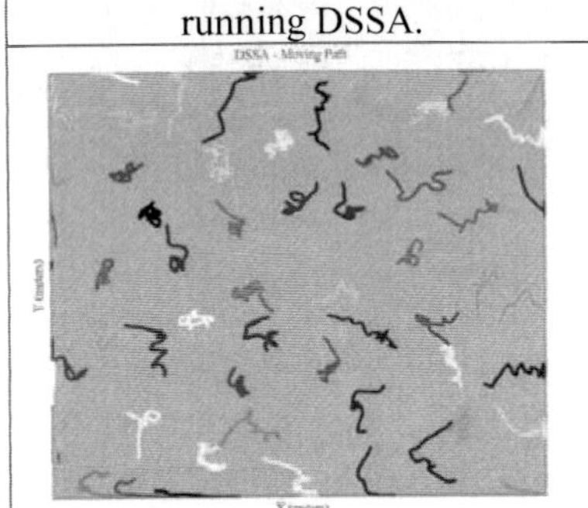	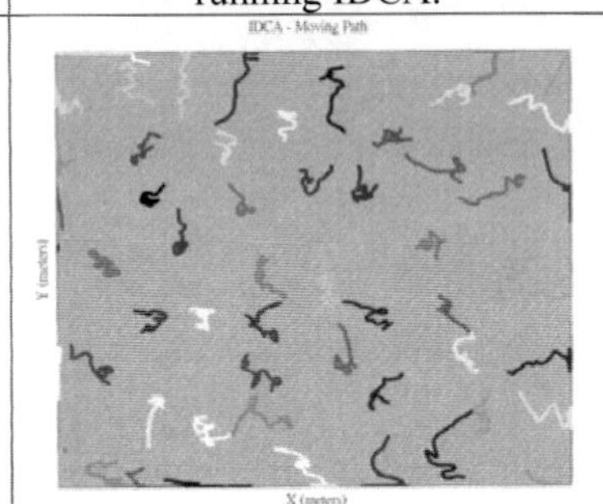	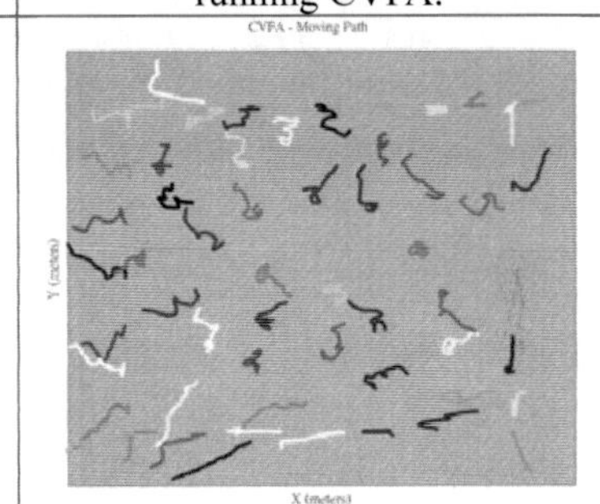
60 sensor-node movements after running DSSA.	60 sensor-node movements after running IDCA.	60 sensor-node movements after running CVFA.
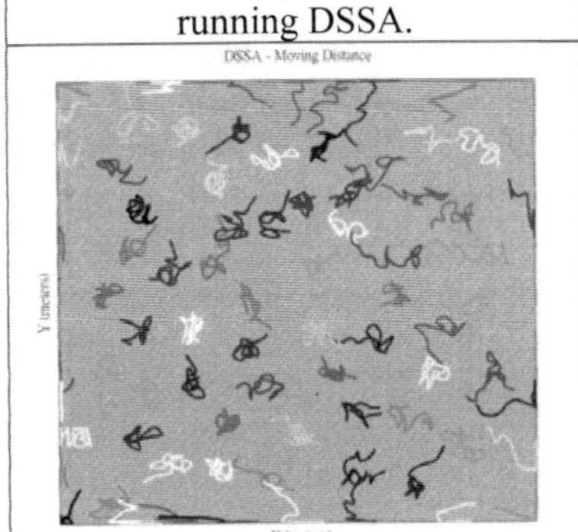	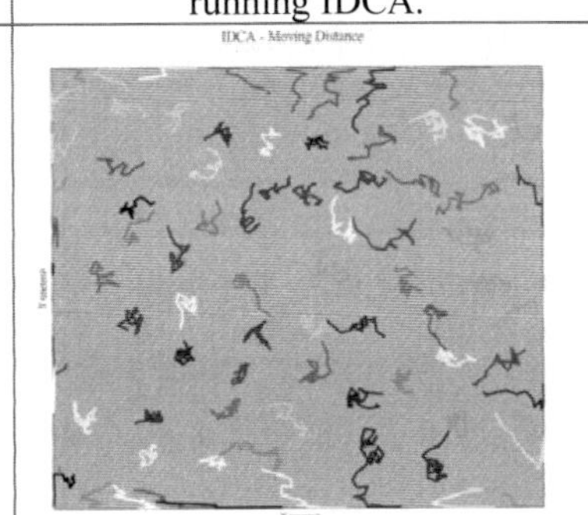	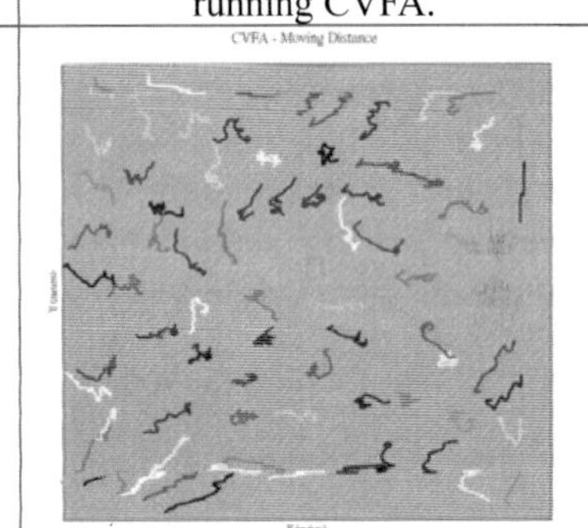
90 sensor-node movements after running DSSA.	90 sensor-node movements after running IDCA.	90 sensor-node movements after running CVFA.
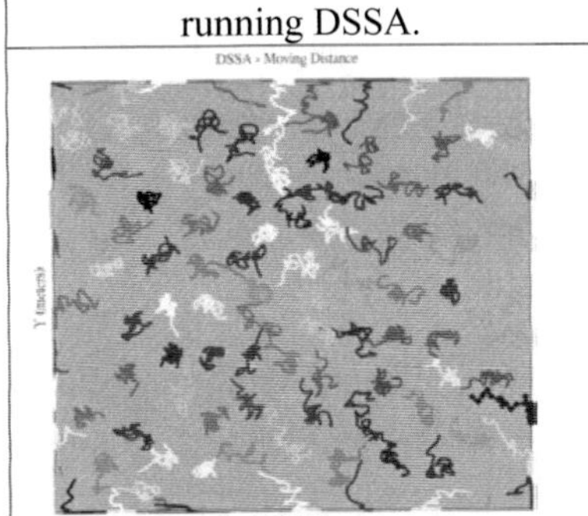	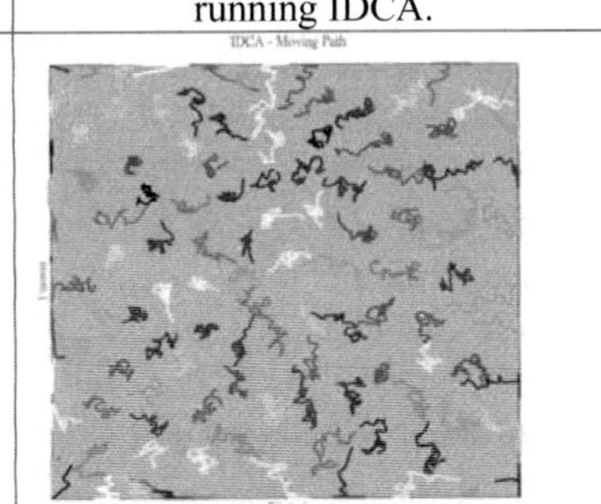	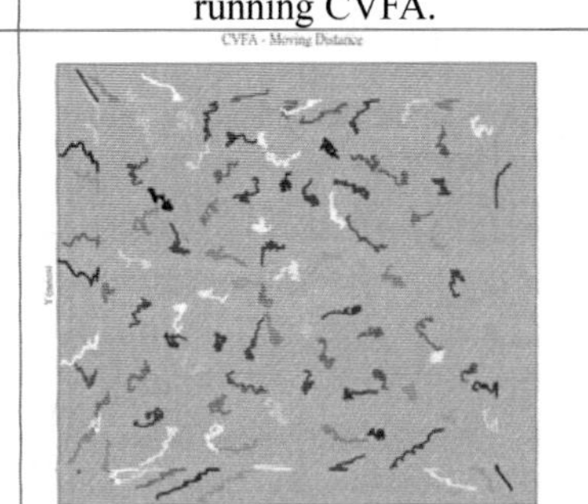
120 sensor-node movements after running DSSA.	120 sensor-node movements after running IDCA.	120 sensor-node movements after running CVFA.

Figure 10. The sensor-node movements.

The movement may be much more expensive than communication and computation in terms of energy [5]. Therefore, the core problem of energy saving is that how to decrease the unnecessary movements of mobile sensors, since movement is the dominant factor in energy consumption. In addition, the average moving distance by each node is related to the energy required for its movement. So, the expected moving distance is important for the estimation of energy required when each node has a limited energy supply. Fig. 10 shows the moving path of all sensors after running these three algorithms in cases of 30, 60, 90 and 120 nodes.

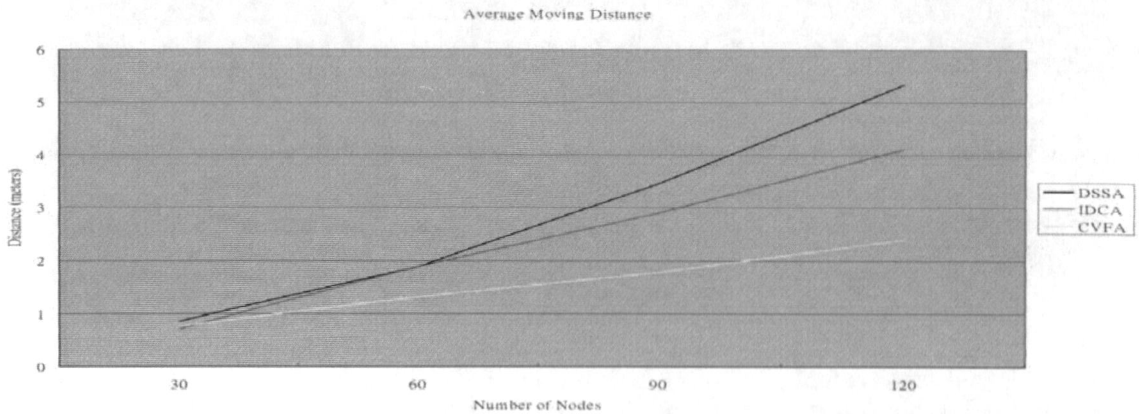

Figure 11. The comparison of average moving distance.

In Fig. 11, we can see that CVFA has the least movement among these algorithms. The reason is as follow. CVFA adds inter-cluster communication that can decrease channel contentions, to acquire global information to compute more precise target location than DSSA or IDCA, and actual movement only occurs when sensors determine their final locations. On the other side, the mobile sensors may move in a zig-zag way, if DSSA or IDCA utilizes narrow range to acquire local information. However, the mobile sensors may also move in a zig-zag way, if DSSA or IDCA utilizes wide range to acquire global information. The channel contention problem is the major reason. When there are a large number of mobile sensors within a wide communication range, channel contention occurs more often and these sensors have to move more than once, because channel contention results in data lost so as to their target locations would be computed incorrect. Comparison between DSSA and IDCA, IDCA has less movement since sensors that located high density move less for energy efficiency. This is the same result illustrated in [5].

4.3. Energy Consumption

Energy consumption in the deployment procedure comes from two major sources, mechanical movement and radio communication. We use the moving distance as the metric for the energy consumption in mechanical movement and the amount of message as the metric for the energy consumption in communication. We assume the energy consumption of movement is 30J to move one meter, it is calculated from the data of the mobile sensor platform [2]. In addition, we also assume that the energy consumption of intra-cluster communication and inter-cluster communication in CVFA is 100.4nJ and 101.6nJ to send one message, respectively. These consumptions are calculated from the first order radio model in [26]. Note that the default cR in DSSA and IDCA is the same with inter-cluster cR, so the energy consumption of communication is also 101.6nJ to send one message.

Fig. 12 shows that CVFA has the least energy consumption of movement, and that is the same result in Fig. 11. However, CVFA has the most energy consumption of communication in Fig. 13, because adding the inter-cluster communication and cluster management. To consider both factors in Fig. 14, we conclude that CVFA is much more energy efficient because the mechanical movement consumes much more energy than radio communication.

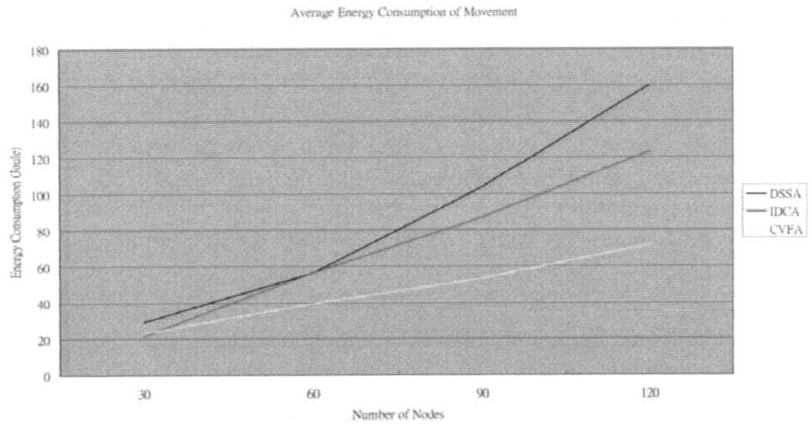

Figure 12. The comparison of average energy consumption of movement.

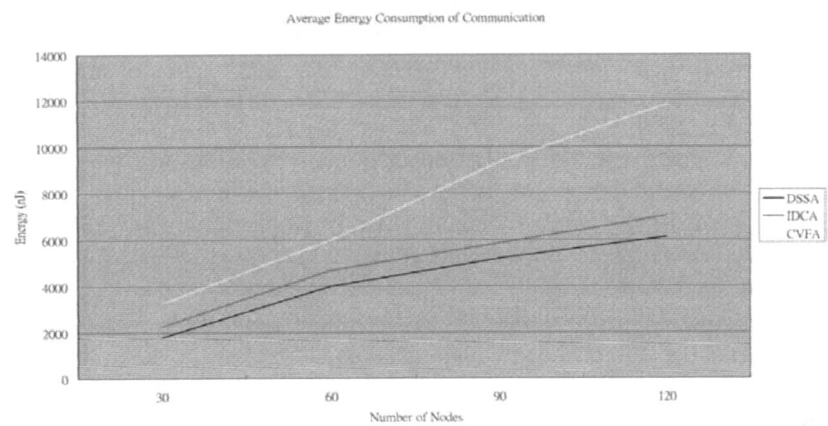

Figure 13. The comparison of average energy consumption of communication.

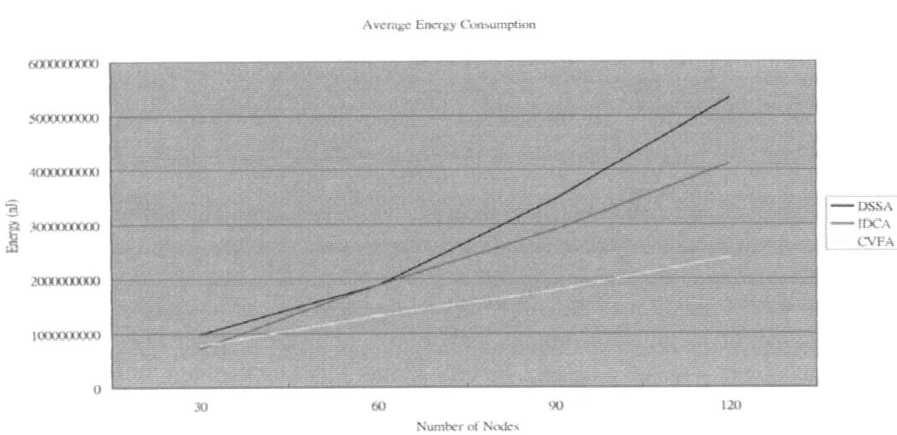

Figure 14. The comparison of average energy consumption.

5. Conclusion

This paper raised the issue of deploying mobile sensor network in an uncertain environment. A deployment region needs to be covered by clusters grouped form a specific number of nodes with limited sensing and communication range. We designed a distributed virtual force algorithm to increase energy-efficiency of sensors in the deployment of clustered WSN. In addition, we proposed a synergistic combination of intra-cluster and inter-cluster deployment scheme. Through aggregation of global information, localized coordination and control ability of clusters, our method can decrease unnecessary node movement and the probability of channel contentions. The performance of our method is determined by the percentage of region covered, the average moving distance required for deployment, and uniformity of the networks. By experiment, under high density region of deployment, our method achieves less average moving distance and more energy efficiency in comparison with previous studies. In future work, we would consider to study how to improve the performance and management of clustering, and how to deploy hierarchical clustered WSN in heterogeneous networks [31, 32]. On the other hand, the relation between cluster-size and neighborhood-size probably influences the degree of saving energy. It is well worth studying the optimal deployment.

References

[1] A.H. Anisi, G. Abdul-Salaam, M.Y.I. Idris, A.W.D. Wahab, Ahmedy I. Energy harvesting and battery power based routing in wireless sensor networks, Wireless Networks, 23(1), 249-266, 2017.
[2] K. Sohrabi, J. Gao, V. Ailawadhi, and G.J. Pottie, Protocols for self-organization of a wireless sensor network, IEEE Personal Communication, Vol. 7, pp. 16-27, 2000.
[3] G.J. Pottie, and W.J. Kaiser, Wireless integrated network sensors, Communications of the ACM, 43(5), pp. 51-58, 2000.
[4] S. Guo, L. He, Y. Gu, B. Jiang, B. He, Opportunistic Flooding in Low-Duty-Cycle Wireless Sensor, Networks with Unreliable, Transactions on Computers, 63(11), 2787-2802, 2014.
[5] N. Heo and P.K. Varshney, Energy-efficient deployment of Intelligent Mobile sensor networks, IEEE Transactions on Systems, Man and Cybernetics, Part A: Systems and Humans, Vol. 35, No. 1, pp. 78-92, 2005.
[6] S. Tilak, N. Abu-Ghazaleh, and W. Heinzelman, Infrastructure tradeoffs for sensor networks, In Proceedings of the 1st ACM international workshop on Wireless sensor networks and applications, pp. 49-58, 2002.
[7] T. Clouqueur, V. Phipatanasuphorn, P. Ramanathan, and K. Saluja, Sensor deployment strategy for target detection, In Proceedings of the 1st ACM international workshop on Wireless sensor networks and applications, pp. 42-48, 2002.
[8] E.H. Callaway Jr, Wireless Sensor Networks: Architectures and Protocols, CRC press, 2003.
[9] I. F. Akyildiz, W. Su, Y. Sankarasubramaniam, and E. Cayirci, A survey on sensor networks, IEEE Communications Magazine, Vol. 40, pp. 102-114, 2002.
[10] D., Suh, K. Jeon, S. Chang, J. Kim, Auto-localized multimedia platform based on a modular Cyber Physical System aligned in a two-dimensional grid, Cluster Computing, 18(4), 1449-1464, 2015.
[11] A. Howard, M.J. Matarić, and G.S. Sukhatme, Mobile sensor network deployment using potential fields: A distributed, scalable solution to the area coverage problem, In Distributed Autonomous Robotic Systems, Vol. 5, pp. 299-308, 2002.
[12] A. Howard, M.J. Matarić, and G.S. Sukhatme, An incremental self-deployment algorithm for mobile sensor networks, Autonomous Robots, Vol. 13, No. 2, pp. 113-126, 2002.
[13] X.Y. Li, P.J. Wan, and O. Frieder, Coverage in wireless ad hoc sensor networks, IEEE Transactions on Computers, Vol. 52, No. 6, pp. 753-763, 2003.
[14] A.F. Winfield, Distributed sensing and data collection via broken ad hoc wireless connected networks of mobile robots, In Distributed Autonomous Robotic Systems, Vol. 4, 2000, pp. 273-282.
[15] S.S. Dhillon and K. Chakrabarty, Sensor placement for effective coverage and surveillance in distributed sensor networks, IEEE, Vol. 3, 2003, pp. 1609-1614.
[16] Y. Zou and K. Chakrabarty, Sensor deployment and target localization in distributed sensor networks, ACM Transactions on Embedded Computing Systems, Vol. 3, No. 1, pp. 61-91,

2004.
[17] V. Kawadia and P.R. Kumar, Power control and clustering in ad hoc networks, In INFOCOM, Vol. 1, 2003, pp. 459-469.
[18] A. Abuarqoub, and H. Al-Bashar, Dynamic clustering and management of mobile wireless sensor networks, Computer Networks, vol. 117, pp. 62-75, 2017.
[19] M. Tubaishat and S. Madria, Sensor networks: an overview, Potentials, Vol. 22, No. 2, pp. 20-23, 2003.
[20] K. Lalitha, R. Thangarajan, S.K. Udgata, C. Poongodi,, A.P. Sahu, GCCR: An Efficient Grid Based Clustering and Combinational Routing in Wireless Sensor Networks, Wireless Personal Communication, 97(1), 1075-1095, 2017.
[21] O. Younis and S. Fahmy, HEED: a hybrid, energy-efficient, distributed clustering approach for ad hoc sensor networks, IEEE Transactions on Mobile Computing, Vol. 3, No. 4, pp. 366-379, 2004.
[22] M. Guerra, D. Efimov, G. Zheng, W. Perruquetti, Finite-time obstacle avoidance for unicycle-like robot subject to additive input disturbances, Autonomous Robots, 41(1), 19-30, 2017.
[23] L. Tong, Q. Zhao, and S. Adireddy, Sensor networks with mobile agents, In IEEE Military Communications Conference, Vol. 1, 2003, pp. 688-693.
[24] H.G. Wen, Y.H. Qi, Z.B. Weng, A Multiband Dual-Polarized Omnidirectional Antenna for 2G/3G/LTE Applications, 17(2), 331-335, 2017.
[25] J.H. Qin, W.M. Fu, H.J. Gao, W.X. Zheng, Distributed k -Means Algorithm and Fuzzy c – Means Algorithm for Sensor Networks Based on Multiagent Consensus Theory, IEEE Transactions on Cybernetics, 47(3), 772-783, 2016.
[26] S.P. Singh, S.C. Sharma, A Survey on Cluster Based Routing Protocols in Wireless Sensor, Networks, Procedia computer science, 45, 687-695, 2015.
[27] P. Rodriguez, S.M. Tan, and C. Gkantsidis, On the feasibility of commercial, legal P2P content distribution, ACM SIGCOMM Computer Communication Review, Vol. 36, No. 1, pp. 75-78, 2006.
[28] L.H. Yen and C. Yu, Link probability, network coverage, and related properties of wireless ad hoc networks, In IEEE International Conference on Mobile Ad-hoc and Sensor Systems, 2004, pp. 525-527.
[29] M.C. Su and C.H. Chou, A modified version of the K-means algorithm with a distance based on cluster symmetry, IEEE Transactions on Pattern Analysis & Machine Intelligence, Vol. 6, pp. 674-680, 2001.
[30] A. Kupcsik, and G. Neumann, Model-based contextual policy search for data-efficient generalization of robot skills, Artificial Intelligence, 247, 415-439, 2017.
[31] K. Zheng, H. Meng, L. Hou, K. Lin, and L. Hu, Dynamic offloading schemes for mobile cloud computing services in heterogeneous networks, Journal of Internet Technology, Vol 16, No. 2, pp. 335-342, 2015.
[32] W.T. Cho, Y.W. Ma, and Y.M. Huang, A smart socket-based multiple home appliance recognition approach over IoT architecture, Journal of Internet Technology, Vol. 16, No. 7, pp. 1227-1238, 2015.

IOP Conf. Series: Journal of Physics: Conf. Series **1064** (2018) 012068 doi:10.1088/1742-6596/1064/1/012068

Numerical investigation on vorticity of rectangular turbulent jet enhanced by plasma

Liang Li, Xueke Che, Tikai Zheng, Chuan Chen, Wangsheng Nie and Xiuqian Li *

Space Engineering University, Beijing, China

*Corresponding author e-mail:lixiuqian123@126.com

Abstract. A three-dimensional plasma excitation model based on the phenomenological model was proposed and used as an excitation source of the jet. The large - eddy simulation method was used to simulate the flow field of a rectangular turbulent jet. The position deviation of the large eddy structures in the simulated flow field is within 8% compared with the experimental data. The model can accurately reflect the vortices distribution of the flow field and has a good accuracy. The variation trend of jet vortices is surfaces and vortices field is studied under different excitation voltages, which proves that the stream wise vortex is the key to the influence of the plasma exciter on the jet. The maximum value of velocity and vortices in the flow field enlarge with the voltage increasing while the turbulent position of the jet advances significantly, and the vortex shedding strength increases. The jet mixing characteristics are significantly enhanced.

1. Introduction

Rectangular jet is widely used in industrial production, chemical industry and aerospace propulsion system, which has the characteristics of easy turbulent and fast diffusion rate, thus can effectively enhance the mixing efficiency of the fluid, improve the fuel premix efficiency and realize the improvement of combustion efficiency [1]. Scholars at home and abroad have carried out a large number of theoretical, experimental and simulation studies on the development and flow characteristics of the rectangular turbulent jets.Kaoru et al. [2] used large eddy simulation (LES) model to simulate rectangular jet, the role of large-scale vortex structure in jet development and momentum transfer in different Reynolds numbers is studied, and it is proved that large-scale vortex structure is the main reason for the instability of jet shear layer; Th. Pandas et al. [3] The experimental research on rectangular bounded jets is carried out, and the internal flow field of the rectangular turbulent jet is measured by using the hot-wire anemometer (HWA) probe array, and the experimental results verify the conclusion of [2]. Yang et al. [4] studied the development process of jets with different nozzle shapes, it is found that the distribution of the stream wise vortices and the span wise vortices in the flow field of the rectangular jet have important influence on the jet development and mixing characteristics, and different mixing effects of the rectangular jet can be achieved by changing the outlet Reynolds number and the aspect ratio (AR) of the nozzle.

Active flow control is an important means to realize accurate jet control and adapt to different working conditions. The jet control can be realized by using mechanical adjusting device [5],

Content from this work may be used under the terms of the Creative Commons Attribution 3.0 licence. Any further distribution of this work must maintain attribution to the author(s) and the title of the work, journal citation and DOI.
Published under licence by IOP Publishing Ltd

controlling jet [6] and magnetic field and so on. Surface dielectric barrier discharge (SDBD) plasma flow control is a promising active flow control method with small additional mass, simple structure and convenient adjustment [7]. SDBD plasma flow control is realized mainly by breakdown the surface air on the electrode using high voltage to form a plasma atmosphere, and then use the potential difference between the electrodes to induce the nearby airflow to produce the jet, thus controlling the flow field. At home and abroad, a large number of theoretical and experimental studies have been carried out on the mechanism of its action.

Simulation is an important means to reveal the control mechanism of SDBD plasma flow control. The phenomenological model does not describe the plasma discharge process in detail, but only gives the effect of the plasma on the gas. It has the characteristics of simple and reliable calculation, and can simulate the surface dielectric barrier discharge induced jet in macroscopic view. Therefore, it plays an important role in the simulation of plasma flow control [8]. For the phenomenological model of SDBD discharge, domestic and foreign scholars have done a lot of research. Suzan Y. B et al [9] proposed a phenomenological model of plasma discharge based on electric field force. Compared with the experimental data, the simulation accuracy of plate induced discharge is higher. In China, Xu eke Chee, Qinghai Chen [10,11] have improved the model, which has been applied to the numerical simulation in the flow environment of adjacent space propeller and low Reynolds number airfoil, which can restrain stall and prevent boundary layer separation.

Based on the previous research, the development of rectangular turbulent jet is modeled and simulated by using three-dimensional SDBD phenomenological model and large eddy simulation, and the flow field structure and jet mixing characteristics under different excitation voltages are investigated. It is helpful to reveal the mechanism and optimal control mode of SDBD plasma flow control.

2. Numerical model

2.1. Phenomenological model

The detailed derivation process of the SDBD plasma phenomenological model under two-dimensional incompressible conditions is given in [9] and it will not be repeated here. During the discharge process, the external potential ϕ is determined by the excitation power source. After the exposed electrode is connected to the high voltage power supply, its boundary conditions can be expressed as:

$$\varphi(t)=\varphi^{\max} f(t) \tag{1}$$

Where $\varphi^{\max}$ the magnitude of the applied voltage and its value is is half of the peak-to-peak value of the excitation voltage VIP-p. In the pulse discharge mode, the voltage change function over time is:

$$f(t)=\begin{cases}\sin(2\pi wt) & nT_m \le t \le (n+D)T_m \\ 0 & (n+D)T_m \le t \le (n+1)T_m\end{cases} \quad n\in N \tag{2}$$

Since the implanted electrode is a ground electrode, its potential is 0; the gradient of the potential at other outer boundaries $\partial\phi/\partial n=0$. In the discharge process, the air ionization region is between the exposed electrode and the implanted electrode, and the charge is mainly distributed above the implanted electrode, and the charge density of other outer boundaries $\rho_c=0$. The charge distribution above the implanted electrode conforms to the Gaussian distribution law:

$$G(x)=\exp\left[-(x-\mu)^2/(2\sigma^2)\right] \tag{3}$$

Where μ denotes the value of x when the function obtains the maximum value, and σ denotes the attenuation coefficient. The change in charge distribution over time of the charge distribution over the implanted electrode over time, $\rho_{c,w}$ can be expressed as:

$$\rho_{c,w}(x,t) = \rho_c^{\max} G(x) f(t) \quad (4)$$

The boundary conditions of the electric potential and the electric charge in the discharge of the SDBD plasma have been obtained. The specific boundary conditions are set as shown in Figure 1:

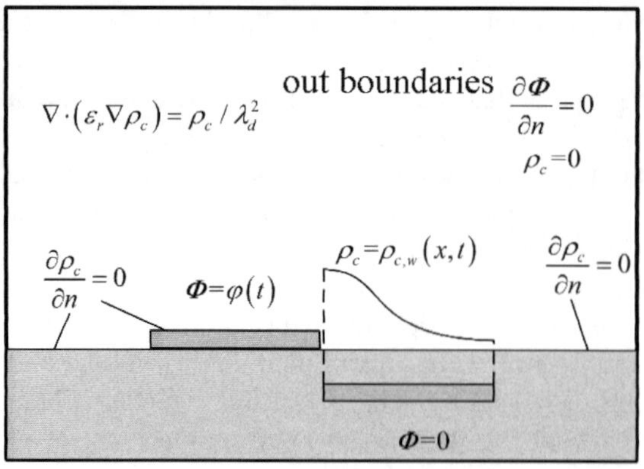

Figure 1. SDBD plasma phenomenological model boundary conditions schematic.

2.2. Numerical method
The vortex structures play an important role in the process of jet instability. The large-scale vortex structures continuously transfer momentum from the main stream area to the adjacent area of the jet, resulting in the constant attenuation of the centreline velocity. Momentum exchanging increases with the entrainment of the scrolling effect happened between the jet boundary layer and the surrounding fluid. The large vortex structure breaks in the course of development, thereby forming a large number of small vortex structures at positions far from the jet's centreline, further dissipating the remaining momentum. Because the scale of the vortex structure varies greatly, the focus of turbulence numerical calculation is how to deal with the relationship between the calculation speed and the calculation scale. Direct numerical simulation (DNS) can solve the structure of various scales in the flow field by directly solving the N-S equations. The precision of DNS is very high but it consumes computational resources. The Reynolds-averaged N-S method (RANS) can calculate quickly but cannot effectively reflected the vortex structures that play an important role in the turbulent flow field. The large eddy simulation method divides the vortex structures into two types: large eddy and small eddy, by filtering function. It is considered that the large vortex has anisotropy and plays a major role in the flow field thus it is solved directly; meanwhile, the small vortex has isotropic and mainly play a role of dissipating momentum in the flow field, so can establish a model to reflect its impact on large-scale movements. LES can effectively reflect the flow field structure while saving a lot of computing resources [12].

The platform with the existed code is used, and phenomenological model accesses the inner surface of the jet nozzle in the model by c. The calculation uses the Smagorinsky-Lilly model, in which the sub lattice turbulent viscosity μ_t is:

$$\mu_t = \rho L_s^2 \left| \overline{S} \right| \quad (5)$$

Where,

$$\left|\overline{S}\right| = \sqrt{2\overline{S}_{ij}\overline{S}_{ij}} \tag{6}$$

$$L_s = \min(\kappa d, C_s V^{\frac{1}{3}}) \tag{7}$$

and L_s is the sub lattice mixing length, κ is the Karman constant, d is the closest distance from the location to the wall, C_s is the Smagorinsky constant, and V is the volume of the calculation unit.

2.3. Mesh and boundary conditions

The geometry of the calculation area and the central profile along the stream wise direction, span wise direction and lateral is showed in Figure 2(a) ~ (d). The shape of the nozzle is rectangular, the geometric dimension is width b=27mm, height h=3mm, aspect ratio (AR) is 9; the excitation electrode is placed on the lower surface of the jet outlet, the width of exposed electrode is L1=5mm, and the width of the implanted electrode L2 = 9 mm, electrode gap d = 0 mm. The size of the flow body area is 150 mm×80 mm×54 mm, and the mesh is encrypted around the centerline of the jet to accurately capture changes in the jet shear layer.

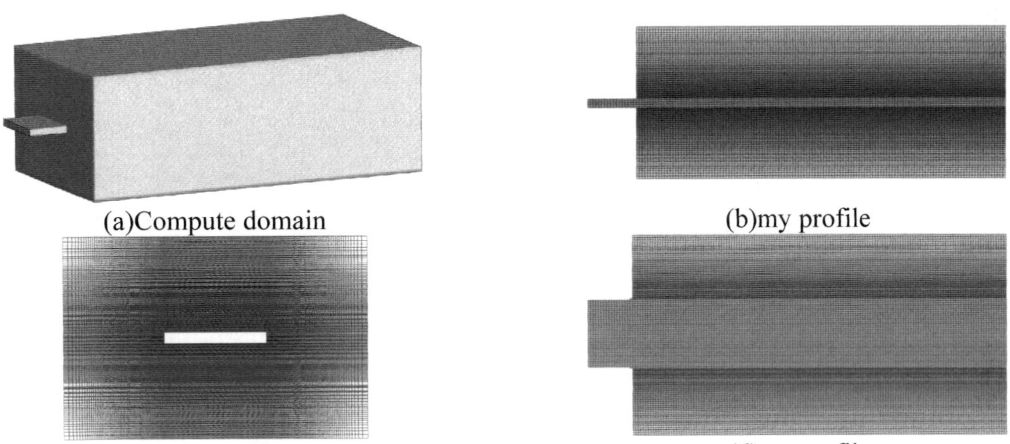

(a)Compute domain (b)my profile

(c)is profile (d)oz. profile

Figure 2. Compute regions and grids.

The inlet boundary conditions were set as flow inlet, its pressure is 1.0 tam, and the outlet was set as pressure outlets, the back pressure of it is 80.0 kappa. The nozzle and wall surfaces were set as non-slip boundaries while the experimental environment pressure was 80.0 kappa. Check out Table 1 for other details. The computing platform is a Think Station P710 tower workstation with two Intel® Xeon E5-2620V4 processors, and the physical core count of a single processor is8 (16 threads) and the clock speed is 2.1 GHz. The size of running memory is 32 Bit, it took about 156 hours to complete calculating a single jet flow for 1.0 s.

Table 1. Flow calculation parameters

Operating pressure p/Pa	Temperature T/K	Mass flow in m/(kg.s^{-1})	Density ρ/(kg.m^{-3})	Viscosity μ/(Pays)	Re
80000	288	3.611	1.18	1.78*10^{-5}	1457

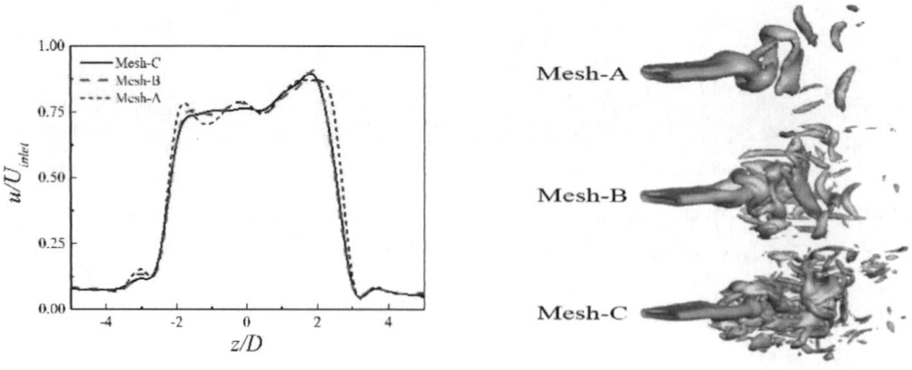

(a)Mean velocity distribution (y=0, x/D=7)

(b)Is surface of Vortices(372s-1)

Figure 3. Results of different mesh densities.

Three alternative sets of meshes, namely Mesh-A (377486 nodes), Mesh-B (822996 nodes), and Mesh-C (1543975 nodes) were prepared to perform mesh-independence tests. Figure 3(a) shows the distribution along the z-axis of the centreline velocity of the profile at x = 7D, which is dimensionless processed using the maximum inlet velocity for comparison. As can be seen from the figure, the velocity distributions calculated by the three sets of grids are basically the same, and all of them can reflect the fluctuation trend of the rectangular turbulent jets at this position and there is only small fluctuations in the numerical values. Considering the transient irregular nature of turbulent jets, the average velocity distribution cannot effectively reflect the influence of the density of nodes on the merits of the calculation results. Figure 3(b) shows the vortices is-surfaces with different grids at a flow time of 1.0 s. Mesh-A only shows a small number of large-scale span wise vortexes near the nozzles and a partial stream wise vortex, while the calculation results of Mesh-B and Mesh-C contain a large number of small-scale vortices with more detail. Considering that Mesh-C has more meshes, the computation speed is slower, and Mesh-B has reflected enough details of the flow field, so it is finally selected for calculation.

3. Model Verification

Using the experimental data that has been obtained by this research group, verify the model built before.

The experiment used a PIV system in a low-pressure cabin with a pressure of 80.0 kappa to complete the shooting of the central section in the z-direction of the rectangular jet. The inlet pressure of the jet was 1 tam, and the jet was formed by the differential pressure inside and outside the chamber. The jet nozzle size and electrode arrangement are consistent with the simulation model. The electrode configuration is counter-flow, that is, the direction of the induced jet flow generated when the electrode is working is opposite to that of the jet mainstream. The tracer particles adopt white smoke particles generated from the combustion of cigarette cake. The PIV system is mainly composed of a ND: YAG double pulse laser, a synchronous controller, a Kodak scientific-grade chip CCD camera (with a resolution of 2048×2048 ox/in), a computer and an image processing software. The laser repetition frequency is 5 Hz. The images obtained by PIV imaging are processed using Micro Vic V3.1.1 calculation software with a spatial resolution of 2.62 mm. The excitation power supply adopts the HFHV30-1 high-frequency high-voltage power supply developed by the Institute of Electrical Engineering of the Chinese Academy of Sciences. The operation modes are continuous and pulsed. In pulse mode, the operating frequency can be adjusted between 10-3000Hz and the duty cycle can be adjusted from 10% to 90%. The voltage measurement uses the Agilent N2271B high-voltage probe, the current measurement uses the Pearson current coil 6595, and the oscilloscope is the Agilent DSO3024A. This article uses the pulse mode of operation; the power carrier frequency is 10 kHz.

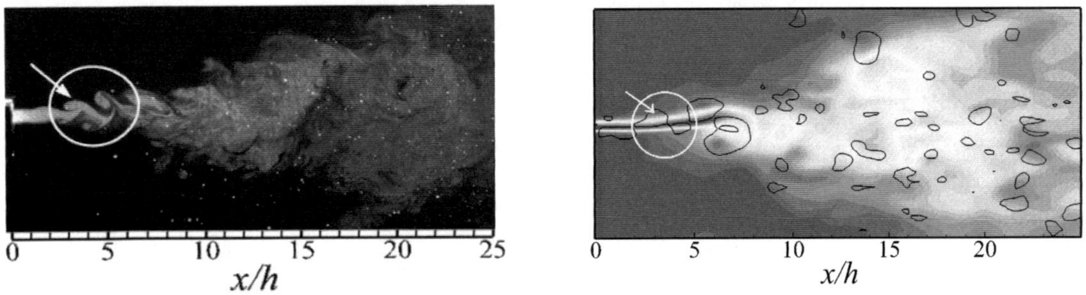

(a) Experiment graph (b) Simulation velocity field

Figure 4. Experiment and simulation flow field (VIP-p=6.0kV, f=50Hz, Duty=50%).

The Q-criterion is used to identify the vortex structure in the simulation flow field. For the incompressible two-dimensional flow field, the expression is [13]:

$$Q=\frac{1}{2}\left(\|\boldsymbol{\Omega}\|^2-\|\boldsymbol{S}\|^2\right) \tag{8}$$

$\boldsymbol{\Omega}$ Is the rotational velocity tensor in the rotational motion, S is the strain tensor in the non-rotating motion, and the expressions are respectively:

$$\boldsymbol{\Omega}=\frac{1}{2}\left[\nabla \boldsymbol{U}-(\nabla \boldsymbol{U})^T\right] \tag{9}$$

$$\boldsymbol{S}=\frac{1}{2}\left[\nabla \boldsymbol{U}+(\nabla \boldsymbol{U})^T\right] \tag{10}$$

For a two-dimensional flow field in the middle section, the Q criterion can be simplified to:

$$Q=\frac{\partial u}{\partial x}\frac{\partial v}{\partial y}-\frac{\partial u}{\partial y}\frac{\partial v}{\partial x}-\frac{1}{2}\left(\frac{\partial u}{\partial x}+\frac{\partial v}{\partial y}\right)^2 \tag{11}$$

Calculate the Q value in the flow field, and select the appropriate threshold to filter the small-scale structure. The black solid line represents the flow in the area where the vortex dominates.

Comparing the two images, it can be seen that the jets obtained by the experiment and simulation have a smooth section of a certain length near the outlet, and the length is about 3 times the width of the nozzle. After that, the mainstream began to presents a small degree of instability. The instability of the shear layer led to vortex rings on both sides of the mainstream; the occurrence of large vortices resulted in aggravating entrainment of jets and surrounding fluids. As the jet develops further, the width in the short axis direction increases rapidly, and the rectangular jet enters the plane jet section. At the same time, it can also be seen that after the jet is ejected from the outlet, its free surface relaxes faster and decays rapidly along the flow direction (Figure 4(b)), resulting in a large velocity gradient. The existence of this disturbance makes the stability of the jet compromised. Disturbance accelerated its spread [14]. The first two sets of vortex pair centers defining the start of jet instability are the comparison standards. The comparisons under other conditions are shown in Table 2. The comparison results show that the numerical simulation method captures large-scale structures in the jet flow field. Better, the method chosen for the calculation model is appropriate and the calculation result is reliable.

Table 2. Position of eddies in experiment and simulation.

Conditions	First eddy position(x/h)		Second eddy position(x/h)	
	Experiment	Simulation	Experiment	Simulation
Plasma off	6.0	6.0	8.0	10.0
4kV	5.0	5.2	7.0	8.0
8kV	3.8	3.8	5.0	5.5

4. Influence of voltage on flow field

The established SDBD flow control model was used to study the effect of different excitation voltages on the disturbance of a rectangular turbulent jet flow field. The electrode configuration of the simulation model uses the inverse configuration [15], that is, the direction of the induced jet generated by the electrode is opposite to that of the mainstream flow of the jet (Figure 5). When the reverse-type jet works, the reversed induced jet will cause the shear layer of the jet to destabilize, and the induced vortices formed above the electrode wall will then incorporate a part of the momentum into the main stream, increasing the exit Reynolds number, thus achieve the promotion effect.

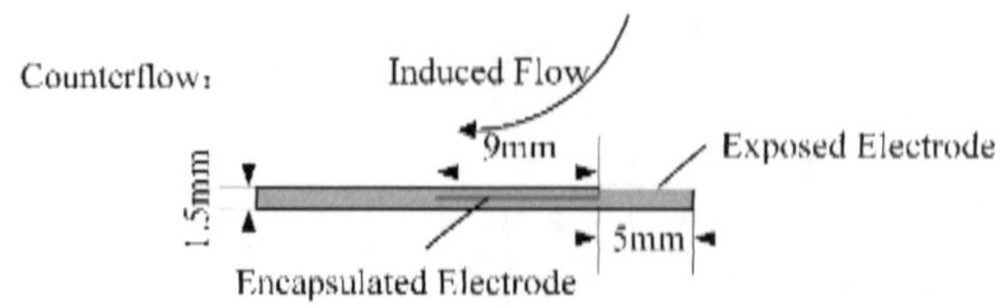

Figure 5. Inverse configuration of the exciter.

Figure 6 shows the vortices is surface of the jet flow under different excitation voltages. For comparison, all is surface data have a vortex value of 372 s-1. The vortices is surface was smooth before x/h<17.0 without excitation, indicating that the jet shear layer was relatively stable and there was no significant disturbance; after this position, a pair of large The scale vortex structure appears, which is also the sign of the instability of the jet, after that the large vortex structure begins to distort and break, accompanied by the vortex shedding and the direction of the vortex ring. After the excitation is applied, the vortices is surface undergoes significant deformation. The first is that the jet instability position is significantly advanced, and the number of vortex structures in the destabilizing position increases. On the one hand, the SDBD plasma exciter induces a part of the vortex structure. Due to the induced jet perturbation, the mainstream of the jet itself begins to destabilize and the vortex structure increases. With the increasing of the excitation voltage, two significant changes occur in the vortices profile. First, the deformation strength increases. When the excitation voltage is 4 kV, the vortex profile with a stable initial emission phase has only a small part of the protruding structure. At 8Kv, the vortices profile at the initial stage of destabilization has a tendency to disengage from the mainstream, the position that the large vortex break is ahead. Under low excitation voltage, the vortex shedding at x/h=20.0 is smaller than the profile value, but in high voltage, the eddy shedding position is about 5h ahead than low voltage, which proves that the excitation voltage has a significant gain effect on the control effect.

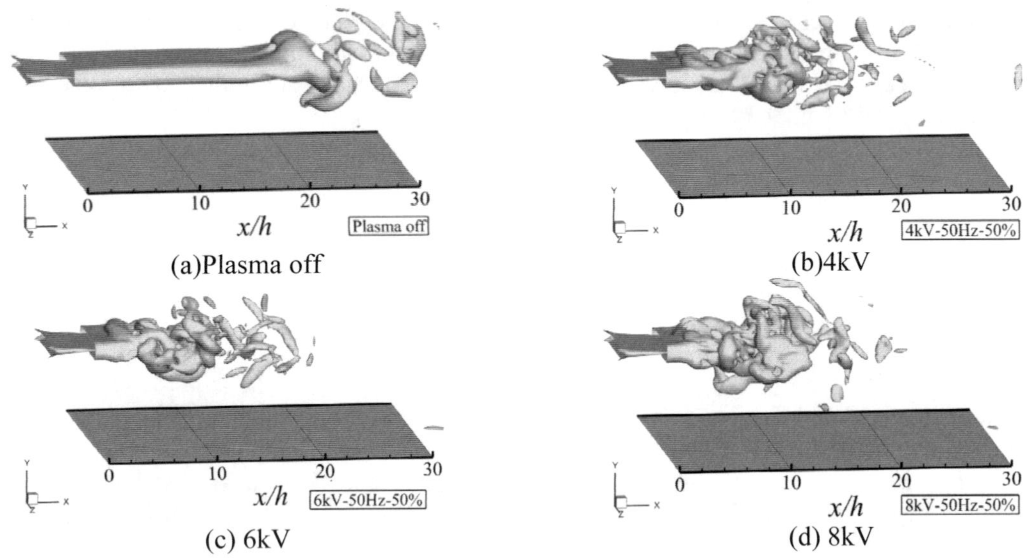

Figure 6. Vortices is-surface in different voltages.

Figure 7 shows the vortices distribution of the profile y=0 under four operating conditions. The ratio of exit velocity U and nozzle height h is used for dimensionless processing. Without excitation, the initial vortices is concentrated and mainly distributed in the jet shear layer, but the scale is small and has no great influence on the shear layer. With the increasing of entrainment, the natural vortex structure appears in the position around x/h=23.0. At this time, the consumption of large vortices to the mainstream velocity is very obvious, and the large vortices develop to a certain extent and begin to distort and break. A large number of small vortices are formed by the shedding vortex structure, which accelerates the momentum dissipation velocity at the back of the flow field. With the increasing of the excitation voltage the maximum vortices in the flow field increases and the center of the jet is disturbed more obviously. It is proved that the induced vortex plays an important role in the initial evolution of the jet.

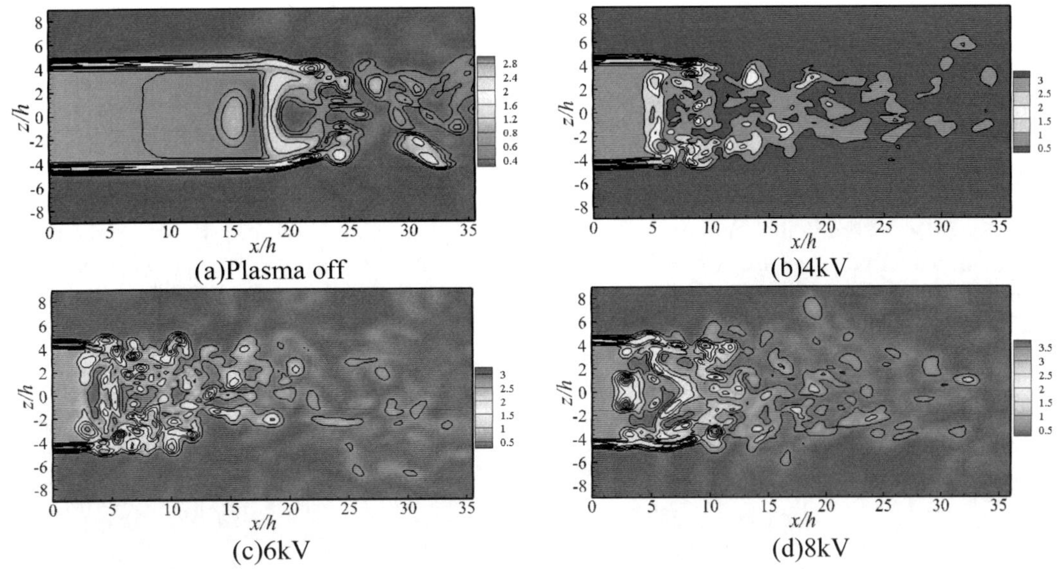

Figure 7. Vortices distribution in different excitation voltage flow fields.

Through the experimental measurements of the initial stage of jet development (especially before the start of the instability), Pandas [3] et al found that the vortex which plays a leading role in the jet development process can be divided into two types according to its direction: the stream wise vortex is and the span wise vortex in, the expression is:

$$\Omega_s = \frac{D}{U}\left(\frac{\partial w}{\partial y} - \frac{\partial v}{\partial z}\right) \tag{12}$$

$$\Omega_n = \frac{D}{U}\sqrt{\left(\frac{\partial u}{\partial z} - \frac{\partial w}{\partial x}\right)^2 + \left(\frac{\partial v}{\partial x} - \frac{\partial u}{\partial y}\right)^2} \tag{13}$$

The exit velocity U and nozzle equivalent diameter D are used to dimensionless the two vortices. It is considered that the span wise vortex in the jet is the most obvious, while the stream wise vortex is relatively concealed, so it is difficult to observe by means of observation. The strength of the span wise vortex is higher relatively, but the stream wise vortex plays a key role in promoting the distortion and fragmentation of the span wise vortex [17]. The conflict between the two vortices increase the momentum of the jet in the early development and promote turbulence.

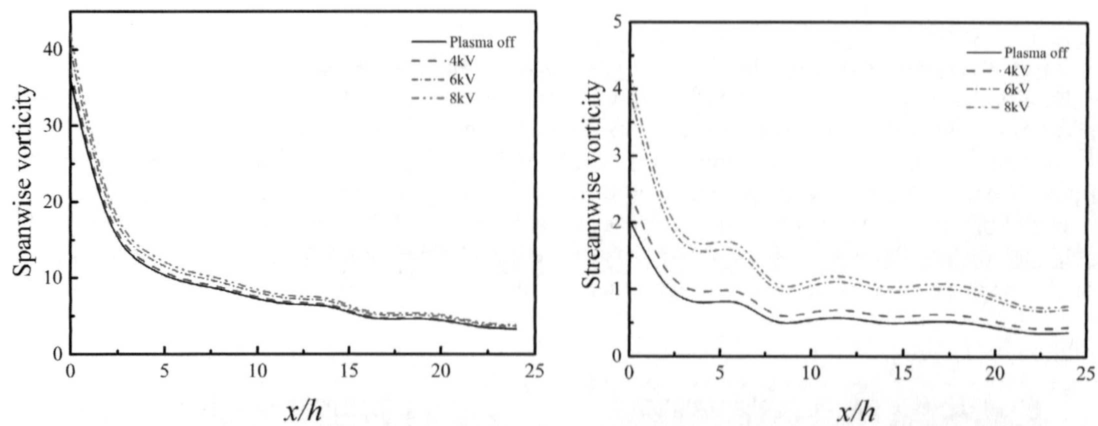

(a)Span wise vortices variation trend (b) Stream wise vortices variation trend

Figure 8. Variation trend of span wise vortices and stream wise vortices in different voltage.

In order to further investigate the influence of stream wise vortex and span wise vortex on jet flow in the initial stage of jet development, the variation of stream wise vortex and span wise vortex in the range of x < 13D(x<24h) in different excitation voltage was extracted (Figure8). It can be seen that the strength of the span wise vortex is obviously larger than that of the stream wise vortex by about an order of magnitude, and the two vortices have a sharp downward trend near the exit, which is consistent with the results obtained in [4]. The difference is that the stream wise vortex fluctuates to a certain extent, and the maximum value of the stream wise vortex increases obviously with the increase of the excitation voltage amplitude of variation. At the same time, the amplitude of variation of the span wise vortex is very small, which proves that the SDBD plasma exciter indirectly affects the stronger span wise vortex mainly by influencing the stream wise vortex, and promotes the turbulence of the jet by promoting its distortion and fragmentation.

5. Conclusion

By establishing the phenomenological model of SDBD plasma, the development of rectangular turbulent jet under different excitation voltages was simulated by using the method of large eddy simulation. The calculation model is in good agreement with the experimental data, and the accuracy is high. The analysis of the calculated results further explains the process of the SDBD plasma exciter acting on the jet. It is proved that the variation of the vortex structure, especially the stream wise vortex, is an important part of the excitation process. Further theoretical basis is provided for the study of plasma flow control.

Acknowledgements

The paper is financially supported by the national nature and science foundation of China (code: 51777214, 91441123).

References

[1] Hu H, Saga T, Kobayashi T, et al. Mixing Process in a Lobed Jet Flow. Aiwa Journal, 2012, 40(7):1339-1345.

[2] Kaoru Iwamoto, Nobuhide Kasagi, Yuji Suzuki. Dynamical Roles of Large-Scale Structures in Turbulent Channel Flow. WCCM VI in conjunction with APCOM'04, 2004, Beijing, China.

[3] Pandas T, Schwab R, Pollard A. The role of vortices in the near field development of sharp-edged, rectangular, wall jets. International Journal of Heat & Fluid Flow, 2017, 67.

[4] Yang X, Long X, Yao X. Numerical investigation on the mixing process in a steam ejector with different nozzle structures. International Journal of Thermal Sciences, 2012, 56(2):95-106.

[5] Davis M. R. Variable control of jet decay. AIAA Journal, 1982, 20(5):606-609.

[6] Raman G. Using Controlled Unsteady Fluid Mass Ad-dition. AIAA Journal, 1997, 35(4):647-656.

[7] Nie W S, Cheng Y F, Che X K. Research progress of dielectric barrier plasma flow control. Advances in Mechanics, 2012,42(6):722-734.

[8] Shao T, Yan P. Atmospheric pressure gas discharge and its plasma application. Science Press, 2015.

[9] Suzen Y. B., Huang P. G., Jacob J.D., et al. AIAA. Numerical Simulations of Plasma Based Flow Control Applications. Aiaa Fluid Dynamics Conference and Exhibit. 2005.

[10] Chen Q Y, Tian X H, Jiang J W, et al. Propeller Plasma Flow Control Enhancement Experiment. Journal of Aerospace Power, 2016, 31(5):1205-1211.

[11] Che X K, Nie W S, Zhou P H, et al. Study on the improvement of plasma volume force image and model. National Fluid Mechanics Conference, Nanjing. October 20-23, 2016.

[12] Zhao P H, Ye T H, Zhu Y M, et al. Large Eddy Simulation of Turbulent Flow in a Circular Jet [J]. Journal of Engineering Thermophysics, 2012, 33(3):529-532.

[13] Zhang Z, Che X K, Nie W S, et al. Study of vortex in flow fields induced by surface dielectric barrier discharge actuator at low pressure based on Q criterion. Plasma Science & Technology, 2018, 20(1):014006.

[14] Kozato Y, Kikuchi S, Imao S, et al. Flow control of a rectangular jet by DBD plasma actuators [J]. International Journal of Heat & Fluid Flow, 2016, 81(825):14-00695-14-00695.

[15] Cheng Yufeng, Che Xueke, Nie Wangheng. Numeri -cal study on propeller flow -separation control by DBD-plasma aerodynamic actuation. IEEE Transactions On Plasma Scinence, 2013, 41 (4): 892-898

[16] Jung D, Gamard S, George W K. Downstream evolution of the most energetic modes in a turbulent axisymmetric jet at high Reynolds number. Part 1. The near-field region. Journal of Fluid Mechanics, 2004, 514(514):173-204.

[17] Pescini E, Marra F, Giorgi M G D, et al. Investigation of the boundary layer characteristics for assessing the DBD plasma actuator control of the separated flow at low Reynolds numbers. Experimental Thermal & Fluid Science, 2017, 81:482-498.

IOP Conf. Series: Journal of Physics: Conf. Series **1064** (2018) 012069 doi:10.1088/1742-6596/1064/1/012069

Study on the influence of turbulence intensity on vortex signal

Zhisen Liu*, **Fugui Huang**

College of Mechanical Engineering & Automation, Huaqiao University, Fujian, China

*Corresponding author e-mail: albertliuhao@126.com

Abstract. Numerical simulation software FLUENT was used to compare the turbulence intensity of the fluid at the outlet of the linear shrinkage tube and curvilinear shrinkage tube. Through the detection points behind the bluff body in the measurement segment, the static pressure distribution is obtained, and the frequency and intensity of the vortex street signal are analysed. The results show that the fluid turbulence intensity in the curvilinear contraction tube is small, and the vortex signal intensity of the measurement segment is higher and the noise signal is less, which is favourable to the measurement of the vortex signal.

1. Introduction

Vortex flow meter, as a new-type velocity flow meter, has been widely used in industrial production. The theoretical shrinkage ratio of vortex street flowmeter can reach 100:1, but it can only reach 15:1 in practical applications. The reason is that vortex street flowmeter has a higher bottom boundary [1]. From the principle of Karman vortex signal generation, the vortex sensor output signal is weak at low velocity. Experts and scholars have found that vortex street signals tend to be aliased and even submerged by noise signals at low flow velocities, resulting in low signal-to-noise ratios for vortex street signals. To improve the performance of vortex flow meters at low velocity, the simplest method is to increase the fluid velocity. The solution adopted in industry is to add linear shrink tube in front of the vortex flow meter.

According to the Fluid Continuity Theorem, the linear shrink tube can significantly increase the fluid velocity, but it has been found in long-term use that the structure increases the fluid turbulence while increasing the fluid velocity, which in turn causes measurement error of the vortex flow meter. Then, a curvilinear shrink tube appeared, and to some extent improved the problem of the linear shrink tube increasing the fluid turbulence, which improved the measurement characteristics of the vortex flow meter. However, there are few theoretical and experimental studies on the Influence of fluid turbulence on vortex signal.

This article will use the computational fluid dynamics simulation software to model the straight pipe, the linear contraction pipe, the curve contraction pipe three kinds of different fluid pipes, takes the straight pipe as the reference, studies and analyses the change of fluid turbulence and the influence of turbulence on vortex signal in linear shrinkage tube, the curve shrinkage tube. The result indicates that low turbulence intensity has a positive effect on vortex signal.

2. Turbulence formation theory

Reynolds experiments show that the fluid has two flow patterns: laminar flow and turbulent flow; the two flow states not only have different fluid particle trajectories, but also their internal structures are

Content from this work may be used under the terms of the Creative Commons Attribution 3.0 licence. Any further distribution of this work must maintain attribution to the author(s) and the title of the work, journal citation and DOI.
Published under licence by IOP Publishing Ltd

completely different. However, as the flow velocity changes, the two flow states can be convert to each other. Taking the transition from laminar flow to turbulent flow as an example, we analyse the conditions and reasons for its transformation.

From Reynolds experiments, the indispensable conditions for the laminar flow to turn into a turbulent flow is that the formation of a vortex and the vortex are separated from the original stratum and incorporated into the adjacent flow stratum. An important prerequisite for vortex formation is fluid viscosity. The distribution of flow velocity on the actual fluid flow cross-section is not uniform, thus promoting internal friction shear stress between the flow layers. For a certain stratum, the shear stress exerted on it by the faster flow layer is in the forward flow direction, and the shear stress exerted on it by the slower flow layer is in the reverse flow direction. Therefore, the shear stress experienced by the selected flow layer tends to promote the generation of the vortex.

When the vortex formed, the velocity distribution near the vortex changes. The rotation direction of the vortex is the same as the direction of the faster flow layer and is opposite to that of the slower flow layer. This leads to an increase in the velocity, and decrease in the pressure of the faster flow layer and a decrease in the velocity, an increase in the pressure of the slower flow layer. As a result, there is a pressure difference between the two sides of the vortex, forming a lateral force. This force will prompt the vortex to disengage from the original stratum and into the faster flow zone. When the vortex body departs from the original stratum and enters the new stratum, the laminar flow becomes turbulent flow [2].

In the shrinkage tube, the flow near the tube wall is greatly increased due to compression, and the fluid velocity at the centre of the tube changes slowly. Shrink tubes increase the difference of flow velocity in different fluid layers. The flow velocity in the linear shrink tube is always increased and the vortex body is always formed. At the outlet of the straight shrink tube, the turbulence intensity of the fluid is maximized. The flow velocity in the curvilinear shrinkage tube increases slowly, then increases rapidly, and then slowly increases. In the whole process, the vortex body most easily forms when the flow velocity increases rapidly, and the fluid turbulence intensity reaches the maximum in this section. The outlet of the curvilinear shrinkage tube is relatively flat and the turbulence intensity of the fluid at the outlet is relatively small.

3. The simulation experiment

3.1. Modelling and Meshing

This paper uses modelling software ICEM CFD to establish three models [3], as shown in Figure 1. The linear tube (Fig. 1a) has a length of 600mm and a diameter of 50mm; the linear shrinkage tube (Fig. 1b) has a length of 600mm, a large diameter of 100mm, a small diameter of 50mm, a shrinkage ratio of 4; and a curvilinear shrinkage tube (Fig. 1c). has a length of 600mm, a large diameter of 100mm, a small diameter of 50mm, and the shrinkage ratio is 4. Trapezoidal bluff body is used in all three types of pipes. The length of the trapezoidal bluff body is 16mm, the width of the inflow plane is 14mm, and the entrance distance is 360mm. To ensure the same velocity in the measuring section, the inlet velocity of the linear shrinkage tube and the curvilinear shrinkage tube is set to 0.4m/s 1.8m/s, 1.5m/s; the inlet velocity of the linear tube is set to 1.6m/s, 3.2m/s, 6.0m/s.

The wall curve of the curvilinear shrinkage tube is the Witozinsky curve, and its equation is as follows:

$$r = \frac{R_2}{\sqrt{1-\left[1-\left(\frac{R_2}{R_1}\right)^2\right]\frac{\left(1-\frac{3x^2}{a^2}\right)^2}{\left(1+\frac{x^2}{a^2}\right)^3}}} \tag{1}$$

In the formula, R_1 is the inlet radius of the shrinkage tube, R_2 is the outlet radius of the shrinkage tube, r is the cross-sectional radius of the shrinkage tube at the axial distance x, $a = \sqrt{3}L$, and L is the length of the shrinkage tube.

The ICEM CFD software is used to divide the established model into grids. Because the model structure is simple, the structured grid is used to divide the model. After meshing, the model is imported into FLUENT for calculation.

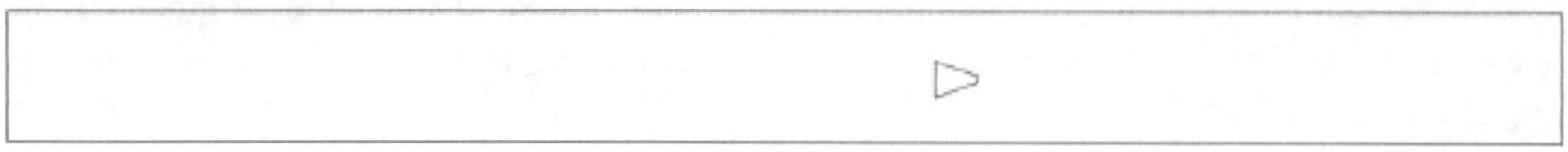

(a)

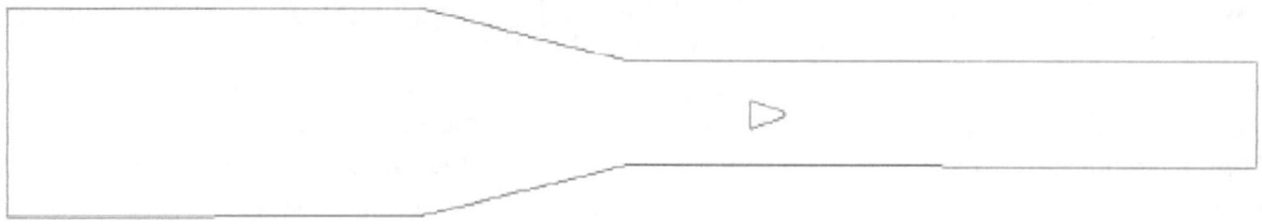

(b)

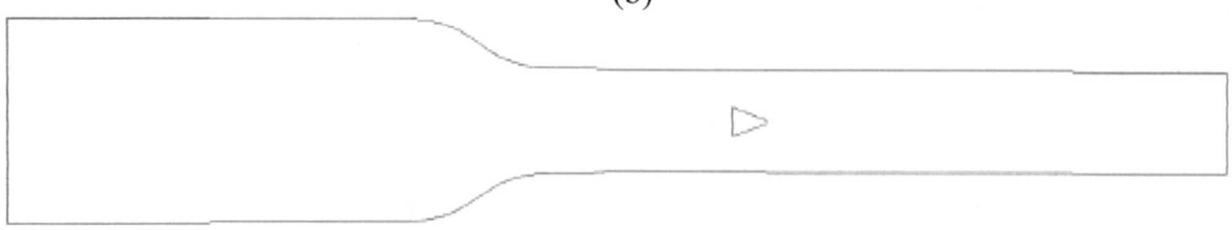

(c)

Figure 1. Three different types of fluid pipes.

3.2. Calculation Models and Algorithm Selection

The results obtained by different numerical models in FLUENT differ in the accuracy and speed of calculations, and appropriate calculation methods should be selected for specific problems to obtain accurate results [4].

This paper selects the RNG k-ε turbulence model. This model is an improved model of the standard k-ε model, and its improvement measures mainly include the following aspects:

(1) An additional term is added to the ε equation, making it more accurate when calculating flow fields with large velocity gradients.

(2) The model considers the rotation effect, so the calculation accuracy for strong rotational flow is also improved.

(3) The model adds an analytical formula for calculating the turbulent Brandt number.

(4) The model can calculate the low Reynolds number effect after proper treatment of the near wall area.

The above measures make the model more suitable for calculating turbulence, and its calculation accuracy and speed are greatly improved.

To control the calculation process and improve the accuracy of the calculation, it is necessary to set the corresponding value in the solver. The settings mainly include algorithms and discrete formats.

In this paper, the SIMPLE form of pressure-velocity correlation algorithm is selected. SIMPLE has good stability and high precision and is used for the calculation of steady flow. The second-order upwind style in the discrete format is selected. The second-order upwind style can be seen as a special case of the flow field variable in the control point of the upstream grid unit. It retains the first and second terms of the Taylor series, and the calculation accuracy is higher.

Considering the calculation accuracy and the calculation speed, the fluid flow field of three different pipeline structures was simulated and analysed using RNG k-ε turbulence model, SIMPLE pressure-velocity coupling algorithm and second-order upwind discrete format.

IOP Conf. Series: Journal of Physics: Conf. Series **1064** (2018) 012069 doi:10.1088/1742-6596/1064/1/012069

4. Analysis of simulation results

4.1. Comparison of Vortex signal strength

The detection point behind the bluff body in the measurement segment records the periodic variation of the pressure. The vortex frequency measurement is performed by using the Fourier Transform (FFT). The following figures show the pressure change curve and frequency diagram of the post-body detection points of the line tube, curvilinear shrinkage tube, and line shrinkage tube at different velocity.

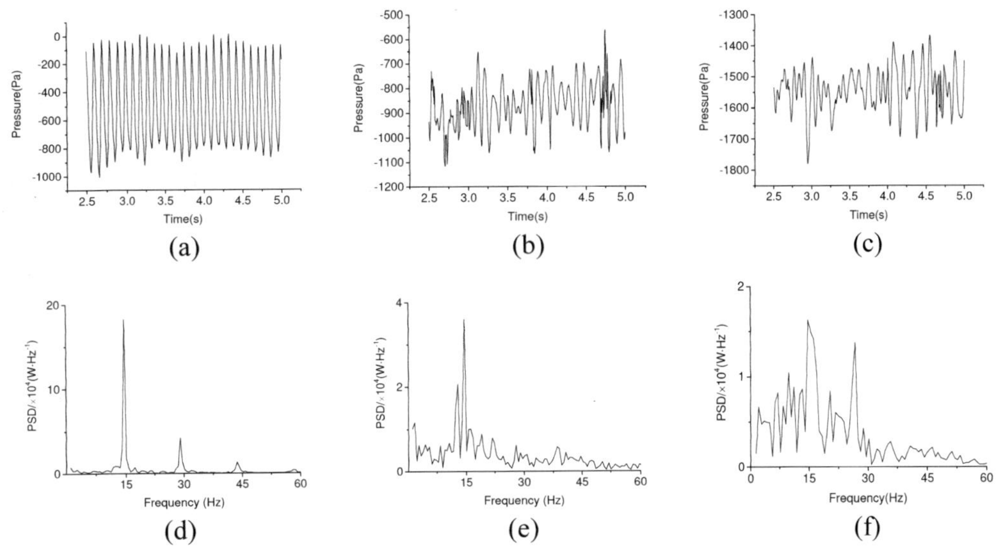

Figure 2. Pressure curve and frequency diagram when velocity is 1.6m/s.

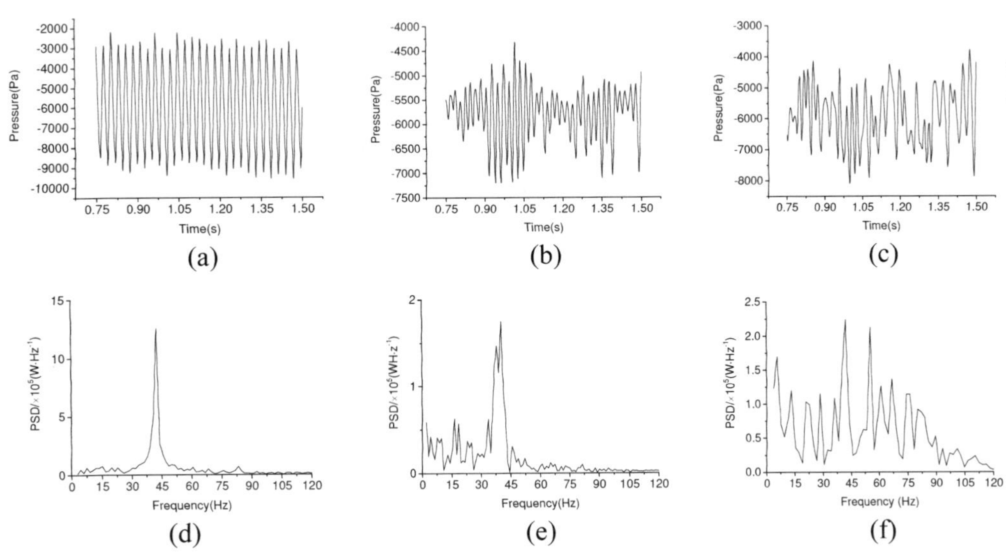

Figure 3. Pressure curve and frequency diagram when velocity is 3.2m/s.

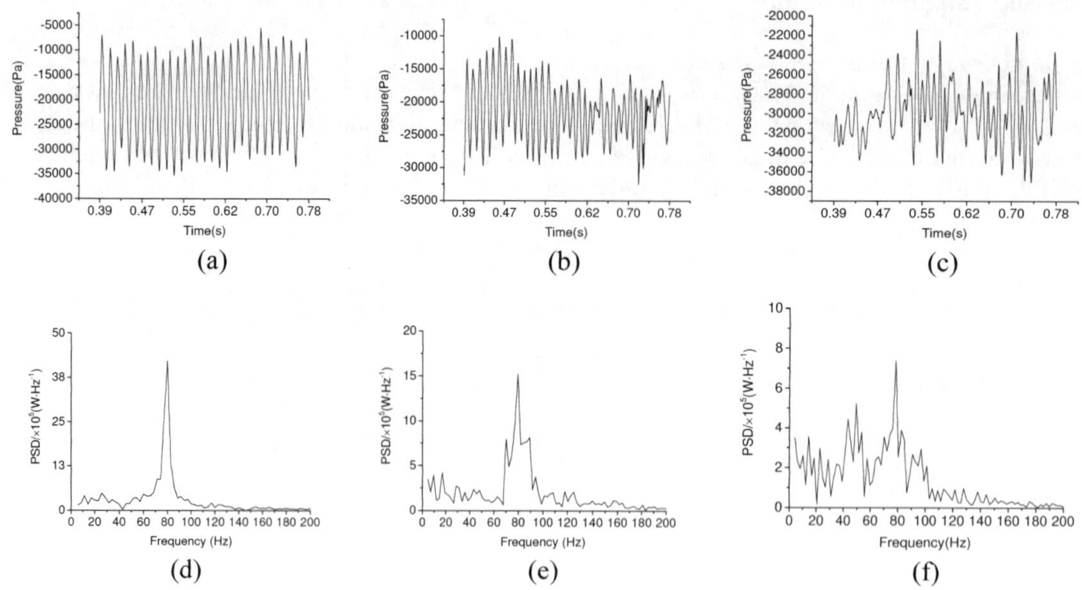

Figure 4. Pressure curve and frequency diagram when velocity is 6.0m/s.

It can be seen from the time domain waveform that the vortex signal is a sine wave, which reflects the vibration characteristics of the fluid caused by the periodic vortex. Due to the change of the pipeline structure, the pressure value at the detection point deviated from the sine law, and the pressure value of linear shrinkage pipe deviated the most. The signal peak-to-peak value corresponds to the magnitude of the pressure fluctuation generated by the vortex. The larger the peak-to-peak value, the stronger the corresponding vortex signal. The comparison shows that the vortex signal intensity is the highest when the flow velocity is 6.0 m/s. The peak value in the frequency domain map corresponds to the frequency of vortex signal, and as the velocity increases, the frequency value also increases accordingly. At the same velocity, the spectral density of the curvilinear tube is greater. Among the three types of pipes, the frequency components of the linear shrink tube are complex and there are many noise signals, which is also the reason for the poor measurement characteristics of the vortex flow meter in actual measurement.

4.2. The comparison of turbulence intensity

Turbulence is an eigenvalue that reflects the turbulence intensity and is the ratio of the root mean square of the pulsation velocity to the corresponding mean velocity [5]. The greater the turbulence intensity, the more vortices are present inside the fluid and the vortex signal collected in the measurement section will have a lot of noise signals. Therefore, low turbulence intensity has a positive effect on the collection and processing of vortex signals. Based on the principle of turbulent flow formation and experience in practical application, the turbulence intensity of the fluid in the linear shrinkage tube at the same velocity is higher than that of the curvilinear shrinkage tube. At different velocity, the turbulent intensity of the linear shrinkage tube and the curvilinear shrinkage tube are almost the same. As shown in the figure below.

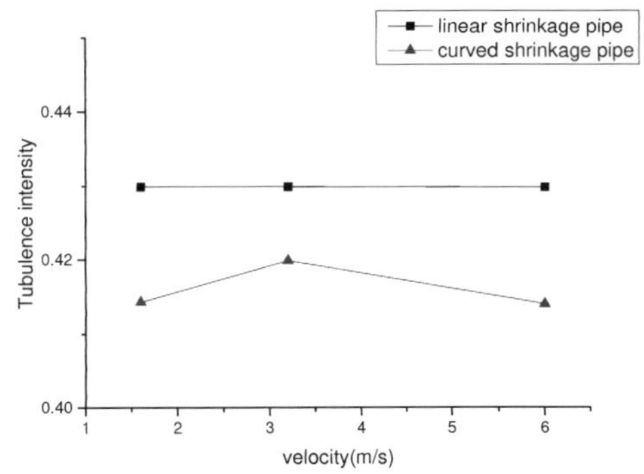

Figure 5. Turbulence intensity at different velocity.

The figure shows that the turbulence intensity in the line shrinkage tube at the same velocity is greater. At different velocity, turbulence intensity changes little. It shows that the turbulence intensity is more affected by the structure of the shrinkage tube, and the turbulence intensity of the fluid in the curvilinear shrinkage tube is smaller than the turbulence intensity of the fluid in the linear shrinkage tube.

5. Conclusion

In this paper, FLUENT simulation software is used to simulate the flow field of straight pipe, straight shrink pipe and curved shrink pipe. The following conclusions are obtained:

(1) At the same flow velocity, the frequency of the vortex street signal is the same; the frequency and intensity of the vortex street signal increase with the increase of the flow velocity.

(2) The turbulence of the fluid in the linear shrink tube is greater than that of the curved shrink tube. At different velocity, the degree of turbulence in the fluid did not differ significantly. Turbulence is more affected by the structure of the shrink tube.

(3) The fluid in the linear contraction tube structure is more disordered, and there are more noise signals in the measurement section, which are consistent with the phenomenon in practical application.

According to the above simulation results, compared to the straight shrink tube, the fluid turbulence in the curve shrink tube is smaller, the noise signal in the measurement section can be reduced, the vortex signal is easier to measure, and the static characteristics of the vortex flow meter are improved.

Acknowledgments

This work was financially supported by Subsidized Project for Cultivating Postgraduates' Innovative Ability in Scientific Research of Huaqiao University.

References

[1] Pankanin G L. The vortex flowmeter: various methods of investigating phenomena[J]. Measurement Science & Technology, 2005, 16(3):R1.

[2] Guo W D A, Guo W Z A, Zhang L B. Experimental Study On Tubulence Intensity Of Flow Around Semicircle Cylinder Pier In Rectangular Channel[J]. Journal of Shenyang Agricultural University, 2009, 40(4):471-474.

[3] Ghaoud T, Clarke D W. Modelling and tracking a vortex flow-meter signal[J]. Flow Measurement & Instrumentation, 2002, 13(3):103-117.

[4] Bovand M, Valipour M S, Eiamsa-Ard S, et al. Numerical analysis for curved vortex tube optimization [J]. International Communications in Heat & Mass Transfer, 2014, 50:98-107.

[5] Luo J, Bing-Qian L I, Jian L I. The Analysis of Coherent Structure in Round Jet Based on Different Contraction Curve in Contraction Section[J]. Modular Machine Tool & Automatic Manufacturing Technique, 2016.

IOP Conf. Series: Journal of Physics: Conf. Series **1064** (2018) 012070 doi:10.1088/1742-6596/1064/1/012070

Bayesian Maximum Entropy Applied to Fluid Measurement

Mengyuan Wang [a], Shi Liu [b,*] and Yubo Liang [c]

School of control and computer engineering, North China Electric Power University, Beijing, China

[a]15652913080@163.com, *,[b]Corresponding author e-mail: liushi@ncepu.edu.cn, [c]lyb941002@163.com

Abstract. To alleviate the ill-posedness and non-linearity of electrical capacitance tomography, Bayesian maximum entropy (BME) mapping method is used to mix another source of information. They are some observation points, which are hard data. They can be seen as error free. Three two-phase flow patterns are preset to verify the method. Results show that the BME prediction maps are more accuracy than ECT reconstructed image. BME images have more clear edges and those sizes are more similar to the original maps. Two quality assessments demonstrate that BME maps has a lower relative error(RE) and higher correlation coefficient(CC), which verify the effectiveness of the BME mapping method in fluid measurement.

1. Introduction

Two-phases flow is commonly exist in many industrial processes such as chemical, nuclear and petroleum industries [1]. Process tomography (PT) is widely used for the measurement of two-phase flow characteristics [2]. The PT technique has the advantages of being a non-invasive measurement, having a fast response and being of low cost. However, due to the ill-posedness and non-linearity of the inverse problem, the accuracy of reconstructed images is low [3].

To improve the accuracy of PT images (ECT image is used in this paper), some observation points are mixing by BME. BME theory was presented by Christakos in the 1990s [4]. It initiates the era of modern geostatistics. BME provides a systematic and rigorous approach for integrating various knowledge bases into spatiotemporal analysis. In this paper, one sources of information is the ECT image, which represents soft data. The other one is observation points, which represents hard data. It may be seen as error free.

2. Method

2.1. Bayesian Maximum Entropy

Based on dispersed observations with different levels of uncertainties, statistical spatial data fusion is becoming a popular tool for spatial prediction [5]. Similar to that employed by Douaik et al [6], a Bayesian spatial prediction approach is adopt based on the Bayesian maximum entropy(BME) [4]. As a spatial prediction method, BME can incorporate a wide variety of information with different uncertainties.

Content from this work may be used under the terms of the Creative Commons Attribution 3.0 licence. Any further distribution of this work must maintain attribution to the author(s) and the title of the work, journal citation and DOI.
Published under licence by IOP Publishing Ltd

Using BME, the target flow pattern map is considered as a random variable. A specific realization of the random variable is $\chi_{map} = (\chi_{hard}, \chi_{soft}, \chi_k)$, where: χ_{hard} is the hard data (accurate); χ_{soft} is the soft data (vague); χ_k is the data will be estimated through BME. $\mathbf{P}_{hard}$, $\mathbf{P}_{soft}$ and $\mathbf{P}_k$ are the locations of χ_{hard}, χ_{soft} and χ_k, respectively. To maximum the information about the target random variable over a certain area, three stages are introduced:

1. The first or prior stage. An initial probability distribution will be generated across space on the general knowledge base (G). Physical laws as well as other sources of information may be included as mean trend and the covariance function for the target parameter.

2. The second stage. The site-specific knowledge available is organized into hard and soft data, which is denoted by S.

3. The third of integration (posterior) stage. The site-specific knowledge S is assimilated with G. The complete probability law at each estimation site will be contained in the final prediction.

When receiving the paper, we assume that the corresponding authors grant us the copyright to use the paper for the book or journal in question. Should authors use tables or figures from other Publications, they must ask the corresponding publishers to grant them the right to publish this material in their paper.

The mathematical formulation of the above stages is as follows. Consider a parameter such as electrical capacitance X (s). It is constrained by G as

$$G\left(s_{map}\right) = \hat{g}_\alpha\left(x_{map}\right), \alpha = 1,2,...,N_C \tag{1}$$

$$\hat{g}_\alpha\left(x_{map}\right) = \int g_\alpha\left(\chi_{map}\right) f_G\left(\chi_{map}\right) d\chi_{map} \tag{2}$$

Where: x_{map} is a vector which indicates the variables associated with X(s) at a set s_{map} of mapping sites. $g_\alpha(\chi_{map})$ Are a set of functions of χ_{map}. Its expectation is known from G [7]. In our case, G consist of the mean

$$mX(s) = E\left[X(s)\right] \tag{3}$$

And covariance function

$$C(s,s') = E\{[X(s) - mX(s)][X(s') - mX(s')]\} \tag{4}$$

Where: *E* [.] indicates statistical expectation. $f_G(\chi_{map})$ Is the multivariate probability distribution function (pdf) based on G, which represents the prior knowledge. G is used to maximize the expected amount of information, which is represented by Shannon's entropy *ENT*:

$$Ent = -\int f_G\left(\chi_{map}\right) log\left[f_G\left(\chi_{map}\right)\right] d\chi_{map} \tag{5}$$

Equation (5) may be maximized by introducing the Lagrange multiplier μ_α:

$$L\left[f_G\left(\chi_{map}\right)\right] = -\int f_G\left(\chi_{map}\right) log\left[f_G\left(\chi_{map}\right)\right] d\chi_{map} - \sum \mu_\alpha\left[\int g_\alpha\left(\chi_{map}\right) f_G\left(\chi_{map}\right) d\chi_{map} - E\left[g_\alpha\left(\chi_{map}\right)\right]\right] \tag{6}$$

The values of μ_α are obtained by solving equation (5) and equation (6). Solving the system of equation (5) leads to

$$f_G\left(\chi_{map}\right)=A\,exp\left[\sum\mu_\alpha g_\alpha\left(\chi_{map}\right)\right] \tag{7}$$

Where: A is a normalization constant [8]. Knowing μ_α and hence the G-based prior pdf (equation (6)) means that the first stage has completed.

At the second stage, the site specific knowledge S is organized as hard and soft data.

At the third stage, to obtain the total knowledge $K=G\cup S$, the prior pdf is updated with the site-specific knowledge S. It is represented as the BME pdf given by:

$$f_K\left(\chi^{\kappa}\right)=\left(BA\right)^{-1}\int exp\left[\sum\mu_\alpha g_\alpha\left(\chi_{map}\right)\right]d\chi_{soft} \tag{8}$$

Where: B is a normalization coefficient. The BME pdf is non-Gaussian in general. It contains spatial dependency and accuracy of the data. The soft data used in this paper is in the form of pdfs. Serre provide the following solution for the pdf-type,

$$f_K\left(\chi_k\left|\chi_{hard},f_s\left(\chi_{soft}\right)\right.\right)=\frac{\int f_G\left(\chi_k,\chi_{hard},\chi_{soft}\right)f_S\left(\chi_{soft}\right)d\chi_{soft}}{\int f_G\left(\chi_{hard},\chi_{soft}\right)f_S\left(\chi_{soft}\right)d\chi_{soft}} \tag{9}$$

2.2. *Quality Assessments*

In this paper, images are analyzed in two aspects: relative error and correlation coefficient, with:

Relative error (RE),

$$\delta=\frac{\sum_{i=1}^{n}\left(\hat{z}_i-z_i\right)}{\sum_{i=1}^{n}z_i}\times100\% \tag{10}$$

And Correlation coefficient (CC),

$$r=\frac{\sum_{i=1}^{n}\left(\hat{z}_i-\bar{\hat{z}}_i\right)\times\left(z_i-\bar{z}\right)}{\sqrt{\sum_{i=1}^{n}\left(\hat{z}_i-\bar{\hat{z}}_i\right)^2\sum_{i=1}^{n}\left(z_i-\bar{z}\right)^2}}\times100\% \tag{11}$$

Where: $\hat{z}$ is the predicted image using BDF. z is the original image reconstructed by ECT. A smaller relative error means the image suffers from lower noise effects. The image which is more similar to the preset one has a larger correlation coefficient.

3. Result and Discussion

To verify the effectiveness of BME mapping method, three typical two-phase flow patterns are preset. They are the core flow, the scattered flow and the sedimented flow. They are widely existed in industry process. There are 6400 pixels each map. The hard data is 100 observation points, which distributed evenly in the measurement area. They are error free. The other source of information is the image reconstructed by Landweber method. It is an interval reconstruction algorithm, which can reconstruct a higher accuracy image than Linear Back Projection (LBP). The ECT image is soft data, which expressed in terms of pdf. The BME algorithm is realized in the SEKS-GUI software libraries.

As Fig. 1 shows, the BME estimated map has a more accuracy edges. The image reconstructed by ECT has a smooth edge. As shown in Fig. 2, the shape of object can be recognized after BME prediction. The object in ECT image is rough although the true one is square. In Fig. 3, the two objects stick together. The objects in the BME prediction map are distributed clearly.

The quality assessment result is shown in Table 1. It indicates that the prediction one has a lower RE and a high CC. It shows that the prediction one has a low error and more relative to the original one, which is expected in fluid measurement.

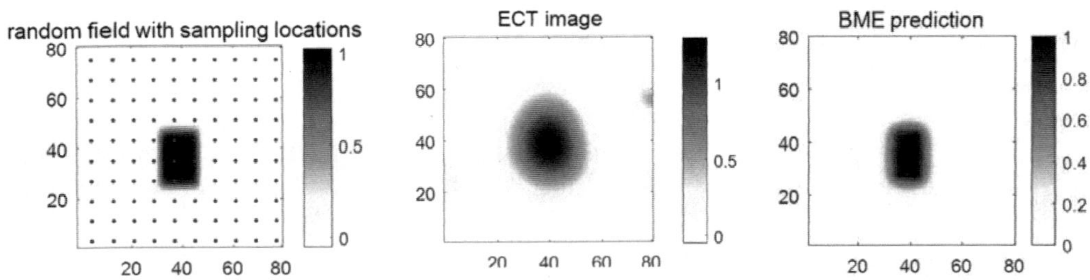

Figure 1. Prediction results of Scattered Flow.

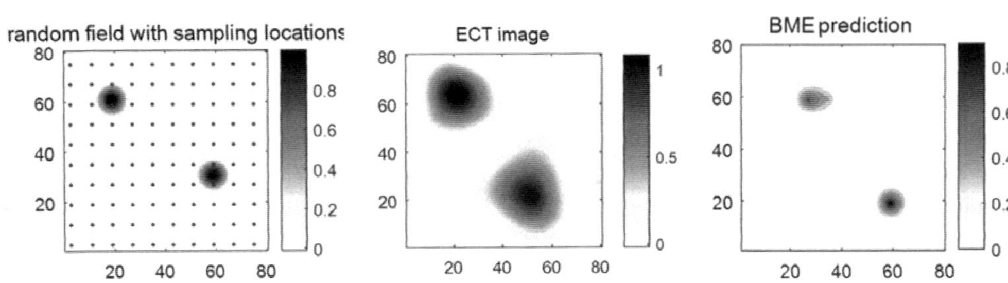

Figure 2. Prediction results of Core Flow.

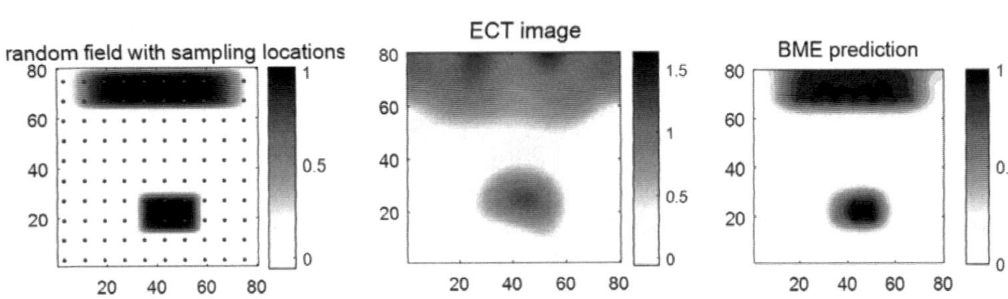

Figure 3. Prediction results of Sedimented Flow.

Table 1. Quality Assessment.

Flow Pattern	Core Flow		Scattered Flow		Sedimented Flow	
	ECT	BME	ECT	BME	ECT	BME
Relative Error (%)	33.23	5.82	36.75	15.84	24.77	11.37
Correlation Error (%)	82.85	96.88	80.95	92.84	83.01	92.63

4. Conclusion

In this paper, to improve the quality of images reconstructed by ECT, the BME mapping method is applied to the fluid measurement. As an important character of flow, flow pattern is predicted with BME. Three typical two-phase flow patterns (i.e. the core flow, the scattered flow and the sedimented flow) are preset to test the method. As the auxiliary source of information, some observation points are taken. Result shows that the prediction one has a more clearly edge. The size and shape of object in the BME map is more nearly to the true one. The objects which stick together in the ECT image can be distributed successfully after fusion. Two quality assessments also show the effectiveness of the method. The fusion map has a low error and more relative to the original one. The effectiveness of BME in fluid measurement is verified in this paper.

Acknowledgments
The authors wish to extend their gratitude to NSFC for sponsoring the research (No 61571189) and to the State Administration of Foreign Experts Affairs for supporting the 111 project (ref: B13009).

References

[1] Polansky, Jiri, and M. Wang. "Proper Orthogonal Decomposition as a technique for identifying two-phase flow pattern based on electrical impedance tomography." Flow Measurement & Instrumentation 53 (2016): 126 - 132.
[2] Liu, Xiao, et al. "An extreme learning machine combined with Landweber iteration algorithm for the inverse problem of electrical capacitance tomography ☆." Flow Measurement & Instrumentation 45.5 (2015): 348 - 356.
[3] York, Trevor. "Status of electrical tomography in industrial applications." Process Imaging for Automatic Control International Society for Optics and Photonics, 2001: 608 - 619.
[4] Christakos, George. "A Bayesian/maximum-entropy view to the spatial estimation problem." Mathematical Geology 22.7 (1990): 763 - 777.
[5] Jafargandomi, Arash, and A. Binley. "A Bayesian trans-dimensional approach for the fusion of multiple geophysical datasets." Journal of Applied Geophysics 96.9 (2013): 38 - 54.
[6] Douaik, A., et al. "Space-time mapping of soil salinity using probabilistic bayesian maximum entropy." Stochastic Environmental Research & Risk Assessment 18.4 (2004): 219 - 227.
[7] Serre, M. L., and G. Christakos. "Modern geostatistics: computational BME analysis in the light of uncertain physical knowledge – the Equus Beds study." Stochastic Environmental Research & Risk Assessment 13. 1 - 2 (1999): 1 - 26.
[8] Christakos George Et. Modern Spatiotemporal Geostatistics. Modern spatiotemporal geostatistics /. Oxford University Press, 2000: 1760 - 1765.

IOP Conf. Series: Journal of Physics: Conf. Series **1064** (2018) 012071 doi:10.1088/1742-6596/1064/1/012071

Study on Flutter Stability Control Technology of Blade Flexible Machining for Airfoil Fan

YunSheng Chen[1], Xinghua Lu[2,*]

[1]Guangzhou Huali Science and Technology Vocational College, Guangzhou, China
[2]Huali College Guangdong University of Technology, Guangdong Guangzhou, China

*Corresponding author e-mail: xhlu@gdut.edu.cn

Abstract. The blade of airfoil wind turbine is affected by cross-section aerodynamic and airfoil aerodynamic damping in flexible machining, which results in flutter in the process of machining, affects machining accuracy. A flutter stability control method based on modal parameter identification and aerodynamic damping coupling compensation for flexible machining of airfoil fan sheet is proposed. The flutter mechanics analysis model for flexible machining of airfoil fan sheet is established by using multi-body system dynamics analysis method. The stability control constraint parameters of blade machining flutter are analyzed. The flutter modal parameters of flexible machining are identified by finite element dynamic analysis method with blade flapping, shimmy and section aerodynamic characteristic parameters are taken as the optimization objectives. The steady state error coupling compensation method is used to control the aerodynamic damping of blade machining and the continuous feedback drive method is used to carry out the adaptive calibration of machining parameters to realize the control optimization of the blade flexible machining of the fan. The simulation results show that the method is used to design the blade flexibility of the airfoil fan, which reduces the flutter effect and improves the machining accuracy and stability control ability.

1. Introduction

Airfoil fan blades are widely used in aero-engines, wind turbines, compressors, high pressure compressors and other mechanical devices as the main components of fan. In the machining of the blade of the airfoil fan, because of the complicated composition of the blade, the bending degree of the component and the component composition of the connecting part, different processing techniques are needed in the process of machining processing. The processing technology of airfoil fan blade involves various technological processes such as NC milling, cutting, forging and grinding. In the whole machining process of airfoil fan blade, due to various kinds of machining excess (such as torsion, front and tail edge shape, contour, etc.) the presence of the position degree leads to flutter easily occurring in the blade processing of the airfoil fan. The machining flutter of fan blade comes from the coupling of aerodynamic force, elastic force and inertial force, which makes it difficult to achieve equilibrium state in the process of machining, resulting in large elastic deformation and vibration.

In order to overcome the machining flutter of the blade of the airfoil fan, the flexible machining method is used to control the flutter and stability, to control the machining accuracy of the blade and to

Content from this work may be used under the terms of the Creative Commons Attribution 3.0 licence. Any further distribution of this work must maintain attribution to the author(s) and the title of the work, journal citation and DOI.
Published under licence by IOP Publishing Ltd

improve the engine performance. The flutter stability control of blade flexible machining of airfoil fan is taken based on the stability adjustment of machining overshoot and the optimization design of machining parameters. Some typical machining flutter control methods have been studied by domestic and foreign scholars, among which, in reference [4], a blade processing method for airfoil fan based on cutting teeth with five blade method is proposed. According to the aerodynamic parameters of the blade section of the fan, the torsional angle is adjusted, and the adjustment parameters of the blade machining machine are studied by numerical method based on the characteristic analysis of tooth surface reference point, the double spiral method is used to optimize the blade machining design to improve the machining performance. However, the anti-interference ability of the method in flutter control is not strong, and the calculation cost of processing parameters is high. In the paper [5], an optimal control method for the parameters of spiral moving cutting tool in fan blade machining based on the analysis of gear tooth contact is proposed. According to the spatial geometry relation of arc bevel gear pair, the coupling control module of blade machining is established based on the sensitivity analysis of machining overshoot type, the problem control ability of machining process can be improved by restraining the influence of overshoot error. This method has the problem that the initial sensitivity of parameters is large and the oscillation amplitude of surge margin is high. In reference [6], the relationship between blade torsion and its power consumption is studied. It is found that the torsion of blade shape causes the increase of air flow angle and the decrease of compressor efficiency, and the error correction of blade processing is carried out according to this principle. The overshoot error of the leading edge profile is reduced and the machining accuracy of the blade is improved, this method cannot effectively suppress the flutter of the blade of the airfoil fan and the contour deviation is large.

Aiming at the above problems, in order to restrain flutter of blade machining of airfoil fan, a flutter stability control method for flexible machining of airfoil fan blade based on modal parameter identification and aerodynamic damping coupling compensation is proposed. The flutter mechanics analysis model for flexible machining of airfoil fan blade is established by using multi-body system dynamics analysis method. The stability control constraint parameter of blade machining is analysed, and the blade branching is carried out. The flutter modal parameters of flexible machining are identified by the finite element dynamic analysis method and the aerodynamic damping control of blade machining is carried out by using the steady-state error coupling compensation method. The continuous feedback drive method is used to calibrate the machining parameters to realize the control and optimization of the flexible machining of fan blades. Finally, simulation tests are carried out to demonstrate the superiority of this method in restraining machining errors and improving the flutter stability control ability of flexible machining.

2. Flutter mechanics analysis model for flexible machining of airfoil fan sheet

2.1. Control constraint parameters for stability control of flexible processing

In order to control flutter stability of airfoil fan blade in flexible machining, a control constraint parameter model is first constructed. The blade of airfoil fan is a fan of a multistage aero-engine in this paper. The mechanical parameters of the fan are analysed and numerically calculated by Numeca Fine/Turbo software. The numerical simulation results and test results of the fan under the condition of zero angle of attack and imported Mach number is about 0.69, the model is analyzed in the geometric model of numerical verification. The aerodynamic damping coefficient of airfoil and its distribution model with geometric parameters of airfoil are constructed by taking static pressure coefficient, tip clearance, tail edge angle, torsion and contour as control constraint parameters. Combined with the aerodynamic damping analysis model of wind turbine blade stall, the aero-elastic coupling multi-body dynamics model of the blade is established, and the aero-elastic response of the blade under various wind speeds is obtained by solving the equation. Combined with the theory of modal superposition, that is, the vibration response of blade is mainly superposed by several main modal responses in front of the blade, the following contact measurement system for flutter stability control in flexible machining of fan is established, which is mainly composed by the mechanical structure of the measuring machine, numerical control system and industrial control computer IPC. According to the above design principle, starting from the profile change caused by blade fouling, the basic geometric parameter model for flutter stability control is defined in Table 1.

Table 1. Basic geometric parameter model

Parameter name	Initial value
Position error	2.34
Blade curvature	12.65
Machining radians	4.78
Curvature difference	11.09
Concave curvature	12.45
Crossed axis angle	6.54
Mean pressure angle	9.12
Mean cone distance	8.34
Node cone angle	6.98
Root cone angle	2.34
Large end modulus	4.56
Midpoint helical angle	21.23

According to the constraint parameter model in Table 1, the flutter control object model for flexible machining of airfoil fan blade is established, and the SA turbulence model for spiral cone blade machining is defined as follows:

$$\begin{aligned} &X(n)=[x_1(n),x_2(n),...,x_m(n)]^T \\ &W_j(n)=[w_1(n),w_2(n),...,w_m(n)]^T, j<m \\ &a_j(n)=[a_1(n),a_2(n),...,a_m(n)]^T, j<m \\ &Y_{j-1}(n)=[y_1(n),y_2(n),...,y_m(n)]^T, j<m \end{aligned} \tag{1}$$

The dynamic response characteristics of blades are analyzed. The geometric characteristics of double helical machining are analyzed by using the method of geometric parameter analysis of flexible blades.

2.2. Multi-body dynamics analysis of flexible blade machining

Combined with the structural dynamic mode of the whole blade, the dynamic analysis of the blade flexible machining is carried out, and the parameters of flutter control are adjusted adaptively. According to the spatial geometric relation of the cutting tooth machining, the relative matrix R under the stall condition of the blade is defined as follows:

$$R=X(n)X^T(n)=\begin{bmatrix} x_1(n)x_1(n) & x_1(n)x_2(n) & ... & x_1(n)x_m(n) \\ x_2(n)x_1(n) & x_2(n)x_2(n) & ... & x_2(n)x_m(n) \\ \mathrm{M} & \mathrm{M} & & \mathrm{M} \\ x_m(n)x_1(n) & x_m(n)x_2(n) & ... & x_m(n)x_m(n) \end{bmatrix} \tag{2}$$

Based on the "super cell" model, the dynamic geometric characteristic parameters of blade machining are sorted, and the size distribution of the eigenvalues can be expressed as follows:

$$\lambda_1>\lambda_2>...>\lambda_{j-1}>\lambda_j>...>\lambda_m \tag{3}$$

According to the spiral motion coefficient of the blade machining, the second order control parameter model of the machine tool is obtained as follows:

$$\sum_{k=1}^{m}[\theta_{jk}(n+1)-\theta_{jk}(n)]q_k=\sum_{k=1}^{m}\eta[\lambda_k-\delta_j^2(n)]\theta_{jk}(n)q_k+\eta\sum_{k=1}^{j-1}\lambda_k a_{jk}(n)q_k \tag{4}$$

According to the vibration direction of blade flutter, the aerodynamic damping parameters are adaptively adjusted and feedback control is carried out to improve the flutter stability control ability of flexible machining.

3. Flutter stability control optimization of blade flexible machining

3.1. Identification of flutter mode parameters in flexible machining

Based on the analysis of control constraint parameters and flutter dynamics, the flutter stability control is optimized. In this paper, a flutter stability control method based on modal parameter identification and aerodynamic damping coupling compensation for flexible machining of airfoil fan blade is proposed. Adaptive geometric calibration of machine tool adjustment parameters is carried out by continuous feedback driving method. The similar geometric characteristic equations are described as follows:

$$\begin{cases} W_i^T(x)X(n)=X^T(n)w_j(n) \\ a_j^T(n)Y_{j-1}(n)=Y_{j-1}^T a_j(n) \end{cases} \tag{5}$$

The uncertain characteristic analysis of over-difference variable of fan blade machining is carried out. The aerodynamic damping characteristic equation of blade machining is established as follows:

$$w_j(n+1)=w_j(n)+\eta[y_j(n)X(n)-y_j^2(n)w_j(n)]=w_j(n)+\eta[X(n)X^T(n)w_j(n)+X(n)X^T(n)Q^T a_j(n)-y_j^2(n)w_j(n)] \tag{6}$$

The flutter modal parameters of flexible machining are controlled by O4H topology with the aerodynamic characteristic parameters of blade and section as the optimization objective. The grid structure model is shown in Figure 1.

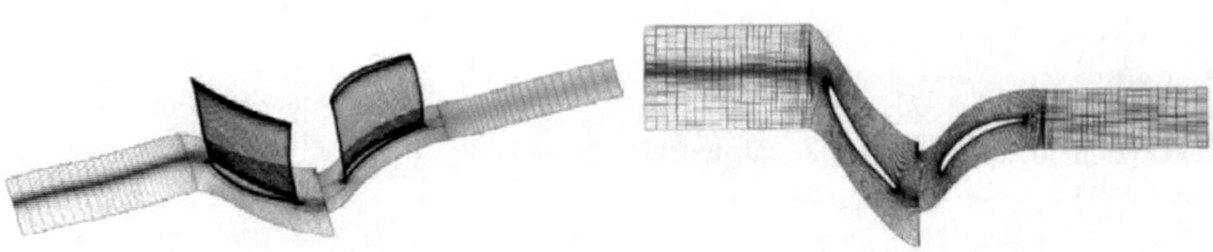

(a) 3D mesh model of blade (b) 2D mesh model of blade

Figure 1. Schematic diagram of head type machining grid for airfoil blade

According to the grid model shown in figure 1, the flutter modal parameters of flexible machining are identified with the finite element dynamic analysis method, and the position degree is selected. The contour and torsion are controlled by the input explanatory variables, and the geometric variation of the overpass is obtained by the analysis of the blade profile bias:

$$x_i(n)=\sum_{j=1}^{M}h_{ij}(n)^T s_j(n)+v_i(n) \tag{7}$$

$$y_j(n) = \sum_{i=1}^{P} f_{ij}(n)^T \boldsymbol{x}_i(n) \tag{8}$$

Wherein, h_{ij} represents the geometric parameters of each section, and f_{ij} represents the sensitivity characteristic. Thus, the flutter modal parameter identification output of blade flexible machining of airfoil fan is obtained:

$$f_{ij}(n+1) = f_{ij}(n) + \mu_{MCMA} \frac{\partial J_{MCMA}(n)}{\partial f_{ij}(n)} \tag{9}$$

Wherein, μ_{MCMA} represents the axial position degree, the step length of the super difference parameter is determined as follows:

$$f_{ij}(n) = [f_{ij}^{(0)}(n), f_{ij}^{(1)}(n), \Lambda\ , f_{ij}^{(L-1)}(n)]^T \tag{10}$$

The explanatory variables of the flutter control for the blade processing is:

$$\frac{\partial J_{MCMA}(n)}{\partial f_{ij}(n)} = \frac{1}{4}\left(\sum_{j=1}^{M} \frac{\partial J_{R,j}(n)}{\partial f_{ij}(n)} + \sum_{j=1}^{M} \frac{\partial J_{I,j}(n)}{\partial f_{ij}(n)}\right) \tag{11}$$

Partial derivation of real parts is obtained:

$$\frac{\partial J_{R,j}(n)}{\partial f_{ij}(n)} = \frac{\partial (\left|y_{R,j}(n)\right|^2 - R_{2,R})^2}{\partial f_{ij}(n)} = 4(\left|y_{R,j}(n)\right|^2 - R_{2,R}) \times y_{R,j}(n)^* \times x_j(n) \tag{12}$$

With the identification of flutter modal parameters of flexible machining and fusion treatment, the compensation ability of steady state error of fan blade machining is improved.

3.2. Adaptive calibration of fan blade flexible machining parameters

The steady-state error coupling compensation method is used to control the aerodynamic damping of blade machining. The output error function is obtained as follows:

$$e_{R,j} = (\left|y_{R,j}(n)\right|^2 - R_{2,R}) \times y_{R,j}(n)^* \tag{13}$$

Combined with the Lyapunov derivation method, the virtual part error function of flutter stability control of fan blade is obtained by removing the disturbance of lift and drag characteristics:

$$e_{I,j} = (\left|y_{I,j}(n)\right|^2 - R_{2,I}) \times y_{I,j}(n)^* \tag{14}$$

The steady-state error coupling compensation method is used to control the aerodynamic damping of blade machining. The control iterative equation is described as follows:

$$f_{ij}(n+1)=f_{ij}(n)+\mu_{MCMA}\times\left(\left(\left|y_{R,j}(n)\right|^2-R_{2,R}\right)y_{R,j}(n)^*+\boldsymbol{j}\left(\left|y_{I,j}(n)\right|^2-R_{2,I}\right)y_{I,j}(n)^*\right)\times x_j(n) \quad (15)$$

The weighting coefficient is modified, and then the modified aerodynamic parameters are input into the equalizer. The coupling response of the aerodynamic load along the whole blade is obtained as follows:

$$\tilde{\boldsymbol{x}}(n)=\boldsymbol{x}(n)-\tilde{\boldsymbol{h}}_j(n)*\hat{s}(n) \quad (16)$$

Wherein, $\tilde{\boldsymbol{h}}_j(n)*\hat{s}(n)$ represents the autocorrelation function with $\boldsymbol{y}_j(n)$, and adopts the continuous feedback driving method for automatic feedback and calibration of machining parameters, then:

$$E[\tilde{X}(n)y_j^*(n-k)]=0 \quad (17)$$

Wherein, $k=K-1$, $\tilde{X}(n)$ is the Toeplitz matrix of $\tilde{\boldsymbol{x}}(n)$. According to the above algorithm design, the flutter control optimization of fan blade flexible machining is realized.

4. Test and analysis of simulation experiment

In order to verify the performance of this method in the control of flutter stability in flexible machining of fan blades, simulation experiments are carried out. The blade model of 5 MW wind turbine is used as the research object, rotation rate of impeller rigid body relative to the center Line of wind Turbine is 12.1r/min, the curvature radius of blade torsional deformation is 8mm, the rated wind speed is 14m/s, the blade aerodynamic load is 2400KN, the aerodynamic damping is 800KN, the blade structure mass is 100kg, the surge margin is 0.96, and the blade height section bias is 0.1mm.The flutter control convergence distribution model under different aerodynamic damping ratio of blade machining is obtained by simulation experiment of fan blade flexible machining as shown in Fig. 2.

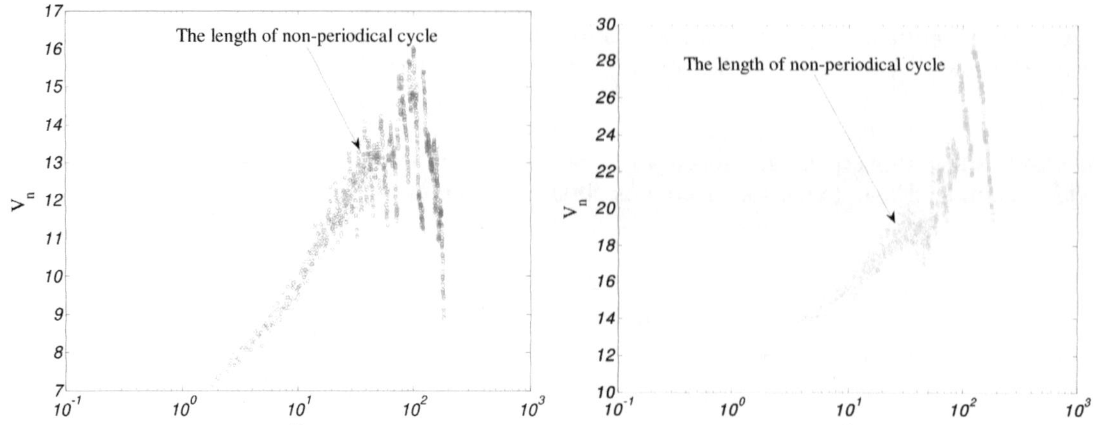

(a) Aerodynamic damping ratio $R=0.25$ (b) Aerodynamic damping ratio $R=0.32$

Figure 2. Flutter control distribution curve of blade flexible machining

Figure 2 shows that the flutter control of the blade of airfoil fan by using this method is more stable, and the flutter suppression ability is stronger. The total pressure ratio is increased by about 0.015 and the wake strength of the profile changes by 0.1 mm. it is obviously larger than the prototype, the Aero elastic damping at each section of the rigid body center of the blade was effectively controlled, and the flutter stability of the whole blade is better. In order to compare the performance, the flexible machining control is carried out by different methods, and the machining precision is tested. The comparison results are shown in Fig. 3 and the analytical figure 3 shows that this method has higher accuracy and better performance.

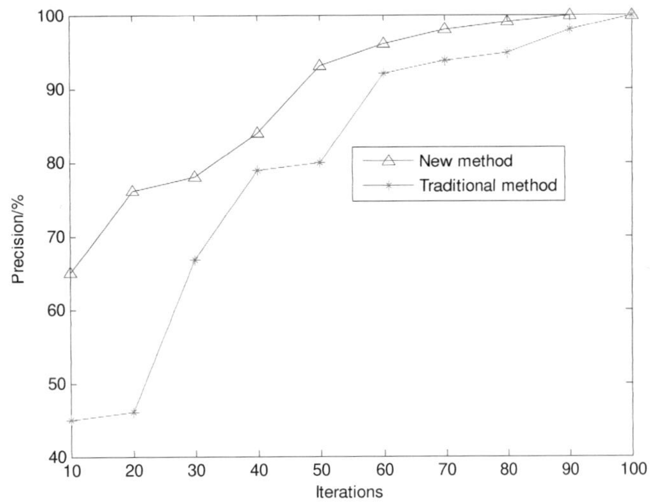

Figure 3. Machining accuracy comparison

5. Conclusion

In order to overcome machining flutter of airfoil fan blade, flexible machining method is adopted to suppress flutter and control stability. In this paper, a flexible machining method based on modal parameter identification and aerodynamic damping coupling compensation is proposed for flexible machining of airfoil fan blade. The vibration stability control method is used to establish the flutter mechanics analysis model of the airfoil blade flexible machining by using the multi-body system dynamics analysis method. The stability control constraint parameter of the blade machining flutter is analyzed and the blade is wielded. The flutter modal parameters of flexible machining are identified by means of finite element dynamic analysis method, and the aerodynamic damping control of blade machining is carried out by using the steady-state error coupling compensation method. The continuous feedback drive method is used to calibrate the machining parameters to realize the control and optimization of the flexible machining of fan blades. The results show that this method can effectively control the flutter of the blade of the airfoil fan, and the output stability is good, and the machining accuracy is improved.

References

[1] CHEN Weiwei, YAN Junfan, LOU Yunge. Design of automatic function intelligent measuring instrument for shaft parts [J]. Machinery, 2016 (4): 76 - 79.

[2] WU Baohai, LIANG Mancang, ZHANG Ying, LUO Ming. Tool Selection of Multi-axis Machining for Channel Parts with Sculptured Surface. Journal of Mechanical Engineering, 2018, 54 (3): 117 - 124.

[3] GENG L, ZHANG Y F,FUH J Y H. A neural network based approach to 5-axis tool-path length estimation for optimal multi-cutter selection [J]. Computer-Aided Design and Applications, 2011, 8 (2): 301 – 313.

[4] DING X, FUH J,LEE K. Interference detection for 3-axis mold machining [J]. Computer-Aided Design, 2001, 33 (8: 561 - 569.

[5] LASEMI A, XUE Deyi, GU Peihua. Recent development in CNC machining of freeform surfaces: A state-of-the-art review[J]. Computer-Aided Design, 2010, 42 (7): 641 - 654.

[6] MCGOVERN S M, GUPTA S M. 2-opt heuristic for the disassembly line balancing problem [C]// Proceedings of the SPIE International Conference on Environmentally Conscious Manufacturing III, 2004: 71 - 84.
[7] ALTEKIN F T, KANDILLER L, OZDEMIREL N E. Profit-oriented disassembly-line balancing [J]. International Journal of Production Research,2008, 46 (10): 2675 - 2693.
[8] PENG Xiongbin, GONG Guofang, LIAO Xiangping, WU Weiqiang, WANG Hui, LOU Haiyang. Modeling and Model Identification of Micro-position-control Hydraulic System. Journal of Mechanical Engineering, 2017, 53 (22): 206 - 211.
[9] MA Zikui, CHEN Wenhua. Friction Torque Calculation Method of Ball Bearings Based on Rolling Creepage Theory. Journal of Mechanical Engineering, 2017, 53 (22): 219 - 224.
[10] CHEN W, MA Z,GAO L,et al. Quasi-static analysis of thrust-loaded angular contact ball bearings part I:Theoretical formulation [J]. Chinese Journal of Mechanical Engineering, 2012, 25 (1): 71 - 80.
[11] ZHAO Chunjiang, WANG Jianmei, HUANG Qingxue. Modification of raceway control theory on the dynamic equations group of high-speed ball bearings [J]. Transactions of the Chinese Society for Agricultural Machinery, 2009, 40 (5): 199 - 202.
[12] SU Bing, ZHANG Shuo,YANG Boyuan. Experiment for sliding friction characteristics of bearing steel GCr15 [J]. Bearing, 2015 (2): 39 - 42.

Optimization Research on Lift-Type Vertical Axis Wind Turbine Airfoil by CFD

Hao Cao*, Xiaowen Wu, Huisheng Ye, Sheng Hu, Ling Lu and Jiwen Peng

State Grid Hunan Electric Power Corporation Limited Research Institute, Changsha, 410007, China

*Corresponding author e-mail: caohao45@126.com

Abstract. The blade airfoil has an important effect on the performance of a lift-type vertical axis wind turbine (VAWT). In this paper, a research has been carried out based on CFD numerical technology to compare the effects of 4 kinds of S series airfoil (S1012, S1016, S1046, S1048) and traditional airfoil NACA0015 on lift-type VAWT. The turbulence model is used *k-ω* SST model. The simulation results indicate that: The airfoil NACA0015 has the highest torque coefficient and power coefficient. In terms of S series airfoils, the airfoil S1012 has the best performance and airfoil S1016 performs worst. The airfoil S1016 has the largest static torque coefficient. And the airfoil which has higher power coefficient would have a lower static torque coefficient for S series airfoil. The static torque coefficient of airfoil NACA0015 is lower than airfoil S1016 and S1046.

1. Introduction

In recent years, the continuous consumption of fossil energy brought serious environmental pollution, and it has become the biggest threat to human development. The wind, as a kind of clean and sustainable energy, has great prospects for development. The global wind power generation had reached 468.99 GW by 2016 [1]. The lift-type VAWT is a normal wind power equipment. Compared to the horizontal axis wind turbine (HAWT), it has the following advantages: without upwind devices, no requirement for flow direction, simpler structure, simpler installation, simpler maintenance and lower cost.

The blade airfoil has an important effect on the performance of lift-type VAWT. A suitable airfoil can effectively improve the wind energy utilization rate of wind turbine. At present, the airfoil NACA0015 is one of the most widely used airfoils in lift-type VAWTs [2]. In recent years, many new airfoils have gradually attracted the attention of investigators. Mohammed M.H. compared the performances of 20 airfoils such as NACA0010, NACA0015, and NACA63418 and so on using Realizable *k-ω* turbulence model [3]. His results shows that the airfoil S1046 has the maximum wind energy efficiency 0.4051, 10.87% higher than the traditional NACA airfoil. Singh M.A. studied the self-initiation of H-type VAWT with airfoil S1210 and the effect of the solidity on it by experimental method [4]. The result shows that the maximum power coefficient is reached to 0.32 when the solidity is 1 and the wind speed is 5.7 m/s, it means the wind turbine can have higher static and dynamic torque at the same time. Consequently, it can be seen that the use of S series airfoil has good effect on improving the performance of VAWT.

Content from this work may be used under the terms of the Creative Commons Attribution 3.0 licence. Any further distribution of this work must maintain attribution to the author(s) and the title of the work, journal citation and DOI.
Published under licence by IOP Publishing Ltd

In this paper, a research has been carried out based on CFD numerical technology to compare the effects of 4 kinds of S series airfoil (S1012, S1016, S1046, S1048) and traditional airfoil NACA0015 on lift-type VAWT. The purpose is to get more suitable airfoil for lift-type VAWT.

2. CFD Model

In this paper, the flow field characteristics and wind energy utilization efficiency of lift-type VAWTs with different airfoils are analyzed by solving the unsteady NS equation, based on the *k-ω* SST turbulence model. And the coupling calculation of pressure and velocity is carried out using SIMPLE algorithm.

2.1. Airfoil Parameter Setting

Figure 1 shows the 5 kinds of airfoil studied in this paper. Obviously, airfoil S1012 is thinner than S1016, while S1046 is thicker than S1048. Comparison between airfoil S1016 and airfoil S1046 shows that the former is thicker than the latter in the middle near the tip of the blade, also the former's overall outline is more symmetrical. Compared to the airfoil NACA0015, the tail edges of S series airfoil are sharper. A 3-blade lift-type VAMT with diameter of 2.5 m is selected in this paper, the blade height is 3.5 m, airfoil chord length is 0.25 m. so the depth-width ratio (AR) is 14 and the solidity is 0.6.

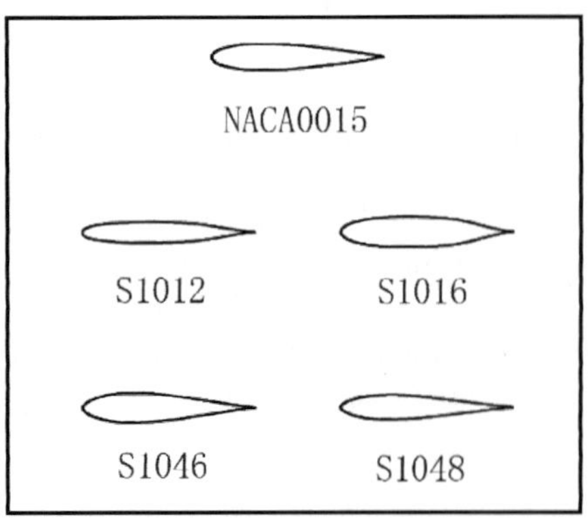

Figure 1. Diagrammatic sketch of 5 airfoils

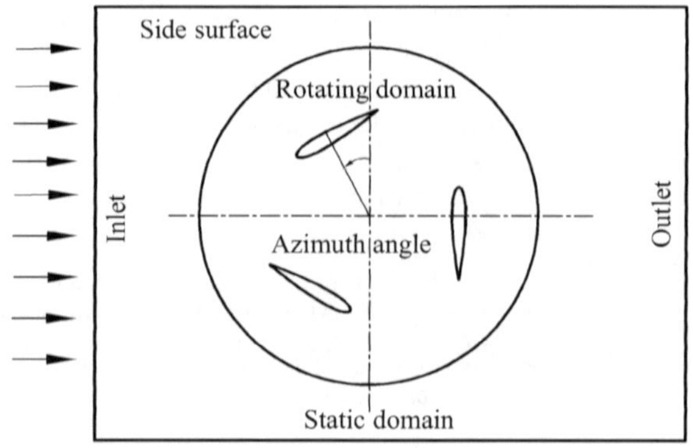

Figure 2. Two-dimensional model of lift-type VAMT

2.2. Calculation Grid Setting

A two-dimensional model is established according to the characteristics of lift-type VAMT, as Figure 2 shown. The calculation region take a square with a side length of 20D (D is the diameter of wind turbine impeller), and the impeller is located in the center of calculation region to ensure that the air flow can be fully developed when the wind turbine is in operation, and the influence of the boundary on the flow field around the turbine impeller can be minimized to obtain accurate results. The whole calculation region is divided into rotating domain and static domain. During the progress of calculation, the rotating domain revolves around the center of rotation over time.

The calculation grid in this paper is mainly constructed by triangular unstructured meshes. In order to make the calculation results more accurate, the grid near the surface of the airfoil is divided by quadrilateral meshes. The height of the first layer quadrilateral mesh on the surface of the airfoil is 0.01mm. There are 15 layers quadrilateral meshes and the thickness of each layer increases in proportion. Figure 3 shows that the meshes around the airfoil are locally encrypted according to the characteristics of the flow field. And Figure 4 shows that arc processing is done on the tail edge of airfoil to ensure mesh quality. At the same time, the grid density outside the rotating region will gradually decrease along the direction away from the turbine impeller to save calculation time.

IOP Conf. Series: Journal of Physics: Conf. Series **1064** (2018) 012072 doi:10.1088/1742-6596/1064/1/012072

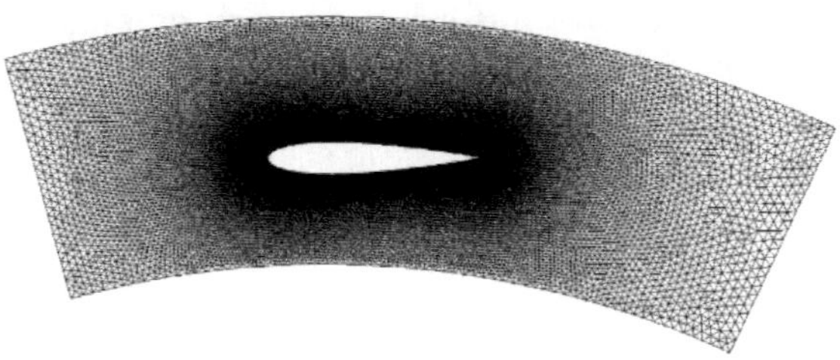

Figure 3. Airfoil peripheral area grid

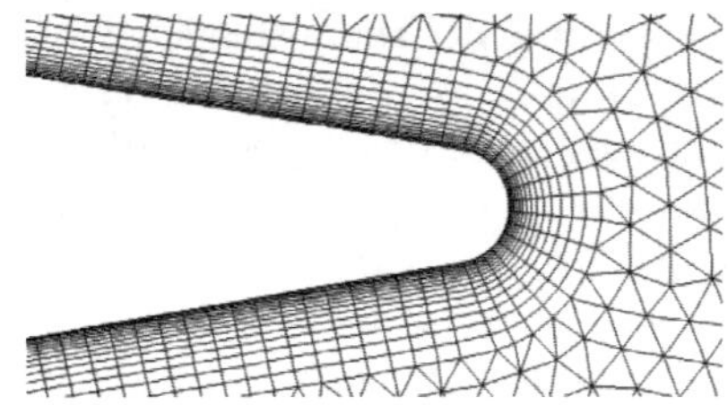

Figure 4. Airfoil tail edge grid

2.3. Turbulence Model Selection

At present, the turbulence models used to simulate the running state of VAWT are mainly as follows: Standard k-ε model, RNG k-ε model, Realizable k-ε model, k-ω SST model and Spalart-Allmaras model. The accuracy of Spalart-Allmaras model is the lowest one in these models because it is a kind of one-equation model. Francesco Balduzzi [13] compared Standard k-ε model, RNG k-ε model and k-ω SST model, his result shows that the turbulence models based on ε are difficult to converge and the k-ω SST model has the best stability and reliability, it is also consistent with the experimental results. In this paper, k-ω SST model is used to simulate the running condition of VAWT, and the enhanced wall function is used to improve the accuracy of calculation in boundary layer.

3. Calculation Result Analysis

In this paper, the input wind flow speed is set to 10 m/s, and the operating conditions of the wind turbine under 7 tip speed ratios(0.5, 1, 1.5, 2, 2.5, 3, 4) are simulated and calculated respectively. Figure 5 and Figure 6 show the curves of torque coefficient (C_t) and power coefficient (C_m) of different airfoils with tip speed ratios (TSR) respectively. It can be seen that the maximum C_t and C_m of the airfoil NACA0015 are greater than the coefficients of S series airfoil. During the S series airfoil, the maximum power coefficient of airfoil S1012 is the highest, and the maximum power coefficient of airfoil S1016 is the smallest. All of these airfoils can work normally under TSR=4.

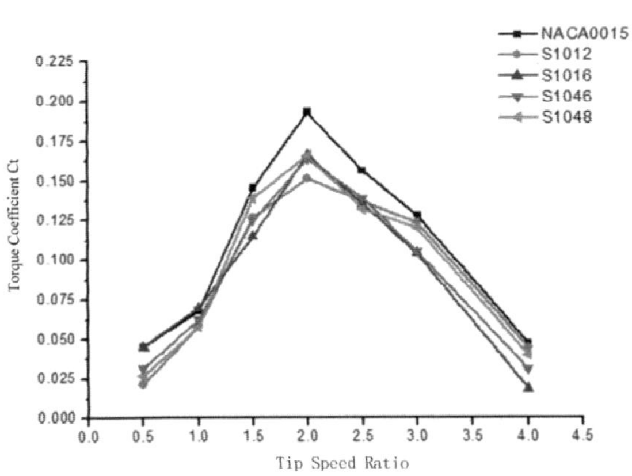

Figure 5. Curve of C_t of different airfoil with TSR

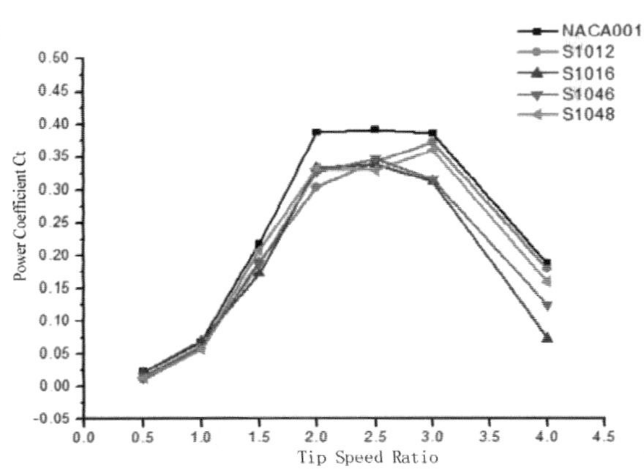

Figure 6. Curve of C_m of different airfoil with TSR

As shown in Figure 5, the torque coefficient C_t of the 5 airfoils have the same trend with TSR. With the increase of TSR, the C_t increase first and then decrease, an obvious peak occurs when TSR=2.5. As shown in Figure 6, the maximum power coefficient C_m of these airfoils increase first and then decrease with the increase of TSR. The C_m of airfoil S1012 and S1048 both have an obvious peak when TSR=3, and the C_m of airfoil NACA0015, S1016, S1046 have an obscure peak when TSR=2.5, the C_m of NACA0015 are almost the same when TSR=2, 2.5, 3.

Figure 7 shows the instantaneous torque coefficients of five airfoils at different azimuths under their optimal TSR. The instantaneous torque coefficients of the five airfoils are higher at the azimuth angle of 110°, 210°, 340°and lower at the azimuth angle of 40°, 160°, and 290°. Figure 7 also shows that the instantaneous toque coefficient of airfoil NACA0015 is in the outer position under most azimuths, and its value is quite different from that of airfoil S1012 and S1048. For the S series airfoils, although the instantaneous torque coefficients of the airfoils S1012 and S1048 are not in the outer circle under most azimuth angles, the optimum TSR is 3 and their instantaneous torque coefficients are not different from those of the airfoil S1016 and S1046. The optimum TSR of S1016 and S1046 is 2.5. Consequently, the optimum C_m of airfoil NACA0015 is the highest, and the C_m of airfoil S1012 and S1048 are bigger than that of other S series airfoil.

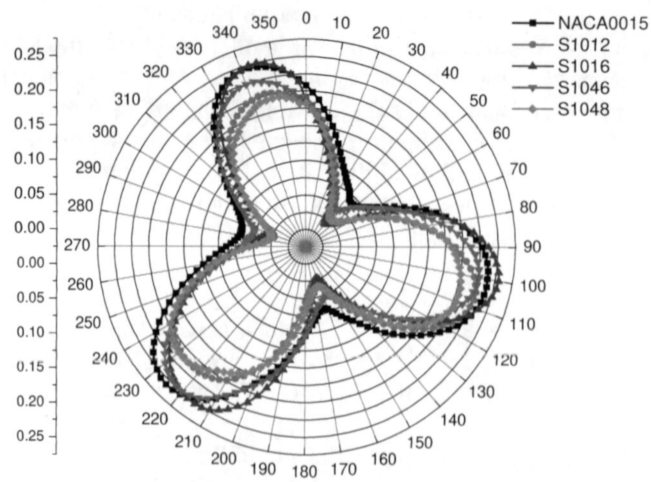

Figure 7. Instantaneous torque coefficient of 5 airfoils at different azimuth angles under optimum TSR

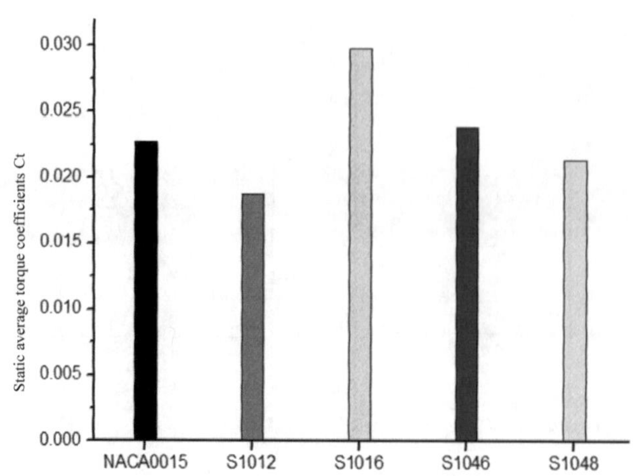

Figure 8. Static average torque coefficients of different airfoils

Figure 8 shows the static average torque coefficients of VAWT with different airfoils. The static average torque coefficient of airfoil S1016 is the highest, it means its automatic start performance is the best, followed by airfoil S1046, while airfoil NACA0015 is in the third place. The static average torque coefficient of the airfoil with larger maximum power coefficient in S series airfoil is smaller.

IOP Conf. Series: Journal of Physics: Conf. Series **1064** (2018) 012072 doi:10.1088/1742-6596/1064/1/012072

4. Conclusion

The performances of the 4 S series airfoils studied in this paper are not as good as the performance of traditional airfoil NACA0015. The maximum torque coefficient and maximum power coefficient of the airfoil NACA0015 are both higher than those of the other 4 S series airfoils. The specific conclusions are as follows:

1) For the torque coefficient, the optimum TSR of 5 airfoils is 2.5, while for the power coefficient, the optimal TSR of the airfoil S1012 and S1048 is 3, and that of the other 3 airfoils is 2.5;

2) All 5 airfoils work properly when the TSR reaches 4;

3) NACA0015 has the maximum power coefficient and torque coefficient;

4) In S series airfoil, the maximum power coefficient of airfoil S1012 is the largest, and the maximum power coefficient of airfoil S1016 is the smallest;

5) The static average torque coefficient of the airfoil S1016 is the highest, and it also have better automatic starting performance. The automatic starting performance of the airfoil NACA0015 is in the middle position;

6) In the S series airfoil, the smaller the maximum power coefficient, the smaller the static average torque coefficient of the airfoil, also the less automatic starting ability.

Compared with the result of Moharmed M.H. [3], there are some discrepancies in the conclusion of this paper. The main reasons for the difference include the different solidities of wind turbine model, different turbulence models and different meshing methods. In addition, the use of new airfoils can enhance VAWT's performance, such as automatic strartup capability. Therefore, it is necessary to continue to strengthen the research on new airfoil in the future.

References

[1] Kumar, Y., et al., Wind energy: Trends and enabling technologies. Renewable and Sustainable Energy Reviews, 2016.53: p. 209-224.

[2] Balduzzi, F., et al., Feasibility analysis of a Darrieus vertical-axis wind turbine installation in the rooftop of a building. Applied Energy, 2012. 97 (3): p. 921-929.

[3] Mohamed, M.H., Performance investigation of H-rotor Darrieus turbine with new airfoil shapes. Energy, 2012. 47 (1): p. 522-530.

[4] Singh, M.A., A. Biswas, and R.D. Misra, Investigation of self-starting and high rotor solidity on the performance of a three S1210 blade H-type Darrieus rotor Renewable Energy, 2015. 76: p. 381-387.

[5] Maître, T., E. Amet, and C. Pellone, Modeling of the flow in a Darrieus water turbine: Wall grid refinement analysis and comparison with experiments. Renewable Energy, 2013. 51 (2): p. 497-512.

IOP Conf. Series: Journal of Physics: Conf. Series **1064** (2018) 012073 doi:10.1088/1742-6596/1064/1/012073

Research on Numerical Wave-making Method Based on Viscous Flow Wave

Jianmin Liu, Qian Sun *, Lei Jiang

School of Marine Engineering, Jimei University, Xiamen, Fujian, China.

*Corresponding author e-mail: sunqian1226@163.com

Abstract. In this paper, the numerical method on physics-based push plate wave is studied. The wave tank is numerically simulated based on the N-S equation of viscous fluid and the standard k- ε turbulence model. The volume of fluid (VOF) is used to capture the free surface, and the wave plate is oscillated by the dynamic mesh technique. In addition, the CFD commercial software Fluent and its secondary development function are employed to establish a 3D numerical regular wave tank. After the analysis of the results, it can be found that the imitating physical push-plate wave created in this paper is of high accuracy and good stability.

1. Introduction

Wave, which is one of the most important driving forces in the marine environment, is closely related to the performance of ships. Recently the research on a numerical wave tank designed to achieve the practical engineering application has become the focus of international offshore engineering field. However, since the development of the wave theory has been stuck in the stage of semi-empirical formula and ideal fluid hypothesis, as well as the high cost of the experimental physical tank and the long cycle, the numerical wave tank emerges as the third research method. The numerical wave tank is designed to simulate the various functions of the experimental physical tank through a computer simulation, so as to complete the scientific research and engineering tasks. In the investigation, perfecting and improving the various boundary techniques of the numerical wave tank is the basis for applying numerical wave tanks to practical engineering problems. With the rapid development of computers, CFD has been widely used, and many commercial fluid dynamics softwares such as CFX, STAR-CD, and FLUENT have appeared. Although there are still some deficiencies, with the rapid development of CFD, they will play more and more important roles in the field of marine engineering.

In this paper, the numerical wave tank is simulated by solving the N-S equation of the viscous fluid. The VOF method is employed to capture the free surface and the secondary development of the fluent software is performed. The dynamic grid technology is adopted to achieve the making of push-plate wave. In the end, the three-dimensional wave field of push plate wave making is simulated.

Content from this work may be used under the terms of the Creative Commons Attribution 3.0 licence. Any further distribution of this work must maintain attribution to the author(s) and the title of the work, journal citation and DOI.
Published under licence by IOP Publishing Ltd

IOP Conf. Series: Journal of Physics: Conf. Series **1064** (2018) 012073 doi:10.1088/1742-6596/1064/1/012073

2. Mathematical Model of the Wave Tank

2.1. Basic control equation

The instantaneous incompressible fluid control equation is averaged over time, thus, the continuous equation of time average is obtained. The time-averaged Navier-Stokes equation, ie., the Reynolds average N-S equation is written as:

$$\frac{\partial(\rho u_i)}{\partial t}+\frac{\partial(\rho u_i u_j)}{\partial x_j}=\rho f_i-\frac{\partial p}{\partial x_i}+\frac{\partial}{\partial x_j}\left(\mu\frac{\partial u_i}{\partial x_i}\right)+[-\frac{\partial(\rho\overline{u_i^{'}u_j^{'}})}{\partial x_j}] \tag{1}$$

Where $-\rho\overline{u_i^{'}u_j^{'}}$ is the Reynolds stress.

2.2. Turbulence model

The standard two-equation model is chosen in the study, which is as follows:

$$\frac{\partial}{\partial t}(\rho\varepsilon)+\nabla\cdot(\rho\varepsilon\mathbf{u})=\nabla\cdot[(\mu+\frac{\mu_t}{\sigma_\varepsilon})\nabla\varepsilon]+C_{1\varepsilon}\frac{\varepsilon}{k}+(G_k+C_{3\varepsilon}G_b)-C_{2\varepsilon}\rho\frac{\varepsilon^2}{k}+S_\varepsilon \tag{2}$$

$$\frac{\partial}{\partial t}(\rho k)+\nabla\cdot(\rho k\mathbf{u})=\nabla\cdot[(\mu+\frac{\mu_t}{\sigma_k})\nabla k]+G_k+G_b-\rho\varepsilon-Y_M+S_k \tag{3}$$

In the equation, G_k represents the turbulent kinetic energy generated by the laminar velocity gradient, G_b is the turbulent kinetic energy generated by buoyancy, Y_M is the fluctuation caused by the diffusion in compressible turbulence, C_μ, $C_{1\varepsilon}$, $C_{2\varepsilon}$, $C_{3\varepsilon}$ are constants, σ_k and σ_ε are the PRANDTL numbers of turbulence in k and ε equations, S_k and S_ε are user-defined. μ_t represents the turbulence speed, which is determined by the formula $\mu_t=\rho C_\mu\frac{k^2}{\varepsilon}$.

2.3. VOF Method

VOF (Volume of Fluid) is put forward by Hirt and Nichols, based on the particle tracking method (MAC). It is to determine the motion interface and its changes by taking the volume fraction α_q of each phase in the fluid elements. It is apt for various incompatible mixed fluids. The variable α_q is the volume ratio of phases in the calculation unit. In the calculation unit, the sum of the volume fractions of all the phases is equal to 1, that is, $\sum\alpha_q=1$. By solving the continuous equations of the phase volume fractions, the free surface between phases is tracked.

3. Wave making

Wave making is the key technology of numerical wave tanks. The schematic diagram of the push-plate wave maker is shown in Fig. 1. The coordinate system is established with the Z axis overlapping the initial position of the wave plate, the Z axis positively upwards, and the X axis coinciding with the still water surface. Assuming that the wave plate is doing small amplitude harmonic movement, h is the depth of the tank. The wave plate is doing reciprocating piston movement with the amplitude Ap. Thus, waves are generated by the wave plate, where H is the wave height, L is the wavelength, and T the period.

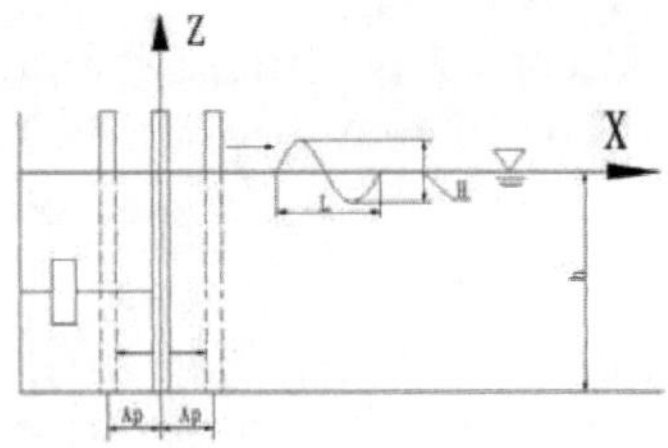

Figure 1. Push-plate wave-maker.

The displacement of the wave plate movement can be expressed as:

$$X_P = A_P \sin(wt) \tag{4}$$

Where X_P the displacement of the wave plate motion is, A_P is its amplitude, and w is the circle frequency, t is time.

Assuming the water is an incompressible and non-viscous fluid, then the velocity potential function of the flow field in the wave groove satisfies the Laplace equation:

$$\frac{\partial^2\phi}{\partial x^2}+\frac{\partial^2\phi}{\partial z^2}=0 \tag{5}$$

Meanwhile, the velocity potential satisfies the following boundary conditions.

(1) Water surface conditions:

$$\left.\frac{\partial\phi}{\partial t}\right|_{z=0}+g\eta=0 \tag{6}$$

$$-\left.\frac{\partial\phi}{\partial z}\right|_{z=0}=\frac{\partial\eta}{\partial t} \tag{7}$$

(2) Underwater conditions: Because the bottom of the tank is waterproof, the vertical speed of the water quality point is zero, so

$$\left.\frac{\partial\phi}{\partial z}\right|_{z=-h}=0 \tag{8}$$

(3) Body surface condition. The water quality velocity on the wave plate is equal to the flat motion speed of the wave plate. Assuming that the displacement and speed of the wave plate are both small, the following boundary conditions can be obtained approximately as follows:

$$\left.\frac{\partial\phi}{\partial x}\right|_{x=0}=A_P w\cos(wt) \tag{9}$$

The above equation is solved with the discrete variable method. According to the corresponding initial and boundary conditions, the velocity potential function of the flow field is:

$$\phi(x,z,t)=\frac{g}{w}C_0Z_0(kz)\sin(kx-wt)+\frac{g}{w}\sum_{n=1}^{n=\infty}C_nZ_n(k_nz)\cos(wt)e^{-k_nx} \tag{10}$$

With $Z_0(kz) = \cosh[k(z+h)]/\cosh(kh)$ and $Z_n(k_n z) = \cos[k(z+h)]/\cosh(k_n h)$. In the formula, the second item is a transient wave, $k = 2\pi / L$ is the wave number. k k_n And the angular frequency meet the dispersive relation: $w^2 = kg\tanh(kh)$, $w^2 = -k_n g\tanh(k_n h)$. C_0 And C_n are $C_0 = \dfrac{2[\cosh(2kh)-1]}{\sinh(2kh)+2kh}$, $C_n = \dfrac{2[\cos(2k_n h)-1]}{\sin(2k_n h)+2k_n h}$ $(n = 1,2,\cdots\cdots)$.

From equation (6), we can derive the wave height at the average position of the wave plate in motion: $\eta_0 = -\dfrac{1}{g}\dfrac{\partial\phi}{\partial t}\Big|_{z=0}$. $\eta_0 = C_0 A_P \cos(kx - wt) + \sum_{n=1}^{n=\infty} C_n A_p \sin(wt) e^{-k_n x}$. The first item on the right is the traveling wave, and the second item is the standing wave produced by the motion of the wave plate.

With formula (4) and the secondary development of Fluent, the motion law of the wave plate is defined. By adopting UDF function DEFINE_CG_MOTION (M_push, dt, cg_vel, cg_omega, time, dtime), the program is written for the movement of push board.

4. Numerical Computation and Analysis of Simulation Results

In the study, the 3D numerical wave tank model simulation of physical is established with the pre-processing software ICEM-CFD. The pressure inlet on top of the wave tank is set to be the boundary condition. The outlet is Outlet-vent, and Loss Coefficient is 1. The appropriate parameters are set on Open Channel, the wave board is set using UDF, and other boundary conditions are set as wall boundary.

The wave plate adopts three-dimensional dynamic grid technology for wave making. Figure 2 shows the partially enlarged view of the dynamic mesh. The finite volume method is employed for the fluid control equations and turbulence models to perform discreteness (FVM). Pressure Semi-Implicit Segmentation Technology is applied for the coupling of pressure speed, and the geometric reconstruction method for capturing the free liquid surface. At the initial time (t=0) on top of the free liquid surface is all air, with a density of 1.225kg /m3. The lower part of the free fluid surface is full of water with a density of 998.2kg/m3.

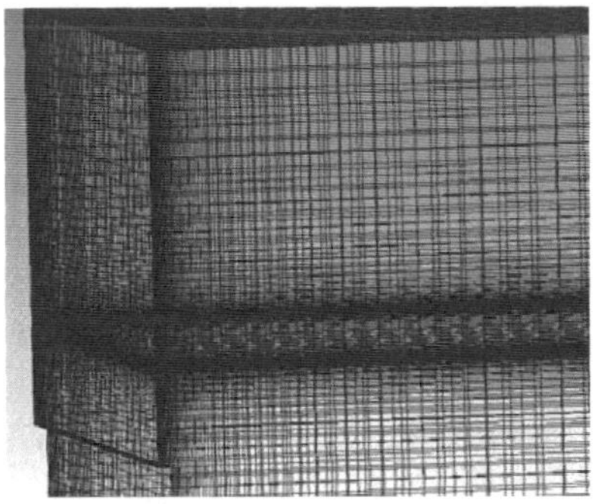

Figure 2. Partial enlargement of the wave plate grid.

The target wavelength is set as 8m and the period as 2.2678s.with the push plate the three-dimensional waveforms with different wave heights of 0.2m, 0.3m, and 0.4m are created. Wavegauges are put in different positions in the numerical wave tank to collect the wave height time history at these locations.

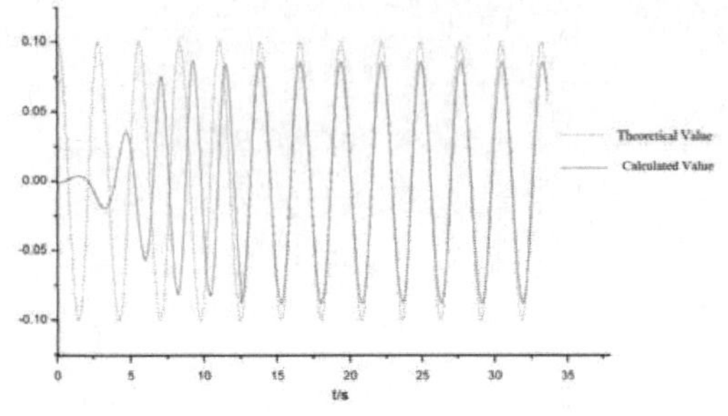

X=1.0L

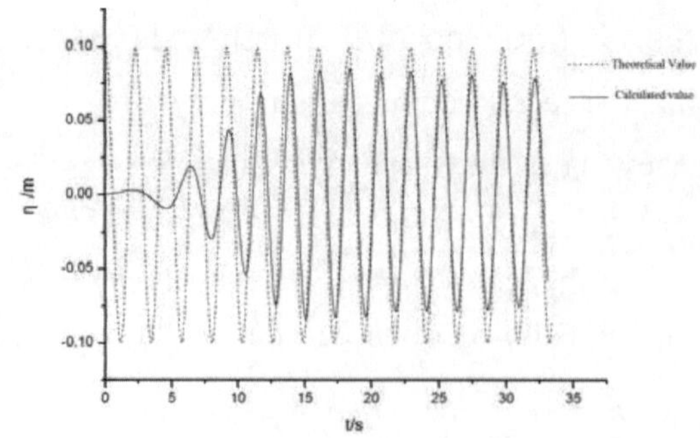

X=2L

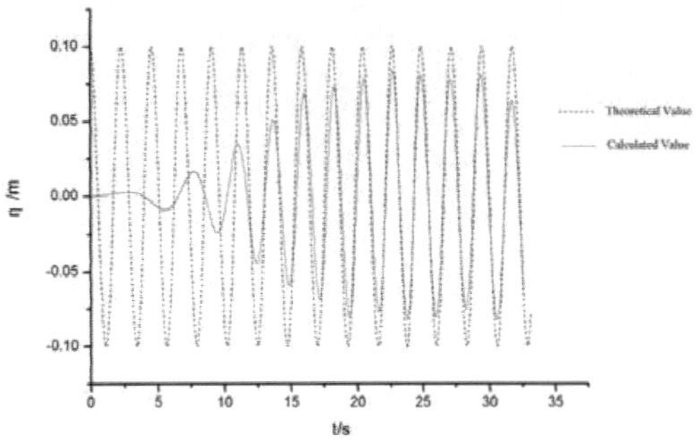

3.0L

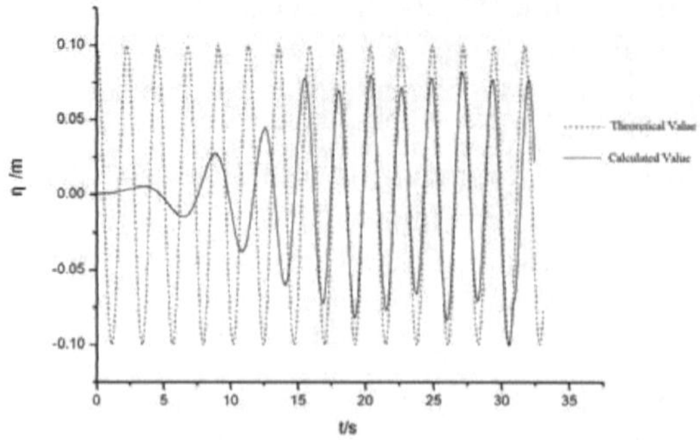

X=4.0L

Figure 3. Comparison of the calculation values of wave surface with time and theoretical solutions.

Figure 3 shows history chart of the wave height over time when the horizontal distances from the wave plate are respectively 1, 2, 3, or 4 times of the wavelength. From the figure, it can be seen that as time goes on, the waveform tends to be stable. The longer away from the push wave plate the horizontal distance is, the longer the development of the waveform needs, and the slowlier the stable wave forms. As a result, the wave height is smaller than the theoretical value. This is because the wave transmission takes time to accumulate and the wave energy is constantly attenuated during the transmission due to the effect of the viscosity. After some periods, when the energy the wavefield acquired is equivalent to the energy dissipated, the wavefield eventually reaches stable. Due to the effect of the viscous dissipation, the smallest is the average of wave height where the distance from the wave plate is 4 times the wavelength.

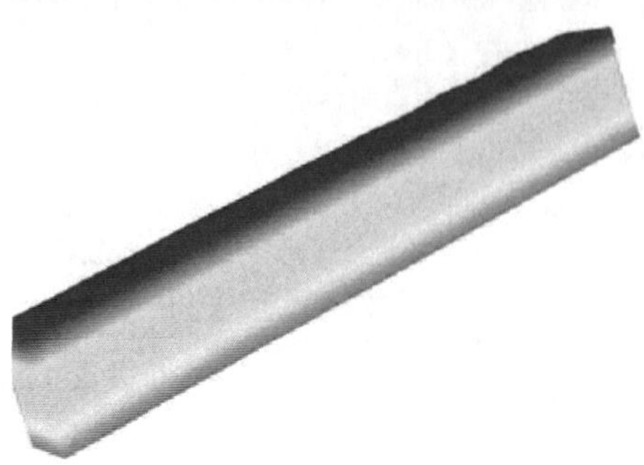

Figure 4. Distribution diagram of pressure field.

Figure 4 is the distribution of the pressure field, which shows the pressure gradually increases along the depth of the water. The free liquid surface is an isobaric surface, which corresponds with the theoretical result. Figures 5 and 6 indicate the instantaneous speed vector. As can be seen, the further away from the wave surface the fluid particle is, the smaller its velocity turns.

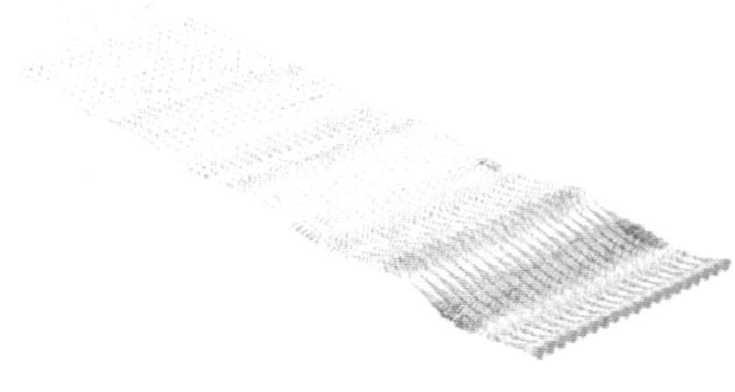

Figure 5. Vector diagram of instantaneous speed.

Figure 6. Vector diagram of the velocity near the free liquid surface

It can be seen from Table 1 that the waveform formed by the wave plate is of high stability and quality. The error of wavelength numerically simulated and theoretically adopted is 0.9063%, and the period error is 0.8863%. The wave heights recorded by the wave height meter at different positions are shown in Figure 7. It can be seen that the wave height decays faster where the distance from the wave plate is 2 times the wavelength, and then gradually stabilizes until a stable wave field is formed.

Table 1. Simulation results of regular waves by wave plate.

	Wavelength /m	Period /s
Theoretical Value	8.0000	2.2678
Numerical Value	7.9275	2.2879
Relative Error /%	0.9063	0.8863

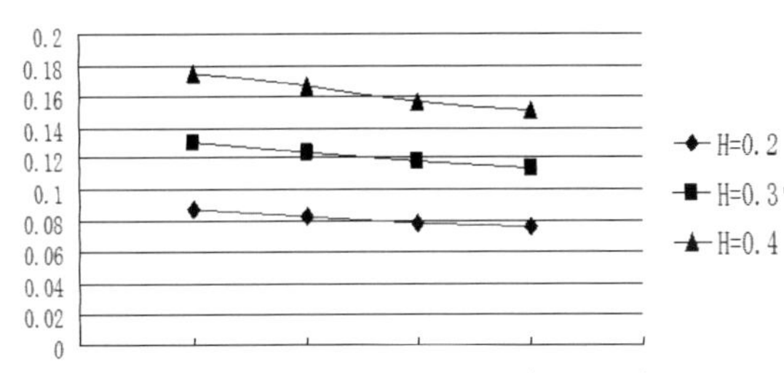

Figure 7. Attenuation curves of different wave height

5. Conclusion

In this paper, with FLUENT software as a platform, using VOF method, the software Fluent is secondary development to achieve the motion of wave plate. By solving the standard turbulence

IOP Conf. Series: Journal of Physics: Conf. Series **1064** (2018) 012073 doi:10.1088/1742-6596/1064/1/012073

equation the numerical simulation of the wave-making in the wave tank is achieved. From the results, we can see that the quality of the wave made by wave plate is of high quality and stability. This study can provide technology support for the numerical wave-making with CFD calculations of ships.

Acknowledgments
This work was financially supported by Fujian Provinial Natural Science Foundation, Project No. 2015J01586; Project No.2017J01485.

References
[1] S.Z. Li, Study on 2-D numerical wave tank based on the software fluent, Ms. Dissertation, . Harbin Institute of Technology, Harbin, China, 2006.
[2] C.S. Wu, D.X. Zhu, M. Gu, Computation of hydrodynamic forces for a ship in regular heading waves by a viscous numerical tank, J. of Ship Mechanics, 2008, 12 (2) 168-179.
[3] A.H. Clement, Benchmark test cases for numerical wave absorption, Proceedings of the Ninth International Offshore and Polar Engineering Conference. Brist, France, 2009, pp.266-289.
[4] C.W. Hirt, B.D. Niehols, Volume of Fluid(VOF) Method for the Dynamies of Free Boundaries. J. of Computational physics, 1981, 39(l) pp.201-225.
[5] F.J. Wang, Computational Fluid Dynamics Analysis CFD Software Principle and Application, Tsinghua University Press, Beijing, 2004.
[6] X.T. Wang, Theory of Absorption Wave Machine, Ms. Dissertation, Dalian University of Technology, Dalian, China, 2002.

Size-dependent vibration of multi-span micropipes conveying fluid based on modified couple stress theory

Yuzhen Zhao*, **Jiaquan Deng, Yongshou Liu and Wei Liu**

Department of Engineering Mechanics, North-western Polytechnical University, Xi'an, China.

*Corresponding author e-mail: zhaoyz1989@mail.nwpu.edu.cn

Abstract. In this article, the size-dependent free vibration and stability of multi-span micropipes conveying fluid are studied. To take the size effect into account, the modified couple stress theory is used to derive the governing equation. And the dynamic stiffness method (DSM) is adopted to determine the natural frequencies. Subsequently, some examples are presented to investigate the effects of fluid velocity and material length scale parameter on stability of the multi-span micropipes conveying fluid. It's found that the natural frequencies and the critical velocities obtained by the modified couple stress theory are higher than those predicted by the classical theory.

1. Introduction

Being the simplest form of fluid structure interaction problem, dynamics of pipes conveying fluid has attracted enormous attentions from researchers due to the rich and interesting dynamic behavior. Numerous works have been carried out on the vibration and stability of pipes conveying fluid [1, 2]. Although various aspects of this problem have been studied in the past half-century, most of the publications are limited to the model of single-span pipe conveying fluid. But in real working environment, the piping systems are always arbitrary multi-span pipes, and the works concerning the arbitrary multi-span pipe conveying fluid are quite limited [3,4]. As pointed out by Wu and Shih [3], the main reason is that the classical mode method is easily applicable only to the special cases where the functions for the mode shapes are obtainable, such as single-span or the periodical multi-span pipes. Finite element method (FEM) may be the most versatile method for the dynamics of multi-span pipes conveying fluid, but it is laborious when the number of element is very large [5].

With the quick growth in science and engineering, fluid-conveying pipes have been made in microscale and applied in micro-electromechanical systems (MEMS), fluid storage, fluid transport and drug deliver [6,7]. Therefore, understanding the dynamic behavior of micropipes conveying fluid is currently a subject of great interest. However, the dynamic behavior of microstructures has been proved to be size dependent and the classical continuum theories are unable to capture the size effect of microstructures. To overcome this problem, a size-dependent nonclassical continuum theory, modified couple stress theory [8], has been proposed to capture the size effect. And the modified couple stress theory has been widely utilized by many researchers to study the dynamic behavior of microstructures [9, 10].

In this work, the size-dependent free vibration and stability of a multi-span micropipe conveying fluid are investigated. The modified couple stress theory and the Hamilton's principle are used to drive

Content from this work may be used under the terms of the Creative Commons Attribution 3.0 licence. Any further distribution of this work must maintain attribution to the author(s) and the title of the work, journal citation and DOI.
Published under licence by IOP Publishing Ltd

the governing equation. The natural frequencies and the critical velocities are determined by using the dynamic stiffness method (DSM). The effects of fluid velocity and material length scale parameter on dynamic behavior of multi-span micropipes conveying fluid are discussed.

2. Modified couple stress theory and governing equation

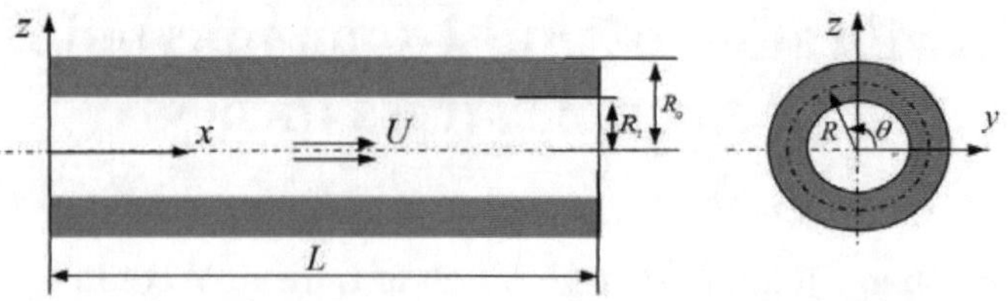

Figure 1. Geometric model of a micropipe conveying fluid.

A micropipe with mean radius *R*, length *L* and fluid velocity *U* is depicted in Figure 1. Displacement components in x-, y-, and z-directions denote as u, v and w, respectively. According to the modified couple stress theory, the strain energy density is a function of both strain tensor (conjugated with stress tensor) and curvature tensor (conjugated with couple stress tensor) [8]. For a linear elastic material occupying region Ω, the strain energy U_m can be expressed as:

$$U_m = \frac{1}{2}\int_{\Omega}\left(\sigma_{ij}\varepsilon_{ij} + m_{ij}\chi_{ij}\right)dv \qquad (i,j=1,2,3) \tag{1}$$

In which, σ_{ij} is the stress tensor, ε_{ij} is the strain tensor. The deviatoric part of couple stress tensor m_{ij} , and the symmetric curvature tensor χ_{ij} are given by [10]

$$m_{ij} = 2l^2 G\chi_{ij}, \qquad \chi_{ij} = \left(\theta_{i,j} + \theta_{j,i}\right)/2 \tag{2}$$

Where δ_{ij} is the Kronecker's delta function, l is material length scale parameter and G is shear modulus. θ_i is the rotation vector and it can be written as:

$$\theta_i = curl\left(u_i\right)/2 \tag{3}$$

Where u_i is the displacement vector. According to Euler-Bernoulli beam theory, the displacement components are given by

$$u = -z\psi\left(x,t\right), \qquad v = 0, \qquad w = w\left(x,t\right) \tag{4}$$

Where ψ is the rotation of pipe cross-section In view of the small deformation assumption, the following approximation can be applied

$$\psi\left(x\right) \approx w\left(x\right)' \tag{5}$$

where the prime represents differentiation with respect to *x*.
Based on Euler-Bernoulli beam theory, the strains can be obtained

$$\varepsilon_{xx} = -zw'', \qquad \varepsilon_{yy} = \varepsilon_{zz} = \varepsilon_{xy} = \varepsilon_{yz} = \varepsilon_{zx} = 0 \tag{6}$$

Accordingly, the stresses can be written as

$$\sigma_{xx} = -Ezw'', \qquad \sigma_{yy} = \sigma_{zz} = \sigma_{xy} = \sigma_{yz} = \sigma_{zx} = 0 \tag{7}$$

It should be mentioned that the Poisson effect is neglected to facilitate the formulation of a simple beam theory in above equations [10].

Combination Eqs. (3)-(5), yields

$$\theta_y = -w', \qquad \theta_x = \theta_z = 0 \tag{8}$$

Substituting Eq. (8) to Eq. (2), yields

$$\chi_{xy} = -w''/2, \qquad \chi_{xx} = \chi_{yy} = \chi_{zz} = \chi_{yz} = \chi_{zx} = 0 \tag{9}$$

Substitution of Eq. (9) to Eq. (2) gives

$$m_{xy} = -Gl^2 w'', \qquad m_{xx} = m_{yy} = m_{zz} = m_{yz} = m_{zx} = 0 \tag{10}$$

Substituting Eqs. (6)-(7) and (9)-(10) to Eq. (1), one obtains

$$U_m = \frac{1}{2}\int_L \left(EI + GA_p l^2\right)\left(w''\right)^2 dx \tag{11}$$

Where, I is the moment of inertia.

The kinetic energies of pipe and fluid can be expressed as [1]

$$T_p = \frac{m}{2}\int_L \left(\dot{w}\right)^2 dx, \qquad T_f = \frac{M_f}{2}\int_L \left[\left(\dot{w} + Uw'\right)^2 + U^2\right] dx \tag{12}$$

where the dot represents differentiation with respect to t.

According to Paidoussis [1], Hamilton's principle for both ends supported pipe conveying fluid can be written as

$$\int_{t_1}^{t_2} \delta\left(T_p + T_f - U_m\right) dt \tag{13}$$

Substituting Eqs. (11) and (12) into Eq. (13), and applying the variational techniques to Eq. (13), the governing equation of the micropipes conveying fluid can be obtained

$$\left(EI + GAl^2\right) w'''' + M_f U^2 w'' + 2M_f U\dot{w}' + \left(M_f + m\right)\ddot{w} = 0 \tag{14}$$

Compared with governing equation of classical pipes conveying fluid [1], the flexural rigidity is replaced by the equivalent flexural rigidity $EI+GAl^2$.

3. Application of DSM to multi-span pipes conveying fluid

In this section, the DSM is utilized to determine the natural frequencies of multi-span pipes conveying fluid. The general solution for Eq. (14) can be written as

$$w(x,t) = \hat{w}(x,\omega)e^{i\omega t} \tag{15}$$

Where ω is the circular frequency and $i=\sqrt{-1}$.
Substituting Eq. (15) into Eq. (14), yields

$$\left(EI+GAl^2\right)\hat{w}''''+M_fU^2\hat{w}''+i2\omega M_fU\hat{w}'-\omega^2\left(M_f+m\right)\hat{w}=0 \tag{16}$$

Then, the displacement in frequency domain can be set as

$$\hat{w}=ce^{ikx} \tag{17}$$

By substituting Eq. (16) to Eq. (17), one obtains

$$\left(EI+GAl^2\right)k^4-M_fU^2k^2-2\omega M_fUk-\omega^2\left(M_f+m\right)=0 \tag{18}$$

The transverse displacement in frequency domain can be rewritten as

$$\hat{w}=\sum_{j=1}^{4}w_je^{ik_jx} \tag{19}$$

Accordingly, the rotation, moment and shear force in frequency domain can be obtained

$$\hat{\theta}=\sum_{j=1}^{4}ik_jw_je^{ik_jx},\hat{M}=\sum_{j=1}^{4}\left(EI+GAl^2\right)k_j^2w_je^{ik_jx},\hat{Q}=\sum_{j=1}^{4}\left(EI+GAl^2\right)ik_j^3w_je^{ik_jx} \tag{20}$$

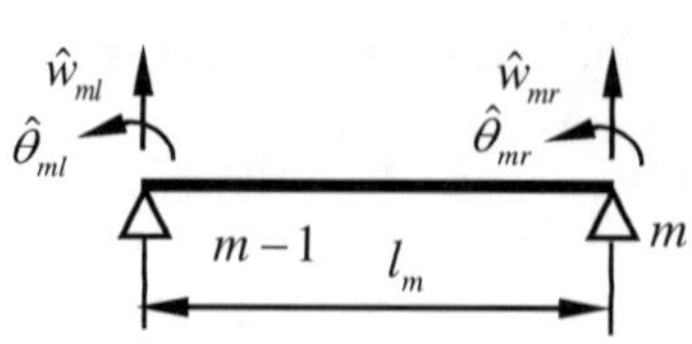

Figure 2. Displacements of the *m*th span element.

The *m*th span element of a multi-span micropipe is shown in Figure 2. The nodal displacements and forces in local coordinate have the following relationships

$$\hat{w}_{ml}=\hat{w}(0),\quad \hat{\theta}_{ml}=\hat{\theta}(0),\quad \hat{w}_{mr}=\hat{w}(l_m),\quad \hat{\theta}_{mr}=\hat{\theta}(l_m) \tag{21}$$

$$\hat{M}_{ml}=-\hat{M}(0),\quad \hat{Q}_{ml}=-\hat{Q}(0),\quad \hat{M}_{mr}=\hat{M}(l_m),\quad \hat{Q}_{mr}=\hat{Q}(l_m) \tag{22}$$

where, the subscripts l and r represent the left and right ends of the element, respectively. And the subscript m indicates the mth span element. According to Eqs. (19) and (20), the nodal displacements and nodal forces can be expressed as

$$\mathbf{D}_m=\mathbf{Y}_m(\omega)\mathbf{w}_m,\qquad \mathbf{F}_m=\mathbf{X}_m(\omega)\mathbf{w}_m \tag{23}$$

The detailed expressions can be written as

$$\begin{Bmatrix} \hat{w}_{ml} \\ \hat{\theta}_{ml} \\ \hat{w}_{mr} \\ \hat{\theta}_{mr} \end{Bmatrix} = \begin{bmatrix} 1 & 1 & 1 & 1 \\ \lambda_1 & \lambda_2 & \lambda_3 & \lambda_4 \\ e^{ik_1 l_m} & e^{ik_2 l_m} & e^{ik_3 l_m} & e^{ik_4 l_m} \\ \lambda_1 e^{ik_1 l_m} & \lambda_2 e^{ik_2 l_m} & \lambda_3 e^{ik_3 l_m} & \lambda_4 e^{ik_4 l_m} \end{bmatrix} \cdot \begin{Bmatrix} w_1 \\ w_2 \\ w_3 \\ w_4 \end{Bmatrix}, \quad \begin{Bmatrix} -\hat{Q}_{ml} \\ -\hat{M}_{ml} \\ \hat{Q}_{mr} \\ \hat{M}_{mr} \end{Bmatrix} = \begin{bmatrix} -\gamma_1 & -\gamma_2 & -\gamma_3 & -\gamma_4 \\ -\beta_1 & -\beta_2 & -\beta_3 & -\beta_4 \\ \gamma_1 e^{ik_1 l_m} & \gamma_2 e^{ik_2 l_m} & \gamma_3 e^{ik_3 l_m} & \gamma_4 e^{ik_4 l_m} \\ \beta_1 e^{ik_1 l_m} & \beta_2 e^{ik_2 l_m} & \beta_3 e^{ik_3 l_m} & \beta_4 e^{ik_4 l_m} \end{bmatrix} \begin{Bmatrix} w_1 \\ w_2 \\ w_3 \\ w_4 \end{Bmatrix} \quad (24)$$

Where, $\lambda_j = ik_j$, $\gamma_j = \left(EI + GAl^2\right)ik_j^3$, $\beta_j = \left(EI + GAl^2\right)k_j^2$ (j=1, 2, 3, 4). Based on Eq. (23), we can get

$$\mathbf{F}_m = \mathbf{K}_m(\omega)\mathbf{D}_m \quad (25)$$

Where $\mathbf{K}_m = \mathbf{X}_m \mathbf{Y}_m^{-1}$ is the dynamic stiffness matrix of the mth span element. The global dynamic stiffness matrix can be assembled by the local dynamic stiffness matrices. And the relationship between the nodal forces and nodal displacements for multi-span micropipes can be written as

$$\mathbf{F}_g = \mathbf{K}_g(\omega)\mathbf{D}_g \quad (26)$$

where the subscript g represents the global coordinate system. After applying the boundary conditions to Eq. (26), the dynamic stiffness matrix $\mathbf{K}_g$ can be used to calculate the natural frequencies. And it has been demonstrated that the DSM is valid and efficient for dynamic analysis of multi-span pipes conveying fluid [4].

4. Numerical results and discussions

For analysis convenience, two non-dimensional quantities are introduced [1]

$$u = \left(M_f / EI\right)^{1/2} LU, \qquad \varpi = \left[\left(M_f + m\right)/EI\right]^{1/2} L^2\omega \quad (27)$$

Where u is dimensionless fluid velocity and ϖ is dimensionless natural frequency.

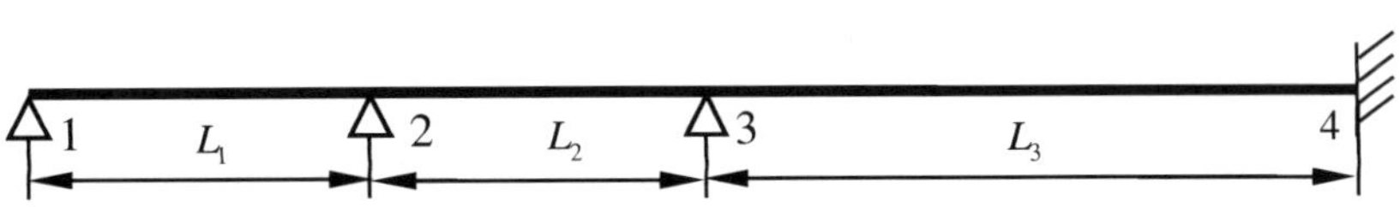

Figure 3. A three-span micropipe conveying fluid.

A typical three-span micropipe conveying fluid is plotted in Figure 3. The material properties of the pipe are E=70 GPa, ν =0.23, ρ_p=2700 kg/m3, ρ_f=1000 kg/m3 and the length scale parameter l is assumed to be 15 μm [10]. The other geometrical parameters are L3=2L1=2L2, Do/Di=0.9, and L/Do=40. Di and Do represent the inner and outer diameters. L is the length of the whole piping system.

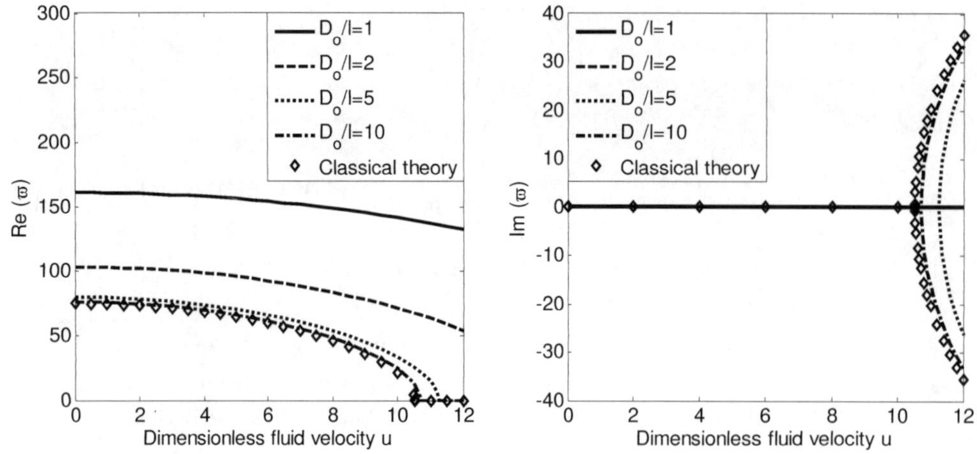

Figure 4. Fundamental natural frequencies versus fluid velocity u with different D_o/l.

Table 1. Comparison of dimensionless critical velocities for different D_o/l.

Critical velocity	D_o/l =1	D_o/l =2	D_o/l =5	D_o/l =10	Classical
u_c	22.54	14.49	11.25	10.71	10.57

To investigate the effects of material length scale parameter and fluid velocity on stability of micropipes conveying fluid, the fundamental natural frequency as a function of fluid velocity u with different Do/l is depicted in Figure 4. The critical velocities for different Do/l are listed in Table 1. It should be mentioned that the real component of natural frequency, re (ϖ), is the oscillation frequency, while the imaginary component of natural frequency, Im (ϖ), is related to exponential growth or decay. It can be seen that the real components decrease with increasing fluid velocity u. When the real part of fundamental natural frequency falls to zero, the conservative piping system loses its stability due to the divergence. And the corresponding velocity is called the critical velocity uc. Obviously, it can be seen from Figure 4 that the natural frequencies predicted by the modified couple stress theory are generally higher than those determined by the classical theory. It's also found that by increasing Do/l, the results predicted by the modified couple stress theory get closer to those obtained by the classical theory. For example, the dimensionless critical velocity for Do/l =1 is 22.54. When the parameter Do/l =10, the dimensionless critical velocity decreases to 10.71 which has very closed to 10.57 predicted by the classical theory. Therefore, the results in Figure 4 and Table 1 demonstrate that the effects of material length scale parameter on the natural frequencies and critical velocities of micropipes conveying fluid are significant, especially when the diameter of micropipe is comparable to the material length scale parameter. However, for a higher value of (Do/l >10), natural frequencies and critical velocities obtained by the modified couple stress theory converge to the classical ones.

5. Conclusion

In this paper, based on the modified couple stress theory, the governing equation of micropipes conveying fluid is derived by the Hamilton's principle. The natural frequencies are determined by the DSM. It is found that the effects of material length scale parameter on natural frequencies and critical velocities are significant. And it makes the micropipe conveying fluid more stable, especially when the diameter of the micropipe is comparable to the material length scale parameter.

IOP Conf. Series: Journal of Physics: Conf. Series **1064** (2018) 012074 doi:10.1088/1742-6596/1064/1/012074

References

[1] M.P. Païdoussis, Fluid-Structure Interactions Slender Structures and Axial Flow Volume 1, 2nd ed., Academic Press, London, 2014.

[2] S.J. Li, B.W. Karney, G.M. Liu, FSI research in pipeline systems-A review of the literature, J. Fluid. Struct. 57 (2015) 277-297.

[3] J.S. Wu, P.Y. Shih, The dynamic analysis of multispan fluid-conveying pipe subject to external load, J. Sound. Vib. 229 (2001) 201-215.

[4] B.H. Li, H.S. Gao, H.B. Zhai, Y.S. Liu, Z.F. Yue, Free vibration analysis of multi-span pipe conveying fluid with dynamic stiffness method, Nucl. Eng. Des. 241 (2011) 666-671.

[5] ZHANG Shunfeng, ZHANG Sheng,WANG Qiang, et al, Transient Properties of Hydraulic Operated Directional Valve Based on Fluid-structure Interaction, Journal of Ordnance Equipment Engineering. 9 (2017) 151-155.(in Chinese).

[6] FAN Weimin, LIU Shuangjie, WANG Lishuang, et al. Application of MEMS Accelerometer in Antiaircraft Missile Fuze, Journal of Ordnance Equipment Engineering. 3 (2018) 75-82.(in Chinese).

[7] W. Xia, L. Wang, Microfluid-induced vibration and stability of structures modeled as microscale pipes conveying fluid based on non-classical Timoshenko beam theory, Microfluid. Nanofluid. 9 (2010) 955-962.

[8] F. Yang, A. C. M. Chong, D.C.C. Lam, P. Tong, Couple stress based strain gradient theory for elasticity, Int. J. Solids. Struct. 39 (2002) 2731-2743.

[9] M. Salamat-talab, A. Nateghi, J. Torabi, Static and dynamic analysis of third-order shear deformation FG micro beam based on modified couple stress theory, Int. J. Mech. Sci. 57 (2012) 63-73.

[10] L. Wang, Size-dependent vibration characteristics of fluid-conveying microtubes, J. Fluid. Struct. 26 (2010) 675-684.

An adaptive variational procedure for a two-phase model with variable density and viscosity

Min Yang*

School of Sciences, Hohai University, Nanjing, China.

*Corresponding author e-mail: njhzm@sina.com

Abstract. In this paper we present an adaptive variational procedure to solve a two-phase field model with variable density, viscosity. The model is a couple system that consists of incompressible Navier-Stokes equations and Allen-Cahn equation. In the scheme the projection method based on pressure is used to deal with the Navier-Stokes equations and stabilization approach is used for the Allen-Cahn equation. By some subtle explicit-implicit treatments, the proposed scheme is energy stable. In particular the adaptive refinement/coarsening is carried out by evaluating the residual error indicators on the error estimates of the Allen-Cahn equation. Some numerical examples are performed to show the accuracy and efficiency.

1. Introduction

The phase field model is a tool to resolve the motion of free interfaces between two distinct fluid components. Its origin can be traced back to Rayleigh [1] and Waals [2]. It was first formulated to study material mixtures and later adopted as a technique to resolve motion of the interface between different material components in material mixtures [3-5]. The most advantage of the phase-field approach is that the nonlinear partial differential equations satisfy thermodynamics consistent energy dissipation law. So we can carry out mathematical analysis and numerical schemes which can satisfy the corresponding discrete energy dissipation laws.

But most of the analysis of the two-phase model have been restricted to the matched density case. In this paper, we extend the previous study to a two-phase field model with variable density and viscosity in which the Allen-Cahn equation is adopted as the transport equation for the phase variable.

We first investigate the energy stability of the proposed scheme on an unstructured grid over a domain of complex geometry. The proposed procedure aims at reducing the error introduced by the mesh adaptivity while solving the coupled Navier-Stokes and Allen-Cahn equations. The effectiveness of the adaptive algorithm is assessed and several numerical experiments are performed. This article makes a significant contribution to the development of a stable, robust and general nonlinear adaptive variational procedure for solving the coupled Navier-Stokes and Allen-Cahn equations to model two-phase model via phase-field modeling.

The paper is organized as follows. We describe a Navier-Stokes Allen-Cahn coupled model in Section 2. In Section 3 we describe an energy stable scheme and discrete energy laware derived and fully-discrete variational formulation is given. Finally, some numerical experiments are presented in Section 4 to assess the effectiveness and robustness of the proposed adaptive scheme.

Content from this work may be used under the terms of the Creative Commons Attribution 3.0 licence. Any further distribution of this work must maintain attribution to the author(s) and the title of the work, journal citation and DOI.
Published under licence by IOP Publishing Ltd

2. A Navier-Stokes Allen-Cahn coupled model

In this paper we consider a two-phase incompressible fluid with variable density and viscosity in a physical domain denoted by Ω with boundary Γ. It is showed that this problem can be described by the following coupled system of the Allen-Cahn and Navier-Stokes equations:

Incompressible Navier-Stokes equations for hydrodynamics:

$$\begin{cases} \rho[\frac{\partial u}{\partial t}+(u\cdot\nabla)u]=\nabla\cdot[\eta D(u)]-\nabla p+B\mu\nabla\phi \quad in \quad \Omega \\ \nabla\cdot u=0 \quad in \quad \Omega \\ u=0 \quad on \quad \Gamma \\ u|_{t=0}=u_0 \quad in \quad \Omega \end{cases} \tag{1}$$

Allen-Cahn equation for the dynamics of phase variable:

$$\begin{cases} \frac{\partial\phi}{\partial t}+u\cdot\nabla\phi=-M\mu \quad in \quad \Omega \\ \mu=-\varepsilon\Delta\phi+f(\phi) \quad in \quad \Omega \\ \partial_n\mu=0,\partial_n\phi=0 \quad on \quad \Gamma \\ \phi|_{t=0}=\phi_0 \quad in \quad \Omega \end{cases} \tag{2}$$

To preserve the volume fraction, we add the Lagrangian multiplier $\xi(t)$ corresponding to the constant volume constraint as follows.

$$\mu=-\varepsilon\Delta\phi+f(\phi)+\xi(t), \quad \frac{d}{dt}\int_\Omega \phi dx=0$$

We can easily get that $\xi(t)=\frac{1}{|\Omega|}\int_\Omega f(\phi(x,t))dx.$ in the above system, p is the pressure, $\eta D(u)=\eta(\nabla u+\nabla u^T)$ denotes the viscous part of the stress tensor, ϕ is the phase-field variable and μ is the chemical potential. The function $f(\phi)=F'(\phi)$, with $F(\phi)$ being the Ginzburg-Landau bulk potential. More precisely, F is defined as $F(\phi)=\frac{1}{4\varepsilon}(\phi^2-1)^2$. And ∇ denotes the gradient operator, n is the outward normal direction on Γ, scalar operator $\partial_n=n\cdot\nabla$ is the partial derivative along direction n, density $\rho=\frac{1+\phi}{2}\rho_1+\frac{1-\phi}{2}\rho_2$, viscosity $\eta=\frac{1+\phi}{2}\eta_1+\frac{1-\phi}{2}\eta_2$, where $(\cdot)_i$ denotes the value of its argument on the phase i of the argument. B denotes the strength of the capillary force comparing the Newtonian fluid stress, M is the phenomenological mobility coefficient, ε is the ratio between interface thickness and domain size.

3. Numerical energy stable scheme

In this section, we adopt the stabilization method with the adaptive method. The unconditional stability of the stabilization method requires that the second derivative of $F(\phi)$ to be bounded. Thanks to a truncated $F(\phi)$ with quadratic growth at infinity can also guarantee the boundless of ϕ in Allen-Cahn equation. More precisely, we replace $F(\phi)$ by $\hat{F}(\phi)$:

$$\hat{F}(\phi)=\begin{cases}\frac{1}{2\varepsilon}(\phi+1)^2, & if \phi<-1,\\ \frac{1}{4\varepsilon}(\phi^2-1)^2, & if\ -1\le\phi\le 1,\\ \frac{1}{2\varepsilon}(\phi-1)^2, & if \phi>1\end{cases} \tag{3}$$

Then setting $\hat{f}(\phi)=\hat{F}(\phi)$, and $L_1=\max_{\phi\in R}|\hat{f}'(\phi)|=\frac{2}{\varepsilon}$. Let $\delta t>0$ be a time discretization step and suppose u^n, ϕ^n and p^n are given, where the superscript n on variables denotes approximations of corresponding variables at time $n\delta t$. Assuming s_1 and s_2 are two positive stabilizing coefficients to be determined. The initial conditions ϕ^0, $p^{-1}=p^0, u^0$ are given and set $\overline{\rho}=\min(\rho_1,\rho_2)$. Now we are ready to present the energy stable scheme:

Step 1: We first solve ϕ^{n+1} and μ^{n+1} from

$$\mu^{n+1}=-\varepsilon\Delta\phi^{n+1}+\hat{f}(\phi^n)+s_1(\phi^{n+1}-\phi^n)+\xi^n(t), \tag{4}$$

$$\frac{\phi^{n+1}-\phi^n}{\delta t}+u_*^n\cdot\nabla\phi^n=-M\mu^{n+1}, \tag{5}$$

with boundary condition $\partial_n\mu^{n+1}=0$, where $u_*^n=u^n-\delta t\frac{B}{\rho^n}\mu^{n+1}\nabla\phi^n$.

Step 2: Then we solve u^{n+1} from

$$\rho^{n+1}=\frac{1+\phi^{n+1}}{2}\rho_1+\frac{1-\phi^{n+1}}{2}\rho_2,\ \eta^{n+1}=\frac{1+\phi^{n+1}}{2}\eta_1+\frac{1-\phi^{n+1}}{2}\eta_2, \tag{6}$$

$$\frac{\frac{1}{2}(\rho^{n+1}+\rho^n)u^{n+1}-\rho^n u_*^n)}{\delta t}+\rho^{n+1}(u^n\cdot\nabla)u^{n+1}+\frac{1}{2}(\nabla\cdot(\rho^{n+1}u^n)u^{n+1})$$

$$=\nabla\cdot(\eta^{n+1}D(u^{n+1}))-\nabla(2p^n-p^{n-1}) \tag{7}$$

with the boundary condition $u^{n+1}=0$

Step 3: In the last step, we get p^{n+1} from

$$\Delta(p^{n+1}-p^n)=\frac{\overline{\rho}}{\delta t}\nabla\cdot u^{n+1} \tag{8}$$

With the boundary condition $\partial_n p^{n+1}=0$.

Then we can get the following finite element method and give the weak form. In a standard finite element method the mesh is fitted to the boundary or interpolates the boundary to some suitable order. Let T_h be a curvilinear triangulation of Ω, such that $\Omega\subset T_h$, but $T_h\not\subset\Omega$. For all triangles $K\in T_h$ holds $K\cap\Omega\neq\phi$ and we define the domain covered by T_h and $\Omega_T:=\bigcup_{K\in T_h}K$. Let h_K be the diameter of K which is less or equal to the grid-parameter h and h: $=\max_{K\in T_h}h_K>0$. Associated with T_h we have the finite element spaces of continuous piece wise polynomial functions

$$V_h = \{v_h \in H^1(\Omega_T): v_h|_K \in P_1(K), \forall K \in T_h\};$$
$$U_h = \{v_h \in H^1(\Omega_T): v_h|_K \in P_2(K), \forall K \in T_h\}$$

Where $P_k(K)$ denotes the space of polynomials of degree less than or equal to k on the element K.

Then the discrete finite element form can be stated as: first, find $(\mu_h^{n+1}, \phi_h^{n+1}) \in V_h \times V_h$, such that for

Any $(\omega_h, \varphi_h) \in V_h \times V_h$

$$(\frac{\phi_h^{n+1}}{\delta t}, \omega_h) + (M\mu_h^{n+1}, \omega_h) = (\frac{\phi_h^n}{\delta t}, \omega_h) - (u_{h*}^n \cdot \nabla \phi_h^n, \omega_h)$$

$$(\varepsilon \nabla \phi_h^{n+1}, \nabla \varphi_h) + s_1(\phi_h^{n+1}, \varphi_h) - (\mu_h^{n+1}, \varphi_h) = (s_1\phi_h^n - \hat{f}(\phi_h^n) - \xi^n(t), \varphi_h)$$

Second, from (6) we get ρ^{n+1}, η^{n+1}, then we find $u_h^{n+1} \in U_h$, such that for any $v_h \in U_h$

$$(\frac{\frac{1}{2}(\rho^{n+1} + \rho^n)u_h^{n+1}}{\delta t} + \rho^{n+1}(u_h^n \cdot \nabla)u_h^{n+1} + \frac{1}{2}(\nabla \cdot (\rho^{n+1}u_h^n)u_h^{n+1}), v_h)$$

$$+ (\eta^{n+1} D(u_h^{n+1}), \nabla v_h) = (\frac{\rho^n u_{h*}^n}{\delta t} - \nabla(2p_h^n - p_h^{n-1}), v_h)$$

Last, find $p_h^n \in V_h$, such that for any $q_h \in V_h$

$$(\nabla p_h^{n+1}, \ \nabla q_h) = (\nabla p_h^n, \nabla q_h) + \frac{\overline{\rho}}{\delta t}(\nabla \cdot u^{n+1}, q_h)$$

The above finite element formulation of the Navier-Stokes and the Allen-Cahn equations is simply the differential operator multiplied by the weighting function and integrated by parts as appropriate, which tends to minimize the residual of the equations in a chosen set of weighting functions. When the solution of the underlying differential equation is smooth, as measured relative to the differential equation on the given mesh, the variational error tends to be small at convergence. If the solution exhibits oscillations near sharp gradients due to dominant convection and reaction effects on an under-resolved mesh, the residual of the equation has a significantly large value. This residual needs to be minimized to obtain an accurate and close-to-converged solution, defined by the user-defined tolerance limit. Therefore, the mesh has to be refined in those areas to capture the sharp gradients in the solution. This suggests using the residual of the differential equation as an indicator of the error for the adaptive mesh algorithm. For the quantification of the error on which our adaptive algorithm will be based, the residual error estimates are derived for the Galerkin discretization of the Allen-Cahn equation as in [7].

4. A numerical experiment

Let $\Omega = [0,1] \times [0,1]$ and the periodic boundary conditions are imposed on all the boundaries. The initial condition for this problem is considered as:

$$\phi(x, y, 0) = 1 + \tanh(\frac{R_1 - \sqrt{(x-0.25)^2 + (y-0.25)^2}}{\sqrt{2\varepsilon}}) + \tanh(\frac{R_2 - \sqrt{(x-0.75)^2 + (y-0.75)^2}}{\sqrt{2\varepsilon}})$$

Where R1 = 0.1 and R2 = 0.15 are the radii of the two circles centered at (0.35, 0.35) and (0.57, 0.57), respectively. The interface thickness parameter is chosen as ε = 0.005. The results on 512×256 are taken as the reference solutions and $\delta t = 0.0005$. the results regarding the error analysis are showed in Table 1.

Table 1. $t = 0.06$, L^2 error and convergence rate

h	u_x	Rate	u_y	Rate	ϕ	Rate	p	Rate
0.2022	4.6331e-02	-	1.2891e-02	-	2.8933e-01	-	1.2488e-02	-
0.1011	3.4722e-03	2.50	2.2548e-03	2.92	7.7651e-02	2.01	6.3244e-03	1.61
0.0505	4.1561e-04	2.71	3.6763e-04	2.83	2.4552e-02	2.24	2.4419e-03	1.82

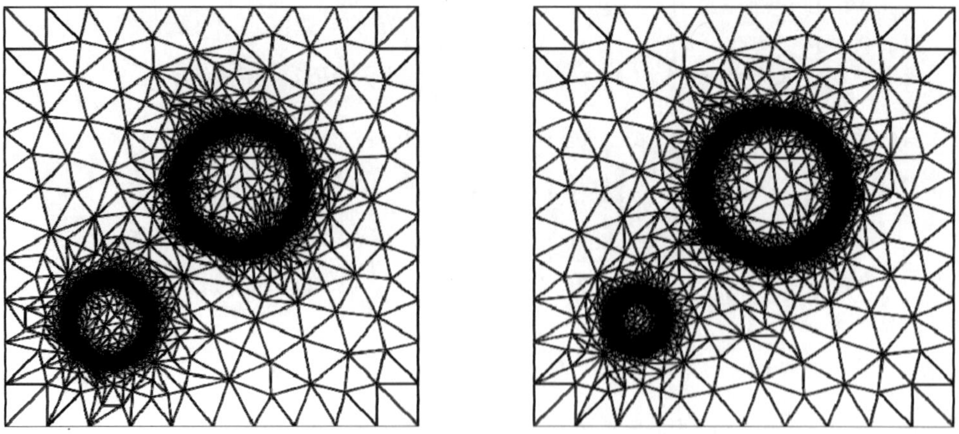

Figure 1. The adaptive mesh at t=0, t=400

5. Conclusion

In this paper, a stable, robust and general adaptive variational partitioned procedure has been proposed to solve the coupled Navier-Stokes and Allen-Cahn equations for two-phase fluid flows. The simplicity of the refinement/coarsening avoids complicated tree-type data structures, thus providing the ease of implementation. Several test problems are considered to demonstrate the various aspects of the proposed adaptive variational scheme.

Acknowledgments

This work was supported by the Fundamental Research Funds for the Central Universities (Grants No.2013B07314).

References

[1] L. Rayleigh, On the theory of surface forces-II. Compressible fluids, Philos. Mag. 33 (1892).
[2] J. van der Waals, The thermodynamic theory of capillarity under the hypothesis of a continuous density variation, J. Stat. Phys. 20 (1893) 197-244.
[3] J.W. Cahn, J.E. Hilliard, Free energy of a nonuniform system. I. Interfacial free energy, J. Chem. Phys. 28 (1958) 258-267.
[4] L.Q. Chen, Y. Wang, The continuum field approach to modeling microstructural evolution, JOM 48 (1996) 13-18.
[5] Jingzhi Zhu, Longqing Chen, Jie Shen, Veena Tikare, Coarsening kinetics from a variable-mobility Cahn–Hilliard equation: application of a semi-implicit Fourier spectral method, Phys. Rev. E 60 (1999) 3564.
[6] J. Shen, X. Yang, Q. Wang, Mass and volume conservation in phase field models for binary fluids, Commun. Comput. Phys. 13 (2013) 1045-1065.
[7] V.Joshi, R.K.Jaiman, An adaptive variational procedure for the conservative and positivity preserving Allen-Cahn phase-field model. J. Comput. Phys.366(2018) 478-504.

IOP Conf. Series: Journal of Physics: Conf. Series **1064** (2018) 012076 doi:10.1088/1742-6596/1064/1/012076

Flow characteristic prediction for liquid-liquid system in a stirred tank based on RNG k-ε model

Bin Qin, Chao Chen, Mengxiao Yu, Xin Li and Zhaoyou Zhu*

College of Chemical Engineering, Qingdao University of Science and Technology, Qingdao, China.

*Corresponding author e-mail: huagongyl@163.com

Abstract. The RNG k-ε model was adopted to calculate liquid–liquid turbulent flow in stirred tanks based on Eulerian–Eulerian model. The radial and tangential velocities of numerical calculation were compared with that of reported experiment. The results show that the calculation results are in satisfactory agreement with the experimental data. The complex liquid-liquid flow characteristics in the stirred tank were successfully predicted by the RNG k-ε model which means that the established calculation model can be an alternative method with inexpensive computing costs and high computational accuracy.

1. Introduction

Liquid-liquid stir is widely used in chemical industry for polymerization, extraction, dissolution, etc. The main purpose of stir is to promote the mass transfer and heat transfer of the two phases. The product quality, yield and economy of the processes are significantly affected by the efficiencies of mass/heat transfer in stirred tank. Therefore, it is of great practical significance to study the flow characteristics of the fluids in the stir process.

There are very few experimental studies on liquid-liquid stir because of the high cost of the experiment and the difficulty of the experimental operation. Svensson and Rasmuson [1, 2] studied the effects of disperse phase volume percentage on liquid-liquid flow characteristics with laser Doppler anemometry (LAD) technique and Particle Image Velocimetry (PIV) technique. The flow density, viscosity, and the fluid flow characteristics change with the dispersed phase percentage changes. Laurenzi et al. [3] studied the two-phase turbulent flow fields of a liquid–liquid system in stirred tank which measured by means of PIV technique. The dispersion conditions at different impeller speeds and the flow characteristics of continuous and disperse phases were studied. Cheng et al. [4] applied the planar laser induced fluorescence (PLIF) method to reveal the difference of macro-mixing behavior between the dispersed phase and the continuous phase of liquid-liquid system.

The difficulties of studying the industrial-scaled stir process with experimental methods are focused on the predicted accuracy and safety. Most actual production data cannot be obtained easily by amplifying the experimental data. Hence, it is necessary to develop computational fluid dynamics (CFD) based methods, which are cost-effective and capable of predicting fluid flow characteristics accurately through numerical simulation. Jaworski and Dudczak [5] studied the influence of the size and number of probes on mixing indices in stirred tanks with a CFD method based on standard k-ε model. Kumaresan and Joshi [6] used CFD methods with sliding mesh approach to study the effect of

Content from this work may be used under the terms of the Creative Commons Attribution 3.0 licence. Any further distribution of this work must maintain attribution to the author(s) and the title of the work, journal citation and DOI.
Published under licence by IOP Publishing Ltd

impeller design on the flow pattern and mixing time for a set of axial flow impellers, and compared the results predicted by CFD methods with the data measured by LAD. Cheng et al. [7] calculated the liquid–liquid flow field in a baffled turbulent liquid–liquid stirred reactor with Rushton turbine by the isotropic standard k–e model and the anisotropic explicit algebraic stress model (EASM) based on Eulerian–Eulerian model. Van den Akker [8] used the anisotropic Large Eddy Simulations (LES) approach to reproduce the local and transient flow conditions in turbulent mixing processes. The LES approach was regarded to be able to simulate the transient process accurately while it requires a large amount of calculation.

The purpose of this work is to study a liquid-liquid system flow characteristics by a CFD approach based on RNG k-ε model and Eulerian-Eulerian model. Then the root mean square (RMS) velocities, radial and tangential velocities calculated by the CFD approach were compared with the reported experimental data [2]. The validity of the approach was verified and the results of the simulation were discussed in detail.

2. Modeling

2.1. Physical model

A standard stirred tank with a six-blade Rushton turbine was used in this work. Four baffles were set in the stirred tank as shown in Fig. 1. The dimension of the stirred tank is shown in Table 1.

Table 1. The dimension of the stirred tank.

Item	Value	Item	Value
Tank diameter (T)	140mm	Blade width (W_b)	D/5
Liquid depth (H)	140mm	Blade length (L)	D/4
Baffle width (W)	T/12	Blade thickness (T_b)	D/50
Impeller diameter (D)	T/3	Disc diameter (D_d)	6D/10
Impeller height (H_i)	T/3	Disc thickness (T_d)	D/50

NaI solution and silicon oil were chosen as continuous aqueous phase and dispersed hydrophobic phase for the liquid-liquid system, respectively. The density and the viscosity of the NaI solution were about 1.34 g/cm3 and 1.4 mPas. The density and the viscosity of the silicon oil were about 0.94 g/cm3 and 11.0 mPas. The silicon oil volume fraction was 7%. The impeller speed was 540 rpm.

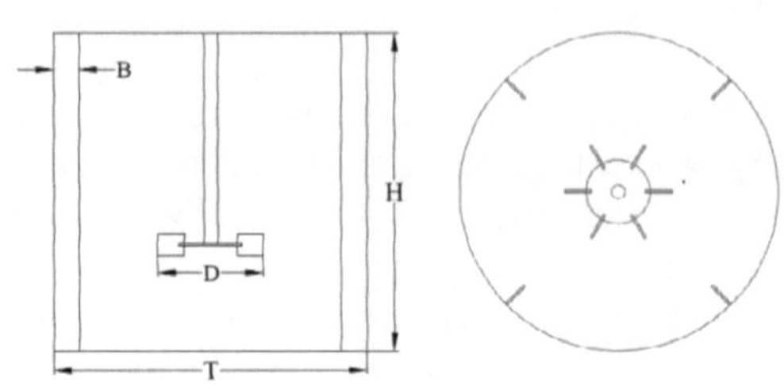

Figure 1. Geometric construction of the stirred tank.

2.2. Mathematical models

A high quality 3D grid of stirred tank was established by the ANSYS ICEM software. The fine grid with 3,778,105 elements was employed in the simulations. The calculation domain is divided into two parts: the center rotation dynamic domain containing the blades with structured grid and the surrounding static domain with unstructured grid.

The reliability of numerical simulation plays an important role in the study of flow field. Fluent 17.0 was used for numerical simulation in this work. The gradient discretization method was Least Squares Cell Based. The momentum discretization method, the volume fraction discretization method,

Turbulent Kinetic Energy discretization method and Turbulent Dissipation Rate discretization method were all discretized by First Order Upwind scheme.

2.2.1. Governing equations. Eulerian-Eulerian model was adopted to calculate the liquid-liquid flow and the phase interactions in the stirred tank. Governing equations are shown as Eq.1 and Eq.2:

$$\frac{\partial}{\partial t}(\alpha_i \rho_i) + \nabla \cdot (\alpha_i \rho_i \vec{u_i}) = 0 \tag{1}$$

$$\frac{\partial}{\partial t}(\alpha_i \rho_i \vec{u_i}) + \nabla \cdot (\alpha_i \rho_i \vec{u_i} \vec{u_i}) = -\alpha_i \nabla p + \nabla \cdot \vec{\tau_i} + \alpha_i \rho_i g + \vec{F_i} \tag{2}$$

Where $\vec{u}$ is velocity, α is volume fraction, ρ is density, g is gravity, p is pressure, $\vec{\tau_i}$ is stress tensor. The $\vec{F}$ is the integration of the forces of phase interactions. The drag force was mainly considered [9] and schiller-naumann model was adopted in this work.

2.2.2. Turbulence models. RNG k-ε model was built for fluid turbulence characteristics and widely used to study stir process. It is found that the calculation accuracy of the model can be satisfied with the calculation requirements and the calculation cost is low [10]. The transport equations for the RNG k-ε model are expressed as Eq.3 and Eq.4:

$$\frac{\partial}{\partial t}(\rho k) + \frac{\partial}{\partial x_i}(\rho k u_i) = \frac{\partial}{\partial x_j}(\alpha_k \mu_{eff} \frac{\partial k}{\partial x_j}) + G_k + G_b - \rho\varepsilon - Y_M + S_k \tag{3}$$

$$\frac{\partial}{\partial t}(\rho\varepsilon) + \frac{\partial}{\partial x_i}(\rho\varepsilon u_i) = \frac{\partial}{\partial x_j}(\alpha_\varepsilon \mu_{eff} \frac{\partial \varepsilon}{\partial x_j}) + C_{1\varepsilon}\frac{\varepsilon}{k}(G_k + C_{3\varepsilon} G_b) - C_{2\varepsilon}\rho\frac{\varepsilon^2}{k} - R_\varepsilon + S_\varepsilon \tag{4}$$

The G_k represents the generation of turbulence kinetic energy due to the mean velocity gradients and it is calculated by Eq.5:

$$G_k = -\rho \overline{u_i u_j} \frac{\partial u_j}{\partial x_j} \tag{5}$$

The G_b represents the generation of turbulence kinetic energy due to buoyancy and it is calculated by Eq.6:

$$G_b = \beta g_i \frac{\mu_t}{\Pr_t} \frac{\partial T}{\partial x_i} \tag{6}$$

The Y_M represents the contribution of the fluctuating dilatation in compressible turbulence to the overall dissipation rate and it is calculated by Eq.7:

$$Y_M = 2\rho\varepsilon M_t^2 \tag{7}$$

The quantities αk and αε are the inverse effective Prandtl numbers for k and ε. C1ε and C2ε are model constants, where C1ε = 1.42 and C2ε = 1.68.

3. Results

3.1. RMS velocities

In order to study the flow characteristics of the liquid-liquid multiphase system in a stirred tank, the RMS velocities of the continuous phase were calculated by a CFD approach based on RNG k-ε model and Eulerian-Eulerian model. The comparison between simulation results and experimental results reported by Svensson and Rasmuson [2] were shown in Fig. 2.

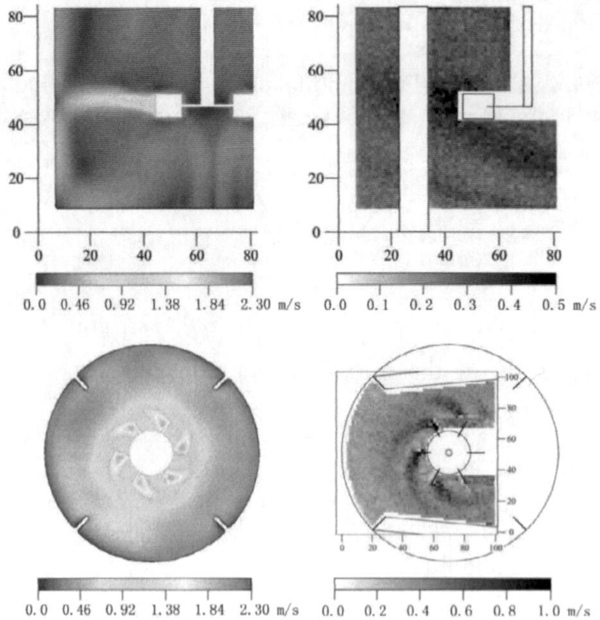

Figure 2. Comparison of simulation results (left) and experimental measurements (right) for RMS velocities of the continuous phase.

It can be seen that the simulation results were in good agreement with experimental results which means that the CFD approach predicted the RMS velocity distribution well. The velocity at impeller disc height was higher than that in other areas, and the further away from the impeller, the smaller the velocity. Moreover, there are a large vortex occurred in the tank which reflect that the efficiency of mixing is good. The locus of the vortex center predicted by the CFD approach and measured by PIV technique can be seen clearly in the vertical planet in Figure 2, which explained that the RNG k-ε model can predict the vortex center in liquid-liquid stir accurately.

3.2. Tangential and radial velocity

The tangential and radial velocities predicted by simulation and measured by experiment were compared. It can be seen in Fig. 3 that the CFD approach can give exact prediction of the velocities.

The data predicted by simulation was in good agreement with experimental data. The tangential velocity at impeller disc height predicted by simulation near the impeller was not as good as far from the impeller. The tangential velocity was in better agreement than the radial velocity with experimental data. It is can be seen in Fig. 3 clearly that the radial velocity and the tangential velocity at impeller disc height were greatly affected by the change of the depth from wall. The absolute values of the radial velocity and the tangential velocity at impeller disc height increase significantly with the increase of the depth. It shows that the closer to the impeller, the larger the kinetic energy of the fluid. The radial velocity and the tangential velocity at 10 mm above the tank bottom were relatively stable with the change of the depth.

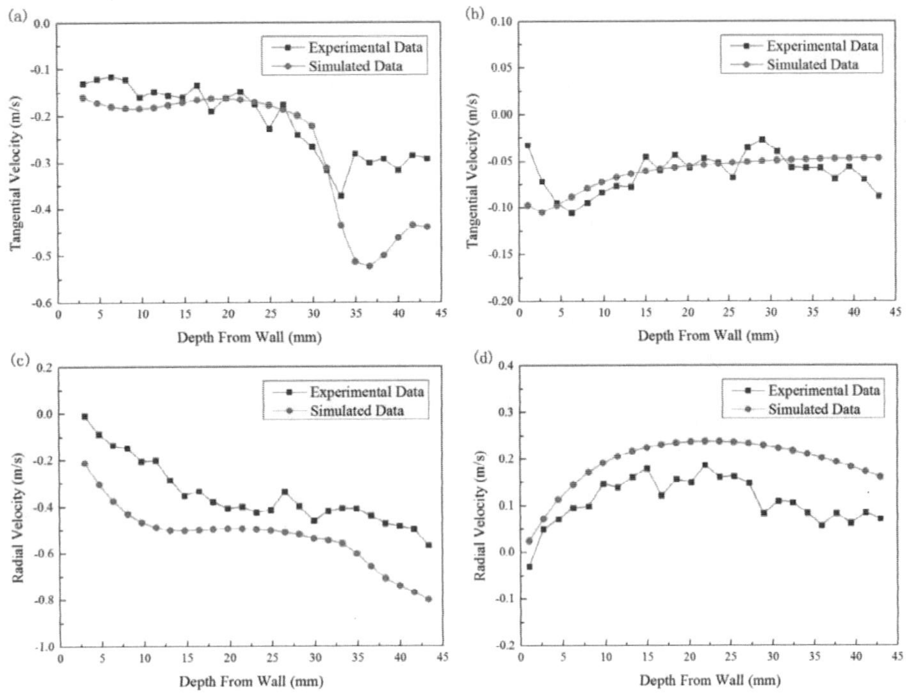

Figure 3. Comparison of simulation predictions and experimental data of radial velocities and tangential velocities for the continuous phase at:

impeller disc height (a, c); 10 mm above bottom(b, d).

4. Conclusion

Liquid-liquid stir in the stirred tank with a six-blade Rushton turbine was simulated by a CFD approach based RNG k-ε model and Eulerian–Eulerian model in this work. The data predicted by simulation were compared with the data measured by experiment. The RNG k-ε model gave reasonable predictions of the RMS velocity, the tangential velocity and the radial velocity. The overall trends of the simulation results are in good agreement with the experimental data. The vortex center in liquid-liquid stir can be predicted accurately with the RNG k-ε model. The radial velocity and the tangential velocity at impeller disc height increase significantly with the increase of the depth from wall. Through the calculation data and the experimental data, it can be concluded that the fluid near the impeller has larger kinetic energy and it is greatly affected by the impeller.

References

[1] F.J.E. Svensson, A. Rasmuson, LDA-measurements in a stirred tank with a liquid-liquid system at high volume percentage dispersed phase, Chem. Eng. Technol. 27 (2004) 335-339.

[2] F.J.E. Svensson, A. Rasmuson, PIV measurements in a liquid–liquid system at volum percentages up to 10% dispersed phase, Exp. Fluids. 41 (2006) 917–931.

[3] F. Laurenzi, M. Coroneo, G. Montante, A. Paglianti, F. Magelli, Experimental and computational analysis of immiscible liquid–liquid dispersions in stirred vessels, Chem. Eng. Res. Des. 87 (2009) 507–514.

[4] D. Cheng, X. Feng, J. Cheng, C. Yang, Z. Mao, Experimental study on the dispersed phase macro-mixing in an immiscible liquid–liquid stirred reactor, Chem. Eng. Sci. 126 (2015) 196–203.

[5] Z. Jaworskia, J. Dudczaka, CFD modelling of turbulent macromixing instirred tanks. Effect of the probe size and number onmixing indices, Couput. Chem. Eng. 22 (1998) S293-S298.

[6] T. Kumaresan, J.B. Joshi, Effect of impeller design on the flow pattern and mixing in stirred tanks, Chem. Eng. J. 115 (2006) 173–193.
[7] D. Cheng, X. Feng, C. Yang, Z. Mao, Modelling and experimental investigation of micromixing of single-feed semi-batch precipitation in a liquid–liquid stirred reactor, Chem. Eng. J. 293 (2016) 291–301.
[8] H.E.A.Van den Akker, The Details of Turbulent Mixing Process and their Simulation, Adv. Chem. Eng. 31 (2006) 151-229.
[9] M.Ljungqvist, A.Rasmuson, Numerical simulation of the two phase flow in an axially stirred reactor, Chem. Eng. Res. Des. 79 (2001) 533-546
[10] Z. Hao, J. Xu, H. Bie, Z. Zhou, Numerical simulation of the effects of baffle on flow field in a stirred tank, Adv. Mater. Res. 732-733 (2013) 432-435.

IOP Conf. Series: Journal of Physics: Conf. Series **1064** (2018) 012077 doi:10.1088/1742-6596/1064/1/012077

Analysis of Morphological Characteristics of Rotating Flow Field Using Proper Orthogonal Decomposition

Weimin Wu[1, a], Jianyao Yao[1, b] and Jingcheng Liu[2, c]

[1] College of Aerospace Engineering, Chongqing University, Chongqing 400044, China.
[2] Chongqing University of Science and Technology, Chongqing 401331, China.

[a]771700967@qq.com, [b] yaojianyao@cqu.edu.cn, [c] liujingcheng1980@126.com

Abstract. We use discrete proper orthogonal decomposition (POD) to investigate three-dimensional rotating flows to better understand the physical mechanism for the details of rotating flow. The results show that use the POD method to study the complicated engineering rotation flow is feasible, especially for the morphological characteristics of trail vortices in such modal spaces. Additionally, the logical architecture of this implementation using POD method can also be applied to other complicated engineering problems.

Keywords: complicated rotating flow, POD method, modal analysis, spectrum characteristics, characteristic plane.

1. Introduction

For the study of complex rotational flow in engineering applications, the traditional CFD method lacks multiple levels of further analysis capabilities. The POD method can not only perform multi-level analysis in orthogonal space, but also quantify its frequency characteristics. Therefore, it is very necessary to use the POD method to further analyze the complex rotation flow problem. In general [1], the POD method can be divided into two categories: continuous POD and discrete POD. For most practical applications, an analytical solution is difficult to obtain, so continuous POD is seldom used in engineering research. Thus, the discrete POD has already been widely applied for model reduction of large-scale CFD problems [2]. Sirovich had made a significant modification of discrete POD to improve its numerical stability and efficiency [3]. Ruelle et al. [4] found that the low-order model formed from the POD basis function is more accurate than when formed by basic functions of other types; for instance, trigonometric functions, Chebyshev polynomials, etc. Kou et al. [5] discussed the transonic buffeting of OAT15A airfoil based on POD and dynamic mode decomposition. The results show that, when using POD to reconstruct flow fields, the overall error is relatively small and the dynamic mode decomposition (DMD) method can more precisely characterize shock discontinuities.

Although POD is already widely used for conventional flow problems, its application to rotating flow is still quite limited. Christensen et al. [6] used POD to study unsteady rotating flow by using a rotating cap inside a cylindrical vessel. The results show that the first transition from steady to oscillatory motion is a supercritical Hopf bifurcation. The numerical solution obtained by using POD is consistent with experimental data. Sanghi et al. [7] investigated the thermally driven two-dimensional flow of air

Content from this work may be used under the terms of the Creative Commons Attribution 3.0 licence. Any further distribution of this work must maintain attribution to the author(s) and the title of the work, journal citation and DOI.
Published under licence by IOP Publishing Ltd

IOP Conf. Series: Journal of Physics: Conf. Series **1064** (2018) 012077 doi:10.1088/1742-6596/1064/1/012077

in a horizontal rotating cylinder and showed that, with less than 20 modes, low-dimensional models accurately reproduce the system dynamics both qualitatively and quantitatively. However, both above-mentioned investigations did not apply POD to the flow pattern in the rotating state and related research on applying POD to rotating flows is sparse.

Thus, we discuss herein the application of POD to modal analysis for the complex engineering problem, in which the related spectrum features are also analyzed. The relevant theories and the application of POD are introduced in Sections 2, and the implementation POD for rotating flows using CFD is given in Section 3. Section 4 presents a 3D numerical example of an impeller of a wind turbine to demonstrate the application of POD. Finally, Section 5 gives the conclusions.

2. Discrete proper orthogonal decomposition for rotating-flow field

For the discrete POD used herein, the sample vectors of the flow parameters for transient solutions at time t_m (i.e., a snapshot) are

$$f(\mathbf{x},t_m)=\left[f(x_1,t_m)\quad f(x_2,t_m)\quad\cdots\quad f(x_n,t_m)\quad\cdots\quad f(x_N,t_m)\right]^T, \tag{1}$$

Where N is the total number of discrete points in the spatial domain.

The important sample matrix for applying POD to a rotating-flow field is comprised of several appropriate sample vectors,

$$\mathbf{F}=\left[f(\mathbf{x},t_1)\quad f(\mathbf{x},t_2)\quad\cdots\quad f(\mathbf{x},t_m)\quad\cdots\quad f(\mathbf{x},t_M)\right], \tag{2}$$

Where $\mathbf{F}$ denotes the sample matrices of flow parameters, which could be velocities, densities, pressures, etc. The quantity M is the total number of instantaneous times; in other words, the total number of snapshots.

To obtain the POD modes for the rotating-flow field, the covariance matrix is first calculated as follows [3]:

$$\mathbf{C}=\mathbf{F}^T\mathbf{F}. \tag{3}$$

Note that $\mathbf{C}$ is a real symmetric matrix and has non-negative eigenvalues. Therefore, the eigenproblem of such a matrix has the form

$$\mathbf{C}\mathbf{G}^{[r]}=\lambda_r\mathbf{G}^{[r]}, \tag{4}$$

Where $\mathbf{G}^{[r]}$ is the r-th eigenvector for the specific flow parameter, in other words, the vector of time-varying coefficients of the rotating-flow modes. The quantity λ_r is the eigenvalue corresponding to the r-th-order POD mode of the specific flow parameter:

$$\mathbf{G}^{[r]}=\left[g^r(t_1)\quad g^r(t_2)\quad\cdots\quad g^r(t_m)\quad\cdots\quad g^r(t_M)\right]^T. \tag{5}$$

The rth-order POD mode of the swirling flow field can be defined by

$$f_r(\mathbf{x})=\frac{1}{M\lambda_r}\sum_{m=1}^{M}g^{[r]}(t_m)f(\mathbf{x},t_m), \tag{6}$$

IOP Conf. Series: Journal of Physics: Conf. Series **1064** (2018) 012077 doi:10.1088/1742-6596/1064/1/012077

Where all eigenvalues can be listed in descending order according to the information-content contribution to the flow field. Therefore, the dominant flow modes in the rotating-flow field can be extracted by using the first several POD modes, which describe the essential features of the flow.

3. Implement proper orthogonal decomposition for rotating flows

The process of the implementation of POD in a rotating-flow field for complex engineering problems contains the following steps:

(1). Carry out a CFD simulation of unsteady flow and assemble the CFD results of snapshots in a time-ordered sequence in the following format,

$$\begin{bmatrix} x_1(t_m) & y_1(t_m) & z_1(t_m) & f_1^1(t_m) & f_1^2(t_m) & \cdots & f_1^D(t_m) \\ x_2(t_m) & y_2(t_m) & z_2(t_m) & f_2^1(t_m) & f_2^2(t_m) & \cdots & f_2^D(t_m) \\ \vdots & \vdots & \vdots & \vdots & \vdots & \vdots & \vdots \\ x_l(t_m) & y_l(t_m) & z_l(t_m) & f_l^1(t_m) & f_l^2(t_m) & \cdots & f_l^D(t_m) \\ \vdots & \vdots & \vdots & \vdots & \vdots & \vdots & \vdots \\ x_N(t_m) & y_N(t_m) & z_N(t_m) & f_N^1(t_m) & f_N^2(t_m) & \cdots & f_N^D(t_m) \end{bmatrix}, \tag{7}$$

Where the D refers to the number of parameter types considered in the POD analysis.

(2). To reduce the computational cost of POD, use the flow-parameter field on the characteristic plane to form the equivalent field instead of the original snapshots. The reduced format of the substituted snapshots is given by

$$\begin{bmatrix} x_1'(t_m) & y_1'(t_m) & z_1'(t_m) & f_1'^1(t_m) & f_1'^2(t_m) & \cdots & f_1'^D(t_m) \\ x_2'(t_m) & y_2'(t_m) & z_2'(t_m) & f_2'^1(t_m) & f_2'^2(t_m) & \cdots & f_2'^D(t_m) \\ \vdots & \vdots & \vdots & \vdots & \vdots & \vdots & \vdots \\ x_l'(t_m) & y_l'(t_m) & z_l'(t_m) & f_l'^1(t_m) & f_l'^2(t_m) & \cdots & f_l'^D(t_m) \\ \vdots & \vdots & \vdots & \vdots & \vdots & \vdots & \vdots \\ x_{N'}'(t_m) & y_{N'}'(t_m) & z_{N'}'(t_m) & f_{N'}'^1(t_m) & f_{N'}'^2(t_m) & \cdots & f_{N'}'^D(t_m) \end{bmatrix}, \tag{8}$$

Where N' is the number of discrete points on the characteristic plane, with $N' \ll N$. Once assembled on the characteristic plane, these discrete points should convey sufficient physical information. The flow field is visualized on the characteristic plane by projection using linear interpolation.

(3). From the substituted snapshots, generate the sample matrix, which is called the subsample matrix, and calculate the POD modes and corresponding time coefficients by using Eq. (5). In addition, obtain the spectrum characteristics of such time coefficients using the fast Fourier transform (FFT).

4. Numerical example of three-dimensional impeller of wind turbine

The radius of the overall cylindrical solution domain is 3.5 m, which is five times that of the impeller blade. To better capture the vortex structures, a nonuniformly distributed mixed mesh is used to discretize the solution domain (see Fig. 1). A dense structured prism grid is used for the downstream vortex-dominated region in the center of the solution domain, and a relatively sparse tetrahedral mesh is used for the outer domain. The total number of cells is approximately 3.89 M.

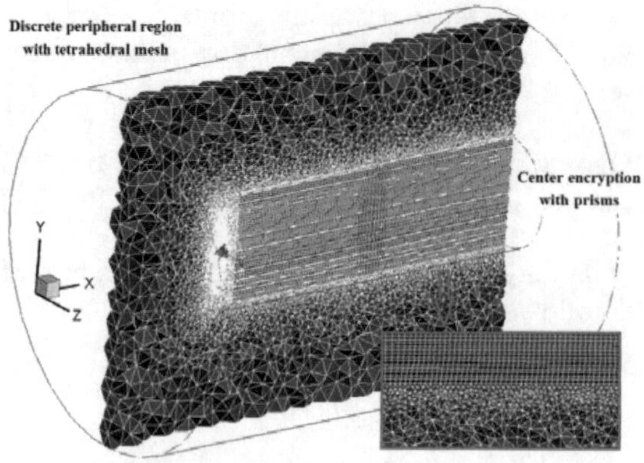

FIG. 1 Mesh layout for CFD impeller model.

The sliding-mesh method is used to exchange flow-flux data between the body-fitted rotating area of the impeller and the peripheral stationary region. The unsteady turbulent incompressible flows are simulated by using the LES method with the sub-grid model of Smagorinsky and the SIMPLEC algorithm. The velocity-inlet and pressure-outlet boundary conditions are applied on the left and right surfaces of the solution domain. The inlet velocity is 10 m/s along the positive X axis, and the rotational velocity of the impeller is 750 rev/min. The time step in the transient-flow calculation is 2.22×10^{-4} s, which corresponds to 1° per step.

When the transient flow reaches a stable periodicity, 60 snapshots for POD analysis are chosen at equal time intervals during a single revolution of the impeller. Next, the reduced-substituted snapshots can be obtained from the characteristic plane of the CFD model at $Z=0$ position. According to the spiral structure of the tip and central vortices in the downstream zone and of the mesh resolution of the solution domain, the radial velocity is clearly more influenced by vortices than by the velocities in the other two directions. Therefore, these POD modes of the first 6 orders in the characteristic plane of the radial velocities shall be further investigated and are displayed in Fig. 2.

Additionally, the corresponding characteristics of the frequency spectrum of the time coefficient of the first 9 modes are also brought out by using a fast Fourier transform to better reveal the physical mechanism of the rotating flow around the wind turbine, as shown in Fig. 3.

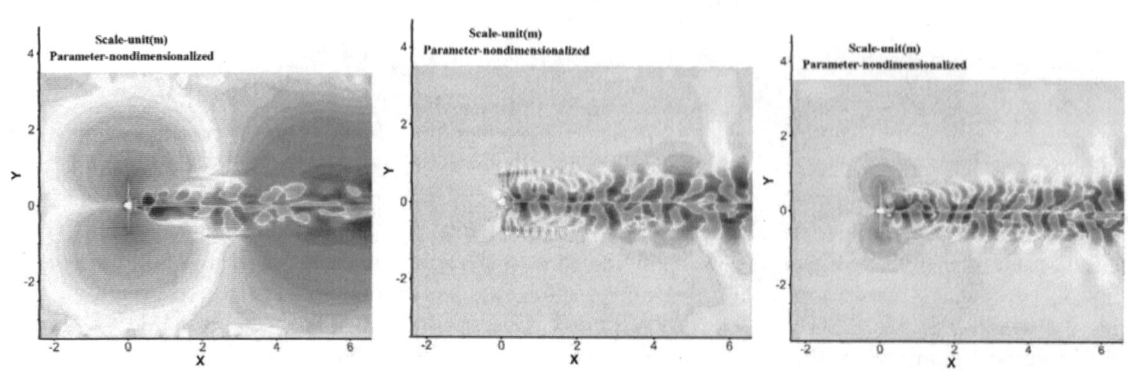

(a) 1-order POD mode. (b) 2-order POD mode. (c) 3-order POD mode.

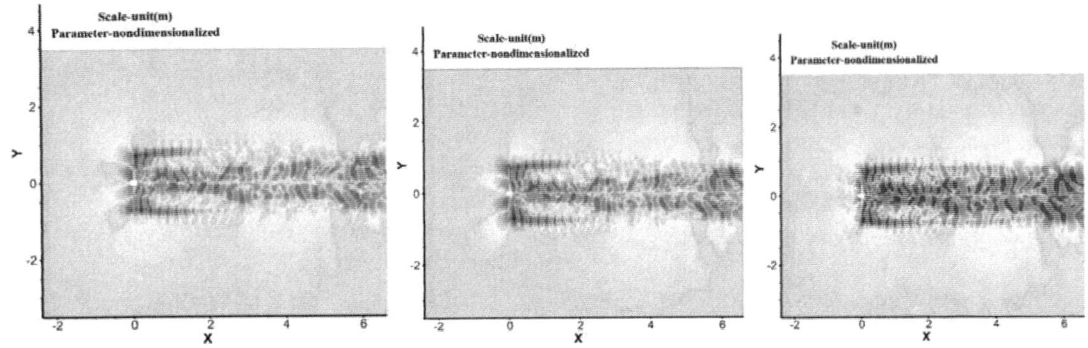

(d) 4-order POD mode. (e) 5-order POD mode. (f) 6-order POD mode.

FIG. 2 First 6 POD modes of radial velocity on characteristic plane.

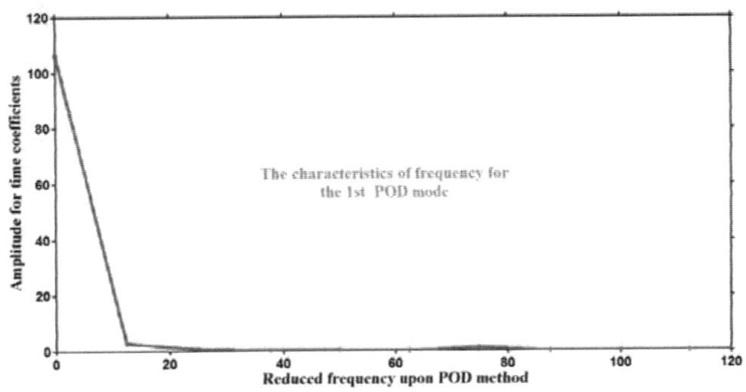

(a) First POD mode.

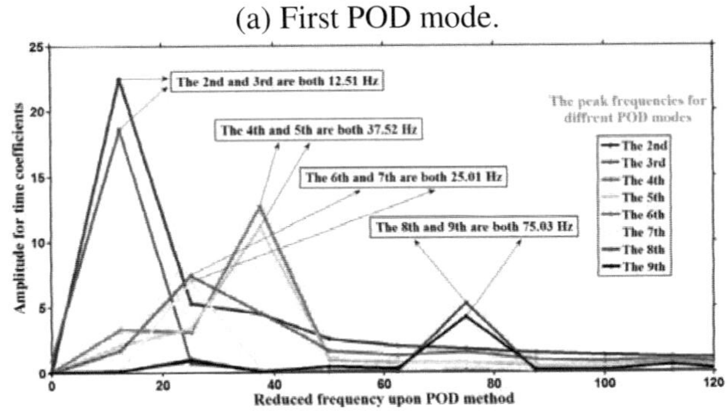

(b) POD modes for 2 to 9 orders.

FIG. 3 Frequency spectrum of time coefficients of first nine modes.

The following observations may be made based on the spatial distributions of the POD modes and the frequency spectrum of the time coefficients:

(1) As shown in Fig. 6, the peak frequency of the first-order mode is zero, indicating the rigid-body displacement or the average value of the rotating flow. The corresponding POD-mode distribution is shown in Fig. 5(a), in which the spatial symmetry of the distribution over a large range is well displayed, except for regions of the central vortex and at the boundaries of the CFD model.

(2) The peak frequencies of the second- and third-order POD modes shown in Figs. 5(b) and 5(c) are both 12.51 Hz, which complies with the component of fundamental frequency of impeller rotation and also shows that the central vortex is the dominant cause of the spatiotemporal periodicity. In these flow

patterns, the spatiotemporal periodicity is well presented mainly for the central vortex along the forward direction of the X axis.

(3) The spatiotemporal periodicity of the fourth- and fifth-order modes shown in Figs. 5(d) and 5(e) is mainly caused by the tip vortex. Because the impeller has three blades, the corresponding peak frequencies of these two modes are both 37.52 Hz, which is the integrated fundamental rotating frequency in the near wake.

(4) In the sixth mode, the tip and central vortices fuse [see Fig. 5(f)], and the spatiotemporal periodicity for vortices is well expressed.

5. Conclusion

This paper presents the theory and implementation of discrete POD for modal analysis in rotating-flow fields. The spectral characteristics of the time coefficient of the flow modality have also been discussed based on the characteristic plane. For unsteady flows, the overwhelming computational cost of POD analysis is incurred in computing snapshots and other calculations of POD analysis are negligible for such complicated wake of wind turbine.

According to the computational results, it is feasible to use the POD method to study the complicated engineering rotation flow. In the low-order flow modes, the displacement of the rigid body of the rotating flow, the fundamental frequency component, the integrated fundamental frequency, the frequency multiplication of the fundamental frequency, etc. have been correspondingly presented, especially for the morphological characteristics of trail vortices in such modal spaces. Therefore, this study can help to further understand the intrinsic physical mechanism of complex rotational flow, and the logical architecture of this implementation using POD method can be applied to other complicated engineering problems, such as turbomachinery, high-energy disc brakes, rotary bioreactors in biomedicine, self-circulation radiators for computers, etc.

References

[1] D. Rempfer and H. F. Fasel, "Evolution of three-dimensional coherent structures in a flat-plate boundary layer," J. Fluid Mech. 260, 257 (2006).

[2] Z. Wang, I. Akhtar and J. Borggaard, et al., "Proper orthogonal decomposition closure models for turbulent flows: A numerical comparison," Computer Methods in Applied Mechanics & Engineering. 10, 237 (2012).

[3] L. Sirovich, "Turbulence and the dynamics of coherent structures part I: Coherent structures," Quarterly of Applied Mathematics. 45, 561 (1987).

[4] D. Ruelle and F. Takens, "On the nature of turbulence," Communications in Mathematical Physics. 20, 167 (1978).

[5] J. Kou, W. Zhang and C. Gao, "Modal Analysis of Transonic Buffet Based on POD and DMD Techniques," Hangkong Xuebao/ acta Aeronautica Et Astronautica Sinica. 37, 2679 (2016).

[6] E. A. Christensen, J. N. Sørensen and M. Brøns, et al., "Low-dimensional representations of early transition in rotating fluid flow," Theoretical & Computational Fluid Dynamics. 5, 259 (1993).

[7] S. Sanghi and N. Hasan, "Proper orthogonal decomposition and its applications," Asia-Pacific Journal of Chemical Engineering. 6, 120 (2015).